IF	Initiation factor
K_M	Michaelis constant
LDL	Low-density lipoprotein
Mb	Myoglobin
NAD^+	Nicotinamide adenine dinucleotide (oxidized form)
NADH	Nicotinamide adenine dinucleotide (reduced form)
$NADP^+$	Nicotinamide adenine dinucleotide phosphate (oxidized form)
NADPH	Nicotinamide adenine dinucleotide phosphate (reduced form)
P_i	Phosphate ion
PAGE	Polyacrylamide gel electrophoresis
PCR	Polymerase chain reaction
PEP	Phosphoenolpyruvate
PIP_2	Phosphatidylinositol *bis*phosphate
PKU	Phenylketonuria
Pol	DNA polymerase
PP_i	Pyrophosphate ion
PRPP	Phosphoribosylpyrophosphate
PS	Photosystem
RF	Release factor
RFLPs	Restriction-fragment-length polymorphisms
RNA	Ribonucleic acid
RNase	Ribonuclease
mRNA	Messenger RNA
rRNA	Ribosomal RNA
tRNA	Transfer RNA
snRNP	Small nuclear ribounuclear protein
S	Svedberg unit
SCID	Severe combined immune deficiency
SSB	Single-strand binding protein
SV40	Simian virus 40
T	Thymine
TDP	Thymidine diphosphate
TMP	Thymidine monophosphate
TTP	Thymidine triphosphate
U	Uracil
UDP	Uridine diphosphate
UMP	Uridine monophosphate
UTP	Uridine triphosphate
V_{max}	Maximal velocity

BIOCHEMISTRY

Biochemistry

Second Edition

Mary K. Campbell

Mount Holyoke College

Illustrations by J/B Woolsey Associates with

art contributions by Irving Geis

SAUNDERS COLLEGE PUBLISHING
HARCOURT BRACE COLLEGE PUBLISHERS

Philadelphia   Fort Worth   San Diego   New York   Orlando
San Antonio   Toronto   Montreal   London   Sydney   Tokyo

Text Typeface: New Baskerville
Compositor: Black Dot Graphics
Acquisitions Editor: John Vondeling
Developmental Editor: Sandra Kiselica
Managing Editor: Carol Field
Senior Project Editor: Margaret Mary Anderson
Copy Editor: Mary Patton
Manager of Art & Design: Carol Bleistine
Art Director: Jennifer Dunn
Art & Design Coordinator: Sue Kinney
Text Designer: Rebecca Lemna
Layout Artist: Caslon, Inc.
Cover Designer: Jennifer Dunn
Text Artwork: J/B Woolsey Associates and Irving Geis
Director of EDP: Tim Frelick
Senior Production Manager: Charlene Squibb
Marketing Manager: Marjorie Waldron

Cover credit: *Inset picture:* ©Dr. Gopal Murti/Science Photo Library, Photo Researchers, Inc. (False-color transmission electron micrograph (TEM) of the bacterium *Escherichia coli* surrounded by its DNA. The bacterium was treated with an enzyme to weaken its cell wall and then placed in water, causing its DNA to be ejected. The DNA is visible as the gold-colored fibrous mass lying around the green bacterial shell.) *Background picture:* ©Michael Freeman/ Phototake. (Peptide chain.)

Frontispiece credit: ©Comstock Inc. Color alteration of frontispiece image and ghosted image done by Bill Neide/Lehigh Press.

Printed in the United States of America

BIOCHEMISTRY, Second Edition
ISBN 0-03-001872-2

Library of Congress Catalog Card Number: 94-066999

567890123 069 9876543

To everyone,

especially my students,

who made this book possible in every way.

Mary K. Campbell

Author

Mary K. Campbell is a professor of chemistry at Mount Holyoke College, where she frequently teaches the one-semester biochemistry course and advises undergraduates working on biochemical research projects. She received her Ph.D. from Indiana University and did postdoctoral work in biophysical chemistry at Johns Hopkins University. Professor Campbell's research interests are in the area of the physical chemistry of biomolecules, specifically, spectroscopic studies of protein–nucleic acid interactions.

Mary Campbell can be found frequently hiking the Appalachian trail with her Bernese mountain dogs, Lolly and Jake.

Irving Geis

Contributor

Irving Geis is well known for his lucid visualizations of molecular structures, particularly proteins and nucleic acids. These have appeared in *Scientific American* for the past thirty years and in major chemistry, biology, and biochemistry textbooks. He is a co-author with R.E. Dickerson, Director of the Molecular Biology Institute of UCLA, of *Chemistry, Matter and the Universe; The Structure and Action of Proteins* and *Hemoglobin: Structure, Function, Evolution and Pathology.*

In addition to drawing, painting, and writing, Irving Geis is a frequent lecturer at universities and medical schools on protein structure and function.

A recent Guggenheim fellowship made possible the assembly and cataloging of his drawings and paintings into The Geis Archives of molecular structure.

We are thankful to Irving Geis for contributing Figures 4.2, 4.3, 4.5, 4.12(a), 4.20, 4.23, Chapter Openers 5 and 6, 6.9, 6.19, 8.2, 9.12, 9.20, and the unnumbered figure on p. 650.

Preface

This text is intended for students in any field of science or engineering who want a one-semester introduction to biochemistry but who do not intend to be biochemistry majors. My main goal in writing this book is to make biochemistry as clear and interesting as possible and to familiarize all science students with the major aspects of biochemistry. For students of biology, chemistry, physics, geology, nutrition, and agriculture, biochemistry impacts greatly on the content of their fields, especially in the areas of medicine and biotechnology. For engineers, studying biochemistry is especially important for those who hope to enter a career in biomedical engineering or some form of biotechnology.

Students who will use this text are at an intermediate level in their studies. A beginning biology course, general chemistry, and at least one semester of organic chemistry are assumed as preparation.

SPECIAL FEATURES

New To This Edition

The order of topics has been changed. The chapters on protein structure and enzyme behavior occupy an earlier position, as does the one on nucleic acid structure. A new Chapter 7 on nucleic acids and biotechnology covers DNA sequencing, cloning, the polymerase chain reaction, gene mapping, and genetic engineering. Coverage of cancer and AIDS has been broadened significantly, including several new sections in Chapters 18 and 20. Control of carbohydrate metabolism (Chapter 14) is treated in greater depth, including hormonal control. More material has been included on plant biochemistry, particularly where there are agricultural applications. Coverage of nutrition, particularly in Chapter 18, has been expanded and updated.

Visual Impact

One of the most distinctive features of this text is its visual impact. Its extensive four-color art program includes artwork by Irving Geis and by John and Bette Woolsey. The illustrations convey meaning so powerfully that many have become, or are certain to become, standard presentations in the field.

Chapter Overviews

Originally written by Irving Geis for the first edition and revised by the author for this edition, the introductory paragraphs serve as overviews for each chapter. These chapter-opening paragraphs tie together the material from previous chapters with the topics to be discussed. They serve as building blocks for new ideas.

Interviews

There are six interviews with prominent scientists that open the parts of this book. These outstanding chemists talk about a broad range of topics, including their education and their research. The interviews represent a brief look into both their professional and personal lives and are included to encourage students to consider a career in the sciences.

Boxes

Boxes on special topics, frequently ones with clinical implications such as cancer and AIDS, highlight points of particular interest to students.

Practice Sessions

New to this edition are practice sessions, designed to give students experience in problem solving. The topics chosen for this treatment are those with which students usually have the most difficulty.

Summaries

Each chapter closes with a concise summary, a broad selection of exercises, and an annotated bibliography.

Interchapters

Two flexible interchapters explain the experimental methods for determining protein structure and the anabolism of nitrogen-containing compounds. These interchapters follow Chapters 4 and 17.

Glossary and Answers

The book ends with an answer section, a glossary of important terms and concepts, and a detailed index.

ORGANIZATION

Because biochemistry is a multidisciplinary science, the first task in presenting it to students of widely different backgrounds is to put it in context. Part I provides the necessary background and connects biochemistry to the other sciences. Part II focuses on the structure and dynamics of important cellular components. Metabolism is covered in Part III. Part IV of the book is devoted to the flow of genetic information.

Some topics are discussed several times; an example is the control of carbohydrate metabolism. The second (or subsequent) discussion makes use of information that students have learned since the first discussion and builds on what they have already learned. It is particularly useful to return to a topic after students have had time to assimilate and reflect on it.

In Part I, an introduction, two chapters relate biochemistry to other fields of science. Chapter 1 deals with some of the less obvious relationships, such as the connections of biochemistry with physics, astronomy, and geology, mostly in the context of the origins of life. This chapter goes on to the more readily apparent linkage of biochemistry with biology, especially with respect to the distinction between prokaryotes and eukaryotes, as well as the role of organelles in eukaryotic cells. Chapter 2 builds on material familiar from general chemistry, such as buffers and the solvent properties of water, emphasizing the biochemical point of view toward such material. Functional groups on organic molecules are discussed, again focusing on their role in biochemistry.

Part II, on the structure of cellular components, focuses on the structure and dynamics of proteins and membranes in addition to giving an introduction to some aspects of molecular biology. Chapters 3 through 5 deal with amino acids, peptides, and the structure and action of proteins. Interchapter A, found between Chapters 4 and 5, describes experimental methods for determining protein structure. Chapters 6 and 7 deal with the structure of nucleic acids and the methods that biotechnology uses to manipulate DNA. Chapter 8 treats the structure of membranes and their lipid components.

Part III, on metabolism, opens with Chapter 9 on chemical principles that provide some unifying themes. Thermodynamic concepts learned earlier in general chemistry are applied to specific biochemical topics such as coupled reactions and hydrophobic interactions. In addition, this chapter explicitly makes the connection between metabolism and electron transfer (oxidation-reduction) reactions. Coenzymes are introduced in this chapter and are discussed in later chapters in the context of the reactions in which they play a role. Glycolysis and the citric acid cycle are treated in Chapters 11 and 12, followed by the electron transport chain and oxidative phosphorylation in Chapter 13. Glycogen metabolism, gluconeogenesis, and the pentose phosphate pathway provide bases for treating control mechanisms in carbohydrate metabolism in Chapter 14. In Chapter 16 photosynthesis rounds out the discussion of carbohydrate metabolism. The catabolic and anabolic aspects of lipid metabolism are dealt with in Chapter 15. The metabolism of nitrogen-containing compounds such as amino acids, porphyrins, and nucleobases are treated in Chapter 17. Interchapter B details the anabolism of these important nitrogen-containing comopunds. A summary chapter (Chapter 18) gives an integrated look at metabolism, including a treatment of hormones and second messengers. The overall look at metabolism includes a brief discussion of nutrition and a somewhat longer one of the immune system.

The final section, Chapters 19 and 20 (Part IV), deals with molecular biology. In the first of these two chapters, the replication of DNA and the translation of the genetic message in RNA provide the focus of discussion.

The ultimate translation of the genetic message in the synthesis of proteins is the subject of the final chapter.

This text attempts to give an overview of important topics of interest to biochemists and to show how the remarkable recent progress of biochemistry impinges on other scientists. The length is intended to allow instructors a choice of favorite topics but not to be overwhelming for the limited amount of time available in one semester.

ALTERNATIVE TEACHING OPTIONS

The order in which individual chapters are covered can be changed to suit the needs of specific groups of students. The portions of Chapter 9 that deal with thermodynamics can be covered before discussion of protein structure, an option preferred by some instructors. The two pairs of chapters that deal with molecular biology (Chapters 6 and 7, and 19 and 20) can be combined as a unit. The whole molecular biology unit can precede metabolism or can follow it, depending on the instructor's preference.

SUPPLEMENTS

This text is accompanied by the following supplements:

- CAMPBELL'S COMPANION AND PROBLEMS BOOK by William M. Scovell (Bowling Green State University) accompanies and complements the text, with the objective of helping students gain a more comprehensive understanding of biochemistry.

  1. Each chapter begins with an introductory paragraph outlining the major topics discussed in the text.
  2. Each chapter contains Learning Objectives to help focus attention on important concepts.
  3. The heart of the book is the additional problems with detailed explanations and answers. These are intended to develop a fuller understanding of general concepts, and in some cases, focus on important details of a structure, reaction, mechanism, metabolic cycle, or pathway. In addition, some problems go beyond the text material to provide a glimpse of rapidly evolving areas.
  4. A section is included that reviews many of the important organic reactions underlying the anabolic and catabolic reactions in biochemistry. This section provides a foundation for clearly realizing that biochemistry is simply "chemistry in living systems."
  5. A number of important topics that are currently experiencing rapid if not explosive development are "SPOTLIGHTED" to point out their role and impact on today's burgeoning understanding of the molecular basis of biology.

- INSTRUCTOR'S MANUAL by Mary Campbell. Includes chapter summaries, lecture outlines, answers to all exercises, and a bank of 25 multiple-choice questions for each chapter.

- OVERHEAD TRANSPARENCIES. One hundred full-color figures from the text.

● COMPUTERIZED TEST BANK. Compatible with IBM and Macintosh computers, it contains 25 multiple-choice questions for each chapter.

Monthly updates of **UVA IMAGES,** a collection of macromolecular structures with explanatory text, is available on Internet by accessing the computer facility at the University of Virginia. Users have the ability to manipulate the images in any dimension.

SAUNDERS CHEMISTRY VIDEODISC VERSION 3 MULTIMEDIA PACKAGE includes still images from the text, as well as hundreds from other Saunders chemistry texts. The disc can be operated via a computer, a bar code reader, or a hand-controlled keypad. It also features molecular simulations and demonstrations.

LECTURE ACTIVE™ SOFTWARE enables instructors to customize their lectures with the Videodisc. Available for both IBM and Macintosh computers.

ACKNOWLEDGMENTS

The help of many others made this book possible. A grant from the Dreyfus Foundation made possible the experimental introductory course that was the genesis of many of the ideas for this text. My colleagues Edwin Weaver and Francis DeToma gave much of their time and energy in initiating that course. Many others at Mount Holyoke were generous with their support, encouragement, and good ideas, especially Anna Harrison, Lilian Hsu, Dianne Baranowski, Sheila Browne, Janice Smith, Jeffrey Knight, Sue Ellen Frederick Gruber, Peter Gruber, Marilyn Pryor, and Sue Rusiecki. Particular thanks go to Doris Rovetti, the chemistry department secretary, and to Sandy Ward, science librarian.

Many biochemistry students have used and commented on earlier versions of this text. The thought of my own mentors has been an inspiration to me: Walter Moore and Henry Mahler, whose untimely death was such a loss to us all, at Indiana University and Paul O. P. Ts'o at The Johns Hopkins University.

I would like to acknowledge my colleagues who contributed their ideas and critiques of my manuscript:

Ronald Bentley, University of Pittsburgh

Henry Butcher, Loyola College

Daniel Buttlaire, San Francisco State University

Todd Carlson, Grand Valley State University

Rick Cote, University of New Hampshire

Edward Funkhauser, Texas A&M University

John Gores, California State Polytechnic University, San Luis Obispo

Ralph Jacobson, California State Polytechnic University, San Luis Obispo

Jerry Jasinski, Keene State College

Debra Kendall, University of Connecticut

Steven Meinhardt, North Dakota State University

Enrique Meléndez-Hevia, Universidad de La Laguna, Canary Islands

Patrick Mobley, California State Polytechnic University, Pomona

Robert Oberlender, University of the Pacific

Howard Ono, California State University, Fresno

Robert Orr, Delaware Valley College

Thomas Prasthofer, North Carolina State University

Rachel Shireman, University of Florida

Anthony Toste, Southwest Missouri State University

Anthony Tu, Colorado State University

Thomas Waddell, University of Tennessee

Jubran Wakim, Middle Tennessee State University

The efforts of John Vondeling, Sandi Kiselica, and Margaret Mary Anderson at Saunders College Publishing were essential to the production of this book. Jennifer Dunn directed the art and design efforts that have had such magnificent results. I feel privileged that Irving Geis has contributed not only some of his classic illustrations but the original version of the introductory overviews to each chapter. John and Bette Woolsey outdid themselves at every turn with illustrations and turned crude sketches into works of art. Computer ray trace space-filling molecular models, which set a new standard of excellence for computer art, were produced by Leonard Lessin, F.B.P.A., in conjunction with Hans Dijkman, Ph.D., and Waldo Feng. Laurel Anderson, photo researcher, found many splendid photographs, in some cases with considerable effort. I extend my most sincere gratitude to those listed here and to all others to whom I owe the opportunity to do this book. Finally, I thank my family, whose moral support has meant so much to me in the course of my work.

Contents Overview

Roberta F. Colman, p. 2

Carl Djerassi, p. 6

Siegfried Reich, p. 64

Jane Richardson, p. 280

Ponzy Lu, p. 594

Jacqueline K. Barton, p. 597

CONTENTS

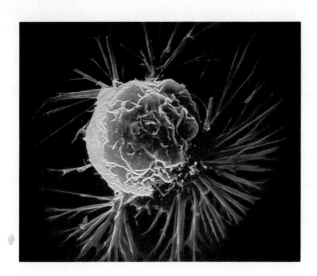

Human breast cancer cell.

Computer-generated model of a globular protein.

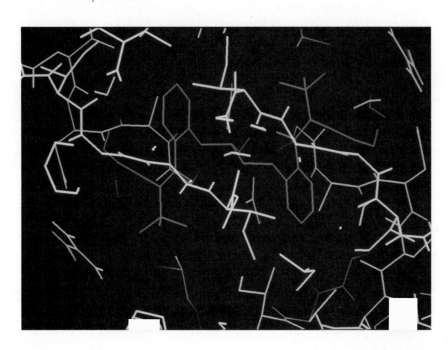

Computer-generated model of an inhibitor of HIV protease.

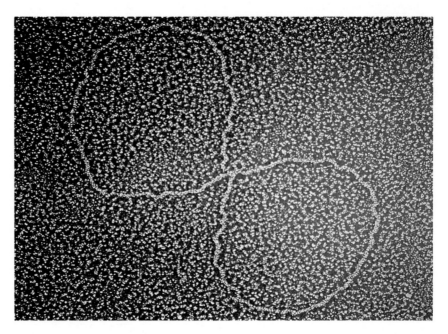

DNA plasmids.

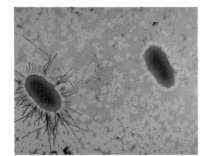

E. coli *bacteria.*

PART III

ENERGETICS AND METABOLISM: CARBOHYDRATES, LIPIDS, AND COMPOUNDS OF NITROGEN   *279*

Interview: Jane Richardson   280

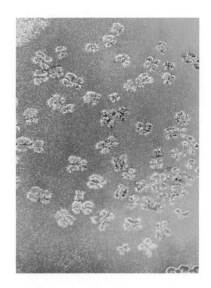

Human chromosomes.

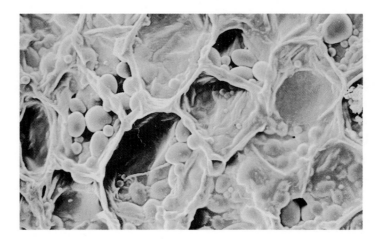

Starch grains in plant cells.

Stored glycogen in liver.

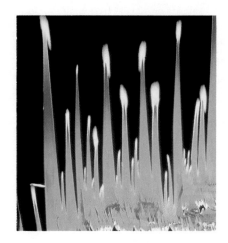

Folic acid crystal.

Mitochondria.

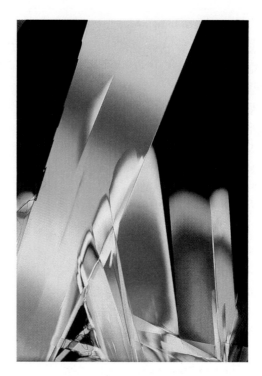

Thiamine crystals.

PART IV

WORKINGS OF THE GENETIC CODE *593*

Interview: Ponzy Lu   594
Interview: Jacqueline K. Barton   597

Crystals of vitamin B-6 photographed under polarized light.

Part I

An Introduction to Biochemistry

Outline

Roberta F. Colman

A colorful carpet from Turkey and a poster of New Guinea adorn the walls of Roberta Colman's office at the University of Delaware's Brown Laboratory. These reminders of travels abroad signify Colman's passion for exploration, which is also evident in her studies of enzymes. A professor in the Department of Chemistry and Biochemistry, Colman earned a bachelor's degree in biochemical sciences from Radcliffe College in 1959 and a Ph.D. in biochemistry from Radcliffe and Harvard University in 1962. She has lectured and published widely, and has held biochemical teaching positions at the medical schools of Washington University and Harvard. Roberta Colman is always aware that her research at the molecular level may have broader implications in medicine, and she has collaborated with medical researchers, including her husband, Robert W. Colman, a professor of medicine at Temple University, with whom she studied proteins of platelets.

How did you become interested in the biochemical sciences?

My first real involvement with the sciences came in high school. I had a really outstanding biology teacher who had a very broad view of science, and he got me excited about learning. I did some research in protozoan metabolism while I was in high school, and that led to an award in the Westinghouse Science Talent Search, which gave me positive feedback. After high school, I wasn't really sure what I wanted to do. I was very interested in journalism and was editor of newspapers in both high school and college. I vacillated for a while and thought I might go into scientific journalism. Eventually I decided that I really wanted to *do* science rather than just write about it. But I guess that background probably helped me anyway, in terms of the writing I do today.

I chose to major in biochemical sciences in college for a couple of reasons. First, it combined interests I had: I liked biology and I liked chemistry. The other reason was that there was a tutorial system at Harvard for certain sciences, including biochemistry. It was an opportunity to meet weekly with an individual faculty member to discuss papers in biochemistry

and scientific books of current interest. This system gave me insight very early into what the field was all about. It really was a very inspirational opportunity. I met throughout college with a biochemist who was a professor at the medical school and eventually did an honors thesis in his laboratory. That gave me an idea of what it was actually like to do scientific research.

What is the value of lab work for students?

I worked in different laboratories every summer during college, and I think that is a very good thing to do. Courses are fine to give you some knowledge base. But if you think you are going to go into scientific research, you really have to experience it on a day-to-day basis.

What was graduate school like?

By the time I finished college I knew I was really excited about biochemistry and I wanted to continue my studies in that area. When I went to graduate school at Harvard, there wasn't a biochemistry department. But there was a committee on biochemistry, which had faculty members from biology and chemistry. I worked in Frank Westheimer's lab in the chemistry department. About half the students in that lab were interested in physical organic chemistry reaction mechanisms, and the other half were interested in biochemical reaction mechanisms. And so there was a lot of interaction.

What was the subject of your dissertation work?

I worked on the mechanism of acetoacetate decarboxylase. Later on, when I was a postdoctoral fellow, I worked on dehydrogenases, also from a mechanistic point of view.

When did you first become a faculty member?

My first faculty position was at Washington University. In choosing my first independent project, I decided I shouldn't continue on a project that was the same as that of any of my mentors. I really wanted to establish my independence. I thought that since I knew a lot about decarboxylase mechanisms and I knew a fair amount about dehydrogenase reactions, I could work on isocitrate dehydrogenase — which is actually both a dehydrogenase and a decarboxylase. So that was a considered decision, because it involved something that built on what I knew before but at the same time it was also different, and the enzyme was significant because it catalyzed a reaction central to metabolic pathways.

When you said you wanted to establish your independence, what did you mean by that?

I wanted to take a project and demonstrate that I could tackle it in a new way — a way that made use of my training but was distinct from what I had done before. People starting out

their careers have to be aware of the need to differentiate themselves from their teachers.

You started in this field before the explosion of interest in biochemistry and biotechnology. What has changed over the years?

Well, for one thing, I think the general public is more aware of the possibilities for biochemistry, and I think they may be a little more wary about it. The number of scientists has increased enormously, and the research funding has not always kept up with that. But there is also something different. I think most biochemists when I was starting were in academia, because that was the place to carry out basic research. I never really considered working in industry at that time. Now there are many fine opportunities to carry out interesting and even basic research in pharmaceutical and chemical companies.

What is it about a university setting that has appealed to you for so long?

It is very exciting to work with students at various levels — undergraduates, graduate students, and postdoc-

toral fellows. It keeps you continually fresh and alive when you are around people just starting in their careers. They tend to be very enthusiastic about trying new things. So I like that. There is also the freedom to explore your own ideas and to keep working on a project for as long as you find that it is fruitful and you are learning new things, and for as long as you can continue to get funding.

Tell us about your current research interests.

What I do is identify and investigate the role of amino acid residues within active and regulatory sites of enzymes. One of the approaches that we have taken is to chemically modify proteins by designing compounds that are structurally very much like natural coenzymes. But for our newly designed compounds, we also tack on a reactive functional group. This whole approach is known as affinity labeling.

How does affinity labeling work?

Affinity labeling is a way of tagging a specific part of an enzyme by making use of its natural affinity for its substrate. What you do is trick the en-

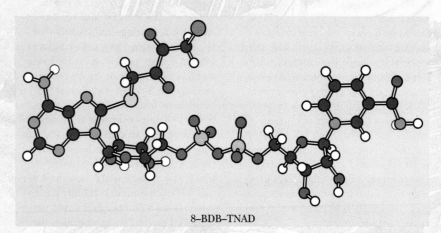

8–BDB–TNAD

The structure of 8-(4-bromo-2,3-dioxobutylthio)-NAD, a compound used by Dr. Colman in her affinity labeling studies of dehydrogenases.

3

zyme into binding with a compound that is very, very similar to a normal metabolite. The compound has great affinity for an enzyme because it looks like the natural substrate. But the compound also has that reactive arm, which will covalently and irreversibly react with the protein somewhere in the vicinity of that binding site. The result is that the compound not only binds to a specific site on the enzyme, but it also permanently labels that site for us to do further studies.

What does tagging involve?

Tagging can be done in a number of ways — with radioactivity, with fluorescence, or with colored molecules or chromophores.

What is the advantage of affinity labeling?

This approach allows us to localize parts of the enzyme, which gives us a handle on identifying either important regulatory sites or active sites. You can follow that up by studying the kinetics and binding properties of the modified protein, and by identifying amino acids at the binding site. Once you have identified an amino acid residue, you can postulate a role for it in binding or catalysis. You can test the postulate by changing the amino acid by site-directed mutagenesis. If it is an amino acid residue that serves as a unique catalyst at an active site, then when you change the residue, the enzyme should be inactivated. Alternatively, when you mutate the amino acid, only the ability of the enzyme to bind substrate may change when you mutate the amino acid. So, by identifying the amino acid by affinity labeling, we have located promising targets for site-directed mutagenesis. This process allows us to refine our ideas about the role the amino acid plays in the enzyme's action.

What would be the ultimate goal of your research?

We are looking for the active sites on enzymes, and the ultimate goal would be to design drugs that could be used to inhibit these enzymes. If there is a crystal structure of the enzyme, then you can examine that part of the molecule in three dimensions and locate regions close to that active site. You can also carry out modeling experiments to find out what inhibitors might fit into that site. We initially were working exclusively with nucleotide binding enzymes, and we looked very extensively at dehydrogenases and kinases. So most of our work was focused on designing nucleotide derivatives. More recently we have been extending this approach to other classes of enzymes, and we have been working very extensively now on glutathione S-transferase.

Why focus on that particular enzyme?

Glutathione S-transferase is an enzyme known to be very important as a detoxification enzyme. From a biochemical perspective, there are two aspects to this. For one thing, glutathione S-transferase may be viewed as a possible defense against carcinogenesis. Obviously, that could be viewed as a very positive effect. But the other thing is that, because it is a detoxifying agent, the enzyme also has the effect of inactivating compounds that are used to treat cancer. For example, when a cancer patient is treated with an alkylating agent in chemotherapy, glutathione S-transferase levels tend to increase in order to detoxify and eliminate the compound that is being used for treatment. In our studies, the aim is to provide basic knowledge about the active sites of these enzymes, which may allow for the design of very specific compounds to inhibit the enzyme. If such compounds could be introduced during chemotherapy, the result might be to prolong the effectiveness of the chemotherapeutic agents.

As someone who is both a professor and researcher, what are the values of teaching and lab work for you?

Well, when you teach, that includes not only lectures but more informal teaching of students in the laboratory. In such work, you continually have to reexamine the things you thought you knew. People ask questions that can be very probing. Also, teaching requires you to learn new material in order to present it to students. And those things may be in areas that you would normally consider outside your own field. Having to read about them in order to present them in lecture, you often encounter new approaches that feed into your own research. I have gotten very good ideas just like that. Teaching also allows you to bring the excitement of research into the classroom. If you are working at the forefront of research, you can tell students about the latest developments, and it gives them a sense of excitement — that this is not a field all wrapped up, that there is much to do. I enjoy this kind of interaction.

Just as your professors did for you, you have helped launch graduate students on their own careers.

I have had students who have taken positions in different areas. One is an assistant professor in a department of pathology, another an associate professor in a department of chemistry. I've had students go to pharmacology departments and one who is in a medicine department. There is quite a number who are working in pharmaceutical companies, and some who are in research institutes. I keep pretty close track of my students and my post-docs. I feel they are my academic "children."

It seems people can go in many different directions with these basic skills in biochemistry.

Yes, I think that's very important to recognize, that this is really an interdisciplinary area. That is more and more true today. One has to know biology and molecular biology and biophysics, and to appreciate NMR spectroscopy, x-ray crystallography, and molecular modeling. And you have to be familiar with experimental, synthetic organic chemistry and enzymology and enzyme kinetics. We are trying to train our students in an interdisciplinary way as much as possible. You can't be an expert in all these areas, but at least you can gain enough knowledge in each area to know what's available — and, if you can't do it all, you know where to look for collaborators.

You have been very widely published, especially in scholarly journals. Why is that important in science?

If you don't write up your work, then it might as well not be done. I mean writing is really part of the science. The major importance of your research is how it will be applied and will influence research projects in other laboratories and in other fields. You do experiments and study in one area, but your hope is that this is going to have influence in many other areas. To do that, it has to get communicated to the world. Journal articles and review articles are part of that process. So is giving seminars in a variety of places. I think that many of the areas we have worked in — particularly affinity labeling — have gained importance in their applications in other laboratories. For example, we have synthesized nucleotide analogues, which have widespread use in many different laboratories. When we describe this work, we usually publish in

detail how one would synthesize these compounds. If we can make these syntheses known and carry out prototypes of reactions, then we hope these procedures will be used in other systems in other laboratories.

Do you believe opportunities have changed for women in biochemistry since you started out?

I think things have changed to some extent, but not entirely. When I first took chemistry at Harvard, I was in a class where there were about 300 students, and 5 were women. Now there is probably an equal balance between men and women in classes.

There are some people who say the whole problem is solved and there is really equal access of women to science and no prejudice at all. I don't think that's quite true. But I get upset when I hear other people say that things are so difficult that women have to work twice as hard and that if a woman wants to get ahead in science she cannot have a family. It is not necessary to make that choice. Women can have both a family and a career in the sciences. I've been married for many years, and I have children who are now grown up. It takes planning, it takes a cooperative family, but women certainly have the abilities and can be very successful.

Is there any strategy you follow in identifying meaningful subjects for research?

One thing is that I have tried to constantly think about how our work may be applied to treating various diseases. I keep track of important diseases with metabolic consequences and of possible medical applications of our research. Although our work may not appear to have relevance today or tomorrow, it may in the long run. I survey the medical literature. And I do

try to collaborate with people in various fields in medicine. We are now involved in a collaboration with a neurologist, for example. The reason for that joint activity is that I think there may be a defect in one of the enzymes we are working on — glutamate dehydrogenase — that could be a factor in certain types of neurological diseases. That is the kind of application that would require me to find a collaborator, because I do not have access to patients, and I would need the medical researcher's expertise to be able to relate a defect that we see in the lab to its effect in the disease process.

Besides teaching and research, do you have any special hobbies or interests?

Well, probably my major hobby is travel. I love to travel, and if I know I am going to a particular place, I will read about it for a whole year before I go. Sometimes my travel is associated with meetings. I gave a talk in Pakistan a year ago and was able to travel to many places in Pakistan. More recently, I was an invited speaker at a meeting of the Federation of Asian and Oceanic Biochemists, which was held in Taiwan. That was a meeting of biochemists from the whole region, and I got to meet people from such places as Thailand and Japan. These were people whom I had not met before but whose work I have followed. Since I was in that part of the world, I also went on to Viet Nam, Laos, and Cambodia. That was very exciting. I had always wanted to visit Angkor Wat.

Interview

Beverly March

Carl Djerassi

arl Djerassi, Professor of Chemistry at Stanford University, is a man of many achievements in science and in letters. Born in Vienna in 1923, he was forced to leave Europe in 1939 in the face of the Nazi occupation. Although he had not completed high school, he was allowed to enroll at Newark Junior College in New Jersey. He also attended Tarkio College in Missouri and then Kenyon College, where he graduated *summa cum laude* in 1942.

Still short of his nineteenth birthday, Djerassi joined the CIBA Corporation, where he and Charles Huttrer synthesized pyribenzamine, one of the first two antihistamines (drugs that act to control allergies). He left CIBA for the University of Wisconsin, where he received his Ph.D. in 1945 at the age of 22. For the next seven years he worked first at CIBA and then at Syntex, S.A., a fledgling pharmaceutical corporation then located in Mexico City. There he was part of the research team that first synthesized cortisone, an important hormone, and later synthesized norethindrone, the first oral contraceptive and one that is still widely used all over the world.

In 1952 Dr. Djerassi moved to Wayne University (now Wayne State) in Detroit, as associate professor, and in 1959 he accepted a professorship at Stanford University, where he has remained ever since. Both at Wayne and at Stanford he maintained an association with Syntex, serving as president of Syntex Research from 1968 to 1972. In 1968 he helped to found the Zoecon Corporation, of which he also became president, and which pioneered the development of hormonal methods of insect control that do not rely on conventional insecticides.

Dr. Djerassi has always been involved in steroid chemistry, but his early interest in synthesis gave way to an interest in the equally important area of the structural determination of naturally occurring steroids, of which there are many thousands. When he began his career, determining the structure of a single, naturally occurring steroid was a difficult process that often took years.

In the 1950s, instrumental methods of structure determination slowly began to replace the laborious, older chemical methods. Dr. Djerassi was one of the pioneers in developing and using techniques such as optical rotatory dispersion, circular dichroism, and mass spectrometry to determine the structures of steroids and other organic compounds. Because of the work of Dr. Djerassi and other pioneers, most unknown structures now can be determined in a few days or less.

Dr. Djerassi has had a prolific career, publishing more than 1000 research papers and seven books dealing with steroids, alkaloids, antibiotics, lipids, and terpenoids, as well as with chemical applications of computer artificial intelligence. His many research awards include the National Medal of Science (1973), the National Medal of Technology (1991), the first Wolf Prize in Chemistry (1978), the Perkin Medal of the Society for Chemical Industry (1975), and the Priestley medal, the highest award given by the American Chemical Society (1992). In 1978 he was inducted into the National Inventors' Hall of Fame.

In recent years Dr. Djerassi, without abandoning his interest in chemistry, has turned to literature. He has published a collection of short stories, two novels (*Cantor's Dilemma* and *The Bourlaxi Gambit*), a collection of poems, and two autobiographies, one titled *Steroids Made It Possible* and the other, intended for a more general audience, called *The Pill, Pygmy Chimps, and Degas's Horse.* Under the auspices of the Djerassi Foundation, he has established an artists' colony near Woodside, California, that provides residences and studio space for visual, literary, choreographic, and musical artists.

Why did you choose a career in chemistry?

I really didn't choose it; it sort of happened to me. I was the child of two physicians, and it was always assumed that I would go to medical school and become a practicing physician. When I arrived in this country at the age of 16 I had taken no chemistry and had not even graduated from high school. Fortunately, I got straight into a junior college in New Jersey and began a pre-med curriculum that included first-year chemistry. I had a first-rate teacher there, named Nathan Washton, who got me interested in chemistry. After one semester at Tarkio College in Missouri (the alma mater of Wallace Carothers, the inventor of nylon), I went to Kenyon College, a small (at that time) college in Ohio, that had only two faculty members in chemistry. Chemistry classes were very small, but I got a first-class education, and that is where I decided to become a chemist rather than go to medical school.

Could you tell us about your work in developing antihistamines?

When I graduated from Kenyon I needed to earn some money, so I got a job as a junior chemist at CIBA Pharmaceutical Corporation in New Jersey. At that time the company became interested in antihistamines, and I was one of only two chemists working on this project. So, despite my youth, I was involved in the synthesis of one of the first two antihistamines produced in this country, pyribenzamine. This compound, which was synthesized during the first year after I graduated from college, turned out to make a significant contribution in the treatment of allergies.

What got you interested in the chemistry of steroids?

While working at CIBA, I began taking classes at NYU and Brooklyn Polytechnic Institute at night, to get an advanced degree, but commuting from New Jersey after a day of work was murderous. After a year of night school, I decided I would go to graduate school full-time. CIBA was involved in steroid projects, and even though I was working on antihistamines, I started reading books on steroids, especially Louis Fieser's *Natural Products Related to Phenanthrene,* a superb book, which turned me on to steroid chemistry. When I went full-time to the University of Wisconsin, I was prepared to work in this field. Fortunately, Wisconsin had two young assistant professors in this area, and I did my Ph.D. with one of them, A. L. Wilds, on the conversion of androgens to estrogens, which at that time was a tough problem.

How did you get the idea to try to synthesize an oral contraceptive?

We did not set out with that objective. Our goal was to develop an orally active progestational hormone — in other words, a compound that would mimic the biological properties of progesterone. At that time progesterone was clinically used for menstrual disorders and infertility, but there were ideas about using it as a contraceptive, because it is progesterone that naturally stops further ovulation after an ovum is fertilized. However, progesterone itself is not active by mouth, and daily injections would be needed. By this time I was associate director of chemical research at Syntex Corporation in Mexico City. By combining ideas discovered by previous investigators, we set out to synthesize a steroid that would not only be active by mouth but would also have enhanced progestational activity. This compound was 19-nor-17α-ethynyltestosterone (norethindrone), whose synthesis we completed on October 15, 1951. It was first tested for menstrual disorders and fertility problems and then as an oral contraceptive. Forty years after its synthesis it is still the active ingredient of about a third of all the oral contraceptives used throughout the world.

How do you feel about the social impact of your contraceptive work?

If I could do it over again, there is no question that I would proceed to do so. By now about 13 to 14 million women in the U.S. and 50 to 60 million in the world use the pill, making it the most widely used method of reversible birth control. Population growth is probably the biggest problem facing us in the world, assuming that the possibility of nuclear warfare is now greatly diminished, and the widespread use of oral contraceptives helps in controlling that. I think the development of these contraceptives is one of the most important contributions that chemistry has made to society. New drugs cure diseases of individuals during their lifetimes, but the use of contraceptives has implications for generations, because if you do not control the production of offspring, you do not control future generations. Furthermore, these compounds have had an enormous impact on women, empowering them to be in control of their own fertility.

Do you think there will be further research in oral contraception — for example, the development of a pill for men?

I think there will be very little fundamental new research in that area over the next couple of decades, and a pill for men is completely out of the ques-

7

tion for the next 15 to 20 years. The reason is that the pharmaceutical industry has turned its back on this field for a number of very complex reasons, not only in the U.S. but in other industrialized countries as well. The only really new approach in chemical birth control in the last 25 years is the French "morning-after pill," RU486 — at present used almost exclusively as an early abortifacient, although it can also be employed as a method of birth control.

Tell us about your work on insect control.

Conceptually there was a relationship between our work on oral contraceptives and insect control. In a way, you could say that steroid oral contraceptives were true biorational methods of human birth control, since progesterone — our conceptional lead compound — is really nature's contraceptive. That was a model on which insect control could be based. At this time (the late 1960s) I was in charge of research at Syntex in addition to being a professor at Stanford University. Governments and the public realized that conventional methods of insect control — largely spraying with chlorinated hydrocarbons such as DDT — were damaging the environment. DDT and similar compounds were being banned, and a new approach was needed. We formed a new company called Zoecon to try to synthesize insect-controlling steroids. In the 1960s, a juvenile hormone, based on a sesquiterpene skeleton, had been discovered, and we decided to focus on it. Insects pass through a juvenile stage controlled by the juvenile hormone, whose production is later shut off by another hormone so that the insect can then mature. Our biorational approach was to synthesize an artificial juvenile hormone that would continue to be applied to immature insects, so that the insect would never reach the stage at which it could reproduce. This turned out to be a new biorational approach to control-

ling mosquitoes, fleas, cockroaches, and other insects that do their damage as adults, and was approved by the Environmental Protection Agency for public use.

You have been a leader in both the synthetic and structure determination areas of organic chemistry. Which of these activities gave you more satisfaction?

In my early days — the 50s and 60s — undoubtedly structure determination was more satisfying than synthesis, especially because I was involved in the development of new physical methods for identifying chemical structures. However, these methods have now made structural determination much more routine and automatic. For example, mass spectrometry enables us to identify compounds on a micro scale, which was unheard of before.

When you embarked on a research project, what proportions of your driving force were provided by pure intellectual curiosity, consideration of potential benefits to society, and material rewards?

In my own work, material rewards never played any initial role, because they always came much later. To a large extent it was initially pure intellectual curiosity, but very soon I started to look for the potential benefits to society. My interest has always been in the biological areas, and I was never interested in war-related research. I never had the slightest question about the societal appropriateness of the work I was doing.

What is your opinion about the interplay between academia and industry in developing new inventions, including chemical inventions?

I have always been connected simultaneously with a university and with

chemical companies. This made me a much better professor, because I became aware of the many steps needed to take a laboratory discovery up to practical realization. Conversely, I was a much better industrial research director because of what I learned in the university. For example, should scientists be concerned with societal implications of their work? Obviously, yes, but it is not easy for a scientist who stays only in the ivory tower. You ought to have a responsibility for taking your work a step further. There should, however, be guidelines to prevent actual or potential conflicts of interest.

Do you think that scientists should be advocates for social or political positions?

Scientists should be advocates, in that every intelligent person should be, provided he or she is well informed. Just because you are a scientist does not mean that you are well informed on any particular issue, but when scientists deal with their own specialties, of course they should. Also, scientists should be interested in the societal implications of their work, just as nonscientists should be informed about certain areas of science, simply because they have such an enormous impact on their daily lives.

You have trained many students in chemical research. On the whole, have they made many contributions to society after they left your laboratories?

Yes. I've trained somewhere between three and four hundred graduate students and postdoctoral fellows in universities, and many others in industry as well. Training these students is probably the biggest professional contribution that one makes. These are very intimate relationships, comparable to those between parents and children. The mentor becomes a role model for the student.

Biochemistry and the Organization of Cells

Scanning electron microscope image of a human breast cancer cell.

Complex living organisms originate with simple light elements. Carbon, hydrogen, and oxygen combine to make up many biomolecules, such as carbohydrates. The addition of nitrogen makes possible the amino acids that combine to form proteins. In turn, added phosphorus provides the ingredients for making DNA. Thus, there occurs a "building up" from atoms to small molecular units to large biomolecules such as proteins and the nucleic acids, DNA and RNA. A collection of interacting molecules, encased in a suitable membrane, becomes a cell—the basic unit of life. Every cell has a central core of the hereditary material DNA, which contains the information to make the complete organism. In one-celled prokaryotes, such as bacteria, the nuclear material is loosely organized. Plant and animal cells (called eukaryotes) are more highly organized, with the nucleus enclosed in a separate membrane. Compartments specialized for particular functions are characteristic of eukaryotic cells. In plants, photosynthesis takes place in chloroplasts: light energy is converted to chemical energy and stored as carbohydrates. In the mitochondria of animal cells, the stored energy of carbohydrates is recovered through respiration, a process in which carbohydrates are oxidized to carbon dioxide and water.

1.1
SOME BASIC THEMES

Living organisms, and even the individual cells of which they are composed, are enormously complex and diverse. Nevertheless, certain unifying features are common to all things that live. All make use of the same types of *biomolecules,* and all use energy. As a result, organisms can be studied via the methods of chemistry and physics. The belief in "vital forces" (forces thought to exist only in living organisms) held by 19th-century biologists has long since given way to awareness of an underlying unity throughout the natural world.

Disciplines that appear to be unrelated to biochemistry can provide answers to important biochemical questions. An example is the discovery, made by physicists in the early 20th century, that x-rays can be diffracted by crystals. The resultant experimental method of x-ray diffraction led to the elucidation of the three-dimensional structures of molecules as complex as proteins and nucleic acids. Biochemistry is a field that draws on many disciplines, and this multidisciplinary nature allows it to use results from many sciences to answer questions about the *molecular nature of life processes.* Enormously important applications of this kind of knowledge are made in medically related fields; an understanding of health and disease at the molecular level leads to more effective treatment of illnesses of all sorts.

The activities within a cell are analogous to the transportation system of a city. The cars, buses, and taxis correspond to the molecules involved in reactions (or series of reactions) within a cell. The routes traveled by these vehicles are likewise comparable to the reactions that occur in the life of the cell. Note particularly that many vehicles travel more than one route—for instance, cars and taxis can go anywhere—whereas other, more specialized modes of transportation such as subways and streetcars are confined to single paths. Similarly, some molecules play multiple roles, whereas others take part only in specific series of reactions. Also, *all the routes operate simultaneously;* we shall see that this is true of the many reactions within a cell.

To continue the comparison, the transportation system of a large city has more kinds of transportation than a smaller one. Whereas a small city may have only cars, buses, and taxis, a large city may have all of these plus others, such as streetcars or subways. Analogously, some reactions are found in all cells and others are found only in specific kinds of cells. Also, more structural features are found in the larger, more complex cells of larger organisms than in the simpler cells of organisms such as bacteria.

An inevitable consequence of this complexity is the large quantity of terminology needed to describe it; learning considerable new vocabulary is part of the study of biochemistry. You will also see many cross-references in this book, a reflection of the many connections among the processes that take place in the cell.

The fundamental similarity of cells of all types makes it interesting and illuminating to speculate on the origins of life. Even the structures of comparatively small biomolecules consist of several parts. Large biomolecules such as proteins and nucleic acids have complex structures, and living cells are enormously more complex. Even so, *both molecules and cells must have arisen ultimately from very simple molecules* such as water, methane, carbon dioxide, ammonia, nitrogen, and hydrogen (Figure 1.1). In turn, these simple molecules must have arisen from atoms. The way in which the universe itself, and the atoms of which it is composed, came to be is a topic of great interest to astrophysicists as well as other scientists.

We have raised a number of issues here, and they touch on many scientific disciplines. Narrowed down, however, they will prove useful for the purposes of this text. A brief look at certain theories of the origins of life will identify some central questions in biochemistry as well as the connections between biochemistry and other sciences.

1.2
ORIGINS OF LIFE

The Earth and Its Age

To date, we are aware of only one planet that supports life: our own. The earth and its waters are universally understood to be the source and mainstay of life as we know it. A natural first question is how the earth, along with the universe of which it is a part, came to be.

Currently, the most widely accepted cosmological theory for the origin of the universe is the *big bang,* a cataclysmic explosion. According to big bang cosmology, all the matter in the universe was originally confined to a comparatively small volume of space. As a result of a tremendous explosion, this "primordial fireball" started to expand with great force. Immediately after the big bang the universe was extremely hot, on the order of 15 billion (15×10^9) K. (Kelvin temperatures are written without a degree symbol.) The average temperature of the universe has been decreasing ever since as a result of expansion, and the lower temperatures have permitted the formation of stars and planets.

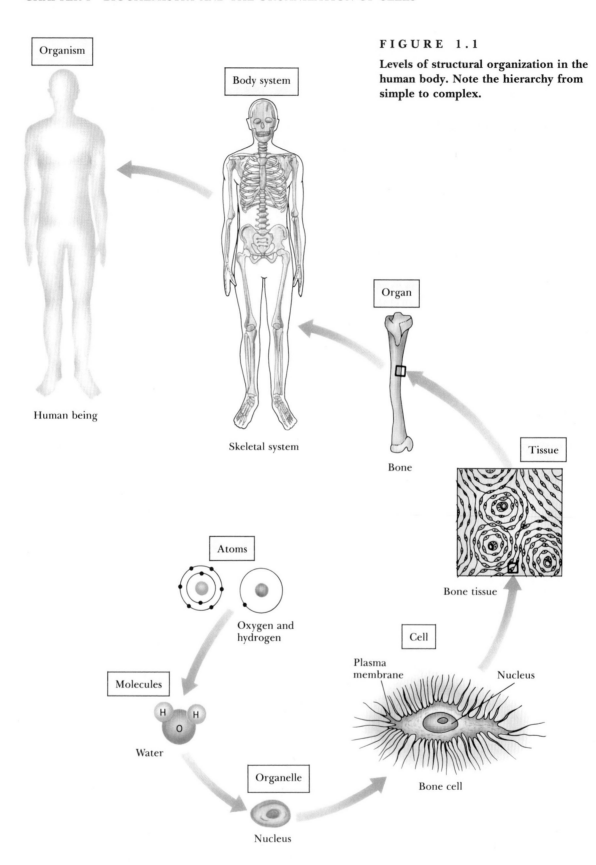

FIGURE 1.1

Levels of structural organization in the human body. Note the hierarchy from simple to complex.

Organism

Human being

Body system

Skeletal system

Organ

Bone

Tissue

Bone tissue

Atoms

Oxygen and hydrogen

Molecules

Water

Cell

Plasma membrane

Nucleus

Bone cell

Organelle

Nucleus

The Hubble space telescope was designed to provide images about seven times clearer than those obtained with ground-based telescopes. It is expected to provide information about precursors of biomolecules in outer space.

In its earliest stages, the universe had a fairly simple composition. Hydrogen, helium, and some lithium were present, having been formed in the original big bang explosion. The rest of the chemical elements are thought to have been formed in three ways: (1) by the thermonuclear reactions that normally take place in stars, (2) in the explosions of stars, and (3) by the action of cosmic rays outside the stars since the formation of the galaxy. The process by which the elements are formed in stars is a topic of interest to chemists as well as to astrophysicists. For our purposes it is noteworthy that the most abundant isotopes of biologically important elements such as carbon, oxygen, nitrogen, phosphorus, and sulfur have *particularly stable nuclei.* These elements were produced by nuclear reactions in first-generation stars, the original stars produced after the beginning of

T A B L E 1 . 1 **Relative Abundance of Important Elements***

ELEMENT	ABUNDANCE IN ORGANISMS	ABUNDANCE IN UNIVERSE
Hydrogen	80–250	10,000,000
Carbon	1000	1,000
Nitrogen	60–300	1,600
Oxygen	500–800	5,000
Sodium	10–20	12
Magnesium	2–8	200
Phosphorus	8–50	3
Sulfur	4–20	80
Potassium	6–40	0.6
Calcium	25–50	10
Manganese	0.25–0.8	1.6
Iron	0.25–0.8	100
Zinc	0.1–0.4	0.12

*Each abundance is given as the number of atoms relative to a thousand atoms of carbon.

FIGURE 1.2

Conditions on early earth would have been inhospitable for most of today's life forms. Very little or no free oxygen (O_2) existed. Volcanoes erupted, spewing gases, and violent thunderstorms produced torrential rainfall that covered the earth. The green arrow indicates the formation of biomolecules from simple precursors.

the universe (Table 1.1). Many first-generation stars were destroyed by explosions called *supernovas,* and their stellar material was recycled to produce second-generation stars such as our own sun, along with our solar system.

Radioactive dating, which makes use of the decay of unstable nuclei, indicates that the age of the earth (and the rest of the solar system) is 4 to 5 billion ($4-5 \times 10^9$) years. The atmosphere of the early earth was very different from the one we live in, and probably went through several stages before reaching its current composition. The most important difference is that, according to most theories of the origins of the earth, very little or no free oxygen (O_2) existed in the early stages (Figure 1.2). The early earth was

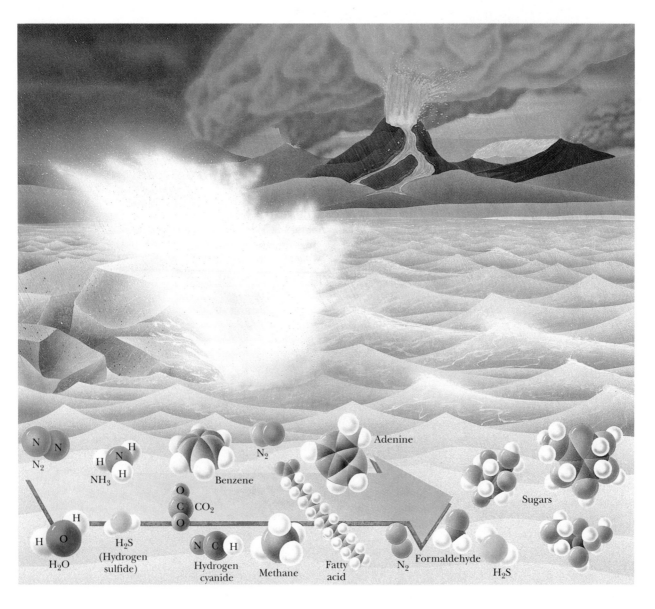

constantly irradiated with ultraviolet light from the sun, since there was no ozone (O_3) layer to block it. Under these conditions, the chemical reactions that produced simple biomolecules took place.

The gases usually postulated to have been present in the atmosphere of the early earth include NH_3, H_2S, CO, CO_2, CH_4, N_2, H_2, and (in both liquid and vapor forms) H_2O. However, there is no universal agreement on the relative amounts of these components from which biomolecules ultimately arose. Many of the earlier theories of the origin of life postulated CH_4 as the carbon source, but more recent studies have shown that appreciable amounts of CO_2 must have existed in the atmosphere at least 3.8 billion (3.8×10^9) years ago. This conclusion is based on geological evidence: the earliest known rocks are 3.8 billion years old, and they are carbonates, which arise from CO_2. Any NH_3 originally present must have dissolved in the oceans, leaving N_2 in the atmosphere as the nitrogen source for formation of proteins and nucleic acids.

Biomolecules

Experiments have been performed in which the simple compounds of the early atmosphere were allowed to react under the varied sets of conditions that might have been present on the early earth. The results of such experiments indicate that these simple compounds react *abiotically* or, as the word indicates (*a,* ''not,'' and *bios,* ''life''), in the absence of life, to give rise to biologically important compounds such as the components of proteins and nucleic acids. According to one theory, reactions such as these took place in the earth's early oceans; other researchers postulate that such reactions occurred on the surfaces of clay particles that were present on the early earth. It is certainly true that mineral substances similar to clay can serve as catalysts in many types of reactions. Both theories have their proponents, and more research will be needed to answer the many questions that remain.

Living cells are assemblages that include very large molecules, such as proteins, nucleic acids, and polysaccharides; these are larger by many powers of ten than the smaller molecules from which they are built. Hundreds or thousands of these smaller molecules, or **monomers,** can be linked to produce macromolecules, which are also called **polymers.** In present-day cells, amino acids (the monomers) combine by polymerization to give **proteins,** and nucleotides (also monomers) combine to give **nucleic acids;** the polymerization of sugar monomers produces polysaccharides. Polymerization experiments with amino acids carried out under early earth conditions have produced *proteinoids,* which are proteinlike polymers. Similar experiments have been done on the abiotic polymerization of nucleotides and sugars.

A polymer in which all the monomers have the same chemical identity is called a *homopolymer.* Since all the monomers are the same, the order in which they are linked does not affect the nature of the polymer. In contrast, different types of monomers can be linked to form a *heteropolymer,* in which the order of monomer units does not necessarily match that of another polymer of the same size and overall composition. Thus, in this case the order in which the monomers are linked can make an important difference

FIGURE 1.3

(a) A homopolymer: all monomers are the same.
(b) Two heteropolymers with monomers linked in different orders. It is possible for two heteropolymers to have the same number and kind of monomer units and to have different properties, if the linking orders differ. These two heteropolymers are different molecules.

(a)
$$X—X—X—X—X—X—X—X—X—X \cdots X$$
$$1 \quad 2 \quad 3 \quad 4 \quad 5 \quad 6 \quad 7 \quad 8 \quad 9 \quad 10 \cdots n$$

X represents a monomer

(b)
$$X—Z—V—Y—W—Z—W—X—Y—V \cdots Z$$
$$1 \quad 2 \quad 3 \quad 4 \quad 5 \quad 6 \quad 7 \quad 8 \quad 9 \quad 10 \cdots n$$

$$V—W—X—Y—Z—X—V—Y—W—Z \cdots X$$
$$1 \quad 2 \quad 3 \quad 4 \quad 5 \quad 6 \quad 7 \quad 8 \quad 9 \quad 10 \cdots n$$

V, W, X, Y, Z represent different monomers

in the properties of the molecule (Figure 1.3). For example, the several types of amino acids and nucleotides can easily be distinguished from one another. The sequence of amino acids determines the properties of proteins. The genetic code lies in the sequence of monomer units in nucleic acids. In polysaccharides, however, the order of monomers does not usually have an extremely important effect on the properties of the polymer, nor does the order of the monomers carry any genetic information. (Other aspects of the *linkage* between monomers are important in polysaccharides, as we shall see when we discuss carbohydrates in Chapter 10).

The ribbon structure of a protein, formed by the linking of amino acids with covalent bonds. The order of amino acids determines the three-dimensional folding of the protein.

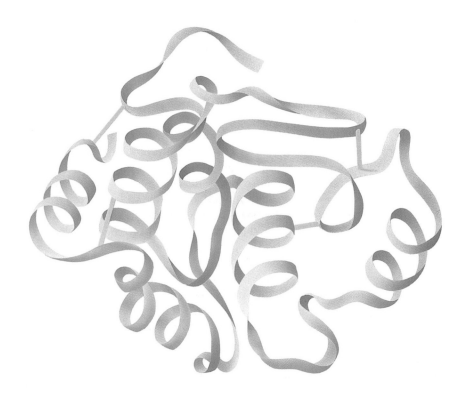

The effect of monomer sequence on the properties of polymers can be illustrated by another example. Proteinoids are artificially synthesized polymers of amino acids, and their properties can be compared with those of true proteins. Although some evidence exists that the order of amino acids in artificially synthesized proteinoids is not completely random — a certain order is preferred — there is no definite amino acid sequence. In contrast, *a well-established, unique amino acid sequence exists for each protein produced by present-day cells.* The enormous variety of possible proteins results from the fact that they are polymers and therefore can have any possible amino acid at each position in their structure. Proteinoids and the class of proteins called *enzymes* display **catalytic activity,** which means that they increase the rates of chemical reactions compared to uncatalyzed reactions. As might be expected, enzymes are far more effective than proteinoids at increasing reaction rates.

The specific amino acids present and their sequences ultimately determine the properties of all types of proteins, including enzymes. The order in which individual amino acids are linked determines the three-dimensional structure of the protein, called its *conformation,* which in turn determines its function and mode of action. One of the most important functions of proteins is catalysis, and the catalytic effectiveness of a given enzyme depends on its amino acid sequence.

In present-day cells, the sequence of amino acids in proteins is determined by the sequence of nucleotides in nucleic acids. The process by which this genetic information is transferred is very complex. *DNA,* one of the nucleic acids, serves as the coding material. The **genetic code** is the means by which the information for the structure and function of all living things is passed from one generation to the next. The workings of the genetic code are no longer completely mysterious, but they are far from completely understood. Theories of the origins of life have the greatest difficulty explaining the development of a coding system, and new insights in this area could throw some light on the present-day genetic code.

Molecules to Cells

One of the most puzzling questions about the origin of life is which came first — catalytic activity (associated with proteins) or coding (associated with nucleic acids). Another way of saying this is: Which came first — proteins or nucleic acids? There are proponents of both viewpoints, but both require modification in light of the discovery that *RNA,* another nucleic acid, is capable of catalyzing its own further processing. Until this discovery, catalytic activity was associated exclusively with proteins. RNA, rather than DNA, is now considered by many scientists to have been the original coding material, and it still serves this function in some viruses. The idea of catalysis and coding both occurring in one molecule will certainly provide a point of departure for more research on the origins of life. (See the article by Cech listed in the bibliography at the end of this chapter.)

According to the theory that proteins came first, aggregates of proteinoids formed on the early earth, probably in the oceans or at their edges. These aggregates took up other abiotically produced precursors of biomolecules to become *protocells,* the precursors of true cells. Model systems for

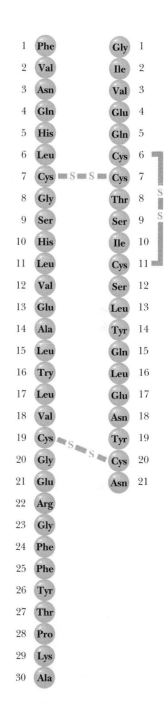

The primary structure of insulin, showing the amino acid sequence. A protein with the same number and kinds of amino acids in a different order is a different protein.

FIGURE 1.4

Proteinoid microspheres, a type of protobiont, are tiny spheres (1 – 2 μm in diameter) that exhibit some of the properties of life.

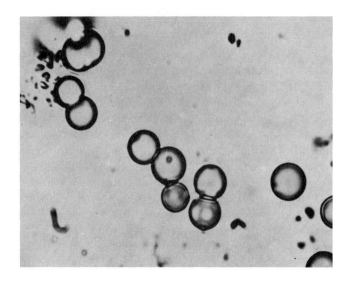

protocells have been devised by several researchers. In one model, artificially synthesized proteinoids are induced to aggregate, forming structures called **microspheres.** *Proteinoid* microspheres are spherical in shape, as the name implies, and in a given sample they are approximately uniform in diameter (Figure 1.4). Such microspheres are certainly not cells, but they provide a model for protocells. Microspheres prepared from proteinoids with catalytic activity exhibit the same catalytic activity as the proteinoids. Furthermore, it is possible to construct such aggregates with more than one type of catalytic activity as a model for primitive cells.

In this proteins-first theory, the rise of a coding system is a crucial point but the one with the most tenuous support from experimental models. It is more difficult to produce models for nucleic acids, especially for DNA, under abiotic conditions than it is to produce proteinoids. Some protonucleic acids have been produced under abiotic conditions, however. The incorporation of protonucleic acids into microspheres to afford protocell models with an effective coding system continues to be a major challenge for researchers in this field.

According to the theory that nucleic acids arose before proteins, the appearance of a form of RNA capable of coding for its own replication was the pivotal point in the origin of life. Experiments have been performed under a variety of conditions to test whether RNA can be polymerized from monomers in the absence of the factors usually required in the reaction mixture, such factors including a preexisting RNA to be copied or an enzyme to catalyze the process. (These experiments were carried out before the discovery, as mentioned, that some currently existing RNAs can catalyze their own processing once they have been formed by other means.) It has been shown that RNA can be produced in the absence of *one* of these factors, but replication in the absence of *both* factors has yet to be demonstrated. "Naked" RNA — that is, RNA not associated with any other substance — does not exist independently. Even a virus, the simplest entity capable of replicating itself, consists of a nucleic acid core, which may be

RNA or DNA, and a protein coat. Many unanswered questions remain about the role of RNA in the origin of life, but clearly the role must be important.

Recently attempts have been made to combine these two approaches into a *double-origin theory*. According to this line of thought, the development of metabolic pathways and the development of a coding system came about separately, and the combination of the two produced life as we know it. The rise of aggregates of molecules capable of catalyzing metabolic reactions was one origin of life, and the rise of a nucleic acid–based coding system was another origin. A double-origin theory acknowledges the importance of coding and also answers the objection based on instability of unprotected RNA.

The theory that life began on clay particles, mentioned earlier, is a form of double-origin theory. According to this point of view, coding arose first, but the coding material was the surface of naturally occurring clay. The pattern of ions on the clay surface is thought to have served as the code, and the process of crystal growth to have been responsible for replication. Simple molecules, and then protein enzymes, arose on the clay surface, eventually giving rise to aggregates similar to the microspheres already described. At some later date, the rise of RNA provided a far more efficient coding system than clay, and RNA-based cells replaced clay-based cells. This scenario assumes that time is not a limiting factor in the process.

At this writing none of the theories of the origin of life is definitely established, and none is definitely disproved. The topic is still under active investigation.

1.3
PROKARYOTES AND EUKARYOTES: DIFFERENCES IN LEVELS OF ORGANIZATION

All cells, prokaryotic and eukaryotic, contain DNA. The total DNA of a cell is called the **genome.** Individual units of heredity, controlling individual traits, are **genes.**

The earliest true cells that evolved must have been very simple, having the minimum apparatus necessary for life processes. The types of organisms living today that probably most resemble the earliest cells are the **prokaryotes.** This word, of Greek derivation (*karyon,* "kernel, nut"), literally means "before the nucleus." Prokaryotes include *bacteria* and *cyanobacteria.* (Cyanobacteria were formerly called blue-green algae; as the newer name indicates, they are more closely related to bacteria.) Prokaryotes are single-celled organisms, but groups of them can exist in association, forming colonies with some differentiation of cellular functions.

Eukaryotes are more complex organisms and can be multicellular or single-celled. The word "eukaryote" means "true nucleus." A well-defined nucleus, set off from the rest of the cell by a membrane, is one of the chief features distinguishing a eukaryote from a prokaryote. A growing body of fossil evidence indicates that eukaryotes evolved from prokaryotes about 1.5 billion (1.5×10^9) years ago, about 2 billion years after life first appeared on earth. Examples of single-celled eukaryotes include yeasts and *Paramecium* (an organism frequently discussed in beginning biology courses); all multi-cellular organisms (such as animals and plants) are eukaryotes. As might be expected, eukaryotic cells are more complex and usually much larger than prokaryotic cells. The diameter of a typical prokaryotic cell is on the order of 1 to 3 μm ($1-3 \times 10^{-6}$ m), whereas that of a typical eukaryotic cell is about 10 to 100 μm. The distinction between prokaryotes and eukaryotes is so basic that it is now a key point in the classification of living organisms; it is far more important than the distinction between plants and animals.

The main difference between prokaryotic and eukaryotic cells is the existence of organelles, especially the nucleus, in eukaryotes. An **organelle** is a part of the cell that has a distinct function; it is surrounded by its own membrane within the cell. In contrast, the structure of a prokaryotic cell is relatively simple, lacking membrane-bounded organelles. Like a eukaryotic cell, however, a prokaryotic cell has a cell membrane, or plasma membrane, separating it from the outside world; this is the only membrane found in the prokaryotic cell. In both prokaryotes and eukaryotes, the cell membrane consists of a double layer (bilayer) of lipid molecules with a variety of proteins embedded in it.

Organelles have specific functions. A typical eukaryotic cell has a *nucleus* with a nuclear membrane. *Mitochondria* (respiratory organelles) and an internal membrane system known as the *endoplasmic reticulum* are also common to all eukaryotic cells. Oxidation reactions take place in eukaryotic mitochondria. In prokaryotes, similar reactions occur on the plasma membrane. *Ribosomes* (particles consisting of RNA and protein), which are the sites of protein synthesis in all living organisms, are frequently bound to the endoplasmic reticulum in eukaryotes. In prokaryotes, ribosomes are found free in the cytosol. A distinction can be made between the cytoplasm and the cytosol. *Cytoplasm* refers to the portion of the cell outside the nucleus, and the *cytosol* is the soluble portion of the cell that lies outside the membrane-bounded organelles. *Chloroplasts,* photosynthetic organelles, are found in plant cells in which photosynthesis takes place. In prokaryotes that are capable of photosynthesis, the reactions take place in laminar arrays called *chromatophores* rather than in chloroplasts.

Table 1.2 summarizes the basic differences between the two cell types.

T A B L E 1 . 2   A Comparison of Prokaryotes and Eukaryotes

ORGANELLE	PROKARYOTES	EUKARYOTES
Nucleus	No definite nucleus; DNA present but not separate from rest of cell	Present
Cell membrane (plasma membrane)	Present	Present
Mitochondria	None; enzymes for oxidation reactions situated on plasma membrane	Present
Endoplasmic reticulum	None	Present
Ribosomes	Present	Present
Chloroplasts	None; photosynthesis localized in chromatophores	Present in green plants

1.4
PROKARYOTIC CELLS

Although no well-defined nucleus is present in prokaryotes, the DNA of the cell is concentrated in one region called the **nuclear region.** This part of the cell directs the workings of the cell very much as the eukaryotic nucleus does. The DNA of prokaryotes is not complexed with proteins as is the DNA of eukaryotes. In general, there is only.a single closed circular molecule of DNA in a prokaryote. This circle of DNA, which is the genome, is attached to the cell membrane. Before a prokaryotic cell divides, the DNA replicates itself and both DNA circles are bound to the plasma membrane. The cell then divides, and each of the two daughter cells receives one copy of the DNA (Figure 1.5).

In a prokaryotic cell the cytosol (the portion of the cell outside the nuclear region) frequently has a slightly granular appearance because of the presence of **ribosomes.** These consist of RNA and protein and thus are also called *ribonucleoprotein particles;* they are the sites of protein synthesis in all organisms. The presence of ribosomes is the main feature of prokaryotic cytosol. (Membrane-bounded organelles, characteristic of eukaryotes, are not found in prokaryotes.)

Every cell is separated from the outside world by a **cell membrane,** or plasma membrane, an assemblage of lipid molecules and proteins. In addition to the cell membrane and external to it, a prokaryotic bacterial cell has a **cell wall** made up mostly of polysaccharide material, a feature it shares with eukaryotic plant cells. The chemical natures of prokaryotic and eukaryotic cell walls differ somewhat, but a common feature is that the polymerization of sugars produces the polysaccharides found in both. Since the cell wall is made up of rigid material, it presumably serves as protection for the cell.

FIGURE 1.5

A typical prokaryote: the bacterium *Escherichia coli* (magnified 100,000×). Division into two cells is nearly complete.

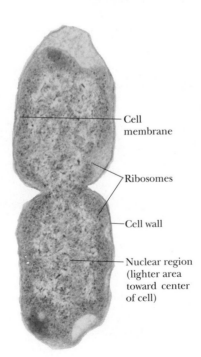

Cell membrane

Ribosomes

Cell wall

Nuclear region (lighter area toward center of cell)

1.5
EUKARYOTIC CELLS

Multicellular plants and animals are both eukaryotes, but obvious differences exist between them. These differences are reflected on the cellular level. Plant cells, like bacteria, have cell walls. A plant cell wall is mostly the polysaccharide cellulose, giving the cell its shape and mechanical stability. Chloroplasts, the photosynthetic organelles, are found in green plants.

Animal cells have neither cell walls nor chloroplasts. Figure 1.6 shows some of the important differences among a typical plant cell, a typical animal cell, and a prokaryote.

FIGURE 1.6

A comparison of (a) a typical animal cell, (b) a typical plant cell, and (c) a prokaryotic cell.

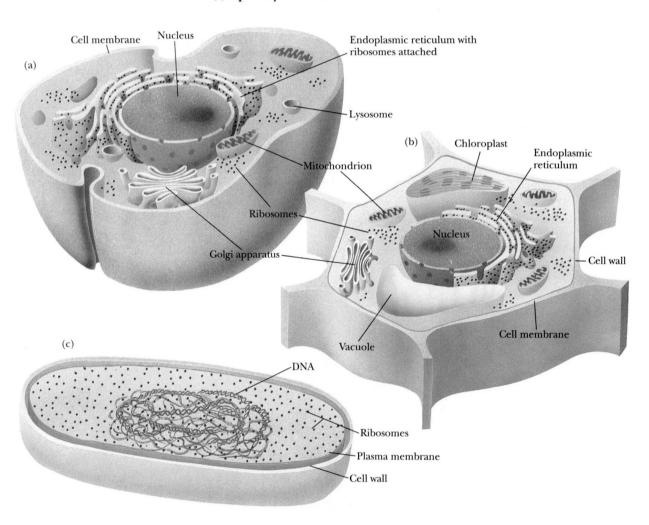

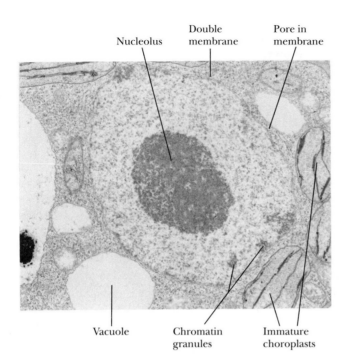

Nucleolus   Double membrane   Pore in membrane

Vacuole   Chromatin granules   Immature choroplasts

FIGURE 1.7

The nucleus of a tobacco leaf cell (magnified 15,000×).

Important Organelles

The **nucleus** is perhaps the most important eukaryotic organelle. A typical nucleus exhibits several important structural features (Figure 1.7). It is bounded by a *nuclear double membrane*. One of its prominent features is the **nucleolus,** which is rich in RNA. The RNA of a cell (with the exception of the small amount produced in such organelles as mitochondria and chloroplasts) is synthesized on a DNA template in the nucleolus for export to the cytoplasm through pores in the nuclear membrane. This RNA is ultimately destined for the ribosomes. Also visible in the nucleus, frequently near the nuclear membrane, is **chromatin,** an aggregate of DNA and protein. The eukaryotic genome (its total cellular DNA) is duplicated before cell division takes place, as in prokaryotes. In eukaryotes, both copies of DNA, which are to be distributed between the daughter cells, are associated with protein. When a cell is about to divide, the loosely organized strands of chromatin become tightly coiled and the **chromosomes** can be seen under an electron or light microscope. The genes, responsible for the transmission of inherited traits, are part of the DNA found in each chromosome.

A second very important eukaryotic organelle is the **mitochondrion,** which, like the nucleus, has a double membrane (Figure 1.8). The outer membrane has a fairly smooth surface, but the inner membrane exhibits many folds called *cristae.* The space within the inner membrane is called the *matrix.* Oxidation processes that occur in mitochondria yield energy for the cell. Most of the enzymes responsible for these important reactions are on the inner mitochondrial membrane. Other enzymes needed for oxidation reactions, as well as DNA different from that in the nucleus, are found in the

DNA CLUES TO THE RIDDLE OF THE ROMANOVS

One of the most tantalizing mysteries of the 20th century involves the circumstances surrounding the execution of the last tsar of Russia, Nicholas II, and his family. Seven members of the Romanov family — the Tsar, the Tsarina, their four daughters, and one son — faced a firing squad in July 1918, accompanied by their personal physician and three servants. In 1991 a grave containing nine skeletons was discovered near the site of the execution. DNA analysis of the bone cells in those skeletons provided important clues for identification. A key fact used in such analysis is that nuclear DNA is inherited from both parents, whereas mitochondrial DNA is inherited from the mother only.

Comparison of nuclear DNA from all nine skeletons showed that five of them were those of related individuals: a father, a mother, and three daughters. The other four skeletons were those of unrelated individuals, presumably the doctor and the three servants. Positive identification of the five Romanovs was achieved by comparison of mitochondrial DNA from their bone cells with that of living individuals related through the maternal line. The bones of the mother and the three daughters contained mitochondrial DNA that matched a sample taken from the blood of Prince Philip of England, whose maternal grandmother was the Tsarina's sister. Similarly, mitochondrial DNA from the bones of the Tsar matched that taken from his living maternal relatives. Five family members were thus identified with a high degree of certainty, but no remains were found to provide a clue to the fates of the son and the remaining daughter.

The son, the Tsarevich Alexei, did not have a realistic chance of surviving shooting by firing squad. He suffered from hemophilia, an inherited defect in the blood clotting mechanism (see Section 5.11). A hemophiliac can bleed to death from a small cut or bruise that would not trouble a healthy person. It is safe to assume that Alexei died quickly, but what happened to his remains is still a mystery. The fate of the fourth daughter is in question; a healthy young

Tsar Nicholas II, Tsarina Alexandra, and three of their five children on the cover of a 1901 magazine.

woman could have recovered from even severe gunshot wounds. This was precisely the claim of the woman who lived most of her life under the name Anna Anderson — that she was the youngest Romanov daughter, Grand Duchess Anastasia, and that she had survived her wounds and had eventually made her way out of Russia. Anna Anderson died in 1984, but a tissue sample taken when she underwent surgery near the end of her life is still in storage. Supporters of her claim want this sample subjected to comparative DNA analysis. The results of such an analysis could provide a definite answer to the puzzle.

internal mitochondrial matrix. Mitochondria also contain ribosomes similar to those found in bacteria. Mitochondria are approximately the size of many bacteria, typically about 1 μm in diameter and 2 to 8 μm in length, and it is theorized that they may have arisen from the absorption of aerobic bacteria by larger host cells.

Outer membrane   Inner membrane

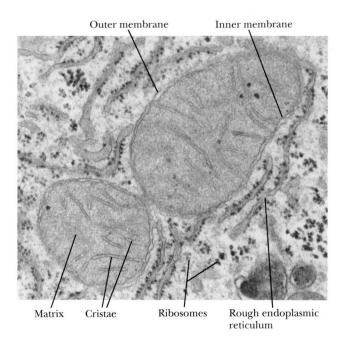

Matrix   Cristae   Ribosomes   Rough endoplasmic
reticulum

FIGURE 1.8
**Mouse liver mitochondria
(magnified 50,000×).**

The **endoplasmic reticulum (ER)** is part of a continuous single-membrane system throughout the cell; the membrane doubles back on itself to give the appearance of a double membrane in electron micrographs. The endoplasmic reticulum is attached to the cell membrane and to the nuclear membrane. It occurs in two forms, rough and smooth. The *rough endoplasmic reticulum* is studded with ribosomes bound to the membrane (Figure 1.9). Ribosomes (which can also be found free in the cytosol) are the sites of protein synthesis in all organisms. The *smooth endoplasmic reticulum* does not have ribosomes bound to it.

FIGURE 1.9
**Rough endoplasmic reticulum
from mouse liver cells (magnified
50,000×).**

Mitochondria

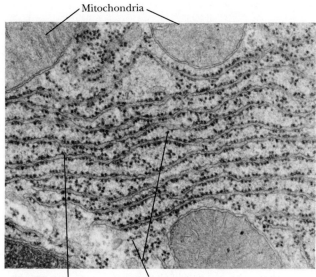

"Double" membranes   Ribosomes
(formed by doubling back of single membranes)

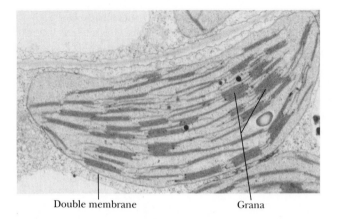

Double membrane   Grana

Chloroplasts are important organelles found only in green plants. They have double membranes and are relatively large, typically up to 2 μm in diameter and 5 to 10 μm in length. The photosynthetic apparatus is found in specialized structures called *grana* (singular *granum*), membranous bodies stacked within the chloroplast. Grana are easily seen through an electron microscope (Figure 1.10). Chloroplasts, like mitochondria, contain a characteristic DNA different from that found in the nucleus. Chloroplasts also contain ribosomes similar to those found in bacteria.

Other Organelles and Cellular Constituents

Membranes are important in the structures of some less well understood organelles. One, the **Golgi apparatus,** is a membrane-bounded organelle (with a single membrane) that is separate from the endoplasmic reticulum but frequently found close to the smooth endoplasmic reticulum. It is an assemblage of flattened vesicles or sacs (Figure 1.11). The Golgi apparatus is involved in secretion of proteins from the cell, but it also occurs in cells in which the primary function is not protein secretion. It additionally appears

FIGURE 1.11

Golgi apparatus from the green
alga *Dunaliella tertiolecta*
(magnified 59,000×).

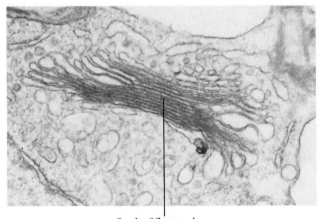

Stack of flattened
membranous vesicles

to be involved in the metabolism of sugars. In particular, it is the site in the cell in which sugars are linked to other cellular components, such as proteins. The function of this organelle is still a subject of research.

Other organelles in eukaryotes are similar to the Golgi apparatus in that they are bounded by single, smooth membranes and have specialized functions. **Lysosomes,** for example, are membrane-bounded sacs containing enzymes that could cause considerable damage to the cell if they were not separated from the lipids, proteins, or nucleic acids that they attack. Inside the lysosome these enzymes break down target molecules, usually from outside sources, as a step in processing nutrients for the cell. **Peroxisomes** are similar to lysosomes; their principal characteristic is that they contain enzymes involved in the metabolism of hydrogen peroxide (H_2O_2), which is toxic. The enzyme *catalase,* which occurs in peroxisomes, catalyzes the conversion of H_2O_2 to H_2O and O_2. **Glyoxysomes** are found in plant cells only. They contain the enzymes that catalyze the *glyoxylate cycle,* a pathway that converts lipids to carbohydrate with glyoxylic acid as an intermediate.

The **cytosol** was long considered to be nothing more than a viscous liquid, but recent studies by electron microscopy have revealed that this part of the cell has some internal organization. The organelles are held in place by a lattice of fine strands that seem to consist mostly of protein. This **microtrabecular lattice,** or **cytoskeleton,** is connected to all organelles (Figure 1.12). Many questions remain about its nature.

FIGURE 1.12

The microtrabecular lattice. (a) This network of filaments, also called the cytoskeleton, pervades the cytosol. Some filaments, called microtubules, are known to consist of the protein tubulin. Organelles such as mitochondria are attached to the filaments. (b) An electron micrograph of the microtrabecular lattice (magnified 87,450×).

(a)

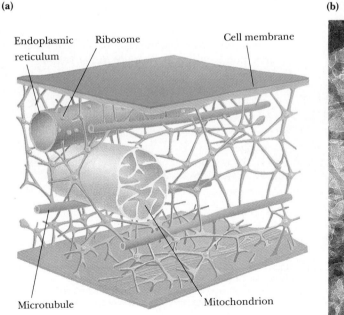

(b)

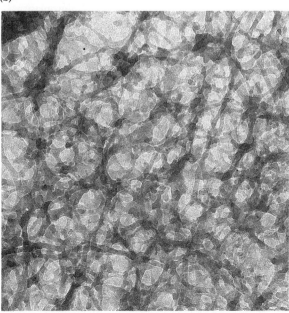

The cell membrane of eukaryotes serves to separate the cell from the outside world. It consists of a double layer of lipids, with several types of proteins embedded in the lipid matrix. Some of the proteins transport specific substances across the membrane barrier. Transport can take place in both directions, with substances useful to the cell being taken in and others being exported out.

Plant cells, but not animal cells, have cell walls external to the plasma membrane. The cellulose that makes up plant cell walls is a major component of plant material; wood, cotton, linen, and most types of paper are mainly cellulose. Also present in plant cells are large central **vacuoles,** sacs surrounded by single membranes. Although vacuoles sometimes appear in animal cells, those in plants are larger and more prominent. They tend to increase in number and size as the plant cell ages. An important function of vacuoles is to isolate waste substances that are toxic to the plant and are produced in greater amounts than the plant can secrete to the environment. These waste products may be unpalatable or even poisonous enough to discourage herbivores (plant-eating organisms) from ingesting them, and may thus provide some protection for the plant.

Table 1.3 summarizes organelles and their functions.

T A B L E   1 . 3   A Summary of Organelles and Their Functions

ORGANELLE	FUNCTION
Nucleus	Location of main genome; site of most DNA and RNA synthesis
Mitochondrion	Site of energy-yielding oxidation reactions; has its own DNA
Chloroplast	Site of photosynthesis in green plants and algae; has its own DNA
Endoplasmic reticulum	Continuous membrane throughout the cell; rough part studded with *ribosomes (the sites of protein synthesis)**
Golgi apparatus	Series of flattened membranes; involved in secretion of proteins from cells and in reactions that link sugars to other cellular components
Lysosomes	Membrane-bounded sacs containing hydrolytic enzymes
Peroxisomes	Sacs that contain enzymes involved in the metabolism of hydrogen peroxide
Cell membrane	Separates the cell contents from the outside world; contents include organelles (held in place by the *microtrabecular lattice**) and the *cytosol*
Cell wall	Rigid exterior layer of plant cells
Central vacuole	Membrane-bounded sac (plant cells)

*Since an organelle is defined as a portion of a cell bounded by a membrane, ribosomes are not, strictly speaking, organelles. Smooth endoplasmic reticulum does not have ribosomes attached, and ribosomes also occur free in the cytosol. The definition of organelle also affects discussion of the cell membrane, cytosol, and microtrabecular lattice.

1.6
THE FIVE-KINGDOM CLASSIFICATION SYSTEM

The original biological classification scheme, established in the 18th century, divided all organisms into two kingdoms: plants and animals. In this scheme, plants are organisms that obtain food directly from the sun and animals are organisms that move about to search for food. It was discovered that some organisms, bacteria in particular, do not have an obvious relationship to either kingdom. It has also become clear that the fundamental division of living organisms is actually not between plants and animals, but between prokaryotes and eukaryotes. In the 20th century, classification schemes have been introduced that divide living organisms into more than the two traditional kingdoms. The five-kingdom system (Figure 1.13) takes into account the differences between prokaryotes and eukaryotes and also provides classifications for eukaryotes that appear to be neither plants nor animals.

The kingdom **Monera** consists of only prokaryotic organisms. Bacteria and cyanobacteria are members of this kingdom. The other four kingdoms

FIGURE   1.13

The five-kingdom classification scheme.

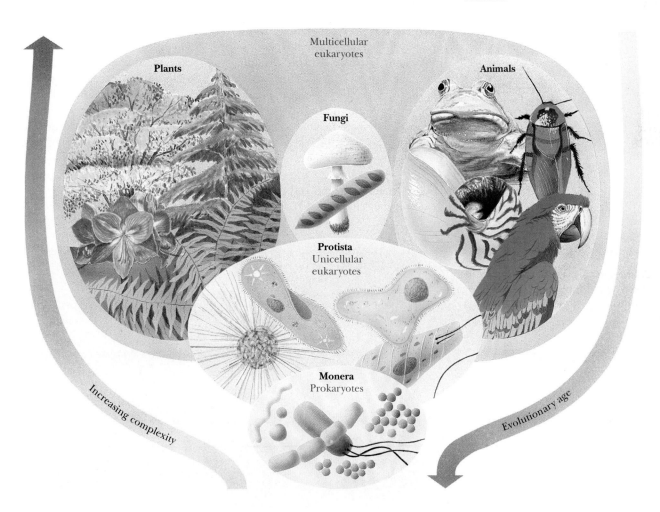

are made up of eukaryotic organisms. The kingdom **Protista** includes unicellular organisms such as yeast, *Euglena, Volvox, Amoeba,* and *Paramecium.* Some protists, such as *Volvox,* form colonies. There is some question among biologists about whether any multicellular organisms should be included in this kingdom. Most biologists do not classify multicellular organisms as protists, but the question continues to be discussed. The three kingdoms that consist mainly of multicellular eukaryotes (with a few unicellular eukaryotes) are **Fungi, Plantae,** and **Animalia.** The fungi include molds and mushrooms. Fungi, plants, and animals must have evolved from simpler eukaryotic ancestors, but the major evolutionary change was the development of eukaryotes from prokaryotes.

1.7
ARE THERE LIFE FORMS OTHER THAN PROKARYOTES AND EUKARYOTES?

A group of organisms exists that can be classified as bacteria on the basis of lacking a well-defined nucleus, but differs from both prokaryotes and eukaryotes in several important ways. These organisms are called *Archaebacteria* (early bacteria) to distinguish them from *Eubacteria* (true bacteria), since they are very primitive organisms. Most of the differences between archaebacteria and other organisms are biochemical features, such as the molecular structures of the cell walls, membranes, and some types of RNA. (The article by C. R. Woese listed in the bibliography at the end of this chapter makes biochemical comparisons between archaebacteria and other life forms.) Some biologists prefer a three-domain classification scheme — *Eubacteria, Archaea* (archaebacteria), and *Eukarya* (eukaryotes) — to the five-kingdom classification. The basis for this preference is the emphasis on biochemistry as the basis for classification.

There are three groups of archaebacteria — methanogens, halophiles, and thermacidophiles — all of which live in extreme environments. *Methanogens* are strict anaerobes that produce methane (CH_4) from carbon dioxide

FIGURE 1.14

A hot spring. Some bacteria can thrive even in these inhospitable conditions.

(CO_2) and hydrogen (H_2). *Halophiles* require very high salt concentrations, such as those found in the Dead Sea, for growth. *Thermacidophiles* require high temperatures and acid conditions for growth; typical conditions are 80 to 90°C and pH 2. These growth requirements may have resulted from adaptations to harsh conditions on the early earth. It will be useful to keep these points in mind as we discuss pathways by which eukaryotes may have evolved from prokaryotes (Figure 1.14).

1.8
COMMON GROUND FOR ALL CELLS

The complexity of eukaryotes raises many questions about how such cells arose from simpler progenitors. Symbiosis plays a large role in current theories of the rise of eukaryotes; the symbiotic association between two organisms is seen as giving rise to a new organism that combines characteristics of both the original ones. The type of symbiosis called *mutualism* is a relationship that benefits both species involved, as opposed to *parasitic* symbiosis, in which one species gains and the other is harmed. A classic example of mutualism (although it has been questioned from time to time) is the lichen, which consists of a fungus and an alga. The fungus provides water and protection for the alga; the alga is photosynthetic and provides food for both partners. Another example is the root nodule system formed by a leguminous plant, such as alfalfa or beans, and anaerobic nitrogen-fixing bacteria (Figure 1.15). The plant gains useful compounds of nitrogen, and the bacteria are protected from oxygen that is harmful to them. Still another example of mutualistic symbiosis, of great practical interest, is the relationship between humans and bacteria, such as *Escherichia coli,* that live in the intestinal tract. The bacteria receive nutrients and protection from their environment. In return, they aid our digestive process; without intestinal bacteria, we would soon develop dysentery and other intestinal disorders. These bacteria are also a source of certain vitamins for us, since they can synthesize the vitamins and we cannot do so.

FIGURE 1.15

Leguminous plants live symbiotically with nitrogen-fixing bacteria in their root systems.

These stromatolites at Shark Bay in Western Australia are approximately 2000 years old. The formations are composed of mats of cyanobacteria and minerals such as calcium carbonate. Some fossil stromatolites are 3.5 billion years old. The fossils indicate that cyanobacteria played a role in the development of life on earth.

In hereditary symbiosis, a larger host cell contains a genetically determined number of smaller organisms. An example is the protist *Cyanophora paradoxa,* a eukaryotic host that contains a genetically determined number of cyanobacteria (blue-green algae). This relationship is an example of **endosymbiosis,** since the cyanobacteria are contained within the host organism. The cyanobacteria are aerobic prokaryotes and are capable of photosynthesis. The host cell gains the products of photosynthesis; in return, the cyanobacteria are protected from the environment and still have access to oxygen and sunlight because of the host's small size. This arrangement can be considered a model for the origin of chloroplasts. In this model, with the passage of many generations cyanobacteria would have gradually lost the ability to exist independently and would have become organelles within a new and more complex type of cell. Such a process may well have given rise to chloroplasts, which are not capable of independent existence. Their autonomous DNA and ribosomal protein-synthesizing apparatus can no longer meet all their needs, but the very fact that these organelles have their own DNA and are capable of protein synthesis suggests that they may have existed as independent organisms in the distant past.

A similar model can be proposed for the origin of mitochondria. Consider this scenario. A large anaerobic host cell assimilates a number of smaller aerobic bacteria. The larger cell protects the smaller ones and provides them with nutrients. As in the example we used for the development of chloroplasts, the smaller cells still have access to oxygen. The larger cell is not itself capable of aerobic oxidation of nutrients, but some of the end-products of its anaerobic oxidation can be further oxidized by the more efficient aerobic metabolism of the smaller cells. As a result, the larger cell can get more energy out of a given amount of food than it could without the bacteria. In time the two associated organisms evolve to form a new aerobic organism, which contains mitochondria derived from the original aerobic bacteria.

The fact that both mitochondria and chloroplasts have their own DNA, which is different from the DNA found in the nucleus of the cell, is an important piece of biochemical evidence in favor of this model. Additionally, both mitochondria and chloroplasts have their own apparatus for synthesis of RNA and proteins. The genetic code in mitochondria differs from that found in the nucleus, which supports the idea of an independent origin. Thus, the remains of these systems for synthesis of RNA and protein could reflect the organelles' former existence as free-living cells. It is reasonable to conclude that large unicellular organisms that assimilated aerobic bacteria went on to evolve mitochondria from the bacteria and eventually gave rise to animals. Other types of unicellular organisms assimilated both aerobic bacteria and cyanobacteria and evolved both mitochondria and chloroplasts; they eventually gave rise to green plants.

These proposed connections between prokaryotes and eukaryotes are not established with complete certainty, and they leave a number of questions. Still, they provide an interesting frame of reference from which to consider the reactions that take place in cells.

SUMMARY

Biochemistry is a multidisciplinary field that addresses questions about the molecular nature of life processes. The fundamental biochemical similarities observed in all living organisms have engendered speculation about the origins of life.

It has been shown that important biomolecules can be produced under abiotic (nonliving) conditions from simple compounds postulated to have been present in the atmosphere of the early earth. These simple biomolecules can polymerize, also under abiotic conditions, to give rise to compounds resembling proteins and nucleic acids.

All cellular activity depends on the presence of catalysts, which increase the rates of chemical reactions, and on the genetic code, which directs the synthesis of the catalysts. In present-day cells, catalytic activity is associated with proteins, and transmission of the genetic code is associated with nucleic acids, particularly with DNA. Both these functions may once have been carried out by a single biomolecule, RNA. It has been postulated that RNA was the original coding material, and it has recently been shown to have catalytic activity as well.

Two main cell types occur in organisms. In *prokaryotes* the cell lacks a well-defined nucleus and an internal membrane; it has only a nuclear region, the portion of the cell that contains DNA, and a cell membrane that separates it from the outside world. The other principal feature of a prokaryotic cell's interior is the presence of ribosomes, the site of protein synthesis.

In contrast, a *eukaryotic* cell has a well-defined nucleus, internal membranes as well as a cell membrane, and a considerably more complex internal structure. In eukaryotes the nucleus is separated from the rest of the cell by a double membrane. Eukaryotic DNA in the nucleus is associated with proteins, which is not the case in prokaryotes. There is a continuous membrane system, called the endoplasmic reticulum, throughout the cell. Eukaryotic ribosomes are frequently bound to the endoplasmic reticulum, but some are also free in the cytosol. Membrane-bounded organelles are characteristic of eukaryotic cells. Two of the most important are mitochondria, the sites of energy-yielding reactions, and chloroplasts, the sites of photosynthesis.

E X E R C I S E S

1. State why the following terms are important in biochemistry: polymer, protein, nucleic acid, catalysis, genetic code.
2. What do you consider to be the strong and weak points of theories of the origin of life? Pay special attention to theories that proteins arose first and to theories that nucleic acids arose first.
3. Comment on RNA's roles in catalysis and coding in theories of the origin of life.
4. Do you consider it a reasonable conjecture that cells could have arisen as bare cytoplasm without a cell membrane?
5. List five differences between prokaryotes and eukaryotes.
6. Draw an idealized animal cell and identify the parts by name and function.
7. Draw an idealized plant cell and identify the parts by name and function.
8. What are the differences between the photosynthetic apparatus of green plants and photosynthetic bacteria?
9. Which organelles are bounded by a double membrane?
10. Which organelles contain DNA?
11. Which organelles are the sites of energy-yielding reactions?
12. Do the sites of protein synthesis differ in prokaryotes and eukaryotes?
13. State how the following organelles differ from each other in terms of structure and function: Golgi apparatus, lysosomes, peroxisomes, glyoxysomes. How do they resemble each other?
14. Assume that a scientist claims to have discovered mitochondria in bacteria. Is such a claim likely to prove valid?

A N N O T A T E D   B I B L I O G R A P H Y

Research progress is very rapid in biochemistry, and the literature in the field is vast and growing. Many books appear each year, and a large number of primary research journals and review journals report on original research. References to this body of literature are provided at the end of each chapter. A particularly useful reference is *Scientific American;* its articles include general overviews of the topics discussed. *Trends in Biochemical Sciences* and *Science* (a journal published weekly by the American Association for the Advancement of Science) are more advanced and can serve as primary sources of information about a given topic. *Science* has excellent illustrations.

Adams, R. L. P., ed. *J. N. Davidson's The Biochemistry of the Nucleic Acids.* 11th ed. New York: Academic Press, 1992. [A classic introduction to the subject.]

Alberts, B., D. Bray, J. Lewis, M. Raff, K. Roberts, and J. D. Watson. *Molecular Biology of the Cell.* 3rd ed. New York: Garland Publishing, 1994. [A particularly well-written and well-illustrated textbook of cell biology.]

Allen, R. D. The Microtubule as an Intracellular Engine. *Sci. Amer.* **256** (2), 42–49 (1987). [The role of the microtrabecular lattice and microtubules in the motion of organelles is discussed.]

Cairns-Smith, A. G. The First Organisms. *Sci. Amer.* **252** (6), 90–100 (1985). [A presentation of the point of view that the earliest life processes took place in clay rather than in the "primordial soup" of the early oceans.]

Cairns-Smith, A. G. *Genetic Takeover and the Mineral Origins of Life.* Cambridge, England: Cambridge Univ. Press, 1982. [A presentation of the idea that life began in clay.]

Cech, T. R. RNA as an Enzyme. *Sci. Amer.* **255** (5), 64–75 (1986). [A discussion of the ways in which RNA can cut and splice itself.]

Dyson, F. *Origins of Life.* Cambridge, England: Cambridge Univ. Press, 1985. [A comparison of theories of the origin of life.]

Eigen, M., W. Gardiner, P. Schuster, and R. Winkler-Oswatitsch. The Origin of Genetic Information. *Sci. Amer.* **244** (4), 88–118 (1981). [A presentation of the case for RNA as the original coding material.]

Heidemann, S. A New Twist on Integrins and the Cytoskeleton. *Science* **260,** 1080–1081 (1993). [A report on new developments in the study of the mechanical properties of the cytoskeleton.]

Hoffman, M. Researchers Find Organism They Can Really Relate To. *Science* **257,** 32 (1992). [A short account of eocytes, whose ribosomal proteins are more closely related to those of eukaryotes than to those of other bacteria.]

Hinkle, P. C., and R. E. McCarty. How Cells Make ATP. *Sci. Amer.* **238** (3), 104–123 (1978). [The role of chloroplasts and mitochondria in generating energy for the cell; an old but particularly good article.]

Horgan, J. In the Beginning. . . . *Sci. Amer.* **264** (2), 116–125 (1991). [A report on new developments in the study of the origin of life.]

Kabnick, K., and D. Peattie. *Giardia:* A Missing Link Between Prokaryotes and Eukaryotes. *Amer. Scientist* **79,** 34–43 (1991). [A description of an organism whose cells resemble those of both prokaryotes and eukaryotes.]

Knoll, A. The Early Evolution of Eukaryotes: A Geological Perspective. *Science* **256,** 622–627 (1992). [A comparison of biological and geological evidence on the subject.]

Lewin, R. RNA Catalysis Gives Fresh Perspective on the Origin of Life. *Science* **231,** 545–546 (1986). [A consideration of RNA splicing in terms of catalysis and coding.]

Lewin, R. No Genome Barriers to Promiscuous DNA. *Science* **224,** 970–971 (1984). [A discussion of the movement of DNA between mitochondrial, chloroplast, and nuclear genomes.]

Lipsky, N. G., and R. E. Pagano. A Vital Stain for the Golgi Apparatus. *Science* **228,** 745–747 (1985). [A description of the first staining method for the Golgi apparatus in living cells.]

Margulis, L. *Early Life*. Boston: Science Books Intl., 1982. [A clear discussion of the origin and development of cells.]

Pool, R. Pushing the Envelope of Life. *Science* **247,** 158–160 (1990). [A *Research News* article describing the extreme conditions under which archaebacteria can flourish.]

Rothman, J. E. The Compartmental Organization of the Golgi Apparatus. *Sci. Amer.* **253** (3), 74–89 (1985). [A description of the functions of the Golgi apparatus.]

Vidal, G. The Oldest Eukaryotic Cells. *Sci. Amer.* **250** (2), 48–57 (1984). [A description of fossil evidence for the rise of eukaryotes.]

Waldrop, M. Goodbye to the Warm Little Pond? *Science* **250,** 1078–1079 (1990). [Facts and theories on the role of meteorite impacts on the early earth in the origin and development of life.]

Weber, K., and M. Osborn. The Molecules of the Cell Matrix. *Sci. Amer.* **253** (4), 100–120 (1985). [An extensive description of the cytoskeleton.]

Woese, C. R. Archaebacteria. *Sci. Amer.* **244** (6), 98–122 (1981). [A detailed description of the biochemical differences between archaebacteria and other types of organ,cisms.]

Chapter 2

Water: The Solvent for Biochemical Reactions

A striking example of water in both the solid and liquid states. Life as we know it depends on the properties of this simple molecule.

Virtually all of the chemical reactions of the cell involve water. They are the reactions of organic chemistry, using the same functional groups and operating in a cellular environment. Life has evolved around the special properties of water. Important structural considerations follow from the nature of the water molecule, which has a partial positive charge on each of its hydrogen atoms and a partial negative charge on its oxygen atom. This allows the water molecule to associate with four others of its kind. Four hydrogen bonds can be formed, pointing to the corners of a tetrahedron. Hydrogen bonds are important everywhere in biomolecular structures; they stitch together parts of protein chains of enzymes as well as hold together the two complementary chains of the DNA double helix. Another unique property of water is its role in the control of acidity within the cell. A cell's survival depends on strict control of its internal pH.

2.1
THE POLAR NATURE OF THE WATER MOLECULE

Water is the principal component of most cells. The geometry of the water molecule and its properties as a solvent play major roles in determining the properties of living systems.

When electrons are shared between atoms in a chemical bond, they need not be shared equally. Bonds with unequal sharing of electrons are referred to as **polar.** The tendency of an atom to attract electrons to itself in a chemical bond (i.e., to become negative) is called **electronegativity.** Atoms of the same element, of course, share electrons equally in a bond — that is, have equal electronegativity — but different elements do not necessarily have the same electronegativity. Fluorine, oxygen, and nitrogen are all highly electronegative — more so than carbon and hydrogen (Table 2.1).

In the O—H bond in water, oxygen is more electronegative than hydrogen, so it has a larger share of the electrons. This difference in electronegativity between oxygen and hydrogen gives rise to *partial* positive and negative charge, usually pictured as δ^+ and δ^-, respectively (Figure 2.1). The O—H bond is thus a polar bond. In situations where the electronegativity difference is quite small, such as in the C—H bond in methane (CH_4), the sharing of electrons in the bond is very nearly equal, and the bond is essentially **nonpolar.**

TABLE 2.1   **Electronegativities of Selected Elements**

ELEMENT	ELECTRONEGATIVITY*
Fluorine	4.0
Oxygen	3.5
Nitrogen	3.0
Carbon	2.5
Hydrogen	2.1

*Electronegativity values are relative and are chosen to be positive numbers ranging from less than 1 for some metals to 4 for fluorine.

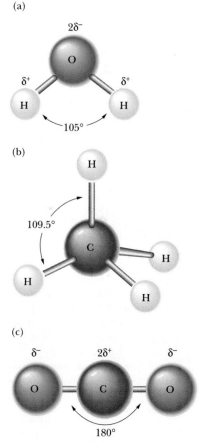

(a)

(b)

(c)

FIGURE 2.1

(a) The structure of the water molecule, showing the polar bonds. (b) The methane molecule, with its nonpolar bonds. (c) Carbon dioxide has polar bonds in opposite directions and is a nonpolar molecule.

It is possible for a molecule to have polar bonds but still be nonpolar because of its geometry. Carbon dioxide is an example. The two C—O bonds are polar, but because the CO_2 molecule is linear, the attraction of the oxygen for the electrons in one bond is cancelled out by the equal and opposite attraction for the electrons by the oxygen on the other side of the molecule.

$$\overset{\delta^-}{O} = \overset{2\delta^+}{C} = \overset{\delta^-}{O}$$

Water is a bent molecule with a bond angle of 105°, and the uneven sharing of electrons in the two bonds is not cancelled out as in CO_2 (Figure 2.1). The result is that the bonding electrons are more likely to be found at the oxygen end of the molecule than at the hydrogen end. Molecules with positive and negative ends are called **dipoles.**

Solvent Properties of Water

The polar nature of water largely determines its solvent properties. *Ionic* compounds with full charges, such as potassium chloride (KCl), and *polar* compounds with partial charges, such as ethyl alcohol (C_2H_5OH) or acetone (($CH_3)_2C{=}O$), tend to dissolve in water (Figure 2.2). The underlying physical principle is electrostatic attraction between unlike charges. The negative end of a water dipole attracts a positive ion or the positive end of another dipole. The positive end of a water molecule attracts a negative ion or the negative end of another dipole. The aggregate of unlike charges, held in proximity to one another because of electrostatic attraction, has a lower energy than would be possible if this interaction could not take place. The

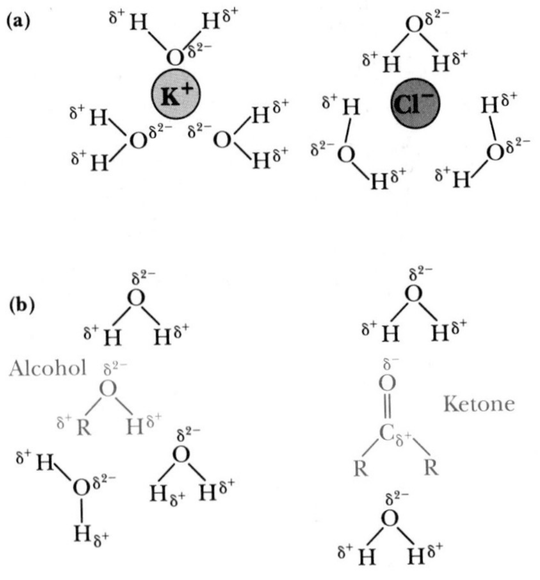

FIGURE 2.2

Ion–dipole and dipole–dipole interactions help ionic and polar compounds dissolve in water. (a) Ion–dipole interactions with water. (b) Dipole–dipole interactions of polar compounds with water. The examples shown here are an alcohol (ROH) and a ketone ($R_2C{=}O$).

lowering of energy makes the system more stable and more likely to exist. These *ion–dipole* and *dipole–dipole* interactions are similar to the interactions between water molecules themselves in terms of the quantities of energy involved. Examples of polar compounds that dissolve easily in water are small organic molecules containing one or more electronegative atoms (such as oxygen or nitrogen), including alcohols, amines, and carboxylic acids. The attraction between the dipoles of these molecules and the water dipole makes them tend to dissolve. Ionic and polar substances are referred to as **hydrophilic** ("water-loving," from the Greek) because of this tendency.

Hydrocarbons (compounds that contain only carbon and hydrogen) are nonpolar. The favorable ion–dipole and dipole–dipole interactions responsible for the solubility of ionic and polar compounds do not occur for nonpolar compounds, and so these compounds tend not to dissolve in water. The interactions between nonpolar molecules and water molecules are weaker than dipolar interactions. The permanent dipole of the water molecule can induce a temporary dipole in the nonpolar molecule by distorting the spatial arrangements of the electrons in its bonds. Electrostatic attraction is possible between the induced dipole of the nonpolar molecule and the permanent dipole of the water molecule (a *dipole–induced dipole* interaction), but it is not as strong as that between permanent dipoles. Hence, its consequent lowering of energy is less than that produced by the attraction of water molecules for one another. The association of nonpolar molecules with water is far less likely to occur than the association of water molecules with themselves.

A full discussion of why nonpolar substances are insoluble in water requires the thermodynamic arguments that we shall develop in Chapter 9. However, the points made here about intermolecular interactions will be useful background information for that discussion. For the moment it is enough to know that it is less favorable thermodynamically for water molecules to be associated with nonpolar molecules than with other water molecules. As a result, nonpolar molecules do not dissolve in water and are referred to as **hydrophobic** ("water-hating"). (Hydrocarbons, in particular, tend to sequester themselves from an aqueous environment.) A nonpolar solid leaves undissolved material in water. A nonpolar liquid forms a two-layer system with water; an example is an oil slick.

Table 2.2 lists some examples of hydrophobic and hydrophilic substances.

T A B L E   2 . 2   Examples of Hydrophobic and Hydrophilic Substances

HYDROPHILIC	HYDROPHOBIC
Polar covalent compounds (e.g., alcohols such as C_2H_5OH [ethanol] and ketones such as $(CH_3)_2C{=}O$ [acetone])	Nonpolar covalent compounds (e.g., hydrocarbons such as C_6H_{14} [hexane])
Sugars	Fatty acids, cholesterol
Ionic compounds (e.g., KCl)	
Amino acids, phosphate esters	

FIGURE 2.3

(a) A sodium salt of fatty acid with an ionized polar head and a nonpolar tail. (b) Formation of a micelle, with the ionized polar groups in contact with the water and the nonpolar parts of the molecule protected from contact with water.

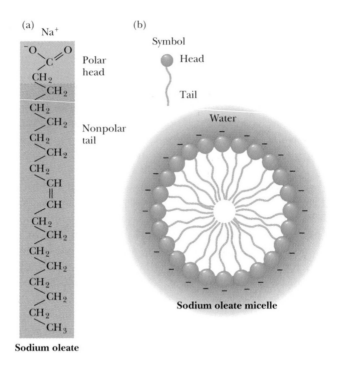

Sodium oleate

Sodium oleate micelle

It is possible for a molecule to have both polar (hydrophilic) and nonpolar (hydrophobic) portions. Substances of this type are called **amphiphilic** (from the Greek *amphi,* "on both ends," and *philic,* "loving") because one part of the molecule tends to dissolve in water and another part in a nonpolar environment. A long-chain fatty acid having a polar carboxylic acid group and a long nonpolar hydrocarbon portion is a prime example of an amphiphilic substance. The carboxylic acid group, the "head" group, contains two oxygen atoms in addition to carbon and hydrogen; it is very polar and can form a carboxylate anion. The rest of the molecule, the "tail," contains only carbon and hydrogen and is thus nonpolar. A compound such as this in the presence of water tends to form structures called **micelles,** in which the polar head groups are in contact with the aqueous environment and the nonpolar tails are sequestered from the water (Figure 2.3).

Interactions between nonpolar molecules themselves are very weak and depend on the attraction between short-lived temporary dipoles and the dipoles they induce. In a large sample of nonpolar molecules there will always be some molecules with these temporary dipoles, which are caused by a momentary clumping of bonding electrons at one end of the molecule. A temporary dipole can induce another dipole in a neighboring molecule in the same way that a permanent dipole does. The interaction energy is low because the association is so short-lived. It is called a **van der Waals bond.**

The arrangement of molecules in cells strongly depends on the molecules' polarity, as we saw with micelles. Reactivity is also important, and it is largely that of the molecules' functional groups. Here is an area of study where organic chemistry has a great deal to contribute to biochemistry.

2.2

THE IMPORTANCE OF FUNCTIONAL GROUPS IN BIOCHEMISTRY

Organic chemistry is the study of compounds of carbon and hydrogen and their derivatives. Since the cellular apparatus of living organisms is made up of carbon compounds, biomolecules are part of the subject matter of organic chemistry. Additionally, many carbon compounds exist that are not found in any organism, and many topics of importance to organic chemistry have little connection with living things.

Until the early part of the 19th century, there was a widely held belief in "vital forces," forces unique to living things. A part of this theory was the idea that the compounds found in living organisms could not be produced in the laboratory. The critical experiment that disproved this belief was performed by the German chemist Friedrich Wöhler in 1828. Wöhler synthesized urea, a well-known waste product of animal metabolism, from ammonium cyanate, a compound obtained from mineral (i.e., nonliving) sources.

$$NH_4OCN \longrightarrow H_2NCONH_2$$

Ammonium   Urea
  Cyanate

It has since been shown that any compound that occurs in a living organism can be synthesized in the laboratory, although in many cases the synthesis represents a considerable challenge to even the most skilled organic chemist.

The reactions of biomolecules can be described by the methods of organic chemistry, and one of the most useful of those methods is the classification of compounds according to **functional groups.** *The reactions of molecules are the reactions of the functional groups.* Table 2.3 lists some

TABLE 2.3 **Functional Groups of Biochemical Importance**

CLASS OF COMPOUND	GENERAL STRUCTURE	CHARACTERISTIC FUNCTIONAL GROUP	NAME OF FUNCTIONAL GROUP	EXAMPLE
Alkenes	$RCH{=}CH_2$ $RCH{=}CHR$ $R_2C{=}CHR$ $R_2C{=}CR_2$	$C{=}C$	Double bond	$CH_2{=}CH_2$
Alcohols	ROH	—OH	Hydroxyl group	CH_3CH_2OH
Ethers	ROR	—O—	Ether linkage	CH_3OCH_3
Amines	RNH_2 R_2NH R_3N	$-N\diagup$	Amino group	CH_3NH_2
Thiols	RSH	—SH	Sulfhydryl group	CH_3SH
Aldehydes	$R-\overset{\displaystyle O}{\overset{\|}{C}}-H$	$-\overset{\displaystyle O}{\overset{\|}{C}}-$	Carbonyl group	$CH_3\overset{\displaystyle O}{\overset{\|}{C}}H$

*The symbol R refers to any carbon-containing group. When there are several R groups in the same molecule, they may be different groups or they may be the same.

continued

T A B L E   2 . 3 *continued*

CLASS OF COMPOUND	GENERAL STRUCTURE	CHARACTERISTIC FUNCTIONAL GROUP	NAME OF FUNCTIONAL GROUP	EXAMPLE
Ketones	$R-\overset{\overset{\displaystyle O}{\|\|}}{C}-R$	$-\overset{\overset{\displaystyle O}{\|\|}}{C}-$	Carbonyl group	$CH_3\overset{\overset{\displaystyle O}{\|\|}}{C}CH_3$
Carboxylic acids	$R-\overset{\overset{\displaystyle O}{\|\|}}{C}-OH$	$-\overset{\overset{\displaystyle O}{\|\|}}{C}-OH$	Carboxyl group	$CH_3\overset{\overset{\displaystyle O}{\|\|}}{C}OH$
Esters	$R-\overset{\overset{\displaystyle O}{\|\|}}{C}-OR$	$-\overset{\overset{\displaystyle O}{\|\|}}{C}-OR$	Ester group	$CH_3\overset{\overset{\displaystyle O}{\|\|}}{C}OCH_3$
Amides	$R-\overset{\overset{\displaystyle O}{\|\|}}{C}-NR_2$	$-\overset{\overset{\displaystyle O}{\|\|}}{C}-N\big<$	Amide group	$CH_3\overset{\overset{\displaystyle O}{\|\|}}{C}N(CH_3)_2$
	$R-\overset{\overset{\displaystyle O}{\|\|}}{C}-NHR$			
	$R-\overset{\overset{\displaystyle O}{\|\|}}{C}-NH_2$			

*The symbol R refers to any carbon-containing group. When there are several R groups in the same molecule, they may be different groups or they may be the same.

biologically important functional groups. Some other groups, which are of vital importance to organic chemists, are missing from the table because molecules containing these groups, such as alkyl halides and acyl chlorides, do not have any particular applicability in biochemistry. Conversely, carbon-containing derivatives of phosphoric acid are little mentioned in beginning courses on organic chemistry, but esters and anhydrides of phosphoric acid (Figure 2.4) are of vital importance in biochemistry. Adenosine triphosphate, or ATP, contains both ester and anhydride linkages involving phosphoric acid.

We shall discuss the reactions of the functional groups when we consider compounds in which they occur. Important classes of biomolecules have characteristic functional groups that determine their reactions.

2.3
THE HYDROGEN BOND

In addition to the interactions discussed in Section 2.1, there is another important type of noncovalent interaction, **hydrogen bonding.** Hydrogen bonding is of electrostatic origin and can be considered a special case of dipole–dipole interaction. When hydrogen is covalently bonded to an electronegative atom such as oxygen, nitrogen, or fluorine, it has a partial positive charge due to the polar bond, a situation that does not occur when hydrogen is covalently bonded to carbon. This partial positive charge on hydrogen can interact with an unshared (nonbonding) pair of electrons (a source of negative charge) on another electronegative atom. All three atoms lie in a straight line, forming a hydrogen bond. This arrangement allows for the greatest possible partial positive charge on the hydrogen and conse-

(a.1)

$$\text{HO}-\overset{\displaystyle O}{\underset{\displaystyle OH}{\overset{\displaystyle \|}{P}}}-\text{OH} + \text{HO}-\text{R} \longrightarrow \text{HO}-\overset{\displaystyle O}{\underset{\displaystyle OH}{\overset{\displaystyle \|}{P}}}-\text{O}-\text{R}$$

 H₂O

Phosphoric acid   Alcohol   Phosphoric acid ester

(a.2)

(b.1)

$$\text{HO}-\overset{O}{\underset{OH}{\overset{\|}{P}}}-\text{OH} + \text{HO}-\overset{O}{\underset{OH}{\overset{\|}{P}}}-\text{OH} \longrightarrow \text{HO}-\overset{O}{\underset{OH}{\overset{\|}{P}}}-\text{O}-\overset{O}{\underset{OH}{\overset{\|}{P}}}-\text{OH}$$

 H₂O

Anhydride

(a.3)

(c)

$$\text{HO}-\overset{O}{\underset{OH}{\overset{\|}{P}}}-\text{O}-\overset{O}{\underset{OH}{\overset{\|}{P}}}-\text{O}-\overset{O}{\underset{OH}{\overset{\|}{P}}}-\text{O}-\cdots$$

Ester

Anhydride

ATP

(b.2)

FIGURE 2.4

ATP and the reactions for its formation. (a.1) Reaction of phosphoric acid with a hydroxyl group to form an ester, which contains a P—O—R linkage. (Phosphoric acid is shown in its nonionized form in this figure.) (a.2) Space-filling model of phosphoric acid. (a.3) Space-filling model of the methyl ester of phosphoric acid. (b.1) Reaction of two molecules of phosphoric acid to form an anhydride, which contains a P—O—P linkage. (b.2) Space-filling model of the anhydride of phosphoric acid. (c) The structure of ATP (*a*denosine *tri*phosphate), showing two anhydride linkages and one ester linkage.

quently for the strongest possible interaction with the unshared pair of electrons on the second electronegative atom (Figure 2.5). The electronegative atom to which the hydrogen is covalently bonded is called the *hydrogen-bond donor,* and the electronegative atom that contributes the unshared pair of electrons to the interaction is the *hydrogen-bond acceptor.* (The hydrogen is not covalently bonded to the acceptor.)

Linear {
    H—O
    |
    H   (hydrogen bond donor)
    ⋮
    O   (hydrogen bond acceptor)
}

Nonlinear {
    O
   H   H
    ⋮
    O
}

FIGURE 2.5

A comparison of linear and nonlinear hydrogen bonds. Nonlinear bonds are weaker than bonds in which all three atoms lie in a straight line.

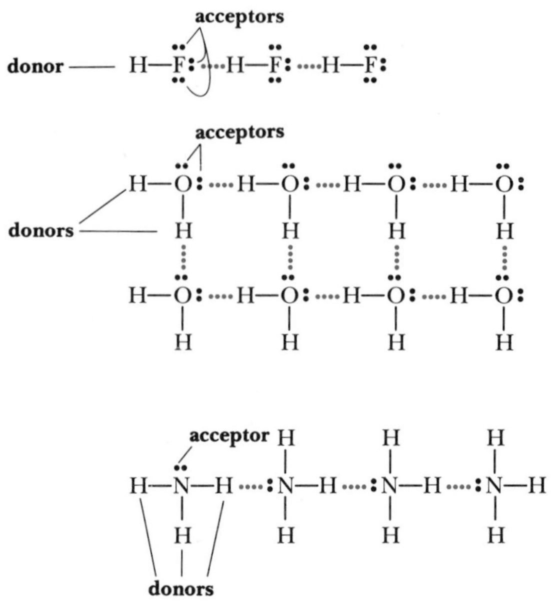

FIGURE 2.6

A comparison of the numbers of hydrogen-bonding sites in HF, H_2O, and NH_3. (Actual geometries are not shown.) Each HF molecule has one hydrogen bond donor and three hydrogen bond acceptors. Each H_2O molecule has two donors and two acceptors. Each NH_3 molecule has three donors and one acceptor.

A consideration of the hydrogen bonding situations in HF, H_2O, and NH_3 can yield some useful insights. Figure 2.6 shows that water constitutes an optimum situation in terms of the number of hydrogen bonds that each molecule can form. Water has two hydrogens to enter into hydrogen bonds, and two unshared pairs of electrons on the oxygen to which other water molecules can be hydrogen bonded. Each water molecule is involved in four hydrogen bonds—as donor in two and as acceptor in two. Hydrogen fluoride has only one hydrogen to enter into a hydrogen bond as a donor, but it has three unshared pairs of electrons on the fluorine that could bond to other hydrogens. Ammonia has three hydrogens to donate to a hydrogen bond but only one unshared pair of electrons, on the nitrogen.

The geometric arrangement of hydrogen-bonded water molecules has important implications for the properties of water as a solvent. The bond angle in water is 105°, as was shown in Figure 2.1, and the angle between the unshared pairs of electrons is similar. The result is a tetrahedral arrangement of water molecules. Liquid water consists of hydrogen-bonded arrays that resemble ice crystals; each of these arrays can contain up to 100 water molecules. The hydrogen bonding between water molecules can be seen more clearly in the regular lattice structure of the ice crystal (Figure 2.7). There are several differences, however, between hydrogen-bonded arrays of this type in liquid water and the structure of ice crystals. In liquid water, hydrogen bonds are constantly breaking and new ones forming, with some molecules breaking off and others joining the cluster. A cluster can break up and re-form in 10^{-10} to 10^{-11} seconds in water at 25°C. An ice crystal, in

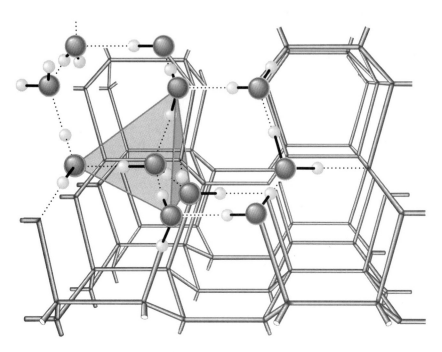

FIGURE 2.7

Tetrahedral hydrogen bonding in H_2O: an array of H_2O molecules in an ice crystal. Each H_2O molecule is hydrogen bonded to four others.

contrast, has a more or less stable arrangement of hydrogen bonds, and of course its number of molecules is many orders of magnitude greater than 100.

Hydrogen bonds are much weaker than normal covalent bonds. Whereas the energy required to break the O—H covalent bond is 460 kJ mol^{-1} (110 kcal mol^{-1}), the energy of hydrogen bonds in water is about 20 kJ mol^{-1} (5 kcal mol^{-1}) (Table 2.4). Even this comparatively small amount of energy is enough to drastically affect the properties of water, especially its

T A B L E 2 . 4 Some Bond Energies

	TYPE OF BOND	ENERGY*	
		(kJ mol^{-1})	*(kcal mol^{-1})*
Covalent bonds	O—H	460	110
(strong)	H—H	416	100
Noncovalent bonds	Hydrogen bond	20	5
(weaker)	Ion–dipole interaction	20	5
	Hydrophobic interaction	4–12	1–3
	Van der Waals bonds	4	1

*Note that two units of energy are used throughout this text. The kilocalorie (kcal) is a commonly used unit in the biochemical literature. The kilojoule (kJ) is an SI unit and will come into wider use as time goes on.

T A B L E   2 . 5   Comparison of Properties of Water, Ammonia, and Methane

SUBSTANCE	MOLECULAR WEIGHT	MELTING POINT (°C)	BOILING POINT (°C)
Water (H_2O)	18.02	0.0	100.0
Ammonia (NH_3)	17.03	−77.7	−33.4
Methane (CH_4)	16.04	−182.5	−161.5

melting point, its boiling point, and its density relative to the density of ice. Both the melting point and the boiling point of water are significantly higher than would be predicted for a molecule of this size (Table 2.5). Other substances of about the same molecular weight, such as methane and ammonia, have much lower melting and boiling points. The forces of attraction between the molecules of these substances are weaker than the attraction between water molecules, which is due to their hydrogen bonds. The energy of this attraction has to be overcome to melt ice or boil water.

Ice has a lower density than liquid water because the fully hydrogen-bonded array in an ice crystal is very open, with a lot of empty space between the molecules. Liquid water is less extensively hydrogen bonded and thus is denser than ice. Most substances contract when they freeze, but the opposite is true of water. Thus, ice cubes and icebergs float, and the expansion resulting from replacement of liquid water by ice crystals can damage surrounding materials. In cold weather, the cooling systems of cars require antifreeze to prevent freezing and expansion of the water, which could crack the engine block. The same principle is used in a method of disrupting cells with several cycles of freezing and thawing.

Hydrogen bonding also plays a role in the behavior of water as a solvent. If a polar solute can serve as a donor or an acceptor of hydrogen bonds, it can form hydrogen bonds with water in addition to being involved in nonspecific dipole–dipole interactions. Figure 2.8 shows some examples. Alcohols, amines, carboxylic acids, and esters, as well as aldehydes and ketones, can all form hydrogen bonds with water, so they are soluble in water. It is difficult to overstate the importance of water to the existence of life on earth, and difficult to imagine life based on another solvent.

F I G U R E   2 . 8

Examples of hydrogen bonding between polar groups and water.

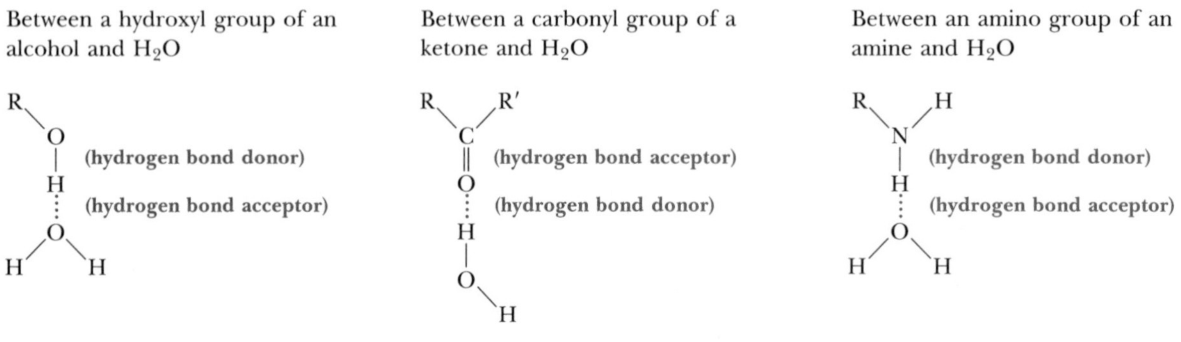

Between a hydroxyl group of an alcohol and H_2O

Between a carbonyl group of a ketone and H_2O

Between an amino group of an amine and H_2O

T A B L E 2 . 6 **Examples of Major Types of Hydrogen Bonds Found in Biologically Important Molecules**

BONDING ARRANGEMENT	MOLECULES WHERE THE BOND OCCURS
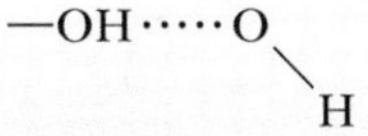 $-OH\cdots O\diagdown H$	H-bond formed in H_2O
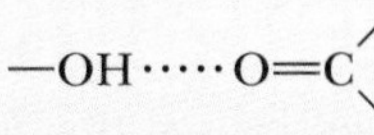 $-OH\cdots O=C\diagup\diagdown$	Bonding of water to other molecules
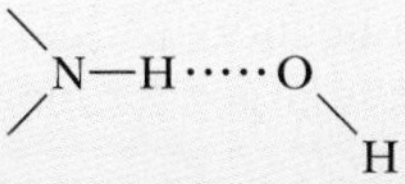 $\diagdown N-H\cdots O\diagdown H$	
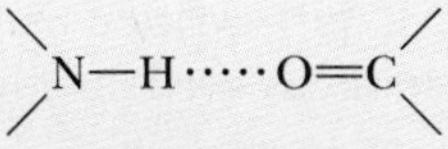 $\diagdown N-H\cdots O=C\diagup\diagdown$	Important in protein and nucleic acid structures
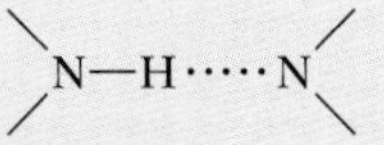 $\diagdown N-H\cdots N\diagup\diagdown$	

Biologically Important Hydrogen Bonds Other Than to Water

Hydrogen bonds have a vital involvement in stabilizing the three-dimensional structures of biologically important molecules, including DNA, RNA, and proteins. The two strands of the double helix in DNA are held together by hydrogen bonds between complementary bases (Chapter 6, Section 6.2). Transfer RNA also has a complex three-dimensional structure stabilized by hydrogen bonds (Section 6.3). Hydrogen bonding in proteins gives rise to two important structures, the α-helix and β-pleated sheet conformations. Both types of conformation are widely encountered in proteins (Chapter 4, Section 4.5). Table 2.6 summarizes some of the most important kinds of hydrogen bonds in biomolecules.

2.4
ACIDS AND BASES

The biochemical behavior of many important compounds depends on their acid–base properties. Acids are defined as proton (hydrogen ion) donors, and bases as proton acceptors. How readily acids or bases lose or gain protons depends on the chemical nature of the compounds involved. The degree of dissociation of acids in water, for example, ranges from essentially complete dissociation for a strong acid to practically no dissociation for a very weak acid, and any intermediate value is possible.

It is useful to derive a numerical measure of the **strength** of an acid, which is the amount of hydrogen ion released when a given amount of acid is dissolved in water. Such an expression, called the **acid dissociation constant,** or K'_a, can be written for any acid, HA, that reacts according to the equation

$$HA \rightleftharpoons H^+ + A^-$$
Acid   Conjugate base

$$K'_a = \frac{[H^+][A^-]}{[HA]}$$

In this expression the square brackets refer to molar concentration, that is, the concentration in moles per liter. For each acid, the quantity K'_a has a fixed numerical value at a given temperature. This value is larger for more completely dissociated acids; *the greater the K'_a, the stronger the acid.*

Strictly speaking, the acid–base reaction we have just written is a proton-transfer reaction in which water acts as a base as well as the solvent.

$$HA(aq) + H_2O(\ell) \rightleftharpoons H_3O^+(aq) + A^-(aq)$$
Acid   Base   Conjugate   Conjugate
                          acid to H_2O base to HA

The notation (aq) refers to solutes in aqueous solution, whereas (ℓ) refers to water in the liquid state. It is well established that there are no "naked protons" (free hydrogen ions) in solution; even the hydronium ion (H_3O^+) is an underestimate of the degree of hydration of hydrogen ion in aqueous solution. All solutes are extensively hydrated in aqueous solution. We will write the short form of equations for acid dissociation in the interests of simplicity, but the role of water should be kept in mind throughout our discussion.

2.5
THE SELF-DISSOCIATION OF WATER
AND THE pH SCALE

The acid–base properties of water play an important part in biological processes because of the central role of water as a solvent. The extent of self-dissociation of water to hydrogen ion and hydroxide ion

$$H_2O \rightleftharpoons H^+ + OH^-$$

is small, but the fact that it takes place determines important properties of many solutes. Both the hydrogen ion (H^+) and the hydroxide ion (OH^-) are associated with several water molecules, as are all ions in aqueous solution, and the water molecule in the equation is itself part of a cluster of such molecules. It is especially important to have a quantitative estimate of the degree of dissociation of water. We can start with the expression

$$K'_a = \frac{[H^+][OH^-]}{[H_2O]}$$

The molar concentration of pure water, $[H_2O]$, is quite large compared to any possible concentrations of solutes and can be considered a constant. (The numerical value is 55.5 M, which can be obtained by dividing the number of grams of water in one liter, 1000, by the molecular weight of water, 18 grams/mole; 1000/18 = 55.5 M.) Thus,

$$K'_a = \frac{[H^+][OH^-]}{55.5}$$

$$K'_a \times 55.5 = [H^+][OH^-] = K_w$$

A new constant, K_w, the **ion product constant for water,** has just been defined, where the concentration of water has been included in its value.

The numerical value of K_w can be determined experimentally by measuring the hydrogen ion concentration of pure water. The hydrogen ion concentration is also equal, by definition, to the hydroxide ion concentration, because water is a monoprotic acid (one that releases a single proton per molecule). At 25°C in *pure* water,

$$[H^+] = 10^{-7} \text{ M} = [OH^-]$$

Thus, at 25°C the numerical value of K_w is given by the expression

$$K_w = [H^+][OH^-] = (10^{-7})(10^{-7}) = 10^{-14}$$

This relationship, which we have derived for pure water, is valid for *any* aqueous solution, whether neutral, acidic, or basic.

The wide range of possible hydrogen ion and hydroxide ion concentrations in aqueous solution makes it desirable to define a quantity that expresses these concentrations more conveniently than by exponential notation. This quantity is called pH and is defined as

$$\text{pH} = -\log_{10}[H^+]$$

with the logarithm taken to the base 10. Note that, because of the logarithms involved, a difference of one pH unit implies a tenfold difference in hydrogen ion concentration, $[H^+]$. The pH values of some typical aqueous samples can be determined by a simple calculation.

_____ P R A C T I C E   S E S S I O N

Since in pure water $[H^+] = 10^{-7}$ M and pH = 7, you should be able to calculate the pH of the following solutions:

1. 10^{-3} M HCl   **2.** 10^{-4} M NaOH

ANSWERS   1. pH = 3   **2.** pH = 10

_____

Pure water with a pH of 7 is neutral, acidic solutions have pH values lower than 7, and basic solutions have pH values higher than 7.

A similar quantity, pK'_a, can be defined by analogy with the definition of pH:

$$pK'_a = -\log_{10}K'_a$$

The pK'_a is another numerical measure of acid strength; the smaller its value, the stronger the acid. This is the reverse of the situation with K'_a, where larger values imply stronger acids (Table 2.7).

TABLE 2.7 Dissociation Constants of Some Acids

ACID	HA	A⁻	K'_a	pK'_a
Pyruvic acid	$CH_3COCOOH$	CH_3COCOO^-	3.16×10^{-3}	2.50
Formic acid	$HCOOH$	$HCOO^-$	1.44×10^{-4}	3.75
Lactic acid	$CH_3CHOHCOOH$	$CH_3CHOHCOO^-$	1.38×10^{-4}	3.86
Benzoic acid	C_6H_5COOH	$C_6H_5COO^-$	6.46×10^{-5}	4.19
Acetic acid	CH_3COOH	CH_3COO^-	1.76×10^{-5}	4.76
Ammonium ion	NH_4^+	NH_3	5.6×10^{-10}	9.25
Oxalic acid (1)	$HOOC-COOH$	$HOOC-COO^-$	5.9×10^{-2}	1.23
Oxalic acid (2)	$HOOC-COO^-$	$^-OOC-COO^-$	6.4×10^{-5}	4.19
Malonic acid (1)	$HOOC-CH_2-COOH$	$HOOC-CH_2-COO^-$	1.49×10^{-3}	2.83
Malonic acid (2)	$HOOC-CH_2-COO^-$	$^-OOC-CH_2-COO^-$	2.03×10^{-6}	5.69
Malic acid (1)	$HOOC-CH_2-CHOH-COOH$	$HOOC-CH_2-CHOH-COO^-$	3.98×10^{-4}	3.40
Malic acid (2)	$HOOC-CH_2-CHOH-COO^-$	$^-OOC-CH_2-CHOH-COO^-$	5.5×10^{-6}	5.26
Succinic acid (1)	$HOOC-CH_2-CH_2-COOH$	$HOOC-CH_2-CH_2-COO^-$	6.17×10^{-5}	4.21
Succinic acid (2)	$HOOC-CH_2-CH_2-COO^-$	$^-OOC-CH_2-CH_2-COO^-$	2.3×10^{-6}	5.63
Carbonic acid (1)	H_2CO_3	HCO_3^-	4.3×10^{-7}	6.37
Carbonic acid (2)	HCO_3^-	CO_3^-	5.6×10^{-11}	10.20
Citric acid (1)	$HOOC-CH_2-C(OH)$ $(COOH)-CH_2-COOH$	$HOOC-CH_2-C(OH)$ $(COOH)-CH_2-COO^-$	8.14×10^{-4}	3.09
Citric acid (2)	$HOOC-CH_2-C(OH)$ $(COOH)-CH_2-COO^-$	$^-OOC-CH_2-C(OH)$ $(COOH)-CH_2-COO^-$	1.78×10^{-5}	4.75
Citric acid (3)	$^-OOC-CH_2-C(OH)$ $(COOH)-CH_2-COO^-$	$^-OOC-CH_2-C(OH)$ $(COO^-)-CH_2-COO^-$	3.9×10^{-6}	5.41
Phosphoric acid (1)	H_3PO_4	$H_2PO_4^-$	7.25×10^{-3}	2.14
Phosphoric acid (2)	$H_2PO_4^-$	HPO_4^{2-}	6.31×10^{-8}	7.20
Phosphoric acid (3)	HPO_4^{2-}	PO_4^{3-}	3.98×10^{-13}	12.40

Monitoring Acidity

There is an equation that connects the K'_a of any weak acid with the pH of a solution containing both that acid and its conjugate base. This relationship has wide use in biochemical practice, especially where it is necessary to control pH for optimum reaction conditions. Some reactions cannot take place if the pH varies from the optimum value. Important biological macromolecules lose activity at extremes of pH. Also, some drastic physiological consequences can result from pH fluctuations in the body (Box 2.1). Section 2.7 has more information about how pH can be controlled. To derive the involved equation, it is first necessary to solve the K'_a equation for the hydrogen ion concentration caused by dissociation of the acid HA.

$$K'_a = \frac{[H^+][A^-]}{[HA]}$$

Solving for $[H^+]$ gives

$$[H^+] = \frac{K'_a[HA]}{[A^-]}$$

Calculate the relative amounts of acetic acid and acetate ion present at the following points when acetic acid is titrated with sodium hydroxide. Also use the Henderson–Hasselbalch equation to calculate the values of the pH at these points. Compare your results with Figure 2.9.

   a. 0.1 equivalent of NaOH is added

   b. 0.3 equivalent of NaOH is added

   c. 0.5 equivalent of NaOH is added

   d. 0.7 equivalent of NaOH is added

   e. 0.9 equivalent of NaOH is added

ANSWERS   a. 90% acetic acid and 10% acetate ion; pH = 3.81   **b.** 70% acetic acid and 30% acetate ion; pH = 4.39   **c.** 50% each of acetic acid and acetate ion; pH = 4.76 (Why?)   **d.** 30% acetic acid and 70% acetate ion; pH = 5.13   **e.** 10% acetic acid and 90% acetic acid; pH = 5.71

release two hydrogen ions and have two K'_as and two pK'_as. The third group is polyprotic acids, which release more than two hydrogen ions. The two examples of polyprotic acids given here, citric acid and phosphoric acid, release three hydrogen ions and have three K'_as and three pK'_as. Amino acids and peptides, the subject of the next chapter, behave as di- and polyprotic acids; we shall see examples of their titration curves when we discuss them.

2.7
BUFFERS

A **buffer solution** consists of a mixture of a weak acid and its conjugate base. Buffer solutions tend to resist a change in pH on addition of moderate amounts of strong acid or base. Let us compare the changes in pH that occur on addition of equal amounts of strong acid or strong base to pure water at pH 7 and to a buffer solution at pH 7. If 1.0 mL of 0.1 M HCl is added to 99.0 mL of pure water, the pH drops drastically. If the same experiment is conducted with 0.1 M NaOH in place of 0.1 M HCl, the pH rises drastically (Figure 2.10).

Calculate the pH value obtained when 1.0 mL of 0.1 M HCl is added to 99.0 mL of pure water. Also, calculate the pH observed when 1.0 mL of 0.1 M NaOH is added to 99.0 mL of pure water. *Hint:* Be sure to take into account the dilution of both acid and base to the final volume of 100 mL.

ANSWERS   Acid added, pH = 3; base added, pH = 11

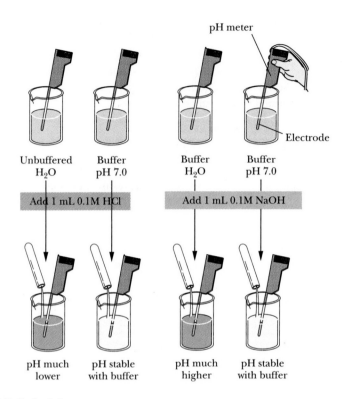

FIGURE 2.10

Buffering. Acid is added to the two beakers on the left. The pH of unbuffered H₂O drops dramatically while that of the buffer remains stable. Base is added to the two beakers on the right. The pH of the unbuffered water rises drastically while that of the buffer remains stable.

The results are different when 99.0 mL of buffer solution is used instead of pure water. A solution that contains the monohydrogen phosphate and dihydrogen phosphate ions, HPO_4^{2-} and $H_2PO_4^-$, in suitable proportions can serve as such a buffer. The Henderson–Hasselbalch equation can be used to calculate the $[HPO_4^{2-}]/[H_2PO_4^-]$ ratio that corresponds to pH 7.0.

PRACTICE   SESSION _____

Convince yourself that the proper ratio for pH 7.00 is 0.63 parts HPO_4^{2-} to 1 part $H_2PO_4^-$ by doing the calculation now.

_____

For purposes of illustration, let us consider a solution in which the concentrations are $[HPO_4^{2-}] = 0.063$ M and $[H_2PO_4^-] = 0.10$ M. If 1.0 mL of 0.10 M HCl is added to 99.0 mL of the buffer, the reaction

$$HPO_4^{2-} + H^+ \rightleftharpoons H_2PO_4^-$$

takes place, and almost all the added H^+ is used up. The concentrations of HPO_4^{2-} and $H_2PO_4^-$ change, and the new concentrations can be calculated.

	CONCENTRATIONS (mol/L)		
	$[HPO_4^{2-}]$	$[H^+]$	$[H_2PO_4^-]$
Before addition of acid	0.063	1×10^{-7}	0.10
Acid added — no reaction yet	0.063	1×10^{-3}	0.10
After acid reacts with HPO_4^{2-}	0.062	To be found	0.101

The new pH can then be calculated using the Henderson–Hasselbalch equation and the phosphate ion concentrations. The appropriate pK'_a is 7.20 (Table 2.7).

$$pH = pK'_a + \log \frac{0.062}{0.101}$$

$$pH = 7.20 + \log \frac{[HPO_4^{2-}]}{[H_2PO_4^-]}$$

The new pH is 6.99, representing a much smaller change than took place in the unbuffered pure water (Figure 2.10). Similarly, if 1.0 mL of 0.1 M NaOH is used, the same reaction takes place as in a titration:

$$H_2PO_4^- + OH^- \rightleftharpoons HPO_4^{2-} + H_2O$$

Almost all the added OH^- is used up, but a small amount remains. Since this buffer is an aqueous solution, it is still true that $K_w = [H^+][OH^-]$. The increase in hydroxide ion concentration implies that the hydrogen ion concentration decreases and that the pH increases. Use the Henderson–Hasselbalch equation to calculate the new pH and convince yourself that the result is pH = 7.01, again a much smaller change in pH than took place in pure water (Figure 2.10). Many biological reactions will not take place unless the pH remains within fairly narrow limits, and as a result buffers have great practical importance in the biochemistry laboratory.

A consideration of titration curves can give insight into the origin of this property (Figure 2.11a). The pH of a sample being titrated changes very slowly in the vicinity of the inflection point of a titration curve. Also, at the

FIGURE 2.11

The relationship between the titration curve and buffering action in $H_2PO_4^-$. (a) The titration curve of $H_2PO_4^-$, showing the buffer region for the $H_2PO_4^-/HPO_4^{2-}$ pair. (b) Relative abundances of $H_2PO_4^-$ and HPO_4^{2-}.

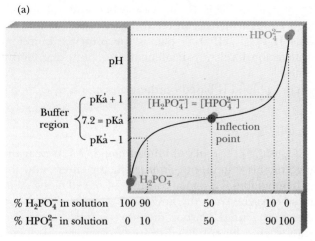

(a)

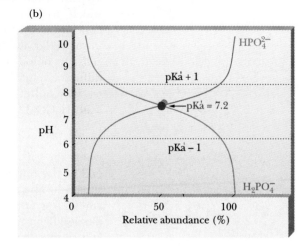

(b)

inflection point half the amount of acid originally present has been converted to the conjugate base. The second stage of ionization of phosphoric acid, $H_2PO_4^- \rightleftharpoons H^+ + HPO_4^{2-}$, was the basis of the buffer just used as an example. The pH at the inflection point of the titration is 7.20, a value numerically equal to the pK'_a of the dihydrogen phosphate ion. At this pH the solution contains equal concentrations of the dihydrogen phosphate and monohydrogen phosphate ions, the acid and base forms.

A buffer solution can maintain the pH at a relatively constant value because of the presence of appreciable amounts of both the acid and its conjugate base. This condition is met at pH values at or near the pK'_a of the acid. If OH^- is added, an appreciable amount of the acid form of the buffer is present in solution to react with the added base. If H^+ is added, there is also an appreciable amount of the basic form of the buffer to react with the added acid.

The $H_2PO_4^-/HPO_4^{2-}$ pair is suitable as a buffer near pH 7.2, and the CH_3COOH/CH_3COO^- pair near pH 4.76. At pH values below the pK_a the acid form predominates, and at pH values above the pK'_a the basic form predominates. An acid-to-base ratio of 10:1 corresponds to a pH lower than the pK'_a by about 1 pH unit, because each unit change in the pH scale corresponds to a tenfold change in the hydrogen ion concentration. An acid-to-base ratio of 1:10 corresponds to a pH about 1 unit higher than the pK'_a. The plateau region in a titration curve, where the pH does not change rapidly, covers a pH range extending approximately 1 pH unit on each side of the pK'_a. Thus, there is a range of about 2 pH units in which the buffer is effective (Figure 2.11b). The condition that a buffer contain appreciable amounts of both a weak acid and its conjugate base applies both to the ratio of the two forms and to the absolute amount of each present in a given solution. If a buffer solution contained a suitable ratio of acid to base, but very low concentrations of both, it would take very little added acid to use up all of the base form, or vice versa. A buffer solution with low concentrations of both acid and base forms is said to have a low **buffering capacity.** A buffer that contains greater amounts of both acid and base has a higher buffering capacity.

Buffer systems in living organisms and in the laboratory are based on many types of compounds. Since physiological pH in most organisms stays around 7, it might be expected that the phosphate buffer system would be widely used in living organisms. This is the case where phosphate ion concentrations are high enough for the buffer to be effective, as in most intracellular fluids. The $H_2PO_4^-/HPO_4^{2-}$ pair is the principal buffer in cells. In blood, phosphate ion levels are inadequate for buffering, and a different system operates.

The buffering system in blood is based on the dissociation of carbonic acid (H_2CO_3),

$$H_2CO_3 \rightleftharpoons H^+ + HCO_3^-$$

where the pK'_a of H_2CO_3 is 6.37. The pH of human blood, 7.4, is near the end of the buffer range of this system, but another factor enters into the situation. The protein hemoglobin is a major constituent of red blood cells, and its function is to carry oxygen from the lungs to the tissues and carbon dioxide from the tissues to the lungs. Carbon dioxide can dissolve in water and in water-based fluids, such as blood. The dissolved carbon dioxide forms carbonic acid, which in turn reacts to produce bicarbonate ion, HCO_3^-:

$$CO_2(g) \rightleftharpoons CO_2(aq)$$

$$CO_2(aq) + H_2O(\ell) \rightleftharpoons H_2CO_3(aq)$$

$$H_2CO_3(aq) \rightleftharpoons H^+(aq) + HCO_3^-(aq)$$

Net equation: $CO_2(g) + H_2O(\ell) \rightleftharpoons H^+(aq) + HCO_3^-(aq)$

At the pH of blood, which is about 1 unit higher than the pK'_a of carbonic acid, most of the dissolved CO_2 is present as HCO_3^-. The CO_2 being transported to the lungs to be expired takes the form of bicarbonate ion, which is bound to the hemoglobin molecule by ionic attraction to a charged group on the protein. There is a direct relationship between the pH of the blood and the pressure of carbon dioxide gas in the lungs.

The phosphate buffer system is common in the laboratory (*in vitro*, outside the living body) as well as in living organisms (*in vivo*). The buffer system based on tris(hydroxymethyl)aminomethane (called TRIS) is also widely used *in vitro*. Other buffers that have come into wide use more recently are **zwitterions,** compounds that each have both a positive charge and a negative charge. Zwitterions are usually considered less likely to interfere with biochemical reactions than some of the earlier buffers (Table 2.8).

TABLE 2.8 Acid and Base Forms of Some Useful Biochemical Buffers

ACID FORM		BASE FORM	pK$_a$
TRIS—H$^+$ (protonated form) $(HOCH_2)_3CNH_3^+$	N—tris[hydroxymethyl]aminomethane (TRIS) $\rightleftharpoons$	TRIS (free amine) $(HOCH_2)_3CNH_2$	8.3
$^-$TES—H$^+$ (zwitterionic form) $(HOCH_2)_3\overset{+}{C}NH_2CH_2CH_2SO_3^-$	N—tris[hydroxymethyl]methyl-2-aminoethane sulfonate (TES) $\rightleftharpoons$	$^-$TES (anionic form) $(HOCH_2)_3CNHCH_2CH_2SO_3^-$	7.55
$^-$HEPES—H$^+$ (zwitterionic form) $HOCH_2CH_2\overset{+}{N}$⟨ ⟩$NCH_2CH_2SO_3^-$ (H)	N—2—hydroxyethylpiperazine-N'-2-ethane sulfonate (HEPES) $\rightleftharpoons$	$^-$HEPES (anionic form) $HOCH_2CH_2N$⟨ ⟩$NCH_2CH_2SO_3^-$	7.55
$^-$MOPS—H$^+$ (zwitterionic form) O⟨ ⟩$^+NCH_2CH_2CH_2SO_3^-$ (H)	3—[N—morpholino]propane-sulfonic acid (MOPS) $\rightleftharpoons$	$^-$MOPS (anionic form) O⟨ ⟩$NCH_2CH_2CH_2SO_3^-$	7.2
$^{2-}$PIPES—H$^+$ (protonated dianion) $^-O_3SCH_2CH_2N$⟨ ⟩$^+NCH_2CH_2SO_3^-$ (H)	Piperazine—N,N'-bis[2-ethanesulfonic acid] (PIPES) $\rightleftharpoons$	$^{2-}$PIPES (dianion) $^-O_3SCH_2CH_2N$⟨ ⟩$NCH_2CH_2SO_3^-$	6.8

Most living systems operate at pH levels close to 7. The pK'_a values of many functional groups, such as the carboxyl and amino groups, are well above or well below this value. As a result, under physiological conditions, many important biomolecules exist as charged species to one extent or another. The practical consequences of this fact are explored in Box 2.1.

SOME PHYSIOLOGICAL CONSEQUENCES OF BLOOD BUFFERING

The process of respiration plays an important role in the buffering of blood. In particular, an increase in H^+ concentration can be dealt with by raising the rate of respiration. Initially, the added hydrogen ion binds to bicarbonate ion, forming carbonic acid.

$$H^+(aq) + HCO_3^-(aq) \rightleftharpoons H_2CO_3(aq)$$

An increased level of carbonic acid raises the levels of dissolved carbon dioxide and, ultimately, gaseous carbon dioxide in the lungs.

$$H_2CO_3(aq) \rightleftharpoons CO_2(aq) + H_2O(\ell)$$
$$CO_2(aq) \rightleftharpoons CO_2(g)$$

A high respiration rate removes this excess carbon dioxide from the lungs, starting a shift in the equilibrium positions of all the foregoing reactions. The removal of gaseous CO_2 decreases the amount of dissolved CO_2. Hydrogen ion reacts with HCO_3^- and in the process lowers the H^+ concentration of the blood back to its original level. In this way blood pH is kept constant.

In contrast, *excessively* deep and rapid breathing, or *hyperventilation*, removes such large amounts of carbon diox-

ide from the lungs that it raises the pH of blood, sometimes to dangerously high levels that bring on weakness and fainting. Sprinters, however, have learned how to make use of the increase in blood pH caused by hyperventilation. Short bursts of strenuous exercise produce high levels of lactic acid in the blood as a result of the breakdown of glucose. The presence of so much lactic acid tends to lower the pH of the blood, but a brief (30-second) period of hyperventilation before a short sprint (say, a 100-meter dash) counteracts the effects of the added lactic acid and maintains the pH balance.

An increase in H^+ in blood can be caused by large amounts of any acid entering the bloodstream. Aspirin, like lactic acid, is an acid, and extreme acidity resulting from the ingestion of large doses of aspirin can cause *aspirin poisoning*. Exposure to *high altitudes* has an effect similar to hyperventilation at sea level. In response to the tenuous atmosphere, the rate of respiration increases. As with hyperventilation, more carbon dioxide is expired from the lungs, ultimately lowering the H^+ level in blood and raising the pH. When a person who normally lives at sea level is suddenly placed at a high elevation, the blood pH rises temporarily, until the person becomes acclimated.

S U M M A R Y

The properties of the water molecule have a direct effect on the behavior of biomolecules. Water is a polar molecule, with partial charges of opposite signs at its ends. There are forces of attraction between the unlike partial charges. In addition, in both the liquid

and solid states water molecules are extensively hydrogen bonded to one another. Polar substances tend to dissolve in water, and nonpolar substances do not. Hydrogen bonding between water and polar solutes takes place in aqueous solutions. The three-dimensional structures of many important biomolecules, including proteins and nucleic acids, are stabilized by hydrogen bonds.

The degree of dissociation of acids in water can be characterized by an acid dissociation constant, K'_a, which gives a numerical indication of the strength of the acid. The self-dissociation of water can be characterized by a similar constant, K_w. Since the hydrogen ion concentration of aqueous solutions can vary by many orders of magnitude, it is desirable to define a quantity, pH, that expresses the concentration of hydrogen ions conveniently. A similar quantity, pK'_a, can be used as an alternative expression for the strength of any acid. The pH of a solution of a weak acid and its conjugate base can be related to the pK'_a of that acid by the Henderson–Hasselbalch equation.

In aqueous solution, the relative concentrations of a weak acid and its conjugate base can be related to the titration curve of that acid. In the region of the titration curve in which the pH changes slowly on addition of acid or base, the acid/base concentration ratio varies within a fairly narrow range (10:1 at one extreme and 1:10 at the other). The tendency to resist a change in pH on addition of relatively small amounts of acid or base is characteristic of buffer solutions. The control of pH by buffers depends on the fact that their compositions reflect the acid/base concentration ratio in the region of the titration curve where there is little change in pH.

Both organic chemistry and biochemistry deal with the reactions of carbon-containing molecules. Both disciplines base their approaches on the behavior of functional groups, but their emphases differ because some functional groups important to organic chemistry do not play a role in biochemistry, and vice versa. Functional groups of importance in biochemistry include carbonyl groups, hydroxyl groups, carboxyl groups, amines, amides, and esters; derivatives of phosphoric acid such as esters and anhydrides are also important.

E X E R C I S E S

1. Rationalize the fact that hydrogen bonding has not been observed between CH_4 molecules.
2. Many properties of acetic acid can be rationalized in terms of a hydrogen-bonded dimer. Propose a structure for such a dimer.
3. Identify the conjugate acids and bases in the following pairs of substances.

   $(CH_3)_3NH^+/(CH_3)_3N$

   $^+H_3N—CH_2COOH/^+H_3N—CH_2—COO^-$

   $^+H_3N—CH_2—COO^-/H_2N—CH_2—COO^-$

   $^-OOC—CH_2—COOH/^-OOC—CH_2—COO^-$

   $^-OOC—CH_2—COOH/HOOC—CH_2—COOH$

4. Calculate the hydrogen ion concentration, $[H^+]$, for each of the following materials.

   Blood plasma, pH 7.4

   Orange juice, pH 3.5

   Human urine, pH 6.2

   Household ammonia, pH 11.5

   Gastric juice, pH 1.8

5. Suggest a suitable buffer range for each of the following substances.

   Lactic acid (pK'_a = 3.86) and its sodium salt

   Acetic acid (pK'_a = 4.76) and its sodium salt

   TRIS (see Table 2.8; pK'_a = 8.3) in its protonated form and its free amine form

   HEPES (see Table 2.8; pK'_a = 7.55) in its zwitterionic form and its anionic form

6. What is the $[CH_3COO^-]/[CH_3COOH]$ ratio in an acetate buffer at pH 5.00?
7. How would you prepare one liter of a 0.05 M phosphate buffer at pH 7.5, using crystalline K_2HPO_4 and a solution of 1 M HCl?
8. The buffer needed for Exercise 7 can also be prepared using crystalline NaH_2PO_4 and a solution of 1 M NaOH. How would you do this?
9. In Section 2.5 we said that at the equivalence point of a titration of acetic acid *essentially all* the acid has been converted to acetate ion. Why do we not say that *all* the acetic acid has been converted to acetate ion?
10. Define buffering capacity. How do the following buffers differ in capacity? How do they differ in pH?

    0.01 M Na_2HPO_4 and 0.01 M NaH_2PO_4

    0.1 M Na_2HPO_4 and 0.1 M NaH_2PO_4

    1.0 M Na_2HPO_4 and 1.0 M NaH_2PO_4

11. Identify the zwitterions in the list of substances in Exercise 3.

12. If you mixed equal volumes of 0.1 M HCl and 0.2 M TRIS (free amine form), is the resulting solution a buffer? Why or why not?

13. The measured pH of a sample of lemon juice is 2.1. Calculate the hydrogen ion concentration.

14. What is the ratio of concentrations of acetate ion and undissociated acetic acid in a solution that has a pH of 5.12?

15. You need to carry out an enzymatic reaction at pH 7.5. A friend suggests a weak acid with a pK'_a of 3.9 as the basis of a buffer. Will this substance and its conjugate base make a suitable buffer? Why or why not?

16. A frequently recommended treatment for hiccups is to hold one's breath. The resulting condition, hypoventilation, causes a buildup of carbon dioxide in the lungs. Predict the effect on the pH of blood.

17. Aspirin is an acid with a pK'_a of 3.5; its structure includes a carboxyl group. To be absorbed into the bloodstream it must pass through the membrane lining the stomach and the small intestine. Electrically neutral molecules can pass through a membrane more easily than can charged molecules. Would you expect more aspirin to be absorbed in the stomach, where the pH of gastric juice is about 1, or in the small intestine, where the pH is about 6? Why?

18. Match each entry in column a with one in column b. (Column a lists the names of some important functional groups, and column b shows their structures.)

Column a	Column b
Amino group	CH_3SH
Carbonyl group (ketone)	$CH_3CH{=}CHCH_3$
	$CH_3CH_2\overset{\displaystyle O}{\overset{\|}{C}}H$
Hydroxyl group	
Carboxyl group	$CH_3CH_2NH_2$
	$CH_3\overset{\displaystyle O}{\overset{\|}{C}}OCH_2CH_3$
Carbonyl group (aldehyde)	
Thiol group	$CH_3CH_2OCH_2CH_3$
	$CH_3\overset{\displaystyle O}{\overset{\|}{C}}CH_3$
Ester linkage	
	$CH_3\overset{\displaystyle O}{\overset{\|}{C}}OH$
Double bond	
Amide linkage	CH_3OH
Ether	$CH_3\overset{\displaystyle O}{\overset{\|}{C}}N(CH_3)_2$

19. Identify the functional groups in the following compounds.

Glucose

A triglyceride

A peptide

Vitamin A

20. A friend who is enthusiastic about health foods and organic gardening asks you whether urea is "organic" or "chemical." How do you reply to this question?

ANNOTATED BIBLIOGRAPHY

Barrow, G. M. *Physical Chemistry for the Life Sciences.* 2nd ed. New York: McGraw-Hill, 1981. [Acid–base reactions are discussed in Chapter 4, with titration curves treated in great detail.]

Fasman, C. D., ed. *Handbook of Biochemistry and Molecular Biology: Physical and Chemical Data Section.* 2 vols. 3rd ed. Cleveland: The Chemical Rubber Company, 1976. [Includes a section on buffers and directions for preparation of buffer solutions (Vol. 1, pp. 353–378). Other sections cover all important types of biomolecules.]

Ferguson, W. J., and N. E. Good. Hydrogen Ion Buffers. *Anal. Biochem.* **104,** 300–310 (1980). [A description of useful zwitterionic buffers.]

Olson, A., and D. Goodsell. Visualizing Biological Molecules. *Sci. Amer.* **268** (6), 62–68 (1993). [An account of how computer graphics can be used to represent molecular structure and properties.]

Pauling, L. *The Nature of the Chemical Bond.* 3rd ed. Ithaca, NY: Cornell University Press, 1960. [A classic. Chapter 12 is devoted to hydrogen bonding.]

Rand, R. Raising Water to New Heights. *Science* **256,** 618 (1992). [A brief perspective on the contribution of hydration to molecular assembly and protein catalysis.]

Simmonds, R. *Chemistry of Biomolecules: An Introduction.* Boca Raton, FL: CRC Press, 1992. [A book that elucidates the structures of biomolecules, their chemical synthesis, and the relationship between biological properties and chemical structure.]

Westheimer, F. H. Why Nature Chose Phosphates. *Science* **235,** 1173–1178 (1987). [A discussion of the importance of phosphate groups in biochemistry, particularly in the backbones of nucleic acids. The author is an eminent organic chemist.]

Westhof, E., ed. *Water and Biological Macromolecules.* Boca Raton, FL: CRC Press, 1993. [A series of articles about the role of water in hydration of biological macromolecules and the forces involved in macromolecular complexation and cell–cell interactions.]

Any current textbook of organic chemistry can be expected to include a discussion of functional groups and their reactions. Most organic chemistry texts have several chapters that discuss carbohydrates, lipids, proteins, and nucleic acids from the point of view of organic chemists.

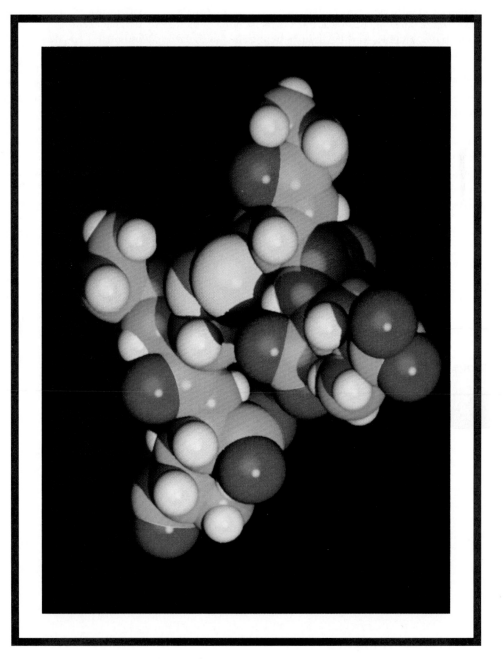

Computer-generated structure of the oxidized form of the peptide glutathione.

Part II

Components of Cells: Structure and Function

Outline

Siegfried Reich

Since childhood, Siegfried Reich has enjoyed working with his hands, taking things apart and putting them together. As a college student, he found expression for that way of solving problems in the laboratory. Today, as senior scientist and project leader for Agouron Pharmaceuticals, Inc., he is still working with his hands — but now with a computer keyboard and mouse. And these days, what he takes apart and puts together are intricate molecular configurations on a computer screen.

Reich received his B.S. in chemistry from San Diego State University in 1982 and went on to earn a Ph.D. in chemistry at the University of California at Irvine in 1986. After two years of postdoctoral work at the University of California at Berkeley, he joined Agouron, which is based in La Jolla, California. It is one of a number of companies created in the 1980s to pursue alternative approaches to drug development. Agouron's specialty is protein structure–based drug design, which combines the disciplines of chemistry, physics, and biology. Siegfried Reich helps design molecules to inhibit the enzymes necessary for the replication of diseases. If these discov-

eries make it through the rigorous drug approval process, they could, in the future, play a role in treatments of maladies such as cancer and AIDS.

Tell us about your research as a graduate student at UC Irvine.

In graduate school, I got into a specific area of synthetic organic chemistry called asymmetric synthesis. Many compounds can take similar, but different, forms — much like a right and a left hand. With asymmetric synthesis, the goal is to synthesize compounds so that you can get one form or the other, selectively. For example, compounds that make up drugs will often exist in different forms. One form may be the desired active component; the other may be inactive or, even worse, toxic. An understanding of asymmetric synthesis helps you to synthesize the correct form.

The aspect of my graduate work I really liked was that it took chemistry from a two-dimensional to a three-dimensional perspective. And it helped me appreciate the relevance to a pharmaceutical setting — to how the three-dimensionality of molecules relates to molecular recognition. In graduate

school we often used Tinker Toy–like scale models of small molecular structures. And in group meetings, we would have a set of these models on the table and discuss why certain reactions were working the way they were and why they gave us certain products. You could look at these models and plan out a synthesis. My thesis project was the synthesis of a natural product, Erythronolide A seco acid, which is a precursor to the antibiotic Erythromycin.

What did you do after graduate school?

I spent two years in postdoctoral research at the University of California at Berkeley, working with Dr. Paul Bartlett, who is very well known for his involvement in the area of bioorganic chemistry. His research group tries to ask biochemical questions by synthesizing certain molecules.

And what kind of questions would those be?

Let me tell you about the specific project I worked on. We were looking at a group of enzymes called proteases, which have the job of clipping or cutting other proteins. We were interested in seeing if we could design a protease inhibitor based on an analysis of the three-dimensional structure of the enzyme. That is how I got my feet wet in this whole concept of using x-ray crystallography — the technology that Agouron is based on. As synthetic chemists we do not solve the crystal structures ourselves; that is done by specialists called crystallographers. But with x-ray crystallography, we can look at a protein in its three-dimensional state and ask how to design a small molecule to interact with that protein, which is how most drugs work.

Using x-ray crystallography must be quite different from using those scale models.

Well, yes and no. The basis is the same. The difference is that now we have moved from a model on your desktop to a computer graphics machine, where you are wearing special stereoptic glasses that allow you to see in three dimensions. So you are doing the same thing — asking questions about molecular recognition, which is what drives drug action. But you are doing it with more technology and a lot greater accuracy.

What got you interested in working for Agouron?

I was excited by the concept of working at a smaller company that was doing very interesting science. I interviewed at a number of large pharmaceutical companies. They were very intriguing places, but most of the larger companies are still based on the old school, which involves screening lots of compounds until you get a hit. Agouron represents a new arm of the pharmaceutical industry: looking at the receptor at the molecular level and asking whether we can design drugs to bind with that receptor. To me that just made a lot more scientific sense, although much is yet to be proven.

What does your job involve at Agouron?

I am part of the Medicinal Chemistry Group at Agouron, which is made up of synthetic organic chemists. Each of us knows how to go into the laboratory and devise and execute a synthesis of a small molecule. What I do is use that background in combination with information I get from crystallographers. I have a graphics machine on my desk, and that allows me to view this enzyme in its three-dimensional state. When I look at the screen, I am looking at the active site of the enzyme, which is like a pocket where its substrate will normally bind. This is where it all happens, pretty much. If you can design a small molecule to complement that active site well enough, then you can shut that enzyme down. That really is the nuts and bolts of what I try to do.

What are the pluses and minuses of trying to *design* new drug molecules, versus *screening* for them?

Well, we can start with a potential minus. Molecular design requires a significant investment in technology. Crystallography is not something you just set up in a month's time. X-ray defractometers, the machines that collect the data, are generally custom-made. And you need to employ a number of crystallographers. The other factor is that crystallography is nothing without crystals. If your particular receptor is not amenable to crystallization, then you have serious problems.

The plus to the drug design approach is that you have an open palette of atoms to use in designing these molecules. You are really limited only by your imagination. With the new technology and with enough manpower behind it, what used to take literally years to solve can now be done in a matter of months. You have an opportunity to come up with things that have never been seen before. That makes the work very interesting. And

it gives you a very strong patent position when a novel active compound is discovered this way.

How do you target a protein for research?

Generally we pick a protein because there is a clear connection to some disease state. A lot hinges upon previous scientific work in the molecular biology and biochemistry arena. Usually, target proteins will have been implicated as crucial to some disease. The other thing that we will certainly take into consideration is the feasibility of getting a crystal.

We start with a theory that there is a connection, that a particular compound will keep cells from proliferating — for example, in cancer. But generally such evidence is obtained in the test tube — *in vitro* evidence that if you shut this enzyme down, cancer cells will cease to proliferate. But that isn't the whole story. The final story is when you can actually shut that enzyme down in the body. And until you have that clinical-type feedback, it's an open question.

Could you give an example?

A good example is the protein that we and other companies are working on right now. That's HIV protease, related to the AIDS virus. For a long time it was clear that the HIV virus needed this enzyme to replicate and thrive, and that was supported by clear evidence in the test tube and cell culture. But only relatively recently has there been evidence from clinical data that administering the compound does have an effect. But even then, you could say we really do not have the final proof. The real proof will be if patients actually have longer, healthier lives after the compound is administered.

You are the group leader of that HIV protease project at your company.

Yes, it's my job to organize it and make sure it is moving at a certain pace. We are at the stage of gearing up to put some of our compounds in patients. But even so, we probably won't have information about clinical efficacy for a year to two years from now.

Aren't there other versions of these types of compounds already under clinical study?

Yes, but there is a problem with currently available products that inhibit HIV protease. Although they are very potent, they are not significantly orally available. A drug like this is a treatment, not a cure, for AIDS. And AIDS patients will probably take such drugs for the rest of their lives, every day. If people have to take this drug intravenously, with a needle, they are going to be a lot more hesitant to take the drug. You want it to be oral.

And what do you hope for the product you are working on?

Our compound has significantly enhanced oral availability. That means you generally do not need to administer as much of the compound orally to achieve a desired effect. If currently

available compounds are given orally, maybe 5 to 10% will actually be absorbed. So a higher overall level of the drug must be maintained just to get that 5 or 10% into the bloodstream.

And when you are working on the HIV protease project, for example, is there more than one solution?

Yes, absolutely. And that gets back to the question about the pros and cons of drug design versus screening. When you sit in front of the screen and view this empty pocket, you ask what three-dimensional small molecule could fit in there. There are going to be a number of solutions to that question. And that is one of the powers of this design approach, in that you are not necessarily biased by any preexisting solution. You really have a full library of atoms that you can use to build into the pocket.

As an organic chemist, what is your role in this process?

I look at the active site on the protein — this pocket — and I try to decide what kind of small molecule will complement that surface, in both its three-dimensional structure and its electronic structure. Then I go ahead

and build that small molecule on the screen. I build it in three dimensions and dock it, and see how well it fits. Where it does not fit as well, I refine and refine. And drawing on my organic chemistry background, I am always thinking in the back of my head about whether what I am working with is something that can actually be made and that will be stable. There is always this practical component. You do not want to have someone sitting in front of a computer screen designing the perfect molecule that will complement everything but that can never be made.

When you have your design, what do you do with it?

Okay, now if I've got a molecule that looks like it fits well, and if that conclusion is supported with computational analysis, I will look into how to synthesize it. If it looks like I can devise a reasonable procedure, I order the necessary chemicals and either I or a co-worker will go ahead and start synthesizing it.

Computer-generated drawings of (a) the HIV protease enzyme and (b) the drug AG-1284 bound in the active site of HIV protease.

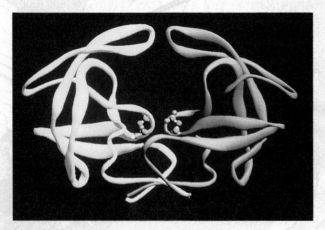

(a)

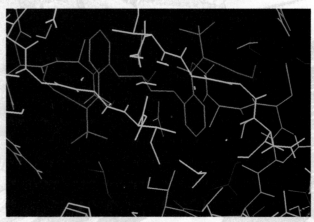

(b)

And once you have the compound made, how do you test it out?

Once we have established that what we have synthesized is the compound we wanted, then we take it to the biochemist, who will conduct an *in vitro* assay. The biochemist begins with the enzyme; then mixes in the inhibitor, the new compound under study; then adds the substrate. The process is watched over a period of time and is measured with a spectrophotometer. If it works, the compound will start to bind to the enzyme and inhibit its activity. Depending on how good your inhibitor is, you will see more or less inhibition.

How often does your inhibitor turn out to be highly active?

Not too often. It's generally a slow process. For example, in the series that we recently developed and described, the HIV protease inhibitor called AG-1284, we started off with a compound in which you could barely see any inhibiting activity. Over the course of a year and a half, we increased its potency over 10,000-fold.

Wow. How do you accomplish that?

In a nutshell, we use protein crystallography as a tool to ask the next question. If we have a compound with borderline activity, the protein will be crystallized again. But this time, it is crystallized with your inhibitor bound to it. So you can get feedback. Where you previously just had this theoretical idea about how this molecule might bind to the active site, now you get to find out how it did bind.

What is the outlook for HIV treatment?

I think everyone understands that the HIV problem is one of the biggest challenges the pharmaceutical industry has ever had. One reason is that this virus mutates, and it's deadly. I think the bottom line to the HIV treatment question will be to administer a number of agents, and that will reduce the chance of mutations cropping up and being a problem.

It must be exciting, working on something like this.

Oh, it is. I have a friend who has AIDS. He is one person. But still, knowing that you can potentially do something that may give a friend some more years in this world is just — you can't put it into words. Also, as a scientist, it is nice to know that, after all those years of schooling, I am doing something that I like to do and that

has such a practical application. Being in industry and having a practical application for your science is very gratifying.

It seems that everything you described requires a multidisciplinary mindset.

Absolutely. None of this would be possible unless we had people from all disciplines — biochemistry, crystallography, organic chemistry, pharmacology. We have regular meetings, and all these people will come together and brainstorm about how we might be able to progress.

What about this field as a future, particularly for careers?

I think it looks very, very good for organic chemistry, due in large part to the explosion of the biotechnology area. It is a natural marriage, having this very powerful field of biotechnology recognizing a requirement for small molecules that can be designed by organic chemists. There is one other thing — the issue of creativity in this job. That is why I do what I do. There is so much room for being creative in a small company, pushing the edge of technology.

Chapter 3

Amino Acids
and Peptides

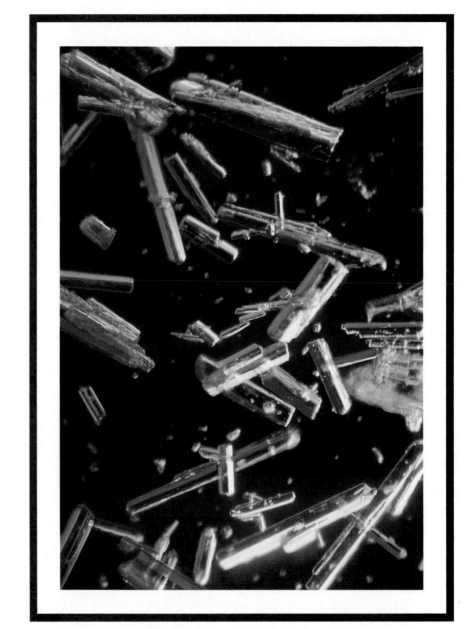

Crystals of glycine, an amino acid, viewed under polarized light.

Proteins are long chains of amino acids linked together by peptide bonds with a positively charged nitrogen-containing amino group at one end and a negatively charged carboxyl group at the other end. Along the chain is a series of different side chains, one for each of the 20 amino acids. A linkage of two amino acids is a dipeptide; three amino acids form a tripeptide. The sequence of the amino acids is of the utmost importance. Glycine–lysine–alanine is a different peptide from alanine–lysine–glycine, with a different chemical significance. (In the same manner, the motto "Talk little, do much" has a different meaning from "Do little, talk much.") For a chain 20 amino acids long there are more than a billion possible sequences; for a protein chain of 100 amino acids there are more possible sequences than there are atoms in the Universe! Literally, the sequence is the message. It determines exactly how the protein will fold up in a three-dimensional conformation to perform its precise biochemical function.

3.1
AMINO ACIDS – THEIR GENERAL FORMULA AND THREE-DIMENSIONAL STRUCTURE

Among all the possible amino acids, only 20 are usually found in proteins. The general structure of amino acids involves an **amino group** and a **carboxyl group,** both of which are bonded to the α carbon (the one next to the carboxyl group). The α carbon is also bonded to a hydrogen and to the **side chain group,** which is represented by the letter R. The R group determines the identity of the particular amino acid (Figure 3.1). The two-dimensional formula shown here can only partially convey the common structure of amino acids, because one of the most important properties of these compounds is their three-dimensional shape, or **stereochemistry.**

Every object has a mirror image. Many pairs of objects that are mirror images can be turned so as to be superimposed on each other; two identical solid-colored coffee mugs are an example. In other cases the mirror-image objects cannot be superimposed on one another but are related to each other as the right hand is to the left. Such nonsuperimposable mirror images are said to be **chiral** (from the Greek *cheir*, "hand"). Many important biomolecules are chiral.

A frequently encountered chiral center in biomolecules is a carbon atom with four different groups bonded to it (Figure 3.2). Such a center occurs in all amino acids except glycine. Glycine has two hydrogen atoms bonded to the α carbon; in other words, the side chain (R group) of glycine is hydrogen. Glycine is not chiral (or, alternatively, is *achiral*) because of this symmetry. In all the other commonly occurring amino acids, the α carbon has four different groups bonded to it, giving rise to two nonsuperimposable mirror-image forms. Figure 3.3 shows perspective drawings of these two possibilities, or **stereoisomers,** for alanine, where the R group is —CH₃. The dashed wedges represent bonds directed away from the observer, and the solid triangles represent bonds directed out of the plane of the paper in the direction of the observer.

The two possible stereoisomers of another chiral compound, L- and D-glyceraldehyde, are shown for comparison with the corresponding forms of

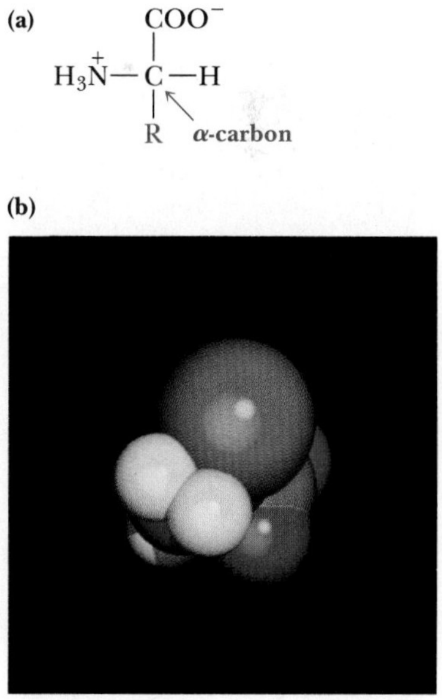

(a)

$$H_3\overset{+}{N} - \underset{\underset{R}{|}}{\overset{\overset{COO^-}{|}}{C}} - H$$

α-carbon

(b)

FIGURE 3.1

(a) The general formula of amino acids, showing the ionic forms that predominate at pH 7. (b) Space-filling model of an amino acid. The large green sphere represents the R group.

(a)

(b)

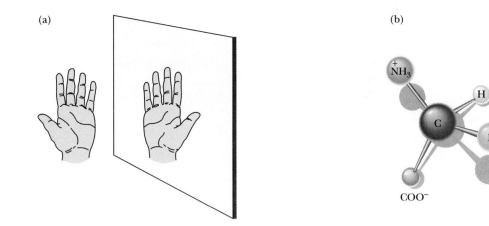

FIGURE 3.2

(a) Left and right hands are mirror images of each other. (b) A pair of mirror-image amino acids. Amino acids found in proteins have the configuration shown on the right side of the mirror.

alanine. These two forms of glyceraldehyde are the basis of the classification of amino acids into L and D forms. The terminology comes from the Latin *laevus* and *dexter,* meaning "left" and "right," respectively. The two stereoisomers of each amino acid are designated as L and **D amino acids** on the basis of their similarity to the glyceraldehyde standard. In the L form of glyceraldehyde, the hydroxyl group is on the left side of the molecule, and in the D form it is on the right side, as shown in perspective in Figure 3.3 (a Fischer projection). In an amino acid, the position of the amino group on the left or right side of the α carbon determines the L or D designation. The amino acids that occur in proteins are all of the L form. Although D amino

(a)

(b)

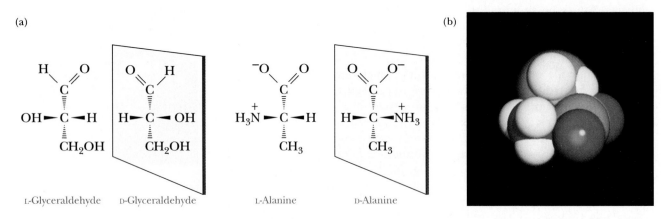

FIGURE 3.3

Stereochemistry of alanine and glyceraldehyde. The amino acids found in proteins have the same chirality as L-glyceraldehyde, which is opposite that of D-glyceraldehyde. (b) Space-filling model of L-alanine.

acids occur in nature, most often in bacterial cell walls and in some antibiotics, they are not found in proteins.

3.2
THE STRUCTURES AND PROPERTIES OF THE INDIVIDUAL AMINO ACIDS

The R groups, and thus the individual amino acids, are classified according to several criteria, two of which are particularly important. The first of these is the *polar or nonpolar nature* of the side chain. The second depends on the presence of an *acidic or basic group* in the side chain. Other useful criteria include the presence of functional groups in the side chains and the nature of those groups.

As mentioned, the side "chain" of the simplest amino acid, glycine, consists of a hydrogen atom, and in this case alone two hydrogen atoms are bonded to the α carbon. In all other amino acids the side chain is larger and more complex (Figure 3.4). We frequently refer to amino acids by three-letter or one-letter abbreviations of their names. Table 3.1 lists these abbreviations.

Group 1 – Amino Acids with Nonpolar Side Chains

One group of amino acids has *nonpolar side chains*. This group consists of alanine, valine, leucine, isoleucine, proline, phenylalanine, tryptophan, and

(Text continues on p. 74)

T A B L E   3 . 1   Names and Abbreviations of the Common Amino Acids

AMINO ACID	THREE-LETTER ABBREVIATION	ONE-LETTER ABBREVIATION
Alanine	Ala	A
Arginine	Arg	R
Asparagine	Asn	N
Aspartic acid	Asp	D
Cysteine	Cys	C
Glutamine	Gln	Q
Glutamic acid	Glu	E
Glycine	Gly	G
Histidine	His	H
Isoleucine	Ile	I
Leucine	Leu	L
Lysine	Lys	K
Methionine	Met	M
Phenylalanine	Phe	F
Proline	Pro	P
Serine	Ser	S
Threonine	Thr	T
Tryptophan	Trp	W
Tyrosine	Tyr	Y
Valine	Val	V

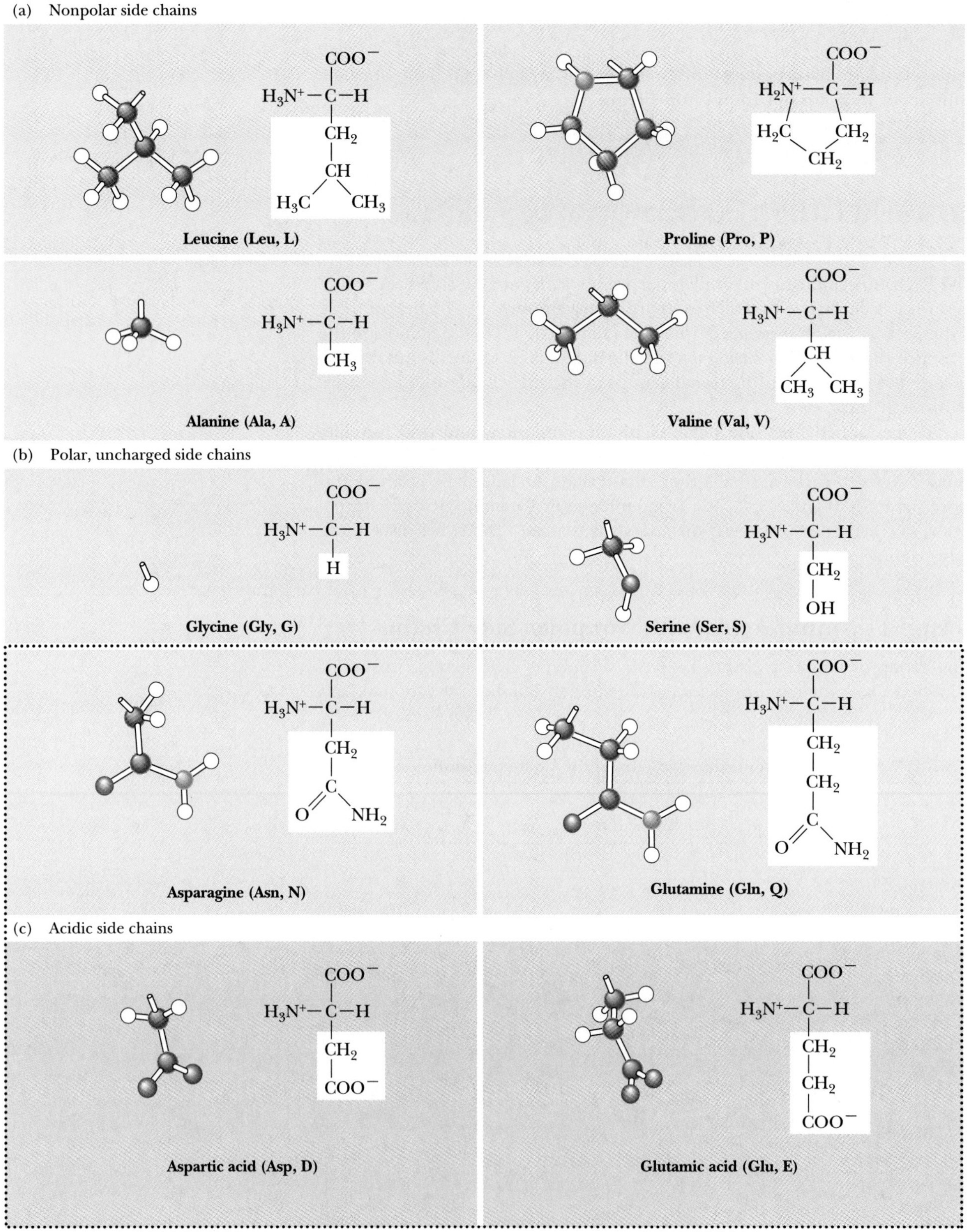

(a) Nonpolar side chains

COO⁻
|
H₃N⁺—C—H
|
CH₂
|
CH
H₃C   CH₃

Leucine (Leu, L)

COO⁻
|
H₂N⁺—C—H
|
H₂C   CH₂
CH₂

Proline (Pro, P)

COO⁻
|
H₃N⁺—C—H
|
CH₃

Alanine (Ala, A)

COO⁻
|
H₃N⁺—C—H
|
CH
CH₃ CH₃

Valine (Val, V)

(b) Polar, uncharged side chains

COO⁻
|
H₃N⁺—C—H
|
H

Glycine (Gly, G)

COO⁻
|
H₃N⁺—C—H
|
CH₂
|
OH

Serine (Ser, S)

COO⁻
|
H₃N⁺—C—H
|
CH₂
|
C
O   NH₂

Asparagine (Asn, N)

COO⁻
|
H₃N⁺—C—H
|
CH₂
|
CH₂
|
C
O   NH₂

Glutamine (Gln, Q)

(c) Acidic side chains

COO⁻
|
H₃N⁺—C—H
|
CH₂
|
COO⁻

Aspartic acid (Asp, D)

COO⁻
|
H₃N⁺—C—H
|
CH₂
|
CH₂
|
COO⁻

Glutamic acid (Glu, E)

FIGURE 3.4

Structures of the amino acids. The 20 amino acids found in proteins are shown in their predominant forms at pH 7. The R groups are shown as ball and stick

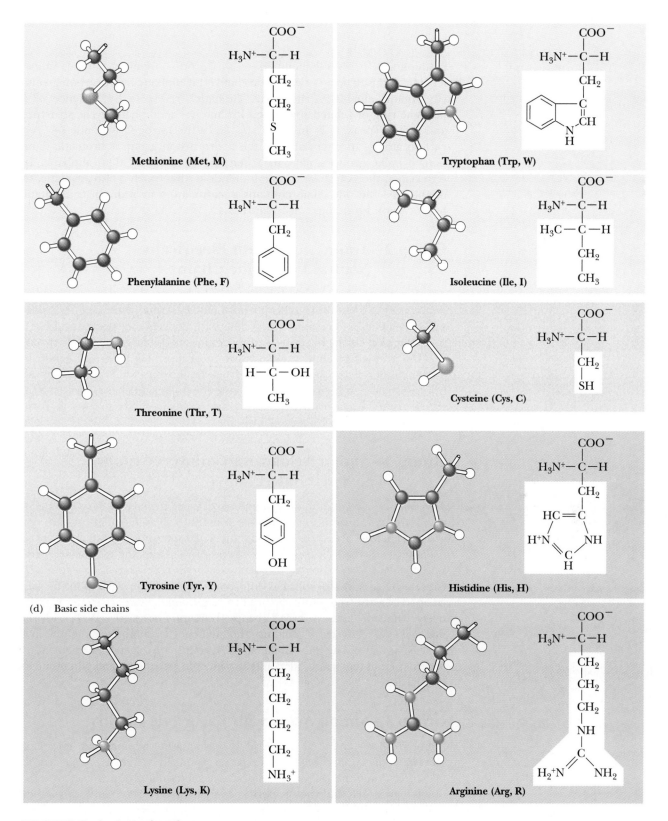

Methionine (Met, M)

$$COO^-$$
$$H_3N^+-C-H$$
$$|$$
$$CH_2$$
$$|$$
$$CH_2$$
$$|$$
$$S$$
$$|$$
$$CH_3$$

Tryptophan (Trp, W)

$$COO^-$$
$$H_3N^+-C-H$$
$$|$$
$$CH_2$$
$$C$$
$$CH$$
$$N$$
$$H$$

Phenylalanine (Phe, F)

$$COO^-$$
$$H_3N^+-C-H$$
$$|$$
$$CH_2$$

Isoleucine (Ile, I)

$$COO^-$$
$$H_3N^+-C-H$$
$$|$$
$$H_3C-C-H$$
$$|$$
$$CH_2$$
$$|$$
$$CH_3$$

Threonine (Thr, T)

$$COO^-$$
$$H_3N^+-C-H$$
$$|$$
$$H-C-OH$$
$$|$$
$$CH_3$$

Cysteine (Cys, C)

$$COO^-$$
$$H_3N^+-C-H$$
$$|$$
$$CH_2$$
$$|$$
$$SH$$

Tyrosine (Tyr, Y)

$$COO^-$$
$$H_3N^+-C-H$$
$$|$$
$$CH_2$$
$$|$$
$$OH$$

Histidine (His, H)

$$COO^-$$
$$H_3N^+-C-H$$
$$|$$
$$CH_2$$
$$HC=C$$
$$H^+N \quad NH$$
$$C$$
$$H$$

(d) Basic side chains

Lysine (Lys, K)

$$COO^-$$
$$H_3N^+-C-H$$
$$|$$
$$CH_2$$
$$|$$
$$CH_2$$
$$|$$
$$CH_2$$
$$|$$
$$CH_2$$
$$|$$
$$NH_3^+$$

Arginine (Arg, R)

$$COO^-$$
$$H_3N^+-C-H$$
$$|$$
$$CH_2$$
$$|$$
$$CH_2$$
$$|$$
$$CH_2$$
$$|$$
$$NH$$
$$|$$
$$C$$
$$H_2^+N \quad NH_2$$

F I G U R E   3 . 4 continued

models and as structures against a white background. The amino acids within the dotted line have carboxyl groups in their side chains, either as free carboxyls or as amides.

methionine. In several members of this group—namely alanine, valine, leucine, and isoleucine—each side chain is an aliphatic hydrocarbon group. (In organic chemistry, the term "aliphatic" refers to the absence of a benzene ring or related structure.) Proline has an aliphatic cyclic structure and strictly speaking is an *imino acid,* since the nitrogen is bonded to two carbon atoms. In phenylalanine the hydrocarbon group is aromatic (contains a cyclic group similar to a benzene ring) rather than aliphatic. In tryptophan the side chain contains an indole ring, which is also aromatic. In methionine the side chain contains a sulfur atom in addition to aliphatic hydrocarbon groupings. (See Figure 3.4.)

Group 2—Amino Acids with Electrically Neutral Polar Side Chains

Another group of amino acids has *polar side chains that are electrically neutral (uncharged) at neutral pH.* This group includes serine, threonine, tyrosine, and cysteine. Glycine is also included here for convenience because it lacks a nonpolar side chain. In serine and threonine the polar group is a hydroxyl (—OH) bonded to aliphatic hydrocarbon groups. The hydroxyl group in tyrosine is bonded to an aromatic hydrocarbon group, which does eventually lose a proton at higher pH. In cysteine the polar side chain consists of an —SH group, which can react with other cysteine —SH groups to form disulfide (—S—S—) bridges in proteins.

Group 3—Amino Acids with Carboxyl Groups in Their Side Chains

Two amino acids, glutamic acid and aspartic acid, have *carboxyl groups in their side chains* in addition to the one present in all amino acids. A carboxyl group can lose a proton, forming the corresponding carboxylate anion (Chapter 2, Section 2.4). Because of the presence of the carboxylate, the side chains of these two amino acids are *negatively charged at neutral pH.* Glutamic acid and aspartic acid differ only by a —CH_2— group in the side chain. They are frequently referred to as glutamate and aspartate, respectively. The side-chain carboxyl groups frequently bond to —NH_2 to form *side-chain amide groups,* yielding the analogous amino acids glutamine and asparagine. The side-chain amide groups are electrically neutral at neutral pH; in that sense, we could also include asparagine and glutamine in Group 2.

Group 4—Amino Acids with Basic Side Chains

There are three amino acids, histidine, lysine, and arginine, that have *basic side chains,* and in all three the side chain is *positively charged at or near neutral pH.* In histidine the pK'_a of the side-chain imidazole group is very close to physiological pH, and the properties of many proteins depend on whether individual histidine residues are or are not charged. In lysine the side-chain amino group is attached to an aliphatic hydrocarbon tail. In arginine the side-chain basic group, the guanidino group, is more complex in structure than the amino group, but it is also bonded to an aliphatic hydrocarbon tail.

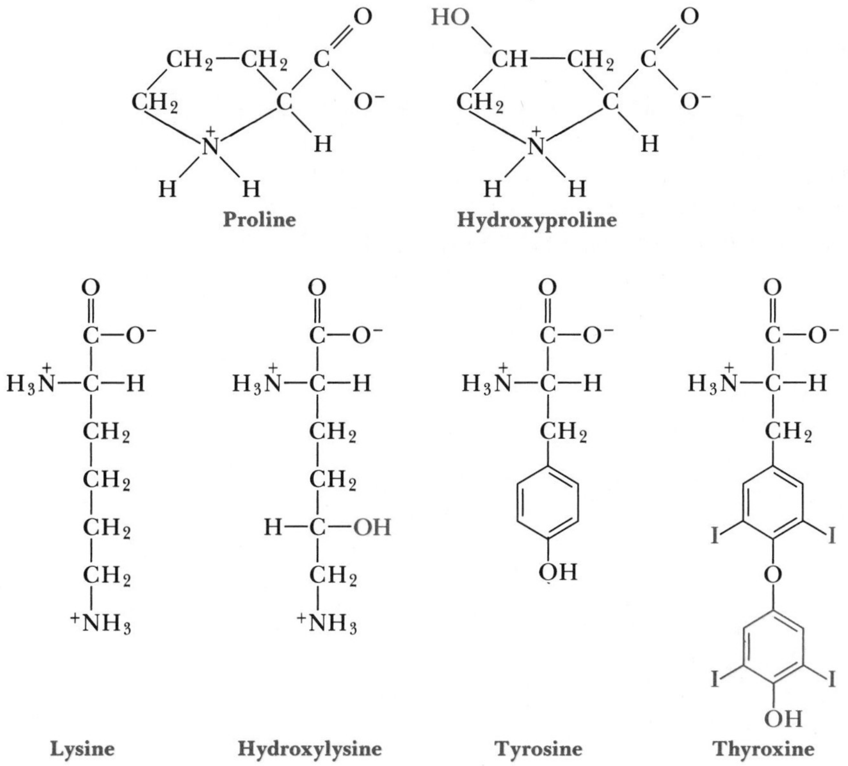

FIGURE 3.5

Structures of hydroxyproline, hydroxylysine, and thyroxine. The structures of the parent amino acids — proline for hydroxyproline, lysine for hydroxylysine, and tyrosine for thyroxine — are included for comparison. All amino acids are shown in their predominant ionic forms at pH 7.

Uncommon Amino Acids

Several other amino acids are known to occur in some, but by no means all, proteins (Figure 3.5). They are derived from the common amino acids and are produced by modification of the parent amino acid after the protein is synthesized by the organism. Hydroxyproline and hydroxylysine differ from the parent amino acids in having hydroxyl groups on their side chains; they are found only in a few connective tissue proteins, such as collagen. Thyroxine differs from tyrosine in having an extra iodine-containing aromatic group on the side chain; it is found only in the thyroid gland, bound to the protein thyroglobulin.

PRACTICE   SESSION

Identify the amino acids with nonpolar side chains in the following group; also pick out the ones with basic side chains.

Alanine, serine, arginine, lysine, leucine, phenylalanine

ANSWERS Nonpolar: alanine, leucine, and phenylalanine; basic: arginine and lysine. Serine is not in either category, since it has a polar side chain.

3.3

TITRATION CURVES OF THE AMINO ACIDS

In a free amino acid the carboxyl group and amino group of the general structure are charged at neutral pH — the carboxylate portion negatively and the amino group positively. Amino acids without charged groups on their side chains exist in neutral solution as zwitterions with no net charge. A zwitterion has equal positive and negative charges; in solution it is electrically neutral.

When an amino acid is titrated, its titration curve indicates the reaction of each functional group with hydrogen ion. In alanine, the carboxyl and amino groups are the two titratable groups. At very low pH alanine has a protonated (and thus uncharged) carboxyl group and a positively charged amino group that is also protonated. Under these conditions the alanine has a net positive charge of 1. As base is added, the carboxyl group loses its proton to become a negatively charged carboxylate group (Figure 3.6a), and the pH of the solution increases. Alanine now has no net charge. As the pH increases still further with addition of more base, the protonated amino group (a weak acid) loses its proton, and the alanine molecule now has a negative charge of 1. The titration curve of alanine is that of a dibasic acid (Figure 3.7a).

In histidine, the imidazole side chain also contributes a titratable group. At very low pH values, the histidine molecule has a net positive charge of 2, because both the imidazole and amino groups have positive charges. As base is added and the pH increases, the carboxyl group loses a proton to become a carboxylate as before, and the histidine now has a positive charge of 1 (Figure 3.6b). As still more base is added, the charged imidazole group loses its proton, and this is the point at which the histidine has no net charge. At still higher values of pH the amino group loses its proton, as was the case with alanine, and the histidine molecule now has a negative charge of 1. The titration curve of histidine is that of a tribasic acid (Figure 3.7b).

Like the acids we discussed in Chapter 2, the amino acids have characteristic values for the K'_as and pK'_as of their titratable groups. The

FIGURE 3.6

Ionization of amino acids. (a) The ionization of alanine (a neutral amino acid). (b) The ionization of histidine (an amino acid with a titratable side chain).

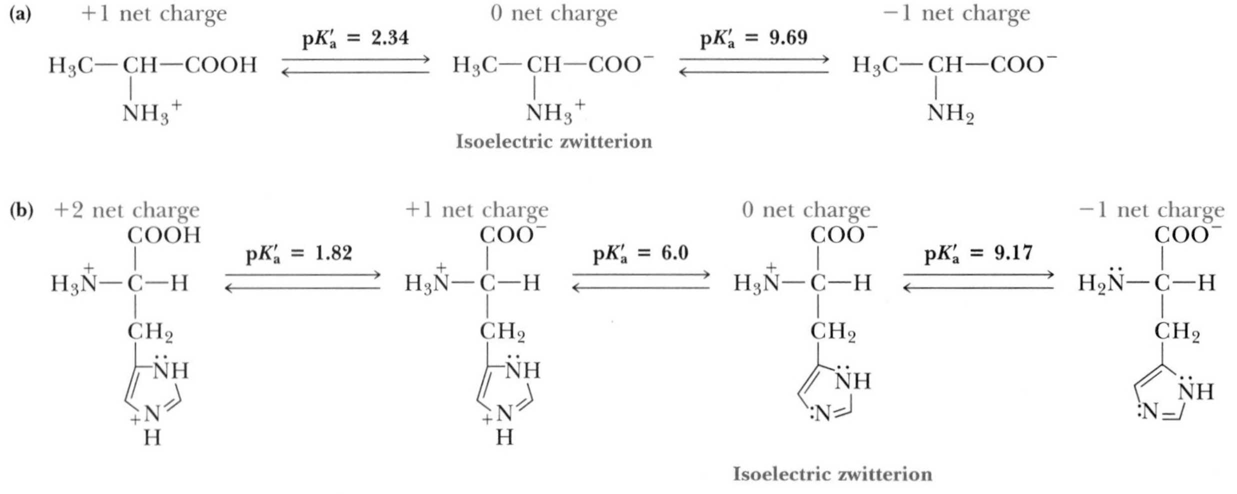

(a)

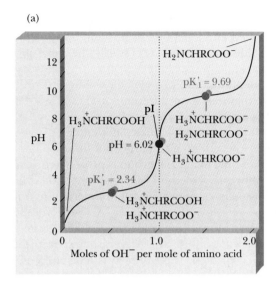

(b)

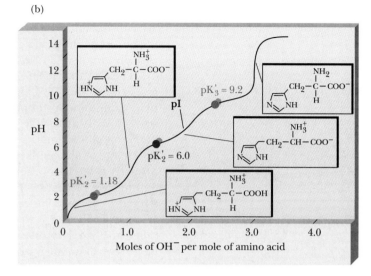

FIGURE 3.7

Titration curves of amino acids.
(a) The titration curve of alanine.
(b) The titration curve of
histidine. The isoelectric pH (pI)
is the value at which positive and
negative changes are the same.
The molecule has no net charge.

pK$'_a$s of α-carboxyl groups are fairly low, around 2. The pK$'_a$s of amino groups are reasonably high, with values ranging from 9 to 10.5. The pK$'_a$s of side-chain groups, including side-chain carboxyl and amino groups, depend on the groups' chemical natures. Table 3.2 lists the pK$'_a$s of the titratable groups of the amino acids. The classification of an amino acid as acidic or basic depends on the pK$'_a$ of the side chain. These groups can still be titrated

TABLE 3.2 pK$'_a$ Values of Common Amino Acids

ACID	α COOH	α NH$_3^+$	RH or RH$^+$
Gly	2.34	9.60	
Ala	2.34	9.69	
Val	2.32	9.62	
Leu	2.36	9.68	
Ile	2.36	9.68	
Ser	2.21	9.15	
Thr	2.63	10.43	
Met	2.28	9.21	
Phe	1.83	9.13	
Trp	2.38	9.39	
Asn	2.02	8.80	
Gln	2.17	9.13	
Pro	1.99	10.60	
Asp	2.09	9.82	3.86*
Glu	2.19	9.67	4.25*
His	1.82	9.17	6.0*
Cys	1.71	10.78	8.33*
Tyr	2.20	9.11	10.07
Lys	2.18	8.95	10.53
Arg	2.17	9.04	12.48

*For these amino acids the R group ionization occurs before the $\alpha-$NH$_3^+$ ionization.

after the amino acid is incorporated into a peptide or protein, but the pK'_a of the titratable group on the side chain is not necessarily the same in a protein as it is in a free amino acid.

The fact that amino acids, peptides, and proteins have different pK'_as gives rise to the possibility that they can have different charges at a given pH. Alanine and histidine, for example, both have net charges of -1 at high pH, above 10; the only charged group is the carboxylate anion. At lower pH, around 5, alanine is a zwitterion with no net charge, but histidine has a net charge of $+1$ at this pH because the imidazole group is protonated. This property is the basis of **electrophoresis,** a method for separating molecules on the basis of charge. This method is extremely useful in determining the structure of proteins and nucleic acids (Interchapter A, Section A.1; Chapter 7, Section 7.2). The pH at which a molecule has no net charge is called the **isoelectric pH** (given the symbol **pI**). At its isoelectric pH a molecule will not migrate in an electric field. This property can be put to use in separation methods.

PRACTICE   SESSION

Which of the following amino acids has a net charge of $+2$ at low pH? Which has a net charge of -2 at high pH?

Aspartic acid, alanine, arginine, glutamic acid, leucine, lysine

ANSWERS   Arginine and lysine have net charges of $+2$ at low pH because of their basic side chains; aspartic acid and glutamic acid have net charges of -2 at high pH because of their carboxylic acid side chains. Alanine and leucine do not fall into either category because they do not have titratable side chains.

3.4
THE PEPTIDE BOND

Individual amino acids can be linked together by formation of covalent bonds. The bond is formed between the α-carboxyl group of one amino acid and the α-amino group of the next one. Water is eliminated in the process, and the linked amino acid **residues** remain after the elimination of water (Figure 3.8a). A bond formed in this way is called a **peptide bond. Peptides** are compounds formed by the linking of small numbers of amino acids, ranging from two to several dozen. In a protein, many amino acids (usually more than a hundred) are linked by peptide bonds to form a **polypeptide chain** (Figure 3.8b). Another name for a compound formed by the reaction between an amino group and a carboxyl group is an *amide,* and the term "amide bond" is a synonym for "peptide bond."

The carbon–nitrogen bond formed when two amino acids are linked in a peptide bond is usually written as a single bond, with one pair of electrons shared between the two atoms. With a simple shift in the position of a pair of electrons, it is quite possible to write this bond as a double bond. This shifting of electrons is well known in organic chemistry and results in **resonance structures,** structures that differ from one another only in the

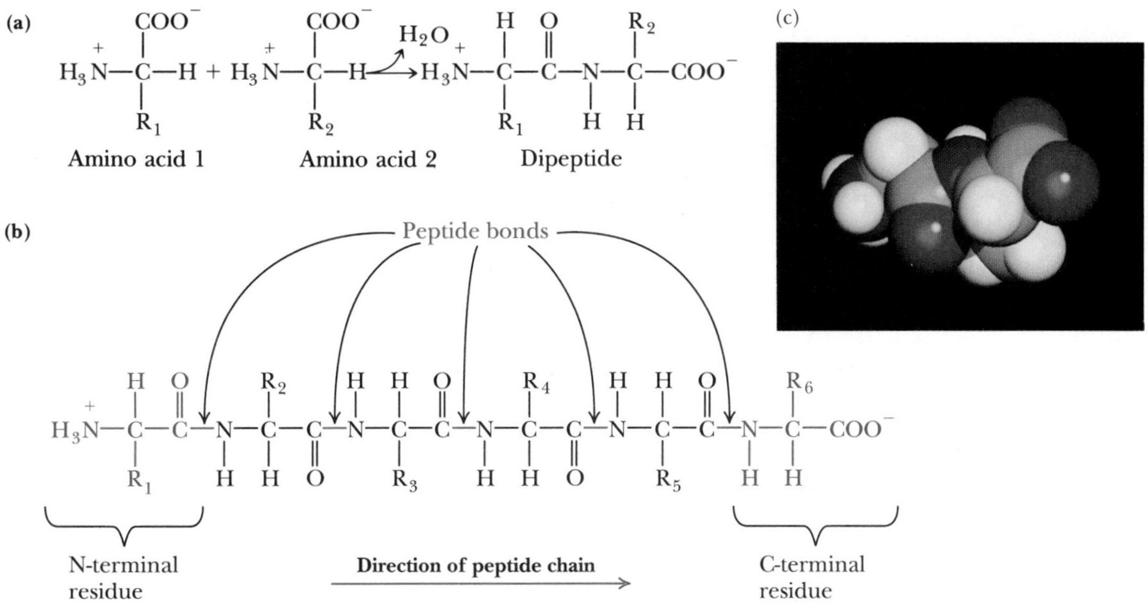

(a)

$$H_3\overset{+}{N}-\underset{R_1}{\overset{\overset{\displaystyle COO^-}{|}}{C}}-H + H_3\overset{+}{N}-\underset{R_2}{\overset{\overset{\displaystyle COO^-}{|}}{C}}-H \xrightarrow{\overset{\displaystyle H_2O}{}} H_3\overset{+}{N}-\underset{R_1}{\overset{\overset{\displaystyle H}{|}}{C}}-\overset{\overset{\displaystyle O}{\|}}{C}-\underset{H}{\overset{\overset{\displaystyle H}{|}}{N}}-\underset{H}{\overset{\overset{\displaystyle R_2}{|}}{C}}-COO^-$$

Amino acid 1   Amino acid 2   Dipeptide

(b)

Peptide bonds

$$H_3\overset{+}{N}-\underset{R_1}{\overset{\overset{\displaystyle H}{|}}{C}}-\overset{\overset{\displaystyle O}{\|}}{C}-\underset{H}{\overset{\overset{\displaystyle H}{|}}{N}}-\underset{H}{\overset{\overset{\displaystyle R_2}{|}}{C}}-\overset{\overset{\displaystyle O}{\|}}{C}-\underset{H}{\overset{\overset{\displaystyle H}{|}}{N}}-\underset{R_3}{\overset{\overset{\displaystyle H}{|}}{C}}-\overset{\overset{\displaystyle O}{\|}}{C}-\underset{H}{\overset{\overset{\displaystyle R_4}{|}}{N}} \cdots$$

N-terminal residue   **Direction of peptide chain** →   C-terminal residue

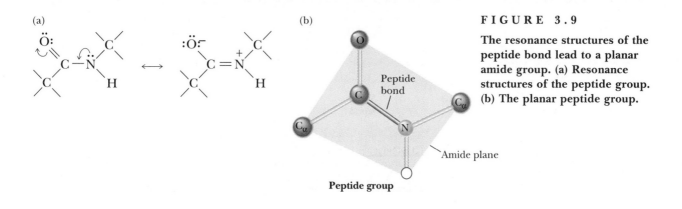

FIGURE 3.8

The peptide bond. (a) Formation of the peptide bond. (b) A small peptide showing the direction of the peptide chain (N-terminal ⟶ C-terminal). (c) Space-filling model of the dipeptide formed by reacting the carboxyl group of glycine with the amino group of alanine.

positioning of electrons. The positions of double and single bonds in one resonance structure are frequently different from their positions in another resonance structure of the same compound. No single resonance structure actually represents the bonding in the compound, but all resonance structures contribute to the bonding situation. The peptide bond can be written as a resonance hybrid of two structures (Figure 3.9), one with a single bond between the carbon and nitrogen and the other with a double bond between the carbon and nitrogen. The peptide bond has *partial double bond character,* and as a result the peptide group that forms the link between the two amino acids is planar. The peptide bond is also stronger than an ordinary single bond because of this resonance stabilization.

FIGURE 3.9

The resonance structures of the peptide bond lead to a planar amide group. (a) Resonance structures of the peptide group. (b) The planar peptide group.

Amide plane

Peptide group

Peptide bond

This structural feature has important implications for the three-dimensional conformations of peptides and proteins. There is free rotation around the bonds between the α carbon of a given amino acid residue and the amino nitrogen and carboxyl carbon of that residue, but there is no significant rotation around the peptide bond. This stereochemical constraint plays an important role in determining how the protein backbone can fold.

3.5
SOME SMALL PEPTIDES OF PHYSIOLOGICAL INTEREST

The simplest possible covalently bonded combination of amino acids is a dipeptide, in which two amino acid residues are linked by a peptide bond. An example of a naturally occurring dipeptide is carnosine, which is found in muscle tissue. This compound, which has the alternative name β-alanyl-L-histidine, has an interesting structural feature. (In the systematic nomenclature of peptides, the **N-terminal** amino acid residue — the one with the free amino group — is given first, then other residues are given as they occur in sequence. The **C-terminal** amino acid residue — the one with the free carboxyl group — is given last.) The N-terminal amino acid residue, β alanine, is structurally different from the α-amino acids we have seen up to now. As the name implies, the amino group is bonded to the second, or β, carbon of the alanine (Figure 3.10). The peptide bond in this dipeptide is formed between the carboxyl group of the β alanine and the amino group of the histidine, which is the C-terminal amino acid. Box 3.1 discusses another dipeptide of some interest.

PRACTICE   SESSION

Write an equation with structures for the formation of a dipeptide formed when alanine reacts with glycine to form a peptide bond. Is there more than one possible product for this reaction?

ANSWER   There are two possible products: alanylglycine, in which alanine is at the N-terminal end and glycine at the C-terminal end, and glycylalanine, in which glycine is at the N-terminal end and alanine at the C-terminal end.

FIGURE 3.10

Structures of carnosine and its component amino acid, β-alanine.

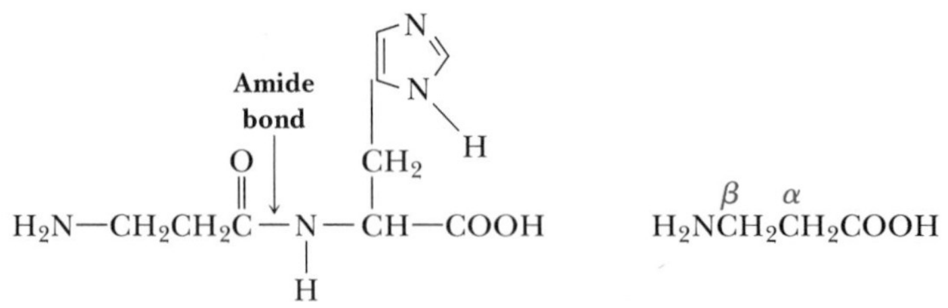

β-Alanyl-L-histidine (carnosine)

β-Alanine

Box 3.1

ASPARTAME, THE SWEET PEPTIDE

The dipeptide L-aspartyl-L-phenylalanine is of considerable commercial importance. The aspartyl residue has a free α-amino group, the N-terminal end of the molecule, and the phenylalanyl residue has a free carboxyl group, the C-terminal end. This dipeptide is about 200 times sweeter than sugar. A methyl ester derivative of this dipeptide is of even greater commercial importance than the dipeptide itself. The derivative has a methyl group at the C-terminal end in an ester linkage to the carboxyl group. The methyl ester derivative is called *aspartame* and is marketed as a sugar substitute under the trade name NutraSweet.

The consumption of common table sugar in the United States is about 100 pounds per person per year. Many people want to curtail their sugar intake in the interest of fighting obesity. Others must limit their sugar intake because of diabetes. One of the commonest ways of doing so is by drinking diet soft drinks. The soft-drink industry is one of the largest markets for aspartame. The use of this sweetener was approved by the U.S. Food and Drug Administration in 1981 after extensive testing, although there is still considerable controversy about its safety. Note that both amino acids have the L configuration. If a D-amino acid is substituted for either amino acid or for both of them, the resulting derivative is bitter rather than sweet.

(a)

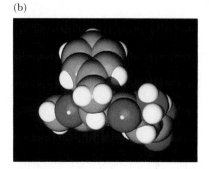

L-Aspartyl-L-phenylalanine (methyl ester)

(b)

(a) **Structure of aspartame. (b) Space-filling model of aspartame.**

Glutathione is a commonly occurring tripeptide; it has considerable physiological importance because it is a scavenger for oxidizing agents. (It is thought that some oxidizing agents are harmful to organisms and play a role in the development of cancer.) In terms of its amino acid composition and bonding order, it is γ-glutamyl-L-cysteinylglycine (Figure 3.11a). Once again the N-terminal amino acid is given first. In this case the γ-carboxyl group of the glutamic acid is involved in the peptide bond; the amino group of the cysteine is bonded to it. The carboxyl group of the cysteine is bonded, in turn, to the amino group of the glycine. The carboxyl group of the glycine forms the other end of the molecule, the C-terminal end. The glutathione molecule shown in Figure 3.11a is the reduced form. It scavenges oxidizing agents by reacting with them. The oxidized form of glutathione is generated from two molecules of the reduced peptide by formation of a disulfide bond between the —SH groups of the two cysteine residues (Figure 3.11b).

Two pentapeptides found in the brain are enkephalins, naturally occurring analgesics. For molecules of this size, abbreviations for the amino

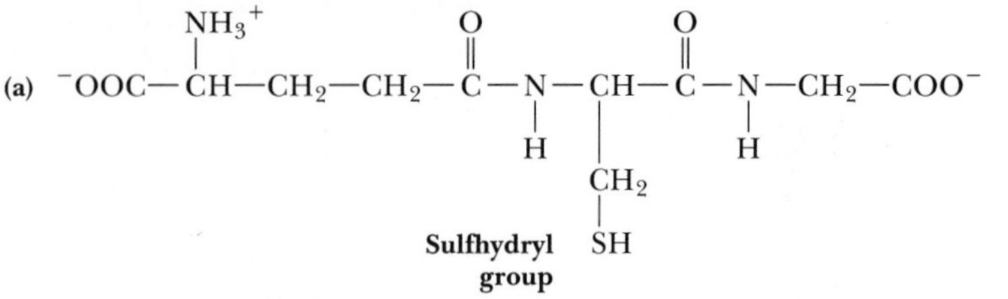

(a)

GSH (Reduced glutathione) (γ-Glu—Cys—SH—Gly)

Sulfhydryl group   SH

(b) 2 GSH $\underset{\substack{\text{+2H +2e}^-\\ \textbf{Reduction}}}{\overset{\substack{\textbf{Oxidation}\\ \text{-2H -2e}^-}}{\rightleftharpoons}}$ GSSG

Reaction of 2 GSH to give GSSG.

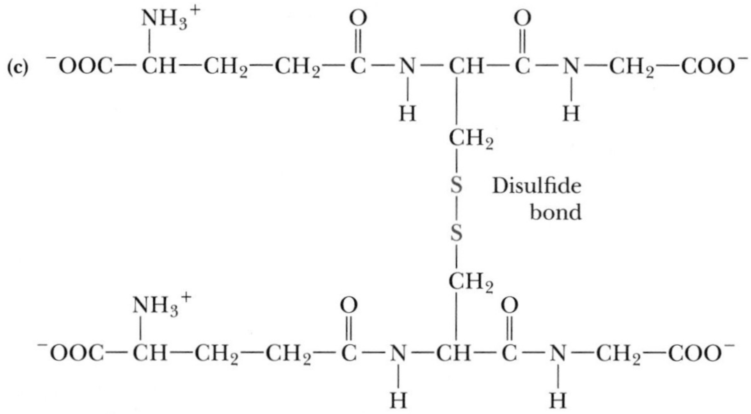

(c)

GSSG (Oxidized glutathione)   (γ-Glu—Cys—Gly)

Disulfide bond

(γ-Glu—Cys—Gly)

FIGURE 3.11

The oxidation and reduction of glutathione.

acids are more convenient than structural formulas. The same notation is used for the amino acid sequence, with the N-terminal amino acid listed first and the C-terminal listed last. The two peptides in question, leucine and methionine enkephalin, differ only in their C-terminal amino acids.

Tyr—Gly—Gly—Phe—Leu (3-letter abbreviations)
Y—G—G—F—L (1-letter abbreviations)

Leucine enkephalin

Tyr—Gly—Gly—Phe—Met
Y—G—G—F—M

Methionine enkephalin

It is thought that the aromatic side chains of tyrosine and phenylalanine in these peptides play a role in their activities. It is also thought that there are similarities between the three-dimensional structures of opiates, such as morphine, and those of the enkephalins. As a result of these structural similarities, opiates bind to the receptors in the brain intended for the enkephalins and thus produce their physiological activities.

Some important peptides have cyclic structures. Two well-known examples with many structural features in common are oxytocin and vasopressin (Figure 3.12). In each there is an —S—S— bond similar to that in the oxidized form of glutathione; the disulfide bond is responsible for the cyclic structure. Each of these contains nine amino acid residues, each has an amide group rather than a free carboxyl group at the C-terminal end, and

(a)   H$_3$N$^+$—Cys^1—Tyr^2—Ile^3

Disulfide bond

S
|
S

Gln^4

Cys^6——Asn^5

Pro^7—Leu^8—Gly^9—C(=O)—NH$_2$

Oxytocin

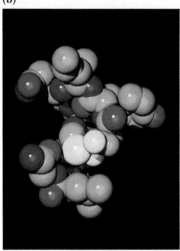

(b)

H$_3$N$^+$—Cys^1—Tyr^2—Phe^3

Disulfide bond

S
|
S

Gln^4

Cys^6——Asn^5

Pro^7—Arg^8—Gly^9—C(=O)—NH$_2$

Vasopressin

FIGURE 3.12

(a) Structures of oxytocin and vasopressin. (b) Space-filling model of oxytocin.

3.2

PEPTIDE HORMONES

Oxytocin induces labor in pregnant women and controls contraction of uterine muscle. During pregnancy the number of receptors for oxytocin in the uterine wall increases. At term the number of receptors for oxytocin is great enough to cause contraction of the smooth muscle of the uterus in the presence of small amounts of oxytocin produced by the body toward the end of pregnancy. The fetus moves toward the cervix of the uterus because of the strength and frequency of the uterine contractions. The cervix stretches, sending nerve impulses to the hypothalamus. When the impulses reach this part of the brain, positive feedback leads to the release of still more oxytocin by the posterior pituitary gland. The presence of more oxytocin leads to stronger contractions of the uterus so that the fetus is forced through the cervix and the baby is born. Oxytocin also plays a role in stimulating the flow of milk in a nursing mother. The process of suckling sends nerve signals to the hypothalamus of the mother's brain. Oxytocin is released and carried by the blood to the mammary glands. The presence of oxytocin causes contraction of smooth muscle in the mammary glands, forcing out the milk that is in them. As suckling continues, more hormone is released, producing still more milk.

Nursing stimulates the release of oxytocin, producing more milk.

Vasopressin plays a role in the control of blood pressure by regulating contraction of smooth muscle. Like oxytocin, vasopressin is released by the action of the hypothalamus on the posterior pituitary and transported by the blood to specific receptors. Vasopressin stimulates reabsorption of water by the kidney, thus having an antidiuretic effect. More water is retained, and the blood pressure increases.

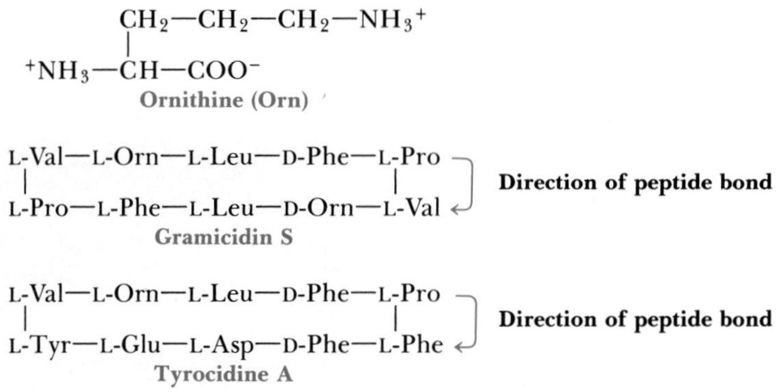

FIGURE 3.13

Structures of ornithine, gramicidin S, and tyrocidine A.

$$CH_2-CH_2-CH_2-NH_3^+$$
$$\quad\quad\quad\quad\mid$$
$$^+NH_3-CH-COO^-$$
Ornithine (Orn)

L-Val—L-Orn—L-Leu—D-Phe—L-Pro ⌐
 │   │   **Direction of peptide bond**
L-Pro—L-Phe—L-Leu—D-Orn—L-Val ⌐
Gramicidin S

L-Val—L-Orn—L-Leu—D-Phe—L-Pro ⌐
 │   │   **Direction of peptide bond**
L-Tyr—L-Glu—L-Asp—D-Phe—L-Phe ⌐
Tyrocidine A

each has a disulfide link between cysteine residues at positions 1 and 6. The difference between these two peptides is that oxytocin has an isoleucine residue at position 3 and a leucine residue at position 8, and vasopressin has a phenylalanine residue at position 3 and an arginine residue at position 8. Both of these peptides have considerable physiological importance as hormones (see Box 3.2).

In some other peptides the cyclic structure is formed by the peptide bonds themselves. Two cyclic decapeptides (peptides containing ten amino acid residues) produced by the bacterium *Bacillus brevis* are interesting examples. Both of these peptides, gramicidin S and tyrocidine A, are antibiotics, and both contain D amino acids as well as the more usual L amino acids (Figure 3.13). In addition, both contain the amino acid ornithine (Orn), which does not occur in proteins but which does play a role as a metabolic intermediate in several common pathways (Chapter 17, Section 17.6).

S U M M A R Y

Amino acids, the monomer units of proteins, have a general structure in common, with an amino group and a carboxyl group bonded to the same carbon atom. The natures of the side chains, referred to as R groups, are the basis of the differences among amino acids.

Except for glycine, amino acids can exist in two forms, designated L and D. These two stereoisomers are nonsuperimposable mirror images. The amino acids found in proteins are of the L form, but some D amino acids occur in nature. A classification scheme for amino acids can be based on the natures of their side chains. Two particularly important criteria are the polar or nonpolar nature of the side chain and the presence of an acidic or basic group in the side chain.

In free amino acids at neutral pH, the carboxylate group is negatively charged and the amino group is positively charged. Amino acids without charged groups on their side chains exist in neutral solution as zwitterions, with no net charge. Titration curves of amino acids indicate the pH ranges in which titratable groups gain or lose a proton. Side chains of amino acids can also contribute titratable groups; the charge (if any) on the side chain must be taken into consideration in determining the net charge on the amino acid.

Peptides are formed by linking the carboxyl group of one amino acid to the amino group of another amino acid in a covalent (amide) bond. Proteins consist of polypeptide chains; the number of amino acids in a protein is usually 100 or more. The peptide group is planar; this stereochemical constraint plays an important role in determining the three-dimensional structures of peptides and proteins. Small peptides, containing two to several dozen amino acid residues, can have marked physiological effects in organisms.

E X E R C I S E S

1. Write equations to show the ionic dissociation reactions of the following amino acids: aspartic acid, valine, histidine, serine, lysine.

2. Predict the ionized forms of the following amino acids at pH 7: glutamic acid, leucine, threonine, histidine, arginine.

3. Based on the information in Table 3.2, is there any amino acid that could serve as a buffer at pH 8? If so, which?

4. Given a peptide with the following amino acid sequence,

    Val—Met—Ser—Ile—Phe—Arg—Cys—Tyr—Leu

    identify the polar amino acids, the aromatic amino acids, and the sulfur-containing amino acids.

5. Identify the charged groups in the peptide shown in Exercise 4 at pH 1 and at pH 7. What is the net charge of this peptide at these two pH values?

6. What are the sequences of all the possible tripeptides that contain the amino acids serine, leucine, and phenylalanine? Use three-letter abbreviations to express your answer.

7. Draw structures of the following amino acids, indicating the charged form that exists at pH 4: histidine, asparagine, tryptophan, proline, and tyrosine.

8. Consider the following peptides:

    Phe—Glu—Ser—Met   and   Val—Trp—Cys—Leu

    Do these peptides have different net charges at pH 1? At pH 7? Indicate the charges at both pH values.

9. Sketch a titration curve for the amino acid cysteine; indicate the pK'_a values for all titratable groups. Also indicate the pH at which this amino acid has no net charge.

10. Sketch a titration curve for aspartic acid; indicate the pK'_a values of all titratable groups. Also indicate the pH range in which the conjugate acid–base pair +1 asp and 0 asp acts as a buffer.

11. Consider the peptides Ser—Glu—Gly—His—Ala and Gly—His—Ala—Glu—Ser. How do these two peptides differ? Would you expect their titration curves to differ? Why or why not?

12. What are the structural differences between the peptide hormones oxytocin and vasopressin? How do they differ in function?

13. How do the oxidized and reduced forms of glutathione differ from one another?

14. Give the amino acid sequence of a peptide that contains one or more D amino acids. What is the biological role of this peptide?

15. What is the stereochemical basis of the observation that D-aspartyl-D-phenylalanine has a bitter taste whereas L-aspartyl-L-phenylalanine is significantly sweeter than sugar?

16. Name and give the structure of an amino acid produced by modification of one of the usual twenty amino acids after protein synthesis. Identify a specific protein in which such an amino acid can be found.

17. Sketch resonance structures for the peptide group, and indicate how these structures contribute to the planar arrangement of this group of atoms.

A N N O T A T E D   B I B L I O G R A P H Y

Barrett, G. C., ed. *Chemistry and Biochemistry of the Amino Acids.* New York: Chapman and Hall, 1985. [Wide coverage of many aspects of the reactions of amino acids.]

Larsson, A., ed. *Functions of Glutathione: Biochemical, Physiological, Toxicological and Chemical Aspects.* New York: Raven Press, 1983. [A collection of articles on the many roles of a ubiquitous peptide.]

McKenna, K. W., and V. Pantic, eds. *Hormonally Active Brain Peptides: Structure and Function.* New York: Plenum Press, 1986. [A discussion of the chemistry of enkephalins and related peptides.]

Siddle, K., and J. C. Hutton. *Peptide Hormone Action — A Practical Approach.* Oxford, England: Oxford University Press, 1990. [A book that concentrates on experimental methods for studying the actions of peptide hormones.]

Steginb, L. D., and L. J. Filer, Jr. *Aspartame — Physiology and Biochemistry.* New York: Marcel Dekker, 1984. [A comprehensive treatment of metabolism, sensory and dietary aspects, preclinical studies, and issues relating to human consumption, including ingestion by phenylketonurics and during pregnancy.]

Wold, F. *In vivo* Chemical Modification of Proteins (Post-Translational Modification). *Ann. Rev. Biochem.* **50,** 788–814 (1981). [A review article on the modified amino acids found in proteins.]

Chapter 4

The Three-Dimensional Structure of Proteins

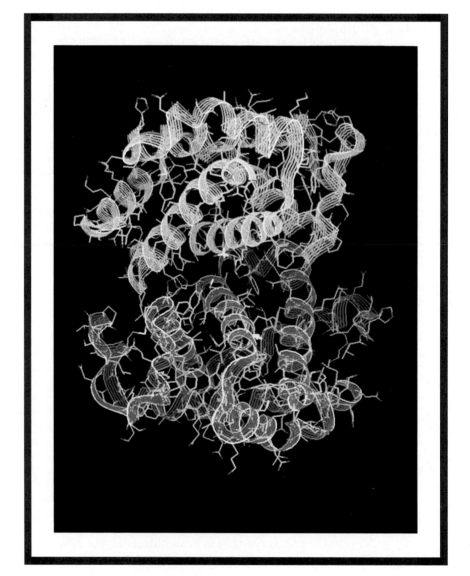

The hemoglobin molecule.

Amino acids joined together form a protein (polypeptide) chain. The repeating units are amide planes containing peptide bonds. These amide planes can twist about their connecting carbon atoms to create the three-dimensional conformations of proteins. Forty years ago, Linus Pauling predicted that linked amino acids could form an α-helix. Years later, his prediction was confirmed when myoglobin, an oxygen-binding protein, was found to be made from Pauling's α-helices. This type of local folding of the protein chain is called *secondary* structure, the linear sequence being the *primary* structure. The conformation of a complete protein chain is its *tertiary* structure. Myoglobin, a molecule that binds oxygen tightly, has a single protein chain. Hemoglobin, a protein with four myoglobinlike subunits fitted together, has a *quaternary* structure. This allows it to change its shape from the *oxy* conformation, when it binds oxygen in the lungs, to the *deoxy* form, when it releases oxygen to working tissues. The discovery of structure–function relationships in hemoglobin led to an understanding of the way in which complex multisubunit enzymes regulate metabolic pathways.

4.1
THE STRUCTURES OF PROTEINS DETERMINE THEIR FUNCTIONS

Levels of Structure in Proteins

Biologically active proteins are polymers consisting of amino acids linked by covalent bonds. Many different possible conformations (three-dimensional structures) are possible for a molecule as large as a protein. Of these many structures, one or, at most, a few have biological activity; these are called the **native conformations.** Most proteins appear to have little regular arrangement of atoms. As a consequence, proteins are frequently described as having large segments of "random structure" (also referred to as random coil or simply coil). The term "random" is really a misnomer, since the same complex structure is found in all molecules of a given protein having the same native conformation. Because of the complexity of proteins, it has become usual to define four levels of structure to attack the problem more efficiently.

Primary structure is the order in which the amino acids are linked together. The peptide H_2N—Leu—Gly—Thr—Val—Arg—Asp—His—COOH has a different primary structure from the peptide H_2N—Val—His—Asp—Leu—Gly—Arg—Thr—COOH, even though both have the same number and kinds of amino acids. Note that the order of amino acids can be written on one line. The primary structure is the one-dimensional first step in specifying the three-dimensional structure of a protein.

Two three-dimensional aspects of a single polypeptide chain, called the secondary and tertiary structures, can be considered separately. **Secondary structure** is the arrangement in space of the atoms in the backbone of the polypeptide chain. The α-helix and β-pleated sheet hydrogen-bonded arrangements (Chapter 2, Section 2.3) are two different types of secondary structure. (The conformations of the side chains of the amino acids are not part of secondary structure.) In very large proteins, the folding of parts of the chain can occur independently of the folding of other parts. Such independently folded portions of proteins are referred to as **domains** or *supersecondary* structure.

Tertiary structure includes the three-dimensional arrangement of all the atoms in the protein, including those in the side chains and in any **prosthetic groups** (groups of atoms other than amino acids).

A protein can consist of multiple polypeptide chains called **subunits.** The arrangement of subunits with respect to one another is the **quaternary structure.** Interaction between subunits is mediated by noncovalent interactions, such as hydrogen bonds and hydrophobic interactions.

The Importance of Primary Structure

We shall discuss secondary structure in more detail in Section 4.3, tertiary structure in Section 4.4, and quaternary structure in Section 4.5.

The amino acid sequence (the primary structure) of a protein determines its three-dimensional structure, which in turn determines its properties. In every protein the correct three-dimensional structure is needed for correct functioning.

One of the most striking demonstrations of the importance of primary structure is found in the hemoglobin associated with **sickle-cell anemia.** In this genetic disease, red blood cells cannot bind oxygen efficiently. The red cells also assume a characteristic sickle shape, giving the disease its name. The sickled cells tend to become trapped in small blood vessels, cutting off circulation and thereby causing organ damage. These drastic consequences stem from a change in one amino acid residue in the sequence of the primary structure.

Considerable research is being done to determine the effects of changes in primary structure on the functions of proteins. It is possible to replace any chosen amino acid residue in a protein with another specific amino acid residue. The conformation of the altered protein, as well as its biological activity, can then be determined. The results of such amino acid substitutions range from negligible effects to complete loss of activity, depending on the protein and the nature of the altered residue.

4.2
PRIMARY STRUCTURE OF PROTEINS

Determining the sequence of amino acids in a protein is a routine, but not trivial, operation. Its several parts must be carried out carefully to obtain accurate results (Figure 4.1).

You may want to read Section A.2 (pp. 125–128) now.

Step 1   The first step in determining the primary structure of a protein is to establish which amino acids are present and in what proportions. It is relatively easy to break a protein down to its component amino acids by heating a solution of the protein in acid, usually 6 M HCl, at 100 to 100°C for 12 to 36 hours. Separation and identification of the products are somewhat more difficult and are best done by an automated instrument known as an *amino acid analyzer.* Interchapter A, on experimental methods, contains a detailed description of its operation. An amino acid analyzer gives both qualitative information about the identities of the amino acids present

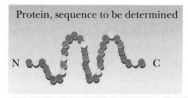

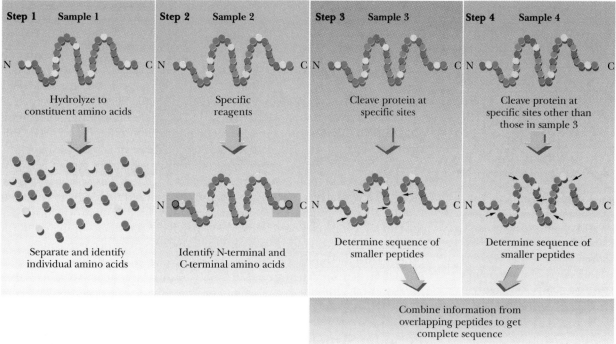

FIGURE 4.1

The strategy for determining the primary structure of a given protein. The amino acid can be determined by four different analyses performed on four separate samples of the same protein.

and quantitative information about the relative amounts of those amino acids.

Step 2 The identities of the N-terminal and C-terminal amino acids in a protein sequence are determined. There are several methods of doing this, both chemical and enzymatic; they are also described in Interchapter A.

When the end-group analysis is complete, it is possible to proceed with steps 3 and 4, the determination of the amino acid sequence. Automated instruments can perform a stepwise modification, cleave the protein, and identify all the amino acids in the polypeptide chain, but the process becomes more difficult as the number of amino acids increases. In most proteins the chain is more than 100 residues long. It is usually necessary to break a long polypeptide chain into fragments of reasonable size for sequencing. The details of the sequencing procedure are described in Interchapter A, Section A.5 (pp. 132–134).

You may want to read Section A.3 (pp. 128–129) now.

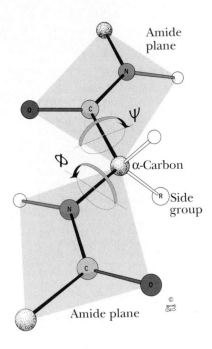

FIGURE 4.2

Definition of the angles that determine the conformation of a polypeptide chain. The rigid planar peptide groups (called "playing cards" in the text) are shaded. The angle of rotation around the C^α—N bond is designated ϕ (phi), and the angle of rotation around the C^α—C bond is designated ψ (psi). These two bonds are the ones around which there is freedom of rotation.

4.3
SECONDARY STRUCTURE OF PROTEINS

The secondary structure of proteins is the arrangement in space of the atoms of the backbone, the polypeptide chain itself. The nature of the bonds in the peptide backbone plays an important role here. Within each amino acid residue are two bonds with reasonably free rotation. They are (1) the bond between the α-carbon and the amino nitrogen of that residue and (2) the bond between the α-carbon and the carboxyl carbon of that residue. The combination of the planar peptide group and the two freely rotating bonds has important implications for the three-dimensional conformations of peptides and proteins. A peptide chain backbone can be visualized as a series of playing cards, each card representing a planar peptide group. The cards are linked at opposite corners by swivels, representing the bonds about which there is considerable freedom of rotation (Figure 4.2). The side chains also play a vital role in determining the three-dimensional shape of a protein, but only the backbone is considered in secondary structure.

The angles Φ (phi) and Ψ (psi), frequently called *Ramachandran angles,* (after their originator, G.N. Ramachandran) are used to designate rotations around the C—N and C—C bonds, respectively. The conformation of a protein backbone can be described by specifying the values of Φ and Ψ for each residue. Two kinds of secondary structures that occur frequently in proteins are the repeating *α-helix* and *β-pleated sheet* (or β-sheet) hydrogen-bonded structures. They are not the only possible secondary structures, but they are by far the most important, and they deserve a closer look.

Periodic Structures in Protein Backbones

The α-helix and β-pleated sheet are periodic structures; their features repeat at regular intervals. The α-helix is rodlike and involves only one polypeptide chain. The β-pleated sheet structure can produce a two-dimensional array and can involve one or more polypeptide chains.

The α-Helix

The α-helix is stabilized by hydrogen bonds within the backbone of a single polypeptide chain. Counting from the N-terminal end, the CO group of each amino acid residue is hydrogen bonded to the NH group of the amino acid four residues away from it in the covalently bonded sequence. The helical conformation allows a linear arrangement of the atoms involved in the hydrogen bonds, which gives the bonds maximum strength and thus makes the helical conformation very stable (Section 2.3). There are 3.6 residues for each turn of the helix, and the *pitch* of the helix (the linear distance between corresponding points on successive turns) is 5.4 Å (Figure 4.3). (The angstrom unit [1 Å = 10^{-8} cm = 10^{-10} m] is convenient for interatomic distances in molecules, but it is not a Système International [SI] unit. Nanometers [1 nm = 10^{-9} m] and picometers [1 pm = 10^{-12} m] are the SI units used for interatomic distances. In SI units the pitch of the α-helix is 0.54 nm, or 540 pm.)

(a)

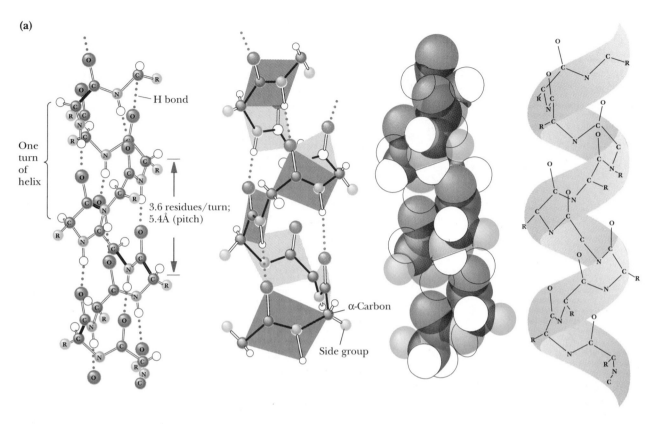

One turn of helix

H bond

3.6 residues/turn; 5.4Å (pitch)

α-Carbon

Side group

(b)

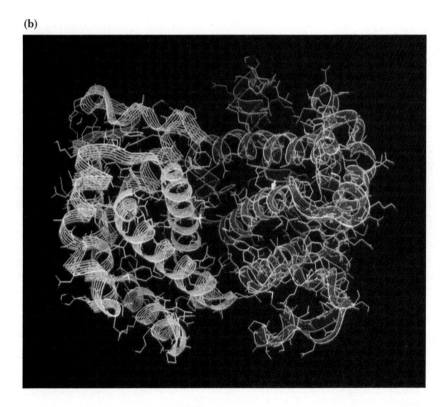

FIGURE 4.3

The α-helix. (a) From left to right, ball-and-stick model of the α-helix, showing terminology; ball-and-stick model with planar peptide groups shaded; computer-generated space-filling model of the α-helix; outline of the α-helix. (b) Model of the protein hemoglobin, showing the helical regions.

Proteins have varying amounts of α-helical structures, varying from a few percent to nearly 100%. Several factors can disrupt the α-helix. The amino acid proline creates a bend in the backbone because of its *cyclic* structure. It cannot fit into the α-helix for two reasons: (1) rotation around the bond between the nitrogen and the α-carbon is severely restricted, and (2) proline's α-amino group cannot participate in intrachain hydrogen bonding. Other localized factors involving the side chains include strong electrostatic repulsion due to the proximity of several charged groups of the same sign, such as groups of positively charged lysine and arginine residues or groups of negatively charged glutamate and aspartate residues. Another possibility is crowding (steric repulsion) due to the proximity of several bulky side chains. In the α-helical conformation all the side chains lie outside the helix; there is not enough room for them in the interior. The β-carbon is just outside the helix, and crowding can occur if it is bonded to two atoms other than hydrogen, as is the case with valine and threonine.

The β-Sheet

The arrangement of atoms in the β-pleated sheet conformation differs markedly from that in the α-helix. The peptide backbone in the β-sheet is almost completely extended. Hydrogen bonds can be formed between different parts of a single chain that is doubled back on itself (**intrachain** bonds) or between different chains (**interchain** bonds). If the peptide chains run in the same direction (that is, if they are all aligned in terms of their N-terminal and C-terminal ends), a **parallel pleated sheet** is formed. When alternating chains run in opposite directions, an **antiparallel pleated sheet** is formed (Figure 4.4). The hydrogen bonding between peptide chains in the

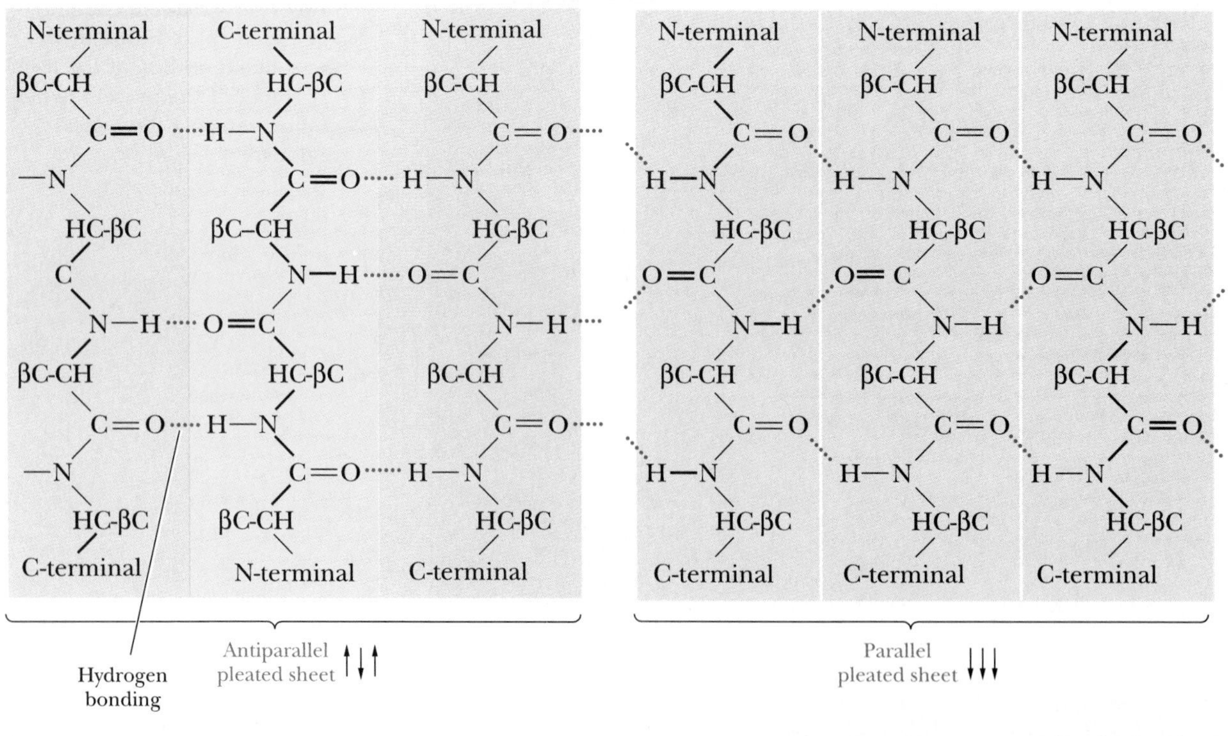

FIGURE 4.4

β-pleated sheet structures. (a) In an antiparallel pleated sheet, the peptide chains run in opposite directions from the N-terminal to the C-terminal ends. (b) In a parallel pleated sheet, the peptide chains run in the same direction.

β-pleated sheet gives rise to a repeated zigzag structure; hence the name "pleated sheet" (Figure 4.5).

α-Helices and β-Sheets in Proteins

The α-helix and β-pleated sheet are combined in many ways as the polypeptide chain folds back on itself in a protein. For steric (spatial) reasons glycine is frequently encountered in **reverse turns,** at which the polypeptide chain changes direction; the single hydrogen of the side chain prevents crowding (Figure 4.6). Because the cyclic structure of proline has the correct geometry for a reverse turn, this amino acid is also frequently encountered in such turns (Figure 4.6c). The combination of α- and β-strands produces a variety of **supersecondary structures** in proteins. The

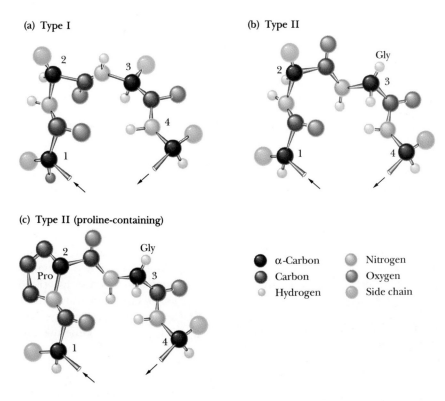

(a) Type I

(b) Type II

(c) Type II (proline-containing)

● α-Carbon	○ Nitrogen
● Carbon	● Oxygen
○ Hydrogen	● Side chain

FIGURE 4.6

Structures of reverse turns. Arrows indicate the directions of the polypeptide chains. (a) A type I reverse turn. In residue 3 the side chain (green) lies outside the loop, and any amino acid can occupy this position. (b) A type II reverse turn. The side chain of residue 3 has been rotated 180° from the position in the type I turn and is now on the inside of the loop. Only the hydrogen side chain of glycine can fit into the space available, so glycine must be the third residue in a type II reverse turn. (c) The five-membered ring of proline has the correct geometry for a reverse turn; this residue normally occurs as the second residue of a reverse turn. The turn shown here is type II, with glycine as the third residue.

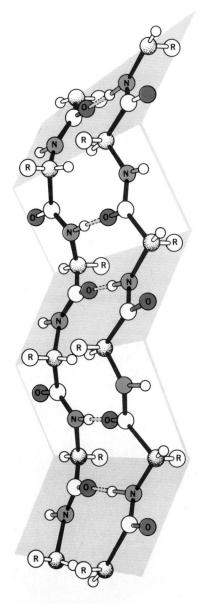

FIGURE 4.5

The three-dimensional form of the antiparallel β-pleated sheet arrangement. The chains do not fold back on each other but are in a fully extended conformation.

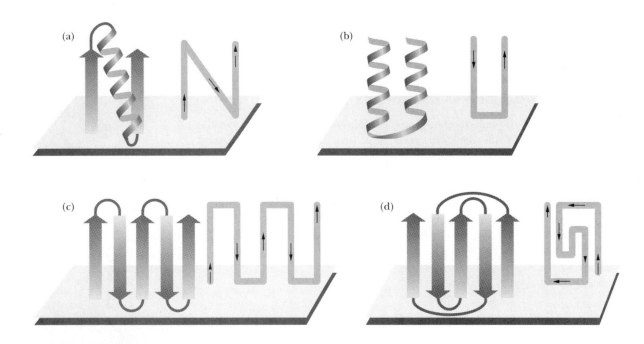

FIGURE 4.7

Schematic diagrams of supersecondary structures. Arrows indicate the directions of the polypeptide chains. (a) A $\beta\alpha\beta$ unit, (b) an $\alpha\alpha$ unit, (c) a β-meander, and (d) the Greek key. The Greek key motif in protein structure resembles the geometric patterns on this ancient Greek vase, giving rise to the name.

commonest feature of this sort is the $\beta\alpha\beta$ **unit,** in which two parallel strands of β-sheet are connected by a stretch of α-helix (Figure 4.7a). An $\alpha\alpha$ **unit** consists of two antiparallel α-helices (Figure 4.7b). In such an arrangement, energetically favorable contacts exist between the side chains in the two stretches of helix. In a β-**meander,** an antiparallel sheet is formed by a series of tight reverse turns connecting stretches of the polypeptide chain (Figure 4.7c). Another kind of antiparallel sheet is formed when the polypeptide chain doubles back on itself in a pattern known as the **Greek key,** named for a decorative design found on pottery from the classical period (Figure 4.7d). When β-sheets are extensive enough they can fold back on themselves, forming a β-**barrel,** a structural feature that occurs in many proteins (Figure 4.8).

The Collagen Triple Helix

Collagen, a component of connective tissue, is the most abundant protein in vertebrates. It is organized in water-insoluble fibers of great strength. A collagen fiber consists of three polypeptide chains wrapped around each other in a ropelike twist, or triple helix. Each of the three chains has, within limits, a repeating sequence of three amino acid residues, X—Pro—Gly or X—Hyp—Gly, where Hyp stands for hydroxyproline, and any amino acid can occupy the first position, designated as X. Quite frequently, X is also

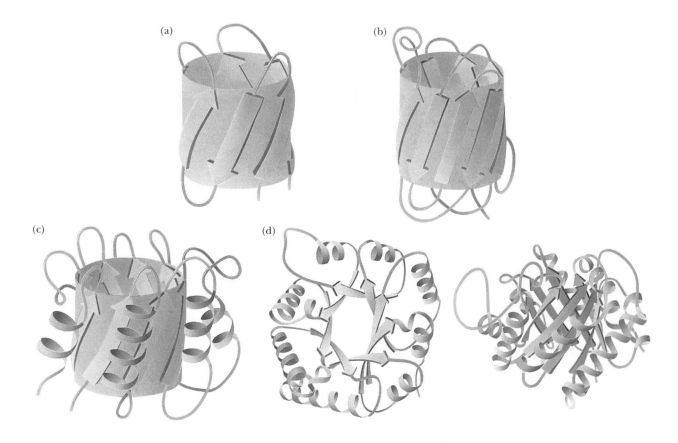

FIGURE 4.8

Some β-barrel arrangements. (a) A linked series of β-meanders. This arrangement occurs in the protein rubredoxin from *Clostridium pasteurianum*. (b) The Greek key pattern occurs in human prealbumin. (c) A β-barrel involving alternating βαβ units. This arrangement occurs in triose phosphate isomerase from chicken muscle. (d) Top and side views of the polypeptide backbone arrangement in triose phosphate isomerase. Note that the α-helical sections lie outside the actual β-barrel.

proline or hydroxyproline. (Hydroxyproline is formed from proline by a specific hydroxylating enzyme after the amino acids are linked together.) Hydroxylysine also occurs in collagen. In the amino acid sequence of collagen, every third position must be occupied by glycine. The triple helix is arranged so that every third residue on each chain is inside the helix. Only glycine is small enough to fit into the space available (Figure 4.9a).

The three individual collagen chains are themselves helices. They are twisted around each other in a superhelical arrangement to form a stiff rod. This triple helical molecule is called **tropocollagen;** it is 300 nm (3000 Å) long and 1.5 nm (15 Å) in diameter. The three strands are held together by hydrogen bonds involving the hydroxyproline and hydroxylysine residues. The molecular weight of the triple-stranded array is about 300,000; each strand contains about 800 amino acid residues (Figure 4.9b). Collagen is

FIGURE 4.9

The collagen triple helix, known as tropocollagen. Each of the three chains of the collagen triple helix has a helical conformation. Each chain also has an amino acid sequence that can be considered a repeat of the same tripeptide, glycine-hydroxyproline-proline. Some variations are possible (see text), but glycine must occur at every third position along the chain, and most of the other residues are proline or hydroxyproline. The three chains are wrapped around each other to form a ropelike structure.

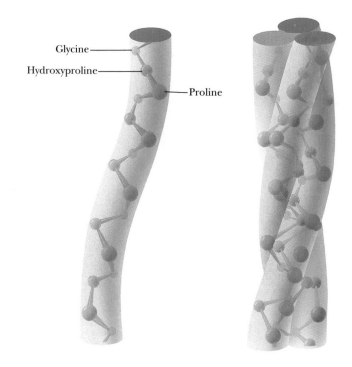

Glycine

Hydroxyproline

Proline

both intramolecularly and intermolecularly linked by covalent bonds formed by reactions of lysine and histidine residues. The amount of cross-linking in a tissue increases with age. That is why meat from older animals is tougher than meat from younger animals.

Collagen in which the proline is not hydroxylated to hydroxyproline to the usual extent is less stable than normal collagen. Symptoms of scurvy, such as bleeding gums and skin discoloration, are the results of fragile collagen. The enzyme that hydroxylates proline and thus maintains the normal state of collagen requires ascorbic acid (vitamin C) to remain active. Scurvy is ultimately caused by a dietary vitamin C deficiency.

Two Types of Protein Conformations: Fibrous and Globular

It is difficult to draw a clear separation between secondary and tertiary structure. The nature of the side chains in a protein (part of the tertiary structure) can influence the folding of the backbone (the secondary structure). A comparison of collagen with silk and wool fibers can be illuminating. Silk fibers consist largely of the protein fibroin, which, like collagen, has a fibrous structure but which, unlike collagen, consists largely of β-sheets. Fibers of wool consist largely of the protein keratin, which is largely α-helical. The amino acids of which collagen, fibroin, and keratin are composed determines which conformation they will adopt, but all are **fibrous** proteins (Figure 4.10a).

(a)

(b)

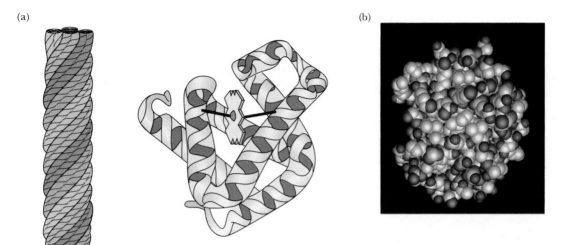

Filament
(four right-hand
twisted protofilaments)

Myoglobin, a globular protein

FIGURE 4.10

A comparison of the shapes of
fibrous and globular proteins (a)
Schematic diagrams of a portion
of a fibrous protein and of a
globular protein. (b) Computer-
generated model of a globular
protein.

In other proteins, the backbone folds back on itself to produce a more or less spherical shape. These are called **globular** proteins (Figure 4.10b), and we shall see many examples of them. Their helical and pleated sheet sections can be arranged so as to bring the ends of the sequence close to each other in three dimensions. Globular proteins, unlike fibrous proteins, are water-soluble and have compact structures; tertiary and quaternary structure are usually associated with them.

4.4
TERTIARY STRUCTURE OF PROTEINS

The tertiary structure of a protein is the three-dimensional arrangement of all the atoms in the molecule. The conformations of the side chains and the positions of any prosthetic groups are parts of the tertiary structure, as is the arrangement of helical and pleated sheet sections with respect to one another. In a fibrous protein such as collagen, the overall shape of which is a long rod, the secondary structure provides most of the information about the tertiary structure as well. The helical backbone of the protein does not fold back on itself, and the only important aspect of the tertiary structure that is not specified by the secondary structure is the arrangement of the atoms of the side chains.

For a globular protein, considerably more information is needed. It is necessary to determine the way in which the helical and pleated sheet sections fold back on each other, in addition to the positions of the side-chain atoms and any prosthetic groups. The interactions between the side chains play an important role in the folding of proteins. The folding pattern frequently brings residues that are widely separated in the amino acid sequence into proximity in the tertiary structure of the native protein.

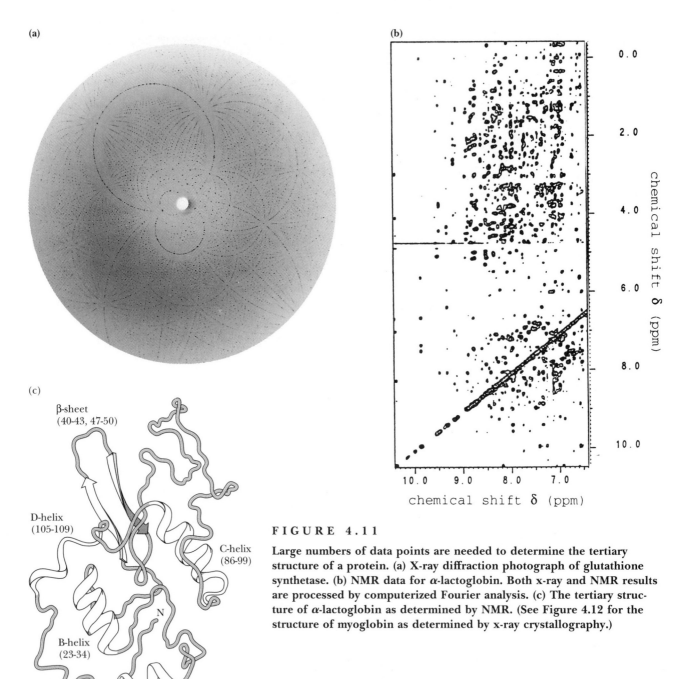

(a)

(b)

chemical shift δ (ppm)

chemical shift δ (ppm)

(c)

β-sheet
(40-43, 47-50)

D-helix
(105-109)

C-helix
(86-99)

N

B-helix
(23-34)

A-helix (5-11)

C

FIGURE 4.11

Large numbers of data points are needed to determine the tertiary structure of a protein. (a) X-ray diffraction photograph of glutathione synthetase. (b) NMR data for α-lactoglobin. Both x-ray and NMR results are processed by computerized Fourier analysis. (c) The tertiary structure of α-lactoglobin as determined by NMR. (See Figure 4.12 for the structure of myoglobin as determined by x-ray crystallography.)

The experimental technique used to determine the tertiary structure of a protein is **x-ray crystallography.** Perfect crystals of proteins can be grown under carefully controlled conditions. In such a crystal, all the individual protein molecules have the same three-dimensional conformation and the same orientation. Crystals of this quality can be formed only from proteins of very high purity.

When a suitably mounted protein crystal is exposed to a beam of x-rays, a *diffraction pattern* is produced on a photographic plate (Figure 4.11a) or a radiation counter. The pattern is produced as a result of scattering of the x-rays by the electrons in each atom in the molecule. The number of electrons in the atom determines the intensity of its scattering of x-rays; heavier atoms scatter more effectively than light atoms. The scattered x-rays from the individual atoms can reinforce each other or cancel each other (set up constructive or destructive interference), giving rise to the characteristic pattern for each type of molecule. A series of diffraction patterns taken from several angles contains the information needed to determine the tertiary structure. The information is extracted from the diffraction patterns through a mathematical analysis known as a *Fourier series.* Many thousands of such calculations are required to determine the structure of a protein, and even though they are performed by computer, the process is a fairly long one. Improving the calculation procedure is a subject of active research. The articles by Hauptmann and by Karle listed in the bibliography at the end of this chapter outline some of the accomplishments in the field.

Another technique that supplements the results of x-ray diffraction has come into wide use in recent years. It is a form of **nuclear magnetic resonance (NMR) spectroscopy.** In this particular application of NMR, called **2-D** (two-dimensional) **NMR,** large collections of data points, taken under slightly different conditions, are subjected to computer analysis (Figure 4.11b). Like x-ray diffraction, this method uses a Fourier series as the mathematical tool for analysis of results. It is similar to x-ray diffraction in another way: it is a long process and requires considerable amounts of computing power. One way in which 2-D NMR differs from x-ray diffraction is that, rather than crystals, it uses protein samples in aqueous solution. This is closer to the state of proteins in cells, and thus it is one of the main advantages of the method.

Myoglobin: An Example of Protein Structure

In many ways, myoglobin, the function of which is oxygen storage in mammalian muscle, is the classic example of a globular protein. We use it here as a case study in tertiary structure (we shall see the tertiary structures of many other proteins in context when we discuss their roles in biochemistry). Myoglobin was the first protein for which the complete tertiary structure (Figure 4.12) was determined by x-ray crystallography. The

FIGURE 4.12

(a) The structure of the myoglobin molecule, showing the peptide backbone and the heme group. The helical segments are designated by the letters A through H. The terms NA and HC indicate the N-terminal and C-terminal ends, respectively. (b) Space-filling model of myoglobin.

(a)

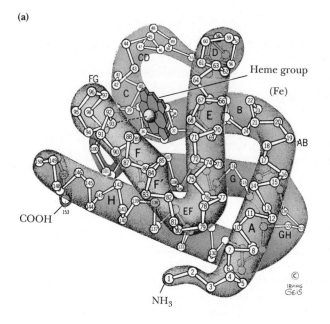

Heme group

(Fe)

COOH

NH₃

(b)

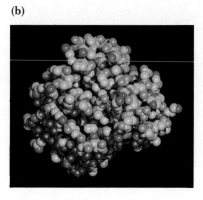

complete myoglobin molecule consists of a single polypeptide chain of 153 amino acid residues and includes a prosthetic group, the **heme** group, which also occurs in hemoglobin. The myoglobin molecule (including the heme group) has a compact structure, with the interior atoms very close to each other. This structure provides examples of many of the forces responsible for the three-dimensional shapes of proteins.

In myoglobin there are eight α-helical regions and no pleated sheet regions. Approximately 75% of the residues in myoglobin are found in these helical regions, which are designated by the letters A through H. Hydrogen bonding in the polypeptide backbone stabilizes the α-helical regions; amino acid side chains are also involved in hydrogen bonds. The polar residues are on the exterior of the molecule. The interior of the protein contains almost exclusively nonpolar amino acid residues. Two polar histidine residues are found in the interior; they are involved in interactions with the heme group and bound oxygen, and thus play an important role in the function of the molecule. The planar heme group fits into a hydrophobic pocket in the protein portion of the molecule and is held in position by hydrophobic attractions between heme's porphyrin ring and the nonpolar side chains of the protein. The presence of the heme group drastically affects the conformation of the polypeptide: the apoprotein (the polypeptide chain alone, without the prosthetic heme group) is not as tightly folded as the complete molecule.

The heme group consists of a metal ion, Fe(II), and an organic part, protoporphyrin IX (Figure 4.13). [The notation Fe(II) is preferred to Fe^{2+} when metal ions occur in complexes.] The porphyrin part consists of four five-membered rings based on the pyrrole structure; these four rings are linked by bridging methine ($-CH=$) groups to form a square planar structure. The Fe(II) ion has six coordination sites and forms six metal-ion complexation bonds. Four of the six sites are occupied by the nitrogen atoms of the four pyrrole-type rings of the porphyrin to give the complete heme group. The presence of the heme group is required for myoglobin to bind oxygen.

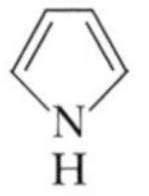

FIGURE 4.13

The structure of the heme group. Four pyrrole rings are linked by bridging groups to form a planar porphyrin ring. Several isomeric porphyrin rings are possible, depending on the nature and arrangement of the side chains. The porphyrin isomer found in heme is protoporphyrin IX. Addition of iron to protoporphyrin IX produces the heme group.

Pyrrole

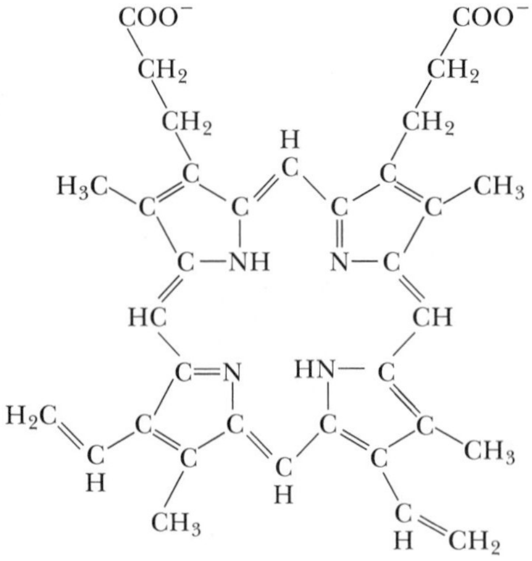

Protoporphyrin IX

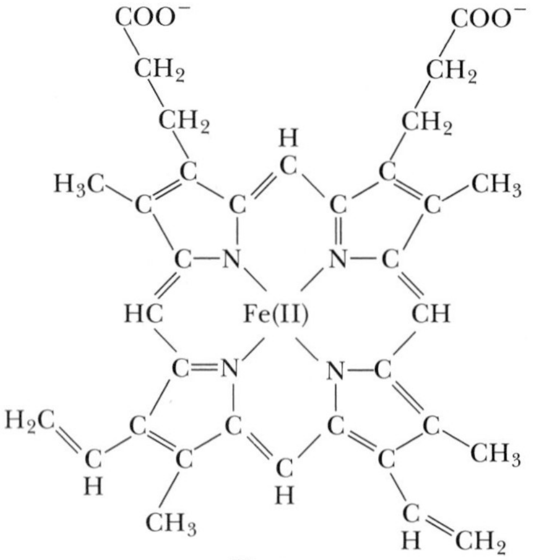

Heme
(Fe-protoporphyrin IX)

(a)

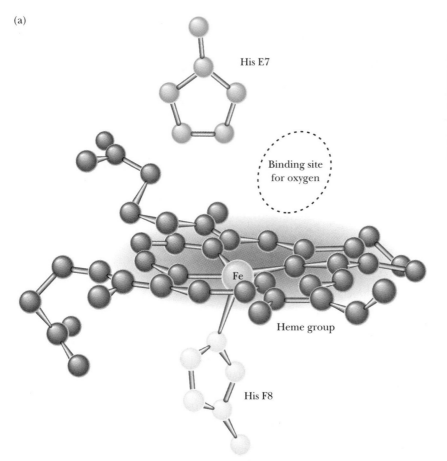

His E7

Binding site
for oxygen

Fe

Heme group

His F8

(b)

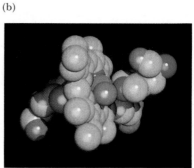

FIGURE 4.14

The oxygen-binding site of myoglobin. (a) The porphyrin ring occupies four of the six coordination sites of the Fe(II). Histidine F8 (His F8) occupies the fifth coordination site of the iron (see text). Oxygen is bound at the sixth coordination site of the iron, and histidine E7 lies close to the oxygen. (b) Space-filling model of the oxygen-binding site of myoglobin. The oxygen is shown in red at the center, the heme group (green) is vertical, and two histidines lie to the left of the heme and the right of the oxygen.

The fifth coordination site of the Fe(II) ion is occupied by one of the nitrogen atoms of the imidazole side chain of histidine residue F8 (the eighth residue in helical segment F). This histidine residue is one of the two in the interior of the molecule. The oxygen is bound at the sixth coordination site of the iron, which remains in the same oxidation state — Fe(II) — whether the oxygen is bound or not. The fifth and sixth coordination sites lie on opposite sides of the plane of the porphyrin ring. The other histidine residue in the interior of the molecule, residue E7 (the seventh residue in helical segment E), lies on the same side of the heme group as the bound oxygen (Figure 4.14). This second histidine is not bound to the iron or to any part of the heme group, but serves to stabilize the binding site for the oxygen. Oxygen does not bind to the isolated heme group. Binding also requires the environment provided by the protein.

Forces Involved in Protein Folding

The primary structure of a protein — the order of amino acids in the polypeptide chain — depends on the formation of peptide bonds, which are covalent. Higher-order levels of structure, such as the conformation of the backbone (secondary structure) and the positions of all the atoms in the protein (tertiary structure), depend on noncovalent interactions; if the protein consists of several subunits, the interaction of the subunits (quaternary structure) also depends on noncovalent interactions. Noncovalent stabilizing forces contribute to the most stable structure for a given protein, the one with the lowest energy.

Several types of hydrogen bonding occur in proteins. *Backbone* hydrogen bonding is a major determinant of secondary structure; hydrogen bonds *between the side chains of amino acids* are also possible in proteins. Nonpolar residues tend to cluster together in the interior of protein molecules as a result of *hydrophobic* interactions. *Electrostatic* attraction between oppositely charged groups, which frequently occurs on the surface of the molecule, results in such groups being close to one another. Several side chains can be *complexed* to a single metal ion. (Metal ions also occur as part of prosthetic groups.)

In addition to these noncovalent interactions, *disulfide bonds* form covalent links between the side chains of amino acids. When such bonds form, they restrict the folding patterns available to polypeptide chains. There are specialized laboratory methods for determining the number and positions of disulfide links in a given protein. Information about the locations of disulfide links can then be combined with knowledge of the primary structure to give the *complete covalent structure* of the protein. Note the subtle difference here: the primary structure is the order of amino acids, whereas the complete covalent structure also specifies the positions of the disulfide bonds (Figure 4.15).

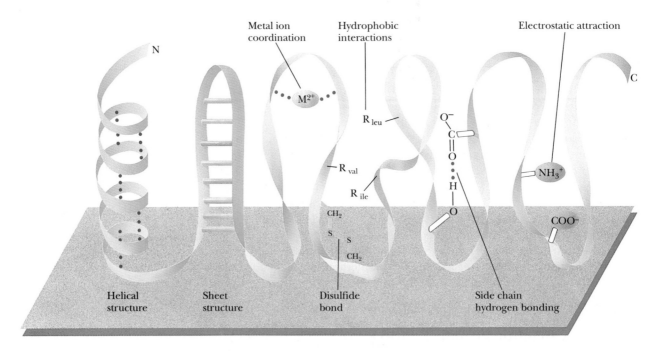

FIGURE 4.15

Forces that stabilize the tertiary structure of proteins. Note that the helical structure and sheet structure are two kinds of backbone hydrogen bonding.

Recall that, as a result of this assortment of stabilizing forces, residues that are far apart in the primary sequence can be close to each other in the three-dimensional structure produced by the folding of the protein. When a polypeptide chain folds back on itself, it can assume a compact globular shape. A different polypeptide chain (or the same chain under different conditions) can assume a rodlike fibrous form.

Not every protein necessarily exhibits all possible structural features. For instance, there are no disulfide bridges in myoglobin and hemoglobin, which are oxygen storage and transport proteins and classic examples of protein structure, but they both contain Fe(II) ions as part of a prosthetic group. In contrast, the enzymes trypsin and chymotrypsin do not contain complexed metal ions, but they do have disulfide bridges. Hydrogen bonds, electrostatic interactions, and hydrophobic interactions occur in most proteins.

The three-dimensional conformation of a protein is the result of the interplay of all the stabilizing forces. It is known, for example, that proline does not fit into an α-helix and that its presence can cause a polypeptide chain to turn a corner, ending an α-helical segment. The presence of proline is not, however, a *requirement* for a turn in a polypeptide chain. Other residues are routinely encountered at bends in polypeptide chains. The segments of proteins at bends in the polypeptide chain and in other portions of the protein that are not involved in helical or pleated sheet structures are the ones frequently referred to as "random" or "random coil." In reality, the forces that stabilize each protein are responsible for its definite conformation.

Denaturation and Refolding

The noncovalent interactions that maintain the three-dimensional structure of a protein are weak, and it is not surprising that they can easily be disrupted. The unfolding of a protein is called **denaturation.** Reduction of disulfide bonds leads to even more extensive unraveling of the tertiary structure. Denaturation and reduction of disulfide bonds are frequently combined when complete disruption of the tertiary structure of proteins is desired. Under proper experimental conditions, the disrupted structure can then be completely recovered. This process of denaturation and refolding is a dramatic demonstration of the relationship between the primary structure of the protein and the forces that determine the tertiary structure.

Proteins can be denatured in several ways. One is *heat.* An increase in temperature favors vibrations within the molecule, and the energy of these vibrations can become great enough to disrupt the tertiary structure. At either high or low *extremes of pH,* at least some of the charges on the protein are missing, and so the electrostatic interactions that would normally stabilize the native, active form of the protein are drastically reduced. This leads to denaturation. The binding of *detergents* such as sodium dodecyl sulfate (SDS) also denatures proteins. Detergents tend to disrupt hydrophobic interactions. If a detergent is charged, it can disrupt electrostatic interactions within the protein, as well. Other reagents such as *urea* and *guanidine hydrochloride* form hydrogen bonds to the protein that are stronger than those within the protein. These two reagents can also disrupt

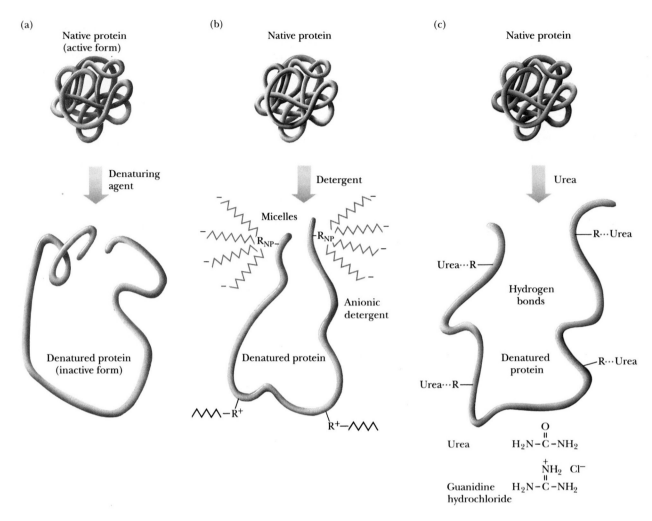

FIGURE 4.16

Denaturation of proteins. (a) The process of denaturation. (b) The mode of action of detergents. R^+ is a positively charged side chain, and R_{NP} is a nonpolar side chain. (c) The mode of action of urea.

hydrophobic interactions in much the same way as detergents (Figure 4.16).

Mercaptoethanol ($HS-CH_2-CH_2-OH$) is frequently used to reduce disulfide bridges to two sulfhydryl groups. Urea is usually added to the reaction mixture to facilitate unfolding of the protein and to increase the accessibility of the disulfides to the reducing agent. If experimental conditions are properly chosen, the native conformation of the protein can be recovered when both mercaptoethanol and urea are removed (Figure 4.17). Experiments of this type provide some of the strongest evidence that the amino acid sequence of the protein contains all the information required for the complete three-dimensional structure. Protein researchers are pursuing with some interest the conditions under which a protein can be denatured—including reduction of disulfides—and the native conformation later recovered.

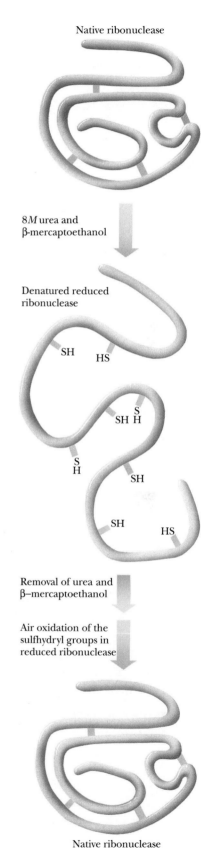

FIGURE 4.17

Denaturation and refolding in ribonuclease. The protein ribonuclease can be completely denatured by the actions of urea and mercaptoethanol. When denaturing conditions are removed, activity is recovered.

4.5
QUATERNARY STRUCTURE OF PROTEINS

Quaternary structure is a property of proteins that consist of more than one polypeptide chain. The number of chains can range from two to more than a dozen, and the chains may be identical or different. Commonly occurring examples are **dimers, trimers,** and **tetramers,** consisting of two, three, and four polypeptide chains, respectively. (The generic term for such a molecule, made up of a small number of subunits, is **oligomer.**) The chains interact with one another noncovalently.

As a result of these noncovalent interactions, subtle changes in structure at one site on a protein molecule may cause drastic changes in properties at a distant site. Proteins that exhibit this property are called **allosteric.** Not all multisubunit proteins exhibit allosteric effects, but many do. (It is also possible for a single polypeptide chain to be an allosteric protein, but this is less usual.)

A classic illustration of quaternary structure of proteins and its effect on properties is a comparison of hemoglobin, an allosteric protein, with myoglobin, which consists of a single polypeptide chain. Keep in mind, however, that interactions between protein subunits occur frequently and can cause the assembly of large structures on the molecular level. For example, the interaction of monomers of the coat protein of a virus produces the viral coat.

Hemoglobin

Hemoglobin is a tetramer, consisting of four polypeptide chains, two α-chains, and two β-chains (Figure 4.18). (In oligomeric proteins, the types of polypeptide chains are designated with Greek letters.) The two α-chains of hemoglobin are identical, as are the two β-chains. The overall structure of hemoglobin is $\alpha_2\beta_2$ in Greek notation. Both the and α- and β-chains of hemoglobin are very similar to the myoglobin chain. The α-chain is 141 residues long and the β-chain is 146 residues long; for comparison, the myoglobin chain is 153 residues long. Lengthy sequences of the α-chain, the β-chain, and myoglobin are **homologous;** that is, most or all of the amino acid residues are in the same positions. The heme group is the same in myoglobin and hemoglobin.

We have already seen that one molecule of myoglobin binds one oxygen molecule. Four molecules of oxygen can bind to one hemoglobin molecule. Both hemoglobin and myoglobin bind oxygen reversibly, but the binding of oxygen to hemoglobin is cooperative, whereas oxygen binding to myoglobin is not. **Cooperative binding** means that when one oxygen molecule is bound, it becomes easier for the next to bind. A graph of the oxygen-

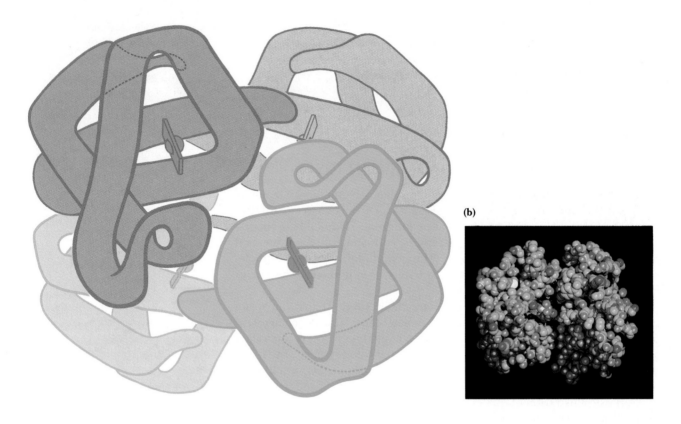

(b)

FIGURE 4.18

The structure of hemoglobin. (a) Hemoglobin ($\alpha_2\beta_2$) is a tetramer consisting of four polypeptide chains, two α-chains, and two β-chains. (b) Space-filling model of hemoglobin.

binding properties of hemoglobin and myoglobin is one of the best ways to illustrate this (Figure 4.19).

When the degree of saturation of myoglobin with oxygen is plotted against oxygen pressure, a steady rise is observed until complete saturation is approached and the curve levels off. The oxygen-binding curve of myoglobin is thus said to be **hyperbolic.** In contrast, the shape of the oxygen-binding curve for hemoglobin is **sigmoidal.** This indicates that the binding of the first oxygen molecule facilitates the binding of the second oxygen, which facilitates the binding of the third, which in turn facilitates the binding of the fourth. This is precisely what is meant by the term "cooperative binding."

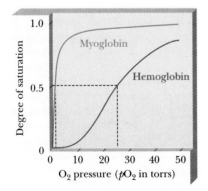

FIGURE 4.19

A comparison of the oxygen-binding behavior of myoglobin and hemoglobin. The oxygen-binding curve of myoglobin is hyperbolic, whereas that of hemoglobin is sigmoidal. Myoglobin is 50% saturated with oxygen at 1 torr partial pressure; hemoglobin does not reach 50% saturation until the partial pressure of oxygen reaches 26 torr.

(a)

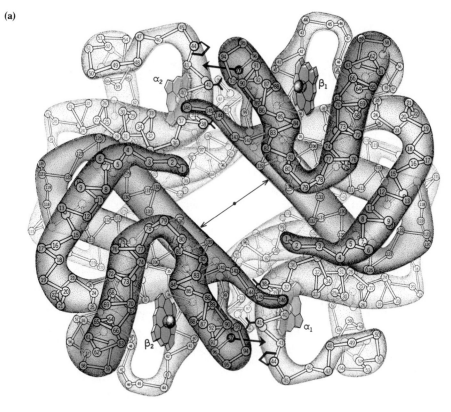

(b)

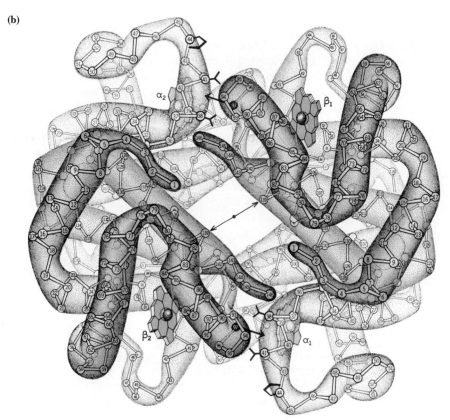

F I G U R E   4 . 2 0

The structures of (a) deoxyhemoglobin and (b) oxyhemoglobin. Note the motions of subunits with respect to one another. There is much less room at the center of oxyhemoglobin.

The two types of behavior are also related to the functions of these proteins. Myoglobin has the function of oxygen *storage* in muscle. It must bind strongly to oxygen at very low pressures, and it is 50% saturated at 1 torr partial pressure of oxygen. (A **torr** is a widely used unit of pressure, but it is not an SI unit. One torr is the pressure exerted by a column of mercury 1 mm high at 0°C. One atmosphere is equal to 760 torr.) The function of hemoglobin is oxygen *transport*, and it must be able both to bind strongly to oxygen and to release oxygen easily, depending upon conditions in the alveoli of lungs (where hemoglobin must bind oxygen for transport to the tissues), which have an oxygen pressure of 100 torr. At this pressure hemoglobin is 100% saturated with oxygen. In the capillaries of active muscles the pressure of oxygen is 20 torr, corresponding to less than 50% saturation of hemoglobin (which occurs at 26 torr). In other words, hemoglobin gives up oxygen easily in capillaries, where the need for oxygen is great.

Structural changes during binding of small molecules are characteristic of allosteric proteins such as hemoglobin. Hemoglobin has different quaternary structures in the bound (oxygenated) and unbound (deoxygenated) forms. The two β-chains are much closer to each other in oxygenated hemoglobin than in deoxygenated hemoglobin. The change is so marked that the two forms of hemoglobin have different crystal structures (Figure 4.20).

Conformational Changes That Accompany Hemoglobin Function

Other ligands are involved in cooperative effects when oxygen binds to hemoglobin. Both H^+ and CO_2, which themselves bind to hemoglobin, affect the affinity of hemoglobin for oxygen by altering the protein's three-dimensional structure in subtle but important ways. This behavior (Figure 4.21) is called the *Bohr effect* after its discoverer, Christian Bohr (the father of physicist Niels Bohr). (The oxygen-binding ability of myoglobin is not affected by the presence of H^+ or CO_2.)

An increase in the concentration of H^+ (that is, a lowering of the pH) reduces the oxygen affinity of hemoglobin. An increase in the concentration

$$HbO_2 + H^+ + CO_2 \underset{\text{Alveoli of lungs}}{\overset{\text{Actively metabolizing tissue (such as muscle)}}{\rightleftharpoons}} O_2 + Hb \overset{CO_2}{\underset{H^+}{<}}$$

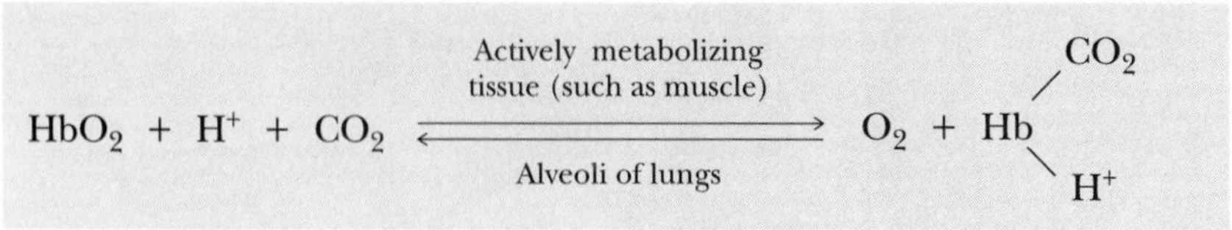

FIGURE 4.21

The general features of the Bohr effect. In actively metabolizing tissue, hemoglobin releases oxygen and binds both CO_2 and H^+. In the lungs, hemoglobin releases both CO_2 and H^+ and binds oxygen.

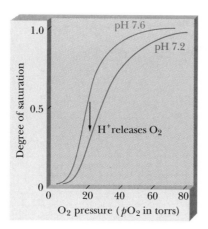

FIGURE 4.22

A comparison of the oxygen-binding capacity of hemoglobin at pH 7.6 and pH 7.2. An increase in the hydrogen ion concentration in blood leads to a release of oxygen from hemoglobin.

of CO_2 also decreases the oxygen-binding capability of hemoglobin, in the following manner (Figure 4.22). The normal pH of blood is 7.4; the pK'_a of H_2CO_3 is 6.35. As a result, about 90% of dissolved CO_2 is present as the bicarbonate ion, HCO_3^-. (The Henderson–Hasselbalch equation can be used to confirm this point. The *in vivo* buffer system involving H_2CO_3 and HCO_3^- in blood was discussed in Chapter 2, Section 2.7.) Some of the bicarbonate ions binds reversibly to hemoglobin. The presence of this group favors the quaternary structure that is characteristic of deoxygenated hemoglobin. Hence, the affinity of hemoglobin for oxygen is lowered.

Hemoglobin's acid–base properties also affect, and are affected by, its oxygen-binding properties. The oxygenated form of hemoglobin is a stronger acid (has a lower pK'_a) than the deoxygenated form. In other words, deoxygenated hemoglobin has a higher affinity for H^+ than does the oxygenated form. Thus, changes in the quaternary structure of hemoglobin can modulate the buffering of blood through the hemoglobin molecule itself.

Table 4.1 summarizes the salient features of the Bohr effect.

In the presence of large amounts of H^+ and CO_2, as in respiring tissue, hemoglobin releases oxygen. The presence of large amounts of oxygen in the lungs reverses the process, causing hemoglobin to bind O_2. The oxygenated hemoglobin can then transport oxygen to the tissues. The process is complex, but it allows for fine tuning of pH as well as levels of CO_2 and O_2.

TABLE 4.1 A Summary of the Bohr Effect

LUNGS	ACTIVELY METABOLIZING MUSCLE
Hemoglobin binds O_2	Hemoglobin releases O_2
Hemoglobin releases H^+, pH increases	Hemoglobin binds H^+, pH decreases

FIGURE   4.23

The binding of BPG to deoxyhemoglobin. Note the electrostatic interactions between the BPG and the protein.

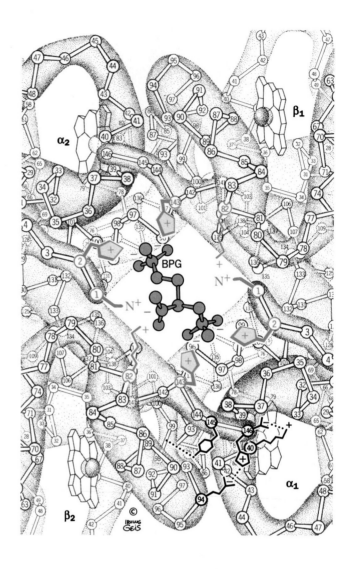

FIGURE   4.24

A comparison of the oxygen-binding properties of hemoglobin in the presence and absence of 2,3-*bis*phosphoglycerate (BPG). Note that the presence of the BPG markedly decreases the oxygen affinity of hemoglobin.

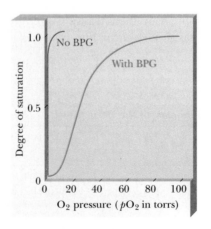

Hemoglobin in blood is also bound to another ligand, 2,3-*bis*phosphoglycerate (BPG), with drastic effects on its oxygen-binding capacity. The binding is electrostatic; specific interactions take place between the negative charges on BPG and positive charges on the protein (Figure 4.23). In the presence of BPG, the partial pressure at which 50% of hemoglobin is bound to oxygen is 26 torr. If BPG were not present in blood, the oxygen-

binding capacity of hemoglobin would be much higher (50% of hemoglobin bound to oxygen at about 1 torr), and little oxygen would be released in the capillaries. "Stripped" hemoglobin, which is isolated from blood and from which the endogenous BPG has been removed, displays this behavior (Figure 4.24).

BPG also plays a role in supplying a growing fetus with oxygen. The fetus obtains oxygen from the mother's bloodstream via the placenta. Fetal hemoglobin (Hb F) has a higher affinity for oxygen than does maternal hemoglobin, allowing for efficient transfer of oxygen from the mother to the fetus (Figure 4.25). Two features of fetal hemoglobin contribute to this higher oxygen-binding capacity. One is the presence of two different polypeptide chains. The subunit structure of Hb F is $\alpha_2\gamma_2$, where the β-chains of Hb A, the usual hemoglobin, have been replaced by the γ-chains

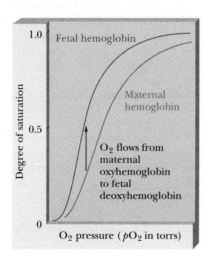

FIGURE 4.25

A comparison of the oxygen-binding capacity of fetal and maternal hemoglobins. Fetal hemoglobin binds less strongly to BPG, and consequently has a greater affinity for oxygen, than does maternal hemoglobin.

ABNORMAL HEMOGLOBINS

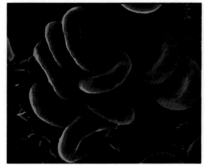
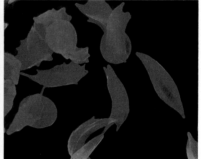

Normal red
blood cells

Sickle
cells

Normal and sickle-cell erythrocytes.

The difference between normal hemoglobin and hemoglobin S (Hb S), the mutant form characteristic of sickle-cell anemia, is the substitution of valine for glutamic acid at position 6 in each of the β-chains. Although sickle-cell hemoglobin is the commonest mutant hemoglobin, it is by no means the only one. More than 400 are known, most of them resulting from substitutions of single amino acids. These altered proteins have provided a great opportunity for studying structure–function relationships in a system consisting of a protein of known structure with a large number of variants, which are also well characterized.

Not all mutant hemoglobins have clinical manifestations. For example, Hb E occurs in as much as 10% of the population of some parts of Southeast Asia. In this variant form, the glutamate at position 26 of the β-chain has been replaced by a lysine. The notation for such a change is Glu B8(26)$\beta \longrightarrow$ Lys, where B8 refers to the location of this residue at position 8 of the B-helix. In Hb E, the change is on the protein's surface, where there is usually little effect on stability. Sickle-cell anemia hemoglobin is a glaring exception to this statement; in this case, alteration of a surface residue changes intermolecular interactions with other subunits. The altered hemoglobin molecules aggregate, leading to observed sickling.

More frequently, when hemoglobin is altered at an internal residue, a marked decrease in the stability of the molecule occurs. Degradation products of such hemoglobins accumulate at the cell membranes of erythrocytes (red blood cells), reducing the membranes' stability. **Hemolytic anemia** arises from the premature cell lysis (disintegration) associated with unstable hemoglobins. One such unstable hemoglo-

bin is Hb Savannah (named for the city in which it was discovered), in which Gly B6(24)β is replaced by Val. There is not enough room for the side chain of the valine between the B-helix and the adjacent E-helix, and consequently the entire structure is disrupted. A similar situation is observed with Hb Bibba [Leu H19(136)$\alpha \longrightarrow$ Pro], in which the proline disrupts the H-helix.

Mutations that affect the binding of the heme group have very noticeable consequences. This is particularly true when such changes stabilize the Fe(III) oxidation state of the heme and thus eliminate binding of oxygen by the defective subunits. **Methemoglobin,** abbreviated Hb M, is the name for hemoglobin in the Fe(III) state. It is brown and is responsible for the color of dried blood and old meat; normal hemoglobin, in which the heme iron is in the Fe(II) oxidation state, is red. Individuals whose blood contains methemoglobin are said to have **methemoglobinemia,** and their blood is chocolate brown. The presence of large concentrations of deoxygenated Hb M in their arterial blood leads to **cyanosis,** characterized by bluish skin. The underlying structural change in Hb M is the substitution of an anionic oxygen ligand for histidine at one of the binding sites of the Fe. In Hb M Iwate [His F8(87)$\alpha \longrightarrow$ Tyr], the tyrosine simply replaces the histidine. In Hb Milwaukee [Val E11(67)$\beta \longrightarrow$ Glu], the glutamate side chain forms an ion pair with the heme iron, stabilizing the Fe(III) oxidation state and preventing the binding of oxygen. Heterozygotes for Hb M (people having one gene for Hb M and one for normal Hb) do not suffer physical disabilities, but there are no recorded cases of persons homozygous for Hb M (having both genes for Hb M). That condition appears to be lethal.

(similar but not identical in structure). The second feature is that Hb F binds less strongly to BPG than does Hb A, because there are fewer positively charged groups of Hb F to stabilize the electrostatic interaction.

Box 4.1 discusses several other variants of hemoglobin whose behavior illustrates the point that changes in the primary structure (the amino acid sequence) of a protein can significantly alter its three-dimensional structure, which in turn influences its chemical properties.

4.6
MUSCLE CONTRACTION: HOW PROTEIN STRUCTURE AFFECTS FUNCTION

Muscle proteins furnish an example of the relationship between structure and function. Changes in the configurations of proteins in muscle cells are ultimately responsible for muscle contraction. Muscle cells are made up of

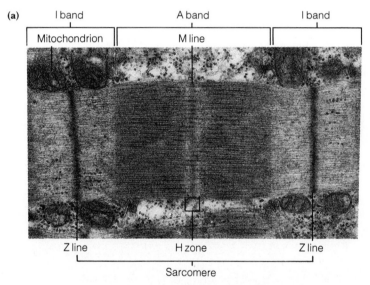

(a)

I band   A band   I band

Mitochondrion   M line

Z line   H zone   Z line

Sarcomere

FIGURE 4.26

The arrangement of filaments in a myofibril. (a) Electron micrograph of a sarcomere. The overlapping thick and thin filaments give rise to the observed light and dark bands. (b) Schematic diagram of a sarcomere.

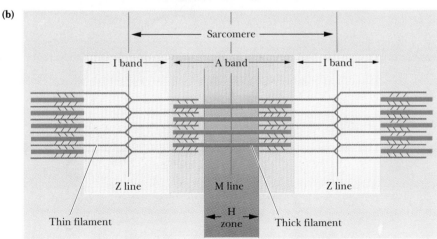

(b)

Sarcomere

I band   A band   I band

Z line   M line   Z line

Thin filament   H zone   Thick filament

fibers called **myofibrils,** which in turn are made up of repeating units called **sarcomeres.** In a sarcomere, thick and thin filaments lie parallel to each other, their ends partially overlapping. The thick filament consists of an aggregate of several molecules of the protein **myosin.** The thin filament consists of an aggregate of the protein **actin.** Other proteins that are involved in muscle contraction occur in the thin filament, but actin and myosin are the most important contributors to the process; we shall concentrate on them.

The presence of thick and thin filaments is responsible for the appearance of the sarcomere, characterized by two alternating bands of different degrees of darkness: a light I band and a heavy A band (Figure 4.26). The thick filament occurs by itself in the comparatively light H zone within the A

FIGURE 4.27

Muscle contraction. (a) In resting muscle, cross-bridges do not form between the thick and thin filaments. ATP becomes attached to the globular head of myosin and is split into ADP and P_i, which remain bound to myosin. (b) In the presence of Ca^{2+}, the ADP and P_i are released. The myosin head binds to actin. (c) The power stroke consists of bending the myosin head at the neck and pulling the filaments past one another. (d) ATP is bound to the myosin head, which is no longer bound to actin, and ATP is split into ADP and P_i. The whole process repeats itself, with myosin binding to the next available binding site on actin.

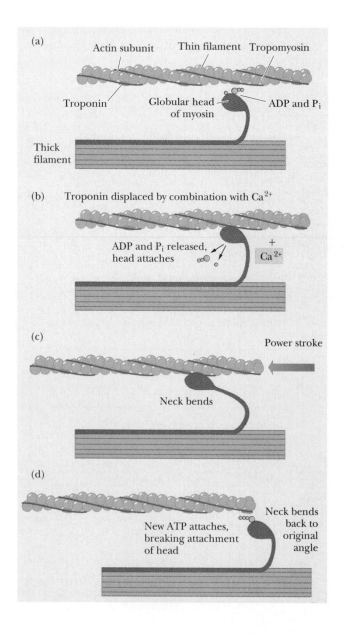

band, and parallel to the thin filament where it overlaps it in the rest of the A band. Within the H zone is the M line, another dark portion representing the overlap of thick filaments. The I band consists of thin filaments only, but it is bisected by the dark Z line, an area rich in proteins other than actin.

In muscle contraction the thick and thin filaments slide past one another. Their lengths do not change, but the amount of overlap increases as the muscle contracts. In the process, the mode of interaction of actin and myosin changes (Figure 4.27). Myosin is principally a fibrous protein, but it has globular portions that appear as projections from the thick filament. The globular portions of myosin interact with actin to form an actomyosin complex (Figure 4.27b). This process requires energy, provided by the hydrolysis of adenosine triphosphate (ATP) to adenosine diphosphate (ADP) and phosphate ion (P_i) (Figure 4.28). When a muscle contracts, each globular region of myosin first tilts and then becomes detached from the site to which it was bound to the actin part of the actomyosin complex, and moves to a new site. The tilting of the globular head of the myosin is considered to provide the **power stroke** for the contraction (Figure 4.27c), a process that makes use of the energy obtained from ATP hydrolysis.

Muscle contraction requires Ca^{2+} as well as energy. Calcium's role in muscle contraction involves the other protein components of the thin filament, troponin and tropomyosin. An increase in the level of Ca^{2+} comes about as a result of the opening of Ca^{2+} channels in the sarcoplasmic reticulum, produced by the action of the neuromuscular junction (Chapter 8, Section 8.7). The high level of Ca^{2+} triggers a conformational change in troponin, a change that is passed along to tropomyosin. The conformational change in tropomyosin, in turn, triggers the tilting and sliding motion of myosin, which is the main feature of muscle contraction. This long series of steps exemplifies the complex network of relationships that exists in all biological processes.

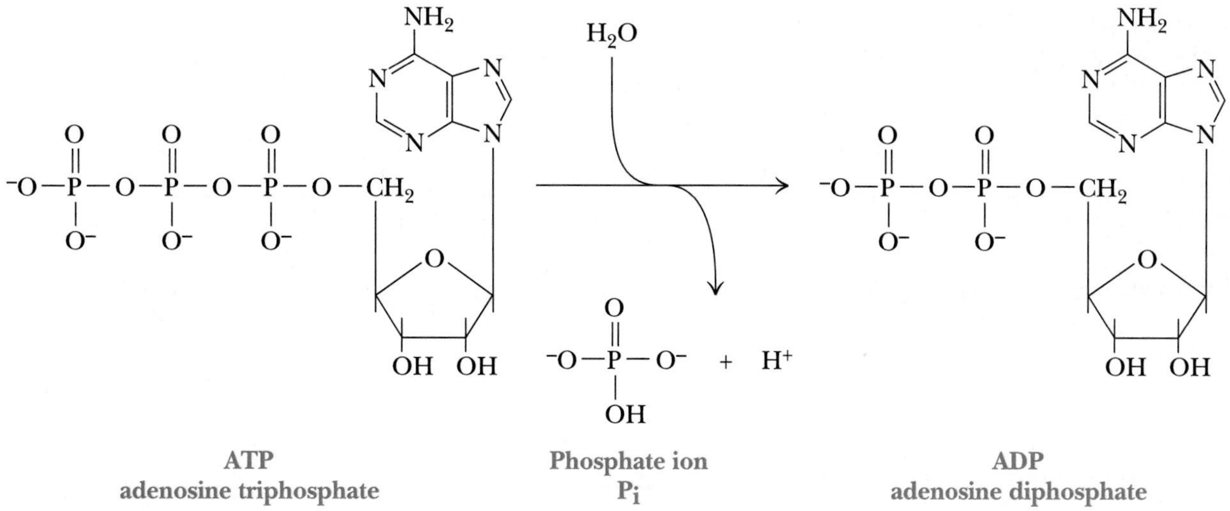

| ATP | Phosphate ion | ADP |
| adenosine triphosphate | P_i | adenosine diphosphate |

FIGURE 4.28

The hydrolysis of ATP to ADP and phosphate ion (P_i) releases energy, which can be used in muscle contraction.

S U M M A R Y

The structure of proteins is complex, with little orderly arrangement of atoms. Many three-dimensional conformations are possible for proteins, but only one, or at most a few, have biological activity; these are called the native conformations. To facilitate structure determination, it is customary to define four levels of organization. Primary structure is the order in which the amino acids are covalently linked together. Secondary structure is the arrangement in space of the atoms in the backbone of the polypeptide chain. Tertiary structure includes the three-dimensional arrangement of *all* the atoms in the protein. Quaternary structure is the arrangement of subunits in multisubunit proteins.

The amino acid sequence (the primary structure) of a protein determines its three-dimensional structure, which in turn determines its properties. A striking example of the importance of primary structure is sickle-cell anemia, a disease caused by a change in one amino acid in each of two of the four chains of hemoglobin. The higher-order (secondary and tertiary) levels of structure depend on noncovalent interactions, including hydrogen bonds, hydrophobic interactions, electrostatic interactions, and complexation of metal ions. The three-dimensional structures of proteins can be completely disrupted and, under proper experimental conditions, completely recovered. This process of denaturation and refolding is a dramatic example of the relationship between the primary structure of the protein and the forces that determine the tertiary structure.

The primary structure of a protein can be determined by chemical methods. Three determinations must be made: (1) which amino acids, and how many of each kind, are found in the protein; (2) which amino acids occur at the N-terminal and C-terminal ends of the molecule; and (3) the sequence of amino acids. The secondary and tertiary structures of a protein can be determined simultaneously by x-ray crystallography. The oxygen storage protein myoglobin was the first protein for which the complete tertiary structure was determined by crystallography.

The individual polypeptide chains of multisubunit proteins interact with one another noncovalently. As a result, subtle changes in structure at one site on the molecule can cause drastic changes in properties at a distant site. Proteins that exhibit this property are referred to as allosteric. The properties of the allosteric protein hemoglobin can be contrasted with those of myoglobin, which is not allosteric. In hemoglobin, an oxygen transport protein, the binding of oxygen is cooperative (as each oxygen is bound, it becomes easier for the next one to bind) and is modulated by such ligands as H^+, CO_2, and BPG (2,3-*bis*phosphoglycerate). The binding of oxygen to myoglobin is not cooperative.

Muscle contraction provides a case study for the relationship between protein structure and function. The process of contraction comes about as a result of the motions of proteins with respect to one another. The motions of the proteins, in turn, depend on changes in their conformations.

E X E R C I S E S

1. Match the following statements about protein structure with the proper levels of organization.
   - (a) Primary structure
   - (b) Secondary structure
   - (c) Tertiary structure
   - (d) Quaternary structure

   - (1) The three-dimensional arrangement of all atoms
   - (2) The order of amino acid residues in the polypeptide chain
   - (3) The interaction between subunits in proteins that consist of more than one polypeptide chain
   - (4) The arrangement in space of the polypeptide backbone

2. A biochemistry student characterizes the process of cooking meat as an exercise in denaturing proteins. Comment on the validity of this remark.

3. You hear the comment that the difference between wool and silk is the difference between helical and pleated sheet structures. Do you consider this a valid point of view? Why or why not?

4. What is the nature of "random" structure in proteins?

5. List five forces that are responsible for maintaining the

correct three-dimensional shapes of proteins. Specify which groups on the protein are involved in each type of interaction.

6. Define denaturation in terms of the effects of secondary, tertiary, and quaternary structure.

7. List two similarities and two differences between hemoglobin and myoglobin.

8. Suggest a way in which the difference between the functions of hemoglobin and myoglobin is reflected in the shapes of their respective oxygen binding curves.

9. In oxygenated hemoglobin, $pK'_a = 6.6$ for the histidines at position 146 on the β-chain. In deoxygenated hemoglobin the pK'_a of these residues is 8.2. How can this piece of information be correlated with the Bohr effect?

10. Suggest an explanation for the observation that covalently modified proteins cannot be denatured reversibly.

11. List some of the differences between the α-helix and β-sheet forms of secondary structure.

12. List some of the possible combinations of α-helices and β-sheets in supersecondary structures.

13. Rationalize the following observations.
    (a) Serine is the amino acid residue that can be replaced with the least effect on protein structure and function.
    (b) Replacement of tryptophan causes the greatest effect on protein structure and function.
    (c) Replacements such as Lys $\longrightarrow$ Arg and Leu $\longrightarrow$ Ile usually have very little effect on protein structure and function.

14. Suggest a reason for the observation that persons with sickle-cell trait sometimes have breathing problems during high-altitude flights.

15. Describe the Bohr effect.

16. Why is proline frequently encountered at the places where the polypeptide chain turns a corner in both myoglobin and hemoglobin?

17. Describe the effect of 2,3-*bis*phosphoglycerate on the binding of oxygen by hemoglobin.

18. How does the oxygen-binding curve of fetal hemoglobin differ from that of adult hemoglobin?

19. List some abnormal hemoglobins and describe how their structures differ from that of normal hemoglobin.

20. Does a fetus homozygous for Hb S have normal Hb F?

21. What is the molecular basis for the observation that blood changes color from red to brown as it dries? What connection does this observation have with abnormal hemoglobins?

A N N O T A T E D   B I B L I O G R A P H Y

Bagshaw, C. R. *Muscle Contraction.* New York: Chapman and Hall, 1982. [A short book about general aspects of muscle contraction.]

Cantor, C. R., and P. R. Schimmel. *Biophysical Chemistry.* San Francisco: W. H. Freeman, 1980. [A multivolume work dealing with properties that are useful for characterizing and separating proteins. Paperback edition available.]

Changeux, J.-P., A. Devillers-Thiery, and P. Chemoulli. Acetylcholine Receptor: An Allosteric Protein. *Science* **225,** 1335–1345 (1984). [A look at the importance of allosteric properties in regulating the action of one of the most important proteins in the nervous system.]

Clore, G., and A. Gronenborn. Structures of Larger Proteins in Solution: Three- and Four-Dimensional Heteronuclear NMR Spectroscopy. *Science* **252,** 1390–1399 (1991). [A review of one of the most powerful experimental methods for determining the complete three-dimensional structure of a protein.]

Dayhoff, M. O., ed. *Atlas of Protein Sequence and Structure.* Washington, DC: National Biomedical Research Foundation, 1978. [A listing of all known amino acid sequences. Updated periodically.]

Dickerson, R. E., and I. Geis. *The Structure and Action of*

Proteins. 2nd ed. Menlo Park, CA: Benjamin-Cummings, 1981. [A well-written and particularly well-illustrated general introduction to protein chemistry.]

Dill, K. Dominant Forces in Protein Folding. *Biochemistry* **29,** 7133–7155 (1990). [A review of the weak interactions that determine protein folding; an advanced treatment.]

Doolittle, R. F. Proteins. *Sci. Amer.* **253** (4), 88–89 (1985). [A well-illustrated discussion of protein structure with emphasis on evolutionary considerations.]

Fermi, G., and M. F. Perutz. *Atlas of Molecular Structures in Biology.* Vol. 2, *Haemoglobin and Myoglobin.* Oxford, England: Clarendon Press, 1981. [A detailed description of the structures of these proteins, with particular emphasis on the results of x-ray crystallography.]

Freifelder, D. *Physical Biochemistry.* 2nd ed. San Francisco: W. H. Freeman, 1982. [An introduction to methods for characterizing and isolating proteins. Paperback edition available.]

Gibbons, A., and M. Hoffman. New 3-D Protein Structures Revealed. *Science* **253,** 382–383 (1991). [Examples of the use of x-ray crystallography to determine protein structure.]

Gierasch, L. M., and J. King, eds. *Protein Folding: Deciphering* the Second Half of the Genetic Code. Waldorf, MD: AAAS Books, 1990. [A collection of articles on recent discoveries about the processes involved in protein folding. Experimental methods for studying protein folding are emphasized.]

Harper, E., and G. Rose. Helix Stop Signals in Proteins and Peptides: The Capping Box. *Biochemistry* **32,** 7605–7609 (1993). [Evidence that a reciprocal backbone–side chain hydrogen bonding interaction can function as a helix stop signal.]

Hauptmann, H. The Direct Methods of X-ray Crystallography. *Science* **233,** 178–183 (1986). [A discussion of improvements in methods of performing the calculations involved in determining protein structure; based on a Nobel Prize address. This article should be read in connection with the one by Karle, and provides an interesting contrast to the articles by Perutz and Kendrew.]

Jaenicke, R. Protein Folding and Protein Association. *Angew. Chem. Int. Ed. Engl.* **23,** 395–413 (1984). [A discussion of a possible "folding code" for protein tertiary structure. The genetic code for the amino acid sequence determines the folding code.]

Jaenicke, R. Protein Folding: Local Structures, Domains, Subunits, and Assemblies. *Biochemistry* **30,** 3147–3161 (1991). [A review article about a topic that has been called "the second half of the genetic code."]

Karle, J. Phase Information from Intensity Data. *Science* **232,** 837–843 (1986). [A Nobel Prize address on the subject of x-ray crystallography. See remarks on the article by Hauptmann.]

Kendrew, J. C. Myoglobin and the Structure of Proteins. *Science* **139,** 1259–1266 (1963). [Based on a Nobel Prize address.]

————. The Three-Dimensional Structure of a Protein Molecule. *Sci. Amer.* **205** (6), 96–111 (1961). [An introduction to determination of protein structure by x-ray crystallography. This article, the preceding one, and the ones by Perutz demonstrate the earliest accomplishments in protein crystallography and are interesting contrasts to the articles by Hauptmann and Karle.]

Leszczynski, J. F., and G. D. Rose. Loops in Globular Proteins: A Novel Category of Secondary Structure. *Science* **234,** 849–855 (1986). [Makes the point that the "random" portions of proteins are not really random.]

Monod, J., J.-P. Changeux, and F. Jacob. Allosteric Proteins and Cellular Control Systems. *J. Mol Biol.* **6,** 306–329 (1963). [The original model for the mode of action allosteric of proteins, and still one of the standard references on the subject.]

O'Shea, E., J. Klemm, P. Kim, and T. Alber. X-ray Structure of the GCN4 Leucine Zipper, A Two-Stranded, Parallel Coiled Coil. *Science* **254,** 539–544 (1991). [Description of a widely distributed and highly important structural motif in proteins.]

Perutz, M. The Hemoglobin Molecule. *Sci. Amer.* **211** (5), 64–76 (1964).

————. The Hemoglobin Molecule and Respiratory Transport. *Sci. Amer.* **239** (6), 92–125 (1978). [The relationship between molecular structure and cooperative binding of oxygen. Also see remarks on the articles by Kendrew.]

Rhodes, D., and A. Klug. Zinc Fingers. *Sci. Amer.* **268** (2) 56–65 (1993). [How the structure of these zinc-containing proteins allows them to play a role in regulating the activity of genes.]

Richards, F. The Protein Folding Problem. *Sci. Amer.* **264** (1), 54–63 (1991). [A description of the practical difficulties involved in a task that is theoretically possible: predicting the three-dimensional structure of a protein from its amino acid sequence.]

Richardson, J. The Anatomy and Taxonomy of Protein Structure. *Adv. Prot. Chem.* **34,** 168–339 (1981). [An extensive review of secondary and tertiary structure, with excellent illustrations.]

Taylor, E. Molecular Muscle. *Science* **261,** 35–36 (1993). [How the structure of the protein myosin allows it to act as a molecular motor in muscle contraction. See also the articles by Rayment *et al.* on pp. 50 and 58 of the same issue.]

Experimental Methods for Determining Protein Structure

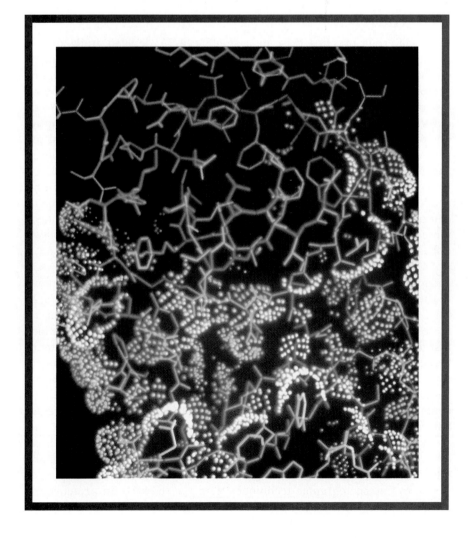

Computer-generated model of the protein pepsin.

Since a cell contains thousands of *different* protein molecules, the task of separating them and determining the structure of a single protein is exceedingly difficult. There are many techniques for characterizing a protein—ranging from strategies for discovering the number and type of its constituent amino acids to those for elucidating its complete amino acid sequence. When a protein has been degraded to its amino acids, they can be identified according to their charge and polarity, by chromatography and electrophoresis. The amino acids at the ends of a protein can be established by chemical labeling. The whole chain can be degraded by specific cleavage to give related peptide fragments. Each peptide can then be degraded one amino acid at a time to discover its sequence. In a final step of structure determination, a complete protein can be subjected to x-ray diffraction analysis to determine its three-dimensional conformation. Before this can be accomplished, however, the protein must be purified by a number of techniques, after which it may be crystallized.

A.1
ISOLATION OF PROTEINS

Many different proteins exist in a single cell. A detailed study of the properties of any one protein requires a homogeneous sample consisting of only one kind of molecule. The separation and isolation, or purification, of proteins constitute an essential first step for experimentation. In general, separation techniques focus on three properties that can be the sources of differences among molecules: size, charge, and polarity. **Chromatography,** one of two common general methods, makes use of differences in all three properties. **Electrophoresis,** the other general method, depends on charge and size. Some forms of these methods are particularly useful for large molecules, and we shall initially concentrate on them. Other forms can be applied to molecules of any size including smaller fragments that are used in determining the amino acid sequence once a protein has been purified; those techniques shall appear in an example later in this chapter, when we discuss protein sequencing.

The word "chromatography" comes from the Greek *chroma,* "color," and *graphein,* "to write"; the technique was first used around the beginning of the 20th century to separate plant pigments with easily visible colors. It has long since been possible to separate colorless compounds, as long as there are methods for detecting them. Chromatography is based on the fact that different compounds can distribute themselves to varying extents between different *phases,* or separable portions of matter. One phase is *stationary* and one is *mobile.* The mobile phase flows over the stationary material and carries the sample to be separated along with it. The components of the sample interact with the stationary phase to different extents. Some components interact relatively strongly with the stationary phase and are therefore carried along more slowly by the mobile phase than are those that interact less strongly. The differing mobilities of the components are the basis of the separation.

Many chromatographic techniques used for research on proteins are forms of **column chromatography,** in which the material that makes up the

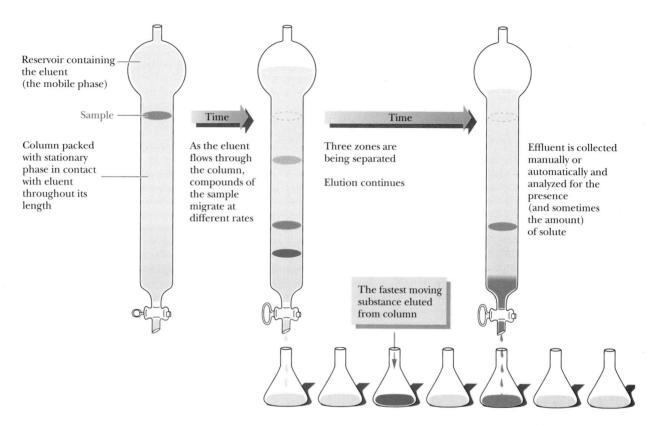

FIGURE A.1

An example of column chromatography.

stationary phase is packed in a column. The sample is a small volume of concentrated solution that is applied to the top of the column; the mobile phase, called the **eluent,** is passed through the column. The sample is diluted by the eluent, and the separation process also increases the volume occupied by the sample. In a successful experiment, all the sample eventually comes off the column. Figure A.1 diagrams an example of column chromatography.

Molecular sieve chromatography separates molecules on the basis of size, making it a useful way to sort proteins of varied molecular weights. It is a form of column chromatography in which the stationary phase (the material used to pack the column) consists of cross-linked gel particles. The gel particles are usually in bead form and consist of one of two kinds of polymers: a carbohydrate polymer (such as dextran or agarose) or polyacrylamide. The cross-linked structure of these polymers produces pores in the material. The extent of cross-linking can be controlled to select a desired pore size.

When a sample is applied to the column, smaller molecules can enter the pores and thus tend to be delayed in their progress down the column, unlike the larger molecules. As a result, the larger molecules are eluted first, followed later by the smaller ones after their escapes from the pores.

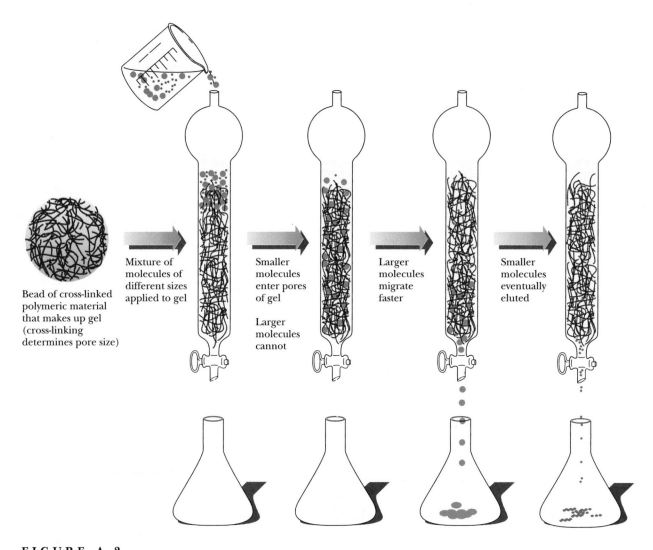

Bead of cross-linked polymeric material that makes up gel (cross-linking determines pore size)

Mixture of molecules of different sizes applied to gel

Smaller molecules enter pores of gel

Larger molecules cannot

Larger molecules migrate faster

Smaller molecules eventually eluted

FIGURE A.2

Schematic representation of molecular sieve chromatography.

Molecular sieve chromatography is represented schematically in Figure A.2. The advantages of this type of chromatography are its convenience as a way to separate molecules on the basis of size and the fact that it can be used to estimate molecular weight by comparison with standard samples.

Affinity chromatography makes use of the specific binding properties of many proteins. It is another form of column chromatography, with some sort of polymeric material used as the stationary phase. The distinguishing feature of affinity chromatography is that the polymer is covalently linked to some compound, called a **substrate,** that binds specifically to the desired protein (Figure A.3). The other proteins in the sample do not bind to the column and can easily be eluted with buffer, while the bound protein remains on the column. The bound protein can then be eluted from the column via the addition of high concentrations of the substrate in soluble form. The protein binds to the substrate in the mobile phase and is recovered from the column. Affinity chromatography is a convenient separation method and has the advantage of producing very pure proteins.

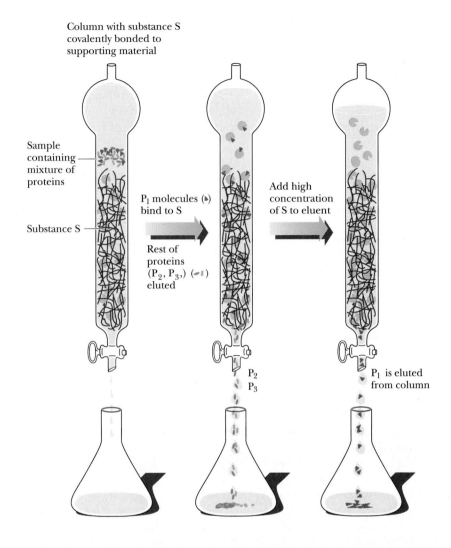

Column with substance S
covalently bonded to
supporting material

Sample
containing
mixture of
proteins

Substance S

P_1 molecules (◗) bind to S

Rest of proteins ($P_2, P_3,$) (◖) eluted

Add high concentration of S to eluent

P_2 P_3

P_1 is eluted from column

FIGURE A.3

The principle of affinity chromatography. In a mixture of proteins, only one (designated P_1) will bind to a substance (S) called the substrate. Ⓢ is the substance S in solution. The binding is reversible: $P_1 + \rightleftharpoons P_1{-}S$.

Electrophoresis is based on the motion of charged particles in an electric field toward an electrode of opposite charge. The separation is based on their mobilities. For our purposes it is enough to know that the motion of a charged molecule in an electric field depends on the ratio of its charge to its mass. A sample is applied to some sort of supporting medium. With the use of electrodes, an electric current is passed through the medium to achieve the desired separation.

Polymeric gels similar to those used in molecular seive chromatography can be used as supporting media for electrophoresis. The gel, usually polyacrylamide, is prepared and cast as a continuous cross-linked matrix (rather than the bead form employed in column chromatography). **Polyacrylamide gel electrophoresis** is a useful technique, and some variations can increase its usefulness (Figure A.4).

In one variation of polyacrylamide gel electrophoresis, the protein sample is treated with the detergent sodium dodecyl sulfate (SDS) before being applied to the gel. The structure of SDS is $CH_3(CH_2)_{10}CH_2O$ $SO_3{}^-Na^+$. The anion binds strongly to proteins via nonspecific adsorption.

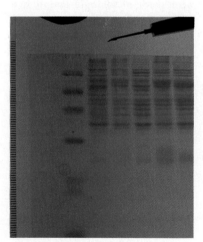

FIGURE A.4

Separation of proteins by gel electrophoresis. Each band seen in the gel represents a different protein. In the SDS-PAGE technique, the sample is treated with detergent before being applied to the gel. In isoelectric focusing, a pH gradient runs the length of the gel.

The larger the protein, the more of the anion it will adsorb. SDS completely denatures proteins, breaking all the noncovalent interactions that determine tertiary and quaternary structure. This means that multisubunit proteins can be analyzed as the component polypeptide chains. All the proteins in a sample have a negative charge as a result of adsorption of anion. In **SDS–polyacrylamide gel electrophoresis** (SDS-PAGE), the acrylamide offers more resistance to large molecules than to small molecules. Because the ratio of charge to mass is approximately the same for all the proteins in the sample, the size of the protein now becomes the determining factor in the separation: small proteins move faster than large ones. Like molecular sieve chromatography, SDS–polyacrylamide gel electrophoresis can be used to estimate the molecular weights of proteins by comparison with standard samples. (A similar electrophoretic technique has wide usefulness in nucleic acid research and is discussed in Chapter 7, Section 7.1.)

Isoelectric focusing is another variation of gel electrophoresis. Since different proteins have different titratable groups, they also have different isoelectric points. Recall (Chapter 3, Section 3.3) that the isoelectric pH (pI) is the pH at which a protein (or amino acid or peptide) has no net charge. At the pI the number of positive charges exactly balances the number of negative charges. In an isoelectric focusing experiment, the gel is prepared with a pH gradient that parallels the electric field gradient. As proteins migrate through the gel under the influence of the electric field, they encounter regions of different pH, and the charge on the protein changes. Eventually each protein reaches the point at which it has no net charge — its isoelectric point — and it no longer migrates. Each protein remains at the position on the gel corresponding to its pI, allowing for an effective method of separation.

An ingenious combination of isoelectric focusing in one dimension and SDS-PAGE in a second dimension, at 90° to the first, allows for enhanced separation (Figure A.5). A sample is applied to a gel, and the electrophoretic separation is run in one dimension by isoelectric focusing. The resulting gel gives a separation similar to that shown in Figure A.4. It is then placed on top of another gel, and the SDS-PAGE separation is run at 90° to the first.

FIGURE A.5

Two-dimensional electrophoresis. (a) The experimental arrangement. The first separation has already been made on a single sample in a narrow polyacrylamide gel, using isoelectric focusing. The result is similar to that in Figure A.4. The gel is then extruded from the tube that contained it and placed on a slab gel for SDS-PAGE electrophoresis. (b) The results of two-dimensional electrophoresis. Each spot is a different protein.

(a)

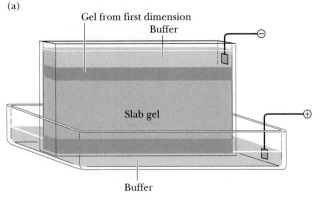

(b)

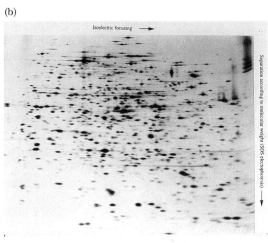

A.2
SEPARATION OF AMINO ACIDS: THE MODE OF OPERATION OF AN AMINO ACID ANALYZER

After a protein has been purified to homogeneity, researchers can determine the abundance of each amino acid in the polypeptide chain. The operation of an **amino acid analyzer** is an example of column chromatography. There are two types of amino acid analyzers; the difference between them lies in the nature of the interaction that takes place between the stationary phase and the components of the sample, providing the basis for separation. In **ion exchange chromatography,** the interaction in question is electrostatic attraction. In the other types, **high-performance liquid chromatography (HPLC),** the separation depends on differences in polarity. (We will address HPLC in Section A.4, when we discuss its role in the separation of peptides in protein sequencing.)

To consider an example of ion exchange chromatography, let us assume that the sample contains a mixture of cations (positively charged ions) and that the stationary phase consists of an anionic substance (negatively charged ions) with counterions (cations) bound to the anionic sites by electrostatic forces. A cation with a charge of +2 in the sample interacts more strongly with an anion in the stationary phase than does a cation with a charge of +1 in the sample. The less highly charged cation is therefore carried along more quickly by the mobile phase than the more highly charged one, and the difference in rates of migration eventually brings about the desired separation.

Synthetic resins are used as ion exchangers in many chromatographic operations. These **ion exchange resins** are cross-linked long-chain polymers, available commercially in bead form. The beads (which are used to pack the column) have diameters on the order of micrometers (1 μm = 10^{-6}m). Each bead can contain as many as several thousand charged groups, all positive or all negative (Figure A.6). The resins making up the beads

FIGURE A.6

The mode of action of cation exchange resins. In the forward reaction the amino acid is bound to the resin, and in the reverse reaction it is released to the mobile phase. The process occurs many times on each bead and along the entire column. The components of the mixture encounter new beads as they move.

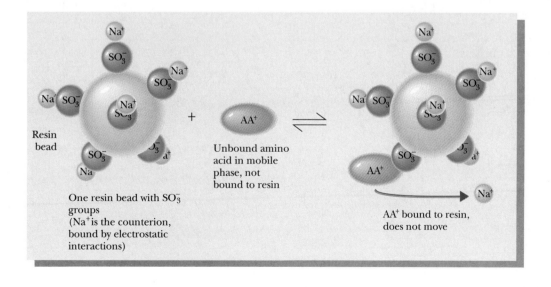

Resin bead

One resin bead with SO_3^- groups (Na^+ is the counterion, bound by electrostatic interactions)

Unbound amino acid in mobile phase, not bound to resin

AA^+ bound to resin, does not move

typically contain negatively charged sulfonate groups, $-SO_3^-$ (the ionized form of the sulfonic acid group, $-SO_3H$). Before the column is packed, the beads are soaked in a solution containing the counterion desired for the start of the experiment — for instance, Na^+. In the column, other positively charged ions can replace the Na^+ through an exchange process. The separation of amino acids takes place on the basis of the extent to which each amino acid is positively charged and the extent to which it can replace the sodium ion as the counterion that balances the negative charge of each sulfonate group of the cation exchange resin.

At the start of a typical separation, the mixture of amino acids is applied to the top of the column, and the flow of eluting buffer is started. As the elution proceeds, some of the Na^+ is replaced by positively charged amino acids, AA^+, which become the counterions. Imagine a sample consisting of three amino acids — aspartic acid, serine, and histidine — dissolved in a buffer of pH 3.25. At this pH the amino groups of all three amino acids are protonated, as is the side-chain imidazole of the histidine. Essentially all the α-carboxyl groups are ionized, and about 20% of the side-chain carboxyl groups of the aspartic acid are ionized (Figure A.7). (Use the Hender-

FIGURE A.7

The separation of amino acids by ion exchange chromatography. (a) Sample components. (b) An example of the separation process by ion exchange.

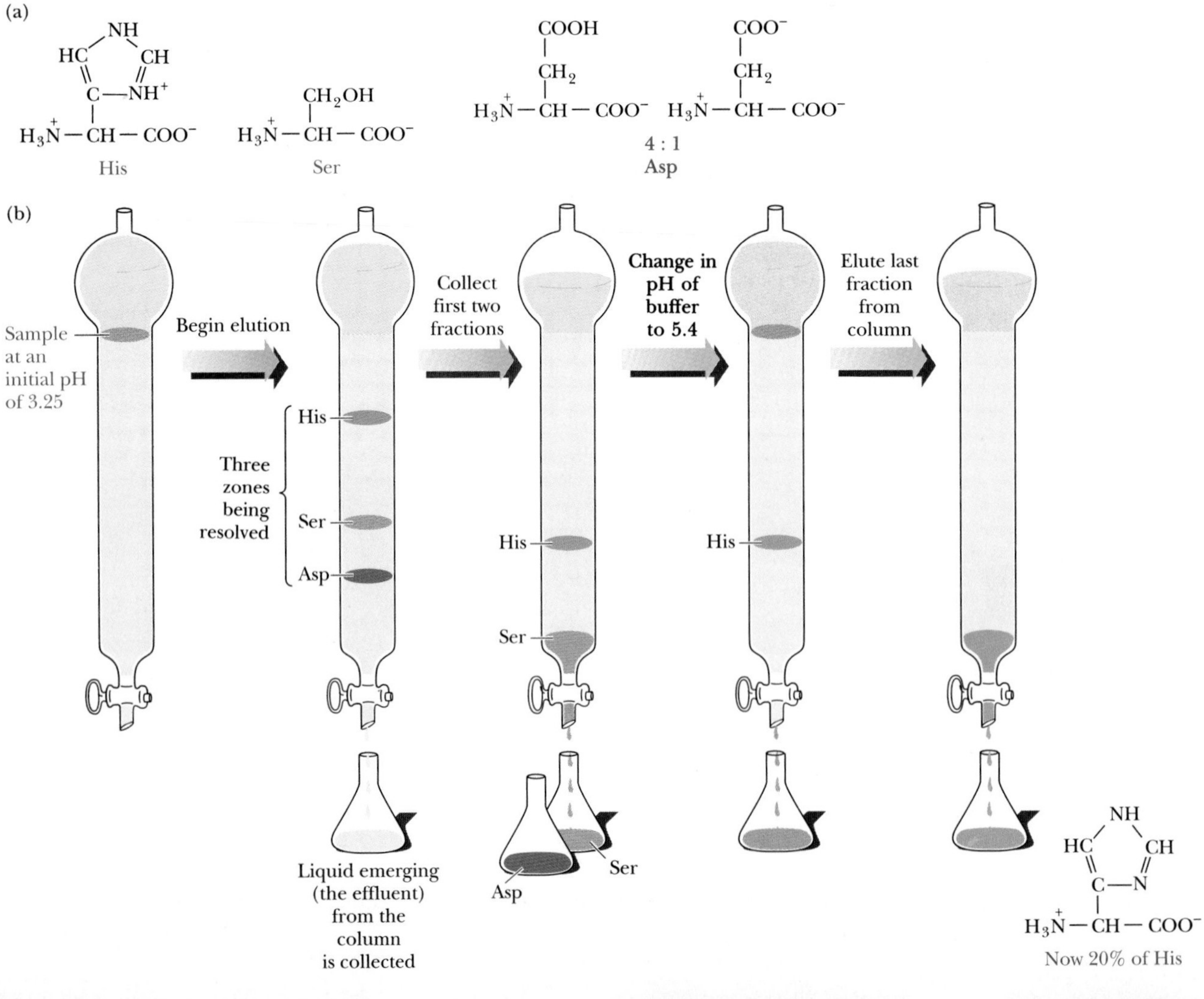

(a)

His   Ser   Asp (4:1)

(b)

Sample at an initial pH of 3.25   Begin elution   Three zones being resolved   His / Ser / Asp   Liquid emerging (the effluent) from the column is collected   Asp   Ser

Collect first two fractions   His / Ser

Change in pH of buffer to 5.4   His

Elute last fraction from column

Now 20% of His

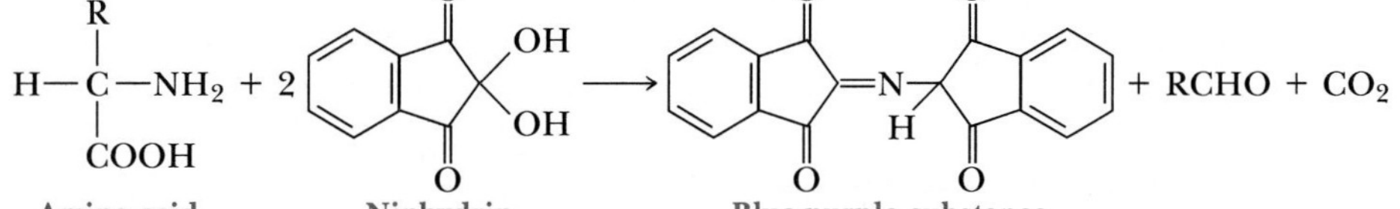

FIGURE A.8

The reaction of amino acids with ninhydrin. Note that the original amino acid is converted to an aldehyde with one carbon fewer than the parent acid. The carboxyl group of the original amino acid is lost as CO_2, and the amino nitrogen appears in the product.

son–Hasselbalch equation to confirm the degree of ionization of the side-chain carboxyl of the aspartic acid.)

In this example the histidine has a net positive charge of 1, the serine is electrically neutral, and the aspartic acid is a 4:1 mixture of electrically neutral molecules and molecules with a net negative charge of 1. The histidine is bound most strongly to the cation exchange resin because of its charge, the serine less strongly, and the aspartic acid least strongly. More eluting buffer is added, and a second stage of exchange takes place in which the positively charged amino acids are replaced by cations from the buffer. The aspartic acid is eluted first, followed by the serine; the histidine remains bound to the resin. The pH of the eluting buffer is raised in stages to facilitate release from the column of amino acids with positively charged R groups. Let us say the pH of the buffer is raised to 5.4. At this value, about 20% of the histidine side chains are deprotonated, and these molecules have no net charge. (Why? Use the Henderson — Hasselbalch equation again to confirm this point.) The histidine is then eluted from the column. Good resolution can be achieved with an amino acid analyzer, and the individual amino acids are well separated from one another.

The individual amino acids can be detected by titration of each fraction of the **eluate** (the solution that has been eluted from the column) with ninhydrin to produce a purple compound (a yellow one in the case of the amino acid proline) (Figure A.8). The absorption of light of any desired wavelength by each fraction collected from the column can then be

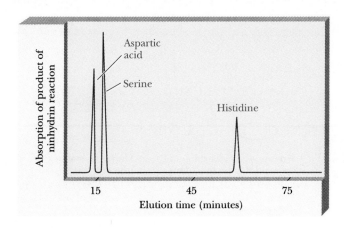

FIGURE A.9

Schematic representation of the results of an amino acid analysis. The sample shown here is a mixture of aspartic acid, serine, and histidine, shown in Figure A.7. In a protein sample there is a peak for each amino acid present. Proper elution conditions ensure good separation of the peaks.

measured. The intensity of the absorption gives information about the relative amount of each amino acid as well as serving as a means of detection. Comparison with standard samples enables accurate determination of the identity and amount of each amino acid. Amino acid analyzers are automated so that eluting buffers can be changed when necessary, and ninhydrin is automatically added to the eluate and the absorption recorded. Figure A.9 shows the results of an amino acid analysis. The area under each peak provides a quantitative estimate of the relative amount of each amino acid. This procedure does not, however, give information about the order of amino acids, which is necessary to determine the primary structure of the protein.

A.3
N-TERMINAL AND C-TERMINAL AMINO ACIDS

The identities of the N-terminal and C-terminal amino acids in a protein sequence can be determined in several ways, both chemical and enzymatic. The well-known methods for identifying the N-terminal amino acid have largely been superseded by advances in sequence determination. The Edman method (Section A.5) for sequencing peptides involves cleavage of residues one at a time, *starting at the N-terminal end.* Consequently, the process of sequencing a peptide automatically provides information about the N-terminal residue. Specialized methods are no longer needed for this aspect of protein structure determination.

The C-terminal amino acid can be determined by chemical methods or by enzymatic cleavage. One chemical method is treatment with **hydrazine,** which reacts with the carbonyl group of each peptide bond. The bond is cleaved, and each amino acid derivative is released as the hydrazide derivative, NH_2—CHR—CO—NH—NH_2, a compound similar to an amide. Since the C-terminal amino acid is not involved in a peptide bond, it remains in the mixture as the only unmodified amino acid. After chromatographic separation and comparison with the standards, the C-terminal amino acid can be identified (Figure A.10).

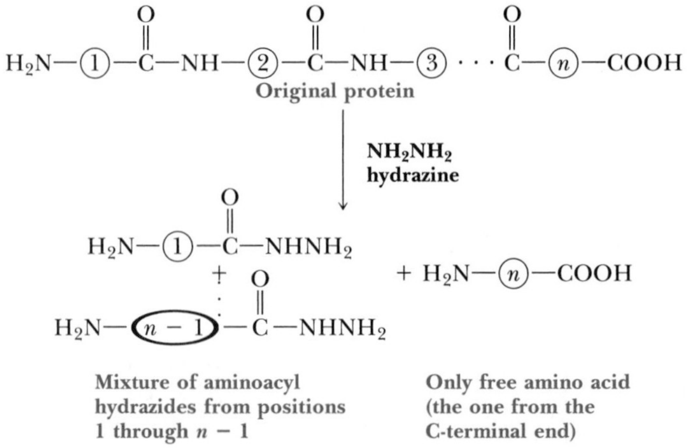

FIGURE A.10

The use of hydrazine to determine C-terminal amino acids. The products of the reaction are separated chromatographically and identified by comparison with standards.

Carboxypeptidases are used for enzymatic determination of the C-terminal amino acid. They are exopeptidases; in this case they cleave polypeptides from the C-terminal end. The difficulty they present is that they continue to digest the polypeptide after the C-terminal residue has been removed. Careful control of reaction conditions is necessary to avoid ambiguous and confusing results.

Determination of the N-terminal and C-terminal residues can indicate whether a given protein consists of one amino acid chain or whether two or even more chains are covalently bonded together. In insulin, for example, there are two N-terminal residues, glycine and phenylalanine, and two C-terminal residues, asparagine and alanine. Insulin consists of two polypeptide chains, designated A and B, held together by two sets of disulfide linkages (Figure A.11). In such a situation it is necessary to be sure that the separation techniques discussed earlier, such as chromatography and electrophoresis, have correctly identified the products of the reaction. Careful experimentation is needed to avoid ambiguous results.

A.4
SEPARATION OF PEPTIDES

In the process of determining the primary structure of a protein, one applies a "divide-and-conquer" strategy by breaking the protein into smaller peptides, separating them, and then determining their sequence. By the Edman sequencing method (Section A.5), it is easier to get reliable results for shorter peptides than for the whole protein at once. In this section we describe the cleavage of proteins into peptides and the separation of those peptides.

Proteins can be cleaved at specific sites by enzymes or by chemical reagents. The enzyme **trypsin** cleaves peptide bonds at amino acids that have positively charged R groups, such as lysine and arginine. The cleavage takes place in such a way that the amino acid with the charged side chain ends up at the C-terminal end of one of the peptides produced by the reaction (Figure A.12a). The C-terminal amino acid of the original protein can be any one of the 20 amino acids and is not necessarily one at which cleavage takes place. A peptide can be automatically identified as the C-terminal end of the original chain if its C-terminal amino acid is not a site of cleavage.

Another enzyme, **chymotrypsin,** cleaves peptide bonds preferentially at aromatic amino acids such as tyrosine, tryptophan, and phenylalanine. The aromatic amino acid ends up at the C-terminal ends of the peptides produced by the reaction (Figure A.12b).

In the case of the chemical reagent **cyanogen bromide** (CN—Br), the sites of cleavage are internal methionine residues. The sulfur of the methionine reacts with the carbon of the cyanogen bromide to produce a homoserine lactone at the C-terminal end of the fragment (Figure A.13).

The cleavage of a protein by any of these reagents produces a mixture of peptides. The use of several such reagents on different samples of a protein to be sequenced produces different mixtures. The sequences of a set of peptides produced by one reagent will overlap the sequences produced by

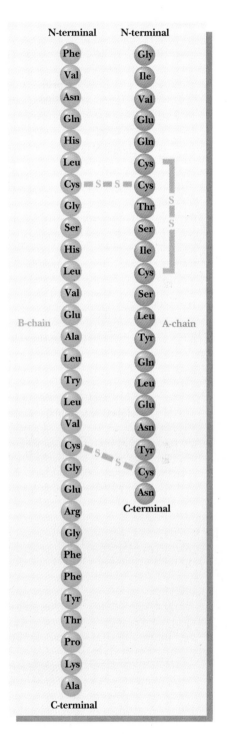

FIGURE A.11

Schematic representation of the two chains of insulin, as determined by end-group analysis.

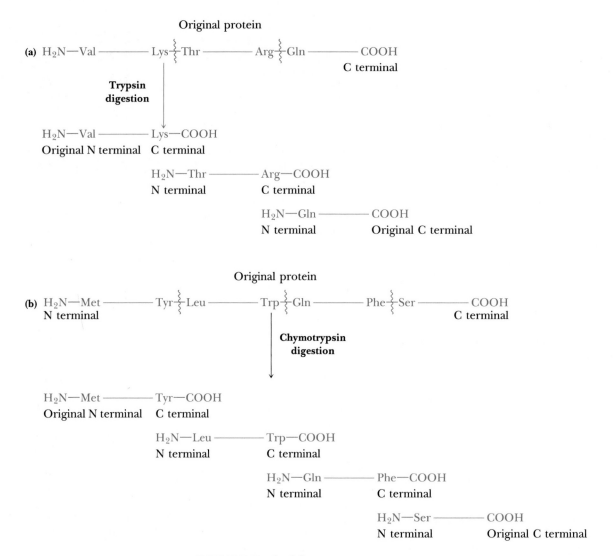

FIGURE A.12

Cleavage of proteins by enzymes. Both trypsin and chymotrypsin are endopeptidases, enzymes that cleave internal peptide bonds. (a) Trypsin hydrolyzes proteins at basic amino acid residues. (b) Chymotrypsin hydrolyzes proteins at aromatic amino acid residues.

another reagent (Figure A.14). This makes it possible to arrange the peptides in the proper order once their own sequences have been determined.

The individual peptides in each mixture must be separated from one another before they can be sequenced. We have already discussed ion exchange column chromatography (Section A.2), a method that depends on differences in electrical charge as the basis for separation of the components of a mixture. Another method, high-performance liquid chromatography

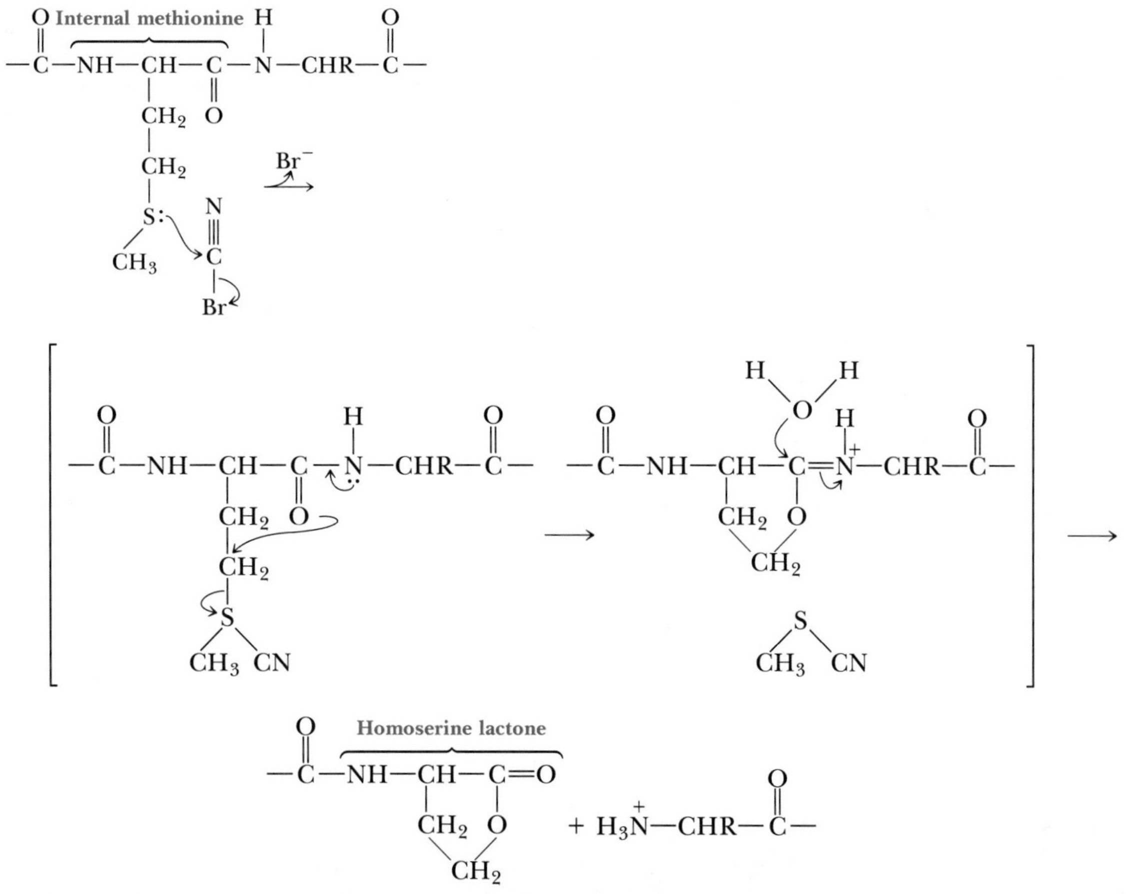

FIGURE A.13

Cleavage of proteins at internal methionine residues by cyanogen bromide.

(HPLC), can be used to separate peptides on the basis of polarity. HPLC is similar to other forms of column chromatography. A distinguishing feature is the long, narrow columns packed with adsorbent particles of considerably smaller diameter than those used in other kinds of column chromatography. HPLC offers a quantitative improvement over earlier forms of chromatography: its columns operate under high pressure, allowing for more efficient separation. There are forms of HPLC based on ion exchange and on

Chymotrypsin $^+H_3N-Leu-Asn-Asp-Phe$
Cyanogen bromide $^+H_3N-Leu-Asn-Asp-Phe-His-Met$
Chymotrypsin $His-Met-Thr-Met-Ala-Trp$
Cyanogen bromide $Thr-Met$
Cyanogen bromide $Ala-Trp-Val-Lys-COO^-$
Chymotrypsin $Val-Lys-COO^-$

Overall sequence $^+H_3N-Leu-Asn-Asp-Phe-His-Met-Thr-Met-Ala-Trp-Val-Lys-COO^-$

FIGURE A.14

Use of overlapping sequences to determine protein sequence. Partial digestion was effected using chymotrypsin and cyanogen bromide.

(a)

Normal-phase chromatography
Polar stationary phase
Less polar mobile phase

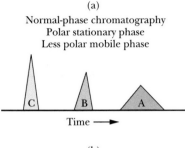

Time ———➤

(b)

Reversed-phase chromatography
Nonpolar stationary phase
Highly polar mobile phase

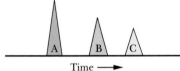

Time ———➤

FIGURE A.15

Chromatographic separations based on differences in polarity. (a) In normal-phase HPLC, the stationary phase is polar. The mobile phase is less polar and elutes the least polar solute first. (b) In reverse-phase HPLC, the stationary phase is nonpolar. The mobile phase is highly polar and elutes the most polar solute first.

molecule size. Here we shall use HPLC as an example of separation based on differences in polarity.

The original scenario for chromatographic separation on the basis of polarity — that is, the scenario for **normal-phase HPLC** — is as follows. The stationary phase is some polar substance (usually aqueous) immobilized on a support (in this case, the small adsorbent particles); the mobile phase is a liquid less polar than water. A polar organic molecule tends to dissolve more easily in a polar liquid than in a nonpolar organic solvent. As a result, when the mobile phase flows over the stationary phase, carrying the sample with it, the relatively more polar peptides tend to partition themselves in the (polar) stationary phase and are carried along more slowly by the (less polar) mobile phase than are the less polar peptides, which have a greater affinity for the mobile phase.

More recently, **reversed-phase HPLC** has been developed. In this variation the stationary phase is nonpolar and the mobile phase is polar. Reversed-phase HPLC gives excellent separations (Figure A.15).

PRACTICE   SESSION

A solution of a peptide of unknown sequence was divided into two samples. One sample was treated with trypsin and the other with chymotrypsin. The smaller peptides obtained by trypsin treatment had the following sequences:

Leu—Ser—Tyr—Ala—Ile—Gln

Asp—Gly—Met—Phe—Val—Lys

The smaller peptides obtained by chymotrypsin treatment had the following sequences:

Val—Lys—Leu—Ser—Tyr

Ala—Ile—Gln

Asp—Gly—Met—Phe

Deduce the sequence of the original peptide.

ANSWER   Asp—Gly—Met—Phe—Val—Lys—Leu—Ser—Tyr—Ala—Ile—Gln

A.5
SEQUENCING OF PEPTIDES: THE EDMAN METHOD

The actual sequencing of each peptide produced by specific cleavage of a protein is accomplished by repeated application of a procedure called **Edman degradation.** The sequence of a peptide containing 10 to 20 residues can easily be determined by this method in about 30 minutes, using as little as 10 picomoles of material. (The amino acid sequences of the individual peptides in Figure A.14 were determined by the Edman method after the peptides were separated from one another.) It then becomes possible to determine the sequence of the entire protein. The overlapping sequences of peptides produced by different reagents provide the key to solving the puzzle. The alignment of like sequences on different peptides makes it possible to deduce the overall sequence.

In the sequencing of a peptide, the Edman reagent, **phenyl isothiocyanate,** reacts with the peptide's N-terminal residue. The modified amino acid can be cleaved off, *leaving the rest of the peptide intact,* and can be detected as the phenylthiohydantoin derivative of the amino acid. The second amino acid of the original peptide can then be treated in the same way, and then the third. With an automated instrument called a **sequencer** (Figure A.16), the process is repeated until the whole peptide is sequenced.

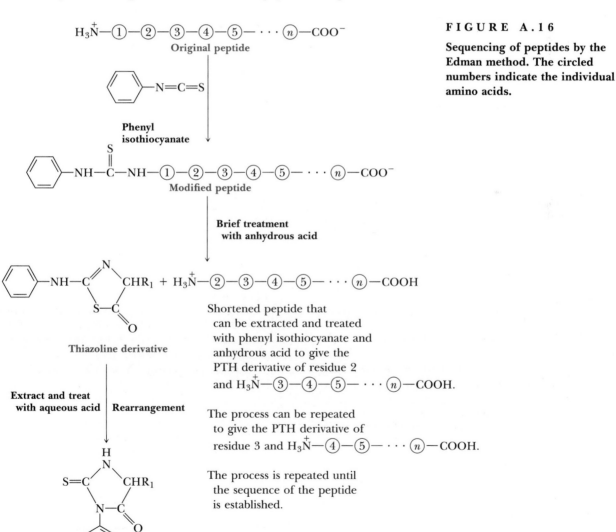

FIGURE A.16

Sequencing of peptides by the Edman method. The circled numbers indicate the individual amino acids.

Shortened peptide that can be extracted and treated with phenyl isothiocyanate and anhydrous acid to give the PTH derivative of residue 2 and $H_3\overset{+}{N}$—③—④—⑤— $\cdots$ n—COOH.

The process can be repeated to give the PTH derivative of residue 3 and $H_3\overset{+}{N}$—④—⑤— $\cdots$ n—COOH.

The process is repeated until the sequence of the peptide is established.

Phenylthiohydantoin (PTH) derivative of N-terminal amino acid, usually identified by HPLC

Another sequencing method makes use of the fact that the amino acid sequence of a protein reflects the base sequence of the DNA in the gene that coded for that protein. Using currently available methods, it is easier to obtain the sequence of the DNA than that of the protein. (See Chapter 7, Sections 7.1 and 7.2, for a discussion of sequencing methods for nucleic acids.) Convenient though this method may be, it does not determine the positions of disulfide bonds or detect amino acids, such as hydroxyproline, that are modified after translation.

S U M M A R Y

Two of the most important methods for separation of amino acids, peptides, and proteins are chromatography and electrophoresis. The varied forms of chromatography rely on differences in charge, in polarity, or in size of the molecules to be separated, depending on the application. In electrophoresis, differences in charge and in size are the criteria for separation. Determination of the N-terminal and C-terminal amino acids of proteins depends on use of these separation methods after the ends of the molecule have been chemically labeled. Selective cleavage of the protein into peptides by enzymatic or chemical hydrolysis produces fragments of manageable size for sequencing. The determination of the amino acid sequence is then accomplished by the Edman method.

E X E R C I S E S

1. An amino acid mixture consisting of lysine, leucine, and glutamic acid is to be separated by ion exchange chromatography, using a cation exchange resin at pH 3.5, with the eluting buffer at the same pH. Which of these amino acids will be eluted from the column first? Will any other treatment be needed to elute one of these amino acids from the column?

2. An amino acid mixture consisting of phenylalanine, glycine, and glutamic acid is to be separated by HPLC. The stationary phase is aqueous and the mobile phase is a solvent less polar than water. Which of these amino acids will move the fastest? Which one will be slowest?

3. In reverse-phase HPLC, the stationary phase is nonpolar and the mobile phase is a polar solvent at neutral pH. Which of the three amino acids in Exercise 2 will move fastest on a reverse-phase HPLC column? Which one will be slowest?

4. Molecular sieve chromatography is a useful method for removing salts such as ammonium sulfate from protein solutions. Describe how such a separation is accomplished.

5. Show by a series of equations (with structures) the first stage of the Edman method applied to a peptide that has leucine as its N-terminal residue.

6. A sample of an unknown peptide was divided into two aliquots. One aliquot was treated with trypsin, and the other with cyanogen bromide. Given the following sequences (N-terminal to C-terminal) of the resulting fragments, deduce the sequence of the original peptide.

Trypsin treatment

Asn—Thr—Trp—Met—Ile—Lys
Gly—Tyr—Met—Gln—Phe
Val—Leu—Gly—Met—Ser—Arg

Cyanogen bromide treatment

Gln—Phe
Val—Leu—Gly—Met
Ile—Lys—Gly—Tyr—Met
Ser—Arg—Asn—Thr—Trp—Met

7. A sample of a peptide of unknown sequence was treated with trypsin; another sample of the same peptide was treated with chymotrypsin. The sequences (N-terminal to C-terminal) of the smaller peptides produced by trypsin digestion were

Met—Val—Ser—Thr—Lys

Val—Ile—Trp—Thr—Leu—Met—Ile

Leu—Phe—Asn—Glu—Ser—Arg

The sequences of the smaller peptides produced by chymotrypsin digestion were

Asn—Glu—Ser—Arg—Val—Ile—Trp

Thr—Leu—Met—Ile

Met—Val—Ser—Thr—Lys—Leu—Phe

Deduce the sequence of the original peptide.

8. How can molecular sieve column chromatography be used to arrive at an estimate of the molecular weight of a protein?

Chapter 5

The Behavior of Proteins: Enzymes

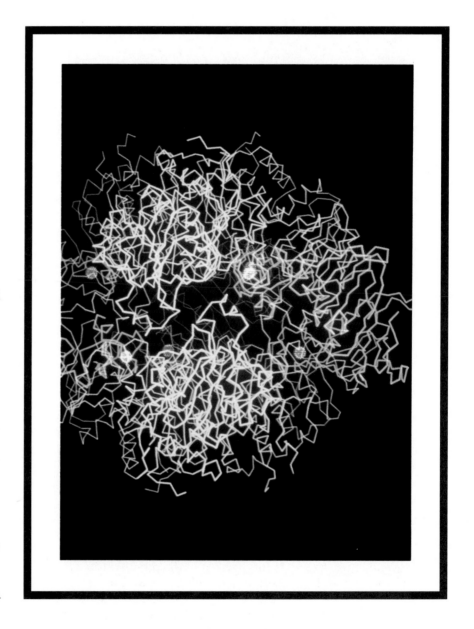

The structure of the enzyme aspartate transcarbamolyase (ATCase) in the T form. Catalytic subunits are shown in yellow and regulatory subunits in green.

Your automobile is powered by the oxidation of the hydrocarbon gasoline to carbon dioxide and water in a controlled explosion within an engine, where hot gases can reach 4000°F. In contrast, the living cell gets its energy by oxidizing the carbohydrate glucose to carbon dioxide and water at a temperature (in humans) of 98.6°F. The secret ingredient in living organisms is *catalysis,* a process performed by protein enzymes. Their three-dimensional architecture gives them exquisite specificity to select the substrate molecules to which they will bind and on which they will operate. Each enzyme has, in fact, a miniature "operating table" where the substrate is momentarily held in a predetermined position so that it can be cut or altered with surgical precision. The scene of the operation, called the active site, is usually a groove, cleft, or cavity on the surface of the protein. Enzyme surgery — cleaving molecules or "stitching" them together — frequently occurs many times (and in some cases many thousands of times) per second. The miracle of life is that a myriad of chemical reactions in the cell is occurring simultaneously with great accuracy and at astonishing speed. Without the proper enzymes to process the food you eat, it might take you 50 years to digest breakfast.

5.1
ENZYMES ARE BIOLOGICAL CATALYSTS

Of all the functions of proteins, the one that is probably most important is **catalysis.** In the absence of catalysis, most reactions in biological systems would take place far too slowly to provide products at an adequate pace for a metabolizing organism. The catalysts that serve this function in organisms are called **enzymes.** With the exception of some RNAs that catalyze their own splicing (described in Chapter 19, Section 19.10), all enzymes are globular proteins. Enzymes are the most efficient catalysts known; they can increase the rate of a reaction by a factor of up to 10^{20} over uncatalyzed reactions. Nonenzymatic catalysts, in contrast, typically enhance the rate of reaction by factors of 10^2 to 10^4. Enzymes are highly specific, even to the point of being able to distinguish stereoisomers of a given compound. In many cases the actions of enzymes are fine-tuned by regulatory processes.

5.2
CATALYSIS: KINETIC VS. THERMODYNAMIC
ASPECTS OF REACTIONS

The rate of a reaction and its thermodynamic spontaneity are two different topics, although they are closely related. This is true of all reactions, whether or not a catalyst is involved. The difference between the energies of the reactants (the initial state) and the energies of the products (the final state) of a reaction is the **standard free energy change,** or $\Delta G°$, for that reaction. The *spontaneity* of a reaction depends on $\Delta G°$ (see Chapter 9, Section 9.2). Enzymes, like all catalysts, speed up reactions. They cannot alter the equilibrium constant or the free energy change. The reaction rate depends

(a)

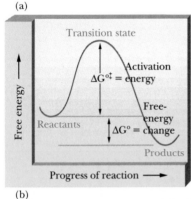

(b)

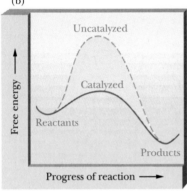

FIGURE 5.1

Activation energy profiles. (a) The activation energy profile for a typical reaction. The reaction shown here is exergonic (energy-releasing). Note the difference between the activation energy ($\Delta G^{\circ\ddagger}$) and the standard free energy of the reaction (ΔG°). (b) A comparison of activation energy profiles for catalyzed and uncatalyzed reactions. The activation energy of the catalyzed reaction is much less than that of the uncatalyzed reaction.

on the **activation energy** ($\Delta G^{\circ\ddagger}$), the energy input required to initiate the reaction. The activation energy for an uncatalyzed reaction is higher than that for a catalyzed reaction; in other words, an uncatalyzed reaction requires more energy to get started, and that is why its rate is slower than that of a catalyzed reaction.

The reaction of glucose and oxygen gas to produce carbon dioxide and water is an example of a reaction that requires a number of enzymatic catalysts:

$$Glucose + 6\ O_2 \longrightarrow 6\ CO_2 + 6\ H_2O$$

This reaction is spontaneous in the thermodynamic sense, because its free energy change is negative ($\Delta G^{\circ} = -2880$ kJ mole^{-1} = -689 kcal mole^{-1}). The energy that must be supplied to start the reaction (which then proceeds with a release of energy) — the activation energy — is conceptually similar to the act of pushing an object to the top of a hill so that it can then slide down the other side.

Activation energy and its relationship to the free energy change of a reaction can best be shown graphically. In Figure 5.1a, the x coordinate shows the extent to which the reaction has taken place, and the y coordinate indicates free energy. The *activation energy profile* shows the intermediate stages of a reaction, those between the initial and final states. Activation energy profiles are essential in the discussion of catalysts. The activation energy directly affects the rate of reaction, and the presence of a catalyst speeds up a reaction by changing the mechanism and thus lowering the activation energy.

Figure 5.1a plots the energies for an exergonic, spontaneous reaction such as the complete oxidation of glucose. At the maximum of the curve connecting the reactants and the products lies the **transition state**, with the needed amount of energy and the correct arrangement of atoms to produce products. Another way of viewing the activation energy is that it is the amount of free energy required to bring the reactants to the transition state.

The analogy of traveling over a mountain pass between two valleys is frequently used in discussions of activation energy profiles. The change in energy corresponds to the change in elevation, and the progress of the reaction corresponds to the distance traveled. The analogue of the transition state is the top of the pass. Considerable effort has gone into elucidating the intermediate stages in reactions of interest to chemists and biochemists, and into determining the pathway or reaction mechanism that lies between the initial and final states. *Reaction dynamics*, the study of the intermediate stages of reaction mechanisms, is currently a very active field of research.

The most important effect of a catalyst on a chemical reaction is apparent from a comparison of the activation energy profiles of the same reaction, catalyzed and uncatalyzed, as shown in Figure 5.1b. The standard free energy change for the reaction, ΔG°, remains unchanged on addition of a catalyst, but the activation energy, $\Delta G^{\circ\ddagger}$, is lowered. In the hill-and-valley analogy, the catalyst is a guide that finds an easier path over the pass between the two valleys. A similar comparison can be made between two routes from San Francisco to Los Angeles. The highest point on Interstate 5

TABLE 5.1   Lowering of the Activation Energy of Hydrogen Peroxide Decomposition by Catalysts

REACTION CONDITIONS	ACTIVATION ENERGY	
	$kJ\ mol^{-1}$	$kcal\ mol^{-1}$
No catalyst	75.2	18.0
Platinum surface	48.9	11.7
Catalase	23.0	5.5

is Tejon Pass (elevation 4400 feet), analogous to the uncatalyzed path. The highest point on U.S. Highway 101 is not much over 1000 feet. Thus, Highway 101 is an easier route, analogous to the catalyzed pathway. The initial and final points of the trip are the same, but the paths between them are different, as are the mechanisms of catalyzed and uncatalyzed reactions. As a result, the rate of the catalyzed reaction is much greater than the rate of the uncatalyzed reaction. Enzymatic catalysts enhance a reaction rate by many powers of ten.

The biochemical reaction in which hydrogen peroxide (H_2O_2) is converted to water and oxygen

$$H_2O_2 \longrightarrow H_2O + O_2$$

provides an example of the effect of catalysts on activation energy. The activation energy of this reaction is lowered if the reaction is allowed to proceed on platinum surfaces, but it is lowered even more by the enzyme catalase. Table 5.1 summarizes the energies involved.

Since it is well known that the rate of a chemical reaction increases with temperature, one might be tempted to assume that this is universally true for biochemical reactions. In fact, increase of reaction rate with temperature occurs only to a limited extent with biochemical reactions. It is helpful to raise the temperature at first, but eventually a point is reached at which heat denaturation of the enzyme (Chapter 4, Section 4.4) plays a part. Above this temperature, adding more heat denatures more enzyme and slows down the reaction.

5.3
ENZYME KINETICS

The *rate of a chemical reaction* is usually expressed in terms of a change in the concentration of a reactant or of a product in a given time interval. Any convenient experimental method can be used to monitor changes in concentration. In a reaction of the form $A + B \longrightarrow P$, the rate of reaction can be expressed in terms of either the rate of disappearance of one of the reactants or the rate of appearance of product. The rate of disappearance of A is $-\Delta[A]/\Delta t$, where Δ symbolizes change, $[A]$ is the concentration of A in moles liter^{-1}, and t is time. Likewise, the rate of disappearance of B is $-\Delta[B]/\Delta t$, and the rate of appearance of P is $\Delta[P]/\Delta t$. The rate of the reaction can be expressed in terms of any of these changes, because the rates of appearance of product and disappearance of reactant are related by the stoichiometric equation for the reaction:

$$\text{Rate} = -\frac{\Delta[A]}{\Delta t} = -\frac{\Delta[B]}{\Delta t} = \frac{\Delta[P]}{\Delta t}$$

The negative signs for the changes in concentration of A and B indicate that A and B are being used up in the reaction, while P is being produced.

It has been established that the rate of a reaction at a given time is proportional to the product of the concentrations of the reactants raised to the appropriate powers,

$$\text{Rate} \propto [A]^f[B]^g$$

or, as an equation,

$$\text{Rate} = k[A]^f[B]^g$$

where k is a proportionality constant called the **rate constant.** The exponents f and g *must be determined experimentally.* They are *not necessarily* equal to the coefficients of the balanced equation, but frequently they are. The square brackets, as usual, denote molar concentration. When the exponents in the rate equation have been determined experimentally, a mechanism for the reaction — a description of the detailed steps along the path between reactants and products — can be proposed.

The exponents in the rate equation are usually small whole numbers such as 1 or 2. (There are also some cases in which the exponent 0 occurs.) The values of the exponents are related to the number of molecules involved in the detailed steps that constitute the mechanism. The **overall order** of a reaction is the sum of all the exponents. If, for example, the rate of a reaction A $\longrightarrow$ B is given by the rate equation

$$\text{Rate} = k[A]^1 \tag{5.1}$$

where k is the rate constant and the exponent for the concentration of A is 1, then the reaction is **first order** with respect to A, and first order overall. The rate of radioactive decay of the widely used tracer isotope phosphorus 32 (^{32}P; atomic weight $= 32$) depends only on the concentration of ^{32}P present; here we have an example of a first-order reaction. Only the ^{32}P atoms are involved in the mechanism of the radioactive decay, which, as an equation, takes the form

$$^{32}P \longrightarrow \text{decay products}$$
$$\text{Rate} = k[^{32}P]^1 = k[^{32}P]$$

If the rate of a reaction A + B $\longrightarrow$ C + D is given by

$$\text{Rate} = k[A]^1[B]^1 \tag{5.2}$$

(where k is the rate constant, the exponent for the concentration of A is 1, and the exponent for the concentration of B is 1), then the reaction is said to be first order with respect to A, first order with respect to B, and **second order** overall. In the reaction of methyl bromide (CH_3Br) with hydroxide ion (OH^-) to give methyl alcohol (CH_3OH) and bromide ion (Br^-), the rate of reaction depends on the concentrations of both reactants:

$$CH_3Br + OH^- \longrightarrow CH_3OH + Br^-$$

$$Rate = k[CH_3Br]^1[OH^-]^1 = k[CH_3Br][OH^-]$$

where k is the rate constant. Both the methyl bromide and the hydroxide ion take part in the reaction mechanism. The reaction of methyl bromide with hydroxide ion is first order with respect to CH_3Br, first order with respect to OH^-, and second order overall.

Many common reactions are first or second order. Once the order of the reaction is determined experimentally, conclusions can be drawn about the mechanism of a reaction.

The possibility exists that exponents in a rate equation may be equal to zero, with the rate for a reaction A $\longrightarrow$ B given by the equation

$$Rate = k[A]^0 = k \tag{5.3}$$

Such a reaction is called **zero order,** and its rate, which is constant, depends not on concentrations of reactants but on other factors such as the presence of catalysts. Enzyme-catalyzed reactions can exhibit zero-order kinetics when the concentrations of reactants are so high that the enzyme is completely saturated with reactant molecules. This point will be discussed in more detail later in this chapter, but for the moment we can consider the situation analogous to a traffic bottleneck where six lanes of cars are trying to cross a two-lane bridge. The rate at which the cars cross is not affected by the number of waiting cars, only by the number of lanes available on the bridge.

5.4
THE TRANSITION STATE IN ENZYMATIC REACTIONS

In an enzyme-catalyzed reaction, the enzyme binds to the **substrate** (one of the reactants) to form a complex. The formation of the complex leads, in turn, to formation of the transition-state species, which then forms the product. The nature of transition states in enzymatic reactions is a large field of research in itself, but some general statements can be made on the subject. A substrate binds to a small portion of the enzyme called the **active site,** frequently situated in a cleft or crevice in the protein and consisting of certain amino acids that are essential for enzymatic activity (Figure 5.2). The catalyzed reaction takes place at the active site, usually in several steps. The first step is the binding of substrate to the enzyme, which occurs because of highly specific interactions between the substrate and the side chains of the amino acids making up the active site. Two important models have been developed to describe the binding process. The first, the **lock-and-key model,** assumes a high degree of similarity between the shape of the substrate and the geometry of the binding site on the enzyme (Figure 5.2a). The substrate binds to a site into which it fits exactly, like a key in a lock or the right piece in a three-dimensional jigsaw puzzle. The second model takes into account the fact that proteins have some three-dimensional flexibility.

FIGURE 5.2

Two models for the binding of a substrate to an enzyme. (a) In the lock-and-key model, the shape of the substrate and the conformation of the active site are complementary to one another. (b) In the induced-fit model, the enzyme undergoes a conformational change on binding to substrate. The shape of the active site becomes complementary to the shape of the substrate only after the substrate binds to the enzyme.

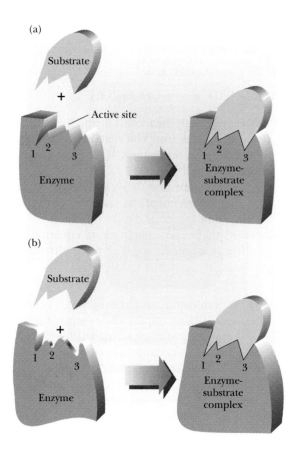

According to this **induced-fit** model, the binding of the substrate induces a conformational change in the enzyme that results in an exact fit once the substrate is bound (Figure 5.2b). The binding site has a different three-dimensional shape before the substrate is bound.

When the substrate is bound and the transition state is formed, the bonds are rearranged. In the transition state, the substrate is bound close to atoms with which it is to react. Furthermore, the substrate is placed in the correct orientation with respect to those atoms. Both effects, proximity and

FIGURE 5.3

Formation of product from substrate (bound to the enzyme), followed by release of the product.

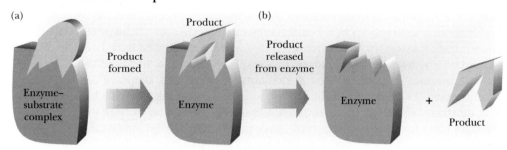

orientation, speed up the reaction. As bonds are broken and new bonds are formed, the substrate is transformed into product. The product is released from the enzyme, which can then catalyze the reaction of more substrate to form more product (Figure 5.3). Each enzyme has its own unique mode of catalysis, which is not surprising in view of enzymes' great specificity. Even so, there are some general modes of catalysis in enzymatic reactions. Two enzymes, chymotrypsin and aspartate transcarbamoylase, are good examples of these general principles.

5.5
TWO EXAMPLES OF
ENZYME-CATALYZED REACTIONS

Chymotrypsin is an enzyme that catalyzes the hydrolysis of peptide bonds, with some specificity for residues containing aromatic side chains. Chymotrypsin cleaves peptide bonds at other sites as well, such as leucine, histidine, and glutamine, but with a lower frequency than at aromatic amino acid residues. It also catalyzes the hydrolysis of ester bonds.

Reactions catalyzed by chymotrypsin

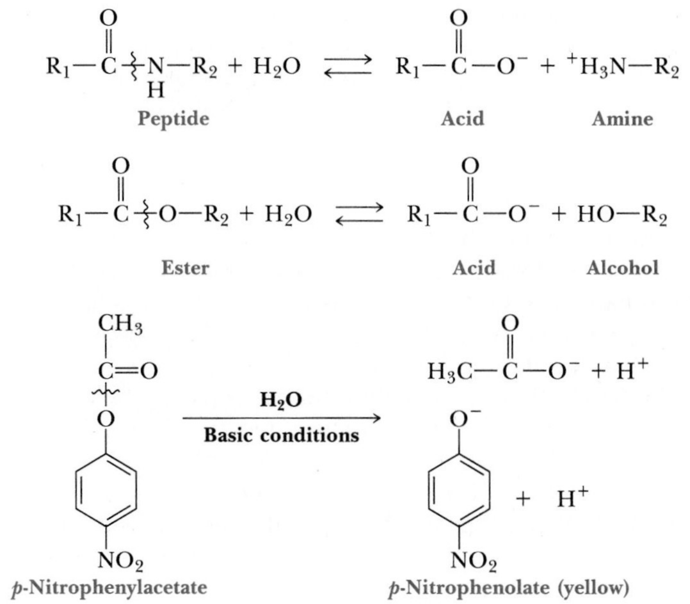

Although ester hydrolysis is not important to the physiological role of chymotrypsin in the digestion of proteins, it is a convenient model system for investigating the enzyme's catalysis of hydrolysis reactions. A usual laboratory procedure is to use *p*-nitrophenyl esters as the substrate and to monitor the progress of the reaction by the appearance of a yellow color in the reaction mixture due to the production of *p*-nitrophenolate ion.

In a typical reaction in which a *p*-nitrophenyl ester is hydrolyzed by chymotrypsin, the experimental rate of the reaction depends on the

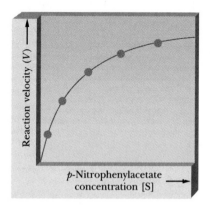

FIGURE 5.4

Dependence of reaction velocity, V, on p-nitrophenylacetate concentration, [S], in a reaction catalyzed by chymotrypsin. The shape of the curve is hyperbolic.

concentration of the substrate, in this case the *p*-nitrophenyl ester. At low substrate concentrations the rate of reaction increases as more substrate is added. At higher substrate concentrations the rate of the reaction changes very little with addition of more substrate, and a maximum rate is reached. When these results are presented in a graph, the curve is *hyperbolic* (Figure 5.4).

Another enzyme-catalyzed reaction is the one catalyzed by **aspartate transcarbamoylase** (ATCase). This reaction is the first step in a pathway leading to the formation of cytidine triphosphate (CTP) and uridine triphosphate (UTP), which are ultimately needed for the biosynthesis of RNA and DNA. In this reaction carbamoyl phosphate reacts with aspartate to produce carbamoyl aspartate and phosphate ion.

Carbamoyl phosphate + Aspartate $\longrightarrow$

$$\text{Carbamoyl aspartate} + HPO_4^{2-}$$

The reaction catalyzed by aspartate transcarbamoylase

The rate of this reaction also depends on substrate concentration — in this case, the concentration of aspartate (the carbamoyl phosphate concentration is kept constant). Experimental results show that once again the rate of the reaction depends on substrate concentration at low and moderate concentrations, and once again a maximum rate is reached at high substrate concentrations. There is, however, one very important difference. For this reaction, a graph showing the dependence of reaction rate on substrate concentration has a *sigmoidal,* rather than hyperbolic shape (Figure 5.5).

The results of experiments on the reaction kinetics of chymotrypsin and aspartate transcarbamoylase are representative of experimental results obtained with many enzymes. The overall kinetic behavior of many enzymes resembles that of chymotrypsin, and other enzymes behave similarly to aspartate transcarbamoylase. We can use this information to draw some general conclusions about the behavior of enzymes.

The comparison between the kinetic behaviors of chymotrypsin and ATCase is reminiscent of the relationship between the oxygen-binding behaviors of myoglobin and hemoglobin, discussed in Chapter 4. ATCase and hemoglobin are allosteric proteins; chymotrypsin and myoglobin are not. (Recall from Section 4.5 that allosteric proteins are the ones in which subtle changes at one site affect structure and function at another site. Cooperative effects, such as the fact that the binding of the first oxygen molecule to hemoglobin makes it easier for other oxygen molecules to bind, are a hallmark of allosteric proteins.) The differences in behavior between allosteric and nonallosteric proteins can be understood in terms of models based on structural differences between the two kinds of proteins. We shall need a model that explains the hyperbolic plot of kinetic data for nonallosteric enzymes, and another model that explains the sigmoidal plot for allosteric enzymes, when we discuss the mechanisms of the many enzyme-

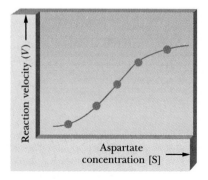

FIGURE 5.5

Dependence of reaction velocity, V, on aspartate concentration, [S], in a reaction catalyzed by aspartate transcarbamoylase. The shape of the curve is sigmoidal.

catalyzed reactions as we encounter them in subsequent chapters. The Michaelis–Menten model is widely used for nonallosteric enzymes, and several models are used for allosteric enzymes.

5.6
THE MICHAELIS–MENTEN APPROACH TO ENZYME KINETICS

A particularly useful model for the kinetics of enzyme-catalyzed reactions was devised in 1913 by Leonor Michaelis and Maud Menten. Theirs is still the basic model for nonallosteric enzymes. It is widely used even though it has undergone many modifications.

A typical reaction might be the conversion of some substrate, S, to a product, P. The stoichiometric equation for the reaction is

$$S \longrightarrow P$$

The mechanism for an enzyme-catalyzed reaction can be summarized in the form

$$E + S \underset{k_{-1}}{\overset{k_1}{\rightleftharpoons}} ES \overset{k_2}{\longrightarrow} E + P \tag{5.4}$$

Note the assumption that the product is not converted to substrate to any appreciable extent. In this equation, k_1 is the rate constant for the formation of the enzyme–substrate complex, ES, from the enzyme, E, and the substrate, S; k_{-1} is the rate constant for the reverse reaction, dissociation of the ES complex to free enzyme and substrate; and k_2 is the rate constant for the conversion of the ES complex to product P and the subsequent release of product from the enzyme. The enzyme appears explicitly in the mechanism, and the concentrations of both free enzyme, E, and enzyme-substrate complex, ES, therefore appear in the rate equations. It is a characteristic of catalysts that they are regenerated at the end of the reaction, and this is true of enzymes.

When we measure the rate (also called the velocity) of an enzymatic reaction at varying substrate concentrations, we see that the rate depends on the substrate concentration, [S]. We can graph our results as in Figure 5.6.

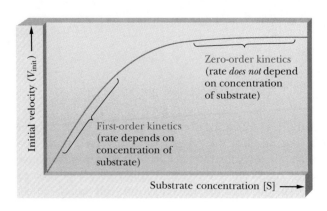

FIGURE 5.6

The rate and the observed kinetics of an enzymatic reaction depend on substrate concentration. The concentration of enzyme, [E], is constant.

In the lower region of the curve the reaction is first order (Section 5.3), implying that the velocity, V, depends on substrate concentration, [S]. In the upper portion of the curve the reaction is zero order; the rate is independent of concentration. The active sites of all enzyme molecules are saturated. The reaction proceeds at its maximum velocity, written V_{max}.

The substrate concentration at which the reaction proceeds at one-half its maximum velocity has a special significance. It is given the symbol K_M, which can be considered an inverse measure of the affinity of the enzyme for the substrate. The lower the K_M, the higher the affinity.

Let us examine the mathematical relationships among the quantities [E], [S], V_{max}, and K_M. The general mechanism of the enzyme-catalyzed reaction involves binding of the enzyme, E, to the substrate to form a complex, ES, which then forms the product. The rate of formation of the enzyme-substrate complex, ES, is

$$\text{Rate of formation} = \frac{\Delta[ES]}{\Delta t} = k_1 \, [E][S] \tag{5.5}$$

where $\Delta[ES]/\Delta t$ means the change in the concentration of the complex, $\Delta[ES]$, during a given time Δt, and k_1 is the rate constant for the formation of the complex.

The complex, ES, breaks down in two reactions, by returning to enzyme and substrate or by giving rise to product and releasing enzyme. The rate of disappearance of complex is the sum of the rates of the two reactions.

$$\text{Rate of breakdown} = -\frac{\Delta[ES]}{\Delta t} = k_1[ES] + k_2[ES] \tag{5.6}$$

The negative sign in the term $\Delta[ES]/\Delta t$ means that the concentration of the complex decreases as the complex breaks down. The term k_{-1} is the rate constant for the dissociation of complex to regenerate enzyme and substrate, and k_2 is the rate constant for the reaction of the complex to give product and enzyme.

Enzymes are capable of processing the substrate very efficiently, and a **steady state** is soon reached in which the rate of formation of the enzyme–substrate complex equals the rate of its breakdown. Very little complex is present, and it turns over rapidly, but its concentration stays the same with time. According to the *steady-state theory*, then, the rate of appearance of the enzyme–substrate complex equals the rate of its disappearance,

$$\frac{\Delta[ES]}{\Delta t} = \frac{-\Delta[ES]}{\Delta t} \tag{5.7}$$

and

$$k_1[E][S] = k_{-1}[ES] + k_2[ES] \tag{5.8}$$

To solve for the concentration of the complex ES, it is necessary to know the concentration of the other species involved in the reaction. The initial concentration of substrate is a known experimental condition and does not change significantly during the initial stages of the reaction. It is much greater than the enzyme concentration. The initial concentration of the enzyme, $[E]_0$, is also known, but a large proportion of it may be involved in the complex. The concentration of free enzyme, $[E]$, is the difference between $[E]_0$, the initial concentration, and $[ES]$, which can be written as an equation:

$$[E] = [E]_0 - [ES] \tag{5.9}$$

Substituting for the concentration of free enzyme, $[E]$, in Equation 5.8,

$$k_1([E]_0 - [ES])[S] = k_{-1}[ES] + k_2[ES] \tag{5.10}$$

Collecting all the rate constants for the individual reactions,

$$\frac{([E]_0 - [ES])\,[S]}{[ES]} = \frac{k_{-1} + k_2}{k_1} = K_M \tag{5.11}$$

where K_M is called the **Michaelis constant.** It is now possible to solve Equation 5.11 for the concentration of enzyme–substrate complex, $[ES]$:

$$\frac{[E]_0[S] - [ES][S]}{[ES]} = K_M$$

$$[E]_0[S] - [ES][S] = K_M[ES]$$

$$[E]_0[S] = [ES](K_M + [S])$$

or

$$[ES] = \frac{[E]_0[S]}{K_M + [S]} \tag{15.12}$$

In the initial stages of the reaction, so little product is present that no reverse reaction of product to complex need be considered. It is the *initial rate* that is usually determined in enzymatic reactions, and this rate depends on the rate of breakdown of the enzyme–substrate complex into product and enzyme. In the Michaelis–Menten model, the initial rate (V_{init}; in some texts the notation is V_0) of formation of product depends only on the rate of the breakdown of the ES complex

$$V_{init} = k_2[ES] \tag{5.13}$$

and on substitution of the expression for $[ES]$ from Equation 5.12:

$$V_{init} = \frac{k_2[E]_0[S]}{K_M + [S]} \tag{5.14}$$

If the substrate concentration is so high that the enzyme is completely saturated with substrate ($[ES] = [E]_0$), the reaction proceeds at its maximum possible rate (V_{max}), and, substituting $[E]_0$ for $[ES]$ in Equation 5.13,

$$V_{init} = V_{max} = k_2[E]_0 \tag{5.15}$$

The original concentration of enzyme is a constant, which means that

$$V_{max} = \text{constant}$$

This expression for V_{max} resembles that for a zero-order reaction, given in Equation 5.3:

$$Rate = k[A]^0 = k$$

Note that the concentration of substrate, [A], appears in Equation 5.3, rather than the concentration of enzyme, [E], as in Equation 5.15. When the enzyme is saturated with substrate, zero-order kinetics with respect to substrate are observed; thus, it is necessary to use comparatively low concentrations of substrate to determine initial rates of enzymatic reactions.

Substituting the expression for V_{max} into Equation 5.14 enables us to relate the initial and maximum rates of an enzymatic reaction:

$$V_{init} = \frac{V_{max}[S]}{K_M + [S]} \tag{5.16}$$

Figure 5.6 shows the effect of increasing substrate concentration on the observed rate. In such an experiment the reaction is run at several substrate concentrations, and the rate is determined by following the disappearance of reactant or the appearance of product by way of any convenient method. In the early stages of the reaction and at low substrate concentrations, first-order kinetics are observed. At higher substrate concentrations, when the enzyme is saturated, the constant reaction rate characteristic of zero-order kinetics is observed.

From now on we shall need to discuss the rate of the reaction in general (V) rather than just the initial rate (V_{init}). The constant rate at saturation is the V_{max} for the enzyme, and the value of V_{max} can be estimated from the graph. The value of K_M can also be estimated from the graph. From Equation 5.16,

Michaelis–Menten equation

$$V = \frac{V_{max}[S]}{K_M + [S]}$$

When experimental conditions are adjusted so that $[S] = K_M$,

$$V = \frac{V_{max}[S]}{[S] + [S]}$$

and

$$V = \frac{V_{max}}{2}$$

In other words, when the rate of the reaction is one-half its maximum value, the substrate concentration is equal to the Michaelis constant (Figure 5.7), and this fact is the basis of the graphical determination of K_M.

The curve that describes the rate of a nonallosteric enzymatic reaction is hyperbolic. It is quite difficult to determine a single point at which the rate

FIGURE 5.7

Graphical determination of V_{max} and K_M from a plot of reaction velocity, V, against substrate concentration, [S]. V_{max} is the constant rate reached when the enzyme is completely saturated with substrate, a value that frequently must be estimated from such a graph.

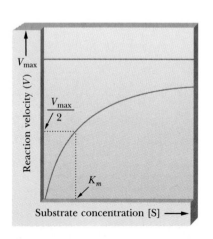

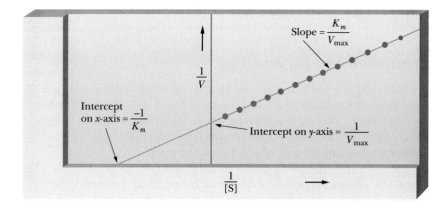

FIGURE 5.8

A Lineweaver–Burk double reciprocal plot of enzyme kinetics. The
reciprocal of reaction velocity,
$1/V$, is plotted against the reciprocal of the substrate concentration, $1/[S]$. The slope of the line
is K_M/V_{max}, and the y-intercept is
$1/V_{max}$. The x-intercept is $-1/K_M$.

levels off. This, in turn, makes it difficult to determine the V_{max} and thus the K_M of the enzyme. It is considerably easier to work with a straight line than a curve. One can transform the equation for a hyperbola (Equation 5.16)

$$V = \frac{V_{max}[S]}{K_M + [S]}$$

into an equation for a straight line by taking the reciprocals of both sides:

$$\frac{1}{V} = \frac{K_M + [S]}{V_{max}[S]}$$

$$\frac{1}{V} = \frac{K_M}{V_{max}[S]} + \frac{[S]}{V_{max}[S]}$$

$$\frac{1}{V} = \frac{K_M}{V_{max}} \cdot \frac{1}{[S]} + \frac{1}{V_{max}} \tag{5.17}$$

The equation now has the form of a straight line, $y = mx + b$, where $1/V$ takes the place of the y coordinate and $1/[S]$ takes the place of the x coordinate. The slope of the line, m, is K_M/V_{max}, and the intercept, b, is $1/V_{max}$. Figure 5.8 presents this information graphically as a **Lineweaver–Burk double-reciprocal plot.** It is usually easier to draw the best straight line through a set of points than to estimate the best fit of points to a curve. There are convenient computer methods for drawing the best straight line through a series of experimental points.

_____ PRACTICE   SESSION

The following data were obtained for an enzymatic reaction. Plot them using the Lineweaver–Burk method, and determine values for K_M and V_{max}. The symbol mM represents millimoles L^{-1}; 1 mM = 1×10^{-3} moles L^{-1}.

SUBSTRATE CONCENTRATION (mM)	VELOCITY (mM sec^{-1})
2.5	0.024
5.0	0.036
10.0	0.053
15.0	0.060
20.0	0.064

ANSWER $V_{max} = 0.0847$ mM sec^{-1}; $K_M = 6.39$ mM

Significance of K_M and V_{max}

We have already seen that when the rate of a reaction, V, is one-half the maximum rate possible, $V = V_{max}/2$, then $K_M = $ [S]. One interpretation of the Michaelis constant, K_M, is that it equals the concentration of substrate at which 50% of the enzyme active sites are occupied by substrate. The Michaelis constant has the units of concentration.

Another interpretation of K_M relies on the assumptions of the original Michaelis–Menten model of enzyme kinetics. Recall Equation 5.4:

$$E + S \underset{k_{-1}}{\overset{k_1}{\rightleftharpoons}} ES \overset{k_2}{\longrightarrow} E + P$$

As before, k_1 is the rate constant for the formation of the enzyme–substrate complex, ES, from the enzyme and substrate; k_{-1} is the rate constant for the reverse reaction, dissociation of the ES complex to free enzyme and substrate; and k_2 is the rate constant for the formation of product, P, and the subsequent release of product from the enzyme. Also recall from Equation 5.11 that

$$K_M = \frac{k_{-1} + k_2}{k_1}$$

Consider the case where the reaction ES $\longrightarrow$ E + S takes place more frequently than ES $\longrightarrow$ E + P. In kinetic terms, this means that the dissociation rate constant k_{-1} is greater than the rate constant for the formation of product, k_2. If k_{-1} is *much* larger than k_2 ($k_{-1} \gg k_2$), as is assumed by the steady-state model, then approximately

$$K_M = \frac{k_{-1}}{k_1}$$

It is informative to compare the expression for the Michaelis constant with the equilibrium constant expression for the dissociation of the ES complex:

$$ES \underset{k_1}{\overset{k_{-1}}{\rightleftharpoons}} E + S$$

T A B L E   5 . 2   Turnover Numbers and K_M for Some Typical Enzymes

ENZYME	FUNCTION	k_{cat} = TURNOVER NUMBER*	K_M†
Acetylcholinesterase	Regenerates an important substance in the transmission of nerve impulses	1.4×10^4	9.5×10^{-5}
Carbonic anhydrase	Catalyzes hydration of CO_2	1.0×10^6	1.2×10^{-2}
Catalase	Catalyzes conversion of H_2O_2 to H_2O and O_2	1.0×10^7	2.5×10^{-2}
Chymotrypsin	A proteolytic enzyme	1.9×10^2	6.6×10^{-4}

*The units of turnover numbers are (moles substrate) (mole enzyme)$^{-1}$ second^{-1}.
†The units of K_M are moles liter^{-1}.

The k's are the rate constants, as before. The equilibrium constant expression is

$$K_{eq} = \frac{[E][S]}{[ES]} = \frac{k_{-1}}{k_1}$$

This expression matches that for K_M and makes the point that, when the assumption that $k_{-1} \gg k_2$ is valid, K_M is simply the dissociation constant for the ES complex. The K_M is a measure of how tightly the substrate is bound to the enzyme. The greater the value of the K_M, the less tightly the substrate is bound to the enzyme.

The V_{max} is a measure of the turnover number of an enzyme, a quantity equal to the catalytic constant, k_{cat}.

$$k_{cat} = V_{max} = \text{turnover number}$$

The **turnover number** is the number of moles of substrate that react to form product per mole of enzyme per unit time. This statement assumes that the enzyme is fully saturated with substrate and thus that the reaction is proceeding at the maximum rate. Table 5.2 lists turnover numbers for typical enzymes. All are *per second.*

Turnover numbers are a particularly dramatic illustration of the efficiency of enzymatic catalysis. Catalase is an example of a particularly efficient enzyme. In Section 5.1 we encountered catalase in its role in converting hydrogen peroxide to water and oxygen. As Table 5.2 indicates, it can transform 10 million moles of substrate to product every second.

5.7
INHIBITION OF ENZYMATIC REACTIONS

An **inhibitor,** as the name implies, is a substance that interferes with the action of an enzyme and slows the rate of a reaction. A good deal of information about enzymatic reactions can be obtained by observing the changes in the reaction caused by the presence of inhibitors. There are two ways in which inhibitors can affect an enzymatic reaction. A *reversible* inhibitor can bind to the enzyme and subsequently be released, leaving the enzyme in its original condition. An *irreversible* inhibitor reacts with the enzyme to produce a protein that is not enzymatically active, and from which the original enzyme cannot be regenerated.

Two major classes of reversible inhibitors can be distinguished on the basis of the sites on the enzyme to which they bind. One class consists of compounds very similar in structure to the substrate. In this case the inhibitor can bind to the active site and block the substrate's access to it. This mode of action is called **competitive inhibition** because the inhibitor competes with the substrate for the active site on the enzyme. The other major class of reversible inhibitors includes any inhibitor that binds to the enzyme at a site other than the active site and, as a result of binding, causes a change in the structure of the enzyme, especially around the active site. The substrate may still be able to bind to the active site, but the enzyme cannot catalyze the reaction as efficiently as it could in the absence of the inhibitor. This mode of action is called **noncompetitive inhibition** (Figure 5.9).

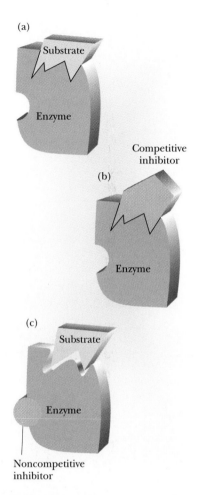

FIGURE 5.9

Modes of action of inhibitors. The distinction between competitive and noncompetitive inhibitors is that a competitive inhibitor prevents binding of the substrate to the enzyme, whereas a noncompetitive inhibitor does not. (a) An enzyme–substrate complex in the absence of inhibitor. (b) A competitive inhibitor binds to the active site; the substrate cannot bind. (c) A noncompetitive inhibitor binds at a site other than the active site. The substrate still binds, but the enzyme cannot catalyze the reaction because of the presence of the bound inhibitor.

The two kinds of inhibition can be distinguished from one another in the laboratory. The reaction is carried out in the presence of inhibitor at several substrate concentrations, and the rates obtained are compared with those of the uninhibited reaction. The differences in the Lineweaver–Burk plots for the inhibited and uninhibited reactions provide the basis for the comparison.

Kinetics of Competitive Inhibition

In the presence of a competitive inhibitor, the slope of the Lineweaver–Burk plot changes, but the intercept does not. The V_{max} is unchanged, but the K_M increases. More substrate is needed to get to a given rate, including V_{max}, in the presence of inhibitor than in its absence. This applies to the specific value $V_{max}/2$ (recall that at $V_{max}/2$, the substrate concentration, [S], equals K_M) (Figure 5.10). Competitive inhibition can be overcome by a sufficiently high substrate concentration.

In the presence of a competitive inhibitor, the equation for an enzymatic reaction becomes

$$EI \xrightleftharpoons{I} E \xrightleftharpoons{S} ES \longrightarrow E+P$$

where EI is the enzyme–inhibitor complex. The dissociation constant for the enzyme–inhibitor complex can be written

$$EI \rightleftharpoons E+I$$

$$K_I = \frac{[E][I]}{[EI]}$$

It can be shown algebraically, although we shall not do it here, that in the presence of inhibitor the value of K_M increases by the factor

$$1 + \frac{[I]}{K_I}$$

FIGURE 5.10

A Lineweaver–Burk double reciprocal plot of enzyme kinetics for competitive inhibition. (a) The binding equilibria to be considered. (b) In the plot of $1/V$ versus $1/[S]$, the red circles represent the presence of a competitive inhibitor, and the blue circles represent the control reaction with no inhibitor present. The value of V_{max} remains the same; the measured value of K_M increases.

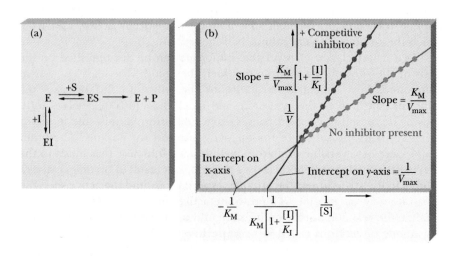

If we substitute $K_M(1 + [I]/K_I)$ for K_M in Equation 5.17, we obtain

$$\frac{1}{V} = \frac{K_M}{V_{max}}\left(1 + \frac{[I]}{K_I}\right) \cdot \frac{1}{[S]} + \frac{1}{V_{max}}$$

$$y = \qquad m \qquad \cdot \ x \ + \ b$$

(5.18)   *Competitive inhibition*

Here the term $1/V$ takes the place of the y coordinate, and the term $1/[S]$ takes the place of the x coordinate, as was the case in Equation 5.17. The intercept $1/V_{max}$, the b term in the equation for a straight line, has not changed from the earlier equation, but the slope K_M/V_{max} in Equation 5.17 has increased by the factor $(1 + [I]/K_I)$. The slope, the m term in the equation for a straight line, is now

$$\frac{K_M}{V_{max}}\left(1 + \frac{[I]}{K_I}\right)$$

accounting for the changes in the slope of the Lineweaver–Burk plot. Note that the y intercept does not change. This algebraic treatment of competitive inhibition agrees with experimental results, validating the model, just as experimental results validate the underlying Michaelis–Menten model for enzyme action.

Box 5.1

Methanol Poisoning: Enzyme Inhibition in Action

Methanol ("wood alcohol") is a toxic substance that can cause blindness at moderate doses and death at higher doses. It is a competitive inhibitor of the enzyme alcohol dehydrogenase, for which the usual substrate is ethanol. When methanol takes the place of ethanol, it reacts to produce formaldehyde (instead of the products of the usual reaction with ethanol). The formaldehyde causes the harmful effects.

A great increase in substrate concentration can overcome competitive inhibition. Hence, part of the medical treatment for methanol poisoning is to administer a large dose of ethanol (ethyl alcohol), enough to cause intoxication under normal circumstances. The added ethanol (the substrate) displaces the methanol (the competitive inhibitor) to reverse the effects of poisoning.

Kinetics of Noncompetitive Inhibition

The kinetic results of noncompetitive inhibition differ from those of competitive inhibition. The Lineweaver–Burk plots for a reaction in the presence and absence of a noncompetitive inhibitor show that both the slope and the y axis intercept change for the inhibited reaction (Figure 5.11). The value of V_{max} decreases, but that of K_M remains the same. Increasing the substrate concentration cannot overcome noncompetitive inhibition.

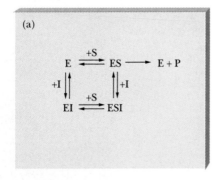

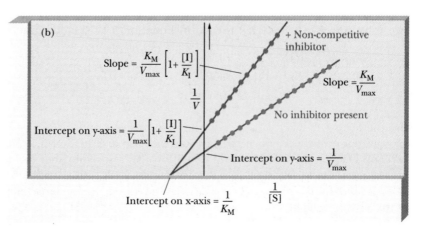

FIGURE 5.11

A Lineweaver–Burk double reciprocal plot of enzyme kinetics for noncompetitive inhibition. (a) The binding equilibria to be considered. (b) A plot of $1/V$ versus $1/[S]$. The red circles represent the presence of a noncompetitive inhibitor, the blue circles represent the control reaction with no inhibitor present. The value of V_{max} decreases; the value of K_M remains the same.

The reaction pathway has become considerably more complicated, and several equilibria have to be considered.

$$E \underset{}{\overset{+S}{\rightleftharpoons}} ES \longrightarrow E + P$$

$$+I \updownarrow \qquad \updownarrow +I$$

$$EI \underset{+S}{\rightleftharpoons} ESI$$

In the presence of a noncompetitive inhibitor, I, the maximum velocity of the reaction, V^I_{max}, has the form (we shall not do the derivation here)

$$V^I_{max} = \frac{V_{max}}{1 + [I]/K_I}$$

where K_I is again the dissociation constant for the enzyme–inhibitor complex, EI. Recall that the maximum rate, V_{max}, appears in the expressions for both the slope and the intercept in the equation for the Lineweaver–Burk plot (Equation 5.17):

$$\frac{1}{V} = \frac{K_M}{V_{max}} \cdot \frac{1}{[S]} + \frac{1}{V_{max}}$$

$$y = m \cdot x + b$$

In noncompetitive inhibition we replace the term V_{max} with the expression for V^I_{max}, to obtain

Noncompetitive inhibition

$$\frac{1}{V} = \frac{K_M}{V_{max}}\left(1 + \frac{[I]}{K_I}\right) \cdot \frac{1}{[S]} + \frac{1}{V_{max}}\left(1 + \frac{[I]}{K_I}\right) \qquad (5.19)$$

$$y = \qquad m \qquad \cdot x + \qquad b$$

The expressions for both the slope and the intercept in the equation for a Lineweaver–Burk plot of an uninhibited reaction have been replaced by more complicated expressions in the equation that describes noncompetitive inhibition. This interpretation is borne out by the observed results.

Using the following data, determine by the Lineweaver–Burk method whether the inhibition of this enzymatic reaction is competitive or noncompetitive.

SUBSTRATE CONCENTRATION (moles L^{-1})	V, NO INHIBITOR (arbitrary units)	V, INHIBITOR PRESENT (same arbitrary units)
0.0292	0.182	0.083
0.0584	0.265	0.119
0.0876	0.311	0.154
0.117	0.330	0.167
0.175	0.372	0.192

ANSWERS Noncompetitive

5.2

ENZYME INHIBITION IN THE TREATMENT OF AIDS

A key strategy in the treatment of acquired immunodeficiency syndrome, or AIDS, has been to develop specific inhibitors that selectively block the actions of enzymes unique to the human immunodeficiency virus (HIV), which causes AIDS. Many laboratories are taking this approach to the development of therapeutic agents.

One newly synthesized compound has shown considerable promise. This substance, designed by Agouron Pharmaceuticals* and designated AG-1284, is a potent inhibitor of HIV protease, an enzyme essential to the production of new virus particles in infected cells. Other inhibitors of HIV protease are known, but AG-1284 binds so strongly to the enzyme that nanomolar (1×10^{-9}M) amounts can produce inhibition. The design and synthesis of this drug were planned with this strong binding in mind.

The structure of HIV protease, including its active site, was known from the results of x-ray crystallography. With this structure in mind, scientists designed and synthesized a compound to bind to the active site; it inhibited the enzyme at micromolar (1×10^{-6}M) concentrations. Improvements were made in the drug design by determining the structures of a series of inhibitors bound to the active site of HIV protease. These structures were also elucidated by x-ray crystallography. This process eventually led to AG-1284, the most potent drug of all the compounds tested.

HIV protease, the enzyme inhibited by AG-1284, is unique to HIV. It catalyzes the processing of viral proteins in an infected cell. Without these proteins, viable virus particles cannot be released to cause further infection. It remains to be seen whether AG-1284 will be safe and effective in a clinical situation, but the approach to its synthesis will certainly be of use in the future.

*See interview with Agouron chemist Siegfried Reich on page 000.

A drawing and a computer-generated model of AG-1284, an inhibitor of HIV protease.

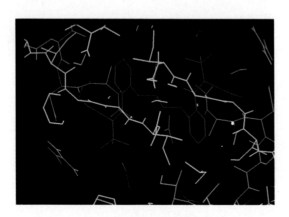

5.8
THE MICHAELIS–MENTEN MODEL DOES NOT DESCRIBE THE BEHAVIOR OF ALLOSTERIC ENZYMES

The behavior of many well-known enzymes can be described quite adequately by the Michaelis–Menten model, but allosteric enzymes behave very differently. Earlier in this chapter we saw that there are similarities between the reaction kinetics of an enzyme such as chymotrypsin, which does not display allosteric behavior, and the binding of oxygen by myoglobin, also an example of nonallosteric behavior. The analogy extends to show the similarity in the kinetic behavior of an allosteric enzyme such as aspartate transcarbamoylase (ATCase) and the binding of oxygen by hemoglobin. Both ATCase and hemoglobin are allosteric proteins; the behaviors of both exhibit cooperative effects caused by subtle changes in quaternary structure. (Recall that *quaternary structure* is the arrangement in space that results from the interaction of subunits through noncovalent forces, and *cooperativity* refers to the fact that the binding of low levels of substrate facilitates the action of the protein at higher levels of substrate, whether the action is catalytic or oxygen-binding.) In addition to displaying cooperative kinetics, allosteric enzymes have a different response to the presence of inhibitors from that of nonallosteric enzymes, as characterized by the Michaelis–Menten model.

Control Mechanisms That Affect Allosteric Enzymes

ATCase catalyzes the first step in a series of reactions with the end product cytidine triphosphate (CTP).

**The reaction catalyzed by ATCase
leads eventually to the production of CTP**

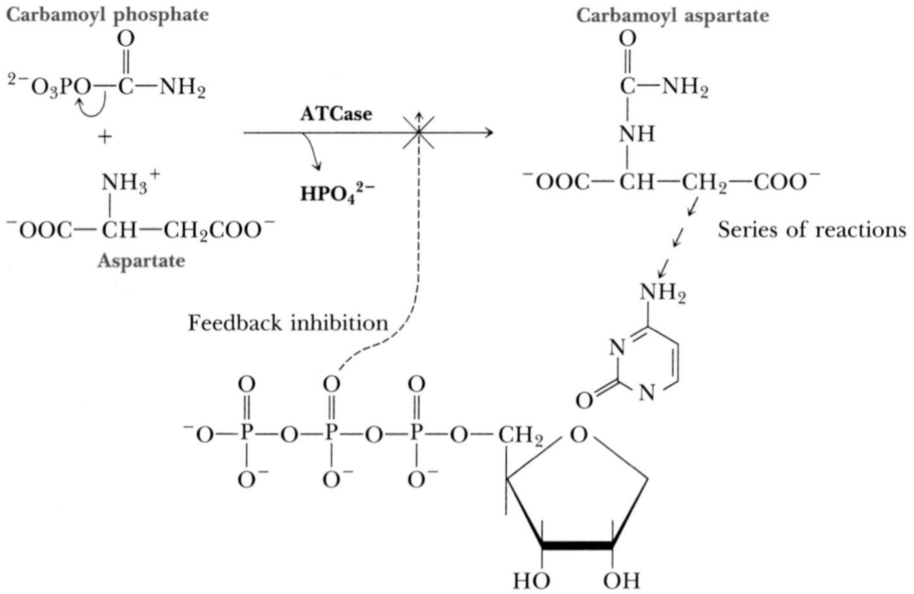

Cytidine triphosphate (CTP)
Allosteric inhibitor of ATCase

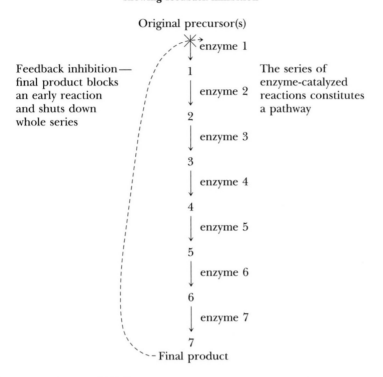

A schematic representation of a pathway
showing feedback inhibition

Original precursor(s)

enzyme 1

Feedback inhibition—
final product blocks
an early reaction
and shuts down
whole series

1

enzyme 2

The series of
enzyme-catalyzed
reactions constitutes
a pathway

2

enzyme 3

3

enzyme 4

4

enzyme 5

5

enzyme 6

6

enzyme 7

7

Final product

FIGURE 5.12

Schematic representation of a
pathway, showing feedback inhibition.

CTP is an inhibitor of ATCase, the enzyme that catalyzes the first reaction in the pathway. This behavior is an example of **feedback inhibition** (also called end-product inhibition), in which the end product of the sequence of reactions inhibits the first reaction in the series (Figure 5.12). Feedback inhibition is an efficient control mechanism because the entire series of reactions can be shut down when an excess of the final product exists; intermediates in the pathway are not formed. Feedback inhibition is a general feature of metabolism and is not confined to allosteric enzymes. However, the observed kinetics of the ATCase reaction, including the mode of inhibition, are typical of allosteric enzymes.

When ATCase catalyzes the condensation of aspartate and carbamoyl phosphate to form carbamoyl aspartate, the graphical representation of the rate as a function of increasing substrate concentration is a sigmoidal curve rather than the hyperbola obtained with nonallosteric enzymes (Figure 5.13a). The sigmoidal curve is indicative of the cooperative behavior of allosteric enzymes. (In this two-substrate reaction, aspartate is the substrate for which the concentration is varied, while the concentration of carbamoyl phosphate is kept constant at high levels.)

Figure 5.13b compares the rate of the uninhibited reaction of ATCase with the reaction rate in the presence of CTP. In the latter case, a sigmoidal curve still describes the rate behavior of the enzyme, but the curve is shifted to higher substrate levels; a higher concentration of aspartate is needed for the enzyme to achieve the same rate of reaction. At high substrate concentrations, the same maximal rate (V_{max}) is observed in the presence and absence of inhibitor. Since, in the Michaelis–Menten scheme, the V_{max} changes when a reaction takes place in the presence of a noncompetitive inhibitor, noncompetitive inhibition cannot be the case here. The same

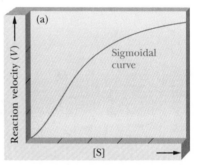

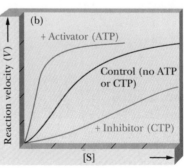

FIGURE 5.13

The kinetics of allosteric enzymes.
(a) Reaction velocity, *V*, as a function of substrate concentration,
[S], in the reaction of ATCase.
The substrate concentration that
varies is that of aspartate. The sigmoidal curve is evidence of cooperative kinetics. (b) The effect of
an activator (ATP) and an inhibitor (CTP) on the kinetics of the
ATCase reaction.

Michaelis–Menten model associates this sort of behavior with competitive inhibition, but that part of the model still does not provide a reasonable picture. Competitive inhibitors bind to the same site as the substrate because they are very similar in structure. The CTP molecule is very *different* in structure from the substrate, aspartate. It is much more likely that CTP is bound to a different site on the ATCase molecule. There is, in fact, experimental evidence for two different binding sites: the enzyme can be modified so that CTP cannot bind, but the binding of aspartate is not affected by this modification.

The situation becomes "curiouser and curiouser" when the ATCase reaction takes place not in the presence of CTP, a pyrimidine nucleoside triphosphate, but in the presence of ATP, a purine nucleoside triphosphate.

Adenosine triphosphate (ATP)
a purine nucleotide;
activator of
ATCase

The structural similarities between CTP and ATP are apparent, but ATP is not a product of the pathway that includes the reaction of ATCase and that produces CTP. Both ATP and CTP are needed for the synthesis of RNA and DNA. The relative proportions of ATP and CTP are specified by the needs of the organism. If there is not enough CTP compared to ATP, the enzyme requires a signal to produce more. In the presence of ATP the rate of the enzymatic reaction is increased at lower levels of aspartate and the shape of the rate curve becomes less sigmoidal and more hyperbolic (Figure 5.13b). In other words, there is less cooperativity in the reaction. The binding site for ATP on the enzyme molecule is the same as that for CTP, which is not surprising in view of their structural similarity, but ATP is an activator rather than an inhibitor like CTP. When CTP is in short supply in an organism, the ATCase reaction is not inhibited, and the binding of ATP increases the activity of the enzyme still more.

The key to allosteric behavior, including cooperativity and modifications of cooperativity, is the existence of multiple forms for the quaternary structures of allosteric proteins. The word "allosteric" is derived from *allo,* "other," and *steric,* "shape," referring to the fact that the possible conformations affect the behavior of the protein. The binding of substrates, inhibitors, and activators changes the quaternary structure of allosteric proteins, and the changes in structure are reflected in behavior. A substance that modifies the quaternary structure, and thus the behavior, of an allosteric protein by binding to it is called an **allosteric effector.** The term "effector" can apply to substrates, inhibitors, or activators. Several models for the behavior of allosteric enzymes have been proposed, and it is worthwhile to compare them.

Let us first define two terms. **Homotropic effects** are allosteric interactions that occur when several identical molecules are bound to a protein. The binding of substrate molecules to different sites on an enzyme, such as the binding of aspartate to ATCase, is an example of a homotropic effect. **Heterotropic effects** are allosteric interactions that occur when different substances (such as inhibitor and substrate) are bound to the protein. In the ATCase reaction, inhibition by CTP and activation by ATP are both heterotropic effects.

5.9
MODELS FOR THE BEHAVIOR
OF ALLOSTERIC ENZYMES

The two principal models for the behavior of allosteric enzymes are the concerted model and the sequential model. They were proposed in 1965 and 1966, respectively, and both are in current use as bases for interpreting experimental results. The concerted model has the advantage of comparative simplicity, and it describes the behavior of some enzyme systems very well. The sequential model sacrifices a certain amount of simplicity for a more realistic picture of the structure and behavior of proteins; it also deals very well with the behavior of some enzyme systems.

The Concerted Model for Allosteric Behavior

In 1965 Jeffries Wyman, Jacques Monod, and Jean-Pierre Changeux proposed the **concerted model** for the behavior of allosteric proteins in a paper that has become a classic in the biochemical literature (it is listed in the bibliography at the end of this chapter). In this picture the protein has two conformations, the R (relaxed) conformation, which binds substrate tightly, and the T (tight, also called taut) conformation, which binds substrate less tightly. The distinguishing feature of this model is that the conformations of *all* subunits change simultaneously. Figure 5.14 shows a hypothetical protein with two subunits. Both subunits change conformation from the inactive T conformation to the active R conformation at the same time; that is, a concerted change of conformation occurs. The binding of the first molecule of substrate to one subunit facilitates the binding of the second substrate molecule to the other subunit. This is exactly what is meant by cooperative binding. In the absence of substrate, the enzyme exists mainly in the T form, in equilibrium with small amounts of the R form. The presence of substrate shifts the equilibrium to produce more of the R form.

In the concerted model, the effects of inhibitors and activators can also be considered in terms of shifting the equilibrium between the T and R forms of the enzyme. The binding of inhibitors to allosteric enzymes is

FIGURE 5.14

The concerted model for allosteric behavior. The T (inactive) and R (active) forms of the enzyme are in equilibrium. The equilibrium lies to the left, in favor of the T form. The cooperative binding of substrate then shifts the equilibrium to the right, in favor of the R form.

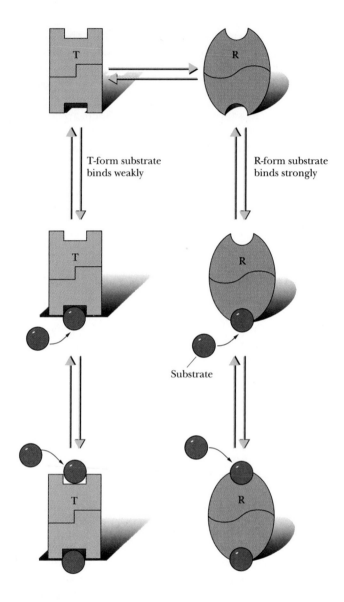

T-form substrate binds weakly

R-form substrate binds strongly

Substrate

cooperative; allosteric inhibitors bind to and stabilize the T form of the enzyme. The binding of activators to allosteric enzymes is cooperative, as well; allosteric activators bind to and stabilize the R form of the enzyme.

When an activator (A) is present, the cooperative binding of A shifts the equilibrium between the T and R forms, with the R form favored (Figure 5.15a). As a result, there is less need for substrate (S) to shift the equilibrium in favor of the R form and less need for cooperativity in the binding of S.

When an inhibitor (I) is present, the cooperative binding of I also shifts the equilibrium between the T and R forms, but this time the T form is favored (Figure 5.15b). More substrate (S) is needed to shift the T-to-R equilibrium in favor of the R form. A greater degree of cooperativity is needed in the binding of S.

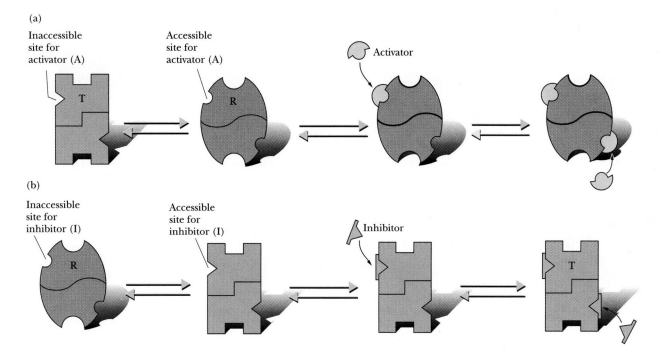

FIGURE 5.15

(a) Concerted model for the cooperative binding of activator A to allosteric enzyme. The T $\rightleftharpoons$ R equilibrium shifts to the right. (b) Concerted model for the cooperative binding of inhibitor I to an allosteric enzyme. The T $\rightleftharpoons$ R equilibrium shifts to the left.

The Sequential Model for Allosteric Behavior

The name Daniel Koshland is associated with the direct **sequential model** of allosteric behavior. The distinguishing feature of this model is that the binding of substrate induces the conformational change from the T form to the R form — the type of behavior postulated by the induced-fit theory of substrate binding. The change from T to R in one subunit favors the same change in the other subunits. This "passing along" of conformational changes from one subunit to another is the form in which cooperative binding is expressed in this model (Figure 5.16a).

In the sequential model, the binding of activators and inhibitors also takes place by the induced-fit mechanism. The conformational change that begins with binding of inhibitor or activator to one subunit affects the conformations of other subunits. The net result is to favor the R state when activator is present and the T form when inhibitor (I) is present (Figure 5.16b). Binding of I to one subunit causes a conformational change so that the T form is even less likely to bind substrate than before. This conformational change is passed along to other subunits, making them also more likely to bind inhibitor and less likely to bind substrate. This is an example of cooperative behavior that leads to more inhibition of the enzyme. Likewise, binding of an activator causes a conformational change that favors substrate binding, and this effect is passed from one subunit to another.

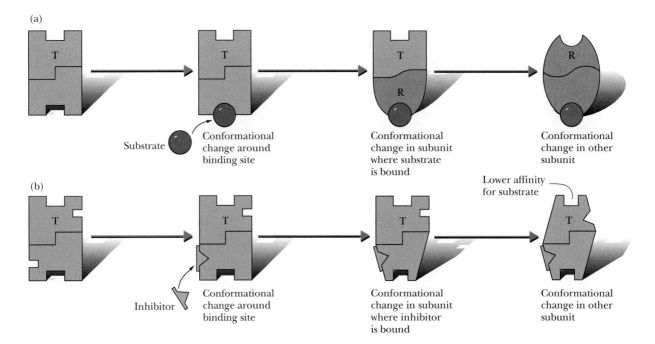

FIGURE 5.16

(a) Sequential model of cooperative binding of substrate S to an allosteric enzyme. (b) Sequential model of cooperative binding of inhibitor I to an allosteric enzyme.

The sequential model for the binding of effectors of all types, including substrates, to allosteric enzymes has a unique feature, another possible type of change in enzyme behavior brought about by the binding of the first molecule. The conformational changes thus induced can make the enzyme less likely to bind more molecules of the same type. This phenomenon, called **negative cooperativity,** has been observed in a few enzymes. One is tyrosyl tRNA synthetase, which plays a role in protein synthesis. In the reaction catalyzed by this enzyme, the amino acid tyrosine forms a covalent bond to a molecule of transfer RNA (tRNA). In subsequent steps the tyrosine is passed along to its place in the sequence of the growing protein. The tyrosyl tRNA synthetase consists of two subunits. Binding of the first molecule of substrate to one of the subunits inhibits binding of a second molecule to the other subunit. The sequential model successfully accounts for the negative cooperativity observed in the behavior of tyrosyl tRNA synthetase. The concerted model makes no provision for negative cooperativity.

5.10
ZYMOGENS: THE BASIS OF ANOTHER TYPE OF CONTROL MECHANISM IN ENZYME ACTION

Allosteric interactions control the behavior of proteins through reversible changes in quaternary structure, but this mechanism, effective though it may be, is not the only one available. A **zymogen,** an inactive precursor of an enzyme, can be irreversibly transformed into an active enzyme by cleavage of covalent bonds.

The proteolytic enzymes trypsin and chymotrypsin provide a classic example of zymogens and their activation. Their inactive precursor molecules, trypsinogen and chymotrypsinogen, respectively, are formed in the pancreas, where they would do damage if they were in an active form. In the small intestine, where their digestive properties are needed, they are activated by cleavage of specific peptide bonds.

Chymotrypsinogen consists of a single polypeptide chain 245 residues long, with five disulfide (—S—S—) bonds. When chymotrypsinogen is secreted into the small intestine, trypsin present in the digestive system cleaves the peptide bond between arginine 15 and isoleucine 16, counting from the N-terminal end of the chymotrypsinogen sequence (Figure 5.17). The cleavage produces active π-chymotrypsin. The 15-residue fragment remains bound to the rest of the protein by a disulfide bond. Although π-chymotrypsin is fully active, it is not the end product of this series of reactions. It acts on itself to remove two dipeptide fragments, producing α-chymotrypsin, which is also fully active. The two dipeptide fragments cleaved off are Ser 14–Arg 15 and Thr 147–Asn 148; the final form of the enzyme, α-chymotrypsin, has three polypeptide chains held together by two of the five original, and still intact, disulfide bonds. (The other three disulfide bonds remain intact as well; they link portions of single polypeptide chains.) When the term "chymotrypsin" is used without specifying the π or the α form, it is the final α form that is meant.

The changes in primary structure that accompany the conversion of chymotrypsinogen to α-chymotrypsin bring about changes in the tertiary structure. The enzyme is active because of its tertiary structure, just as the zymogen is inactive because of its tertiary structure. The three-dimensional structure of chymotrypsin has been determined by x-ray crystallography.

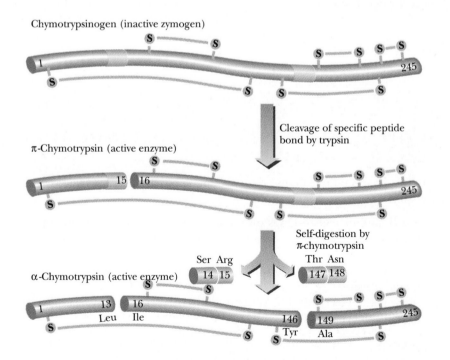

Chymotrypsinogen (inactive zymogen)

Cleavage of specific peptide bond by trypsin

π-Chymotrypsin (active enzyme)

Self-digestion by π-chymotrypsin

Ser Arg   Thr Asn
14   15   147 148

α-Chymotrypsin (active enzyme)

Leu Ile   Tyr Ala

FIGURE 5.17

The activation of chymotrypsinogen.

The protonated amino group of the isoleucine residue exposed by the first cleavage reaction is involved in an ionic bond with the carboxylate side chain of aspartate residue 194. This ionic bond is necessary for the active conformation of the enzyme, since it is near the active site. Chymotrypsinogen lacks this bond; therefore, it does not have the active conformation and cannot bind substrate.

Blood clotting also requires a series of proteolytic activations involving several proteins, particularly the conversions of prothrombin to thrombin and of fibrinogen to fibrin. Blood clotting is a complex process; for this discussion it is sufficient to know that activation of zymogens plays a crucial role. In the final, best characterized step of clot formation, the soluble protein fibrinogen is converted to the insoluble protein fibrin as a result of the cleavage of four peptide bonds. The cleavage occurs as the result of action of the proteolytic enzyme thrombin, which in turn is produced from a zymogen called prothrombin. The conversion of prothrombin to thrombin requires a number of proteins called **clotting factors,** as well as Ca^{2+}.

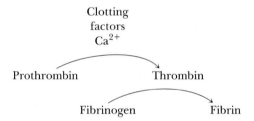

Some of the processes involved in blood clotting.

The early stages of blood clotting consist of an elaborate multistep mechanism that allows for fine tuning of the process but can also cause great problems if something goes wrong with one of the steps. The molecular disease **hemophilia,** for example, is typically caused by a lack of one of the clotting factors. A hemophiliac can bleed to death from a very small cut that would not trouble a normal person.

5.11
ACTIVE-SITE EVENTS IN ENZYMES: A LOOK AT REACTION MECHANISMS

We can ask several questions about the mode of action of an enzyme. Here are some of the most important:

1. Which amino acid residues on the enzyme are in the active site and catalyze the reaction? In other words, which are the essential amino acid residues?
2. What is the spatial relationship of the essential amino acid residues in the active site?
3. What is the mechanism by which the essential amino acid residues catalyze the reaction?

Answers to these questions are available for chymotrypsin, and we shall use its mechanism as an example of enzyme action. Information on well-known systems such as chymotrypsin can lead to general principles that are applicable to all enzymes. Enzymes catalyze chemical reactions in many ways, but all reactions have in common the requirement that some reactive group on the enzyme interact with the substrate. In proteins the α-carboxyl and α-amino groups of the amino acids are no longer free, since they have formed peptide bonds. Thus, the side-chain reactive groups are the ones involved in the action of the enzyme. Hydrocarbon side chains do not contain reactive groups and are not involved in the process. Functional groups that can play a catalytic role include the imidazole group of histidine, the hydroxyl group of serine, the carboxyl side chains of aspartate and glutamate, the sulfhydryl group of cysteine, the amino side chain of lysine, and the phenol group of tyrosine.

Chymotrypsin catalyzes the hydrolysis of peptide bonds adjacent to aromatic amino acid residues in the protein being hydrolyzed; other residues are attacked at lower frequency. In addition, chymotrypsin catalyzes the hydrolysis of esters in model studies in the laboratory. The use of model systems is common in biochemistry, since a model provides the essential features of a reaction in a simple form that is easier to work with than the one found in nature. The amide (peptide) bond and the ester bond are similar enough that the enzyme can accept both types of compounds as substrates. Model systems based on the hydrolysis of esters are frequently used to study the peptide hydrolysis reaction.

A typical model compound is *p*-nitrophenyl acetate, which is hydrolyzed in two stages. The acetyl group is covalently attached to the enzyme at the end of the first stage (Step 1) of the reaction, but the *p*-nitrophenolate ion is released. In the second stage (Step 2) the acyl-enzyme intermediate is hydrolyzed, releasing acetate and regenerating the free enzyme.

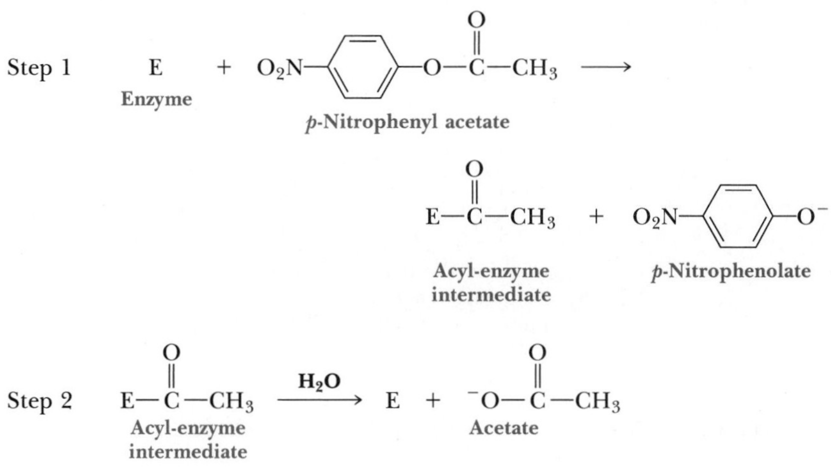

The hydrolysis of *p*-nitrophenyl acetate catalyzed by chymotrypsin.

Determining the Essential Amino Acid Residues

The serine residue at position 195 is required for the activity of chymotrypsin; in this respect chymotrypsin is typical of a class of enzymes known as **serine proteases.** The enzyme is completely inactivated when serine 195 reacts with diisopropylphosphofluoridate (DIPF), forming a covalent bond that links the serine side chain with the DIPF. The formation of covalently modified versions of specific side chains on proteins is called

The labeling of the active serine of chymotrypsin by diisopropylphosphofluoridate (DIPF).

$$Enz—CH_2OH$$

Serine 195

+

$$
\begin{array}{c}
H \\
| \\
H_3C—C—CH_3 \\
| \\
O \\
| \\
F—P{=}O \\
| \\
O \\
| \\
H_3C—C—CH_3 \\
| \\
H
\end{array}
$$

Diisopropylphosphofluoridate
(DIPF)

$\longrightarrow$

$$
\begin{array}{c}
H \\
| \\
H_3C—C—CH_3 \\
| \\
O \\
| \\
Enz—CH_2O—P{=}O \\
| \\
O \\
| \\
H_3C—C—CH_3 \\
| \\
H
\end{array}
$$

Labeled enzyme
(inactive)

+   HF

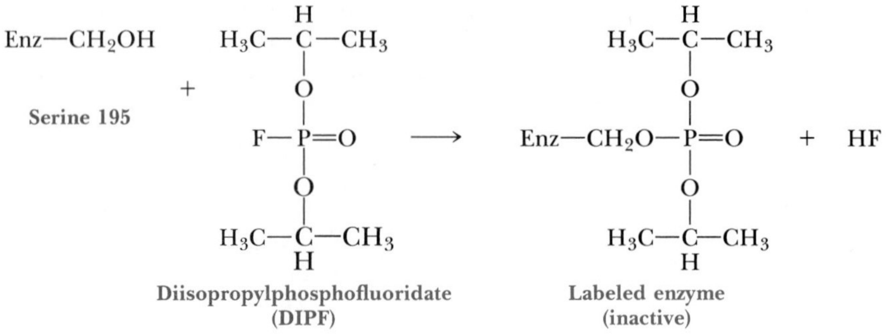

labeling; it is widely used in laboratory studies. The other serine residues of chymotrypsin are far less reactive and are not labeled by DIPF.

Histidine 57 is another essential amino acid residue in chymotrypsin. Chemical labeling again provides the evidence for involvement of this residue in the activity of chymotrypsin. In this case, the reagent used to label the essential amino acid residue is *N*-tosylamido-L-phenylethyl chloromethyl ketone (TPCK), also called tosyl-L-phenylalanine chloromethyl ketone.

The labeling of the active-site histidine of chymotrypsin by TPCK.

(a)   Phenylalanyl moiety chosen because of specificity of chymotrypsin for aromatic amino acid residues

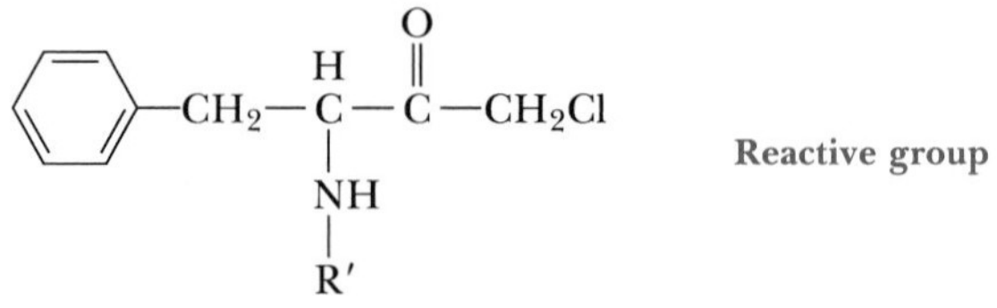

$$
\begin{array}{c}
 & & & O \\
 & & & \| \\
\text{—}CH_2\text{—}\overset{\displaystyle H}{\underset{\displaystyle NH}{C}}\text{—}C\text{—}CH_2Cl \\
 & & & \\
 & & & \\
 & & R'
\end{array}
$$

Reactive group

Structure of N-tosylamido-L-phenylethyl chloromethyl ketone
(TPCK), a labeling reagent for chymotrypsin
(R' represents a tosyl (toluenesulfonyl) group)

(b)

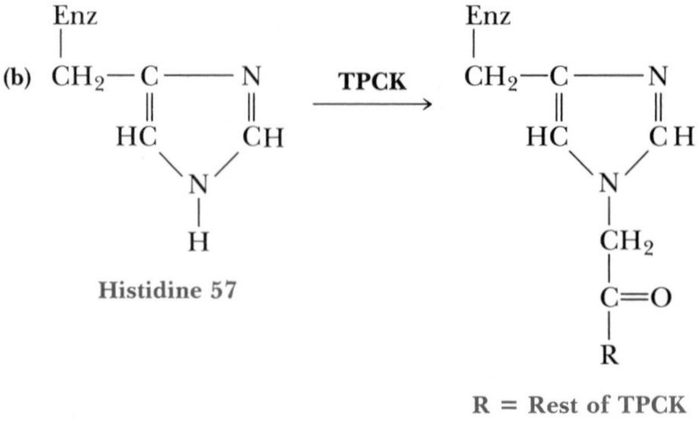

$$
\begin{array}{c}
Enz \\
| \\
CH_2—C{=}N \\
\quad | \quad \quad \| \\
HC \quad \quad CH \\
\quad \backslash \quad / \\
\quad N \\
\quad | \\
\quad H
\end{array}
\xrightarrow{\text{TPCK}}
\begin{array}{c}
Enz \\
| \\
CH_2—C{=}N \\
\quad | \quad \quad \| \\
HC \quad \quad CH \\
\quad \backslash \quad / \\
\quad N \\
\quad | \\
\quad CH_2 \\
\quad | \\
\quad C{=}O \\
\quad | \\
\quad R
\end{array}
$$

Histidine 57

R = Rest of TPCK

The phenylalanine moiety is bound to the enzyme because of the specificity for aromatic amino acid residues at the active site, and the active-site histidine residue reacts because the labeling reagent is similar to the usual substrate.

The Architecture of the Active Site

Both serine 195 and histidine 57 are required for the activity of chymotrypsin; therefore, they must be close to each other in the active site. The determination of the three-dimensional structure of the enzyme by x-ray crystallography provides evidence that the active-site residues do indeed have a close spatial relationship. The folding of the chymotrypsin backbone, mostly in an antiparallel pleated sheet array, positions the essential residues around an active-site pocket (Figure 5.18). Only a few residues are directly involved in the active site, but the whole molecule is necessary to provide the correct three-dimensional arrangement for those critical residues.

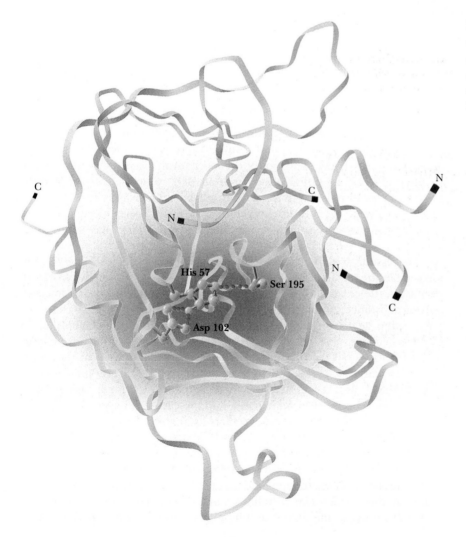

FIGURE 5.18

The tertiary structure of chymotrypsin places the essential amino acid residues close to one another. They are shown in blue and red.

Other important pieces of information about the three-dimensional structure of the active site emerge when a complex is formed between chymotrypsin and a substrate analog. When one such substrate analog, formyl-L-tryptophan,

The structure of formyl-L-tryptophan.

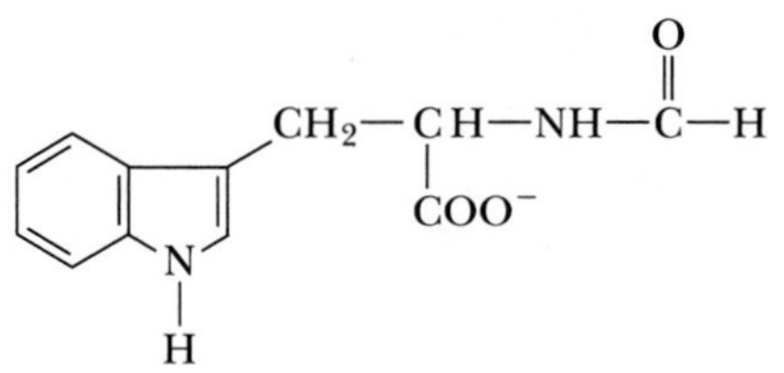

is bound to the enzyme, the tryptophan side chain fits into a hydrophobic pocket near serine 195. This type of binding is not surprising in view of the specificity of the enzyme for aromatic amino acid residues at the cleavage site. The results of x-ray crystallography show, in addition to the binding site for aromatic amino acid side chains of substrate molecules, a definite arrangement of the amino acid side chains that are responsible for the catalytic activity of the enzyme. The residues involved in this arrangement are serine 195 and histidine 57.

The Mechanism of Chymotrypsin Action

Any postulated reaction mechanism must be modified or discarded if it is not consistent with experimental results. There is consensus, but not total agreement, on the main features of the mechanism discussed in this section. At one point it was thought that aspartate 102, which is essential for the activity of chymotrypsin, was involved in a "charge-relay" system with serine 195 and histidine 57. It is now known that the charge-relay mechanism does not apply to chymotrypsin and that the aspartate must play some other role.

The essential amino acid residues, serine 195 and histidine 57, are involved in the mechanism of catalytic action. In the terminology of organic chemistry, the oxygen of the serine side chain is a **nucleophile,** or nucleus-seeking substance. A nucleophile tends to bond to sites of positive charge or polarization (electron-poor sites), in contrast with an **electrophile,** or electron-seeking substance, which tends to bond to sites of negative charge or polarization (electron-rich sites). The nucleophilic oxygen of the serine attacks the carbonyl carbon of the peptide group. The carbon now has four single bonds, and a tetrahedral intermediate is formed; the original —C=O bond becomes a single bond, and the carbonyl oxygen becomes an oxyanion. The acyl-enzyme intermediate is formed from the tetrahedral species (Figure 5.19). The histidine and the amino portion of the original peptide group are involved in this part of the reaction, as the amino group hydrogen bonds to the imidazole portion of the histidine. Note that the imidazole is already protonated and that the proton came from the hydroxyl group of the serine. The histidine behaves as a base in abstracting the proton from the serine; in the terminology of the physical organic chemist,

1st stage reaction

2nd stage reaction

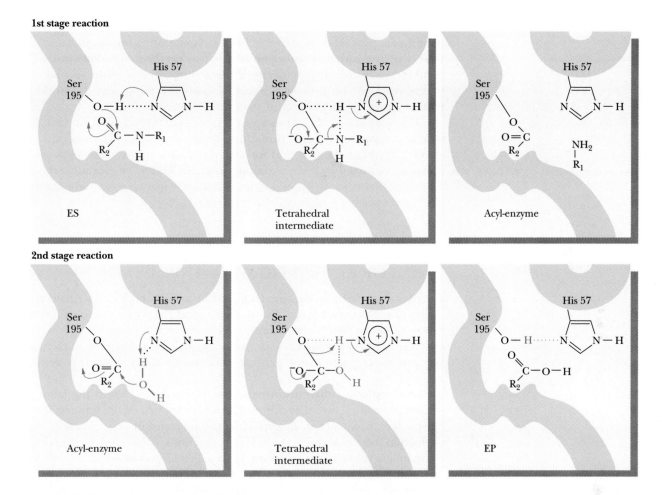

FIGURE 5.19

The mechanism of action of chymotrypsin. In the first stage of the reaction, the nucleophile serine 195 attacks the carbonyl carbon of the substrate. In the second stage, water is the nucleophile that attacks the acyl-enzyme intermediate. Note the involvement of histidine 57 in both stages of the reaction. (From G. Hammes, 1982, *Enzyme Catalysis and Regulation*, Academic Press, New York.)

the histidine acts as a general base catalyst. The carbon–nitrogen bond of the original peptide group breaks, leaving the acyl-enzyme intermediate. The proton abstracted by the histidine has been donated to the leaving amino group. In donating the proton, the histidine has acted as an acid in the breakdown of the tetrahedral intermediate, although it acted as a base in its formation.

In the deacylation phase of the reaction, the last two steps are reversed, with water acting as the attacking nucleophile. In this second phase the water is hydrogen-bonded to the histidine. The oxygen of water now performs the nucleophilic attack on the acyl carbon that came from the original peptide group. Once again a tetrahedral intermediate is formed. In the final step of the reaction, the bond between the serine oxygen and the carbonyl carbon breaks, releasing the product with a carboxyl group where the original peptide group used to be and regenerating the original enzyme. Note that the serine is hydrogen bonded to the histidine. This hydrogen bond increases the nucleophilicity of the serine, whereas in the second part of the reaction the hydrogen bond between the water and the histidine increased the nucleophilicity of the water.

The mechanism of chymotrypsin action is particularly well studied and, in many respects, typical. Numerous types of reaction mechanisms for enzyme action are known, and we shall discuss them in the contexts of the reactions catalyzed by the enzymes in question. To lay the groundwork, it is useful to discuss some general types of catalytic mechanisms and how they affect the specificity of enzymatic reactions.

5.12
TYPES OF CATALYTIC MECHANISMS

The overall mechanism for a reaction may be fairly complex, as we have seen in the case of chymotrypsin, but the individual parts of a complex mechanism can themselves be fairly simple. Concepts such as nucleophilic attack and acid catalysis commonly enter into discussions of enzymatic reactions. We can draw quite a few general conclusions from these two general descriptions.

Nucleophilic substitution reactions play a large role in the study of organic chemistry, and they are excellent illustrations of the importance of kinetic measurements in determining the mechanism of a reaction. The distinction between the S_N1 and S_N2 reactions (first- and second-order nucleophilic substitution reactions, respectively) is a classic example, particularly with optically active starting materials. There are many examples of nucleophilic substitutions in enzymatic reaction mechanisms.

To discuss acid–base catalysis, it is helpful to recall the definitions of acids and bases. In the Brønsted–Lowry definition, an acid is a proton donor and a base is a proton acceptor. The concept of **general acid–base catalysis** depends on donation and acceptance of protons by groups such as imidazole, hydroxyl, carboxyl, sulfhydryl, amino, and phenolic side chains of amino acids; all these functional groups can act as acids or bases. The donation and acceptance of protons gives rise to the bond breaking and reformation that constitute the enzymatic reaction.

A second form of acid–base catalysis is based on another, more general definition of acids and bases. In the Lewis formulation, an acid is an electron-pair acceptor, and a base is an electron-pair donor. Metal ions, including such biologically important ones as Mn^{2+}, Mg^{2+}, and Zn^{2+}, are Lewis acids. Thus, they can play roles in **metal-ion catalysis** (also called Lewis acid–base catalysis). The involvement of Zn^{2+} in the enzymatic activity of carboxypeptidase A is an example of this type of behavior. This enzyme catalyzes the hydrolysis of C-terminal peptide bonds of proteins. The Zn(II), which is required for the activity of the enzyme, is complexed to the imidazole side chains of histidines 69 and 196 and to the carboxylate side chain of glutamate 72. [The notation Zn(II) is used because the zinc ion is involved in a complex.] The zinc ion is also complexed to the substrate.

Imidazole   Carboxylate
Imidazole ◢ ╲ │ ╱
        Zn(II)
         ⋮
         O
         │   ‖
Rest of chain⋯C—C
         │   ╲
                N—CHR—COO⁻
                │
                H

A zinc ion is complexed to three side chains of carboxypeptidase and to a carbonyl group on the substrate.

The type of binding involved in the complex is similar to the binding that links iron to the large ring involved in the heme group. Binding of the substrate to the zinc ion polarizes the carbonyl group, making it susceptible to attack by water and allowing the hydrolysis to proceed more rapidly than it does in the uncatalyzed reaction.

         ╲ │ ╱
         Zn(II)
          ⋮
          O
          │   ‖
Rest of chain— C —C— N —CHR—COO⁻
          │   │
                   H
        O
      H╱ ╲H

          │
          ↓

        O   H
        │   ‖   │
Rest of chain— C —C—O⁻ + H—N⁺—CHR—COO⁻
        │   │
                       H

The reaction catalyzed by carboxy-peptidase A.

A definite connection exists between the concepts of acids and bases and the idea of nucleophiles and their complementary substances, electrophiles. A Lewis acid is an electrophile, and a Lewis base is a nucleophile. Catalysis by enzymes, including their remarkable specificity, is based on these well-known chemical principles operating in a complex environment.

The nature of the active site plays a particularly important role in the specificity of enzymes. An enzyme that displays **absolute specificity,** catalyzing the reaction of one, and only one, substrate to a particular product, is likely to have a fairly rigid active site that is best described by the lock-and-key model of substrate binding. The many enzymes that display **relative specificity,** catalyzing the reactions of structurally related substrates to related products, apparently have more flexibility in their active sites and are better characterized by the induced-fit model of enzyme–substrate binding; chymotrypsin is a good example. Finally, there are **stereospecific** enzymes with specificity in which optical activity plays a role. The binding

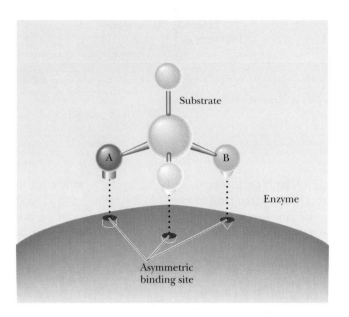

FIGURE 5.20

An asymmetric binding site on an enzyme can distinguish between identical groups such as A and B. Note that the binding site consists of three parts, giving rise to asymmetric binding because one part is different from the other two.

site itself must be asymmetric in this situation (Figure 5.20). If the enzyme is to bind specifically to an optically active substrate, the binding site must have the shape of the substrate and not its mirror image. There are even enzymes that introduce a center of optical activity into the product. The substrate itself is not optically active in this case. There is only one product, which is one of two possible isomers, not a mixture of optical isomers.

5.13
COENZYMES

Coenzymes are nonprotein substances that take part in enzymatic reactions and are regenerated for further reaction. Metal ions frequently play such a role, and they make up one of two important classes of coenzymes. The other important class is a mixed bag of organic compounds; many of them are vitamins or are metabolically related to vitamins.

Because metal ions are Lewis acids (electron-pair acceptors), they can act as Lewis acid–base catalysts. They can also form coordination compounds by behaving as Lewis acids while the groups to which they bind act as Lewis bases. Coordination compounds are an important part of the chemistry of metal ions in biological systems, as shown by Zn(II) in carboxypeptidase and by Fe(II) in hemoglobin. The coordination compounds formed by metal ions tend to have quite specific geometries, which are an aid in positioning the groups involved in a reaction for optimum catalysis.

TABLE 5.3 Coenzymes, Their Reactions, and Vitamin Precursors

COENZYME	REACTION TYPE	VITAMIN PRECURSOR	SEE SECTION
Biotin	CO_2 fixation	Biotin	14.2, 15.6
Coenzyme A	Acyl transfer	Pantothenic acid	9.11, 12.3, 15.6
Flavin coenzymes	Oxidation–reduction	Riboflavin (B_2)	9.9, 12.3
Lipoic acid	Acyl transfer	—	12.3
Nicotinamide adenine coenzymes	Oxidation–reduction	Niacin	9.9, 11.3, 12.3
Pyridoxal phosphate	Transamination	Pyridoxine (B_6)	17.4
Tetrahydrofolic acid	Transfer of one-carbon units	Folic acid	17.4
Thiamine pyrophosphate	Aldehyde transfer	Thiamine (B_1)	11.4, 14.4

Some of the most important organic coenzymes are vitamins and their derivatives, especially B vitamins. Quite a few of them are involved in oxidation–reduction reactions, which provide energy for the organism. Others serve as group-transfer agents in metabolic processes (Table 5.3). We shall see these coenzymes again when we discuss the reactions in which they are involved. For the present we shall investigate one particularly important oxidation–reduction coenzyme and one group-transfer coenzyme.

Nicotinamide adenine dinucleotide (NAD^+) is a coenzyme in many oxidation–reduction reactions. Its structure (Figure 5.21) has three parts: a

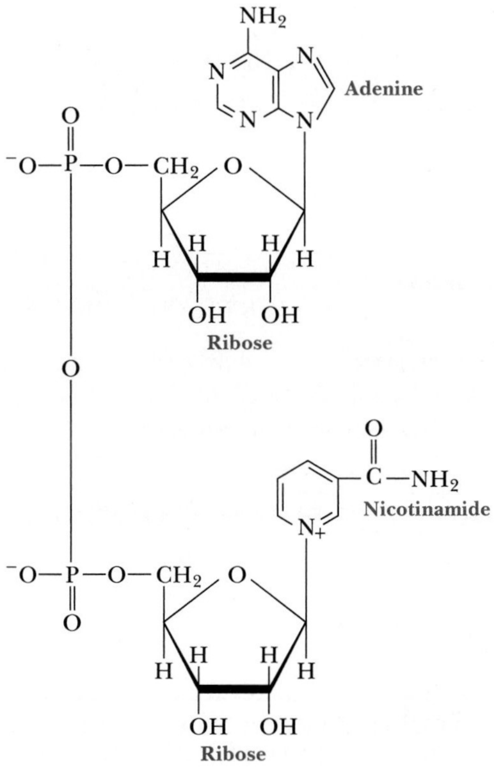

FIGURE 5.21

The structure of nicotinamide adenine dinucleotide (NAD^+).

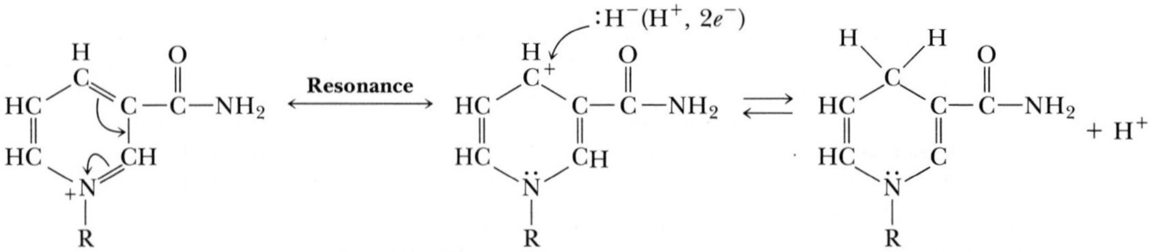

FIGURE 5.22

The role of the nicotinamide ring in oxidation–reduction reactions. R is the rest of the molecule. In reactions of this sort, an H^+ is transferred along with the two electrons.

FIGURE 5.23

Forms of vitamin B_6. The first three structures are vitamin B_6 itself, and the last two structures show the modifications that give rise to the metabolically active coenzyme.

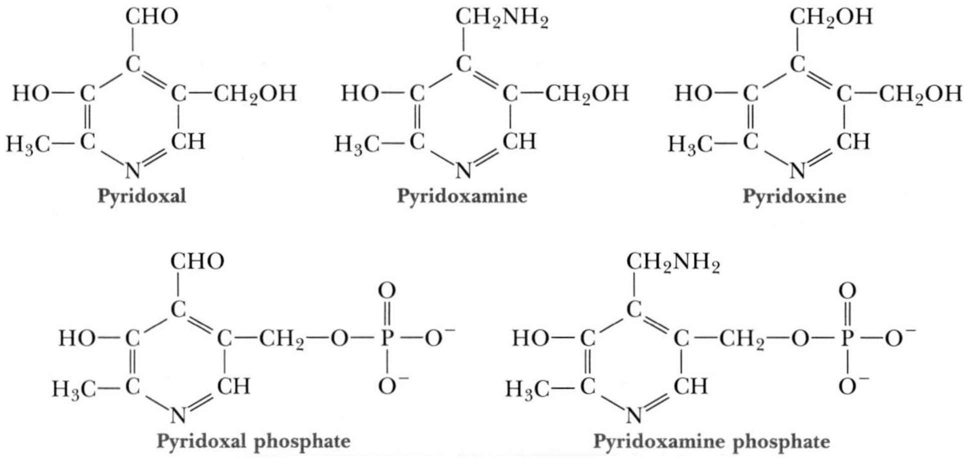

Pyridoxal   Pyridoxamine   Pyridoxine

Pyridoxal phosphate   Pyridoxamine phosphate

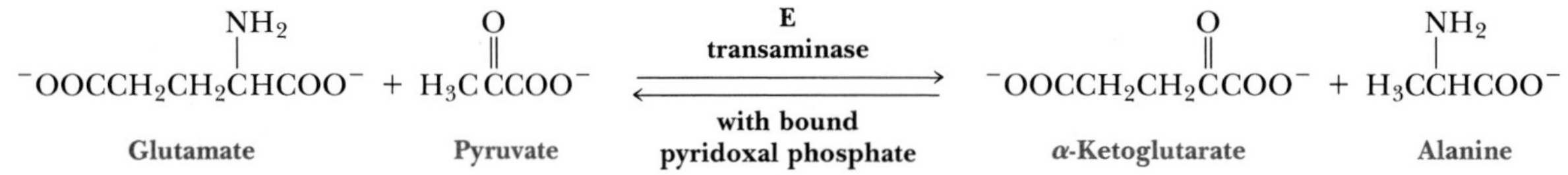

Glutamate   Pyruvate   α-Ketoglutarate   Alanine

This amino (NH_2) group transfer reaction occurs in two stages:

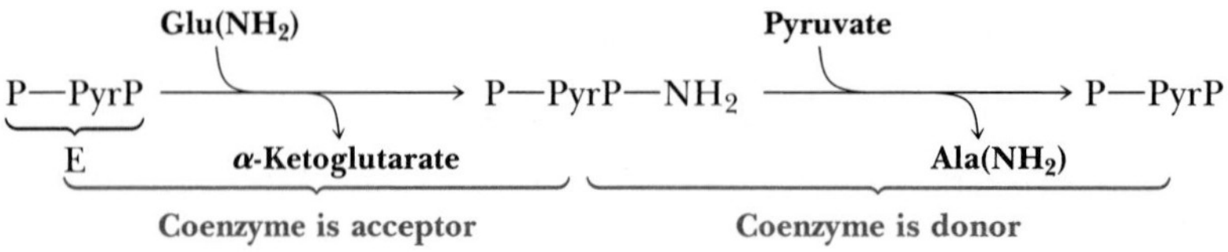

Coenzyme is acceptor   Coenzyme is donor

FIGURE 5.24

The role of pyridoxal phosphate as a coenzyme in a transamination reaction. PyrP is pyridoxal phosphate, P is the apoenzyme (the polypeptide chain alone), and E is the active holoenzyme (polypeptide plus coenzyme).

nicotinamide ring, an adenine ring, and two sugar–phosphate groups linked together. The nicotinamide ring contains the site at which oxidation and reduction reactions occur (Figure 5.22). Nicotinic acid is another name for the vitamin niacin. The adenine–sugar–phosphate portion of the molecule is structurally related to nucleotides.

The B_6 vitamins (pyridoxal, pyridoxamine, and pyridoxine and their phosphorylated forms, which are the coenzymes) are involved in the transfer of amino groups from one molecule to another, an important step in the biosynthesis of amino acids (Figure 5.23). In the reaction, the amino group is transferred from the donor to the coenzyme and then from the coenzyme to the ultimate acceptor (Figure 5.24).

SUMMARY

Probably the most important function of proteins is catalysis. Biological catalysts are called enzymes. With the exception of some RNAs that catalyze their own self-splicing, all enzymes are globular proteins. Enzymes are the most efficient catalysts known. They speed up a reaction by lowering the activation energy, a kinetic parameter. Catalysts do not affect the thermodynamics of the reaction.

The first step in an enzyme-catalyzed reaction is the binding of the enzyme to the substrate to form a complex. The formation of the complex leads to formation of the transition-state species, which in turn forms the product. A substrate binds to a small portion of the enzyme called the active site. Two models have been proposed to describe enzyme–substrate binding: the lock-and-key model, in which there is an exact fit between the enzyme and substrate, and the induced-fit model, in which the enzyme is considered to have conformational flexibility and in which there is an exact fit only when the substrate is bound.

The kinetics of many enzyme-catalyzed reactions can be described by the Michaelis–Menten model. In this model the concept of the steady state, with a constant concentration of the enzyme–substrate complex, plays a vital role.

Inhibitors can give a considerable amount of information about enzymatic reactions. A reversible inhibitor can bind to the enzyme and subsequently be released. An irreversible inhibitor reacts with the enzyme to produce a protein that is not enzymatically active. There are two kinds of reversible inhibitors. Competitive inhibitors bind to the active site and block access of the substrate to the active site. Noncompetitive inhibitors bind to the enzyme at a site other than the active site and cause a change in the structure of the enzyme, especially around the active site, as a result of binding. In the Michaelis–Menten model, competitive inhibitors increase the K_M but leave the V_{max} unchanged; noncompetitive inhibitors change the V_{max} but leave the K_M unchanged.

The Michaelis–Menten model does not describe the behavior of allosteric enzymes. Changes in quaternary structure on binding of substrates, inhibitors, and activators all affect the observed kinetics of such enzymes. In the concerted model for allosteric behavior, the binding of substrate, inhibitor, or activator to one subunit shifts the equilibrium between an active form of the enzyme (which binds substrate strongly) and an inactive form (which does not bind substrate strongly). The conformational change takes place in all subunits at the same time. In the sequential model, the binding of substrate induces the conformational change in one subunit, and the change is subsequently passed along to other subunits. Both models are useful; they may eventually be incorporated in a single, more inclusive model.

Another type of control mechanism in enzyme action is zymogen activation, in which an inactive precursor of an enzyme is transformed into an active enzyme by cleavage of covalent bonds. For example, the proteolytic enzymes trypsin and chymotrypsin arise from the zymogens trypsinogen and chymotrypsinogen, respectively. Similar protein activations take place in blood clotting.

Several questions arise about the events that occur at the active site of an enzyme in the course of a reaction. Some of the most important of these questions address the nature of the essential amino acid residues, their spatial arrangement, and the mechanism of the reaction. Chymotrypsin is a good example of an enzyme for which most of the questions have been answered. Its essential amino acid residues have been determined to be serine 195, histidine 57, and aspartate 102. The complete three-dimensional structure of chymotrypsin, including the architecture of the active site, has been determined by x-ray crystallography. Nucleophilic attack by serine is the main feature of the mechanism, with histidine hydrogen bonded to serine in the course of the reaction. Common organic reaction mechanisms such as nucleophilic substitution and general acid–base catalysis are known to play roles in enzymatic catalysis.

Coenzymes are nonprotein substances that take part in enzymatic reactions and are regenerated for further reaction. Metal ions can serve as coenzymes, frequently by acting as Lewis acids. There are also many organic coenzymes, most of which are vitamins or structurally related to vitamins.

E X E R C I S E S

1. For the reaction of glucose with oxygen to produce carbon dioxide and water,

$$\text{Glucose} + 6\ O_2 \longrightarrow 6\ CO_2 + 6\ H_2O$$

the $\Delta G°$ is -2880 kJ mol^{-1}, a strongly exergonic reaction. However, a sample of glucose can be maintained indefinitely in an oxygen-containing atmosphere. Reconcile these two statements.

2. For the hypothetical reaction

$$3\ A + 2\ B \longrightarrow 2\ C + 3\ D$$

the rate was experimentally determined to be

$$\text{Rate} = k[A]^1[B]^1$$

What is the order of the reaction with respect to A? With respect to B? What is the overall order of the reaction? Suggest how many molecules each of A and B are likely to be involved in the detailed mechanism of the reaction.

3. Distinguish between the lock-and-key and induced-fit models for binding of a substrate to an enzyme.

4. Show graphically the dependence of reaction velocity on substrate concentration for an enzyme that follows Michaelis–Menten kinetics and for an allosteric enzyme.

5. Define steady state, and comment on the relevance of this concept to theories of enzyme reactivity.

6. For an enzyme that displays Michaelis–Menten kinetics, what is the reaction velocity, V (as a percentage of V_{max}), observed at (a) $[S] = K_M$; (b) $[S] = 0.5K_M$; (c) $[S] = 0.1K_M$; (d) $[S] = 2K_M$; (e) $[S] = 10K_M$?

7. How is the turnover number of an enzyme related to V_{max}?

8. How can competitive and noncompetitive inhibition be distinguished in terms of K_M?

9. Draw Lineweaver–Burk plots for the behavior of an enzyme for which the following experimental data are available.

[S] (mM)	V, NO INHIBITOR (mmol min^{-1})	V, INHIBITOR PRESENT (mmol min^{-1})
3.0	4.58	3.66
5.0	6.40	5.12
7.0	7.72	6.18
9.0	8.72	6.98
11.0	9.50	7.60

What are the K_M and V_{max} values for the inhibited and uninhibited reactions? Is the inhibitor competitive or noncompetitive?

10. Distinguish between the molecular mechanisms of competitive and noncompetitive inhibition.

11. What features distinguish enzymes that undergo allosteric control from those that obey the Michaelis–Menten equation?

12. Distinguish between the concerted and sequential models for the behavior of allosteric enzymes.

13. Name three proteins that are subject to the control mechanism of zymogen activation.

14. List three coenzymes and their functions.

15. Is the following statement true or false, and why? "The mechanisms of enzymatic catalysis have nothing in common with those encountered in organic chemistry."

16. An experiment is performed to test a suggested mechanism for an enzyme-catalyzed reaction. The results fit the model exactly (to within experimental error). Do the results prove that the mechanism is correct? Why or why not?

17. What properties of metal ions make them useful coenzymes?

18. Briefly describe the role of nucleophilic catalysis in the mechanism of the chymotrypsin reaction.

19. Explain why cleavage of the bond between arginine 15 and isoleucine 16 of chymotrypsinogen activates the zymogen.

20. An inhibitor that specifically labels chymotrypsin at histidine 57 is N-tosyl-L-phenylalanyl chloromethyl ketone (TPCK). How would you modify the structure of this inhibitor to label the active site of trypsin?

21. The enzyme lactate dehydrogenase catalyzes the reaction

$$\text{Pyruvate} + \text{NADH} + H^+ \longrightarrow \text{lactate} + \text{NAD}^+$$

NADH absorbs light at 340 nm in the near ultraviolet region of the electromagnetic spectrum, but NAD^+ does not. Suggest an experimental method for following the rate of this reaction, assuming that you have available a spectrophotometer capable of measuring light at this wavelength.

22. Determine the values of K_M and V_{max} for an enzymatic reaction, given the following data.

SUBSTRATE CONCENTRATION (moles L^{-1})	VELOCITY (mM min^{-1})
2.500	0.588
1.000	0.500
0.714	0.417
0.526	0.370
0.250	0.256

23. For the following reaction, determine the K_M and whether the inhibition is competitive or noncompetitive.

[S] MOLARITY	V, NO INHIBITOR (arbitrary units)	V, INHIBITOR PRESENT (same arbitrary units)
5×10^{-5}	0.014	0.006
1×10^{-4}	0.026	0.010
5×10^{-4}	0.092	0.040
1.5×10^{-3}	0.136	0.086
2.5×10^{-3}	0.150	0.120
5×10^{-3}	0.165	0.142

A N N O T A T E D   B I B L I O G R A P H Y

Abraham, D., F. Wireko, R. Randad, C. Poyart, J. Kister, B. Bohn, J. Liard, and M. Kunart. Allosteric Modifiers of Hemoglobin. *Biochemistry* **31**, 9141–9149 (1992). [A report on continuing research on allosteric effectors of hemoglobin, with emphasis on treatment of sickle-cell anemia.]

Althaus, I., J. Chou, A. Gonzales, M. Deibel, K. Chou, F. Kezdy, D. Romero, J. Palmer, R. Thomas, P. Aristoff, W. Tarpley, and F. Reusser. Kinetic Studies with the Nonnucleoside HIV-1 Reverse Transcriptase Inhibitor U-88204E. *Biochemistry* **32**, 6548–6554 (1993). [How enzyme kinetics can play a role in AIDS research.]

Bachmair, A., D. Finley, and A. Varshavsky. *In Vivo* Half-Life of a Protein Is a Function of Its Amino Terminal Residue. *Science* **234**, 179–186 (1986). [A particularly striking example of the relationship between structure and stability in proteins.]

Bender, M. L., R. L. Bergeron, and M. Komiyama. *The Bioorganic Chemistry of Enzymatic Catalysis.* New York: J. Wiley, 1984. [A discussion of mechanisms in enzymatic reactions.]

Danishefsky, S. Catalytic Antibodies and Disfavored Reactions. *Science* **259**, 469–470 (1993). [A short review of chemists' use of antibodies as the basis of "tailor-made" catalysts for specific reactions.]

Dressler, D., and H. Potter. *Discovering Enzymes.* New York: Scientific American Press, 1991. [A well-illustrated book that introduces important concepts of enzyme structure and function.]

Dugas, H., and C. Penney. *Bioorganic Chemistry: A Chemical*

Approach to Enzyme Action. New York: Springer-Verlag, 1981. [Discusses model systems as well as enzymes.]

Fersht, A. *Enzyme Structure and Mechanism.* 2nd ed. New York: W. H. Freeman, 1985. [A thorough coverage of enzyme action.]

Gibbs, W. Try, Try Again. *Sci. Amer.* **269** (1), 101–103 (1993). [A report on medical applications of monoclonal antibodies.]

Hammes, G. *Enzyme Catalysis and Regulation.* New York: Academic Press, 1982. [A good basic text on enzyme mechanisms.]

Koshland, D. E. Correlation of Structure and Function of Enzyme Action. *Science* **142,** 1533–1541 (1963). [The definitive statement of the induced-fit theory. Mainly of historic interest, but a classic.]

Kraut, J. How Do Enzymes Work? *Science* **242,** 533–540 (1988). [An advanced discussion of the role of transition states in enzymatic catalysis.]

Lerner, R., S. Benkovic, and P. Schultz. At the Crossroads of Chemistry and Immunology: Catalytic Antibodies. *Science* **252,** 659–667 (1991). [A review of how antibodies can bind to almost any molecule of interest and then catalyze some reaction of that molecule.]

Marcus, R. Skiing the Reaction Rate Slopes. *Science* **256,** 1523–1524 (1992). [A brief, advanced-level look at reaction transition states.]

Monod, J., J.-P. Changeux, and F. Jacob. Allosteric Proteins and Cellular Control Systems. *J. Mol. Biol.* **6,** 306–329 (1963). [The original article on allosterism, describing the concerted model.]

Moore, J. W., and R. G. Pearson. *Kinetics and Mechanism.* 3rd ed. New York: John Wiley Interscience, 1980. [A classic, quite advanced treatment of the use of kinetic data to determine mechanisms.]

Rini, J., U. Schulze-Gahmen, and I. Wilson. Structural Evidence for Induced Fit as a Mechanism for Antibody–Antigen Recognition. *Science* **255,** 959–965 (1992). [The results of structure determination by x-ray crystallography.]

Sigman, D., ed. *The Enzymes.* Vol. 20, *Mechanisms of Catalysis.* San Diego: Academic Press, 1992. [Part of a definitive series on enzymes and their structures and functions.]

Sigman, D., and P. Boyer, eds. *The Enzymes.* Vol. 19, *Mechanisms of Catalysis.* San Diego: Academic Press, 1990. [Part of a definitive series on enzymes and their structures and functions.]

Vallee, B., and D. Auld. New Perspective on Zinc Biochemistry: Cocatalytic Sites in Multi-Zinc Enzymes. *Biochemistry* **32,** 6493–6500 (1993). [A review of the role of zinc and other transition-metal ions in enzymatic catalysis.]

Wong, I., and T. Lohman. Allosteric Effects of Nucleotide Cofactors on *Escherichia coli* Rep Helicase–DNA Binding. *Science* **256,** 350–355 (1992). [The mode of action of an allosteric enzyme that is important in DNA replication.]

Chapter 6

Nucleic Acids: How Structure Conveys Information

A computer-generated image of DNA, looking down the helix axis.

Genes, the hereditary material of the chromosomes, are essentially long stretches of double-helical DNA. In a process mediated by RNA, the other kind of nucleic acid, the sequence of DNA bases in each gene specifies the sequence of amino acids in a single polypeptide (protein) chain. The protein's amino acid sequence, in turn, determines its structure and function. Thus, the genetic code contained in DNA ultimately directs the activities of proteins, the essential machinery of life. Each cell carries in its DNA the instructions for making the complete organism. When the cell divides, each new cell bears a copy of the original DNA. Replication of the hereditary material is made possible by the complementary nature of the DNA bases. Adenine on one strand pairs with thymine on the opposite strand of the double helix. The same is true for the other two bases: guanine on one strand pairs with cytosine on an opposite strand. Thus, one strand of DNA is a template for the other strand. It is now possible to control some aspects of genetic coding. Starting in the 1970s, techniques have been introduced for manipulating DNA by cutting and splicing it in a manner that both mimics and transcends natural processes. These techniques will provide valuable insight into the manner in which proteins interact with DNA molecules to control gene activation and repression.

6.1
LEVELS OF STRUCTURE IN NUCLEIC ACIDS

In Chapter 4 we identified four levels of structure — primary, secondary, tertiary, and quaternary — in proteins. Nucleic acids can be viewed in the same way. The *primary structure* of nucleic acids is the order of bases in the polynucleotide sequence, and the *secondary structure* is three-dimensional conformation of the backbone. The *tertiary structure* is specifically the supercoiling of the molecule.

DNA and RNA are the two kinds of nucleic acids. Important differences between them appear in their secondary and tertiary structures, and so we shall describe these structural features separately for DNA and for RNA. The *quaternary structure* of nucleic acids frequently includes interaction with proteins. One well-known example is the association of RNA and proteins in ribosomes; another is the self-assembly of tobacco mosaic virus, in which the nucleic acid strand winds through a cylinder of coat protein subunits.

6.2
THE COVALENT STRUCTURE OF POLYNUCLEOTIDES

The monomers of nucleic acids are **nucleotides.** An individual nucleotide consists of three parts — a base, a sugar, and a phosphoric acid residue — all of which are covalently bonded together.

The order of bases in the nucleic acids specifies the genetic code. The **nucleic acid bases** (also called **nucleobases**) are of two types, pyrimidines and purines (Figure 6.1). Three **pyrimidine bases,** *cytosine, thymine,* and

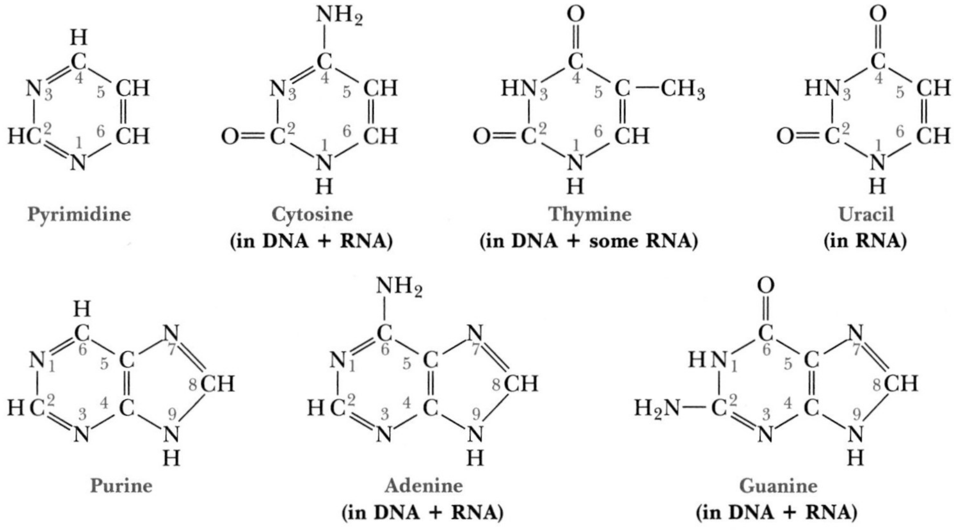

FIGURE 6.1

Structures of the common nucleobases. The structures of pyrimidine and purine are shown for comparison.

uracil, commonly occur. Cytosine is found in both RNA and DNA. Uracil occurs only in RNA. In DNA thymine is substituted for uracil; thymine is also found to a small extent in some forms of RNA. The common **purine bases** are adenine and guanine, both of which are found in both RNA and DNA (Figure 6.1). In addition to these five commonly occurring bases, there are "unusual" bases, with slightly different structures, that are found principally, but not exclusively, in transfer RNA (Figure 6.2). In most cases, but not all, the base is modified by methylation.

A **nucleoside** is a compound that consists of a base and a sugar covalently linked together. It differs from a nucleotide by lacking a phosphate group in its structure. In a nucleoside, a base forms a *glycosidic*

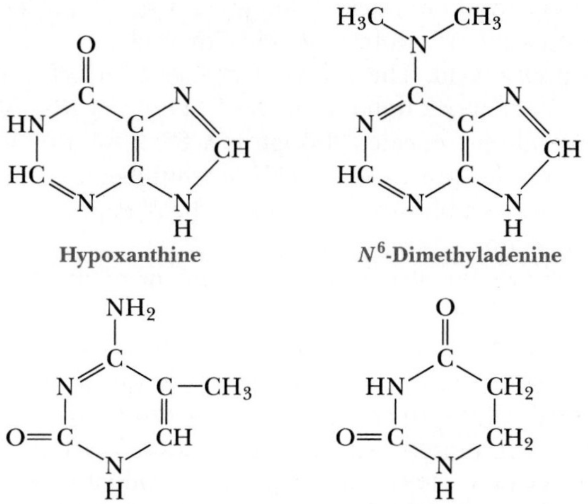

FIGURE 6.2

Structures of some of the less common nucleobases.

FIGURE 6.3

A comparison of the structures of
a ribonucleoside and a deoxyribonucleoside. Nucleosides do not
have a phosphate group in their
structures.

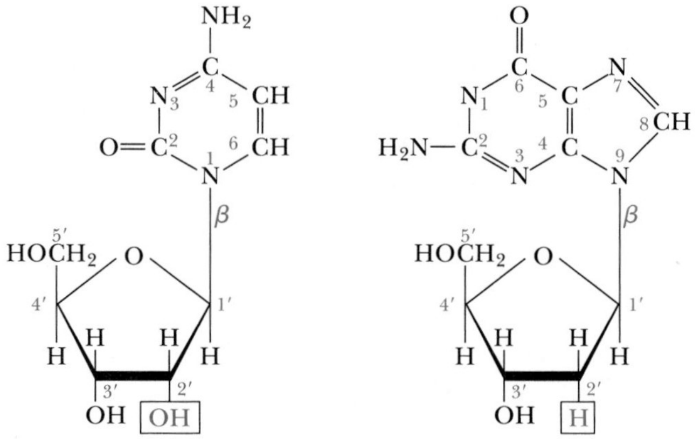

Cytidine
A ribonucleoside

Deoxyguanosine
A deoxyribonucleoside

linkage with the sugar. (Glycosidic linkages and stereochemistry of sugars are discussed in detail in Chapter 10, Section 10.2. For now, it is sufficient to say that a *glycosidic bond* is one that links a sugar and some other moiety.) When the sugar is β-D-ribose, the resulting compound is a **ribonucleoside;** when the sugar is β-D-deoxyribose, the resulting compound is a **deoxyribonucleoside** (Figure 6.3). The glycosidic linkage is from the C-1' carbon of the sugar to the N-1 nitrogen of pyrimidines or to the N-9 nitrogen of purines. (The ring atoms of the base and the carbon atoms of the sugar are both numbered, with the numbers of the sugar atoms primed to prevent confusion.) Note that the sugar is linked to a nitrogen in both cases (an *N*-glycosidic bond).

When phosphoric acid is esterified to one of the hydroxyl groups of the sugar portion of a nucleoside, a nucleotide is formed (Figure 6.4). A nucleotide is named for the parent nucleoside, with the suffix "monophosphate" added; the position of the phosphate ester is specified by the number of the carbon atom at the hydroxyl group to which it is esterified—for instance, adenosine 3'-monophosphate, deoxycytidine 5'-monophosphate. The 5' nucleotides are the ones most commonly encountered in nature.

The polymerization of nucleotides gives rise to nucleic acids. The linkage between monomers in nucleic acids involves formation of two ester bonds by phosphoric acid. The hydroxyl groups to which the phosphoric acid is esterified are those bonded to the 3' and 5' carbons on adjacent residues. The resulting repeated linkage is a 3', 5'-phosphodiester bond. The nucleotide residues of nucleic acids are numbered from the 5' end, which normally carries a phosphate group, to the 3' end, which normally has a free hydroxyl group.

Figure 6.5 shows the structure of a fragment of an RNA chain. The *sugar–phosphate backbone* repeats itself down the length of the chain. The most important feature of the structure of nucleic acids is the identities of the bases, and abbreviated forms of the structure can be written to convey this essential information. In one system of notation, single letters, such as A, G, C, U, and T, represent the individual bases. Vertical lines show the positions of the sugar moieties to which the individual bases are attached, and a diagonal line through the letter "P" represents a phosphodiester

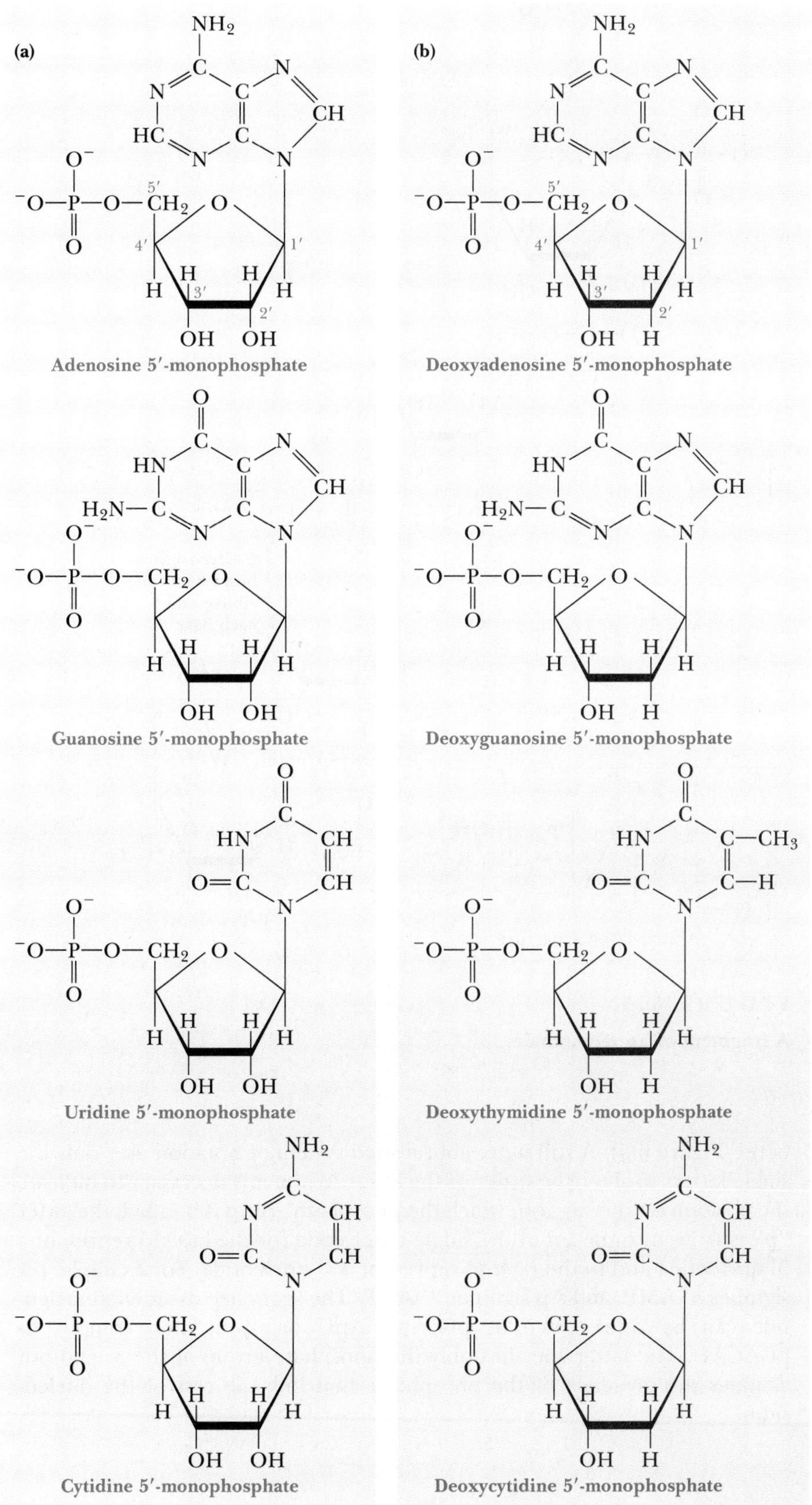

(a)

Adenosine 5'-monophosphate

(b)

Deoxyadenosine 5'-monophosphate

Guanosine 5'-monophosphate

Deoxyguanosine 5'-monophosphate

Uridine 5'-monophosphate

Deoxythymidine 5'-monophosphate

Cytidine 5'-monophosphate

Deoxycytidine 5'-monophosphate

FIGURE 6.4

The structures and names of the commonly occurring nucleotides. Each nucleotide has a phosphate group in its structure. All structures are shown in the forms that exist at pH 7. (a) Ribonucleotides. (b) Deoxyribonucleotides.

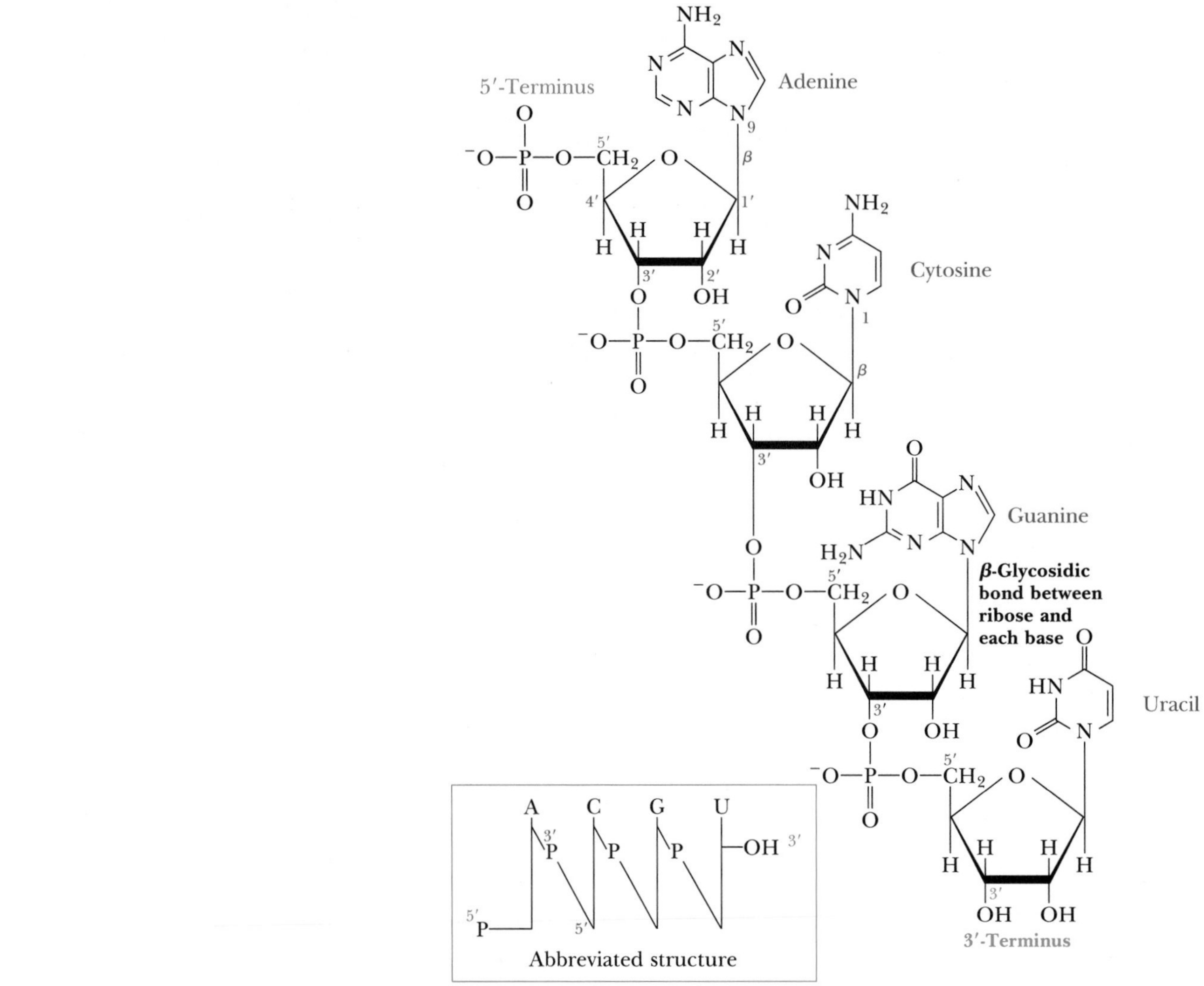

FIGURE 6.5

A fragment of an RNA chain.

bond (Figure 6.5). A still more abbreviated system of notation uses only the single letters to show the order of the bases. When it is necessary to indicate the position on the sugar to which the phosphate group is bonded, the letter "p" is written to the *left* of the single-letter code for the base to represent a 5′ nucleotide and to the *right* to represent a 3′ nucleotide. For example, pA signifies 5′-AMP, and Ap signifies 3′-AMP. The sequence of an oligonucleotide can be represented as pGpApCpApU or, even more simply, as pGACAU. The latter specifies only the phosphate group at the 5′ end but assumes the presence of the phosphates that link the rest of the nucleosides.

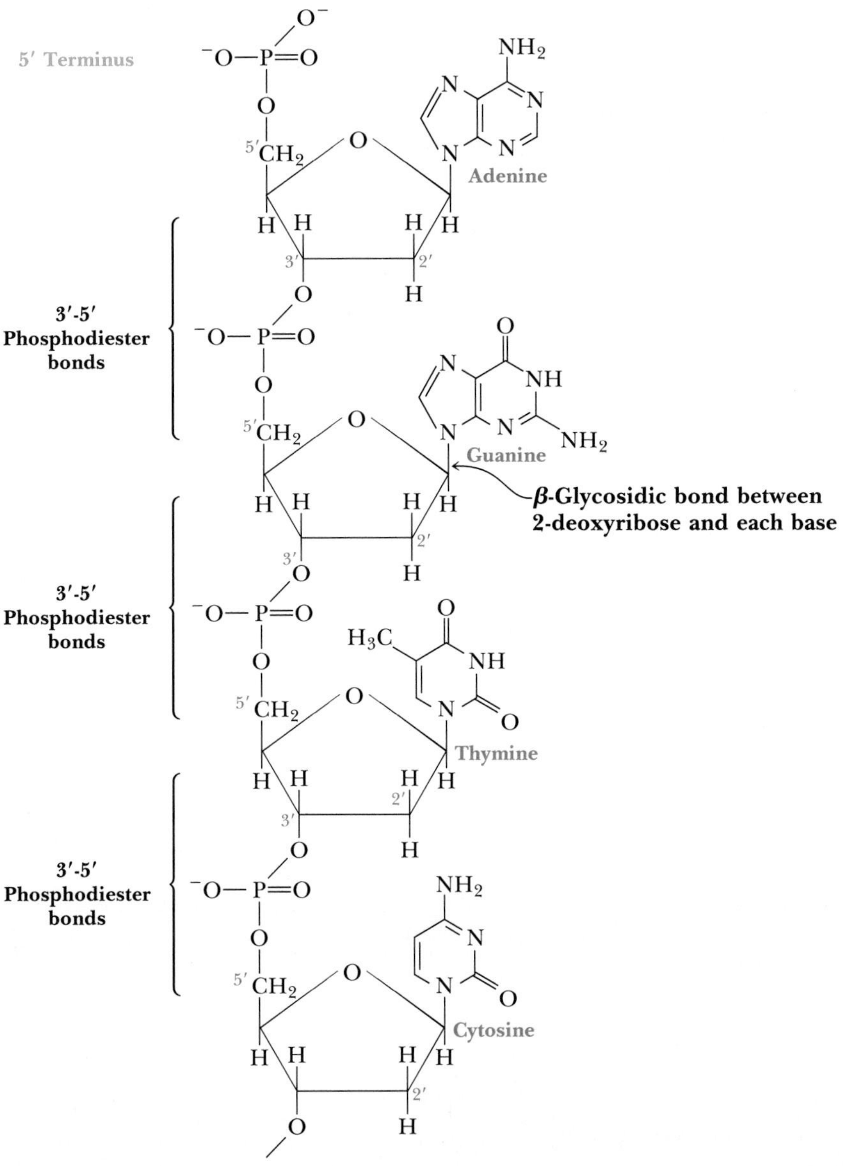

FIGURE 6.6

A portion of a DNA chain.

A portion of a DNA chain differs from the RNA chain just described only in the fact that the sugar is 2′-deoxyribose rather than ribose (Figure 6.6). In abbreviated notation the deoxyribonucleotide is specified in the usual manner. Sometimes a "d" is added to indicate a deoxyribonucleotide residue — e.g., dG is substituted for G — and the deoxy analogue of the ribooligonucleotide in the preceding paragraph would be pd(GACAT).

A nucleoside derivative that is very much in the news is 3'-azido-3'-deoxythymidine (AZT). This compound has shown promise in the treatment of AIDS (acquired immune deficiency syndrome), as has 2'-3'-dideoxyinosine (DDI). Propose a reason for the effectiveness of these two compounds. (*Hint:* How might these two compounds fit into a DNA chain?)

The structure of 3'-azido-3'-deoxythymidine (AZT) and 2',3'-dideoxyinosine (DDI).

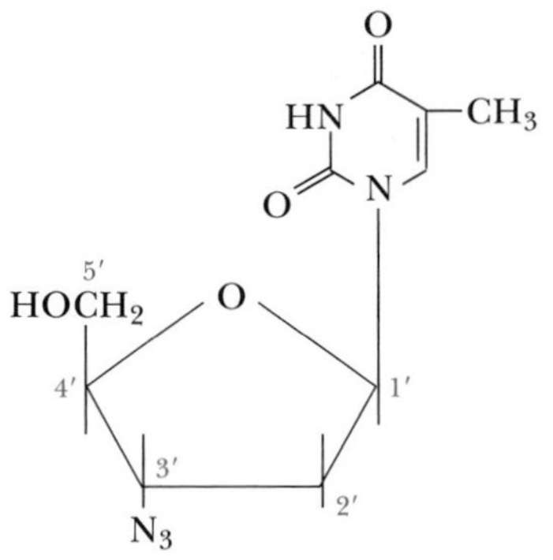

AZT

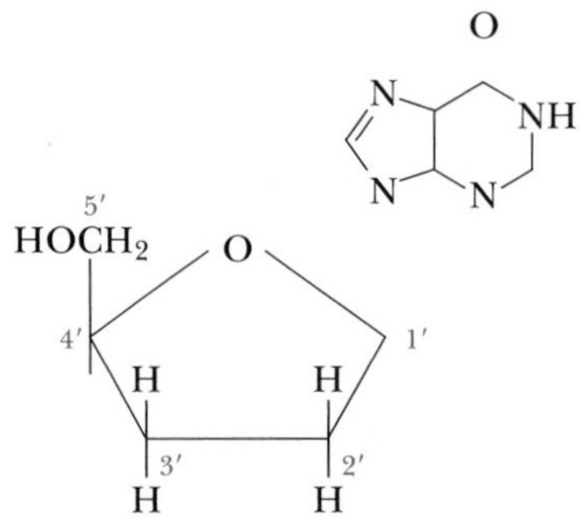

DDI

6.3
THE STRUCTURE OF DNA

Secondary Structure of DNA: The Double Helix

Representations of the double-helical structure of DNA have become common in the popular press as well as in the scientific literature. When this structure was proposed by Watson and Crick in 1953, it touched off a flood of research activity, leading to great advances in molecular biology. The determination of the double-helical structure was based on chemical analysis of DNA base composition and on x-ray diffraction patterns. Both of these lines of evidence were necessary for the conclusion that DNA consists of two polynucleotide chains wrapped around each other to form a helix. Hydrogen bonds between bases on opposite chains determine the alignment of the helix, with the paired bases lying in planes perpendicular to the helix axis. The sugar–phosphate backbone is the outer part of the helix (Figure 6.7). The chains run in antiparallel directions, one 3' to 5' and one 5' to 3'.

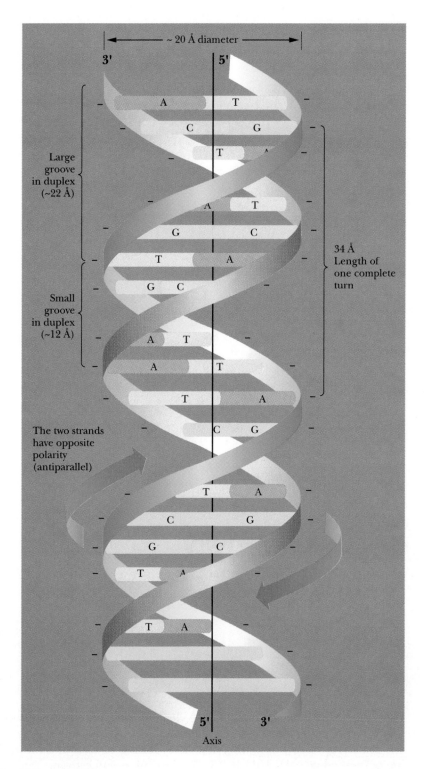

FIGURE 6.7

The double helix. A complete turn of the helix spans ten base pairs, covering a distance of 35 Å (3.4 nm). The individual base pairs are spaced 3.4 Å apart. The places where the strands cross hide base pairs that extend perpendicular to the viewer. The inside diameter is 11 Å, and the outside diameter is 20 Å. Within the cylindrical outline of the double helix are two grooves, a small one and a large one. Both are large enough to accommodate polypeptide chains. The minus signs alongside the strands represent the many negatively charged $-PO_2^-$ — groupings along the entire length of each strand.

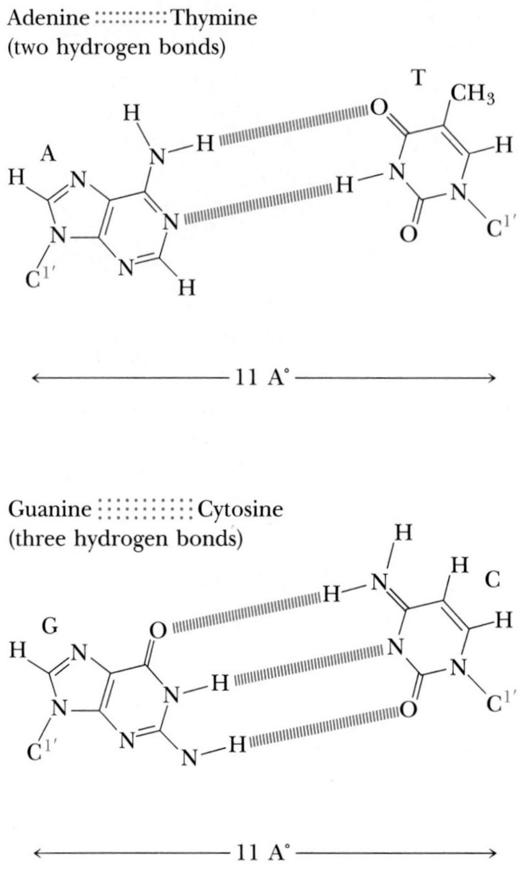

Adenine :::::::::: Thymine
(two hydrogen bonds)

Guanine :::::::::::: Cytosine
(three hydrogen bonds)

FIGURE 6.8

Base pairing. The adenine–thymine (A-T) base pair has two hydrogen bonds, whereas the guanine–cytosine (G-C) base pair has three hydrogen bonds.

The x-ray diffraction pattern of DNA demonstrated the helical structure. The combination of evidence from x-ray diffraction and chemical analysis led to the conclusion that the base pairing is **complementary,** meaning that adenine pairs with thymine, and guanine with cytosine. (Since complementary base pairing occurs along the entire double helix, the two chains are also referred to as *complementary strands*.) By 1953, studies of the base composition of DNA from many species had already shown that, to within experimental error, the mole percentages of adenine and thymine (moles of these substances as percentages of the total moles present) were equal; the same was found to be the case with guanine and cytosine. An adenine–thymine (A-T) base pair has two hydrogen bonds between the bases; a guanine–cytosine (G-C) base pair has three (Figure 6.8).

The inside diameter of the sugar–phosphate backbone of the double helix is about 11 Å (1.1 nm). The distance between the points of attachment of the bases to the two strands of the sugar–phosphate backbone is the same for the two base pairs (A-T and G-C), about 11 Å (1.1 nm), so these base pairs are the right size to fit into the helix. Base pairs other than A-T and G-C are possible, but they do not have the correct dimensions to fit the inside diameter of the double helix (Figure 6.8). The outside diameter of the helix is 20 Å (2 nm). The length of one complete turn of the helix along its axis is 34 Å (3.4 nm), containing ten base pairs. The atoms that make up the two polynucleotide chains of the double helix do not completely fill an imaginary cylinder around the double helix, but leave empty spaces known as grooves. There is a large *major groove* and a smaller *minor groove* in the double helix; both can be sites at which drugs or polypeptides bind to DNA (see Figure 6.7). At neutral, physiological pH, each phosphate group of the backbone carries a negative charge. Positively charged ions such as Na^+ or Mg^{2+} and polypeptides with positively charged side chains are frequently associated with DNA as a result of electrostatic attraction; eukaryotic DNA, for example, is complexed with histones, which are positively charged proteins, in the cell nucleus. The antiparallel relationship of the two polynucleotide chains is an aspect of the complementarity of the two strands (see Figure 6.7).

The form of DNA that we have been discussing so far is called B-DNA. It is thought to be the principal form that occurs in nature. However, other secondary structures can occur, depending on conditions such as the nature of the positive ion associated with the DNA. One of those other forms is A-DNA, which has 11 base pairs for each turn of the helix. Its base pairs are not perpendicular to the helix axis but lie at an angle of about 20° to the perpendicular, like the blades of a propeller (Figure 6.9). An important shared feature of A-DNA and B-DNA is that both are right-handed helices; that is, each helix winds upward in the direction in which the fingers of the right hand curl when the thumb is pointing upward (Figure 6.10).

A variant form of the double helix, Z-DNA, is left-handed; it winds in the direction of the fingers of the left hand (Figure 6.9). Z-DNA is known to occur in nature, most often when there are repeated G-C sequences, but its function has not been determined at this writing. It may play a role in the regulation of gene expression. The Z form of DNA is also a subject of active research among biochemists who want to design drugs that will bind specifically to it.

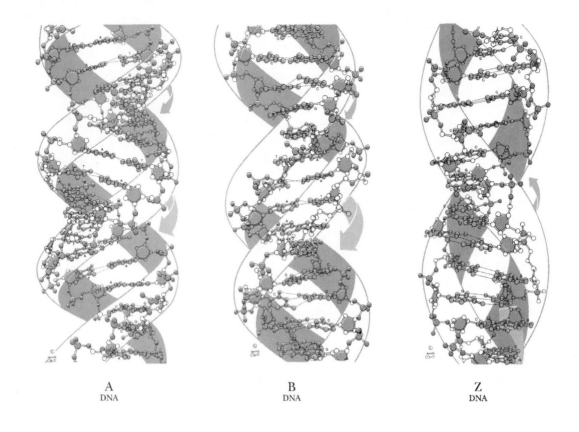

(a)

A
DNA

B
DNA

Z
DNA

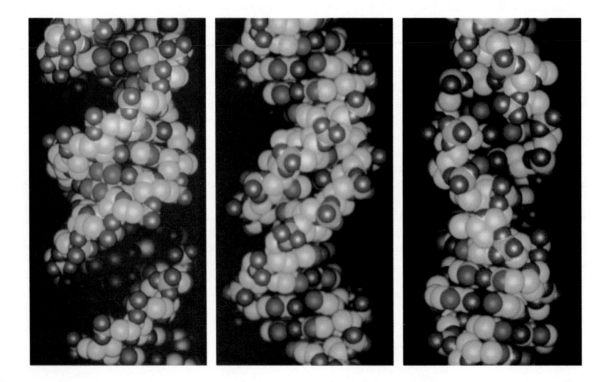

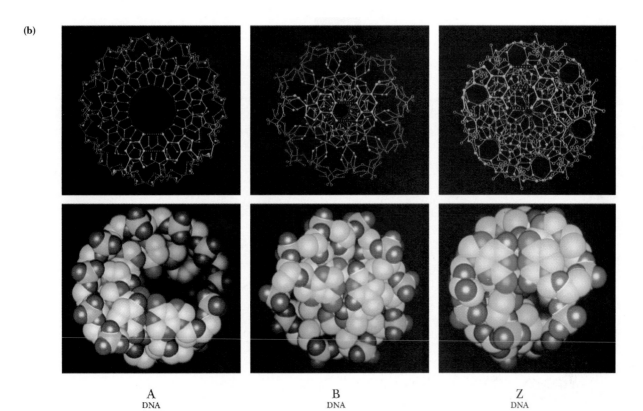

(b)

A
DNA

B
DNA

Z
DNA

FIGURE 6.9

A comparison of the A, B, and Z forms of DNA. Part (a) shows side views of all three, part (b), top views. Both parts include computer-generated space-filling models (bottom). The top half of both parts shows corresponding ball and stick drawings. In the A form, the base pairs have a marked propeller twist with respect to the helix axis. In the B form, the base pairs lie in a plane that is close to perpendicular to the helix axis. Z-DNA is a left-handed helix and in this respect differs from A-DNA and B-DNA, both of which are right-handed helices.

FIGURE 6.10

Right- and left-handed helices are related to each other in the same way as right and left hands.

Box
6.1

TRIPLE-HELICAL DNA: A TOOL FOR DRUG DESIGN

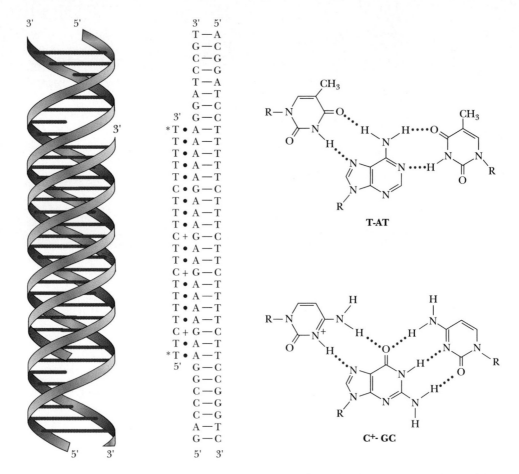

Model of a triple helix (*left*); schematic diagram of a triple helix complex (*middle*). C$^+$ is protonated cytosine. T* indicates the site of attachment of the third helix. (*Right*) The hydrogen bonding scheme for triple helix formation.

Triple-helical DNA was first observed in 1957 in the course of an investigation of synthetic polynucleotides, but for decades it remained a laboratory curiosity. Recent studies have shown that synthetic oligonucleotides (usually about 15 nucleotide residues long) will bind to specific sequences of naturally occurring double-helical DNA. The oligonucleotides are chemically synthesized to have the correct base sequence for specific binding. The oligonucleotide that forms the third strand fits into the major groove of the double helix and forms specific hydrogen bonds. When the third strand is in place, the major groove is inaccessible to proteins that might otherwise bind to that site — specifically, proteins that activate or repress expression of that portion of DNA as a gene. This behavior suggests a possible *in vivo* role for triple helices, especially in

view of the fact that hybrid triplexes with a short RNA strand bound to a DNA double helix are particularly stable. Research on ribozymes has revealed that RNA acts in many roles once thought exclusively reserved for proteins — in this case, roles involved with regulation of genetic expression, similar to the behavior of these synthetic oligonucleotides.

In another aspect of this work, researchers who have studied triple helices have synthesized oligonucleotides with reactive sites that can be positioned in definite places in DNA sequences. Such a reactive site can be used to modify or cleave DNA at a chosen point in a given sequence. This kind of specific cutting of DNA is crucial to recombinant DNA technology and to genetic engineering.

Tertiary Structure of DNA: Supercoiling

The DNA molecule has a length considerably greater than its diameter; it is not completely stiff and can fold back on itself in a manner similar to that of proteins as they fold into their tertiary structures. The double helix we have discussed so far is relaxed, which means that it has no twists in it other than the helical twists themselves. Further twisting and coiling, or *supercoiling,* of the double helix is possible. The first example of supercoiling we shall consider is the case in which DNA is not complexed to proteins. This "naked" DNA occurs in prokaryotes. (In eukaryotes, DNA *is* complexed to proteins of varied types, and the supercoiling pattern is different.)

Supercoiling in Prokaryotic DNA

If the sugar–phosphate backbone of a prokaryotic DNA forms a covalently bonded circle, the structure is still relaxed. Some extra twists are added if the DNA is unwound slightly before the ends are joined to form the circle. A strain is introduced in the molecular structure, and the DNA assumes a new conformation to compensate for the unwinding. If, because of unwinding, a right-handed double helix acquires an extra left-handed helical twist (a supercoil), the circular DNA is said to be *negatively* supercoiled (Figure 6.11a). Under different conditions it is possible to form a right-handed, or *positively,* supercoiled structure in which there is overwinding of the closed-circle double helix. The difference between the positively and negatively supercoiled forms lies in their right- and left-handed natures, which in turn depend on the overwinding or underwinding of the double helix.

Supercoiling has been observed experimentally in naturally occurring DNA. Particularly strong evidence has come from electron micrographs that clearly show coiled structures in circular DNA from a number of different sources, including bacteria, viruses, mitochondria, and chloroplasts (Figure 6.11b). It has been known for some time that prokaryotic DNA is normally circular, but supercoiling is a relatively recent subject of research. Comput-

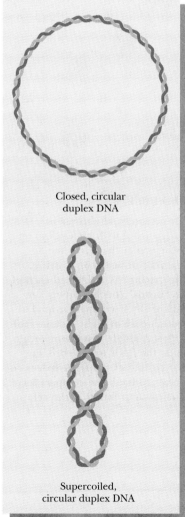

FIGURE 6.11

(a) Schematic representation of circular and supercoiled DNA. (b) Electron micrograph of super-coiled DNA.

(a)

Closed, circular
duplex DNA

Supercoiled,
circular duplex DNA

(b)

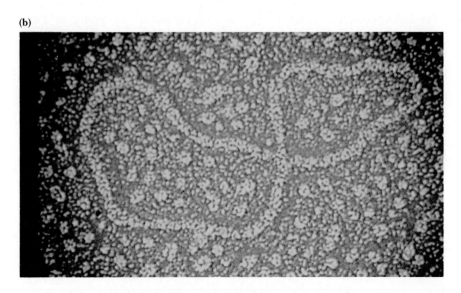

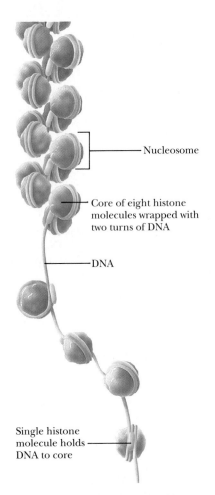

— Nucleosome

— Core of eight histone
molecules wrapped with
two turns of DNA

— DNA

Single histone
molecule holds —
DNA to core

FIGURE 6.12

The structure of chromatin. DNA is associated with histones in an arrangement that gives the appearance of beads on a string. The "string" is DNA, and each of the "beads" (nucleosomes) consists of DNA wrapped around a protein core of eight histone molecules. A single histone molecule holds the DNA to the core. Further coiling of the DNA spacer regions produces the compact form of chromatin found in the cell.

er modeling has helped scientists visualize many aspects of the twisting and knotting of supercoiled DNA, particularly by obtaining "stop-action" images of very fast changes.

Enzymes that affect the supercoiling of DNA have been isolated from a variety of organisms. Naturally occurring circular DNA is negatively supercoiled except during replication, when it becomes positively supercoiled. It is critical for the cell to have control over (regulate) this process. Two classes of enzymes are involved in the regulation, one to relax the supercoil and one to rewind it. **Topoisomerases** are enzymes that hydrolyze a phosphodiester linkage in one strand of the double helix, relax the supercoiling by rotating one strand around the other, and then reseal the break. **DNA gyrases** induce negative supercoiling in relaxed, closed-circular DNA.

Supercoiling in Eukaryotic DNA

The supercoiling of the nuclear DNA of eukaryotes (such as plants and animals) is more complicated than the supercoiling of the circular DNA from prokaryotes. Eukaryotic DNA is complexed with basic proteins that have abundant positively charged side chains at physiological (neutral) pH. Electrostatic attraction between the negatively charged phosphate groups on the DNA and the positively charged groups on the proteins favors the formation of complexes of this sort. The resulting material is called **chromatin.** The supercoiling of chromatin must accommodate the presence of the proteins.

The principal proteins in chromatin are the **histones,** of which there are five main types, called H1, H2A, H2B, H3, and H4. All these proteins contain large numbers of basic amino acid residues, such as lysine and arginine. In the chromatin structure, the DNA is tightly bound to all the types of histone but H1. The H1 protein is comparatively easy to remove from chromatin, but dissociating the other histones from the complex is more difficult. Proteins other than histones are also complexed with the DNA of eukaryotes, but they are neither as abundant nor as well studied as histones.

In electron micrographs, chromatin resembles beads on a string (Figure 6.12). This appearance reflects the molecular composition of the protein–DNA complex. Each "bead" is a **nucleosome** consisting of DNA wrapped around a histone core. This protein core is an octamer, which includes two molecules of each type of histone but H1; the composition of the octamer is $(H2A)_2(H2B)_2(H3)_2(H4)_2$. The "string" portions are called **spacer regions;** they consist of DNA complexed to some H1 histone and nonhistone proteins. As the DNA coils around the histones in the nucleosome, about 200 base pairs are in contact with the proteins; the spacer region is about 30 to 50 base pairs long.

6.4
DENATURATION OF DNA

We have already seen that the hydrogen bonds between base pairs are an important factor in holding the double helix together. The amount of stabilizing energy associated with one hydrogen bond is not great, but there are so many of them that the cumulative effect is considerable. In addition, the stacking of the bases in the native conformation of DNA contributes some stabilization energy. (It is generally thought that the stacking of the

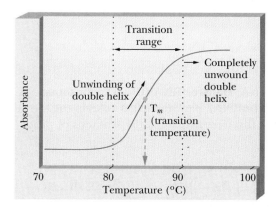

FIGURE 6.13

The experimental determination of DNA denaturation. This is a typical melting curve profile of DNA, depicting the hyperchromic effect observed on heating. The transition (melting) temperature, T_m, is displaced as the percentage of guanine and cytosine (the GC content) increases. (The entire curve would be shifted to the right for a DNA with higher GC content and to the left for a DNA with lower GC content.)

bases, which are aromatic, is due to hydrophobic interactions.) Energy must be added to a sample of DNA to break the hydrogen bonds and disrupt the stacking interactions. This is usually carried out by heating the DNA in solution.

The heat denaturation of DNA, also called **melting,** can be monitored experimentally by observing the absorption of ultraviolet light. The bases absorb light in the 260-nm wavelength region. As the DNA is heated and the strands separated, the wavelength of absorption does not change, but the amount of light absorbed increases (Figure 6.13). This effect is called **hyperchromicity.** It is based on the fact that the bases, which are stacked on top of one another in native DNA, become unstacked as the DNA is denatured. Since the bases interact differently in the stacked and unstacked orientations, their absorbance changes. Heat denaturation is a way to obtain single-stranded DNA (Figure 6.14), which has many uses. Some of them are discussed in Chapter 7.

Under a given set of conditions, there is a characteristic midpoint of the melting curve (the transition temperature, or melting temperature, written T_m) for DNA from each distinct source. The underlying reason for this property is that each type of DNA has a given, well-defined base composition. A G-C base pair has three hydrogen bonds, and an A-T base pair has only two. The higher the percentage of G-C base pairs, the higher the melting temperature of a DNA molecule.

Renaturation of denatured DNA is possible on slow cooling (Figure 6.14). The annealed, separated strands can recombine and form the base pairs responsible for maintaining the double helix.

FIGURE 6.14

The double helix unwinds when DNA is denatured, with eventual separation of strands. The double helix is re-formed on renaturation with slow cooling and annealing.

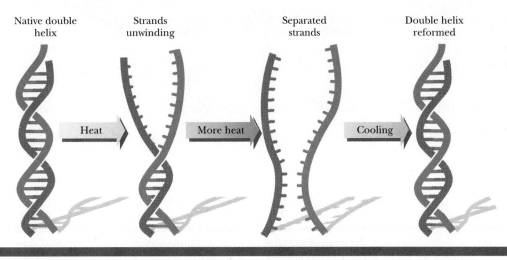

6.5
THE PRINCIPAL KINDS OF RNA
AND THEIR STRUCTURES

Three kinds of RNA, **transfer RNA (tRNA), ribosomal RNA (rRNA),** and **messenger RNA (mRNA),** play important roles in the life processes of cells (Figure 6.15). All three participate in the synthesis of proteins in a series of

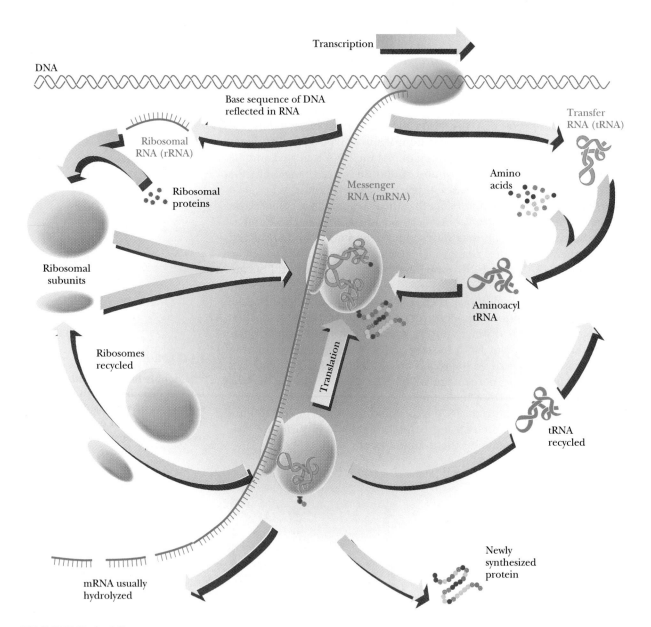

FIGURE 6.15

Flow chart showing the roles of various types of RNA. The synthesis of all types of RNA is directed by the base sequences of DNA. Ribosomal RNA associates with proteins to form ribosomes, the sites of protein synthesis. Transfer RNA binds specifically to amino acids and transports them to the site on the ribosome at which they are incorporated into proteins. Messenger RNA associates with the ribosome and specifies the site of incorporation of each amino acid into the growing protein.

TABLE 6.1 **The Roles of Different Kinds of RNA**

RNA TYPE	SIZE	FUNCTION	EXTENT OF HYDROGEN BONDING
Transfer RNA (tRNA)	Small	Transports amino acids to sites of protein synthesis	Considerable
Ribosomal RNA (rRNA)	Several kinds, variable in size	Combines with proteins to form ribosomes, the sites of protein synthesis	Considerable
Messenger RNA (mRNA)	Variable	Directs amino acid sequence of proteins	None

reactions ultimately directed by the base sequence of the cell's DNA. The base sequences of all types of RNA are determined by that of DNA. The process by which the order of bases is passed from DNA to RNA is called **transcription** (Chapter 19, Section 19.8).

Ribosomes, in which rRNA is associated with proteins, are the sites for assembly of the growing polypeptide chain in protein synthesis. Amino acids are brought to the assembly site covalently bonded to tRNA, as aminoacyl-tRNAs. The order of bases in mRNA specifies the order of amino acids in the growing protein; this process is called **translation** of the genetic message. A sequence of three bases in mRNA directs the incorporation of a particular amino acid into the growing protein chain. (We shall discuss the details of translation in Chapter 20, along with the genetic code, which specifies amino acids during translation.) Table 6.1 summarizes the types of RNA.

Transfer RNA

The smallest of the three important kinds of RNA is tRNA. Dozens of different species of tRNA molecules can be found in every living cell, since there is at least one tRNA that bonds specifically to each of the amino acids that commonly occur in proteins. Frequently there are several tRNA molecules for each amino acid. A RNA is a single-stranded polynucleotide chain, usually about 80 nucleotide residues long and generally having a molecular weight of about 25,000 daltons.

Intrachain hydrogen bonding occurs in tRNA, forming A-U and G-C base pairs similar to those that occur in DNA except for the substitution of uracil for thymine. The molecule folds back on itself in a kind of **cloverleaf structure,** which can be considered the secondary structure of tRNA (Figure 6.16). The hydrogen-bonded portions of the molecule are called *stems,* and the non-hydrogen-bonded portions are *loops.* Some of these loops contain

FIGURE 6.16

The cloverleaf structure of transfer RNA. Double-stranded regions (shown in red) are formed by folding of the molecule and stabilized by hydrogen bonds (| | |) between complementary base pairs. Peripheral loops are shown in yellow. There are three major loops (numbered) and one minor loop of variable size (not numbered).

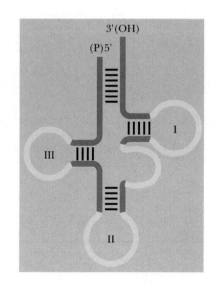

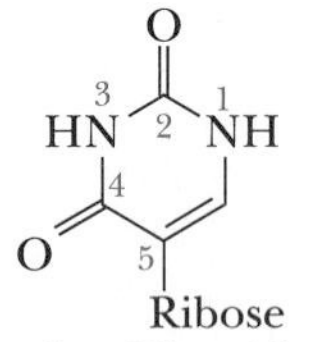

Pseudouridine (ψ)

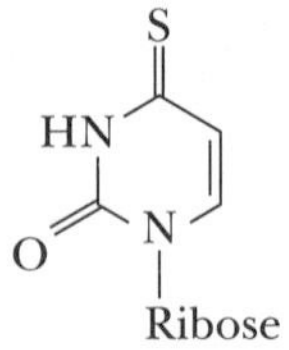

4-Thiouridine

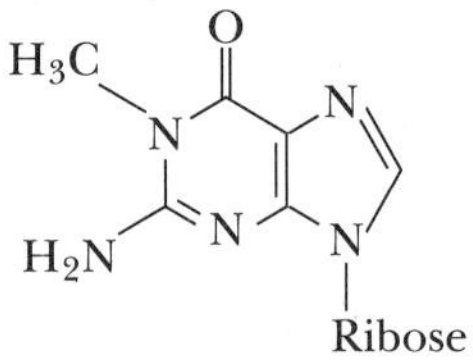

1-Methylguanosine (mG)

FIGURE 6.17

Structures of some modified bases found in transfer RNA. Note that in pseudouridine the pyrimidine is linked to ribose at C-5 rather than the usual N-1.

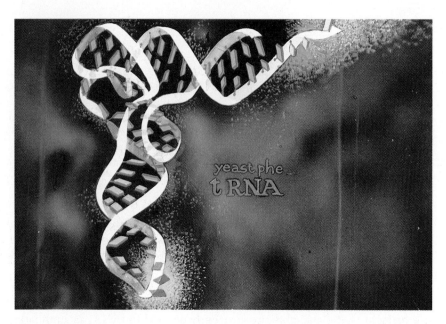

FIGURE 6.18

The three-dimensional structure of transfer RNA. The amino acid is attached at the upper end, shown in blue. The site at which tRNA is bound to mRNA in the course of protein synthesis is shown in green at the bottom of the figure.

modified bases (Figure 6.17). During protein synthesis, both tRNA and mRNA are bound to the ribosome in a definite spatial arrangement that ultimately ensures the correct order of the amino acids in the growing polypeptide chain.

A particular tertiary structure is necessary for tRNA to interact with the enzyme that covalently attaches the amino acid to the 3' end. To produce this tertiary structure, the cloverleaf folds into an L-shaped conformation that has been determined by x-ray diffraction (Figure 6.18).

Ribosomal RNA

In contrast with tRNA, rRNA molecules tend to be quite large, and only a few types of rRNA exist in a cell. Because of the intimate association between rRNA and proteins, a useful approach to understanding the structure of rRNA is to investigate ribosomes themselves.

The RNA portion of a ribosome accounts for 60 to 65% of the total weight, and the protein portion constitutes the remaining 35 to 40% of the weight. Dissociation of ribosomes into their components has proved to be a useful way of studying their structure and properties. A particularly important endeavor has been to determine both the number and the kind of RNA and protein molecules that make up ribosomes. This approach has helped elucidate the role of ribosomes in protein synthesis. In both prokaryotes and eukaryotes, a ribosome consists of two subunits, one larger than the other. In turn, the smaller subunit consists of 1 large RNA

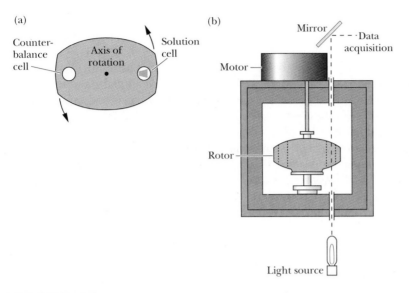

FIGURE 6.19

The analytical ultracentrifuge. (a) Top view of an ultracentrifuge rotor. The solution cell has optical windows; the cell passes through a light path once each revolution. (b) Side view of an ultracentrifuge rotor. The optical measurement taken as the solution cell passes through the light path makes it possible to monitor the motion of sedimenting particles.

molecule and about 20 different proteins; the larger subunit consists of 2 RNA molecules in prokaryotes (3 in eukaryotes) and about 35 different proteins in both prokaryotes and eukaryotes. The subunits are easily dissociated from one another in the laboratory by lowering the Mg^{2+} concentration of the medium. Raising the Mg^{2+} concentration to its original level reverses the process, and active ribosomes can be reconstituted by this method.

A technique called **analytical ultracentrifugation** has proved very useful for monitoring the dissociation and reassociation of ribosomes. Figure 6.19 shows an analytical ultracentrifuge. We need not consider all the details of the method as long as it is clear that the basic principle of the experiment is observation of the motion of ribosomes, RNA, or protein in a centrifuge. Both the size and the shape of a particle determine how fast it will move toward the bottom of the tube. The motion of the particle is characterized by a **sedimentation coefficient,** expressed in **Svedberg units (S),** which are named after Theodor Svedberg, the Swedish scientist who invented the ultracentrifuge. The S value increases with the molecular weight of the sedimenting particle, but it is not directly proportional to it, because the particle's shape also affects its sedimentation rate.

Ribosomes and ribosomal RNA have been studied extensively via sedimentation coefficients. Most research on prokaryotic systems has been done with the bacterium *Escherichia coli,* which we shall use as an example here. An *E. coli* ribosome typically has a sedimentation coefficient of 70S. When an intact 70S bacterial ribosome dissociates, it produces a light 30S subunit and a heavy 50S subunit. Note that the values of sedimentation

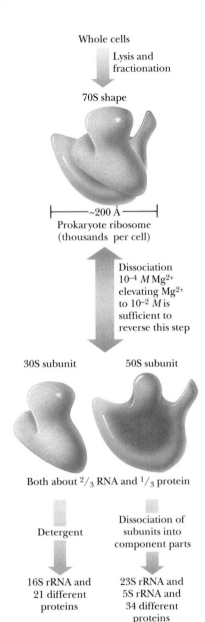

Whole cells

Lysis and
fractionation

70S shape

—~200 Å—

Prokaryote ribosome
(thousands per cell)

Dissociation
10^{-4} M Mg^{2+}
elevating Mg^{2+}
to 10^{-2} M is
sufficient to
reverse this step

30S subunit   50S subunit

Both about $^2/_3$ RNA and $^1/_3$ protein

Detergent

Dissociation of
subunits into
component parts

16S rRNA and
21 different
proteins

23S rRNA and
5S rRNA and
34 different
proteins

FIGURE 6.20

The subunit structure of ribosomes. The individual components can be mixed, producing functional subunits. Reassociation of subunits gives rise to an intact ribosome.

coefficients are not additive, showing the dependence of the S value on the shape of the particle. The 30S subunit contains a 16S rRNA and 21 different proteins. The 50S subunit contains a 5S rRNA, a 23S rRNA, and 34 different proteins (Figure 6.20). For comparison, eukaryotic ribosomes have a sedimentation coefficient of 80S, and the small and large subunits are 40S and 60S, respectively. The small subunit of eukaryotes contains an 18S rRNA, and the large subunit contains three types of rRNA molecules: 5S, 5.8S, and 28S.

The 5S rRNA has been isolated from many different types of bacteria, and the nucleotide sequences have been determined. A typical 5S rRNA is about 120 nucleotide residues long and has a molecular weight of about 40,000 daltons. Some sequences have also been determined for the 16S and 23S rRNA molecules. These larger molecules are about 1500 and 2500 nucleotide residues long, respectively. The molecular weight of 16S rRNA is about 500,000 daltons, and that of 23S rRNA is about 1 million daltons. There appear to be considerable degrees of secondary and tertiary structure in the larger RNA molecules. A secondary structure has been proposed for 16S rRNA (Figure 6.21), and suggestions have been made about the way in which the proteins associate with the RNA to form the 30S subunit. The *self-assembly of ribosomes* takes place in the living cell, and the process can be duplicated in the laboratory. Elucidation of ribosomal structure is an active field of research. The binding of antibiotics to ribosomal subunits so as to prevent self-assembly of the ribosome is one focus of the investigation. The structure of ribosomes is also one of the features used to compare and contrast eukaryotes, eubacteria, and archaebacteria (Chapter 1). (For more information on this subject, see the articles by Lake, especially the review article, listed in the bibliography at the end of this chapter.)

Messenger RNA

The least abundant of the three types of RNA is mRNA. In most cells it constitutes no more than 5 to 10% of the total cellular RNA, whereas tRNA accounts for 10 to 15% and the various types of rRNA compose 75 to 80% of the total. The sequences of bases in mRNA specify the orders of the amino acids in proteins. In rapidly growing cells, many different proteins are needed within a short interval. Fast turnover in protein synthesis becomes essential. Consequently, it is logical that mRNA is formed when it is needed, directs the synthesis of proteins, and then is degraded so that the nucleotides can be recycled. Of the three types of RNA—tRNA, rRNA, and mRNA—mRNA is the one that usually turns over most rapidly in the cell. Both tRNA and rRNA (as well as ribosomes themselves) can be recycled intact for many rounds of protein synthesis.

The sequence of mRNA bases that directs the synthesis of a protein reflects the sequence of DNA bases in the gene that codes for that protein. Messenger RNA molecules are heterogeneous in size, as are the proteins whose sequences they specify. There is probably no intrachain folding in mRNA; it is very likely an open chain. It is also likely that several ribosomes are associated with a single mRNA molecule at some time during the course of protein synthesis.

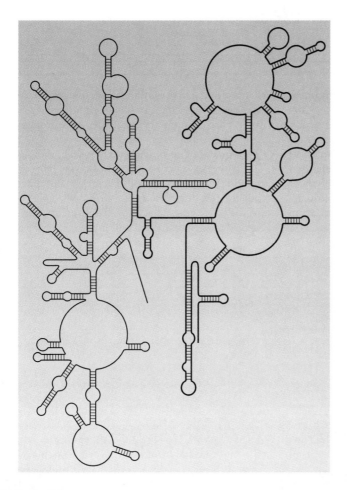

FIGURE 6.21

A proposed secondary structure for 16S rRNA. The intrachain folding pattern includes loops and double-stranded regions.

6.6
STRATEGY FOR DETERMINING THE BASE SEQUENCE OF NUCLEIC ACIDS

We have already seen that the primary structure of a protein determines its secondary and tertiary structures. The same is true of nucleic acids; the nature and order of monomer units determine the properties of the whole molecule. The base pairing in both RNA and DNA depends on a series of complementary bases, whether these bases are on different polynucleotide strands, as in DNA, or on the same strand, as is frequently the case in RNA. The fact that the number of commonly occurring bases in nucleic acids is less than the number of amino acids in proteins has made the determination of the primary structure of nucleic acids a challenging task. The situation is

Box 6.2

RESTRICTION NUCLEASES: "MOLECULAR SCISSORS"

Many enzymes act on nucleic acids. A group of specific enzymes acts in concert to ensure the faithful replication of DNA, and another group directs the transcription of the base sequence of DNA into that of RNA. (We shall need all of Chapter 19 to describe the manner in which these enzymes operate.) Other enzymes, called **nucleases,** catalyze the hydrolysis of the phosphodiester backbones of nucleic acids. Some nucleases are specific for DNA, others for RNA. Cleavage from the ends of the molecule (by exonucleases) is known, as is cleavage in the middle of the chain (by endonucleases). Some enzymes are specific for single-stranded nucleic acids, and others cleave double-stranded ones. One group of nucleases, **restriction nucleases,** has played a crucial role in the development of recombinant DNA technology.

This class of enzymes was discovered in the course of genetic investigations of bacteria and **bacteriophages** (**phages** for short; from the Greek *phagein,* "to eat"), the viruses that infect bacteria. The researchers noted that bacteriophages that grew well in one strain of the bacterial

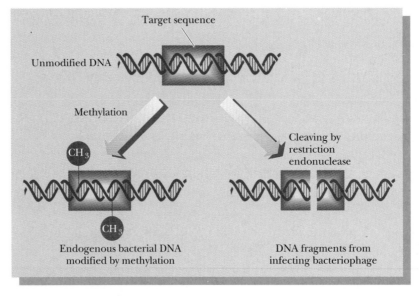

Target sequence

Unmodified DNA

Methylation

CH₃

CH₃

Endogenous bacterial DNA modified by methylation

Cleaving by restriction endonuclease

DNA fragments from infecting bacteriophage

Methylation of endogenous DNA protects it from cleavage by its own restriction endonucleases.

analogous to a jigsaw puzzle in which all the pieces are the same color, which is more challenging than a jigsaw puzzle with a colored pattern that helps to relate the pieces. Sequencing of nucleic acids is now fairly routine, and this relative ease would have amazed the scientists of the 1950s and 1960s.

The methods designed specifically for sequencing DNA depend on its unique properties, particularly on how it is replicated. There are, however, some similarities to the methods for sequencing proteins. For instance, in both processes an early step is the hydrolysis of the macromolecule (protein or DNA) with specific enzymes to produce large chunks of manageable size. In the case of DNA, the enzymes in question are restriction nucleases (Box 6.2). Separation methods for oligo- and polynucleotides are also important. We shall discuss experimental methods for working with them and with nucleic acids in Chapter 7.

RESTRICTION NUCLEASES: "MOLECULAR SCISSORS"

species they infected frequently grew poorly (had *restricted* growth) in another strain of the same species. Further work showed that this phenomenon arises from a subtle difference between the phage DNA and the DNA of the bacterial strain in which phage growth is restricted. This difference is the presence of methylated bases at certain sequence-specific sites in the host DNA and not in the viral DNA.

The growth-restricting host cells contain cleavage enzymes, the restriction nucleases, that produce double-chain breaks at the unmethylated specific sequences in phage DNA; the cells' own corresponding DNA sequences, in which methylated bases occur, are not attacked. These cleavage enzymes consequently degrade DNA from any source *but* the host cell. The most immediate consequence is a slowing of the growth of the phage in that bacterial strain, but it is important for our discussion that DNA from any source can be cleaved by such an enzyme if it contains the target sequence. More than 800 restriction endonucleases have been discovered in a variety of bacterial species. More than 100 specific sequences are recognized by one or more of these enzymes. The following table shows several target sequences.

It is also noteworthy that most of these double-chain cleavages are separated by several base pairs, leaving short single-stranded segments that can base-pair with complementary segments from any DNA that has been treated with the same enzyme. This fact makes it possible to join DNA segments from different organisms; the only requirement is that they have the base sequence recognized by the appropriate restriction nuclease.

ENZYME*	RECOGNITION AND CLEAVAGE SITE
*Bam*HI	5'-GGATCC-3' 3'-CCTAGG-5'
*Eco*RI	5'-GAATTC-3' 3'-CTTAAG-5'
*Hae*III	5'-GGCC-3' 3'-CCGG-5'
*Hin*dIII	5'-AAGCTT-3' 3'-TTCGAA-5'
*Hpa*II	5'-CCGG-3' 3'-GGCC-5'
*Not*I	5'-GCGGCCGC-3' 3'-CGCCGGCG-5'
*Pst*I	5'-CTGCAG-3' 3'-GACGTC-5'

Arrows indicate the phosphodiester bonds cleaved by the restriction nucleases.
*The name of the restriction nuclease consists of a three-letter abbreviation of the bacterial species from which it is derived — for example, *Eco* for *Escherichia coli.*

6.7
RECOMBINANT DNA: THE BASIS OF GENETIC ENGINEERING

DNA molecules containing covalently linked segments derived from two or more DNA sources are called **recombinant DNA.** (Another name for recombinant DNA is **chimeric DNA,** after the chimera, a monster in Greek mythology that has the head of a lion, the body of a goat, and the tail of a serpent.) Producing recombinant DNA requires cutting and splicing two strands of DNA molecules in very specific ways. The cutting is done by restriction enzymes so as to have a convenient way to produce matching end sequences in DNA molecules from different sources. With the end sequences matched up, the splicing is done by a group of enzymes called **DNA ligases.**

Many Restriction Enzymes Produce "Sticky Ends"

Each restriction nuclease hydrolyzes only a specific bond of a specific sequence in DNA. The sequences recognized by restriction enzymes — their sites of action — read the same from left to right as they do from right to left (on the complementary strand). The term for such a sequence is a **palindrome.** ("Able was I ere I saw Elba" and "Madam I'm Adam" are well-known linguistic palindromes.) A typical restriction enzyme called *Eco*RI is isolated from *E. coli* (each restriction nuclease is designated by an abbreviation of the name of the organism in which it occurs). The *Eco*RI site in DNA is 5'-GAATTC-3', where the base sequence on the other strand is 3'-CTTAAG-5'. The sequence from left to right on one strand is the same as the sequence from right to left on the other strand. The phosphodiester bond between G and A is the one hydrolyzed. This same break is made on both strands of the DNA. There are four nucleotide residues — two adenine and two thymine in each strand — between the two breaks on opposite strands, leaving the **sticky ends,** which can still be joined by hydrogen bonding between the complementary bases. With the ends held in place by the hydrogen bonds, the two breaks can then be resealed covalently by the action of DNA ligases (Figure 6.22). If no ligase is present, the ends can

FIGURE 6.22

Hydrolysis of DNA by restriction endonucleases. (a) Separation of ends. (b) Resealing of ends by ligase.

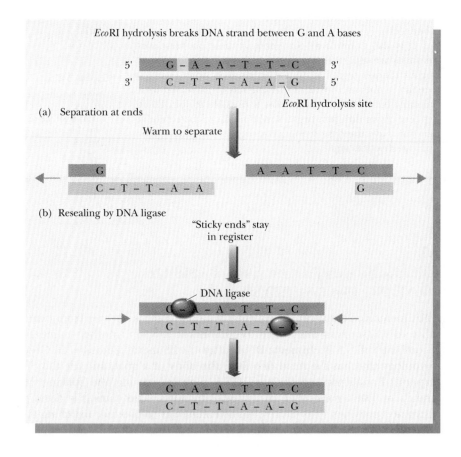

remain separated, and the hydrogen bonding at the sticky ends holds the mocule together until gentle warming or vigorous stirring effects a separation. Blunt-end cuts and cuts with less than absolute specificity as to sequence and site also occur. See Box 6.2 for an example: *Hae*III produces a blunt-end cut.

$$
\begin{array}{ccc}
\text{5'-GG}\overset{\downarrow}{\text{C}}\text{C-3'} & \xrightarrow{\textit{Hae} \text{ III}} & \text{5'-GG CC-3'} \\
\text{3'-CCGG-5'} & & \text{3'-CC GG-5'} \\
\underset{\uparrow}{} & & \text{Blunt-end cut}
\end{array}
$$

DNA from different sources can be treated with the same restriction enzyme, producing identical sequences of overlapping, sticky ends. The overlapping sequences from the two different DNAs can hydrogen bond with each other and then be joined covalently by DNA ligases. Techniques have recently been developed for the attachment of synthetic complementary "linkers" to the ends of DNAs that are to be recombined. This development liberates researchers from dependence on naturally occurring sites for the cut-and-splice process. Other methods have been developed to cut DNA exclusively at chosen sites. This enables researchers to obtain larger pieces of DNA than can be obtained with restriction enzymes.

Some of these techniques involve the use of a triple helix with a reactive site on the chemically synthesized third strand, designed to bind to a specific DNA sequence. For more details, see the articles by Roberts and by Maher in the bibliography at the end of this chapter.

Using "Sticky Ends" to Construct Recombinant DNA

If DNA from two different organisms has recognition sites for the same restriction nuclease, the two kinds of DNA will have the same kind of sticky ends as a result of treatment with that enzyme. If digested samples of DNA from the two sources are mixed, in some cases the sticky ends that anneal to one another will be from different sources. The nicks in the covalent structure can be sealed with DNA ligase, producing recombinant DNA (Figure 6.23).

In the laboratory, when two different kinds of DNA are combined, one of the DNAs is normally from a viral or bacterial source. The virus is usually a bacteriophage; the bacterial DNA typically is derived from a **plasmid,** a circular DNA molecule that is not part of the main circular DNA chromosome of the bacterium. Using DNA from a viral or bacterial source as one of the components of a recombinant DNA enables scientists to take advantage of the rapid growth of viruses and bacteria and obtain greater amounts of the recombinant DNA. This process of making many identical copies of DNA is called **cloning.** Experimental procedures for obtaining and cloning recombinant DNAs are described in Chapter 7, Section 7.3.

Genetic Engineering

In a sense, genetic engineering on an organismal level has been around since humans first started to use selective breeding on plants and animals. This procedure did not deal directly with the molecular nature of genetic material, nor was the appearance of traits under human control. Breeders had to cope with changes that arose spontaneously, and the only choice was

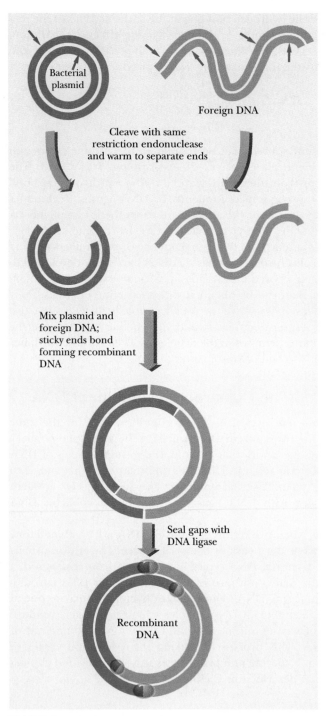

FIGURE 6.23

The methodology for producing recombinant DNA.

(a)

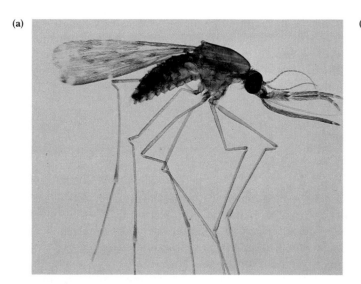

(b)

FIGURE 6.24

(a) Side view of *Anopheles gambiae* mosquito. (b) Two adult female *Anopheles gambiae* mosquitoes (front view). The one on the left is a mutant. Scientists are attempting to produce strains of these mosquitoes, which are unable to transmit malaria to humans, in hopes that they will replace the malaria carriers.

whether to breed for a trait or to let it die out. An understanding of the molecular nature of heredity and the ability to manipulate those molecules in the laboratory have, of course, changed all that.

The practice of selective alteration of organisms for both agricultural and medical purposes has profited greatly from recombinant DNA methods. Genetic engineering of crop plants is an active field of research. Genes for increased yields, frost resistance, and resistance to pests are introduced into commercially important plants such as strawberries, tomatoes, and maize. Similarly, animals of commercial importance—mostly mammals, but including fish—are genetically altered as well. Some variations introduced in animals have medical implications. Mice with altered genetic makeup are used in the research laboratory. In another medically related field, researchers working with insect-borne diseases such as malaria are trying to engineer strains of insects, such as the mosquito *Anopheles gambiae,* that can no longer transmit the infection to humans (Figure 6.24). In all these cases, the focus of research is to introduce *traits that can be inherited* by the descendants of the treated organisms. Chapter 7, Section 7.7, includes a discussion of methods for carrying out these projects.

In the treatment of human genetic disease, however, the aim is not to produce heritable changes. Serious ethical questions arise with the manipulation of human genetics, and consequently the focus of research has been on forms of **gene therapy** in which cells of specific tissues in a living person are altered in a way that alleviates the disease. Examples of diseases that may someday be treated in this way include cystic fibrosis, hemophilia, Duchenne muscular dystrophy, and severe combined immune deficiency (SCID). The last of these is also known as the bubble syndrome, because its victims must live in isolation (in a large "bubble") to avoid infection. The fundamental cause of SCID is a deficiency in the enzyme adenosine deaminase; the treatment, currently in clinical trials, is to introduce into the patient's body new cells that can make this enzyme (see Chapter 7, Section 7).

6.8
DNA RECOMBINATION OCCURS IN NATURE

When recombinant DNA technology was in its early stages in the 1970s, considerable concern arose both about safety and about ethical questions. Some of the ethical questions are still matters of concern. One that has definitely been laid to rest is the question of whether the process of cutting and splicing DNA is an unnatural process. Indeed, DNA recombination is a common part of the natural crossing over of chromosomes. Figure 6.25 shows mechanisms for DNA recombination. There are many, varied reasons for *in vivo* recombination of DNA, two of which are the maintenance of genetic diversity and the repair of damaged DNA.

Two different DNA molecules can recombine to form a third DNA that is itself different from each of the original DNA molecules. This is DNA recombination, or **genetic recombination,** a process in which *regions of homology* between the original DNA molecules play an important role. In other words, there have to be matching stretches of base sequence in the two DNA molecules for the process to take place with maximum efficiency. (The reason for this requirement will become apparent.) Both the original DNA molecules are nicked; the nicks are eventually sealed in a different arrangement. The details of the process can be carried out in a variety of ways, but the manner we shall describe here is typical of genetic recombination. An endonuclease nicks two homologous chromosomes on different DNA molecules. The two nicked strands cross over, and each base pairs with the complementary strand on the other DNA. A suitable ligase seals the nicked ends in their crossed-over form. As a result of unwinding and rewinding of the helix, the branch point — the point at which the strands cross — migrates. During this migration step, two single strands that were originally parts of different DNA molecules form a double helix, because they have complementary base sequences. This fact is the basis of the requirement for sequence homology, mentioned earlier.

The recombined product can also be drawn in an X-shaped form, which will be more useful for future discussion. A rotation of half the crossed-over product gives the untangled chi intermediate, which gets its name from its resemblance to the Greek letter chi (χ). The chi intermediate has two possible fates that depend on the orientation of the two nicks, followed by sealing. The result of one orientation is two DNA molecules, each with one strand that is the same as the original DNA strand and one strand that incorporates parts of both original DNA strands. This product is thus described as **single-strand heterozygous.** The other possible orientation of two nicks, followed by sealing, gives rise to two DNA molecules in which both strands incorporate parts of both original DNA strands. This product is known as **double-strand heterozygous.**

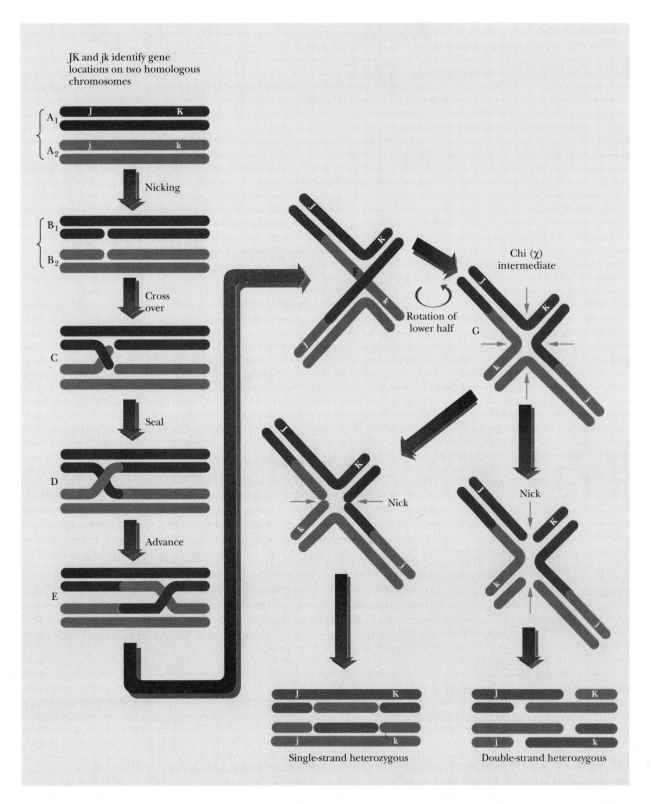

FIGURE 6.25

Mechanisms for DNA recombination.

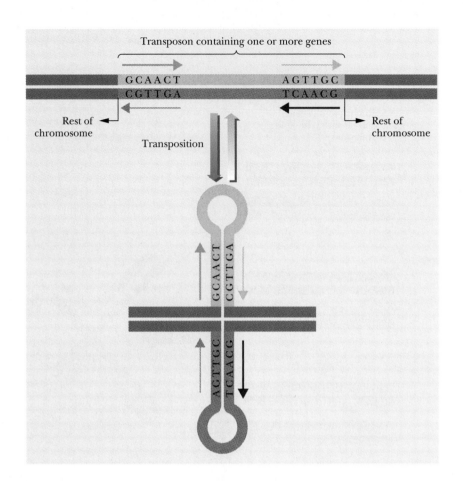

FIGURE 6.26

A possible mechanism for transposition of DNA segments.

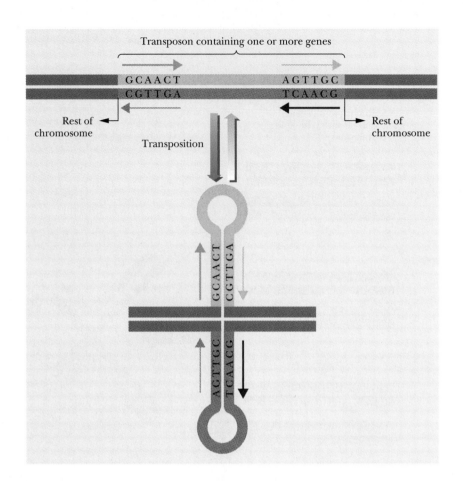

The concept of **transposition of DNA** within a chromosome or from one chromosome to another ("jumping genes") was greeted with general skepticism when it was first proposed, but it is now well established that genes and clusters of genes can change position. Both prokaryotes and eukaryotes are subject to gene transposition. This phenomenon plays a part in the rapid development of mutations in bacteria, including the appearance of antibiotic resistance. The movable segments of DNA are called **transposable elements** (or **transposons**), and they begin and end with an inverted repeat sequence. The existence of the inverted repeat sequence could allow the transposon to loop out from the rest of the DNA, but this has not been established with certainty (Figure 6.26). Many questions remain about DNA transposition, and research is under way to elucidate the details of the process.

SUMMARY

The double helix originally proposed by Watson and Crick is the most striking feature of DNA structure. The two coiled strands run in antiparallel directions and are held together by hydrogen bonds between complementary bases. Adenine pairs with thymine and guanine with cytosine. Supercoiling is a feature of DNA structure in both prokaryotes and eukaryotes.

Eukaryotic DNA is complexed with histones and other basic proteins, whereas prokaryotic DNA occurs in "naked" form, not complexed to proteins.

When DNA is denatured, the double-helical structure breaks down; the progress of this phenomenon can be followed by monitoring the absorption of ultraviolet light. The temperature at which DNA becomes denatured by heat depends on its base composition; higher temperatures are needed to denature DNA rich in G-C base pairs.

The three kinds of RNA—transfer RNA (tRNA), ribosomal RNA (rRNA), and messenger RNA (mRNA)—differ somewhat in structure. Transfer RNA is relatively small, about 80 nucleotides long. It exhibits extensive intrachain hydrogen bonding, represented by a cloverleaf structure. Ribosomal RNA molecules tend to be quite large and are complexed with proteins to form ribosomal subunits. Ribosomal RNA also exhibits extensive internal hydrogen bonding. The sequence of bases in a given mRNA determines the sequence of amino acids in a specified protein. The size of mRNA molecules varies with the size of the protein. No evidence exists for extensive hydrogen bonding in mRNA.

Restriction enzymes also have an important role in recombinant DNA technology. These endonucleases produce short single-stranded stretches, called sticky ends, at the ends of cleaved DNA. The sticky ends provide a way to link DNAs from different sources, even to the extent of inserting eukaryotic DNA into bacterial genomes. DNA recombination occurs in nature and plays a particularly important role in the functioning of transposable elements within the genome.

E X E R C I S E S

1. Why does DNA with a high A-T content have a lower transition temperature (T_m) than DNA with a high G-C content?
2. Which of the following statements is (are) true?
   (a) Bacterial ribosomes consist of 40S and 60S subunits.
   (b) Prokaryotic DNA is normally complexed with protein.
   (c) Prokaryotic DNA normally exists as a closed circle.
   (d) Circular DNA is supercoiled.
3. Binding sites for the interaction of polypeptides and drugs with DNA are found in the major and minor grooves. True or false?
4. Which of the following statements is (are) true?
   (a) The two strands of DNA run parallel from their 5' to their 3' ends.
   (b) An adenine–thymine base pair contains three hydrogen bonds.
   (c) Positively charged counterions are associated with DNA.
   (d) DNA base pairs are always perpendicular to the helix axis.
5. Define the following terms:
   (a) Supercoiling   (c) Topoisomerase
   (b) Positive supercoil   (d) Negative supercoil
6. Briefly describe the structure of chromatin.
7. Sketch a typical cloverleaf structure for transfer RNA. Point out any similarities between the cloverleaf pattern and the proposed structures of ribosomal RNA.
8. Show via an example how a restriction endonuclease produces sticky ends in DNA.
9. Give an example of a palindromic sequence in DNA.
10. What is the role of the chi intermediate in genetic recombination?
11. Define the term "transposon." Outline the role of transposons in gene rearrangement.
12. Would you expect tRNA or mRNA to be more extensively hydrogen bonded? Why?
13. Would you expect mRNA or rRNA to be degraded more quickly in the cell? Why?
14. Would you expect to find adenine–guanine or cytosine–thymine base pairs in DNA? Why?
15. A friend tells you that only four different kinds of bases are found in RNA. What would you say in reply?
16. One of the original structures proposed for DNA had all the phosphate groups positioned at the center of a long fiber. Give a reason why this proposal was rejected.

A N N O T A T E D   B I B L I O G R A P H Y

Adams, R. L. P., J. T. Knowles, and D. P. Leader. *The Biochemistry of the Nucleic Acids.* 11th ed. New York: Chapman and Hall, 1992. [New authors have prepared this edition of a classic text originally written by J. N. Davidson.]

Anderson, W. F., and E. G. Diacumakos. Genetic Engineering in Mammalian Cells. *Sci. Amer.* **245** (1), 106–121 (1981). [A description of recombinant DNA techniques.]

Barton, J. K. Recognizing DNA Structures. *Chem. Eng. News* **66**, 30–42 (1988). [An account of the conformations of

DNA and how they change on drug binding.]

Bauer, W. R., F. H. C. Crick, and J. H. White. Supercoiled DNA. *Sci. Amer.* **243** (1), 118–133 (1980). [A description of the types of circular DNA.]

Berg, P., and M. Singer. *Dealing with Genes: The Language of Heredity.* Mill Valley, CA: University Science Books, 1992. [Two leading biochemists have produced an eminently readable book on molecular genetics; highly recommended.]

Brimacombe, R. The Emerging Three-Dimensional Structure and Function of 16S Ribosomal RNA. *Biochemistry* **27**, 4208–4212 (1988). [A short review.]

Darnell, J. E. The Processing of RNA. *Sci. Amer.* **249** (2), 90–100 (1983). [A discussion of the ways in which RNA is modified after synthesis.]

Felsenfeld, G. DNA. *Sci. Amer.* **253** (4), 58–67 (1985). [A review of the main features of DNA structure.]

Grunstein, M. Histones as Regulators of Genes. *Sci. Amer.* **267** (4), 68–74B (1992). [A report on a regulatory as well as structural role for histones.]

Holley, R. W. The Nucleotide Sequence of a Nucleic Acid. *Sci. Amer.* **214** (2), 30–46 (1966). [A description of the original methods used for sequencing transfer RNA. Of historical interest.]

Kornberg, R. D., and A. Klug. The Nucleosome. *Sci. Amer.* **244** (2), 52–64 (1981). [A discussion of the way in which DNA and histone proteins associate with each other.]

Lake, J. A. Evolving Ribosome Structure: Domains in Archaebacteria, Eubacteria, Eocytes and Eukaryotes. *Ann. Rev. Biochem.* **54**, 507–530 (1985). [A review of the evolutionary implications of ribosome structure.]

Lake, J. A. The Ribosome. *Sci. Amer.* **245** (2), 84–97 (1981). [A look at some of the complexities of ribosome structure.]

Maher, L., B. Wold, and P. Dervan. Inhibition of DNA Binding Proteins by Oligonucleotide-Directed Triple Helix Formation. *Science* **245**, 725–730 (1989). [Triple helices and their role in transcription.]

Maxam, A. M., and W. Gilbert. A New Method for Sequencing DNA. *Proc. Natl. Acad. Sci. USA* **74**, 560–564 (1977). [The original article describing the direct chemical method for sequencing DNA. Mainly of historical interest.]

Moffat, A. Triplex DNA Finally Comes of Age. *Science* **252**, 1374–1375 (1991). [Triple helices as "molecular scissors."]

Nomura, M. The Control of Ribosome Synthesis. *Sci. Amer.* **250** (2), 102–114 (1984). [The assembly of ribosomes.]

Prive, G., U. Heinemann, S. Chandrasegaran, L. Kan, M. Kopa, and R. Dickerson. Helix Geometry, Hydration, and G·A Mismatch in a B-DNA Decamer. *Science* **238**, 498–504 (1987). [How different base sequences give DNA slightly different conformations.]

Rennie, J. DNA's New Twists. *Sci. Amer.* **268** (3), 122–132 (1992). [A description of possible variations on the theme of replication, transcription, and translation.]

Roberts, R., and D. Crothers. Stability and Properties of Double and Triple Helices: Dramatic Effects of RNA or DNA Backbone Composition. *Science* **258**, 1463–1466 (1992). [The nature of the sugar plays an important role in the conformation of the triple helix.]

Roberts, L. New Scissors for Cutting Chromosomes. *Science* **249**, 127 (1990). [A *Research News* article about a method for choosing exactly the right place to cut DNA into large pieces.]

Ross, J. The Turnover of Messenger RNA. *Sci. Amer.* **260** (4), 48–55 (1989). [The regulation of the rate at which messenger RNA is degraded in the cell.]

Saenger, W. *Principles of Nucleic Acid Structure.* New York: Springer-Verlag, 1984. [A fairly advanced treatment of the subject.]

Schleif, R. DNA Binding by Proteins. *Science* **241**, 1182–1187 (1988). [Structural motifs in DNA-binding proteins.]

Schlick, T., and W. Olson. Trefoil Knotting Revealed by Molecular Dynamics Simulations of Supercoiled DNA. *Science* **257**, 1110–1115 (1992). [The application of computer techniques to the study of DNA supercoiling.]

Scovell, W. M. Supercoiled DNA. *J. Chem. Ed.* **63**, 562–565 (1986). [A discussion focused mainly on the topology of circular DNA.]

Sinden, R. A. Supercoiled DNA: Biological Significance. *J. Chem. Ed.* **64**, 294–301 (1987). [The enzymes involved in DNA supercoiling.]

Smith, L. M., J. Z. Sanders, R. J. Kaiser, et al. Fluorescence Detection in Automated DNA Sequence Analysis. *Nature* **321**, 674–679 (1986). [The first report of an automated method for sequencing of DNA.]

Wasserman, S. A., and N. R. Cozzarelli. Biochemical Topology: Applications to DNA Recombination and Replication. *Science* **232**, 951–960 (1986). [A discussion of the ways in which DNA recombination and replication can produce knotted DNA structures.]

Watson, J. D., and F. H. C. Crick. Molecular Structure of Nucleic Acid. A Structure for Deoxyribose Nucleic Acid. *Nature* **171**, 737–738 (1953). [The original article describing the double helix. Of historical interest.]

Westheimer, F. H. Why Nature Chose Phosphates. *Science* **235**, 1173–1178 (1987). [A discussion of the importance of phosphate groups in biochemistry, particularly in the backbones of nucleic acids. The author is an eminent organic chemist.]

Most textbooks of organic chemistry have a chapter on nucleic acids.

Chapter 7

Nucleic Acid Biotechnology Techniques

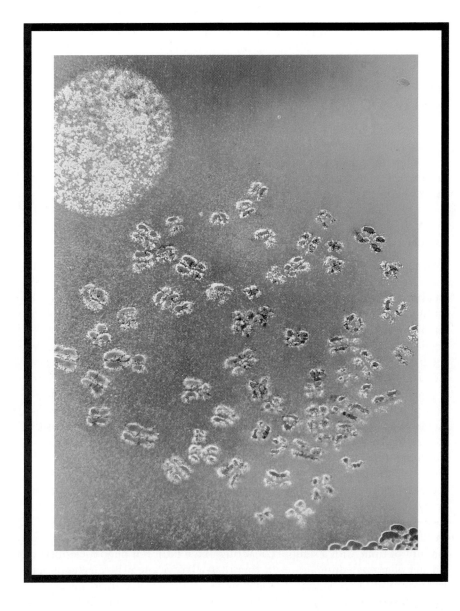

Human chromosomes viewed through a scanning electron microscope.

Methods of manipulating nucleic acids take advantage of their unique properties, particularly those of DNA. Methods of determining the base sequence of DNA make use of the way in which it is replicated. Interactions with specific proteins play a pivotal role in nucleic acid research. The discovery of restriction nucleases made it possible to produce recombinant DNA in the laboratory, mimicking a process that takes place extensively in nature. The ability to pick out specific genes and to increase the available supply of their DNA, whether by cloning or by chain reaction amplification, allows a degree of manipulation of living organisms far beyond that achieved by selective breeding. These methods have made it possible to contemplate the Human Genome Project, in which human DNA is to be completely mapped and the base sequence eventually determined. Medical applications of this technology to genetic diseases are widely discussed in the media. Another application, which potentially touches everyone's life, is agriculture. "Not everybody gets sick, but everybody has to eat." The results of genetic engineering are on the way to your dinner table.

7.1
METHODS FOR WORKING WITH NUCLEIC ACIDS

Experiments on nucleic acids frequently involve extremely small quantities of materials of widely varying molecular size. Two of the prime necessities are to separate the components of a mixture and to detect the presence of nucleic acids; fortunately, powerful methods exist for accomplishing both these goals.

Separation Techniques

Any separation method makes use of differences among the items to be separated. Charge and size are two properties of molecules that are frequently used for separation, and in the case of nucleic acids, size is of paramount importance. One of the most widely used techniques in molecular biology, **polyacrylamide gel electrophoresis (PAGE)** (Interchapter A, Section A.1), makes use of this combination of properties. The charge on the molecules to be separated leads them to move through the gel toward an electrode of opposite electric charge. Nucleic acids and oligonucleotide fragments are negatively charged at neutral pH because of the presence of phosphate groups. When these negatively charged molecules are placed in an electric field between two electrodes, they all migrate toward the positive electrode. In both nucleic acids and oligonucleotides, each nucleotide residue contributes a negative charge from the phosphate to the overall charge of the fragment, but the mass of the nucleic acid or oligonucleotide increases correspondingly. Thus, the ratio of charge to mass remains approximately the same regardless of the size of the molecule in question. As a result, the separation takes place simply on the basis of size and is due to the sieving action of the gel (Interchapter A, Section A.1). In a given amount of time, with a sample consisting of a mixture of oligonucleotides, a smaller oligonucleotide moves farther than a larger one in an electrophoretic separation. The oligonucleotides move in the electric field because of their charge; the distances they move in a given time depend on their sizes.

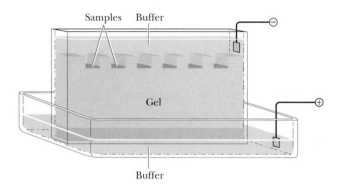

FIGURE 7.1

The experimental setup for gel electrophoresis. The gel is placed in a vertical position. The samples are applied at the top of the gel. When the current is applied, the negatively charged oligonucleotides migrate toward the positive electrode at the bottom.

The experimental design of the separation procedure takes advantage of the fact that the oligonucleotides differ in size. The separation is done with the gel in a vertical position. The negative electrode is at the top of the gel, and the positive electrode is at the bottom. There is room for several samples on each gel. Each sample is loaded at a given place (a distinct track) at the top of the gel, and the current is applied until the separation is complete, with the smallest oligonucleotide products near the bottom of the gel (Figure 7.1).

Detection Methods

The most commonly used method for detecting the separated products is based on radioactive labeling of the sample, usually with the isotope of phosphorus of mass number 32 (^{32}P, spoken as "P thirty-two"). When the labeled oligonucleotides have been separated, the gel is placed in contact with a piece of x-ray film. The radioactively labeled oligonucleotides expose the portions of the film with which they are in contact. When the film is developed, the positions of the labeled substances show up as dark bands. This technique is called **autoradiography,** and the resulting film image is an **autoradiograph** (Figure 7.2). An alternative method to radioactive labeling

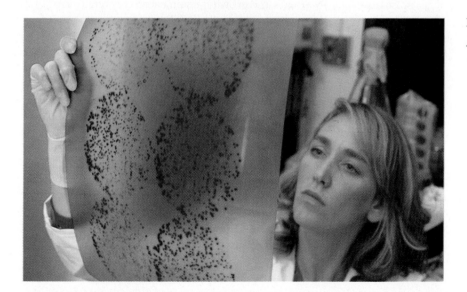

FIGURE 7.2

An example of an autoradiograph.

of oligonucleotides is fluorescent labeling. The fluorescence method has been refined to the point where it can detect amounts of substances measured in nanomoles, and it does not have the hazards associated with radioactivity.

7.2
DETERMINING THE BASE SEQUENCES OF NUCLEIC ACIDS

The method devised by Sanger and Coulson for determining the base sequences of nucleic acids depends on selective interruption of oligonucleotide synthesis. A single-stranded DNA fragment whose sequence is to be determined is used as a template for the synthesis of a complementary strand. The new strand grows from the 5′ to the 3′ end. This unique direction of growth is true for all nucleic acid synthesis (Chapter 19, Sections 19.2, 19.3, and 19.8). The synthesis is interrupted at every possible site in a population of molecules.

The interruption of synthesis depends on the presence of 2′,3′-dideoxyribonucleoside triphosphates (ddNTPs).

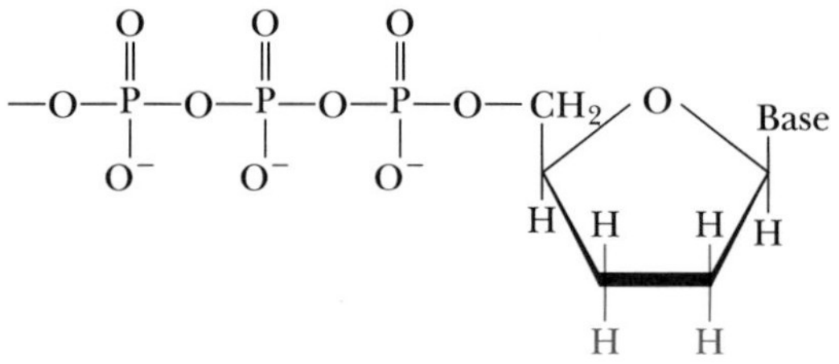

ddNTP

The 3′-hydroxyl group of deoxyribonucleoside triphosphates (the usual monomer unit for DNA synthesis) has been replaced by a hydrogen. These ddNTPs can be incorporated in a growing DNA chain, but they lack a 3′-hydroxyl group to form a bond to another nucleoside triphosphate. The incorporation of a ddNTP into the growing chain causes termination at that point. The presence of small amounts of ddNTPs in a replicating mixture causes random termination of chain growth.

The DNA to be sequenced is mixed with a short oligonucleotide that serves as a primer for synthesis of the complementary strand. The primer is hydrogen bonded to the 3′ end of the DNA to be sequenced. The DNA with primer is divided into four separate reaction mixtures. Each reaction mixture contains all four deoxyribonucleoside triphosphates (dNTPs), one of which is labeled to allow the newly synthesized fragments to be visualized by autoradiography. (In an alternative procedure, the primer can be labeled at the 5′ end with ^{32}P or with a radioactive isotope of sulfur, ^{35}S, which substitutes for one of the oxygens in the phosphate groups of the nucleoside triphosphate at the 5′ end of the primer.) In addition, each of the reaction mixtures contains one of the four ddNTPs. Synthesis of the chain is allowed to proceed in each of the four reaction mixtures. In each mixture, chain termination occurs at all possible sites for that nucleotide.

When gel electrophoresis is performed on each reaction mixture, a band corresponding to each position of chain termination appears. The sequence of the newly formed strand, which is complementary to that of the template DNA, can be "read" directly from the sequencing gel. A variation on this method is to use a single reaction mixture with a different fluorescent label on each of the four ddNTPs. Each fluorescent label can be detected by its characteristic spectrum, requiring only a single gel electrophoresis experiment. The use of fluorescent labels makes it possible to automate DNA sequencing, with the whole process under computer control (Figure 7.3). Commercial kits are available for these sequencing methods.

When RNA is to be sequenced, the method of choice is not to analyze the RNA itself but to use the methods of DNA sequencing on a DNA

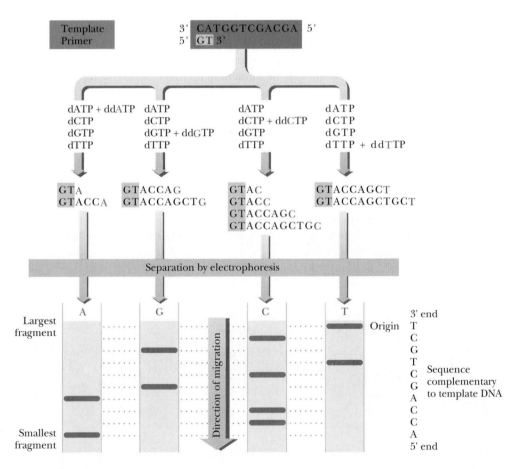

FIGURE 7.3

The Sanger-Coulson method for sequencing DNA. A primer is hydrogen bonded to the 3' end of the DNA to be sequenced. Four reaction mixtures are prepared; each contains the four dNTPs and one of the four possible ddNTPs. In each reaction mixture synthesis takes place, but in a given population of molecules, synthesis is interrupted at every possible site. A mixture of oligonucleotides of varying length is produced. The components of the mixture are separated by gel electrophoresis.

complementary (cDNA) to the RNA in question. The cDNA, in turn, is generated by use of the enzyme reverse transcriptase, which catalyzes the synthesis of DNA from an RNA template.

7.3
RECOMBINANT DNA AND CLONING

The use of restriction enzymes and DNA ligases produces comparatively few molecules of a given recombinant DNA (Chapter 6, Section 6.7), but further experiments with that DNA make it necessary to have a sample of the DNA in question that is large enough to work with and is pure. It is possible to achieve both of these goals by cloning the selected DNA. The term **clone** refers to a genetically identical population, whether of organisms, cells, viruses, or DNA molecules. Every member of the population is derived from a single cell, virus, or DNA molecule. It is particularly easy to see how individual bacteriophages (recall this term from Chapter 6, Box 6.2) and bacterial cells can produce large numbers of progeny. Bacteria grow rapidly, and large populations can be obtained relatively easily under laboratory conditions. Viruses also grow easily. We shall examine each of these examples in turn.

A virus can be considered a genome with a protein coat, usually consisting of many copies of one kind of protein or, at most, a small number of different kinds of proteins. The viral genome can be DNA or RNA, double-stranded or single-stranded. For purposes of this discussion we confine our attention to DNA viruses with double-stranded DNA. In the cloning of bacteriophages, a "lawn" of bacteria covering a Petri dish is infected with the phage. Each individual virus infects a bacterial cell and reproduces, as do its progeny when they infect and destroy other bacterial cells. As the virus multiplies, a clear spot, called a "plaque," appears on the Petri dish, marking the area in which the bacterial cells have been killed (Figure 7.4).

FIGURE 7.4

The cloning of a virus. The progeny of each individual phage (bacterial virus) infect and destroy bacteria on the Petri dish, leaving clear spots known as plaques. Each plaque indicates the presence of a clone.

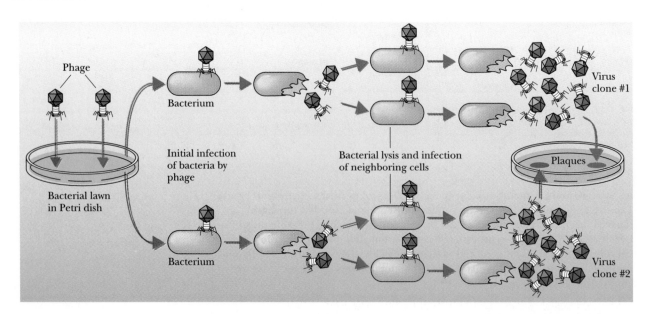

recombination takes place, the gene for tetracycline resistance will no longer be functional. The presence or absence of antibiotic resistance will be useful in detecting the presence of plasmids and whether they have acquired a DNA insert.

When the population of plasmids has been prepared, it is mixed with bacteria that are sensitive to both antibiotics. Cells that do not acquire a plasmid will not grow on a medium that contains either antibiotic (Figure 7.11). Cells that acquire a plasmid without a DNA insert will grow on a medium that contains both antibiotics, because each of these cells has an intact gene for tetracycline resistance. Cells that acquire a plasmid with an insert will be resistant to ampicillin but not to tetracycline. The group of clones that has acquired a plasmid with an insert constitutes the recombinant library. The whole library can be stored for future use, or a single clone can be selected for further study. The process of constructing a DNA library can be quite laborious, leading many researchers to obtain previously constructed libraries from other laboratories. Some journals require that libraries and individual clones that have been discussed in articles published by them should be freely available to other laboratories.

RNA libraries are not constructed and cloned as such. Rather, the RNA of interest (usually mRNA) is used as the template for synthesis of complementary DNA (cDNA) in a reaction catalyzed by reverse transcriptase. The cDNA is incorporated into a vector (Figure 7.12). From this point on, the process of producing a **cDNA library** is virtually identical to that for constructing a genomic DNA library.

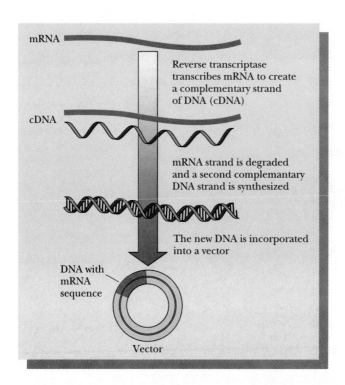

FIGURE 7.12

Reverse transcriptase catalyzes the synthesis of a strand of complementary DNA (cDNA) on a template of mRNA. The cDNA directs the synthesis of a second strand, which is then incorporated into a vector.

Finding an Individual Clone in a DNA Library

Imagine that, after a DNA library has been constructed, someone wants to find a single clone out of the hundreds of thousands, or possibly millions, in the library. This degree of selectivity requires specialized techniques. One of the most useful of these techniques depends on separation and annealing of complementary strands. An imprint of the dish on which the bacterial colonies (or phage plaques) have grown is taken by **blotting.** A nitrocellulose disc is placed on the dish and then removed. Some of each colony or plaque is transferred to the disc, and *the position of each is the same as it was on the dish.* The rest of the original colonies or plaques remain on the dish and can be stored for future use (Figure 7.13).

FIGURE 7.13

Selecting a desired clone from a DNA library. A portion of each clone is transferred to a nitrocellulose disc by blotting. The disc is treated with a denaturing agent to unwind all the DNA. A single-stranded radioactive probe for the desired DNA is added and allowed to anneal. After excess probe is removed, the presence of the desired clone and the presence of bound probe are detected by exposure to x-ray film.

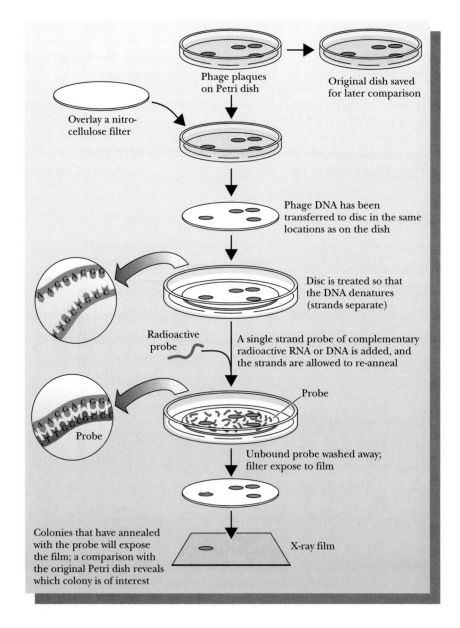

Phage plaques on Petri dish

Original dish saved for later comparison

Overlay a nitrocellulose filter

Phage DNA has been transferred to disc in the same locations as on the dish

Disc is treated so that the DNA denatures (strands separate)

Radioactive probe

A single strand probe of complementary radioactive RNA or DNA is added, and the strands are allowed to re-anneal

Probe

Probe

Unbound probe washed away; filter expose to film

Colonies that have annealed with the probe will expose the film; a comparison with the original Petri dish reveals which colony is of interest

X-ray film

Kary B. Mullis, inventor of the polymerase chain reaction and 1993 Nobel laureate in chemistry.

The nitrocellulose disc is treated with a denaturing agent to unwind all the DNA on it. The next step is to expose the disc to a solution that contains single-stranded DNA (or RNA) that has a sequence complementary to one of the strands in the clone of interest (Figure 7.13). The added single-stranded nucleic acid is called the **probe;** it is tagged in some way, usually with a radioactive label, to ensure detection. The probe anneals to the DNA of interest and only to that DNA. Any excess solution is washed off the nitrocellulose disc, and the disc is placed in contact with x-ray film (Section 7.1). Only those spots on the disc in which some of the probe has annealed to the DNA already there are radioactive, and only those spots expose the x-ray film. The probe causes the desired DNA spot to "**light up.**" Since the original Petri dish has been saved, the desired clone can be picked off the plate and allowed to reproduce.

If the nucleotide sequence of the desired DNA segment is not known and no probe is available, a complication arises. If the gene of interest directs the synthesis of a given protein, one chooses a vector that will allow cloned genes to be transcribed and translated. If the presence of the desired protein can be detected by its function, that serves as the basis for detecting it. Alternatively, radioactively labeled antibodies can be used as a basis for detection.

7.5
THE POLYMERASE CHAIN REACTION

It is possible to increase the amount of a given DNA many times over without cloning that DNA. The method that makes this amplification possible is the **polymerase chain reaction (PCR).** Any chosen DNA can be amplified, and it need not be separated from the rest of the DNA in a sample before the procedure is applied. PCR copies both complementary strands of the desired DNA sequence.

At the start of the process, the two strands are separated by heating, after which short oligonucleotide segments are added in large excess and,

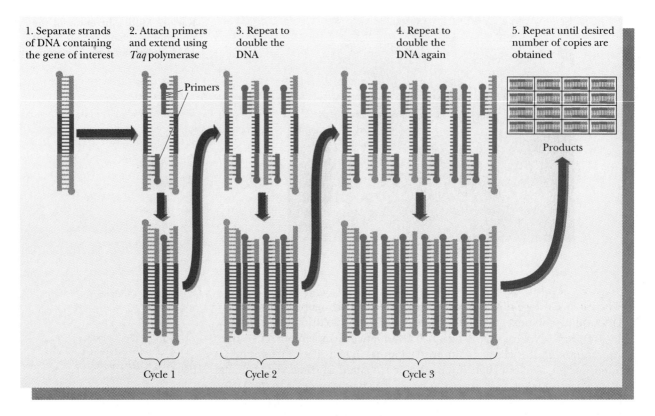

1. Separate strands of DNA containing the gene of interest

2. Attach primers and extend using *Taq* polymerase

Primers

3. Repeat to double the DNA

4. Repeat to double the DNA again

5. Repeat until desired number of copies are obtained

Products

Cycle 1   Cycle 2   Cycle 3

FIGURE 7.14

The polymerase chain reaction (PCR). The amount of DNA in the target area (red) doubles with each cycle. The two strands are separated by heat denaturation. Primers are added, then DNA polymerase catalyzes the addition of nucleotides to produce a complementary strand, duplicating the original DNA. In the second round, the DNA strands are separated and the process is repeated. The target sequence is duplicated again. More than 20 cycles can be completed in an hour, producing millions of copies of the target DNA.

via cooling, are allowed to anneal to the desired DNA. These short segments are complementary to the ends of the DNA chosen for amplification and are there to serve as primers for copying the entire complementary strand in the presence of DNA polymerase and the other factors needed for the synthesis. The two complementary strands grow in the 5′-to-3′ direction (Figure 7.14). This first round doubles the amount of the desired DNA. The process of unwinding the two strands, annealing the initiators, and copying the complementary strand is repeated, bringing about a second doubling of the selected double-stranded DNA. It is not necessary to add more primer, since it is present in large excess. It is also not necessary to add more DNA polymerase. A heat-stable polymerase from *Thermus aquaticus* remains active under the conditions of amplification. The whole process is automated. Control of the temperature to which the strands are heated to separate them is crucial, as is the temperature chosen for annealing the primers. The time allowed for heating and annealing is an important factor, as well.

The amount of DNA continues to double in subsequent rounds of amplification. After about an hour, and 25 to 40 cycles of replication, one obtains millions to hundreds of millions of copies of the desired DNA segment, usually a few hundred to a few thousand base pairs in length (Figure 7.14). Other DNA sequences are not amplified and do not interfere with the reaction or subsequent use of the amplified DNA. Amplification of the amounts of DNA in extremely small samples has made it possible to obtain accurate analyses that were not possible earlier. Forensic applications of the technique have resulted in positive identifications of crime victims and suspects. Even minuscule amounts of ancient DNA, such as those available from Egyptian mummies, can now be researched after amplification.

7.6
GENE MAPPING

A **genetic map** of an organism, also called a linkage map, is the order of genes on a chromosome, an order that can be determined by the methods of classical genetics. We shall use the bacterium *Escherichia coli* as an example.

Classical Genetic Mapping

For decades it has been well known that some bacteria, including *E. coli,* can transfer DNA from one cell to another by a process known as **conjugation.** The cells are in physical contact as DNA is transferred from one to another (Figure 7.15). The longer the cells are in contact, the more DNA can be

FIGURE 7.15

Conjugation in bacteria. The amount of DNA transferred depends on the length of time the bacteria remain in contact. Genes that lie close to each other on the chromosome are transferred together. Genes that lie far apart are not transferred together.

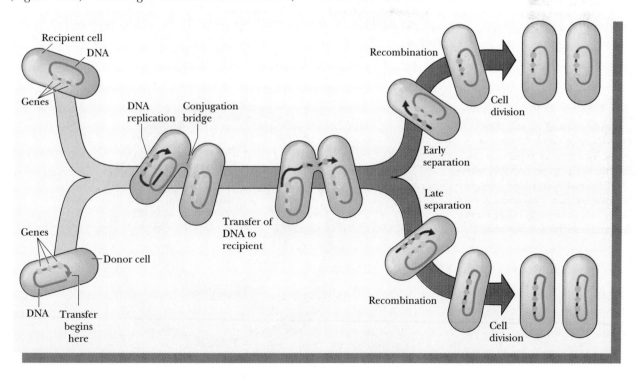

Recipient cell
DNA
Genes
DNA replication
Conjugation bridge
Transfer of DNA to recipient
Recombination
Cell division
Early separation
Late separation
Recombination
Cell division
Genes
Donor cell
DNA Transfer begins here

Conjugation in *E. coli* bacteria.

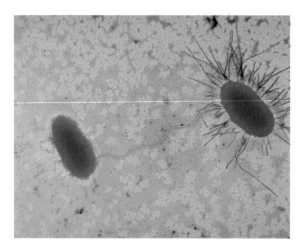

transferred. Genes close to each other on the bacterial genome "travel together" in the course of conjugation. If the flask containing the cells is shaken, the contact is broken and less of the DNA is transferred. The amount of DNA transferred in the process can be detected by use of recipient cells that are missing some genes (usually ones that direct the biosynthesis of some important compound) and testing for the presence of those genes in the recombinants. The order in which genes are transferred can be determined by varying the amount of time allowed for conjugation. Genes that are closer together tend to be linked in the course of recombination and to be transferred together even during short recombination periods. Genes that are farther apart are much less likely to be linked and thus are less likely to be transferred together. This method and variations on it produce the genetic map (Figure 7.16). The construction of the map is based entirely on genetic methods that rely on visible traits and does not deal directly with the nature of the molecules involved. It is also possible to

FIGURE 7.16

A genetic map of *E. coli*. The three-letter symbols represent genes. The arrows represent the directions for gene transfer when two different donors are used in the process of conjugation. Genes that lie close to each other on the map are transferred together.

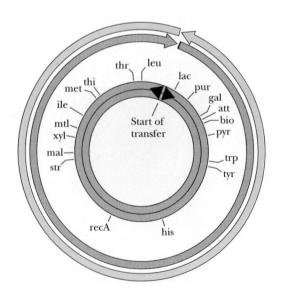

construct a **physical** or **molecular map** of the DNA sequence on chromosomes, using the methods discussed in this chapter. We shall now see what the physical map has in common with the classical genetic map.

Physical Mapping of Chromosomes

Construction of a physical map of chromosomes depends in an important way on the organism involved. In a prokaryote there is normally a single chromosome not separated from the rest of the cell; a prokaryotic chromosome is a circle of double-helical DNA. In a eukaryote there are multiple chromosomes, sequestered in the nucleus; each contains a single, linear DNA double helix. In prokaryotes, genes lie close to each other without intervening DNA. In eukaryotes, genes are separated by long stretches of noncoding DNA sequences along the chromosome. Dealing with all these complexities requires the powerful methods developed for recombinant DNA technology.

Mapping Prokaryotic Chromosomes

The physical map of the *E. coli* genome was obtained by analyzing the restriction fragments that are produced when the entire genome is digested with a restriction nuclease. A particular enzyme was chosen that gives rise to relatively few fragments, which can be separated by gel electrophoresis. The positions of the fragments on the circular genome were determined by comparison with the classical genetic map (Figure 7.17), on which the positions of more than 1000 genes had already been determined at the time this work was done. Cloned segments of these known genes were available as probes in DNA blotting to locate a given gene on a given fragment. Consequently, it has been possible to correlate the physical and genetic maps of *E. coli*. Note that the two kinds of maps parallel each other. Any newly discovered genes can be placed accurately on a restriction fragment as well as on the genetic linkage map.

Genetic and Physical Mapping of Eukaryotic Chromosomes

Mapping genes on the larger and more complex chromosomes of eukaryotes represents a considerable challenge. Classical genetic studies on the fruit fly *Drosophila melanogaster* in the early 20th century determined linkage patterns of genes via methods similar to those just described for bacteria. This work produced detailed maps of genes on chromosomes, but the positions of many more genes were completely unknown. Even less information was available for other eukaryotes, and this situation remained unchanged until recently.

In humans, the first genes to be mapped were those on the X chromosome. Females have two X chromosomes, whereas males have one X chromosome and one Y chromosome. If a gene on the single X chromosome of a male is nonfunctional, there is no chance of the function's being replaced by the corresponding gene on the other X chromosome, as can

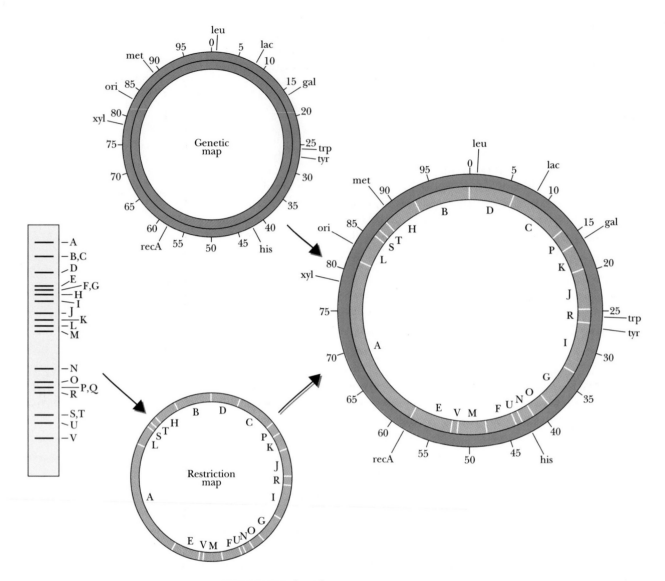

FIGURE 7.17

A comparison of the physical and genetic maps of *E. coli*. The rectangle on the left represents the results of gel electrophoresis of a restriction enzyme digest of the *E. coli* genome. This information is used to construct a map of the physical fragments of the genome and to see how the physical map "lines up" with the genetic map.

happen in females. As a result, some human males exhibit traits that arise from nonfunctional genes on the X chromosome. Examples include red-green color blindness and diseases such as hemophilia and Duchenne muscular dystrophy. Females are very seldom affected by these traits but are carriers of them; a mother passes them on to her sons. Such observations made it possible to locate on the X chromosome the genes involved in the traits (Figure 7.18). Progress in locating human genes on chromosomes

FIGURE 7.18

A schematic diagram of the human X chromosome, showing the locations of genes associated with a number of diseases.

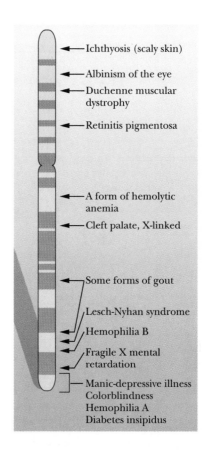

other than the sex chromosomes (X and Y) was slow until recombinant DNA methods became available.

Globin genes are an example of a cluster of genes on an **autosomal** (nonsex) chromosome. The family of genes that encodes the β-chain of hemoglobin consists of a cluster arranged along chromosome 11 in the order in which the genes are expressed in the course of development (Figure 7.19). Five different β-type chains are made from the early embryonic stage to shortly after birth. The form known as ε-globin (epsilon-globin) is made only in the early embryo stage. The two γ-globin (gamma-globin) forms, Gγ and Aγ, are made later in fetal life, while the gene for the epsilon form is switched off. After birth, the genes for δ-globin (delta-globin) and β-globin are expressed, and the genes for the gamma forms are switched off in turn. Note that the map also contains two **pseudogenes,** designated Ψβ. (The Greek letter psi, Ψ, indicates the pseudogene.) Pseudogenes are nonfunctional copies of genes that cannot normally be expressed because of some mutation, frequently a deletion. Pseudogenes occur to varying extents in eukaryotes and are particularly prevalent in mammals.

A similar arrangement appears in the α-globin gene cluster on chromosome 16 (Figure 7.19). The variant of α-globin that is produced in the early embryo is ζ-globin (zeta-globin). This gene is switched off later in development, at which time the α-globin gene is expressed. Note the presence of pseudogenes for both the alpha and zeta forms (Ψα and Ψζ, respectively). The mechanism of regulation of these two gene clusters on different chromosomes is a topic of considerable interest to biologists.

It is not surprising that human globin genes were among the first mammalian genes to be cloned using recombinant DNA methods. There was considerable incentive to develop gene therapies for the genetic diseases that arise from abnormal hemoglobins; it was also comparatively easy to carry out the actual cloning process. Protein synthesis in human red blood cells is largely dedicated to the production of hemoglobin, which accounts for about 90 percent of the protein in these cells. Most of the mRNA in red blood cells is of two kinds, those for the α-globins and those for the β-globins. Using these two readily available mRNAs, it was possible to construct probes to identify the globin genes in cDNA libraries. Each of these genes was contained in a clone that contained stretches of DNA flanking the gene itself. The flanking sequences were cloned by themselves;

FIGURE 7.19

Maps of the gene clusters that code for the variant forms of the α- and β-chains of human hemoglobin. Different α-type and β-type chains are made during embryonic development and after birth (see text). The Greek letter psi (ψ) indicates pseudogenes, which are nonfunctional copies of genes.

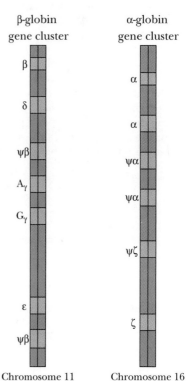

they could then be used to construct probes for segments of DNA farther along the chromosome. This process is called **chromosome walking.** It was repeated until a physical map of the entire gene cluster of the β-globin gene family was obtained.

RFLPs: A Powerful Method for Mapping Large Genomes

In organisms (such as humans) with two sets of chromosomes, a given gene on one chromosome may differ slightly from the corresponding gene on the paired chromosome. In the language of genetics, these genes are **alleles.** When they are the same on the paired chromosomes, the organism is **homozygous** for that gene; when they differ, the organism is **heterozygous.** A difference between alleles, even a change in one base pair, can frequently mean that one allele has a recognition site for a restriction nuclease and the other does not. Restriction fragments of different sizes are obtained on treatment with the nuclease (Figure 7.20); they are called **restriction-fragment-length polymorphisms,** or **RFLPs** (pronounced "riflips") for short. These polymorphisms (a word meaning "many shapes") are analyzed via gel electrophoresis to separate the fragments by size, followed by blotting and the annealing of a probe for a gene being sought.

FIGURE 7.20

A change of one base pair elimi-nates a restriction nuclease cleav-age site. A portion of DNA that codes for normal β-globin has a cleavage site for the restriction nuclease *Dde*I. The corresponding DNA with the sickle-cell mutation does not have this cleavage site. The difference can be detected by electrophoresis, followed by blot-ting and the annealing of a probe for the β-globin gene. (The abbre-viation bp stands for base pairs.)

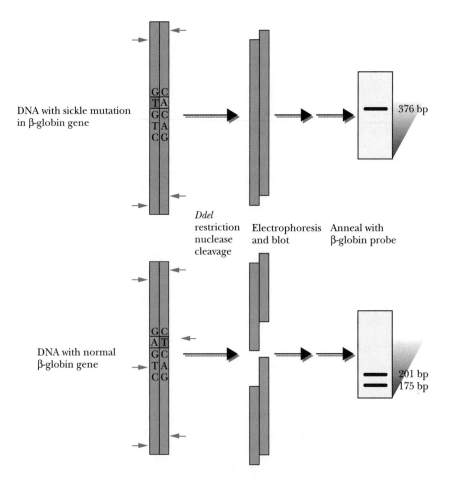

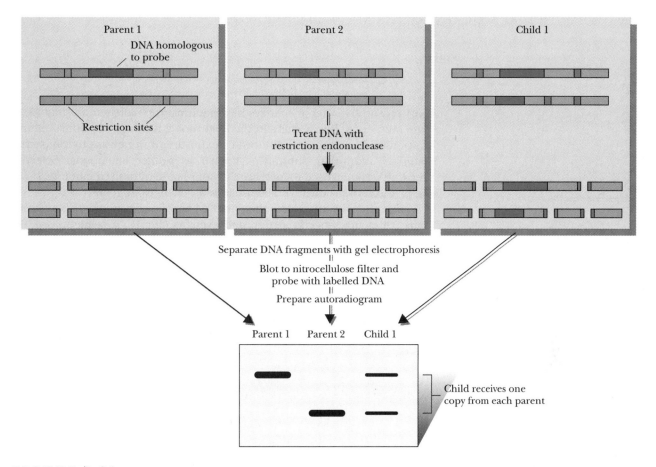

Restriction-fragment-length polymorphisms (RFLPs) can be detected by probes for homologous DNA. Parent 1 has DNA with two restriction sites near the DNA homologous to the probe. Since Parent 1 has the same pattern of restriction sites on both DNA strands, one large restriction fragment will be detected. Parent 2 has three restriction sites on each DNA strand. A single smaller restriction fragment will be detected by the homologous probe for Parent 2. Their child has inherited one copy for each DNA strand from each parent. The child's DNA will produce one fragment of each size.

Research has shown that such polymorphisms are quite common, much more so than the traits such as eye color and inherited diseases that were used for earlier genetic mapping. RFLPs can be used as markers on the genetic linkage map in the same way as mutations in visible traits, because they are inherited in the manner predicted by classical genetics (Figure 7.21). However, since they are more abundant than mutations within coding sequences, they have provided many more markers for detailed genetic mapping. RFLP analysis was used extensively in the process of locating the altered gene that causes cystic fibrosis, a prevalent genetic disease. Once the gene was located on chromosome 7, a series of RFLP markers was used to help map its exact position (Figure 7.22); then the gene was cloned and its protein product characterized. The protein in question is involved in the transport of ions in the lung. This information deepens our insight into the nature of cystic fibrosis and provides approaches to new treatment.

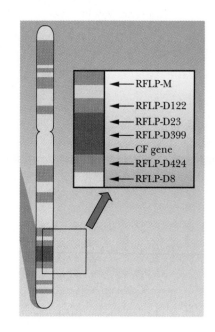

Localization of the gene associated with cystic fibrosis (CF) on human chromosome 7. The RFLP markers are given arbitrary names. The location of the CF gene is given relative to the RFLP markers.

7.7
GENETIC ENGINEERING

Until recently, heritable changes in organisms were solely those that arose from mutations. Researchers in the field took advantage of both spontaneous mutations and those produced by exposure of organisms to radioactive materials and other substances known to induce mutations. Selective breeding was then used to increase the population of desired mutants. It was not possible to produce "custom-tailored" changes in genes.

Since the advent of recombinant DNA technology, it is possible (within limits) to change specific genes, and even specific DNA sequences within

FIGURE 7.23

Synthesis of insulin in humans. The insulin gene is a split gene. The intervening sequence (intron) encodes an RNA transcript that is spliced out of mRNA. Only the portions of the gene called exons are reflected in the base sequence of mRNA. Once protein synthesis takes place, the polypeptide is folded, cut, and spliced. The end product, active insulin, has two polypeptide chains as a result.

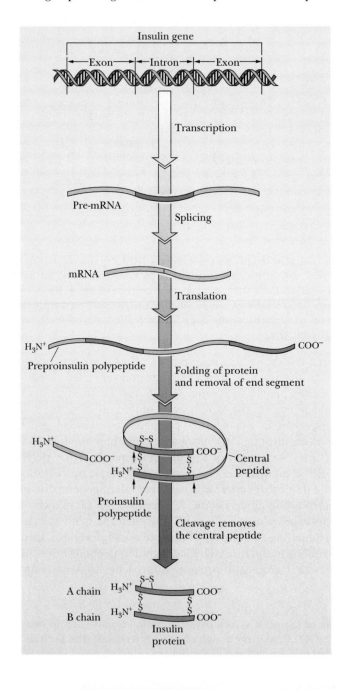

those genes, to alter inherited characteristics of organisms. Bacteria can be altered to produce large amounts of medically and economically important proteins. Animals can be treated to cure or alleviate their genetic diseases, and agriculturally important plants can be made to produce greater crop yields or be given increased resistance to pests.

Bacteria as "Protein Factories"

An application of genetic engineering that is of considerable practical importance is the production of human insulin by *E. coli*. This process is far from straightforward. A significant problem is that the insulin gene is split. It contains an **intron,** a DNA sequence that codes for RNA that will eventually be deleted in the processing of the mRNA that directs the synthesis of the protein (see Chapter 19, Section 19.9). Only the RNA transcribed from DNA sequences called **exons** will appear in mature mRNA (Figure 7.23). Bacteria do not have the cellular apparatus for splicing introns out of RNA transcripts to give functional mRNA. One might think that the problem could be solved by using cDNA obtained from the mRNA for insulin in a reverse transcriptase–catalyzed reaction. The problem here is that the polypeptide encoded by this mRNA contains an end peptide and a central peptide, which is to be removed by further processing in insulin-producing cells to give two polypeptide chains, designated A and B (Figure 7.23).

The approach to this problem is to use two synthetic DNAs, one encoding the A chain of insulin and the other encoding the B chain. These synthetic DNAs are produced in the laboratory using methods that were developed by synthetic organic chemists. Each DNA is inserted into a separate plasmid vector (Figure 7.24). The vectors are taken up by two

FIGURE 7.24

Active human insulin can be produced in bacteria by the use of two separate batches of *E. coli*. Each batch produces one of the two chains, the A chain or the B chain. The two chains are mixed to produce active insulin.

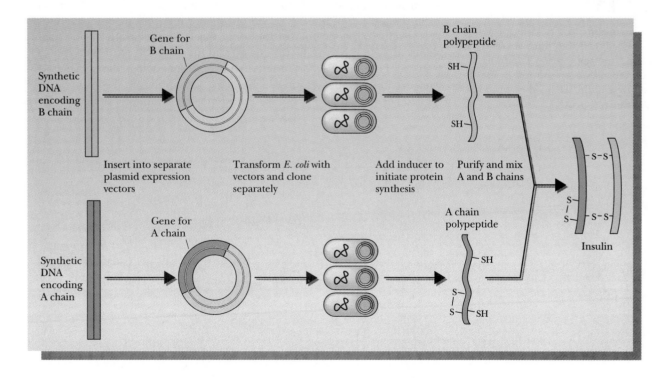

different populations of *E. coli.* The two groups are then cloned separately; each group of bacteria produces one of the two polypeptide chains of insulin. The A and B chains are extracted and mixed, finally producing functional human insulin.

Genetic Engineering in Eukaryotes

When the target organism for genetic engineering is an animal or a plant, one must consider that these are multicellular organisms with multiple kinds of tissues. In bacteria, altering the genetic makeup of a cell implies a change in the whole single-celled organism. In multicellular organisms, one possibility is to change a gene in a specific tissue, one that contains only one kind of differentiated cell. In other words, the change is *somatic,* affecting only the body tissues of the altered organism. In contrast, changes in germ cells (egg and sperm cells), called *germ-line* changes, are passed on to succeeding generations. If germ cells are to be modified, the change must be made at an early stage in development, before the germ cells are sequestered from the rest of the organism. Attempts to produce such changes have succeeded in comparatively few organisms, such as plants, fruit flies, and some other animals such as mice. We shall consider somatic and germ-line alterations in turn.

Gene Therapy

FIGURE 7.25

Gene therapy in bone marrow cells. A cloned gene that directs the synthesis of a missing protein is introduced into bone marrow cells from the mouse. The transformed cells are replaced in the mouse's body, where they produce the desired protein.

Alterations in somatic cells can be used for **gene therapy,** in which a genetic disease is treated by the introduction of a gene for a missing protein whose lack causes the disease. A form of gene therapy that is currently undergoing clinical trials involves the gene for adenosine deaminase, an enzyme involved in purine catabolism (Section 17.9). If this enzyme is missing, dATP builds up in tissues, inhibiting the action of the enzyme ribonucleotide reductase. This results in a deficiency of the other three deoxyribonucleoside triphosphates (dNTPs). The dATP (in excess) and the other three dNTPs (deficient) are precursors for DNA synthesis. This imbalance particularly affects DNA synthesis in lymphocytes, on which much of the immune response depends. Individuals who are homozygous for adenosine deaminase deficiency devel-

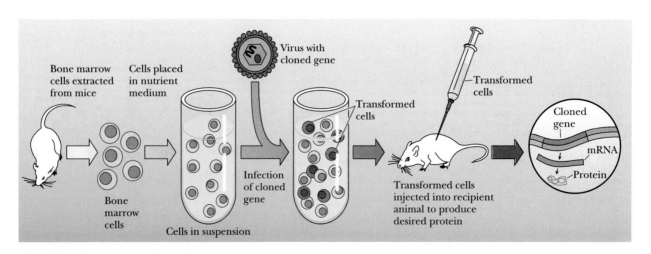

op **severe combined immune deficiency (SCID),** the "bubble boy" syndrome. They are easily prone to infection because of their highly compromised immune systems. The ultimate goal of the planned gene therapy is to take bone marrow cells from affected individuals; introduce the gene for adenosine deaminase into the cells, using a virus as a vector; and then reintroduce the bone marrow cells in the body, where they will produce the desired enzyme. This procedure has been worked out in mice (Figure 7.25). Earlier clinical trials have focused on correction of the gene in mature T cells; the healthy cells are given to the recipient via transfusions. In trials at the National Institutes of Health (NIH), two girls, aged four and nine at the start of treatment, have shown improvement to the extent that they attend regular public schools and have no more than the average number of infections. Administration of bone marrow cells in addition to T cells is the next step; clinical trials of this procedure are under way in Milan, Italy, where the recipient is a boy who was five years old at the start of treatment.

Transgenic Organisms

Germ-line alterations in humans raise serious ethical questions; consequently, no such procedures are planned. Germ-line modifications have been carried out in other eukaryotes, producing genetically altered mice for research, and some commercially important organisms such as crop plants and livestock. Organisms that carry heritable genes introduced by genetic engineering are described as **transgenic;** the introduced gene is called a **transgene.** In mammals such as mice, the method involves microinjection of DNA that carries the gene to be introduced into the nucleus of a fertilized mouse egg (Figure 7.26). The injected DNA integrates itself into the genome. Eggs treated in this manner are injected into a foster mother to develop. When the mice are born, DNA samples are taken from their tails to

FIGURE 7.26

The procedure for producing transgenic mammals. The mice on the left have not received new genes. The mouse on the right has received a new gene, which it can pass on to its offspring. The mouse on the right is a transgene.

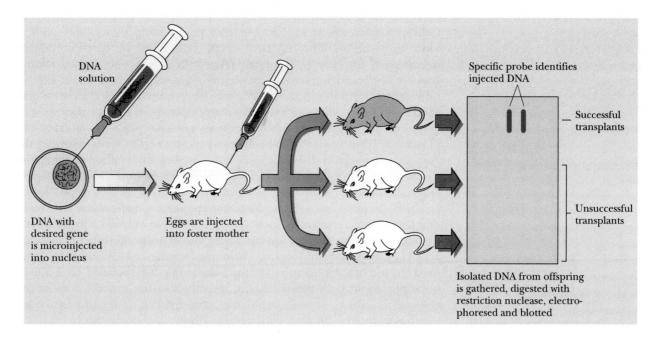

DNA solution

Specific probe identifies injected DNA

Successful transplants

Unsuccessful transplants

DNA with desired gene is microinjected into nucleus

Eggs are injected into foster mother

Isolated DNA from offspring is gathered, digested with restriction nuclease, electrophoresed and blotted

FIGURE 7.27

A transgenic tomato plant. Recombinant DNA methods have produced plants that resist defoliation by caterpillars. Tomatoes with a longer shelf life are another result of this research.

test for incorporation of the injected DNA. The analysis is carried out by subjecting the extracted DNA to restriction nuclease digestion, followed by electrophoresis and blotting (methodology we have seen before). In DNA samples from animals that have incorporated the new gene, testing with a probe for that gene will indicate its presence. If the probe does not bind to any DNA, the animal has not incorporated the new gene.

A vector widely used for genetic engineering in plants is based on bacterial plasmids from the crown gall bacterium, *Agrobacterium tumefaciens*. Cells of this bacterium bind to wounded plant tissue, allowing plasmids to move from the bacterial cells to the plant cells. Some of the plasmid DNA inserts itself into the DNA of the plant cells in the only known natural transfer of genes from a bacterial plasmid to a eukaryotic genome. Expression of plasmid genes in the plant gives rise to a tumor called a crown gall. Whole, healthy plants can grow from gall cells, even though they are not germ cells. (This process, of course, does not take place in animals.) The plants that grow from the gall cells can produce fertile seeds, allowing the gene that has been transferred to be continued in a new strain of the plant. Genes from any desired source can be incorporated into the *A. tumefaciens* plasmid and then transferred to a plant.

This method was used to genetically engineer tomato plants that resist defoliation by caterpillars (Figure 7.27). A gene that encodes a protein toxic to caterpillars was taken from the bacterium *Bacillus thuringensis* to bring about this modification. Work is continuing on other useful modifications of food crops. Many observers of this whole line of research have raised questions about both the safety and the ethics of the process. Laboratory research is subject to strict regulation, and field trials require clearance, which is granted only after considerable review and deliberation.

SUMMARY

Two of the prime necessities for successful experiments with nucleic acids are to separate the components of a mixture and to detect the presence of nucleic acids. Polyacrylamide gel electrophoresis (PAGE) is widely used as a separation method; labeling with radioactive or fluorescent materials allows detection of small amounts of nucleic acids.

Restriction enzymes play a large role in the manipulation of DNA. DNA samples from different sources can be selectively cut and spliced by use of restriction enzymes, producing recombinant DNA. DNA from another source can be introduced into the genome of a virus or a bacterium. In bacteria the foreign DNA is usually introduced into a plasmid, a smaller circular DNA separate from the main bacterial chromosome. The growth of the virus or bacterium also produces large amounts of the other DNA by the process of cloning. This technology has advanced to the point where it is possible to cut the entire genome of a eukaryotic organism into fragments and clone the fragments. The assemblage of cloned fragments is called a DNA library. An alternative method of producing large amounts of a given DNA, called the polymerase chain reaction, depends only on enzymatic reactions and does not require viral or bacterial hosts.

These methods have made it possible to determine the physical locations of genes on chromosomes, supplementing the results of classical genetic mapping. Genetic engineering is on the verge of producing new therapies for inherited disease. Genetically altered organisms such as mice and maize have been engineered, and many more changes are to come.

EXERCISES

1. How does the use of restriction nucleases of different specificities aid in the sequencing of DNA?
2. Define cloning of DNA.
3. A friend who works for a company that sells photographic supplies has noticed that molecular biology laboratories order a lot of x-ray film, and she wonders why. What do you tell your friend?
4. Why is temperature control so important in the polymerase chain reaction?
5. Suppose that you are a prosecuting attorney. How has the introduction of the polymerase chain reaction changed your job?
6. Why is it a large undertaking to construct a DNA library?
7. What role did restriction enzymes play in localizing the gene associated with cystic fibrosis?
8. The genes for both the α and β-globin chains of hemoglobin contain introns; i.e., they are split genes. How would this fact affect your plans if you wanted to introduce the gene for β-globin into a bacterial plasmid and have the bacteria produce β-globin?
9. Outline the methods you would use to produce human growth factor (a substance used in the treatment of dwarfism) in bacteria.

ANNOTATED BIBLIOGRAPHY

Ausubel, F., R. Brent, R. Kingston, D. Moore, J. Seidman, J. Smith, and K. Struhl, eds. *Short Protocols in Molecular Biology,* 2nd ed. New York: John Wiley & Sons, 1992. [A compendium of methods, continually updated.]

Berg, P., and M. Singer. *Dealing with Genes: The Language of Heredity.* Mill Valley, CA: University Science Books, 1992. [Two leading biochemists have produced an eminently readable book on molecular genetics. Highly recommended.]

Bloch, W. A Biochemical Perspective of the Polymerase Chain Reaction. *Biochemistry* **30,** 2735–2747 (1991). [A review article.]

Editors of *Science* and various authors. Molecular Advances in Genetic Disease. *Science* **256,** 766–813 (1992). [An extensive special report on molecular genetics and new treatments for genetic disease.]

Guyer, R., and D. Koshland. The Molecule of the Year. *Science* **246,** 1543–1546 (1989). [A review of the most important scientific developments of the year, including the first award of the Molecule of the Year distinction to the polymerase chain reaction.]

Joyce, G. Directed Molecular Evolution. *Sci. Amer.* **267** (6), 90–97 (1992). [How the polymerase chain reaction can be used to select and amplify chosen molecules out of a larger population.]

Chapter 8

Lipids and Membranes

A cholesterol rainbow formed by adding naphthalene to cholesterol and heating the mixture.

The most striking feature of lipids is their long, oily hydrocarbon chains, which are insoluble in water. As a fatty acid, a lipid contains a carboxyl head group attached to a hydrocarbon "tail." With three long-chain fatty acids, the triacylglycerols (also referred to as fats) are ideal reservoirs for energy storage in the cell. Some lipids have large charged polar heads in addition to their uncharged hydrocarbon tails.

The chief ingredients of biological membranes are the phospholipids. In water they spontaneously form lipid bilayers, with their flexible tails in the hydrophobic interior of the membrane and their polar heads on exterior surfaces in contact with water. About half of the membrane consists of protein molecules associated with the lipid bilayer. Some small molecules can migrate through the membrane from a high concentration on one side to a low concentration on the other side, by simple diffusion. Some proteins form pores that allow specified ions and small molecules to pass through the membrane. Heart muscle cells, which act in close synchrony, are connected by gap junctions — gated tubes that join the cells through their outer membranes. On the surfaces of cells are glycoproteins and lipoproteins that recognize other molecules, as well as receptors that act as gates for the passage of ions and molecules into the cell.

8.1
THE DEFINITION OF A LIPID

Lipids are compounds that occur frequently in nature; they are found in places as diverse as egg yolks and the human nervous system. They are an important component of plant, animal, and microbial membranes. The definition of a lipid is based on solubility. Lipids are marginally soluble (at best) in water and soluble in nonpolar organic solvents such as chloroform and acetone. Fats and oils are typical lipids in terms of their solubility, but that fact does not really define their chemical nature. In terms of chemistry, lipids are a mixed bag of compounds that share some properties based on structural similarities, mainly a preponderance of nonpolar groups.

Classified according to chemical nature, lipids fall into two main groups. One group, which consists of open-chain compounds with polar head groups and long nonpolar tails, includes *fatty acids, triacylglycerols, sphingo-lipids, phosphoacylglycerols,* and *glycolipids.* The second major group consists of fused-ring compounds, the *steroids;* an important representative of this group is cholesterol.

8.2
THE CHEMICAL NATURES OF THE LIPID TYPES

Fatty Acids

A fatty acid has a carboxyl group at the polar end and a hydrocarbon chain at the nonpolar tail. Fatty acids are *amphiphilic* compounds because the carboxyl group is hydrophilic and the hydrocarbon tail is hydrophobic. The carboxyl group can ionize under the proper conditions.

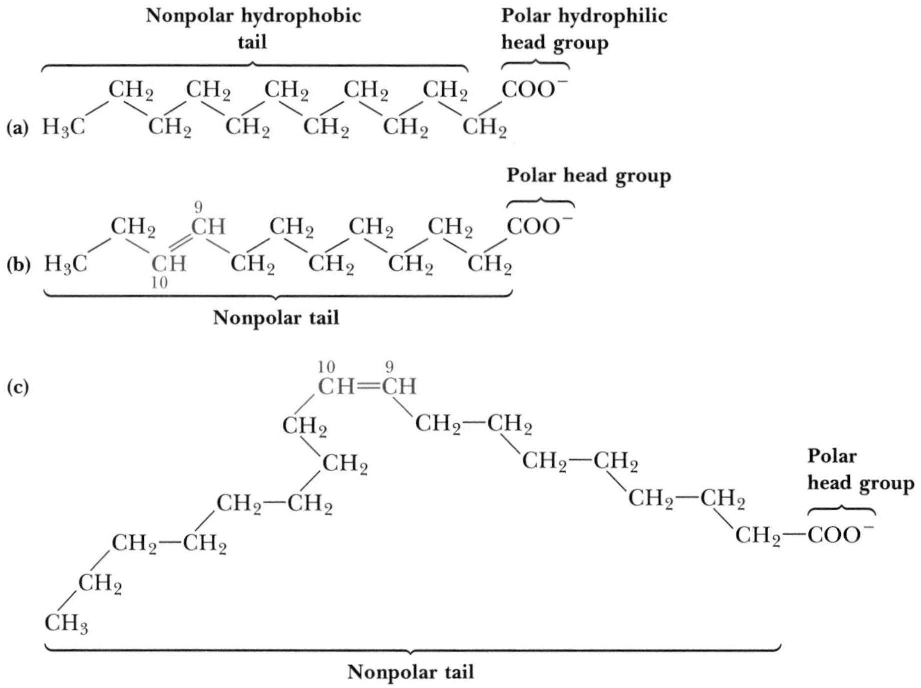

FIGURE 8.1

Structures of representative fatty acids. (a) Dodecanoate, the ionized form of the saturated fatty acid dodecanoic acid (lauric acid). (b) Dodecenoate, the ionized form of the unsaturated fatty acid dodecenoic acid, showing the effect of a *trans* double bond. (c) Oleate, the ionized form of oleic acid, another unsaturated fatty acid. Note that the *cis* double bond introduces a kink in the hydrocarbon chain. The double bonds in both unsaturated fatty acids are at the ninth carbon atom from the carboxyl end.

A fatty acid that occurs in a living system normally contains an even number of carbon atoms, and the hydrocarbon chain is usually unbranched (Figure 8.1). If there are carbon–carbon double bonds in the chain, the fatty acid is **unsaturated;** if there are only single bonds, the fatty acid is **saturated.** Tables 8.1 and 8.2 list a few examples of the two classes. In

T A B L E 8 . 1 Typical Naturally Occurring Saturated Fatty Acids

ACID	NUMBER OF CARBON ATOMS	FORMULA	MELTING POINT (°C)
Lauric	12	$CH_3(CH_2)_{10}CO_2H$	44
Myristic	14	$CH_3(CH_2)_{12}CO_2H$	58
Palmitic	16	$CH_3(CH_2)_{14}CO_2H$	63
Stearic	18	$CH_3(CH_2)_{16}CO_2H$	71
Arachidic	20	$CH_3(CH_2)_{18}CO_2H$	77

TABLE 8.2 **Typical Naturally Occurring Unsaturated Fatty Acids**

ACID	NUMBER OF CARBON ATOMS	DEGREE OF UNSATURATION*	FORMULA	MELTING POINT (°C)
Palmitoleic	16	16:1	$CH_3(CH_2)_5CH{=}CH(CH_2)_7CO_2H$	−0.5
Oleic	18	18:1	$CH_3(CH_2)_7CH{=}CH(CH_2)_7CO_2H$	16
Linoleic	18	18:2	$CH_3(CH_2)_4CH{=}CH(CH_2)CH{=}$ $CH(CH_2)_7CO_2H$	−5
Linolenic	18	18:3	$CH_3(CH_2CH{=}CH)_3(CH_2)_7CO_2H$	−11
Arachidonic	20	20:4	$CH_3(CH_2)_4(CH{=}CHCH_2)_4(CH_2)_2CO_2H$	−50

*Degree of unsaturation refers to the number of double bonds.

unsaturated fatty acids the stereochemistry at the double bond is usually *cis* rather than *trans* (Figure 8.1c). Note that the double bonds are isolated from one another by several singly bonded carbons; fatty acids do not have conjugated double bond systems. The notation used for fatty acids indicates the number of carbon atoms and the number of double bonds. In this system 18:0 denotes an 18-carbon saturated fatty acid (no double bonds), and 18:1 denotes an 18-carbon fatty acid with one double bond. Note that in the unsaturated fatty acids (except arachidonic acid) in Table 8.2 there is a double bond at the ninth carbon atom from the carboxyl end. The position of the double bond results from the way in which unsaturated fatty acids are synthesized in organisms (Chapter 15, Section 15.6). Unsaturated fatty acids have lower melting points than saturated ones. Plant oils are liquid at room temperature because they have higher proportions of unsaturated fatty acids than do animal fats, which tend to be solids.

Fatty acids are rarely found free in nature, but they form parts of many commonly occurring lipids.

Triacylglycerols

Glycerol is a simple compound that contains three hydroxyl groups (Figure 8.2). When all three of the alcohol groups form ester linkages with fatty acids, the resulting compound is a **triacylglycerol;** an older name for this type of compound is **triglyceride.** Note that the three ester groups are the polar part of the molecule, whereas the tails of the fatty acids are nonpolar. It is usual for three different fatty acids to be esterified to the alcohol groups of the same glycerol molecule. Triacylglycerols do not occur as components of membranes (as do other types of lipids), but they accumulate in adipose tissue (primarily fat cells) and provide a means of storing fatty acids, particularly in animals. They serve as concentrated stores of metabolic energy; complete oxidation of fats yields about 9 kcal/g, in contrast with 4 kcal/g for carbohydrates and proteins (see Chapter 18, Section 18.4). Triacylglycerols do not play an important role in plants.

When fatty acids are put to use by an organism, the ester linkages of triacylglycerols are hydrolyzed by enzymes called **lipases.** The same hydrolysis reaction can take place outside organisms, with acids or bases as catalysts. When a base such as sodium or potassium hydroxide is used, the products of

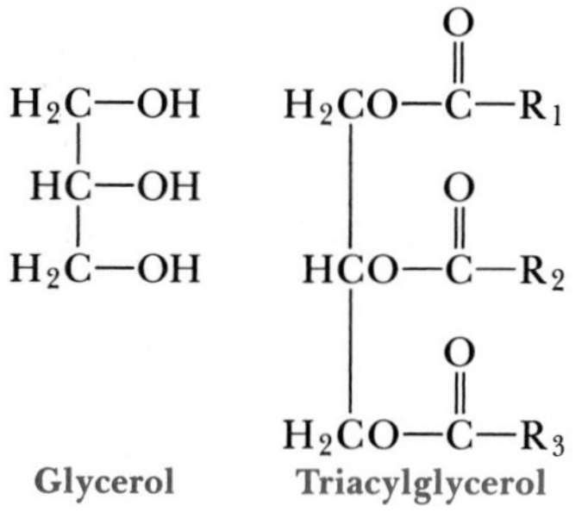

FIGURE 8.2

Structure of a triacyglycerol, with glycerol shown for comparison. R_1, R_2, and R_3 are three different fatty acids, which may be saturated or unsaturated and can occur in any combination.

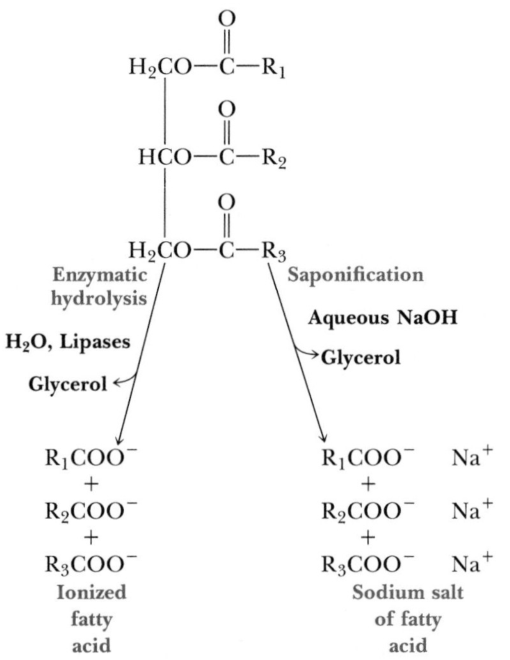

$$H_2CO-\overset{\overset{\displaystyle O}{\|}}{C}-R_1$$

$$HCO-\overset{\overset{\displaystyle O}{\|}}{C}-R_2$$

$$H_2CO-\overset{\overset{\displaystyle O}{\|}}{C}-R_3$$

Enzymatic   Saponification
hydrolysis

H₂O, Lipases   **Aqueous NaOH**

                             Glycerol
Glycerol

R_1COO^- R_1COO^- Na^+
    +   +
R_2COO^- R_2COO^- Na^+
    +   +
R_3COO^- R_3COO^- Na^+
Ionized   Sodium salt
fatty   of fatty
acid   acid

FIGURE 8.3

Hydrolysis of triacyglycerols. The term "saponification" refers to the reaction of a glyceryl ester with sodium or potassium hydroxide to produce a soap, which is the corresponding salt of the long-chain fatty acid.

the reaction, which is called **saponification** (Figure 8.3), are glycerol and the sodium or potassium salts of the fatty acids. These salts are soaps. When soaps are used with hard water, the calcium and magnesium ions in the water react with the fatty acids and form a precipitate — the characteristic scum left on the insides of sinks and bathtubs. The other product of saponification, glycerol, is used in creams and lotions as well as for the manufacture of nitroglycerin.

Phosphoacylglycerols (Phosphoglycerides)

It is possible for one of the alcohol groups of glycerol to be esterified by a phosphoric acid molecule rather than by a carboxylic acid. In such lipid molecules, two fatty acids are also esterified to the glycerol molecule. The resulting compound is called a **phosphatidic acid** (Figure 8.4a). Fatty acids are usually monobasic acids with only one carboxyl group able to form an ester bond, but phosphoric acid is tribasic and thus can form more than one ester linkage. One molecule of phosphoric acid can form ester bonds both to glycerol and to some other alcohol, creating a **phosphatidyl ester** (Figure 8.4b). Both phosphatidic acids and phosphatidyl esters are classed as **phosphoacylglycerols.** The natures of the fatty acids vary widely, as they do in triacylglycerols. As a result, the names of the types of lipids (such as

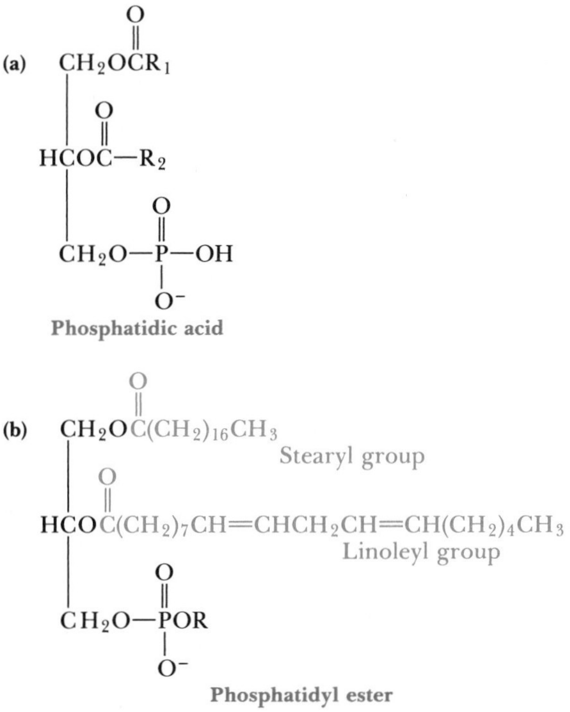

(a)

$$
\begin{array}{l}
\text{CH}_2\text{OC}R_1 \\[2pt]
\text{HCOC}-R_2 \\[2pt]
\text{CH}_2\text{O}-\text{P}-\text{OH} \\
\end{array}
$$

Phosphatidic acid

(b)

$$
\begin{array}{l}
\text{CH}_2\text{OC(CH}_2)_{16}\text{CH}_3 \\
\qquad\qquad\qquad \text{Stearyl group}\\[2pt]
\text{HCOC(CH}_2)_7\text{CH}{=}\text{CHCH}_2\text{CH}{=}\text{CH(CH}_2)_4\text{CH}_3 \\
\qquad\qquad\qquad \text{Linoleyl group}\\[2pt]
\text{CH}_2\text{O}-\text{POR} \\
\end{array}
$$

Phosphatidyl ester

FIGURE 8.4

The molecular architecture of phosphoacylglycerols. (a) A phosphatidic acid, in which glycerol is esterified to phosphoric acid and to two different carboxylic acids. R_1 and R_2 represent the hydrocarbon chains of the two carboxylic acids. (b) A phosphatidyl ester (phosphoacylglycerol). Glycerol is esterified to two carboxylic acids, stearic acid and linoleic acid, as well as to phosphoric acid. Phosphoric acid, in turn, is esterified to a second alcohol, ROH.

triacylglycerols and phosphoacylglycerols) that contain fatty acids must be considered generic names.

The classification of a phosphatidyl ester depends on the nature of the second alcohol esterified to the phosphoric acid. Some of the most important lipids in this class are **phosphatidyl ethanolamine** (cephalin), **phosphatidyl serine, phosphatidyl choline** (lecithin), **phosphatidyl inositol, phosphatidyl glycerol,** and **diphosphatidyl glycerol** (cardiolipin) (Figure 8.5). In each of these types of compounds the natures of the fatty acids in the molecule can vary widely. All these compounds have long, nonpolar, hydrophobic tails and polar, highly hydrophilic head groups and thus are markedly amphiphilic. (We have already seen this characteristic in fatty acids.) In a phosphoacylglycerol the polar head group is charged, since the phosphate group is ionized at neutral pH. There is frequently also a positively charged amino group contributed by an amino alcohol esterified to the phosphoric acid. Phosphoacylglycerols are important components of biological membranes.

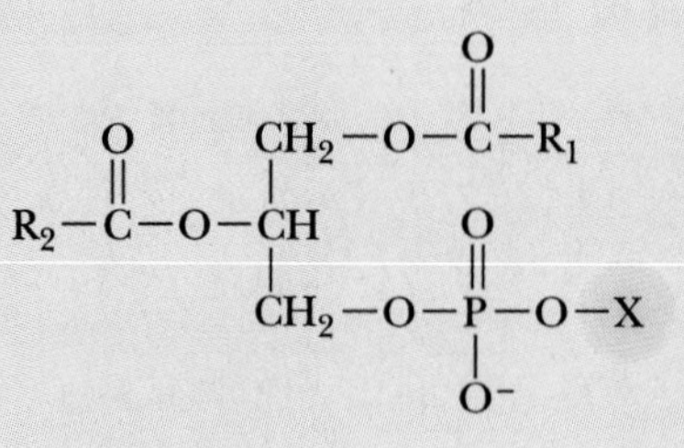

$$R_2-\overset{\overset{\displaystyle O}{\|}}{C}-O-\overset{\displaystyle CH_2-O-\overset{\overset{\displaystyle O}{\|}}{C}-R_1}{\underset{\displaystyle CH_2-O-\overset{\displaystyle O}{\underset{\displaystyle O^-}{P}}-O-X}{CH}}$$

X—OH
is a second
alcohol esterified
to phosphoric acid

Name of X—OH	Formula of —X		Name of Phosphoacylglycerol
Ethanolamine	$-CH_2CH_2NH_3^+$	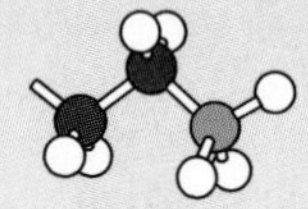	Phosphatidylethanolamine
Choline	$-CH_2CH_2N(CH_3)_3^+$	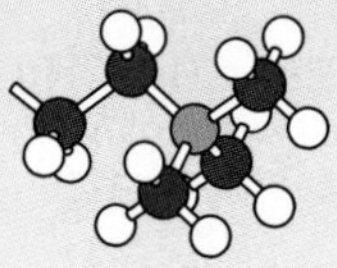	Phosphatidylcholine (lecithin)
Serine	$-CH_2CH(NH_3^+)COO^-$	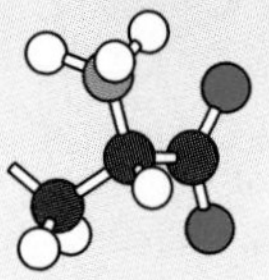	Phosphatidylserine
Inositol	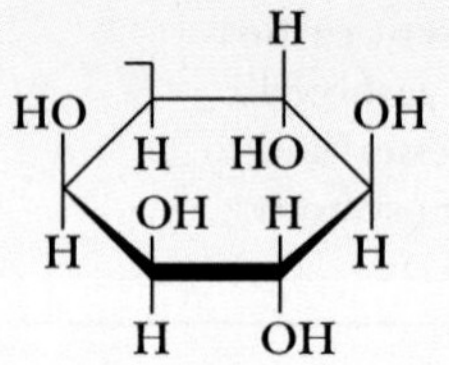	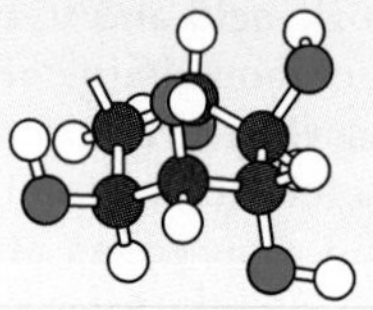	Phosphatidylinositol
Glycerol	$-CH_2CH(OH)CH_2OH$	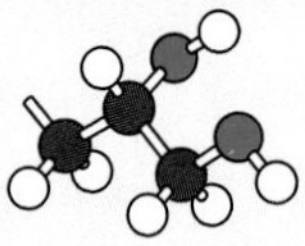	Phosphatidylglycerol
Phosphatidylglycerol	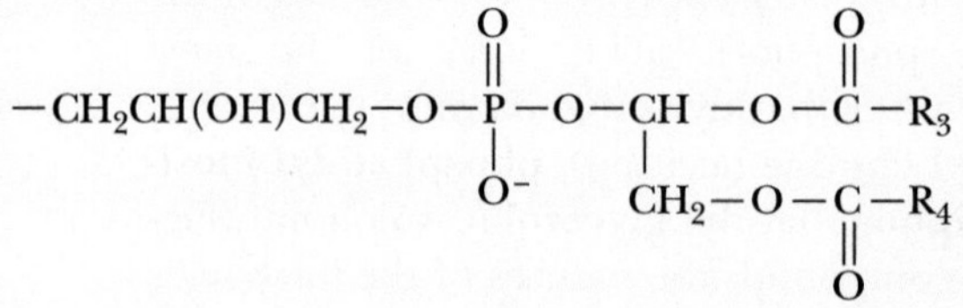 $-CH_2CH(OH)CH_2-O-\overset{\displaystyle O}{\underset{\displaystyle O^-}{P}}-O-\overset{\displaystyle CH}{\underset{\displaystyle CH_2-O-\overset{\displaystyle O}{C}-R_4}{}}-O-\overset{\displaystyle O}{C}-R_3$		Diphosphatidylglycerol (cardiolipin)

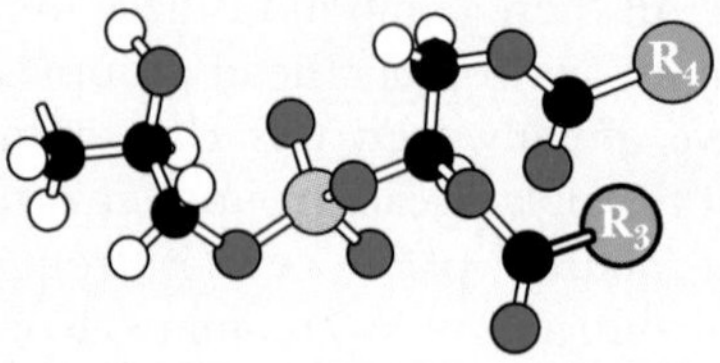

Inositol structure:

HO   OH
   H   HO
   OH   H
H   H
   H   OH

FIGURE 8.5

Structures of some phosphoacylglycerols.

Sphingolipids

Sphingolipids do not contain glycerol, but they do contain the long-chain amino alcohol **sphingosine,** from which this class of compounds takes its name. Sphingolipids are found in both plants and animals; they are particularly abundant in the nervous system. The simplest compounds of this class are the **ceramides,** which consist of one fatty acid linked to the amino group of sphingosine by an amide bond. In **sphingomyelins** the primary alcohol group of sphingosine is esterified to phosphoric acid, which in turn is esterified to another amino alcohol, choline (Figure 8.6). Note the structural similarities between sphingomyelin and other phospholipids. Two long hydrocarbon chains are attached to a backbone that contains alcohol groups. One of the alcohol groups of the backbone is esterified to phosphoric acid. A second alcohol, choline in this case, is also esterified to the phosphoric acid. We have already seen that choline occurs in phosphoacylglycerols as well. Sphingomyelins are amphiphilic; they occur in cell membranes in the nervous system (Box 8.1).

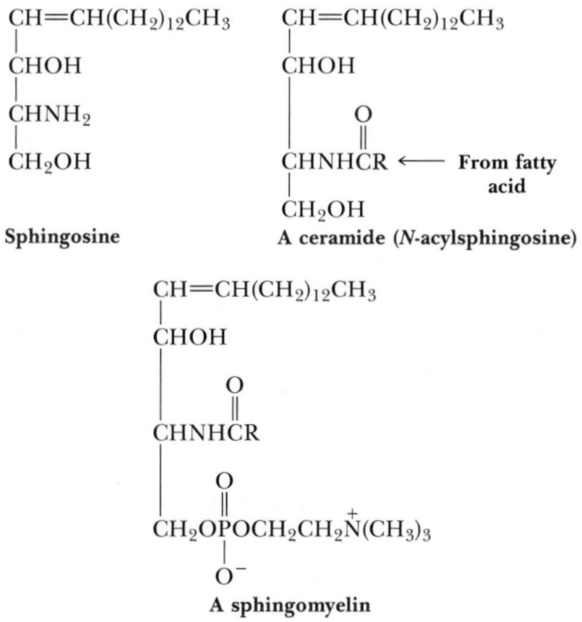

FIGURE 8.6

Structures of some sphingolipids.

MYELIN AND MULTIPLE SCLEROSIS

Annette Funicello enjoyed a successful career in television and films before she was stricken with multiple sclerosis. She started to display the lack of coordination characteristic of the early stages of this disease, causing concern among those who knew her. To end speculations, she announced that she had developed multiple sclerosis.

Myelin is the lipid-rich membrane sheath that surrounds the axons of nerve cells; it has a particularly high content of sphingomyelins. It consists of many layers of plasma membrane that have been wrapped around the nerve cell. Unlike many other types of membranes (Section 8.5), myelin is essentially all lipid bilayer with only a small amount of embedded protein. Its structure, consisting of segments with nodes separating them, promotes rapid transmission of nerve impulses from node to node. Loss of myelin leads to the slowing and eventual cessation of the nerve impulse. In **multiple sclerosis,** a crippling and eventually fatal disease, the myelin sheath is progressively destroyed by **sclerotic plaques**, which affect the brain and spinal cord. These plaques appear to be of autoimmune origin. The progress of the disease is marked by periods of active destruction of myelin interspersed with periods in which no destruction of myelin takes place. Persons affected by multiple sclerosis suffer from weakness, lack of coordination, and speech and vision problems.

Glycolipids

If a carbohydrate is bound to an alcohol group of a lipid by a glycosidic linkage, the resulting compound is a **glycolipid.** Quite frequently, ceramides (see Figure 8.6) are the parent compounds for glycolipids, and the glycosidic bond is formed between the primary alcohol group of the ceramide and a sugar residue. The resulting compound is called a **cerebroside.** In most cases the sugar is glucose or galactose; for example, a glucocerebroside is a cerebroside that contains glucose. As the name indicates, cerebrosides are found in nerve and brain cells, primarily in cell membranes. The carbohydrate portion of these compounds can be very complex.

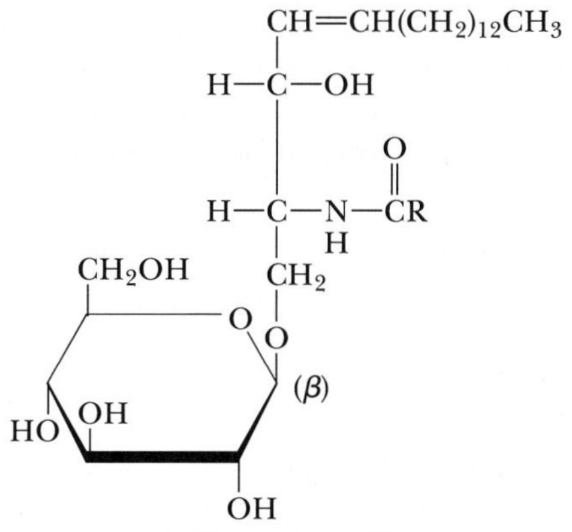

A Glucocerebroside

Steroids

Many compounds of widely differing functions are classified as **steroids** on the basis of the same general structure: a fused-ring system consisting of three six-membered rings (the A, B, and C rings) and one five-membered ring (the D ring). There are many important steroids, including sex hormones. (See Chapter 18, Section 18.5, for descriptions of more steroids of biological importance.) The steroid that is of most interest in our discussion of membranes is **cholesterol** (Figure 8.7). The only hydrophilic group in the cholesterol structure is the single hydroxyl group. As a result, the molecule is highly hydrophobic. Cholesterol is widespread in biological membranes, especially in animals, but it does not occur in prokaryotic cell

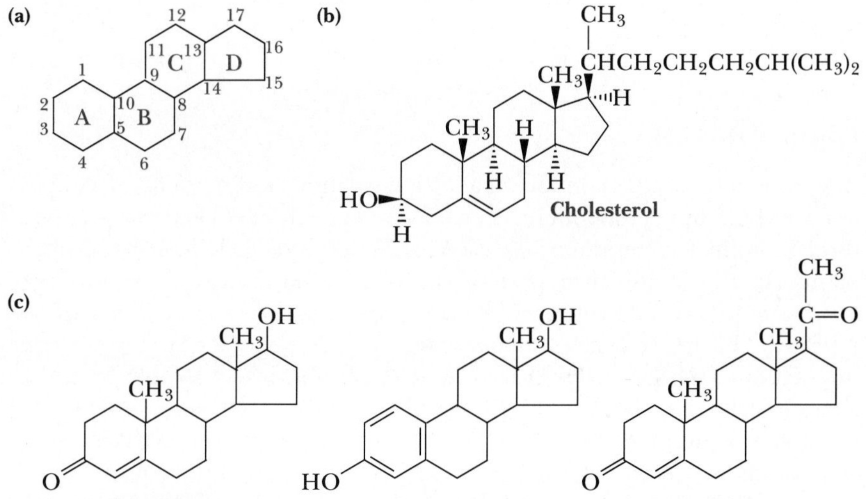

FIGURE 8.7

Structures of some steroids. (a) The fused-ring structure of steroids. (b) Cholesterol. (c) Some steroid sex hormones.

membranes. In spite of its many important biological functions, including its role as a precursor of other steroids and of vitamin D_3, cholesterol is best known for its harmful effects on health. It plays a role in the development of **atherosclerosis,** a condition in which lipid deposits block the blood vessels and lead to heart disease (see Chapter 15, Section 15.8).

8.3
THE NATURE OF BIOLOGICAL MEMBRANES

Every cell has a cell membrane (also called a plasma membrane); eukaryotic cells also have membrane-bounded organelles such as nuclei and mitochondria. The molecular basis of the membrane's structure lies in its lipid and protein components. Now it is time to see how the interaction between the lipid bilayer and membrane proteins determines membrane function. Membranes not only separate cells from the external environment; they also play important roles in transport of specific substances into and out of cells. In addition, a number of important enzymes are found in membranes and depend on that environment for their functions.

Phosphoglycerides are prime examples of amphiphilic molecules, and they are the principal lipid components of membranes. The existence of **lipid bilayers** depends on hydrophobic interactions (Chapter 9, Section 9.6). These bilayers are frequently used as models for biological membranes because they have many features in common with them — such as a hydrophobic interior and an ability to control the transport of small molecules and ions — yet are simpler and easier to work with in the laboratory. The most important difference between lipid bilayers and cell membranes is that the latter contain proteins as well as lipids. The protein component of a membrane can make up 20 to 80% of its total weight. An understanding of membrane structure requires knowledge of how the protein and lipid components contribute to the properties of the membrane.

Lipid Bilayers

Biological membranes contain, in addition to phosphoglycerides, glycolipids as part of the lipid component. Steroids are present in eukaryotes — cholesterol in animal membranes and similar compounds called phytosterols in plants. In the lipid-bilayer part of the membrane (Figure 8.8), the polar head groups are in contact with water, and the nonpolar tails lie in the interior. The whole bilayer arrangement is held together by noncovalent interactions such as van der Waals and hydrophobic interactions (Chapter 2, Section 2.1). The surface of the bilayer is polar and contains charged groups. The nonpolar hydrocarbon interior of the bilayer consists of the saturated and unsaturated chains of fatty acids and the fused ring system of cholesterol. Both the inner and outer layers of the bilayer contain mixtures of lipids, but their compositions differ and can be used to distinguish the inner and outer layers from each other (Figure 8.9). Bulkier molecules tend to occur in the outer layer, and smaller molecules tend to occur in the inner layer.

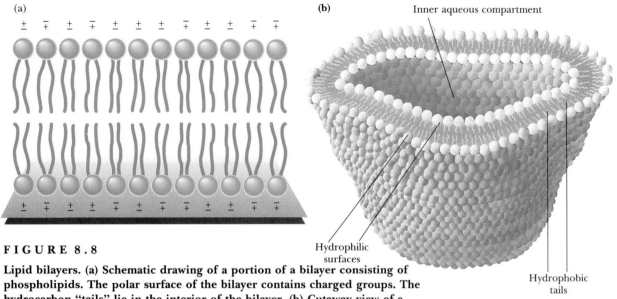

FIGURE 8.8

Lipid bilayers. (a) Schematic drawing of a portion of a bilayer consisting of phospholipids. The polar surface of the bilayer contains charged groups. The hydrocarbon "tails" lie in the interior of the bilayer. (b) Cutaway view of a lipid bilayer vesicle. Note the aqueous inner compartment and the fact that the inner layer is more tightly packed than the outer layer. (From M. S. Bretscher, "The Molecules of the Cell Membrane." *Scientific American*, October 1985, p. 103.)

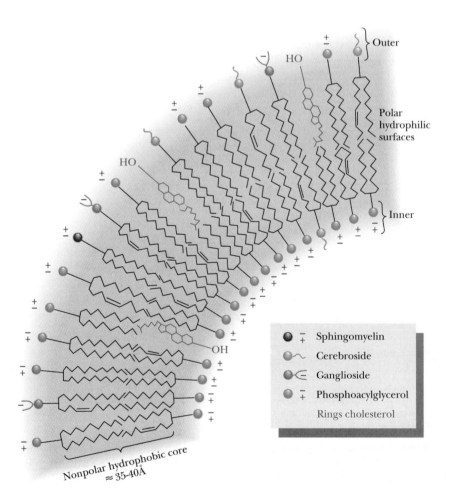

FIGURE 8.9

Lipid bilayer asymmetry. The compositions of the outer and inner layers differ; the concentration of bulky molecules is higher in the outer layer, which has more room.

Legend:
- ⬤ ‾₊ Sphingomyelin
- ● ∼ Cerebroside
- ● ⊂ Ganglioside
- ● ‾₊ Phosphoacylglycerol
- Rings cholesterol

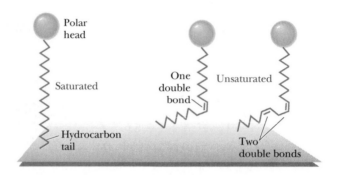

FIGURE 8.10

The effect of double bonds on the conformations of the hydrocarbon tails of fatty acids. Unsaturated fatty acids have kinks in their tail region.

The arrangement of the hydrocarbon interior of the bilayer can be ordered and rigid or disordered and fluid. The bilayer's fluidity depends on its composition. In saturated fatty acids, a linear arrangement of the hydrocarbon chains leads to close packing of the molecules in the bilayer, and thus to rigidity. In unsaturated fatty acids there is a kink in the hydrocarbon chain that does not exist in saturated fatty acids (Figure 8.10). The kinks cause disorder in the packing of the chains, with a more open structure than would be possible for straight saturated chains (Figure 8.11). In turn, the disordered structure due to the presence of unsaturated fatty acids causes greater fluidity in the bilayer. The lipid components of a bilayer are always in motion, to a greater extent in more fluid bilayers and to a lesser extent in more rigid ones.

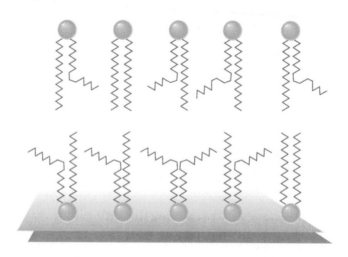

FIGURE 8.11

Schematic drawing of a portion of a highly fluid phospholipid bilayer. The kinks in the unsaturated side chains prevent close packing of the hydrocarbon portions of the phospholipids.

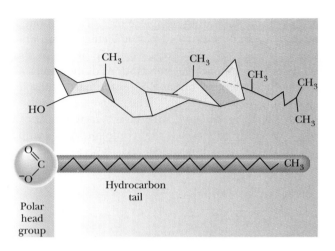

FIGURE 8.12

Stiffening of the lipid bilayer by cholesterol. The presence of cholesterol in a membrane reduces fluidity by stabilizing extended chain conformations of the hydrocarbon tails of fatty aicds, as a result of van der Waals interactions.

The presence of cholesterol may also enhance order and rigidity. The fused-ring structure of cholesterol is itself quite rigid, and the presence of cholesterol stabilizes the extended straight-chain arrangement of saturated fatty acids by van der Waals interactions (Figure 8.12). The lipid portion of a plant membrane has a higher percentage of unsaturated fatty acids, especially polyunsaturated (containing two or more double bonds) fatty acids, than the lipid portion of an animal membrane. Furthermore, the presence of cholesterol is characteristic of animal, rather than plant, membranes; other steroids (phytosterols) are found in plant membranes. As a result, animal membranes are less fluid (more rigid) than plant membranes, and the membranes of prokaryotes, which contain no appreciable amounts of steroids, are the most fluid of all.

With heat, ordered bilayers become less ordered; bilayers that are comparatively disordered become even more disordered. This is a cooperative transition that takes place at a characteristic temperature, like the melting of a crystal (which is also a cooperative transition) (Figure 8.13). The transition temperature is higher for more rigid and ordered membranes than it is for relatively fluid and disordered membranes.

Recall that the distributions of lipids are not the same in the inner and outer portions of the bilayer. Since the bilayer is curved, the molecules of the inner layer are more tightly packed (refer to Figure 8.8). Bulkier molecules such as cerebrosides (see Section 8.2; cerebrosides are glycolipids that contain a sugar moiety in addition to the amino alcohol and a long-chain fatty acid) tend to be located in the outer layer. There is very little

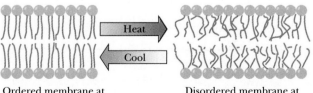

Ordered membrane at lower temperature

Disordered membrane at higher temperature

FIGURE 8.13

Under the influence of heat, a membrane becomes more disordered. This transition takes place at a characteristic temperature, T_m.

tendency for "flip-flop" migration of lipid molecules from one layer of the bilayer to another. Lateral motion of lipid molecules within one of the two layers frequently takes place, however, especially in more fluid bilayers. Several methods exist for monitoring the motions of molecules within a lipid bilayer. These methods depend on labeling some part of the lipid component with an easily detected "tag." The tags are usually fluorescent compounds, which can be detected with high sensitivity. Another kind of labeling method depends on the fact that some nitrogen compounds have unpaired electrons. These compounds are used as labels and can be detected by magnetic measurements.

8.4
MEMBRANE PROTEINS

Proteins in a biological membrane can be associated with the lipid bilayer in either of two ways—as **peripheral proteins** on the surface of the membrane or as **integral proteins** within the lipid bilayer (Figure 8.14). Peripheral proteins are usually bound to the charged head groups of the lipid bilayer by polar or electrostatic interactions, or both. They can be removed by such mild treatment as raising the ionic strength of the medium. The relatively numerous charged particles present in a medium of higher ionic strength undergo more electrostatic interactions with the lipid and with the protein, "swamping out" the comparatively fewer electrostatic interactions between the protein and the lipid.

It is much more difficult to remove integral proteins from membranes. Harsh conditions, such as treatment with detergents or extensive sonication (exposure to ultrasonic vibrations), are usually required. Such measures frequently denature the protein, which often remains bound to lipids in spite of all efforts to obtain it in pure form. The denatured protein is of course inactive, whether or not it remains bound to lipids. Fortunately, nuclear magnetic resonance techniques are now enabling researchers to study proteins of this sort in living tissue. The structural integrity of the whole membrane system appears to be necessary for the activities of most membrane proteins.

FIGURE 8.14

Some types of associations of proteins with membranes. The proteins marked 1, 2, and 4 are integral proteins, and protein 3 is a peripheral protein. Note that the integral proteins can be associated with the lipid bilayer in several ways. Protein 1 transverses the membrane, protein 2 lies entirely within the membrane, and protein 4 projects into the membrane.

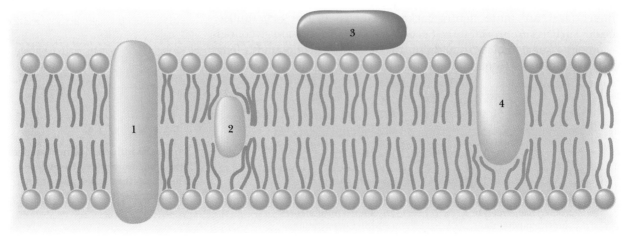

Membrane proteins have a variety of functions. Most, but not all, of the important functions of the membrane as a whole are those of the protein component. **Transport proteins** help move substances in and out of the cell, and **receptor proteins** are important in the uptake of extracellular signals, such as hormones or neurotransmitters, into the cell. In addition, some enzymes are tightly bound to membranes; examples include many of the enzymes responsible for aerobic oxidation reactions, which are found in specific parts of mitochondrial membranes. Some of these enzymes are on the inner surface of the membrane, and some are on the outer surface. There is an uneven distribution of proteins of all types on the inner and outer layers of all cell membranes, just as there is an asymmetric distribution of lipids.

8.5
THE FLUID MOSAIC MODEL OF MEMBRANE STRUCTURE

We have seen that biological membranes have both lipid and protein components. How do these two parts combine to produce a biological membrane? The **fluid mosaic model** is the most widely accepted description of biological membranes at the moment. The term "mosaic" implies that the two components exist side by side without forming some other substance of intermediate nature. No extensive formation of lipid–protein complexes takes place, for example. Instead, the basic structure of biological membranes is that of the lipid bilayer, with the proteins embedded in the bilayer structure (Figure 8.15). The term "fluid mosaic" implies that the same sort of lateral motion occurs in membranes that we have already seen in lipid bilayers. The proteins "float" in the lipid bilayer and can move along the plane of the membrane.

FIGURE 8.15

Fluid mosaic model of membrane structure. Membrane proteins can be seen embedded in the lipid bilayer. (From S. J. Singer, in G. Weismann and R. Claiborne, eds., *Cell Membranes: Biochemistry, Cell Biology, and Pathology,* **New York: HP Pub., 1975, p. 37.)**

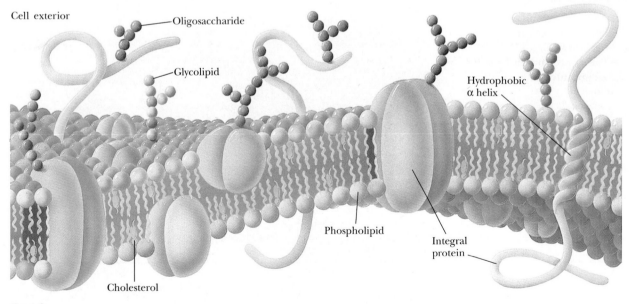

Cell exterior

Oligosaccharide

Glycolipid

Hydrophobic α helix

Phospholipid

Integral protein

Cholesterol

Cytosol

FIGURE 8.16

Replica of a freeze-fractured membrane. In the freeze-fracture technique, the lipid bilayer is split parallel to the surface of the membrane. The hydrocarbon tails of the two layers are separated from each other, and the proteins can be seen as "hills" in the replica shown. In the other layer, seen edge on, there are "valleys" where the proteins were. (From S. J. Singer, in G. Weismann and R. Claiborne, eds., *Cell Membranes: Biochemistry, Cell Biology, and Pathology*, New York: HP Pub., 1975, p. 38.)

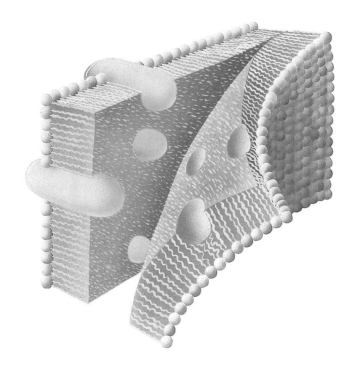

Electron micrographs can be made of membranes that have been frozen and then fractured along the interface between the two layers. The outer layer is removed, exposing the interior of the membrane. The interior has a granular appearance because of the presence of the integral membrane proteins (Figures 8.16 and 8.17).

FIGURE 8.17

Electron micrograph of a freeze-fractured thylakoid membrane of a pea (magnified 110,000×). The grains protruding from the surface are integral membrane proteins.

8.6
MEMBRANE FUNCTION: AN INTRODUCTION

As already mentioned, three important functions take place in or on membranes (in addition to the structural role of membranes as the boundaries and containers of all cells and of the organelles within eukaryotic cells). The first of these functions is **transport.** Membranes are semipermeable barriers to the flow of substances into and out of cells and organelles. Transport through the membrane can involve the lipid bilayer as well as the membrane proteins. The other two important functions primarily involve the membrane proteins. One of these functions is **catalysis.** As we have seen, enzymes can be bound—in some cases very tightly—to membranes, and the enzymatic reaction takes place on the membrane. The third significant function is the **receptor property,** in which proteins bind specific biologically important substances that trigger biochemical responses in the cell.

We shall discuss enzymes bound to membranes in subsequent chapters (especially in our treatment of aerobic oxidation reactions). The other two functions we now consider in turn.

Membrane Transport

The most important question about transport of substances across biological membranes is whether or not the process requires expenditure of energy by the cell. In **passive transport** a substance moves from a region of higher concentration to one of lower concentration. In other words, the movement of the substance is in the same direction as a *concentration gradient,* and no energy is expended by the cell. In **active transport** a substance moves from a region of lower concentration to one of higher concentration (against a concentration gradient), and this process does require expenditure of energy by the cell.

The process of passive transport can be subdivided into two categories, simple diffusion and facilitated diffusion. In **simple diffusion** a molecule moves directly through an opening or pore in the membrane without interacting with another molecule. The transported molecule goes either directly through the lipid bilayer or through an opening in a **channel protein** (Figure 8.18). Small molecules such as water, oxygen, and carbon

FIGURE 8.18

Schematic representation of a channel protein, shown as having four subunits. Both channel proteins, which play a role in simple diffusion, and carrier proteins, which mediate facilitated diffusion, have pores through which substances can pass. The difference between the two is that in facilitated diffusion substances entering the cell bind to the carrier proteins, whereas in simple diffusion the ions or small molecules simply pass through the pore in the channel protein.

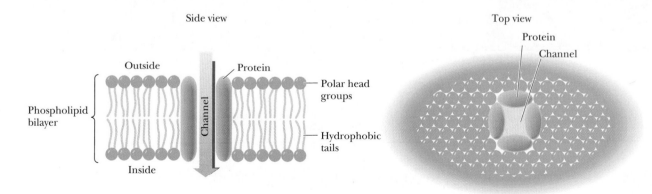

Side view

Outside

Protein

Polar head groups

Channel

Phospholipid bilayer

Hydrophobic tails

Inside

Top view

Protein

Channel

dioxide can pass directly through the lipid bilayer, but larger molecules, especially polar ones, require a channel protein. Ions cannot pass through the lipid bilayer because of their charge, and they also require channel proteins. The sizes and shapes of the openings in channel proteins make them specific for given ions or molecules.

In facilitated diffusion the molecule to be transported across the membrane binds to a **carrier protein** rather than simply passing through, as it would a channel protein in simple diffusion. In both a channel protein and a carrier protein a pore is created by the folding of the backbone and side chains. Many of these proteins have several α-helical portions that span the membrane; in others a β-barrel forms the pore. In one example the helical portion of the protein spans the membrane. The exterior, which is in contact with the lipid bilayer, is hydrophobic, whereas the interior through which ions pass, is hydrophilic.

Active transport requires moving substances against a concentration gradient. The situation is so markedly similar to pumping water uphill that

FIGURE 8.19

The sodium–potassium pump (see text).

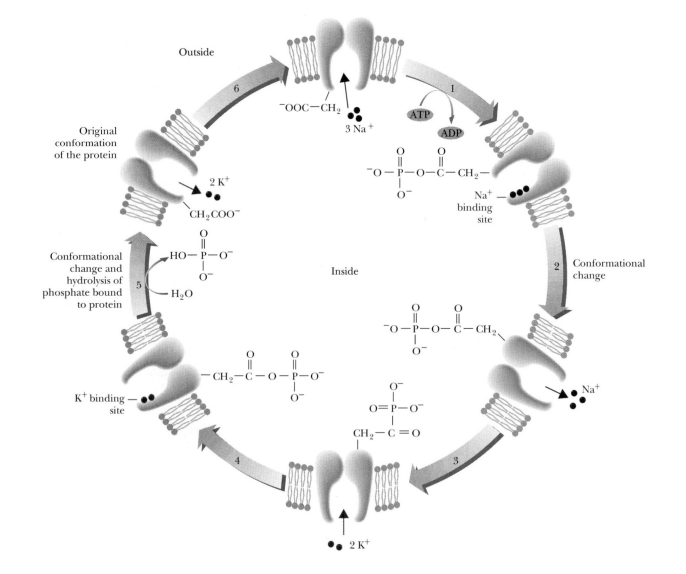

one of the most extensively studied examples of active transport, moving potassium ions into a cell and simultaneously moving sodium ions out of the cell, is referred to as the **sodium–potassium ion pump.**

Under normal circumstances the concentration of K^+ is higher inside a cell than in extracellular fluids ($[K^+]_{inside} > [K^+]_{outside}$), but the concentration of Na^+ is lower inside the cell than out ($[Na^+]_{inside} < [Na^+]_{outside}$). The energy required to move these ions against their gradients comes from an exergonic (energy-releasing) reaction, the hydrolysis of ATP to ADP and P_i (phosphate ion). There can be no transport of ions without hydrolysis of ATP. The same protein appears to serve as both the enzyme that hydrolyzes the ATP (the ATPase) and the transport protein; it consists of several subunits. The reactants and products of this hydrolysis reaction — ATP, ADP, and phosphate ion — remain within the cell, and the phosphate becomes covalently bonded to the transport protein for part of the process.

The Na^+–K^+ pump operates in several steps (Figure 8.19). One subunit of the protein hydrolyzes the ATP and transfers the phosphate group to an aspartate side chain on another subunit (Step 1). Simultaneously, binding of 3 Na^+ from the interior of the cell takes place. The phosphorylation of one subunit causes a conformational change in the protein, which opens a channel or pore through which the 3 Na^+ can be released to the extracellular fluid (Step 2). Outside the cell, 2 K^+ bind to the pump enzyme, which is still phosphorylated (Step 3). Another conformational change occurs when the bond between the enzyme and the phosphate group is hydrolyzed. This second conformational change regenerates the original form of the enzyme and allows the 2 K^+ to enter the cell (Step 5). The pumping process transports three Na^+ ions out of the cell for every two K^+ ions transported into the cell (Figure 8.20).

The operation of the pump can be reversed when there is no K^+ and a high concentration of Na^+ in the extracellular medium; in this case ATP is produced by the phosphorylation of ADP. The actual operation of the Na^+–K^+ pump is not completely understood and may be even more complicated than we now know. There is also a calcium ion (Ca^{2+}) pump,

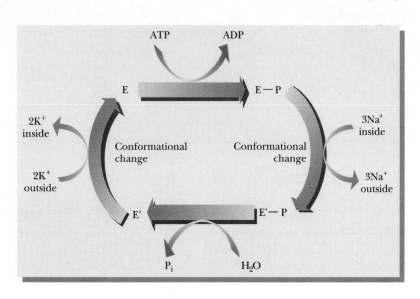

FIGURE 8.20

Another way of looking at the sodium–potassium pump. E is the enzyme (original conformation). E—P is the enzyme with a covalently bound phosphate. E′ is the enzyme after hydrolysis of phosphate (second conformation). E′—P is the enzyme in the second conformation with covalently bound phosphate.

which is a subject of equally active investigation. Unanswered questions about the detailed mechanism of active transport provide opportunities for future research.

Membrane Receptors

The first step in the effects of some biologically active substances is binding of the substance to a protein receptor site on the exterior of the cell. The interaction between receptor proteins and the active substances to which they bind is very similar to enzyme–substrate recognition. There is a requirement for essential functional groups that have the right three-dimensional conformation with respect to one another. The binding site, whether on a receptor or an enzyme, must provide a good fit for the substrate. In receptor binding, as in enzyme behavior, there is a possibility for inhibition of the action of the protein by some sort of "poison" or inhibitor. The study of receptor proteins is less advanced than the study of enzymes, since many receptors are tightly bound integral proteins and their activity depends on the membrane environment. Receptors are often large oligomeric proteins (ones with several subunits), with molecular weights on the order of hundreds of thousands. Also, quite frequently there are very few molecules of the receptor in each cell, adding to the difficulties of isolating and studying this type of protein.

An important type of receptor is that for low-density lipoprotein (LDL), the principal carrier of cholesterol in the bloodstream. LDL is a particle that

FIGURE 8.21

The mode of action of the LDL receptor. A portion of the membrane, with LDL receptor and bound LDL, is taken into the cell as a vesicle. The receptor protein releases LDL and is returned to the cell surface when the vesicle fuses with the membrane. LDL releases cholesterol in the cell. An oversupply of cholesterol inhibits synthesis of the LDL receptor protein. An insufficient number of receptors leads to elevated levels of LDL and cholesterol in the bloodstream. This situation increases the risk of heart attack.

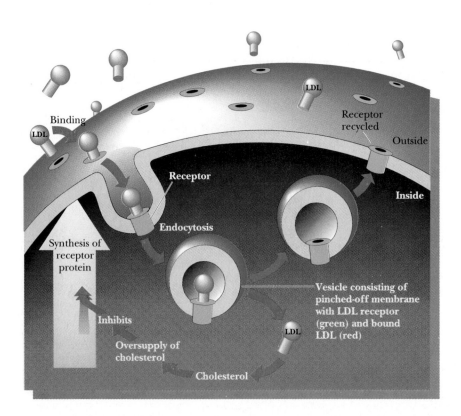

consists of various lipids—in particular, cholesterol and phosphoglycer-ides—as well as a protein. The protein portion of the LDL particle binds to the LDL receptor of a cell. The complex formed between the LDL and the receptor is pinched off into the cell in a process called **endocytosis.** (This important aspect of receptor action is described in detail in the articles by Brown and Goldstein and by Dautry-Varsat and Lodish listed in the bibliography at the end of this chapter.) The receptor protein is then recycled back to the surface of the cell (Figure 8.21). The cholesterol portion of the LDL is used in the cell, but an oversupply of cholesterol causes problems. Excess cholesterol inhibits the synthesis of LDL receptor. If there are too few receptors for LDL, the level of cholesterol in the bloodstream increases. Eventually the excess cholesterol is deposited in the arteries, blocking them severely. This blocking of arteries, called **atheroscle-rosis,** can eventually lead to heart attacks and strokes. In many industrial-ized countries, typical blood cholesterol levels are high, and the incidence of heart attacks and strokes is correspondingly high. (We can say more about this subject after we have seen the pathway by which cholesterol is synthesized in the body, in Chapter 15, Section 15.8.)

8.7
A CASE STUDY IN MEMBRANE BEHAVIOR: THE NEUROMUSCULAR JUNCTION

The complex action of the neuromuscular junction between a nerve cell (a neuron) and a muscle cell provides examples of important aspects of receptor behavior in conjunction with membrane transport. One of the most important features of this behavior is the operation of **gated channels.** The channels in proteins, which serve as avenues for transport into the cell, are not necessarily open all the time. Some channels are open continuously, but others are opened only transiently by gating mechanisms. **Ligand-gated channels** are controlled by the binding of some substance in the extracellu-lar fluid to the receptor; **voltage-gated channels** are controlled by the voltage across the membrane (membrane potential) generated by the concentration gradients of ions (Figure 8.22). (In the normal situation of **membrane polarization,** the external side of the plasma membrane has a positive charge with respect to the internal side.) In the neuromuscular junction a specific sequence of opening and closing of four such gated channels takes place. (We shall use skeletal muscle junctions for our example; the process in smooth muscle differs somewhat.)

   Figure 8.23 shows a resting neuromuscular junction and an activated junction with muscular contraction. Of the four gated channels that play a role in the process, three are voltage-gated and one is ligand-gated. The three voltage-gated channels are:

1. A Ca^{2+} channel in the plasma membrane of the nerve cell (Ca^{2+} enters the nerve cell through this channel)
2. An Na^+ channel in the plasma membrane of the muscle cell (Na^+ enters the muscle cell)
3. A Ca^{2+} channel in the internal membrane system (sarcoplasmic reticu-lum) of the muscle cell (Ca^{2+} flows from the sarcoplasmic reticulum into the cytosol of the muscle cell)

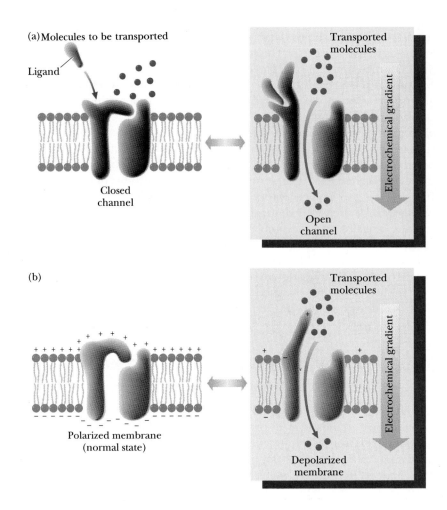

The ligand-gated channel in the plasma membrane of the muscle cell is gated by binding of the neurotransmitter acetylcholine. When open, the acetylcholine-gated cation channel allows the inflow of Na^+ into the muscle

$$CH_3-\overset{\displaystyle O}{\overset{\|}{C}}-O-CH_2CH_2-\overset{+}{N}(CH_3)_3$$
Acetylcholine

cell and the outflow of K^+, both in the directions of their respective concentration gradients.

The activation of the neuromuscular junction is initiated by the nerve impulse. A characteristic of the impulse is a decrease in the membrane potential—that is, depolarization of the membrane. The depolarization leads to the transient opening of the Ca^{2+} channel in the nerve cell. (This is the first of the four gated channels involved in the process.) The Ca^{2+} concentration is about 1000 times higher outside the cell than inside, so Ca^{2+} flows freely into the cell.

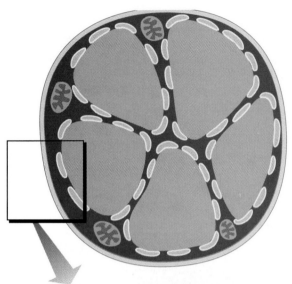

FIGURE 8.23

Comparison of resting and activated neuromuscular junctions. (a) A resting neuromuscular junction. (b) An activated neuromuscular junction causing muscular contraction. The gated channels are numbered in the sequence in which they open.

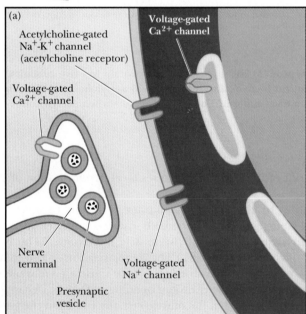

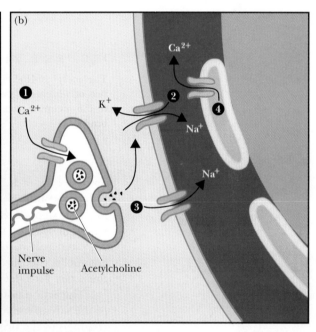

The presence of increased levels of Ca^{2+} inside the nerve cell triggers the release of acetylcholine from membrane-bounded vesicles within the nerve cell. The vesicles fuse with the plasma membrane of the nerve cell, releasing acetylcholine into the space between the nerve cell and the muscle cell. Acetylcholine is taken up by receptors on the muscle cell plasma membrane. These receptors are the second of the four gated channels involved in the operation of the neuromuscular junction; they are acetylcholine-gated channels for Na^+ and K^+. Among the four gated channels that control the action of the neuromuscular junction, this protein is the only receptor with a binding site for a molecule, but the receptor action depends on all four proteins acting in concert. The flow of Na^+ into

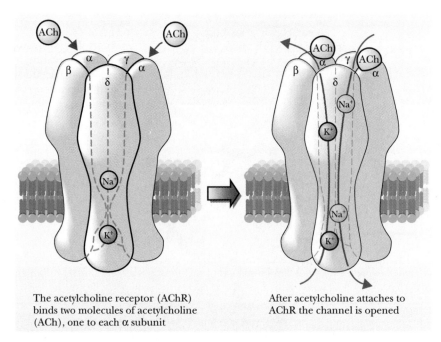

The acetylcholine receptor (AChR) binds two molecules of acetylcholine (ACh), one to each α subunit

After acetylcholine attaches to AChR the channel is opened

FIGURE 8.24

The acetylcholine receptor (AChR) in action. This protein has five subunits, two of which are identical ($\alpha_2\beta\gamma\delta$). Two molecules of acetylcholine (ACh) bind to the receptor, one to each subunit. The channel opens when acetylcholine binds.

and the flow of K^+ out of the muscle cell causes local depolarization of the muscle membrane.

The depolarization of the muscle membrane leads, in turn, to the transient opening of voltage-gated Na^+ channels, causing a greater inflow of Na^+ and still further depolarization of the muscle membrane. (The Na^+ channel is the third of the four channels.) The depolarization — also called the *action potential* — eventually spreads to the whole muscle membrane.

The action potential leads to the transient opening of the Ca^{2+} channels of the sarcoplasmic reticulum within the muscle cell, the fourth of the gated channels in the process. (In cells, the internal membrane system is connected to the plasma membrane.) The outflow of Ca^{2+} from the sarcoplasmic reticulum into the cytosol drastically increases the cytosolic concentration of Ca^{2+}, triggering the Ca^{2+}−dependent process of contraction of the myofibrils within the muscle cell (muscle contraction).

The acetylcholine receptor has been the subject of considerable research. It is a large protein consisting of five subunits, two of which are identical; its composition is designated $\alpha_2\beta\gamma\delta$. It is now known that two molecules of acetylcholine bind to the receptor, one to each α subunit (Figure 8.24). When acetylcholine binds to the receptor, the channel opens. The binding site of each subunit is part of a hydrophilic region of a largely hydrophobic membrane-spanning protein. (See the article by Changeux listed in the bibliography at the end of this chapter.)

8.8
LIPID-SOLUBLE VITAMINS

Some vitamins, having a variety of functions, are of interest in this chapter because they are soluble in lipids. They are themselves hydrophobic, which accounts for their solubility (Table 8.3).

Vitamin A

The extensively unsaturated hydrocarbon **β-carotene** is the precursor of **vitamin A,** which is also known as **retinol.** As the name suggests, β-carotene is abundant in carrots, and it also occurs in other vegetables, particularly the yellow ones. When an organism requires vitamin A, β-carotene is converted to the vitamin (Figure 8.25a).

A derivative of vitamin A plays a crucial role in vision when it is bound to a protein called **opsin.** The cone cells in the retina of the eye contain several types of opsin and are responsible for vision in bright light and for color vision. The rod cells in the retina contain only one type of opsin; they are responsible for vision in dim light. The chemistry of vision has been more extensively studied in rod cells than in cone cells, and we shall discuss events that take place in rod cells.

Vitamin A has an alcohol group that is enzymatically oxidized to an aldehyde group, forming **retinal** (Figure 8.25b). This oxidation in the eye is catalyzed by the enzyme alcohol dehydrogenase. When one is poisoned by methanol (See Chapter 5, Box 5.1), the methanol is oxidized to formaldehyde (HCHO) by alcohol dehydrogenase. Further oxidation of formaldehyde produces formic acid (HCOOH), a highly toxic compound that causes the blindness associated with methanol poisoning.

Two isomeric forms of retinal, involving *cis-trans* isomerization around one of the double bonds, are important in the behavior of this compound *in vivo.* The aldehyde group of retinal forms an imine (also called a Schiff base) with the side-chain amino group of a lysine residue in rod-cell opsin (Figure 8.26, page 270).

The product of the reaction between retinal and opsin is **rhodopsin.** The outer segment of rod cells contains flat membrane-bounded discs, the membrane consisting of about 60% rhodopsin and 40% lipid. (For more details about rhodopsin, see Box 8.2, page 269.)

T A B L E 8 . 3 **Lipid-Soluble Vitamins and Their Functions**

VITAMIN	FUNCTION
Vitamin A	Serves as the site of the primary photochemical reaction in vision
Vitamin D	Regulates calcium (and phosphorus) metabolism
Vitamin E	Serves as an antioxidant; necessary for reproduction in rats, and may be necessary for reproduction in humans
Vitamin K	Has a regulatory function in blood clotting

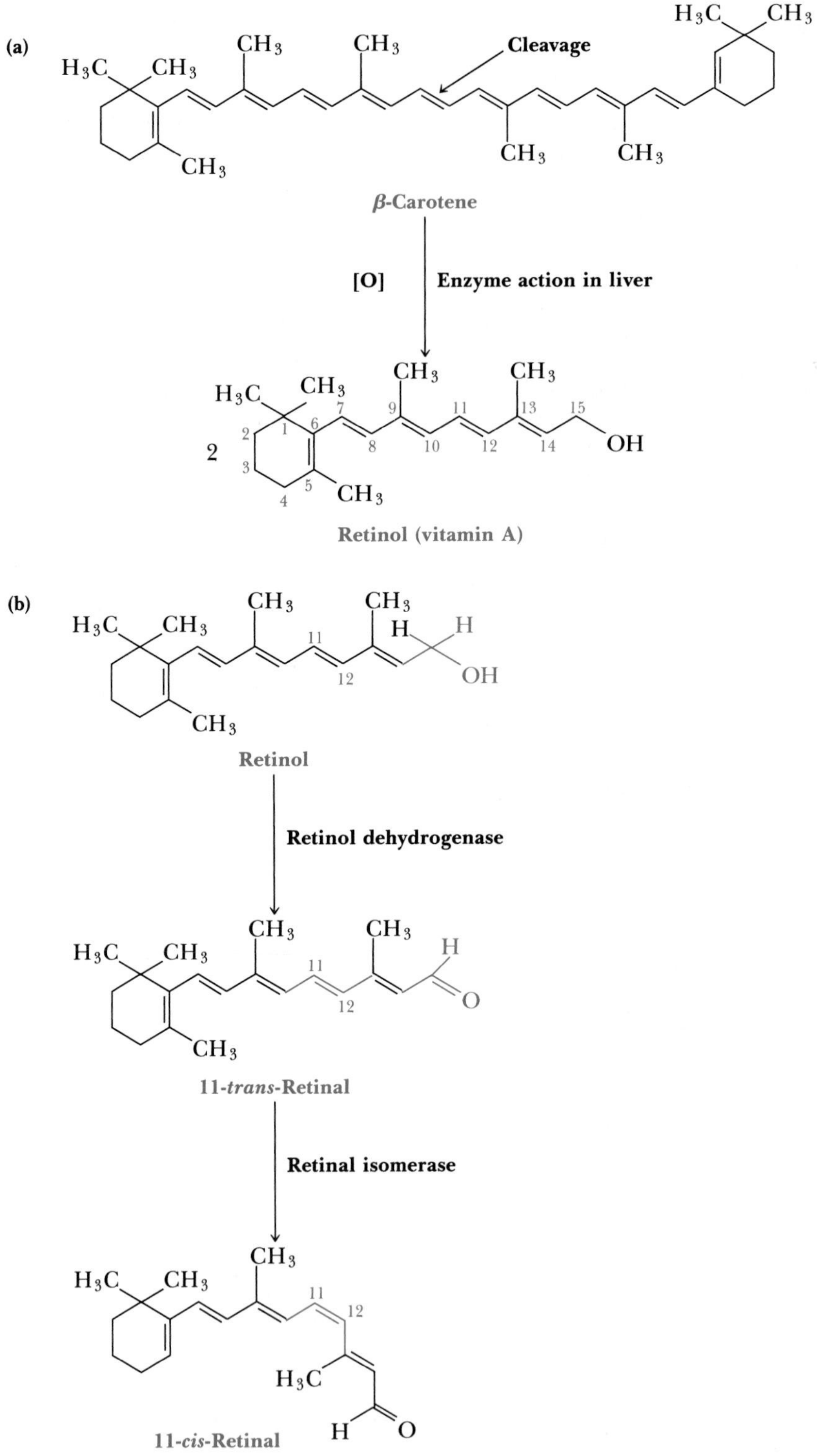

FIGURE 8.25

Reactions of vitamin A. (a) The conversion of β-carotene to vitamin A. (b) The conversion of vitamin A to 11-*cis*-retinal.

Box 8.2

THE CHEMISTRY OF VISION

The primary chemical reaction in vision, the one responsible for generating an impulse in the optic nerve, involves *cis-trans* isomerization around one of the double bonds in the retinal portion of rhodopsin. When rhodopsin is active (that is, when it can respond to visible light), the double bond between carbon atoms 11 and 12 of the retinal (11-*cis*-retinal) has the *cis* orientation. Under the influence of light, an isomerization reaction occurs at this double bond, producing all-*trans*-retinal. Since the all-*trans* form of retinal cannot bind to opsin, all-*trans*-retinal and free opsin are released. As a result of this reaction, an electrical impulse is generated in the optic nerve and transmitted to the brain to be processed as a visual event. The active form of rhodopsin is regenerated by enzymatic isomerization of the all-*trans*-retinal back to the 11-*cis* form and subsequent re-formation of the rhodopsin.

Vitamin A deficiency can have drastic consequences, as would be predicted from its importance in vision. Night blindness and even total blindness can result, especially in children. On the other hand, an excess of vitamin A can have harmful effects such as bone fragility. Lipid-soluble compounds are not excreted as readily as water-soluble substances, and it is possible for excessive amounts of lipid-soluble vitamins to accumulate in adipose tissue.

The primary chemical reaction of vision

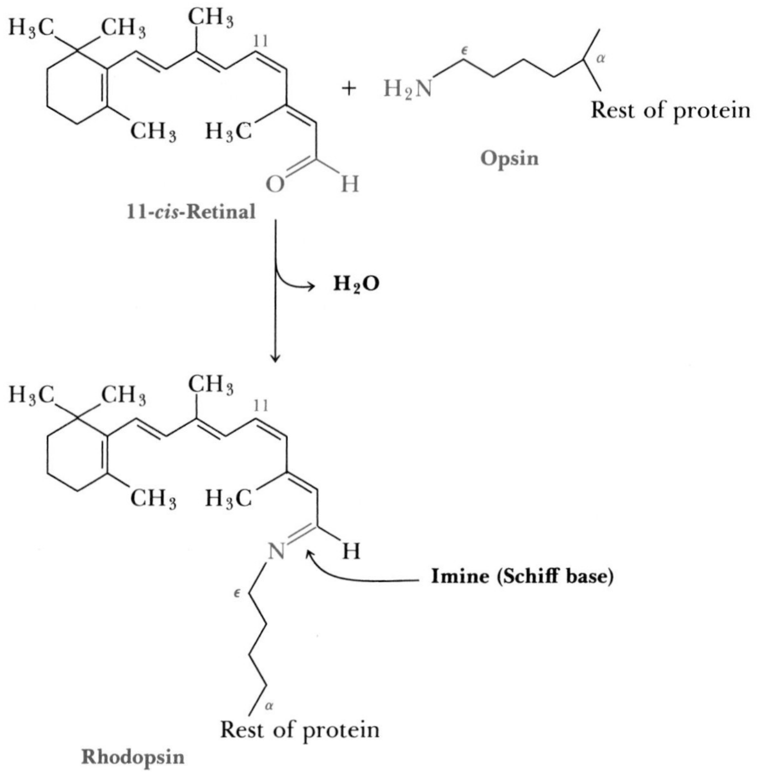

FIGURE 8.26

The formation of rhodopsin from 11-*cis*-retinal and opsin.

Vitamin D

The several forms of **vitamin D** play a major role in regulation of calcium and phosphorus metabolism. One of the most important of these compounds, vitamin D_3 (cholecalciferol), is formed from cholesterol by the action of ultraviolet radiation from the sun. Vitamin D_3 is further processed in the body to form hydroxylated derivatives, which are the metabolically active form of this vitamin (Figure 8.27). The presence of vitamin D_3 leads to increased synthesis of a Ca^{2+}-binding protein, which increases absorption of dietary calcium in the intestines. This process results in calcium uptake by the bones.

A deficiency of vitamin D can lead to **rickets,** a condition in which the bones of growing children become soft, resulting in skeletal deformities. Children, especially infants, have higher requirements for vitamin D than do adults. Milk with vitamin D supplements is available to most children. Adults who are exposed to normal amounts of sunlight do not usually require vitamin D supplements.

Vitamin E

The most active form of **vitamin E** is ***α*-tocopherol.** In rats, vitamin E is required for reproduction and for prevention of the disease **muscular**

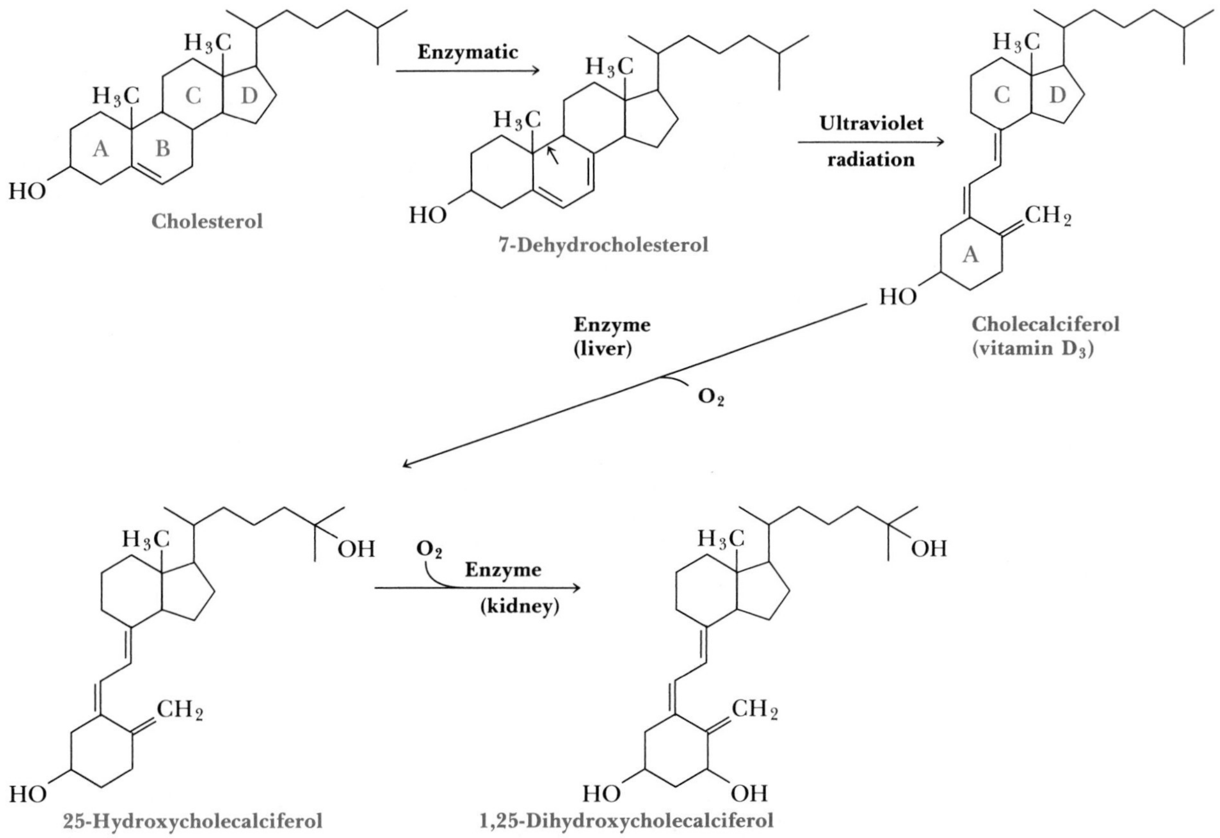

FIGURE 8.27

Reactions of vitamin D. The photochemical cleavage occurs at the bond shown by the arrow; electron rearrangements after the cleavage produce vitamin D₃. The final product, 1,25-dihydroxy-cholecalciferol, is the form of the vitamin that is most active in stimulating the intestinal absorption of calcium and phosphate and in mobilizing calcium for bone development.

dystrophy. It is not definitely known whether this requirement exists in humans. A well-established chemical property of vitamin E is that it is an **antioxidant**—that is, a good reducing agent—so it reacts with oxidizing agents before they can attack other biomolecules. The antioxidant action of vitamin E has been shown to protect important compounds, including vitamin A, from degradation in the laboratory; it probably also serves this function in organisms. Recent research has shown that the interaction of vitamin E with membranes enhances its effectiveness as an antioxidant. Another function of antioxidants such as vitamin E is to react with, and thus remove, the very reactive and highly dangerous substances known as **free radicals.** A free radical has at least one unpaired electron, which accounts for its high degree of reactivity. Free radicals may play a part in the development of cancer and in the aging process.

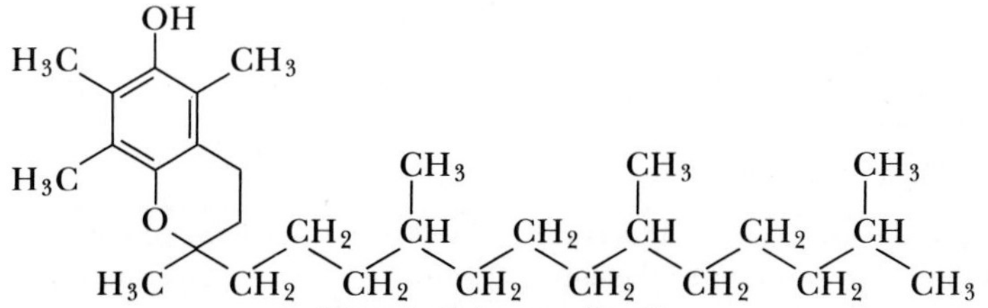

The most active form of vitamin E is *α*-tocopherol.

Vitamin K

The name of **vitamin K** comes from the German *Koagulation,* because this vitamin is an important factor in the blood clotting process. The bicyclic ring system contains two carbonyl groups, the only polar groups on the molecule (Figure 8.28). There is a long unsaturated hydrocarbon side chain that consists of repeating *isoprene* units, the number of which determines the exact form of vitamin K. Several forms of this vitamin can be found in a single organism, and the reason for this variation is not well understood. Vitamin K is not the first vitamin we have encountered that contains isoprene units, but it is the first one in which the number of isoprene units and their degree of saturation make a difference. (See whether you can pick out isoprene-derived portions of the structures of vitamins A and E.) It is also known that the steroids are biosynthetically derived from isoprene units, but the structural relationship is not immediately obvious (Chapter 15, Section 15.8).

The presence of vitamin K is required in the complex process of blood clotting, which involves many steps and many proteins and has stimulated numerous unanswered questions. It is known that vitamin K is required to modify the protein prothrombin. In blood clotting, the side chains of several

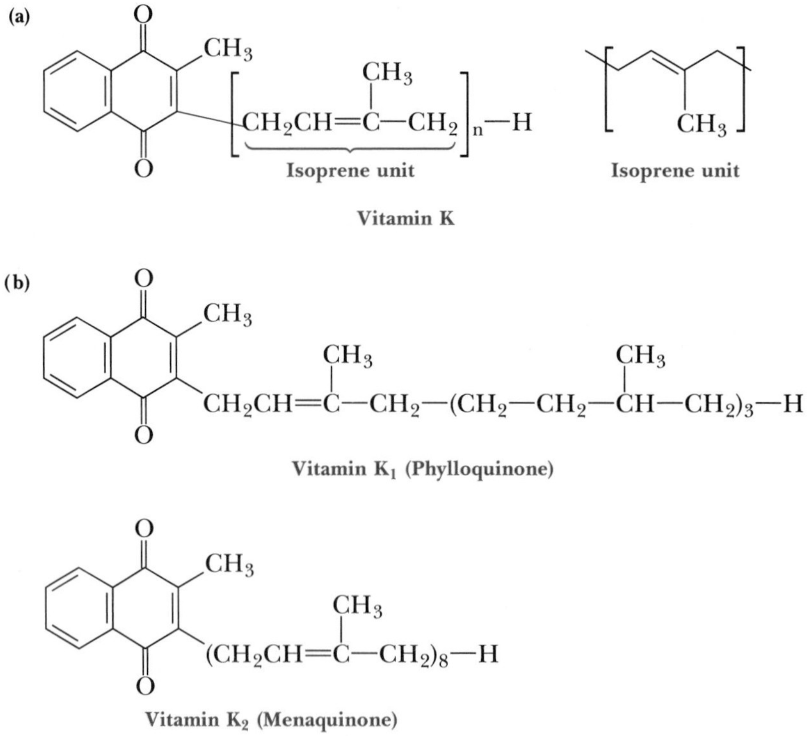

FIGURE 8.28

(a) The general structure of vitamin K, which is required for blood clotting. The value of n is variable but usually <10. (b) Vitamin K_1 has one unsaturated isoprene unit. The rest are saturated. Vitamin K_2 has eight unsaturated isoprene units.

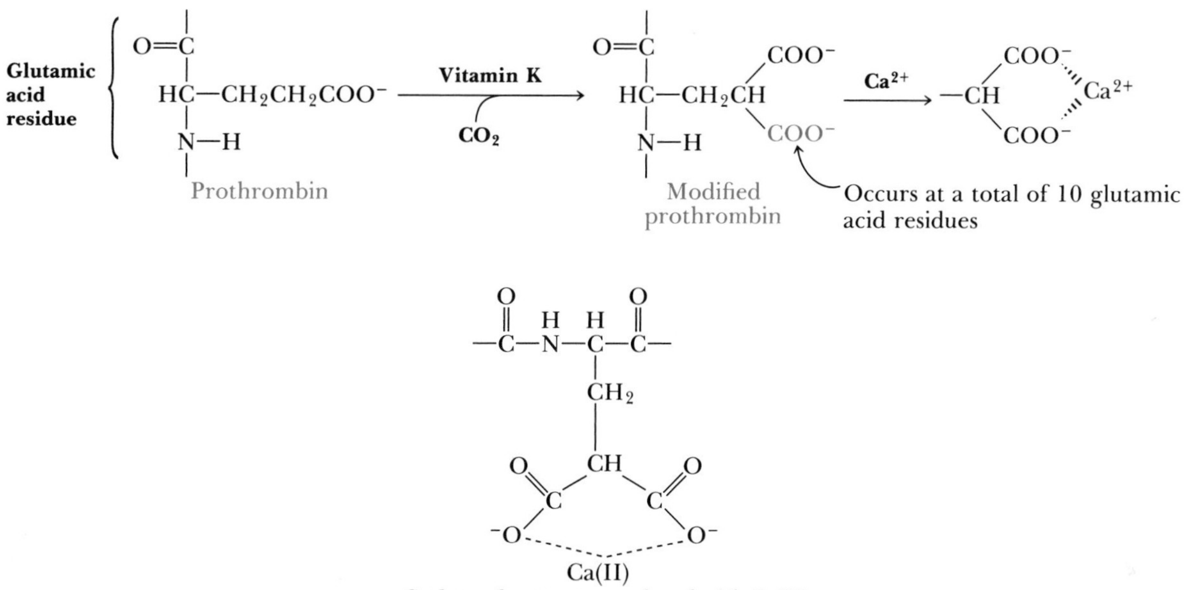

γ-Carboxyglutamate complexed with Ca(II)

FIGURE 8.29

The role of vitamin K in the modification of prothrombin. The detailed structure of the γ-carboxyglutamate at the calcium complexation site is shown below.

glutamate residues of prothrombin are altered by the addition of another carboxyl group. This modification of glutamate produces γ-carboxylglutamate residues (Figure 8.29). The two carboxyl groups in proximity form a **bidendate** ("two teeth") **ligand,** which can bind calcium ion (Ca^{2+}). If prothrombin is not modified in this way, it does not bind Ca^{2+}. Even though there is a lot more to be learned about blood clotting and the role of vitamin K in the process, this point at least is well established, because Ca^{2+} is required for blood clotting.

8.9
PROSTAGLANDINS AND LEUKOTRIENES

A group of compounds derived from fatty acids has a wide range of physiological activities; they are called **prostaglandins** because they were first detected in seminal fluid, which is produced by the prostate gland. It has since been shown that they are widely distributed in a variety of tissues. The metabolic precursor of all prostaglandins is **arachidonic acid,** a fatty acid that contains 20 carbon atoms and four double bonds. The double bonds are not conjugated. The production of the prostaglandins from arachidonic acid takes place in several steps, which are catalyzed by enzymes. The prostaglandins each have a five-membered ring; they differ from one another in the numbers and positions of double bonds and oxygen-containing functional groups (Figure 8.30).

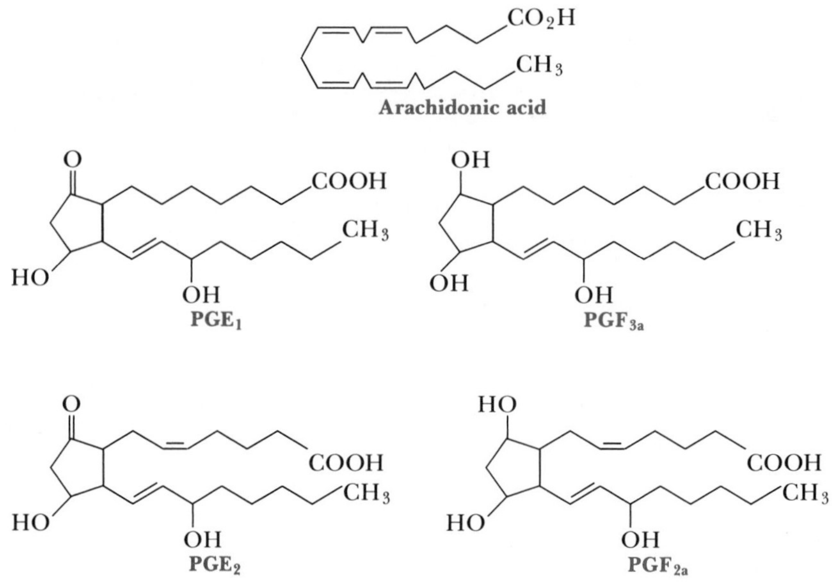

FIGURE 8.30

Arachidonic acid and some prostaglandins.

Elucidation of the structures of prostaglandins and their laboratory synthesis have been topics of great interest to organic chemists, largely because of the many physiological effects of these compounds and their possible usefulness in the pharmaceutical industry. Some of the functions of prostaglandins are control of blood pressure, stimulation of smooth muscle contraction, and induction of inflammation. Aspirin inhibits the synthesis of prostaglandins, a property that accounts for its anti-inflammatory and fever-reducing properties. Cortisone and other steroids also have anti-inflammatory effects because of their inhibition of prostaglandin synthesis.

Prostaglandins are known to inhibit the aggregation of blood platelets. They may thus be of therapeutic value by preventing the formation of blood clots, which can cut off the blood supply to the brain or the heart and cause certain types of strokes and heart attacks. Even if this behavior were the only useful property of prostaglandins, it would justify considerable research effort. Heart attacks and strokes are two of the leading causes of death in industrialized countries.

FIGURE 8.31

Leukotriene C.

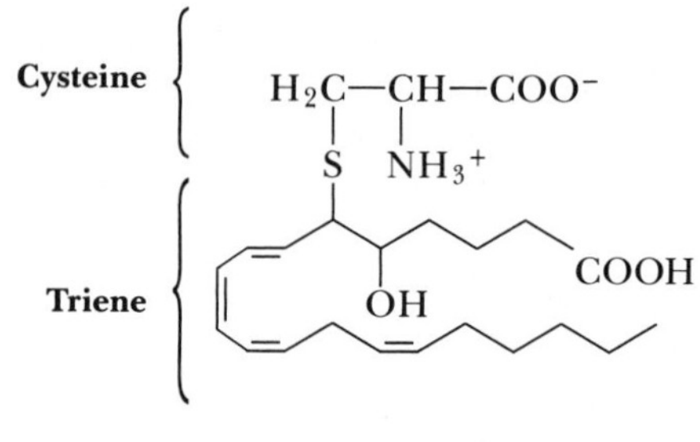

Leukotrienes are compounds that, like prostaglandins, are derived from arachidonic acid. They are found in leukocytes (white blood cells) and have three conjugated double bonds; these two facts account for their name. (Fatty acids and their derivatives do not normally contain conjugated double bonds.) Leukotriene C (Figure 8.31) is a typical member of this group; note the 20 carbon atoms in the carboxylic acid backbone, a feature that relates this compound structurally to arachidonic acid. An important property of leukotrienes is their constriction of smooth muscle, especially in the lungs. Asthma attacks may result from this constricting action, since the synthesis of leukotriene C appears to be facilitated by allergic reactions, such as a reaction to pollen. Drugs that inhibit the synthesis of leukotriene C show promise for the treatment of asthma. Leukotrienes may also have inflammatory properties and may be involved in rheumatoid arthritis.

Research on leukotrienes may provide new treatments for asthma, reducing the need for inhalers such as the one shown in use here.

S U M M A R Y

Lipids are compounds that are insoluble in water and soluble in nonpolar organic solvents. One group of lipids consists of open-chain compounds, each with a polar head group and a long nonpolar tail; this group includes fatty acids, triacylglycerols, phosphoacylglycerols, sphingolipids, and glycolipids. A second major group consists of fused-ring compounds, the steroids. Triacylglycerols are the storage forms of fatty acids, and phosphoacylglycerols are important components of biological membranes, as are sphingolipids and glycolipids.

A biological membrane consists of a lipid part and a protein part. The lipid part is a bilayer, with the polar head groups in contact with the aqueous interior and exterior of the cell, and the nonpolar portions of the lipid in the interior of the membrane. Lateral motion of lipid molecules within one layer of a membrane occurs frequently. The proteins that occur in membranes can be peripheral proteins, found on the surface of the membrane, or integral proteins, which lie within the lipid bilayer. The fluid mosaic model describes the interaction of lipids and proteins in biological membranes. No extensive formation of lipid–protein complexes takes place in this model. The proteins "float" in the lipid bilayer.

Three important functions take place in or on membranes. The first, transport across the membrane, can involve the lipid bilayer as well as the membrane proteins. The second, catalysis, is carried out by enzymes bound to the membrane. Finally, receptor proteins in the membrane bind biologically important substances that trigger a biochemical response in the cell.

The most important question about transport of substances across biological membranes is whether the process requires expenditure of energy by the cell. In passive transport, a substance moves from a region of higher concentration to one of lower concentration, requiring no expenditure of energy by the cell. Active transport requires moving substances against a concentration gradient, a situation similar to pumping water up a hill. Energy, as well as a carrier protein, is required for active transport. The sodium–potassium pump is an example of active transport.

The first step in the effects of some biologically active substances is binding to a protein receptor site on the exterior of the cell. The interaction between receptor proteins and the active substances to which they bind is very similar to enzyme–substrate recognition. The action of a receptor frequently depends on a conformational change in the receptor protein. Receptors can be ligand-gated channel proteins, in which the binding of ligand transiently opens a channel protein through which substances such as ions can flow in the direction of a concentration gradient. The activation of the neuromuscular junction involves the action of such a protein, in which the neurotransmitter acetylcholine is the ligand. The binding of the ligand controls the operation of an $Na^+–K^+$ channel. Ligand-gated proteins also offer an interesting contrast with voltage-gated channel proteins, in which the voltage across the cell membrane, rather than receptor action, controls the operation of the channel.

Lipid-soluble vitamins are hydrophobic, accounting for their solubility properties. A derivative of vitamin A plays a crucial role in vision. Vitamin D controls calcium and phosphorus metabolism, affecting the structural integrity of bones. Vitamin E is known to be an antioxidant; its other metabolic functions are not definitely established. The presence of vitamin K is required in the blood clotting process.

The unsaturated fatty acid, arachidonic acid, is the precursor of prostaglandins and leukotrienes, compounds that have a wide range of physiological activities. Stimulation of smooth muscle contraction and induction of inflammation are common to both classes of compounds. Prostaglandins are also involved in control of blood pressure and inhibition of blood platelet aggregation.

E X E R C I S E S

1. What structural features do a triacylglycerol and a phosphatidyl ethanolamine have in common? How do the structures of these two types of lipids differ?

2. What structural features do a sphingomyelin and a phosphatidyl choline have in common? How do the structures of these two types of lipids differ?

3. Which of the following lipids are *not* found in animal membranes?
   (a) Phosphoglycerides
   (b) Cholesterol
   (c) Triacylglycerols
   (d) Glycolipids
   (e) Sphingolipids

4. Draw the structure of a phosphoacylglycerol that contains glycerol, oleic acid, stearic acid, and choline.

5. You have just isolated a pure lipid that contains only sphingosine and a fatty acid. To what class of lipids does it belong?

6. Write the structural formula for a triacylglycerol, and name the component parts.

7. Write an equation, with structural formulas, for the saponification of the triacylglycerol in Exercise 6.

8. Briefly discuss the structure of myelin and its role in the nervous system.

9. How does the structure of steroids differ from that of the other lipids discussed in this chapter?

10. What is the role in vision of the *cis-trans* isomerization of retinal?

11. Give a reason for the toxicity that can be caused by overdoses of lipid-soluble vitamins.

12. What is the structural relationship between vitamin D_3 and cholesterol?

13. List an important chemical property of vitamin E.

14. List two classes of compounds derived from arachidonic acid; suggest some reasons for the amount of biomedical research devoted to these compounds.

15. In lipid bilayers there is an order–disorder transition similar to the melting of a crystal. In a lipid bilayer in which most of the fatty acids are unsaturated, would you expect this transition to occur at a higher temperature, a lower temperature, or the same temperature as it would in a lipid bilayer in which most of the fatty acids are saturated? Why?

16. A membrane consists of 50% protein by weight and 50% phosphoglycerides by weight. The average molecular weight of the lipids is 800 daltons, and the average molecular weight of the proteins is 50,000 daltons. Calculate the molar ratio of lipid to protein.

17. Which of the following statements is (are) consistent with what is known about membranes?
    (a) A membrane consists of a layer of proteins sandwiched between two layers of lipids.
    (b) The compositions of the inner and outer lipid layers are the same in any individual membrane.
    (c) Membranes contain glycolipids and glycoproteins.
    (d) Lipid bilayers are an important component of membranes.
    (e) Covalent bonding takes place between lipids and proteins in most membranes.

18. Inorganic ions such as K^+, Na^+, Ca^{2+}, and Mg^{2+} do not cross biological membranes by simple diffusion. Suggest a reason.

19. Which statements are consistent with the fluid mosaic model of membranes?
    (a) All membrane proteins are bound to the interior of the membrane.
    (b) Both proteins and lipids undergo transverse ("flip-flop") diffusion from the inside to the outside of the membrane.
    (c) Some proteins and lipids undergo lateral diffusion along the inner or outer surface of the membrane.
    (d) Carbohydrates are covalently bonded to the outside of the membrane.
    (e) The term "mosaic" refers to the arrangement of the lipids alone.

20. Which statements are consistent with the known facts about membrane transport?
    (a) Active transport moves a substance from a region where its concentration is lower to one where its concentration is higher.
    (b) Transport does not involve any pores or channels in membranes.

(c) Transport proteins may be involved in bringing substances into cells.

21. The cell membranes of bacteria grown at 20°C tend to have a higher proportion of unsaturated fatty acids than the membranes of bacteria of the same species grown at 37°C. (In other words, the bacteria grown at 37°C have a higher proportion of saturated fatty acids in their cell membranes.) Suggest a reason for this observation.

22. What types of amino acid residues are likely to be found in the binding site of the LDL receptor? Why?

23. What is the thermodynamic driving force for the formation of phospholipid bilayers?

24. Animals that live in cold climates tend to have higher proportions of polyunsaturated fatty acid residues in their lipids than do animals that live in warm climates. Suggest a reason.

25. Succulent plants from arid regions generally have waxy surface coatings. Suggest why such a coating is valuable for the survival of the plant.

26. In the preparation of sauces that involve mixing water and melted butter, egg yolks are added to prevent separation. How do the egg yolks prevent separation? (*Hint:* Egg yolks are rich in phosphatidylcholine [lecithin].)

A N N O T A T E D   B I B L I O G R A P H Y

Alberts, B., D. Bray, J. Lewis, M. Raff, K. Roberts, and J. D. Watson. *Molecular Biology of the Cell.* 3rd ed. New York: Garland, 1994. [Chapter 6 contains a lucid and well-illustrated description of membrane structure and function from the standpoint of cell biology.]

Bretscher, M. S. The Molecules of the Cell Membrane. *Sci. Amer.* **253** (4), 100–108 (1985). A particularly well-illustrated description of the roles of lipids and proteins in cell membranes.]

Brown, M. S., and J. L. Goldstein. How LDL Receptors Influence Cholesterol and Atherosclerosis. *Sci. Amer.* **251** (5), 58–66 (1984). [A description of lipid metabolism and the role of membrane receptors.]

Brown, M. S., and J. L. Goldstein. A Receptor-Mediated Pathway for Cholesterol Homeostasis. *Science* **232**, 34–47 (1986). [A more recent description of the role of cholesterol in heart disease.]

Chakrin, L. W., and D. M. Bailey. *The Leukotrienes — Chemistry and Biology.* Orlando, FL: Academic Press, 1984. [A collection of articles on the structures and actions of leukotrienes.]

Changeux, J.-P. Chemical Signaling in the Brain. *Sci. Amer.* **269** (5), 58–62 (1993). [A description of the acetylcholine receptor in action.]

Dautry-Varsat, A., and H. F. Lodish. How Receptors Bring Proteins and Particles into Cells. *Sci. Amer.* **250** (5), 52–58 (1984). [A detailed description of endocytosis.]

Dunant, Y., and M. Israel. The Release of Acetylcholine. *Sci. Amer.* **252** (4), 58–66 (1985). [A description of the role of neurotransmitters and receptors for them in the propagation of nerve impulses.]

Hakomori, S. Glycosphingolipids. *Sci. Amer.* **254** (5), 44–53 (1986). [A possible role for this class of cell membrane components in the diagnosis and treatment of cancer.]

Keuhl, F. A., and R. W. Egan. Prostaglandins, Arachidonic Acid and Inflammation. *Science* **210**, 978–984 (1980). [A discussion of both the chemistry of these compounds and their physiological effects.]

Ostro, M. J. Liposomes. *Sci. Amer.* **256** (1), 102–111 (1987). [A description of possible uses of lipid bilayers as vehicles for delivery of drugs to tissues.]

Rasmussen, H. The Cycling of Calcium as an Intracellular Messenger. *Sci. Amer.* **261** (4), 66–73 (1989). [How receptors for calcium ion mediate its intracellular effects.]

Singer, S. J., and G. L. Nicholson. The Fluid Mosaic Model of the Structure of Membranes. *Science* **175**, 720–731 (1972). [The article in which the fluid mosaic model was first introduced.]

Unwin, N., and R. Henderson. The Structure of Proteins in Biological Membranes. *Sci. Amer.* **250** (2), 78–94 (1984). [The results of electron microscopic studies on integral proteins tightly bound in membranes.]

Folic acid crystal.

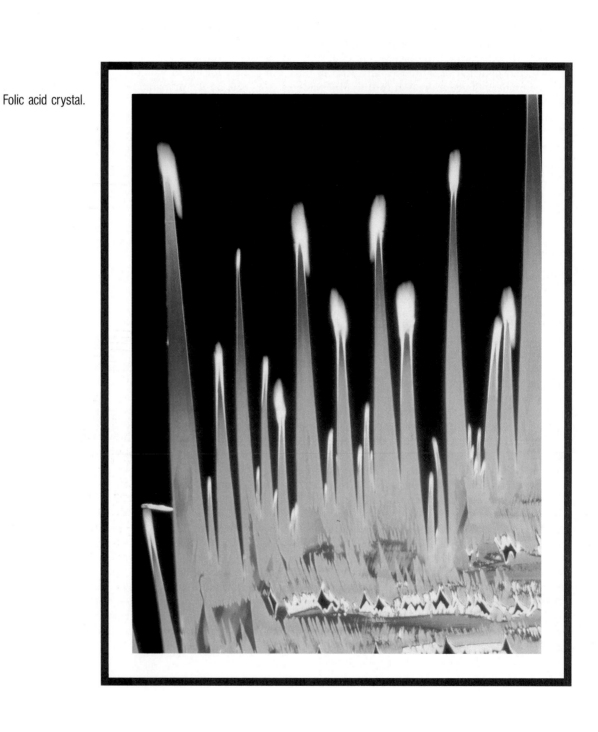

Part III

Energetics and Metabolism: Carbohydrates, Lipids, and Compounds of Nitrogen

Outline

Jane Richardson

With degrees in philosophy and teaching, Professor Jane Richardson has a somewhat unconventional background for a biochemistry researcher. But the MacArthur Fellow and James B. Duke Professor in Duke University's Department of Biochemistry has had a lifelong interest in science and math. Her youthful passion was astronomy, and she built her own telescope. In 1958 she placed third nationally in the Westinghouse Science Talent Search, for devising an ingenious way to track the orbit of the Russian satellite Sputnik through observation.

As an adult, Richardson has applied her keen powers of observation and boundless curiosity to the imaginative study of protein structures. She has regularly worked in partnership with fellow biochemist David C. Richardson, who is also her husband and an avid hiking companion on treks through California's Sierra Nevada.

A widely published contributor to scientific journals and a frequent speaker at scientific forums, Jane Richardson is especially well known for her pioneering drawings of proteins. These striking depictions of three-dimensional protein structures have helped give researchers a common language for envisioning and explaining protein dynamics. Beyond that, the drawings enhance something in which Jane Richardson continues to delight — an aesthetic appreciation for the beauty of nature.

Would you describe what you see as your area of specialization within biochemistry?

What we do is try to understand how proteins fold up to form their three-dimensional structures — what things in the sequence, or in the way parts of the protein interact, are the ones that really determine that it folds up one way rather than another. Now, we are particularly concerned with what lets proteins find a unique structure as opposed to a whole set of similar, about equally favorable, structures.

What led to your interest in this field?

When I went to college at Swarthmore, I was thoroughly taken with philosophy. I had exceptionally good professors and very interesting classes. I was particularly interested in the ideas of the great philosophers — Socrates, Aristotle, and the classic

Europeans. My B.A. is in philosophy, with math and physics as minors. But when I went on to graduate study in philosophy at Harvard, I found that the most active area of the profession seemed to be the part I was least interested in — modern philosophy. I got a master's degree in philosophy and also a master's in education. But what I really enjoyed most at Harvard was taking courses in the botany department. I have found that many of the things I have gotten the most out of were learned either outside class or in classes that I was not supposed to be taking.

What did you do next?

I decided to go back to an earlier interest in the natural sciences. In the early 1960s, about the time my husband and I were finishing college, the first protein structures were being explored. I came to work as a technician at a biochemistry lab at M.I.T, where my husband Dave was doing his research toward a Ph.D. in chemistry. Along with our collaborator, Ted [Edward E.] Hazen, who was a postdoctoral protein chemist, we had to independently rediscover how to do protein crystallography. It was a lot of fun. Working with protein structures has a lot of the same elements as natural history. You are investigating new worlds; it is a bit like going exploring in a tropical jungle.

What led to this original interest in protein structures?

Well, the first thing was undoubtedly precipitated by Chris [Nobel laureate Christian B.] Anfinsen, whom we've worked with over the years. When my husband was starting his graduate work, Chris was trying to find someone to do x-ray crystallography of the structure of the nuclease of *Staphylococcus aureus,* which he saw as a protein that would be particularly good for studying folding. And that was the project we ended up working on when we were first learning how to do crystallography and getting involved in protein structure.

What was the state of the field at the time you started?

When we got started on that project, hemoglobin and myoglobin were the only protein structures that had been solved. By the time we finished, seven or eight years later, we were tied with cytochrome *c* for the tenth structure solved.

What is the value of knowing these structures?

Everything that a protein does—all of its biological activity—depends on bringing the active groups of the various parts together in certain very critical geometries in three-dimensional space. Proteins do essentially everything in biology except store genetic information, and none of those functions would be possible without these precise structures.

Over the years, what have been your main contributions to the field?

We have developed new ideas and techniques, both in ways to research protein structures and in ways to represent and describe protein structures. In research, what Dave and I work on most these days is what is called *de novo* protein design. We were some of the people who started this out years ago. The overall idea is to understand better how a protein three-dimensional structure works and how it is formed. But the main experimental tool that we use is *de novo* design.

And what is the purpose of your new design work?

Our main purpose is to try to understand how the proteins work, how they fold up. Ultimately, the hope is to design new proteins for particular functional purposes.

Could you explain some of the basics of *de novo* protein design?

De novo design is sort of like doing prediction backwards. Instead of working from natural structures, we look at whole categories of structures to find one that is relatively simple, that we think we understand, and that we can try to isolate as a sort of simplest-case paradigm structure. We then use that structure to design a sequence of amino acids that is not related to any natural structures. And then we actually try to make the new structure, either by direct chemical synthesis or by cloning and expression. At that point, you are able to see if it does actually fold up into something that resembles what you expected.

What is the benefit of this work?

The real benefit of this approach, as opposed to modifying natural proteins, is that the natural proteins have a whole history of evolution over millions of years, and they have a lot of things going on in all their interactions with everything in the biological system they are in. There is a lot we do not know about, and any single change you make is fighting against that whole background—and so you are not apt to be able to make small changes in a protein and change it from one structure into something different. We realize at this point that we don't know everything about this problem, and if you are just modifying natural proteins, you don't find the big mistakes, you only find the little ones. But when you essentially design from scratch, you find out some of the things that you hadn't even thought about before.

What lessons have come out of this work?

It turns out that everyone was wrong about which step was going to be the hardest. We thought the hard step would be getting an approximately correct structure—for instance, designing a sequence that really would fold up into a four-helix bundle and go around in one direction rather than the other direction. As it turns out, that part is well within the range of what we know how to do. However, we had also thought that once we had an approximately correct structure the next part would be fairly straightforward. We thought we could make single changes and gradually work our way to a good solid structure. In fact, it turns out that part is the hard part. Designed structures may be approximately right, but they are usually not single unique structures. What we typically end up with is a set of molecules, each one of them with a slightly different structure. They may look similar from a distance, but the details may be different—for instance, how helices interact, or where the side chains really are. These newly designed proteins may take up many different structures that are all about equally favorable. They do not settle down into a unique equilibrium structure the way native proteins do.

What is the explanation?

We think that is because we have mistakes in lots of different places on the inside. We know native proteins can tolerate one or two local mistakes and make up for it. Ours probably have many mistakes, so when we fix one of them individually, it doesn't get any better. Now, we are going back again and looking at the natural proteins to try to figure out how they are accomplishing this, asking what it is that they do in terms of arranging the sequence to fit onto a certain structural framework. We are trying to learn what the principles are that we should be using in order to get over this second-stage problem. Probably the major input in this work is time spent looking at the natural proteins. And thousands of structures are now known. You can't just look at it once and absorb all the information there. So each time something weird happens in one of our designs or we just have a new idea about what we want to do, typically we have to spend quite a bit of time going back and looking at those native structures from this new point of view.

And over the years, you have helped people to visualize protein structures.

Yes, I think the other place where we have been influential has been in representation of proteins — both the hand-drawn ribbon drawings of proteins and computer-graphics methods for visualizing proteins.

And you set a precedent, or style, for how to do this.

Yes, it is kind of a neat idea that lots and lots of people see this whole part of the world through your eyes. Because much of it is subjective. You are always putting your personal ideas into a visual representation, whether it is a hand drawing or a computer-graphics program. In fact, the whole process of scientific research is deciding which as-

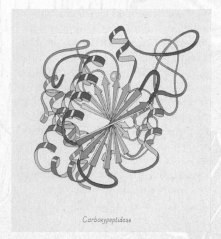 

Carboxypeptidase   *Lactate Dehydrogenase* domain 1

Ribbon diagrams of lactate dehydrogenase and carboxypeptidase.

pects of something complicated are the important aspects to explore and highlight — the ones that really help bring it together and give you insight into what is going on. The whole point is that, although the structure by itself can be complicated and confusing, with drawings and computer graphics you are able to show some order, to simplify.

So this is your special vantage point on the subject of protein structures.

Yes, it is kind of like an equation that pulls together a particular relationship, your theoretical view. In producing any set of drawings, you are emphasizing a relationship you think is informative and insightful, and you are downplaying other things that you think are incidental characteristics. You emphasize a central idea that you think is the important one in these structures. For example, if you are showing a whole set of proteins and your idea is that the shape and organization of the central beta sheet is the main thing — the fundamental organization — of these structures, then you emphasize that aspect in drawings. And you would maybe smooth or simplify other incidental details so that other people will also be able to clearly see this relationship.

How did you originally get involved in hand drawing of proteins?

We were working on a classification system for the tertiary structures of proteins, the overall arrangement of the backbone. And it was very hard to compare the different structures using the kinds of representations that were available at the time. There were drawings that looked a good deal like the ones I ended up doing, but they were done in a variety of different conventions, and when people represented related structures, they often showed them from different viewpoints. Looking at two drawings, you would not be able to tell whether they were really the same subject or not. So, in order to present this information about the structural classification, we had to have a uniform way of representing these things in a way that gave you good clues to identify which beta strands really belong together in one sheet — so you could see the overall shape as a single sheet, rather than just sort of a forest of little arrows that didn't seem to be related to each other. I had to learn how to do these drawings primarily to get our ideas across to other people. But we needed the drawings for ourselves as well, to make sure that we were seeing the right relationships in developing our classification scheme.

Were you trained in art?

Only slightly; I did mechanical drawing in high school. I'm not really that good at drawing things in general. Early on, I spent about two years learning how to do drawings of proteins. And I did an initial set of 100 drawings of different structures in ways so that viewers could see the similarities and differences.

And when you did your drawing, what approach did you use?

The basic drawings were done on top of a plot of the alpha carbon positions of the proteins. I established a standard scale, which made a small protein fit well on an eight-by-ten sheet of paper — so many millimeters per angstrom. I would have a stereo image to look at while I was doing the drawing. I would make drawings first in pencil, then in pen and ink. Later on, I used all sorts of media when I started doing them in color.

Do you still do hand drawings?

Not as much. The computer techniques have improved a lot. There are several quite widely used programs now that make drawings on the computer that look a whole lot like the ones I used to do by hand. And at this point, we have been working more actively on getting new computer-graphics representations that can fill different roles that weren't available before.

What kinds of computer-graphics applications are you working on?

The ones we are working on now span the spectrum. One application is very high end, state of the art, and only doable with supercomputing. What we are trying to develop is a computer graphics model that you can use interactively, changing geometry and moving atoms around and so on. There has been a standard way of doing that for quite a while. But it only acts on the pure geometry of the molecule, and does not pay enough attention to the physics behind protein structures. With currently available technology, you can rotate bonds and connect and disconnect things and move them around, but if you move one atom right on top of another, the computer doesn't know to object. With this new approach, we hope to have a fast enough algorithm so that the energy minimization can be accounted for in real time, while you are moving things. If you have an image of a protein structure on your computer screen and you decide to pull a side chain off in a certain direction and it bumps into another part of the chain, you will be able to see them affect each other like real-life moving objects.

And what about your work at the other end of the spectrum?

Our other really active program is trying to facilitate mass distribution of computer graphics about protein structures. The intent is to make interactive graphics widely available to readers of journals, including graduate or undergraduate students, who can access this information on their Macs or PCs. The approach was originally worked out with a new journal called *Protein Science,* which is published by the Protein Society, and my husband does the programming in this area.

How does this work?

The idea is to interactively let readers look at graphics presentations of subjects covered in the articles. The journal comes out every month with a diskette that has what we call Kinemages, which is short for "kinetic images." It is sort of the next step up from a color stereo image. It has the characteristics of an illustration, in that the authors of the article specify what to show, what to put in and leave out, and what viewpoints are most useful. But then these diskettes also offer the open-endedness of interactive computer graphics. Readers start with what the author wants to show them, but they are not limited to that. They can turn a structure around, add or delete things, look at the back side, measure distances — you know, pull all kinds of information out of it, depending on what they are interested in looking at.

What would you say to people who may be considering a future for themselves in biochemistry?

The area that we have chosen to work in is protein structure. But there are intellectually challenging problems to solve in many areas of biochemistry. There are a lot of places where there are really exciting insights hidden for you to go and find — really rewarding, big ideas to be found. And then the other end of it is the day-to-day details. In our work, looking at these individual protein structures on computer graphics is, ah, just delightful. They are very elegant structures, and being able to go and stare at them and admire them as part of your work is a great fringe benefit and a great deal of fun. It is like an orders-of-magnitude better computer game.

Chapter 9

The Importance of Energy Changes and Electron Transfer in Metabolism

The potential energy of the water at the top of a waterfall is transformed into kinetic energy in spectacular fashion.

Life processes require that molecules taken in as nutrients be torn apart to extract energy and also to provide the building blocks to create new molecules. To maintain a steady state, a living organism needs a constant supply of energy from without to bring order to the constant turmoil within. The energy extraction process takes place in a series of many small steps in which electron donors transfer energy to electron acceptors. These oxidation–reduction reactions are fundamental to the extraction of energy from molecules such as glucose. The principal electron carriers are NADH (the reduced form of nicotinamide adenine dinucleotide) and NAD^+, its oxidized form. NADH is oxidized to NAD^+ when it loses two electrons, and NAD^+ is reduced to NADH when it accepts two electrons. Two electrons from each NADH and two protons join an oxygen atom to form H_2O in the complete oxidation of glucose. Energy generated in this reaction is conserved by transforming "low-energy" ADP to "higher-energy" ATP. The ADP–ATP system is like a very active checking account where deposits and withdrawals are in steady state. The energy from ATP is never used up, only *transferred* in the cell's myriad chemical reactions that require energy.

(a)

(b)

9.1
ENERGY AND CHANGE

Energy can take several forms and can be converted from one form to another. All living organisms require and use energy in varied forms; for example, motion involves mechanical energy, and maintenance of body temperature uses thermal energy. Photosynthesis requires light energy from the sun. Some organisms, such as several species of fish and eels, are striking examples of the use of chemical energy to produce electrical energy. The formation and breakdown of biomolecules involve changes in chemical energy.

Any process that will actually take place is **spontaneous** in the specialized sense used in thermodynamics. The laws of thermodynamics can be used to predict whether any change involving transformations of energy will take place. An example of such a change is a chemical reaction in which covalent bonds are broken and new ones formed. Another example is the formation of noncovalent interactions, such as hydrogen bonds or hydrophobic interactions, when proteins fold to produce their characteristic three-dimensional structures. The tendency of polar and nonpolar substances to exist in separate phases is a reflection of the energies of interaction between the individual molecules — in other words, the thermodynamics of the interaction.

(c)

Three examples of transformations of energy in biological systems. (a) The running horse represents conversion of chemical energy to mechanical energy. (b) The electric fish (*Torpedinidae*) converts chemical energy to electrical energy, and (c) phosphorescent bacteria convert chemical energy into light energy.

J. Willard Gibbs (1839–1903). The symbol G is given to free energy in his honor. His work is the basis of biochemical thermodynamics, and he is considered by some to have been the greatest scientist born in the United States. (The Bettmann Archive)

9.2
THE CRITERION FOR SPONTANEITY

The most useful criterion for predicting the spontaneity of a process is the **free energy,** which has the symbol G. (Strictly speaking, the use of this criterion requires conditions of constant temperature and pressure, which are usual in biochemical thermodynamics.) It is not possible to measure absolute values of energy, only the *changes* in energy that occur during a process. The value of the change in free energy (ΔG, where the symbol Δ indicates change) gives the needed information about the spontaneity of the process under consideration.

The free energy of a system decreases in a spontaneous process, so ΔG is negative ($\Delta G < 0$). Such a process is called **exergonic,** meaning that energy is given off. When the free energy change is positive ($\Delta G > 0$), the process is nonspontaneous (the reverse of such a process is spontaneous). For a nonspontaneous process to occur, energy must be supplied. Nonspontaneous processes are also called **endergonic,** meaning that energy is absorbed. For a process at **equilibrium,** with no net change in either direction, the free energy change is zero ($\Delta G = 0$).

$\Delta G < 0$ Spontaneous exergonic — energy released

$\Delta G = 0$ Equilibrium

$\Delta G < 0$ Nonspontaneous endergonic — energy required

An example of a spontaneous process is the aerobic metabolism of glucose, in which glucose reacts with oxygen to produce carbon dioxide, water, and energy for the organism.

$$\text{Glucose} + 6\ O_2 \longrightarrow 6\ CO_2 + 6\ H_2O \qquad \Delta G < 0$$

An example of a nonspontaneous process is the phosphorylation of ADP (adenosine diphosphate) to give ATP (adenosine triphosphate). This reaction takes place in living organisms because metabolic processes supply energy.

ADP
adenosine diphosphate   phosphate
                                                     ion

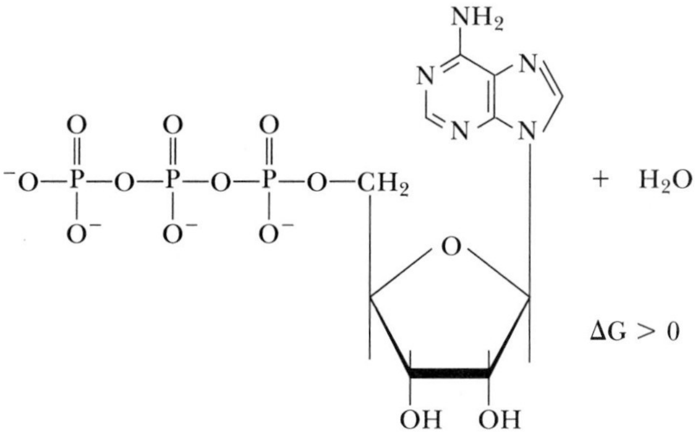

ATP
adenosine triphosphate

9.3
STANDARD STATES AND THE
STANDARD FREE ENERGY CHANGE

We can define **standard conditions** for any process and then use those standard conditions as the basis for comparing reactions. The choice of standard conditions is arbitrary. For a process under standard conditions, all substances involved in the reaction are in their **standard states,** in which case they are also said to be at **unit activity.** For pure solids and pure liquids, the standard state is the pure substance itself. For gases, the standard state is usually taken as a pressure of 1 atmosphere of that gas. For solutes, the standard state is usually taken as 1 molar concentration. Strictly speaking, these definitions for gases and for solutes are approximations, but they are valid for all but the most exacting work.

For any general reaction

$$a\text{A} + b\text{B} \longrightarrow c\text{C} + d\text{D}$$

we can write an equation that relates the free energy change (ΔG) for the reaction under *any* conditions to the free energy change under *standard* conditions ($\Delta G°$); the superscript ° refers to standard conditions. This equation is

$$\Delta G = \Delta G° + 2.303RT \log \frac{[\text{C}]^c[\text{D}]^d}{[\text{A}]^a[\text{B}]^b}$$

In this equation, the square brackets indicate molar concentrations, R is the gas constant (8.31 J mol^{-1}K^{-1}), and T is the absolute temperature. This equation holds under all circumstances; the reaction does not have to be at equilibrium. The value of ΔG under a given set of conditions depends on the value of $\Delta G°$ and on the concentration of reactants and products. In most biochemical reactions it is better to speak of $\Delta G°$, which is independent of concentrations. There is only one $\Delta G°$ for a reaction at a given temperature.

When the reaction is at equilibrium, $\Delta G = 0$, and thus

$$0 = \Delta G^\circ + 2.303RT \log \frac{[C]^c[D]^d}{[A]^a[B]^b}$$

$$\Delta G^\circ = -2.303RT \log \frac{[C]^c[D]^d}{[A]^a[B]^b}$$

The concentrations are now equilibrium concentrations, and this equation can be rewritten

$$\Delta G^\circ = -2.303RT \log K_{eq}$$

where K_{eq} is the equilibrium constant for the reaction. We have here a relationship between the equilibrium concentrations of reactants and products and the standard free energy change. Once we have determined the equilibrium concentrations of reactants by any convenient method, we can calculate the equilibrium constant, K_{eq}. We can then calculate the standard free energy change, ΔG°, from the equilibrium constant.

9.4
A MODIFIED STANDARD STATE FOR BIOCHEMICAL APPLICATIONS

We have just seen that the calculation of standard free energy changes includes the stipulation that all substances be in standard states, which for solutes can be approximated as a concentration of 1 molar. If the hydrogen ion concentration of a solution is 1 molar, the pH is zero. (Recall that the logarithm of 1 to any base is zero.) The interior of a living cell is, in many respects, an aqueous solution of the cellular components, and the pH of such a system is normally in the neutral range. Biochemical reactions in the laboratory are usually carried out in buffers that are also at or near neutral pH. For this reason it is convenient to define for biochemical practice a modified standard state, one that differs from the original standard state only by the change in hydrogen ion concentration from 1 M to 1×10^{-7} M, implying a pH of 7. When free energy changes are calculated on the basis of this modified standard state, they are designated by the symbol $\Delta G^{\circ\prime}$.

Example: Use of Equilibrium Constants To Determine $\Delta G^{\circ\prime}$

Let us assume that the relative concentrations of reactants have been determined for a reaction carried out at pH 7 and 25°C (298 K). Such concentrations can be used to calculate an equilibrium constant, K'_{eq}, which in turn can be used to determine the standard free energy change, $\Delta G^{\circ\prime}$, for the reaction. A typical reaction to which this kind of calculation can be applied is the hydrolysis of ATP at pH 7, yielding ADP, monohydrogen phosphate ion (written as P_i), and H^+ (the reverse of a reaction encountered earlier in this chapter):

$$ATP + H_2O \rightleftharpoons ADP + P_i + H^+$$

$$K'_{eq} = \frac{[ADP][P_i][H^+]}{[ATP]} \qquad pH\ 7,\ 25°C$$

The concentrations of the solutes are used to approximate their activities, and the activity of the water is one. The experimentally determined value for K'_{eq} is 2.23×10^5. Substituting $R = 8.31\ J\ mol^{-1}\ K^{-1}$, $T = 298\ K$, and log $K'_{eq} = 5.348$,

$$\Delta G°' = -2.303RT \log K'_{eq}$$

$$\Delta G°' = -(2.303)(8.31\ J\ mol^{-1}K^{-1})\ (298\ K)\ (5.348)$$

$$\Delta G°' = -3.0500 \times 10^4\ J\ mol^{-1} = -30.5\ kJ\ mol^{-1} = -7.29\ kcal\ mol^{-1}$$

1 kJ = 0.239 kcal

In addition to illustrating the usefulness of a modified standard state for biochemical work, the negative value of $\Delta G°'$ indicates that the reaction of hydrolysis of ATP to ADP is a spontaneous process in which energy is released.

9.5
THERMODYNAMICS AND LIFE

From time to time one encounters the statement that the existence of living things is a violation of the laws of thermodynamics, specifically of the second law. A look at the laws will clarify whether life is thermodynamically possible, and further discussion of thermodynamics will increase our understanding of this important topic.

The laws of thermodynamics can be stated in several ways. According to one formulation, the first law is "You can't win" and the second is "You can't break even." Put less flippantly, the first law states that it is impossible to convert energy from one form to another at greater than 100% efficiency. In other words, the first law of thermodynamics is the law of conservation of energy. The second law states that even 100% efficiency in energy transfer is impossible.

The two laws of thermodynamics can be related to the free energy by means of a well-known equation:

$$\Delta G = \Delta H - T\Delta S$$

In this equation G is the free energy, as before; H stands for the **enthalpy**, and S for the entropy. Discussions of the first law focus on the change in enthalpy, ΔH, which is the **heat of a reaction at constant pressure.** This quantity is relatively easy to measure. Enthalpy changes for many important reactions have been determined and are widely available — for instance, in tables in textbooks of general chemistry. Discussions of the second law focus on changes in entropy, ΔS, a concept that is less easily described and measured than enthalpy. Entropy changes are particularly important in biochemistry.

Ludwig Boltzmann (1844–1906). His equation for entropy in terms of the disorder of the universe was one of his supreme achievements; this equation is carved on his tombstone. (The Bettmann Archive)

One of the most useful definitions of entropy arises from statistical considerations. From a statistical point of view, an increase in the entropy of a system (the substance or substances under consideration) represents an increase in its disorder, or randomness. Books have a higher entropy when they are scattered around the reading room of a library than when they are in their proper places on the shelves. Scattered books are clearly in a more random state than books on shelves. The natural tendency of the universe is in the direction of increasing disorder. Another statement of the second law is this: *in any spontaneous process the entropy of the universe increases* ($\Delta S_{universe} > 0$). (This statement is general and applies to any set of conditions; it is not confined to the special case of constant temperature and pressure, as is the statement that the free energy decreases in a spontaneous process.)

The argument that life is a violation of the second law is based on the observation that any living organism is a highly specific, organized, thoroughly nonrandom arrangement of matter, not at all a reflection of increasing disorder. However, the metabolic processes of organisms involve the breakdown of complex molecules in foodstuffs and the production of wastes and heat. Carbon dioxide and water vapor are given off as gases, with the totally random arrangement characteristic of gases. Living things definitely represent local decreases in entropy: a high degree of order is found in complex molecules such as proteins and nucleic acids, and a still higher degree of order is found in organelles such as nuclei, mitochondria, and chloroplasts. The incorporation of organelles into cells, cells into tissues, and tissues into organisms represents a high degree of order, or low entropy. As we have just seen, though, this order is achieved through increases in entropy elsewhere in the universe. The local decrease in the entropy of the small part of the universe (i.e., the system) associated with a living organism is more than offset by the increase in the entropy of the rest of the universe, the surroundings of the system. The entropy of the universe as a whole increases, as must be true for a spontaneous process.

9.6
HYDROPHOBIC INTERACTIONS: A CASE STUDY IN THERMODYNAMICS (OPTIONAL)

Hydrophobic interactions have important consequences in biochemistry. Large arrays of molecules can take on definite structures as a result of hydrophobic interactions. A phospholipid bilayer is one such array. We have seen (Chapter 2, Section 2.1) that phospholipids are molecules that have polar head groups and long nonpolar tails of hydrocarbon chains. Under suitable conditions, a double-layer arrangement is formed so that the polar head groups of many molecules face the aqueous environment while the nonpolar tails are in contact with each other and are kept away from the aqueous environment. These bilayers form three-dimensional structures called **liposomes** (Figure 9.1). Such structures are useful model systems for biological membranes, which consist of similar bilayers with proteins embedded in them. The very existence of membranes depends on hydrophobic interactions.

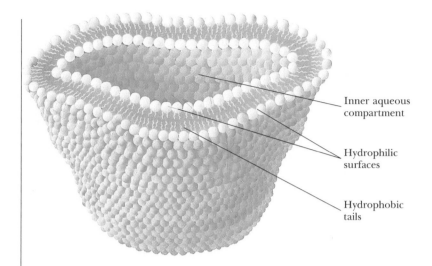

FIGURE 9.1

Schematic diagram of a liposome. This three-dimensional structure is arranged so that hydrophilic head groups of lipids are in contact with the aqueous environment. The hydrophobic tails are in contact with each other and are kept away from the aqueous environment.

Inner aqueous compartment

Hydrophilic surfaces

Hydrophobic tails

Hydrophobic interactions are a major factor in the folding of proteins into the specific three-dimensional structures required for their functioning as enzymes, oxygen carriers, or structural elements. The order of amino acids (i.e., the nature of the side chains) automatically determines the three-dimensional structure of the protein. It is known experimentally that proteins tend to be folded so that the nonpolar hydrophobic side chains are sequestered from water in the interior of the protein, while the polar hydrophilic side chains lie on the exterior of the molecule and are accessible to the aqueous environment (Figure 9.2). What makes hydrophobic interactions favorable?

(a)   (b)

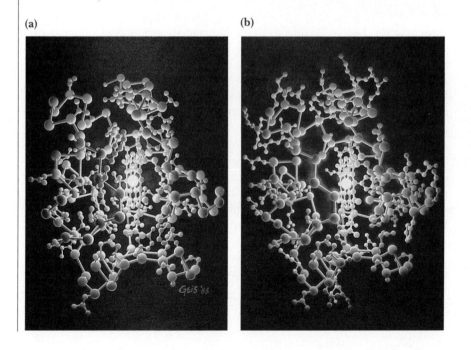

FIGURE 9.2

The three-dimensional structure of the protein cytochrome *c*. (a) The hydrophobic side chains (shown in red) are found in the interior of the molecule. (b) The hydrophilic side chains (shown in green) are found on the exterior of the molecule.

Hydrophobic interactions actually occur and thus are spontaneous processes. The entropy of the universe increases when hydrophobic interactions occur.

$$\Delta S_{universe} > 0$$

We are specifically interested in the polar and nonpolar substances that do not dissolve, rather than the whole universe. We can define the system as the two substances (polar and nonpolar) in which we are interested and the surroundings as the rest of the universe. The entropy change for the whole universe can be divided into the entropy change for the system and the entropy change for the surroundings.

$$\Delta S_{universe} = \Delta S_{system} + \Delta S_{surroundings}$$

The entropy change of the surroundings can be very important, as we saw in our discussion about living things and the second law of thermodynamics. We can also relate $\Delta S_{surroundings}$ to other, easily measured thermodynamic parameters.

It is known (although we shall not derive the equation here) that at constant temperature and pressure

$$\Delta S_{surroundings} = -\frac{\Delta H_{system}}{T}$$

As an example, let us assume that we have tried to mix the liquid hydrocarbon hexane (C_6H_{14}) with water and have obtained not a solution but a two-layer system, one layer of hexane and one of water. The overall process is nonspontaneous, and

$$\Delta S_{sys} + \Delta S_{surr} = \Delta S_{univ} < 0$$

As we have just seen, $\Delta S_{surr} = -\Delta H_{sys}/T$; this enthalpy change is a quantity known as the heat of solution. Some substances either give off heat (the solution is hot to the touch) or take up heat (the solution is cold to the touch) when they dissolve. Such substances are said to have a large heat of solution. Hexane dissolves in ethanol (C_2H_5OH), a liquid that is polar but less polar than water. When a solution of hexane in ethanol is prepared, it is cold to the touch because heat from the surroundings is needed to supply energy for hexane to dissolve in ethanol. In the case of hexane in water, the surroundings cannot supply enough energy to enable hexane to dissolve. The mixing process itself always favors dissolution, since a mixture of two substances has more possible random arrangements than do the separate pure substances. This contribution is known to physical chemists as entropy of mixing. Unfavorable entropy terms enter into the picture if solution formation requires the creation of ordered arrays of solvent — in this case, water.

The heat of solution of nonpolar substances in water, the enthalpy change, is a reflection of the unfavorable entropy change in such a situation.

$$\Delta H_{soln} = \Delta H_{sys}$$

$$-\frac{\Delta H_{soln}}{T} = -\frac{\Delta H_{sys}}{T} = \Delta S_{surr}$$

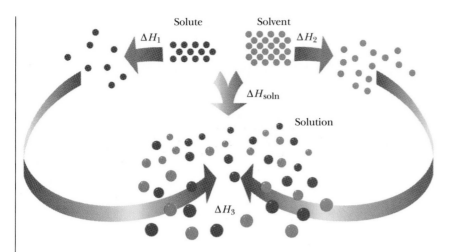

FIGURE 9.3

Enthalpy changes in solution formation. $\Delta H_{soln} = \Delta H_1 + \Delta H_2 + \Delta H_3$, where ΔH_1 is the enthalpy required to overcome interactions between solute molecules, ΔH_2 is the enthalpy required to overcome interactions between solvent molecules (green circles), and ΔH_3 is the enthalpy of interaction between solvent and solute molecules.

The heat of solution, in turn, can be broken down into three parts, all related to the intermolecular interactions discussed in Chapter 2, Section 2.1. The first part (ΔH_1) is the enthalpy required to separate the *solute* molecules from one another by overcoming the interactions between them, and the second part (ΔH_2) is the enthalpy required to separate the *solvent* molecules by overcoming the interactions between them (Figure 9.3). The third part (ΔH_3) is the enthalpy released as a result of the interaction between solute and solvent molecules. The relative magnitudes of these three enthalpy changes determine the overall heat of solution.

$$\Delta H_{soln} = \Delta H_1 + \Delta H_2 + \Delta H_3$$

From the example of an attempt to dissolve hexane in water, we can see that it takes very little energy to separate hexane molecules. They are held together by van der Waals interactions, which are weak. Considerably more energy is needed to overcome the dipole–dipole interactions and hydrogen bonds between water molecules. The energy released as a result of interaction between molecules of water and of hexane is not enough to overcome the interaction between the water molecules themselves (Figure 9.4). In contrast, the heat of solution for dissolving polar solutes in water is favorable.

The only way to achieve a lowering of energy (a favorable enthalpy change) in this situation is to take advantage of the open hydrogen-bonded structure of water. It is possible for a nonpolar molecule to interact with a water molecule that already has four hydrogen bonds and lower the energy, albeit by a small amount. If the water molecule involved in the interaction has fewer than four hydrogen bonds, the nonpolar molecule adjacent to that water molecule takes the place that could be occupied by a water

FIGURE 9.4

Enthalpies of solution (ΔH_{soln}). (a) A favorable ΔH_{soln} (ΔH is negative). An exothermic reaction favors solution formation. (b) An unfavorable ΔH_{soln} (ΔH is positive). An endothermic reaction does not favor solution formation.

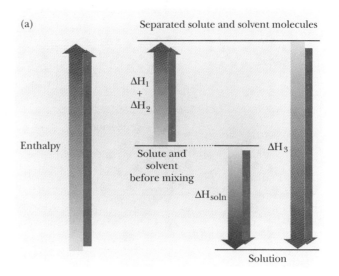

(a)

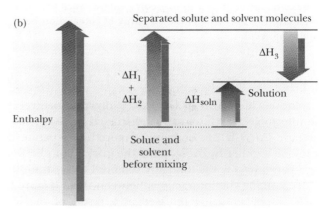

(b)

molecule. In effect, a relatively strong dipole–dipole interaction has been replaced by a weaker interaction, raising the energy.

When a water molecule is already involved in four hydrogen bonds, however, a nonpolar molecule can occupy an empty space in the icelike lattice without replacing a water molecule. Such an arrangement requires a lowering of the entropy of the water, which is unfavorable. The H_2O molecules that have four hydrogen bonds are more like ice than like liquid water. The greater degree of hydrogen bonding introduces a higher degree of order and thus a lower entropy (Figures 9.5 and 9.6). If such an unfavorable entropy change in the water were to take place, it would have to do so as a result of energy being provided by the surroundings, as is the case when hexane dissolves in ethanol. A larger and still more unfavorable entropy change occurs in the surroundings. Enough energy can be supplied from the surroundings, decreasing the entropy of the surroundings, when hexane dissolves in ethanol but not when we try to dissolve hexane in water.

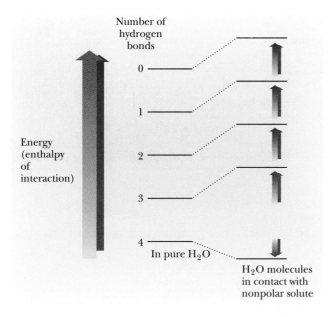

FIGURE 9.5

Enthalpies of interaction between water molecules. The energy of water molecules in pure water depends on the number of hydrogen bonds. Each added hydrogen bond gives a more stable state, lowering the energy by a favorable enthalpy of interaction. It is energetically unfavorable to replace a hydrogen bond with a weaker interaction. Replacement of a hydrogen bond by a dipole-induced dipole interaction with a nonpolar molecule is particularly unfavorable. The only arrangement that is favorable from the standpoint of enthalpy is the addition of the stabilization energy of a dipole-induced dipole interaction to that of four hydrogen bonds around a given water molecule. Such an arrangement, however, is highly unfavorable because of the drastic decrease in the entropy of the water.

The required entropy decrease in the surroundings is too large for the process to take place. Therefore, nonpolar substances do not dissolve in water; rather, nonpolar molecules associate with one another by hydrophobic interactions and sequester themselves from water.

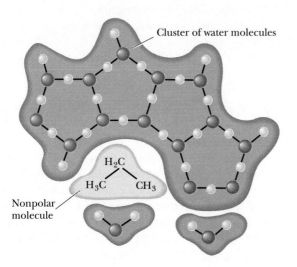

FIGURE 9.6

A nonpolar molecule in the presence of a partial "cage" of hydrogen-bonded water molecules. The water molecules in the interior of the cluster are involved in four hydrogen bonds, which are not shown in this two-dimensional representation. The enthalpy change (ΔH) for placing a nonpolar solute in water is unfavorable except for a small favorable contribution from water molecules with four hydrogen bonds. The entropy change (ΔS) is highly unfavorable, requiring more extensive hydrogen bonding of water (a greater degree of order).

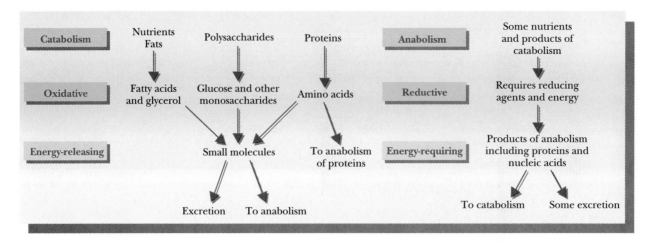

FIGURE 9.7
A comparison of catabolism and anabolism.

9.7
THE NATURE OF METABOLISM

Up to now we have discussed some basic chemical principles and investigated the natures of the molecules of which living cells are composed. We have yet to discuss the chemical reactions of biomolecules themselves, which constitute **metabolism,** the biochemical basis of all life processes. The molecules of carbohydrates, fats, and proteins taken into an organism are processed in a variety of ways (Figure 9.7). The breakdown of larger molecules to smaller ones is called **catabolism.** Small molecules are used as the starting points of a variety of reactions to produce larger and more complex molecules, including proteins and nucleic acids; this process is called **anabolism.** Catabolism and anabolism are separate pathways; they are not simply the reverse of each other.

Catabolism is an oxidative process that releases energy; anabolism is a reductive process that requires energy. We shall need several chapters to explore some of the implications of this statement. In this chapter we discuss oxidation and reduction (electron-transfer reactions) and their relation to the use of energy by living cells.

9.8
THE NATURE OF OXIDATION AND REDUCTION

Oxidation–reduction reactions, also referred to as *redox* reactions, are those in which electrons are transferred from a donor to an acceptor. **Oxidation** is the loss of electrons, and **reduction** is the gain of electrons. The substance that loses electrons (the electron donor) — that is, the one that is oxidized — is called the **reducing agent** or reductant. The substance that gains electrons (the electron acceptor) — the one that is reduced — is called the **oxidizing agent** or oxidant. Both an oxidizing agent and a reducing agent are necessary for the transfer of electrons (an oxidation–reduction reaction) to take place.

An example of an oxidation–reduction reaction is the one that occurs when a strip of metallic zinc is placed in an aqueous solution containing copper ions. Although both zinc and copper ions play roles in life processes, this reaction does not occur in living organisms. However, it is a good place to start our discussion of electron transfer because in this comparatively simple reaction it is fairly easy to follow where the electrons are going (it is not always quite as easy to keep track of the details in other biological redox reactions). The experimental observation is that the zinc metal disappears and zinc ions go into solution, while copper ions are removed from the solution and copper metal is deposited. The equation for this reaction is

$$Zn(s) + Cu^{2+}(aq) \longrightarrow Zn^{2+}(aq) + Cu(s)$$

The notation (s) signifies a solid and (aq) signifies a solute in aqueous solution.

In the reaction between zinc metal and copper ion, the Zn lost two electrons to become Zn^{2+} ion and was oxidized. A separate equation can be written for this part of the overall reaction, and it is called the **half reaction** of oxidation:

$$Zn \longrightarrow Zn^{2+} + 2e^-$$

Zn is the reducing agent (loses electrons, electron donor, is oxidized).

Likewise, the Cu^{2+} ion gained two electrons to form Cu and was reduced. An equation can also be written for this part of the overall reaction and is called the half reaction of reduction.

$$Cu^{2+} + 2e^- \longrightarrow Cu$$

Cu^{2+} is the oxidizing agent (gains electrons, electron acceptor, is reduced).

If the two equations for the half reactions are combined, the result is an equation for the overall reaction:

$$Zn \longrightarrow Zn^{2+} + 2e^- \qquad \text{Oxidation}$$
$$\underline{Cu^{2+} + 2e^- \longrightarrow Cu} \qquad \text{Reduction}$$
$$Zn + Cu^{2+} \longrightarrow Zn^{2+} + Cu \qquad \text{Overall reaction}$$

This reaction is a particularly clear example of electron transfer. It will be useful to keep these basic principles in mind when we examine the flow of electrons in the more complex redox reactions of aerobic metabolism.

9.9
COENZYMES IN BIOLOGICALLY IMPORTANT OXIDATION–REDUCTION REACTIONS

Oxidation–reduction reactions are discussed at length in textbooks of general and inorganic chemistry, but the oxidation of nutrients by living organisms to provide energy requires its own special treatment. The description of redox reactions in terms of oxidation numbers, which is widely practiced with inorganic compounds, can be used to deal with the oxidation of carbon-containing molecules. However, our discussion will be more pictorial and easier to follow if we write equations for the half reactions, then concentrate on the functional groups of the reactants and products and on the number of electrons transferred. An example is the oxidation half reaction for the conversion of ethanol to acetaldehyde.

**The half reaction of oxidation of
ethanol to acetaldehyde**

$$CH_3 - \overset{\overset{\text{H}}{|}}{\underset{\underset{\text{H}}{|}}{C}} \overset{..}{\underset{..}{:O:}} H \rightleftharpoons CH_3 - \overset{\overset{\text{H}}{|}}{C} \overset{..}{::O:} + 2H^+ + 2e^-$$

Ethanol (12 electrons in   Acetaldehyde (10 electrons
groups involved in reaction)   in groups involved in reaction)

Writing the Lewis electron-dot structures for the functional groups involved in the reaction helps us keep track of the electrons being transferred. In the oxidation of ethanol, there are 12 electrons in the part of the ethanol molecule involved in the reaction, and 10 electrons in the corresponding part of the acetaldehyde molecule; two electrons are transferred to an electron acceptor (an oxidizing agent). This type of "bookkeeping" is useful for dealing with biochemical reactions. Many biological oxidation reactions, like this example, are accompanied by the transfer of a proton (H^+). The oxidation half reaction has been written as a reversible reaction because the occurrence of oxidation or reduction depends on the other reagents present.

Another example of an oxidation half reaction is that for the conversion of NADH, the reduced form of nicotinamide adenine dinucleotide, to the oxidized form, NAD^+. This substance is an important coenzyme in many reactions. Figure 9.8 shows the structure of NADH; the nicotinamide portion, the functional group involved in the reaction, is indicated in red. Nicotinamide is a derivative of nicotinic acid (also called niacin), one of the B-complex vitamins (see Chapter 5, Section 5.13). A similar compound is NADPH (for which the oxidized form is $NADP^+$). It differs from NADH by having an additional phosphate group; the site of attachment of this phosphate group to ribose is also indicated in Figure 9.8. To simplify writing the equation for the oxidation of NADH, only the nicotinamide ring is shown explicitly, with the rest of the molecule designated as R. The two electrons that are lost when NADH is converted to NAD^+ can be considered to come from the bond between carbon and the lost hydrogen, with the nitrogen lone-pair electrons becoming involved in a bond.

The half reaction of oxidation of NADH to NAD^+.

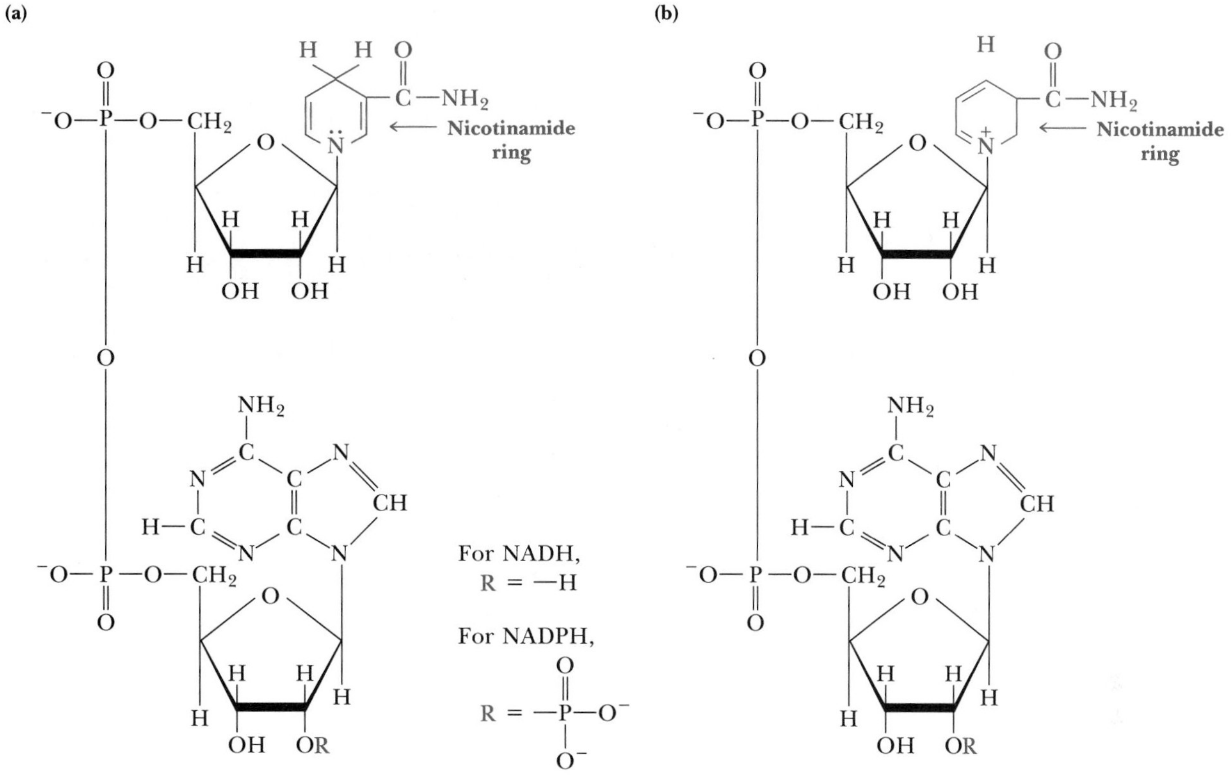

(a) For NADH,
$R = -H$

For NADPH,
$$R = -\overset{\displaystyle O}{\underset{\displaystyle O^-}{\overset{\|}{\underset{|}{P}}}} - O^-$$

FIGURE 9.8

(a) The structure of NADH.
(b) The structure of NAD$^+$, the oxidized form of NADH

The equations for both the reaction of NADH to NAD$^+$ and that of ethanol to acetaldehyde have been written as oxidation half reactions. If ethanol and NADH were mixed in a test tube, no reaction could take place because there would be no electron acceptor. If, however, NADH were mixed with acetaldehyde, which is an oxidized species, a transfer of electrons could take place, producing ethanol and NAD$^+$.

$$NADH \longrightarrow NAD^+ + H^+ + 2e^-$$ Half reaction of oxidation

$$\underline{CH_3CHO + 2H^+ + 2e^- \longrightarrow CH_3CH_2OH}$$ Half reaction of reduction

$$NADH + H^+ + CH_3CHO \longrightarrow NAD^+ + CH_3CH_2OH$$ Overall reaction

$$\underset{\text{Acetaldehyde}}{} \qquad \underset{\text{Ethanol}}{}$$

Such a reaction does take place in some organisms as the last step of alcoholic fermentation. The NADH is oxidized while the acetaldehyde is reduced.

Another important electron acceptor is FAD (flavin adenine dinucleotide) (Figure 9.9), which is the oxidized form of FADH$_2$. The symbol FADH$_2$ explicitly recognizes that protons (hydrogen ions) as well as electrons are accepted by FAD.

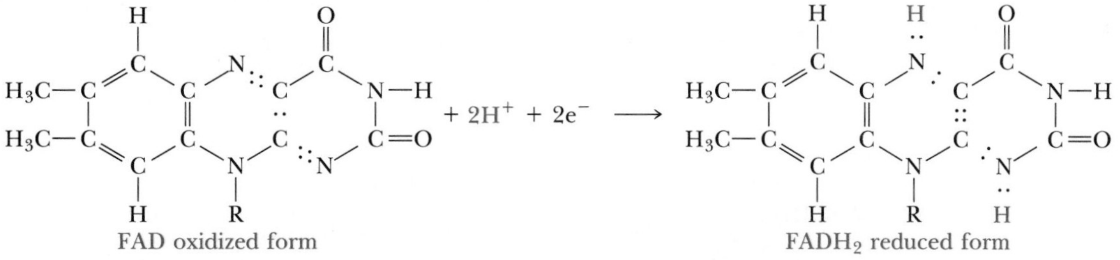

FAD oxidized form $\quad + 2H^+ + 2e^- \longrightarrow \quad$ FADH$_2$ reduced form

The half reaction of reduction of FAD to FADH$_2$.

(a)

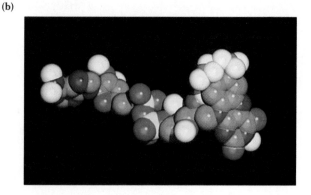

(b)

FIGURE 9.9

(a) The structure of FAD, the oxidized form of *f*lavin *a*denine *d*inucleotide. (b) Space-filling model of FAD.

The structures shown in this equation again point out the electrons that are transferred in the reaction. Several other coenzymes contain the flavin group; they are derived from the vitamin riboflavin (vitamin B_2).

Oxidation of nutrients to provide energy for an organism cannot take place without reduction of some electron acceptor (i.e., an oxidizing agent). The ultimate electron acceptor in aerobic oxidation is oxygen; we shall encounter intermediate electron acceptors as we discuss metabolic processes. Reduction of metabolites plays a significant role in living organisms in anabolic processes. Important biomolecules are synthesized in organisms by many reactions in which a metabolite is reduced while the reduced form of a coenzyme is oxidized.

PRACTICE   SESSION

In the following reactions, identify the substance oxidized, the substance reduced, the oxidizing agent, and the reducing agent.

(a)

Pyruvate + NADH + H$^+$ $\longrightarrow$ lactate + NAD$^+$

(b)

Malate + NAD$^+$ $\longrightarrow$ oxaloacetate + NADH + H$^+$

ANSWERS   In the first reaction pyruvate is reduced and NADH is oxidized; pyruvate is the oxidizing agent and NADH is the reducing agent. In the second reaction malate is oxidized and NAD$^+$ is reduced; NAD$^+$ is the oxidizing agent and malate is the reducing agent.

9.10
COUPLING OF PRODUCTION AND USE OF ENERGY

Another important question about metabolism is: "How is the energy released by the oxidation of nutrients trapped and used?" This energy cannot be used directly; it must be shunted into an easily accessible form of chemical energy. In Section 9.2 we saw that several phosphorus-containing compounds such as ATP can be hydrolyzed easily and that the reaction releases energy. Formation of ATP is intimately linked with the release of energy from oxidation of nutrients. The *coupling* of energy-producing and energy-requiring reactions is a central feature in the metabolism of all organisms.

The phosphorylation of ADP (adenosine diphosphate) to produce ATP (adenosine triphosphate) requires energy, which can be supplied by the oxidation of nutrients.

$$ADP + P_i + H^+ \longrightarrow ATP + H_2O \qquad \Delta G^{\circ\prime} = 30.5 \text{ kJ mol}^{-1}$$
$$= 7.3 \text{ kcal mol}^{-1}$$

or in structural form,

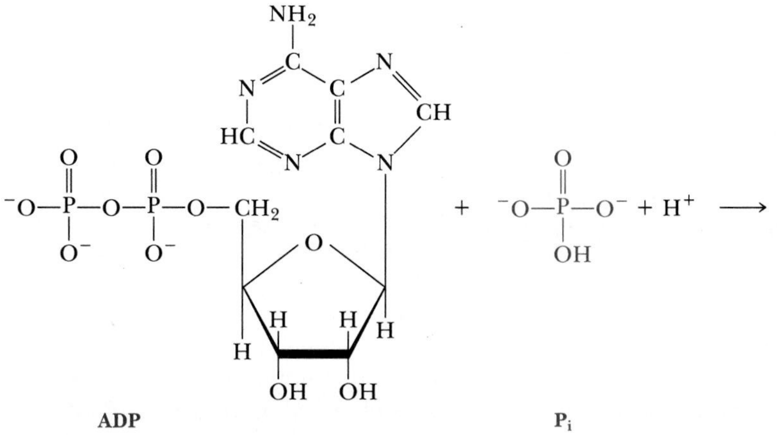

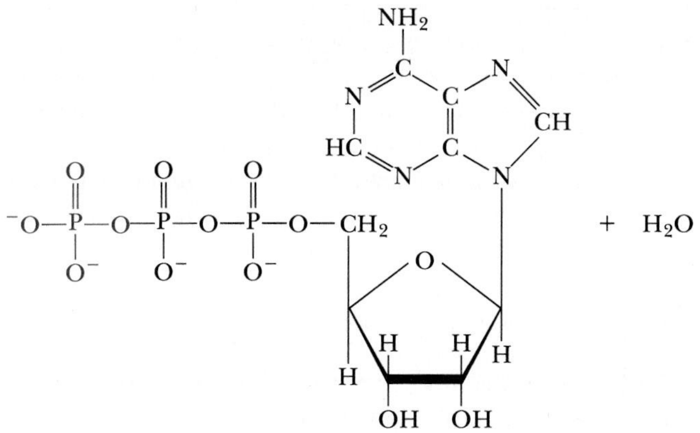

TABLE 9.1   Free Energies of Hydrolysis of Selected Organophosphates

COMPOUND	$\Delta G^{\circ\prime}$ kJ mol^{-1}	$\Delta G^{\circ\prime}$ kcal mol^{-1}
Phosphoenolpyruvate	−61.9	−14.8
Carbamoyl phosphate	−51.4	−12.3
Creatine phosphate	−43.1	−10.3
Acetyl phosphate	−42.2	−10.1
ATP (to ADP)	−30.5	−7.3
Glucose 1-phosphate	−20.9	−5.0
Glucose 6-phosphate	−12.5	−3.0
Glycerol 3-phosphate	−9.7	−2.3

The forms of ADP and ATP shown in the structural equation are in their ionization states for pH 7. The symbol P_i for phosphate ion comes from its name in biochemical jargon, "inorganic phosphate." Note that there are four negative charges on ATP and three on ADP; electrostatic repulsion makes ATP less stable than ADP. Energy must be expended to put an additional negatively charged phosphate group on ADP by forming a covalent bond to the phosphate group being added. The $\Delta G^{\circ\prime}$ for the reaction refers to the usual biochemical convention of pH 7 as the standard state for hydrogen ion (Section 9.4).

The reverse reaction, the hydrolysis of ATP to ADP and phosphate ion, releases 30.5 kJ mol^{-1} (7.3 kcal mol^{-1}) when energy is needed:

$$ATP + H_2O \longrightarrow ADP + P_i + H^+ \qquad \Delta G^{\circ\prime} = -30.5 \text{ kJ mol}^{-1} = -7.3 \text{ kcal mol}^{-1}$$

Nutrients (such as glucose) + O_2

$CO_2 + H_2O$

Catabolism (Oxidative, exergonic)

$ADP + P_i$

ATP

Anabolism (Reductive, endergonic)

Products of anabolism

Precursors

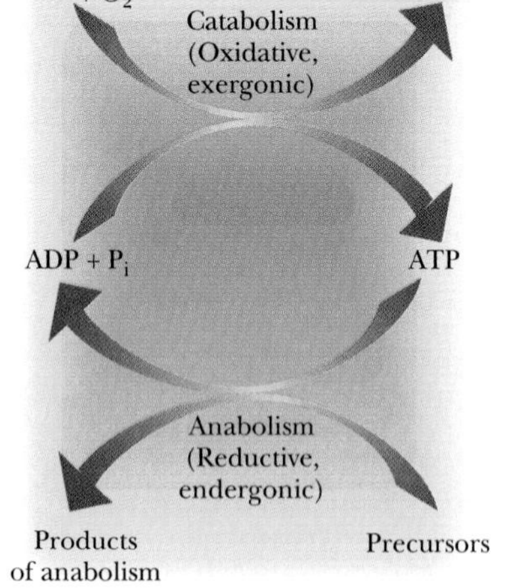

FIGURE 9.10

The role of ATP as energy currency in processes that require energy and processes that use energy.

The bond that is hydrolyzed when this reaction takes place is sometimes called a "high-energy bond," which is shorthand terminology for a reaction in which hydrolysis of a specific bond releases a useful amount of energy. Numerous organophosphate compounds with "high-energy bonds" play roles in metabolism, but ATP is by far the most important (Table 9.1). In some cases the free energy of hydrolysis of organophosphates is higher than that of ATP and thus can drive the phosphorylation of ADP to ATP.

The energy of hydrolysis of ATP is not stored energy, just as an electric current does not represent stored energy. Both ATP and electric current must be produced when they are needed—by organisms or by a power plant, as the case may be. The cycling of ATP and ADP in metabolic processes is a way of shunting energy from its production (by oxidation of nutrients) to its uses (in processes such as biosynthesis of essential compounds or muscle contraction) when it is needed. The oxidation processes take place when the organism needs the energy that can be generated by the hydrolysis of ATP. When chemical energy is stored, it is usually in the form of fats and carbohydrates, which are metabolized as needed. (Small biomolecules such as creatine phosphate can also serve as vehicles for storing chemical energy.) The energy that must be supplied for the many endergonic reactions in life processes comes directly from the hydrolysis of ATP and indirectly from oxidation of nutrients. The latter produces the energy needed to phosphorylate ADP to ATP (Figure 9.10).

Let us examine some biological reactions that release energy and see how some of that energy is used to phosphorylate ADP to ATP. The multistep conversion of glucose to lactate ions is an exergonic and anaerobic process. Two molecules of ADP are phosphorylated to ATP for each molecule of glucose metabolized. The basic reactions are the production of lactate, which is exergonic,

Glucose $\longrightarrow$ 2 lactate ions $\Delta G^{\circ\prime} = -184.5$ kJ mol$^{-1} = -44.1$ kcal mol^{-1}

and the phosphorylation of two moles of ADP for each mole of glucose, which is endergonic.

$\qquad$ 2 ADP + 2 P$_i$ $\longrightarrow$ 2 ATP $\Delta G^{\circ\prime} = 61.0$ kJ $= 14.6$ kcal

(In the interest of simplicity we shall write the equation for phosphorylation of ADP in terms of ADP, P$_i$, and ATP only.) The overall reaction is

$\qquad$ Glucose + 2 ADP + 2 P$_i$ $\longrightarrow$ 2 lactate ions + 2 ATP

$\qquad$ $\Delta G^{\circ\prime}$ overall $= -184.5 + 61.0 = -123.5$ kJ mol$^{-1} = -29.5$ kcal mol^{-1}

Not only can we add the two chemical reactions to obtain an equation for the overall reaction, we can also add the free energy changes for the two reactions to find the overall free energy change. We can do this because the free energy change is a state function; it depends only on the initial and final states of the system under consideration, not on the path between those states. The exergonic reaction provides energy, which drives the endergonic reaction. This phenomenon is called **coupling.** The percentage of the released energy that is used to phosphorylate ADP is the efficiency of energy use in anaerobic metabolism; it is (61.0/184.5) × 100, or about 33%.

The breakdown of glucose goes farther under aerobic conditions than under anaerobic conditions. The end products of aerobic oxidation are six molecules of carbon dioxide and six molecules of water for each molecule of glucose. Up to 38 molecules of ADP can be phosphorylated to ATP when one molecule of glucose is broken down completely to carbon dioxide and water. The exergonic reaction for the complete oxidation of glucose is

$\qquad$ Glucose + 6 O$_2$ $\longrightarrow$ 6 CO$_2$ + 6 H$_2$O

$\qquad$ $\Delta G^{\circ\prime} = -2867$ kJ mol$^{-1} = -685.9$ kcal mol^{-1}

The endergonic reaction for phosphorylation is

$\qquad$ 38 ADP + 38 P$_i$ $\longrightarrow$ 38 ATP $\Delta G^{\circ\prime} = 1159$ kJ $= 277.3$ kcal

The net reaction is

$\qquad$ Glucose + 6 O$_2$ + 38 ADP + 38 P$_i$ $\longrightarrow$ 6 CO$_2$ + 6 H$_2$O + 38 ATP

$\qquad$ $\Delta G^{\circ\prime} = -2867 + 1159 = -1708$ kJ mol$^{-1} = -408.6$ kcal mol^{-1}

(Note that once again we add the two reactions and their respective free energy changes to obtain the overall reaction and its free energy change.) The efficiency of aerobic oxidation of glucose is (1159/2867) × 100, about 40%. More ATP is produced by the coupling process in aerobic oxidation of glucose than by the coupling process in anaerobic oxidation. The hydrolysis of ATP produced by breakdown (aerobic or anaerobic) of glucose can be coupled to endergonic processes, such as muscle contraction in exercise. As any jogger or long-distance swimmer knows, aerobic metabolism involves large quantities of energy, processed in highly efficient fashion. We have now seen two examples of coupling of exergonic and endergonic processes, aerobic oxidation of glucose and anaerobic fermentation of glucose, involving different amounts of energy.

Example: Calculations of Free Energies

We shall use values from Table 9.1 to calculate $\Delta G^{\circ\prime}$ for the reaction shown below and decide whether or not it is spontaneous.

$$\text{ADP + phosphoenolpyruvate} \longrightarrow \text{ATP + pyruvate}$$

From Table 9.1,

Phosphoenolpyruvate + H_2O $\longrightarrow$
$$\text{pyruvate} + P_i \qquad \Delta G^{\circ\prime} = -61.9 \text{ kJ mol}^{-1}$$
$$= -14.8 \text{ kcal mol}^{-1}$$

Also,

$$\text{ATP} + H_2O \longrightarrow \text{ADP} + P_i \qquad\qquad \Delta G^{\circ\prime} = -30.5 \text{ kJ mol}^{-1}$$
$$= -7.3 \text{ kcal mol}^{-1}$$

We want the reverse of the second reaction:

$$\text{ADP} + P_i \longrightarrow \text{ATP} + H_2O \qquad\qquad \Delta G^{\circ\prime} = +30.5 \text{ kJ mol}^{-1}$$
$$= +7.3 \text{ kcal mol}^{-1}$$

We now add the two reactions and their free energy changes:

Phosphoenolpyruvate + H_2O $\longrightarrow$ pyruvate + P_i
$$\underline{\text{ADP} + P_i \longrightarrow \text{ATP} + H_2O}$$
Phosphoenolpyruvate + ADP $\longrightarrow$ pyruvate + ATP

(Net reaction)

$$\Delta G^{\circ\prime} = -61.9 \text{ kJ mol}^{-1} + 30.5 \text{ kJ mol}^{-1} = -31.4 \text{ kJ mol}^{-1}$$

$$\Delta G^{\circ\prime} = -14.8 \text{ kcal mol}^{-1} + 7.3 \text{ kcal mol}^{-1} = -7.5 \text{ kcal mol}^{-1}$$

The reaction is spontaneous.

9.11
METABOLISM PROCEEDS IN STAGES: THE ROLE OF COENZYME A IN ACTIVATION PROCESSES

The metabolic oxidation of glucose that we saw in the last section does not take place in one step. The anaerobic breakdown of glucose requires many steps, and the complete aerobic oxidation of glucose to carbon dioxide and

(a) Thioethanolamine   From pantothenic acid   3′-P-5′-ADP

Functional
sulfhydryl
group

From β-alanine

$$HS{-}CH_2CH_2N{-}\overset{\overset{O}{\parallel}}{C}CH_2CH_2N{-}\overset{\overset{H}{|}}{\underset{\overset{|}{H}}{C}}{-}\overset{\overset{H}{|}}{\underset{\overset{|}{OH}}{C}}{-}\overset{\overset{CH_3}{|}}{\underset{\overset{|}{CH_3}}{C}}{-}CH_2O{-}\overset{\overset{O}{\parallel}}{\underset{\overset{|}{O^-}}{P}}{-}O{-}\overset{\overset{O}{\parallel}}{\underset{\overset{|}{O^-}}{P}}{-}OCH_2$$

Coenzyme A
(CoA-SH)

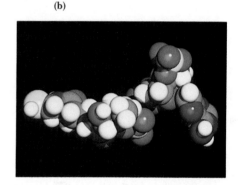

(b)

FIGURE 9.11

(a) The structure of coenzyme A. (b) Space-filling model of coenzyme A.

water has still more steps. One of the most important points about the multistep nature of all metabolic processes, including the oxidation of glucose, is that the many stages allow for efficient production and use of energy. The electrons produced by the oxidation of glucose are passed along to oxygen, the ultimate electron acceptor, by intermediate electron acceptors. Many of the intermediate stages of the oxidation of glucose are coupled to ATP production by phosphorylation of ADP.

A step frequently encountered in metabolism is the process of **activation.** In a reaction of this sort, a metabolite (a component of a metabolic pathway) is bonded to some other molecule such as a coenzyme, and the free energy change for breaking this new bond is negative. In other words, the next reaction in the metabolic pathway is exergonic. For example, if A is the metabolite and it reacts with substance B to give AB, the following series of reactions might take place.

A + coenzyme $\longrightarrow$ A—coenzyme   Activation step

A—coenzyme + B $\longrightarrow$ AB + coenzyme $\Delta G°' < 0$ (exergonic reaction)

The formation of a more reactive substance in this fashion is called *activation.* There are many examples of activation in metabolic processes. We can discuss one of the most useful of them now. It involves forming a covalent bond to a compound known as coenzyme A (CoA).

The structure of CoA is complex. It consists of several smaller components linked together covalently (Figure 9.11). One part is 3′-P-5′-ADP, a derivative of adenosine with phosphate groups esterified to the sugar, as shown in the structure. Another part is derived from the vitamin pantothenic acid, and the part of the molecule involved in activation reactions contains a thiol group. In fact, coenzyme A is frequently written as CoA—SH to emphasize that the thiol group is the reactive portion of the molecule. For example, carboxylic acids form thioester linkages with CoA—SH. The

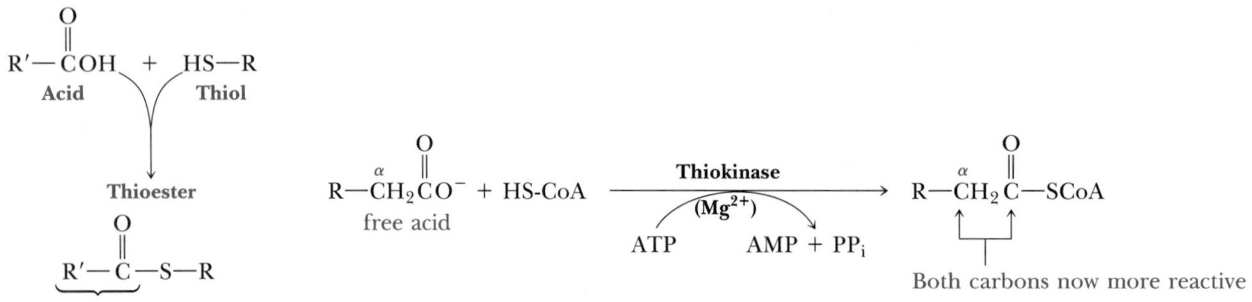

(a) Activated acyl group having increased reactivity.

(b) By "selecting" thioesters over oxyesters, nature has exploited the larger size of the S atom, which contributes less stability to the ester group)

FIGURE 9.12

(a) The general reaction for thioester formation. (b) The formation of a thioester bond between CoA and a carboxylic acid.

metabolically active form of a carboxylic acid is the corresponding acyl-CoA thioester, where the thioester linkage is a "high-energy" bond (Figure 9.12). Acetyl-CoA is a particularly important metabolic intermediate; other acyl-CoA species figure prominently in lipid metabolism.

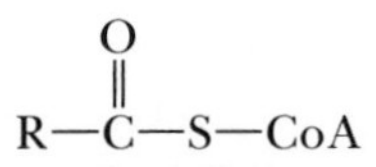

Parenthetically, the important coenzymes we have met in this chapter — NAD^+, $NADP^+$, FAD, and coenzyme A — share an important structural feature: all contain ADP. In $NADP^+$, there is an additional phosphate group at the $2'$ position of the ribose group of ADP. In CoA, the additional phosphate group is at the $3'$ position.

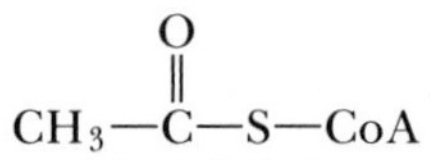

Like catabolism, anabolism proceeds in stages. Unlike catabolism, which releases energy, anabolism requires energy. The ATP produced by catabolism is hydrolyzed to release the needed energy. Reactions in which metabolites are reduced are part of anabolism; they require reducing agents such as NADH, NADPH, and $FADH_2$, all of which are the reduced forms of coenzymes mentioned in this chapter. In their oxidized forms these coenzymes serve as the intermediate oxidizing agents needed in catabolism. In their reduced forms the same coenzymes provide the "reducing power" needed for the anabolic processes of biosynthesis; in this case, the coenzymes act as reducing agents.

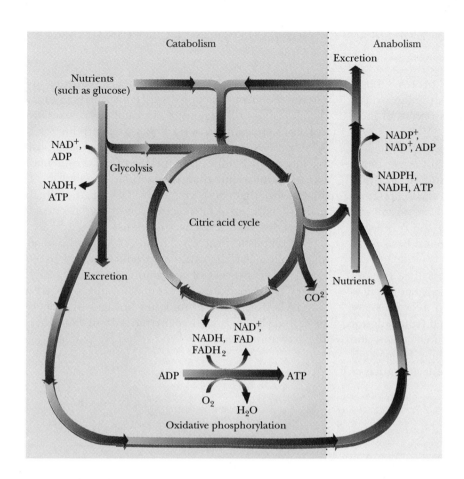

FIGURE 9.13

The role of electron transfer and ATP production in metabolism. NAD$^+$, FAD, and ATP are constantly recycled.

Within the figure:

Catabolism

Anabolism

Excretion

Nutrients (such as glucose)

NAD$^+$, ADP

NADH, ATP

Glycolysis

NADP$^+$, NAD$^+$, ADP

NADPH, NADH, ATP

Citric acid cycle

Excretion

CO^2

Nutrients

NADH, FADH$_2$

NAD$^+$, FAD

ADP → ATP

O$_2$

H$_2$O

Oxidative phosphorylation

We are now in a position to expand on our earlier statements about the natures of anabolism and catabolism. Figure 9.13 is an outline of metabolic pathways that explicitly points out two important features of metabolism: the role of electron transfer and the role of ATP in the release and utilization of energy. Even though this outline is more extended than the one in Figure 9.7, it is still very general. The more important specific pathways have been studied in detail, and some are still the subjects of active research. We shall discuss some of the most important metabolic pathways in the remainder of this text.

SUMMARY

Thermodynamics deals with the changes in energy that determine whether or not a process will take place. In a spontaneous process (one that will take place), the free energy decreases. In a nonspontaneous process (one that will not take place), the free energy increases. The symbol for change in free energy is ΔG. A change in free energy under any set of conditions can be compared to the free energy change under standard conditions ($\Delta G°$). Free energy changes under standard conditions can be related to the equilibrium constant of a reaction by the equation

$$\Delta G° = -2.303RT \log K_{eq}$$

In addition to the free energy, entropy is an important quantity in thermodynamics. The entropy of the universe increases in any spontaneous process. Local decreases in entropy can take place within an overall increase in entropy. Living organisms represent local decreases in entropy.

The reactions of the biomolecules in the cell constitute metabolism. The breakdown of larger molecules to smaller ones is called catabolism. The reaction of small molecules to produce larger and more complex molecules is called anabolism. Catabolism and anabolism are separate pathways, not the reverse of each other. Metabolism is the biochemical basis of all life processes.

Catabolism is an oxidative process that releases energy; anabolism is a reductive process that requires energy. Oxidation–reduction (redox) reactions are those in which electrons are transferred from a donor to an acceptor. Oxidation is the loss of electrons, and reduction is the gain of electrons. Many biologically important redox reactions involve coenzymes such as NADH and FADH$_2$.

The coupling of energy-producing reactions and energy-requiring reactions is a central feature in the metabolism of all organisms. In catabolism, oxidative reactions are coupled to the endergonic production of ATP by phosphorylation of ADP. Aerobic metabolism is a more efficient means of making use of the chemical energy of nutrients than is anaerobic metabolism. In anabolism, the exergonic hydrolysis of the "high-energy" bond of ATP releases the energy needed to drive endergonic reductive reactions.

Metabolism proceeds in stages, and the many stages allow for the efficient production and use of energy. The process of activation, producing "high-energy" intermediates, occurs in many metabolic pathways. The formation of thioester linkages by reaction of carboxylic acids with coenzyme A is an example of the activation process.

E X E R C I S E S

1. For the process

   Nonpolar solute + H$_2$O $\longrightarrow$ solution

   what are the signs of ΔS_{univ}, ΔS_{sys}, and ΔS_{surr}? What is the reason for each answer?
2. Which of the following are spontaneous processes? For each, explain why or why not.
   (a) The hydrolysis of ATP to ADP and P$_i$
   (b) The oxidation of glucose to CO$_2$ and H$_2$O by an organism
   (c) The phosphorylation of ADP to ATP
   (d) The production of glucose and O$_2$ from CO$_2$ and H$_2$O in photosynthesis
3. In which of the following processes does the entropy increase? In each case, explain why it does or does not.
   (a) A bottle of ammonia is opened. The odor of ammonia is soon apparent throughout the room.
   (b) Sodium chloride dissolves in water.
   (c) A protein is completely hydrolyzed to the component amino acids.
4. Which of the following statements is (are) true about the modified standard state for biochemistry? For each, explain why or why not.

   $$[H^+] = 10^{-7}\ M,\ not\ 1\ M$$

   The concentration of any solute is 10^{-7} M.

5. For the hydrolysis of ATP at 25°C (298 K) and pH 7,

   $$ATP + H_2O \longrightarrow ADP + P_i + H^+$$

   the standard free energy of hydrolysis ($\Delta G^{\circ\prime}$) is −30.5 kJ mol^{-1} (− 7.3 kcal mol^{-1}), and the standard enthalpy change ($\Delta H^{\circ\prime}$) is −20.1 kJ mol^{-1} (−4.8 kcal mol^{-1}). Calculate the standard entropy change ($\Delta S^{\circ\prime}$) for the reaction, in both joules and calories. Why is the positive sign of the answer to be expected in view of the nature of the reaction?
6. Which of the following statements are true? For each, explain why or why not.
   (a) An unfavorable entropy change for the water is the most important factor in the insolubility of nonpolar substances in water.
   (b) The entropy of the universe decreases in a spontaneous process.
   (c) The enthalpy change for a reaction is the heat of reaction measured at constant pressure.
   (d) Heat is a reflection of the random motion of molecules.
   (e) An endergonic reaction occurs spontaneously.
7. In most proteins, the majority of the nonpolar side chains are found in the interior of the molecule, while the polar residues tend to be found on the surface. Comment on this observation in light of the material in this chapter.

8. Comment on the statement that the existence of life is a violation of the second law of thermodynamics.

9. Comment on the role of hydrophobic interactions in maintaining the structure of biological membranes.

10. Would you expect the biosynthesis of a protein from the constituent amino acids in an organism to be an exergonic or endergonic process? Give the reason for your answer.

11. Adult humans synthesize large amounts of ATP in the course of a day, but their body weights do not change significantly. In the same time period the structures and compositions of their bodies also do not change appreciably. Explain this apparent contradiction.

12. Would you expect an increase or decrease of entropy to accompany the hydrolysis of phosphatidylcholine to the constituent parts (glycerol, two fatty acids, phosphoric acid, and choline)? Why?

13. The following half reactions play important roles in metabolism.

$$\tfrac{1}{2} O_2 + 2H^+ + 2e^- \longrightarrow H_2O$$

$$NADH + H^+ \longrightarrow NAD^+ + 2H^+ + 2e^-$$

Which of these two is a half reaction of oxidation? Which one is a half reaction of reduction? Write the equation for the overall reaction. Which reagent is the oxidizing agent (electron acceptor)? Which reagent is the reducing agent (electron donor)?

14. All the organophosphate compounds listed in Table 9.1 undergo hydrolysis reactions in the same way as ATP. The following equation illustrates the situation for glucose 1-phosphate.

$$\text{Glucose 1-phosphate} + H_2O \longrightarrow \text{glucose} + P_i$$

$$\Delta G^{\circ\prime} = -20.9 \text{ kJ mol}^{-1}$$

Using the free energy values in Table 9.1, predict whether the following reactions will proceed in the direction written, and calculate the $\Delta G^{\circ\prime}$ for the reaction, assuming that the reactants are initially present in a 1:1 molar ratio.
(a) ATP + creatine $\longrightarrow$ creatine phosphate + ADP
(b) ATP + glycerol $\longrightarrow$ glycerol 3-phosphate + ADP

(c) ATP + pyruvate $\longrightarrow$ phosphoenolpyruvate + ADP
(d) ATP + glucose $\longrightarrow$ glucose 6-phosphate + ADP

15. Short periods of exercise such as sprints are characterized by lactic acid production and the condition known as oxygen debt. Comment on this fact in light of the material discussed in this chapter.

16. Using the data in Table 9.1, calculate the value of $\Delta G^{\circ\prime}$ for the reaction

$$\text{Creatine phosphate} + \text{glycerol} \longrightarrow$$
$$\text{creatine} + \text{glycerol 3-phosphate}$$

(*Hint:* This reaction proceeds in stages. ATP is formed in the first step, and the phosphate group is transferred from ATP to glycerol in the second step.)

17. Calculate the value of $\Delta G^{\circ\prime}$ for the following reaction, using information from Table 9.1.

$$\text{Glucose 1-phosphate} \longrightarrow \text{glucose 6-phosphate}$$

18. Show that the hydrolysis of ATP to AMP and 2 P_i releases the same amount of energy by either of the two following pathways.

   Pathway 1

   $$ATP + H_2O \longrightarrow ADP + P_i$$

   $$ADP + H_2O \longrightarrow AMP + P_i$$

   Pathway 2

   $$ATP + H_2O \longrightarrow AMP + PP_i$$
   $$\text{Pyrophosphate}$$

   $$PP_i + H_2O \longrightarrow 2P_i$$

19. The standard free energy change for the reaction

$$\text{Arginine} + ATP \longrightarrow \text{phosphoarginine} + ADP$$

is +1.7 kJ mol^{-1}. From this information and that in Table 9.1, calculate the $\Delta G^{\circ\prime}$ for the reaction

$$\text{Phosphoarginine} + H_2O \longrightarrow \text{arginine} + P_i$$

20. There is a reaction in carbohydrate metabolism in which glucose 6-phosphate reacts with NADP$^+$ to give 6-phosphoglucono-δ lactone and NADPH (see below). In this reaction, which substance is oxidized, and which is reduced? Which substance is the oxidizing agent, and which is the reducing agent?

glucose-6-phosphate

6-phosphoglucono-δ-lactone

21. There is a reaction in which succinate reacts with FAD to give fumarate and $FADH_2$.

Succinate

Fumarate

In this reaction, which substance is oxidized, and which is reduced? Which substance is the oxidizing agent, and which is the reducing agent?

22. What structural feature do NAD^+, $NADP^+$, and FAD have in common?

A N N O T A T E D   B I B L I O G R A P H Y

Atkins, P. W. *The Second Law.* San Francisco: W. H. Freeman, 1984. [A highly readable nonmathematical discussion of thermodynamics.]

Campbell, J. A. Reversibility and Returnability, or When Can You Return Again? *J. Chem. Ed.* **57,** 345–348 (1980). [A treatment of entropy and its role in spontaneous processes.]

Chang, R. *Physical Chemistry with Applications to Biological Systems.* 2nd ed. New York: Macmillan, 1981. [Chapter 12 contains a detailed treatment of thermodynamics.]

Fasman, G. D., ed. *Handbook of Biochemistry and Molecular Biology.* 3rd ed. Sec. D, *Physical and Chemical Data.* Cleveland: CRC Press, 1976. [Volume 1 contains data on the free energies of hydrolysis of many important compounds, especially organophosphates.]

Harold, F. M. *The Vital Force: A Study of Bioenergetics.* New York: W. H. Freeman, 1986. [Energetic aspects of many important life processes.]

Hinkle, P. C., and R. E. McCarty. How Cells Make ATP. *Sci. Amer.* **238** (3), 104–125 (1978). [Getting old, but a particularly good treatment of energy coupling.]

Prigogine, I., and I. Stengers. *Order Out of Chaos.* Toronto: Bantam Press, 1984. [A comparatively accessible treatment of the thermodynamics of biological systems. Prigogine won the 1977 Nobel Prize in chemistry for his pioneering work on the thermodynamics of complex systems.]

Tanford, C. *The Hydrophobic Effect: Formation of Micelles and Biological Membranes.* 2nd ed. New York: Wiley-Interscience, 1980. [A comprehensive treatment of the topic.]

Most textbooks of general chemistry have several chapters on thermodynamics. Of these, *Chemistry,* 3rd ed. by S. Zumdahl (Lexington, MA: D. C. Heath, 1993) has a particularly good treatment of the thermodynamics of solution formation in Sections 11.2 and 16.3.

Two standard multivolume references cover specific aspects of metabolism in detail. One of these, the third edition of *The Enzymes* (P. D. Boyer, ed., New York: Academic Press), is a series that has been in production since 1970. The other, *Comprehensive Biochemistry* (M. Florkin and E. H. Stotz, eds. New York: Elsevier), has been in production since 1962.

Chapter 10

Carbohydrates

Crystals of glucose, a key molecule in carbohydrate metabolism, viewed under polarized light.

More than half of all the organic carbon on planet Earth is stored in just two carbohydrate molecules — starch and cellulose. Both are polymers of the sugar monomer, *glucose.* The only difference between them is the manner in which the glucose units are joined together. Glucose is made by green plants and stored in starch as the plants' energy reserves. Animals (including humans) have an enzyme that recognizes the helical conformation of starch and can degrade it into its glucose units. Glucose, oxidized to carbon dioxide and water, is our primary energy source. Cellulose, a major component of plant cell walls and of cotton and wood, is a polymer of glucose monomers that all lie in the same plane. We don't possess the enzyme cellulase to break down cellulose, but termites do. A protozoan in their intestines contains cellulase, enabling them to digest wood. We learn this the hard way when wooden houses are damaged by termites. Modified carbohydrates are constituents of bacterial cell walls, where they are cross-linked by short polypeptide chains. By combining specific sugar monomers with amino acids, *glycoproteins* can be formed and act as cell surface markers to be recognized by other biomolecules. This recognition feature is of life-or-death importance in blood transfusions, where compatability of blood types depends on glycoproteins.

When the word "carbohydrate" was coined, it originally referred to compounds of the general formula $C_n(H_2O)_n$. However, only the simple sugars, or **monosaccharides,** fit this formula exactly. The other types of carbohydrates, oligosaccharides and polysaccharides, are based on the monosaccharide units and have slightly different general formulas. **Oligosaccharides** are formed when a few (Greek *oligos*) monosaccharides are linked together; **polysaccharides** are formed when many (Greek *polys*) monosaccharides are bonded together. The reaction that adds monosaccharide units to a growing carbohydrate molecule involves the loss of one H_2O for each new link formed, accounting for the difference in the general formula.

Many commonly encountered carbohydrates are polysaccharides, including *glycogen,* which is found in animals, and *starch* and *cellulose,* which occur in plants. Carbohydrates play a number of important roles in biochemistry. First, they are major energy sources (Chapters 11 through 14 are devoted to carbohydrate metabolism). Second, oligosaccharides play a key role in processes that take place on the surfaces of cells, particularly in cell–cell interactions and immune recognition. In addition, polysaccharides are essential structural components of several classes of organisms. Cellulose is a major component of grass and trees, and other polysaccharides are major components of bacterial cell walls.

10.1
MONOSACCHARIDES: STRUCTURE AND STEREOCHEMISTRY

A monosaccharide can be a polyhydroxy aldehyde (**aldose**) or a polyhydroxy ketone (**ketose**). The simplest monosaccharides contain three carbon atoms and are called trioses (*tri* meaning three). **Glyceraldehyde** is the aldose with

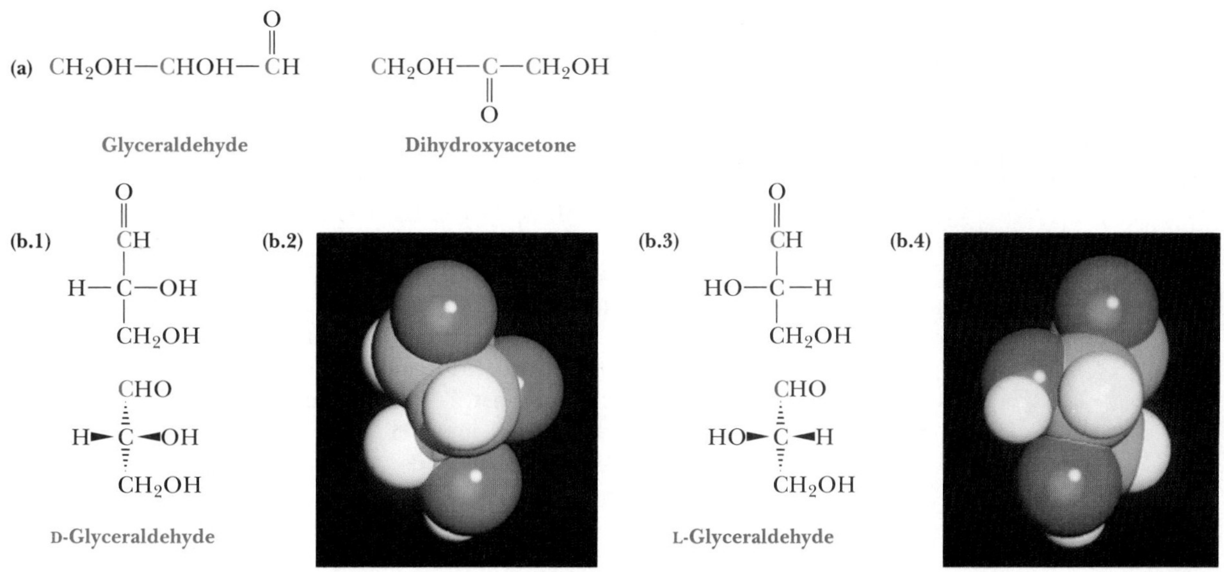

(a) $CH_2OH—CHOH—\overset{\overset{\displaystyle O}{\|}}{C}H$ $CH_2OH—\overset{\underset{\displaystyle O}{\|}}{C}—CH_2OH$

           Glyceraldehyde   Dihydroxyacetone

(b.1)

$$\begin{array}{c} \overset{\overset{\displaystyle O}{\|}}{C}H \\ | \\ H—C—OH \\ | \\ CH_2OH \end{array}$$

$$\begin{array}{c} CHO \\ \vdots \\ H▶C◀OH \\ | \\ CH_2OH \end{array}$$

    D-Glyceraldehyde

(b.2)

(b.3)

$$\begin{array}{c} \overset{\overset{\displaystyle O}{\|}}{C}H \\ | \\ HO—C—H \\ | \\ CH_2OH \end{array}$$

$$\begin{array}{c} CHO \\ \vdots \\ HO▶C◀H \\ | \\ CH_2OH \end{array}$$

    L-Glyceraldehyde

(b.4)

FIGURE 10.1

The structures of the simplest carbohydrates, the trioses. (a) A comparison of glyceraldehyde (an aldotriose) and dihydroxyacetone (a ketotriose). (b.1) The structure of D-glyceraldehyde. (b.2) Space-filling model of D-glyceraldehyde. (b.3) The structure of L-glyceraldehyde. (b.4) Space-filling model of L-glyceraldehyde.

three carbons (an aldotriose), and **dihydroxyacetone** is the ketose with three carbon atoms (a ketotriose). Figure 10.1 shows these molecules. Aldoses with four, five, six, and seven carbon atoms are called aldotetroses, aldopentoses, aldohexoses, and aldoheptoses, respectively. The corresponding ketoses are ketotetroses, ketopentoses, ketohexoses, and ketoheptoses. Six-carbon sugars are the most abundant in nature, but two five-carbon sugars, ribose and deoxyribose, occur in the structures of RNA and DNA, respectively. Four-carbon and seven-carbon sugars play roles in photosynthesis and other metabolic pathways.

We have already seen (Chapter 3, Section 3.1) that some molecules are not superimposable on their mirror images and that these mirror images are **optical isomers** (**stereoisomers**) of each other. A chiral (asymmetric) carbon atom is the usual source of optical isomerism, as was the case with amino acids. The simplest carbohydrate that contains a chiral carbon is glyceraldehyde, which can exist in two isomeric forms that are mirror images of each other (Figure 10.1b). Note that the two forms differ in the position of the hydroxyl group bonded to the central carbon. (Dihydroxyacetone does not contain a chiral carbon atom and does not exist in nonsuperimposable mirror-image forms.) The two forms of glyceraldehyde are designated D-glyceraldehyde and L-glyceraldehyde. Mirror-image stereoisomers are also called **enantiomers,** and D-glyceraldehyde and L-glyceraldehyde are enantiomers of each other. Certain conventions are used for two-dimensional drawings of the three-dimensional structures of stereoisomers. The dashed wedges represent bonds directed away from the viewer, below the plane of the paper, and the solid triangles represent bonds directed oppositely, toward the viewer and out of the plane of the paper. The **configuration** is the three-dimensional arrangement of groups around a chiral carbon atom, and stereoisomers differ from each other in configuration.

The two enantiomers of glyceraldehyde are the only possible stereoisomers of three-carbon sugars, but the possibilities for stereoisomerism increase as the number of carbon atoms increases. To show the structures of

Emil Fischer (1852–1919) was a German-born scientist, who won the Nobel Prize in chemistry in 1902 for his studies on sugars, purine derivatives, and peptides.

the resulting molecules, we need to say more about the convention for a two-dimensional perspective of the molecular structure, which is called the **Fischer projection** method after the German chemist Emil Fischer, who established the structures of many sugars. We shall use some common six-carbon sugars for purposes of illustration. In the Fischer projection, bonds written "vertically" on the two-dimensional paper represent bonds directed *behind* the paper in three dimensions, whereas bonds written "horizontally" represent bonds directed *in front of* the paper in three dimensions. Figure 10.2 shows that the most highly oxidized carbon, in this case the one involved in the aldehyde group, is written at the "top" and is designated carbon 1, or C-1. (In the ketose shown, the ketone group becomes C-2, the carbon atom next to the "top." Most common sugars are aldoses rather than ketoses, so our discussion will focus mainly on aldoses.) The other carbon atoms are numbered in sequence from the "top." The designation of the configuration as L or D depends on the arrangement at the chiral carbon with the highest number. In the cases of both glucose and fructose, this is C-5. In the Fischer projection of the D configuration, the hydroxyl

(a)

		Carbon number
H—C=O	CH$_2$OH	1
H—C—OH	C=O	2
HO—C—H	HO—C—H	3
H—C—OH	H—C—OH	4
H—C—OH	H—C—OH	5
CH$_2$OH	CH$_2$OH	6
D-Glucose (an aldose)	D-Fructose (a ketose)	

(b)

		Carbon number
H—C=O	H—C=O	1
H—C—OH	HO—C—H	2
HO—C—H	H—C—OH	3
H—C—OH	HO—C—H	4
H—C—OH	HO—C—H	5
CH$_2$OH	CH$_2$OH	6
D-Glucose	L-Glucose	

FIGURE 10.2

(a) Examples of an aldose (D-glucose) and a ketose (D-fructose), showing the numbering of carbon atoms. (b) A comparison of the structures of D-glucose and L-glucose.

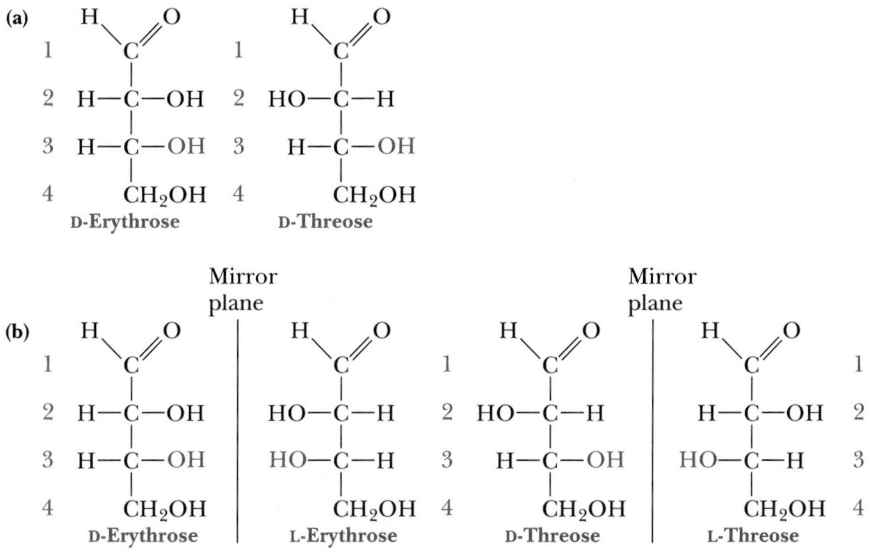

(a)

D-Erythrose   D-Threose

(b)

D-Erythrose   L-Erythrose   D-Threose   L-Threose

FIGURE 10.3

Stereoisomers of an aldotetrose. (a) Diastereomers D-erythrose and D-threose. (b) Enantiomers D- and L-erythrose and D- and L-threose. Carbons are numbered. The designation of D or L depends on the configuration at the highest-numbered carbon atom.

group is on the right of the highest-numbered chiral carbon, whereas in the L configuration the hydroxyl group is on the left of the highest-numbered chiral carbon.

Let us see what happens as another carbon is added to glyceraldehyde to give a four-carbon sugar. In other words, what are the possible stereoisomers for an aldotetrose? The aldotetroses (Figure 10.3) have two chiral carbons, C-2 and C-3, and there are 2^2, or four, possible stereoisomers. Two of the isomers have the D configuration, and two have the L configuration. The two D isomers have the same configuration at C-3, but they differ in configuration (arrangement of the —OH group) at the other chiral carbon, C-2. These two isomers are called D-erythrose and D-threose. They are not superimposable on each other, but neither are they mirror images of each other. Such nonsuperimposable, non-mirror-image stereoisomers are called **diastereomers.** The two L isomers are L-erythrose and L-threose. L-Erythrose is the enantiomer (mirror image) of D-erythrose, and L-threose is the enantiomer of D-threose. L-Threose is a diastereomer of both D- and L-erythrose, and L-erythrose is a diastereomer of both D- and L-threose. Diastereomers that differ from each other in configuration at only one chiral carbon are called **epimers;** D-erythrose and D-threose are epimers.

Aldopentoses have three chiral carbons, and there are 2^3, or eight, possible stereoisomers — four D forms and four L forms. Aldohexoses have four chiral carbons and 2^4, or 16, stereoisomers — eight D forms and eight L forms (Figure 10.4). Some of the possible stereoisomers are much more common in nature than others, and most biochemical discussion centers on

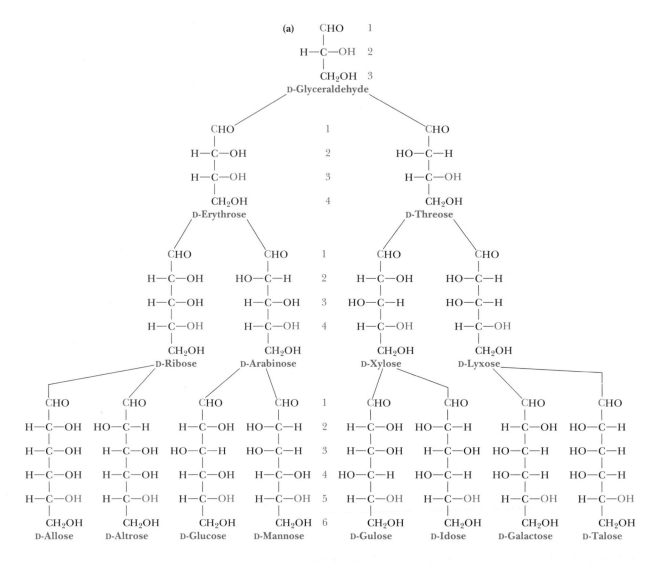

(a)

D-Glyceraldehyde

D-Erythrose   D-Threose

D-Ribose   D-Arabinose   D-Xylose   D-Lyxose

D-Allose   D-Altrose   D-Glucose   D-Mannose   D-Gulose   D-Idose   D-Galactose   D-Talose

(b)

FIGURE 10.4

Stereochemical relationships among monosaccharides. **(a) Aldoses containing from three to six carbon atoms, with the numbering of the carbon atoms shown. Note that the figure shows only half the possible isomers. For each isomer shown there is an enantiomer that is not shown, the L series. (b) The relationship between mirror images is of interest to mathematicians as well as to chemists. Lewis Carroll (C. L. Dodgson), the author of *Alice's Adventures in Wonderland*, was a contemporary of Emil Fischer.**

the common naturally occurring sugars. For example, D sugars rather than L sugars predominate in nature. Most of the sugars we encounter in nature, especially in foods, contain either five or six carbon atoms. We shall discuss D-glucose (an aldohexose) and D-ribose (an aldopentose) far more than many other sugars. Glucose is a ubiquitous energy source, and ribose plays an important role in the structure of nucleic acids.

Cyclic Structures: Anomers

Sugars, especially those with five and six carbon atoms, normally exist as cyclic molecules rather than as the open-chain forms we have shown so far. The cyclization takes place as a result of interaction between the functional groups on distant carbons such as C-1 and C-5 to form a cyclic **hemiacetal** (in aldohexoses). Another possibility (Figure 10.5) is interaction between C-2 and C-5 to form a cyclic **hemiketal** (in ketohexoses). In either case, the

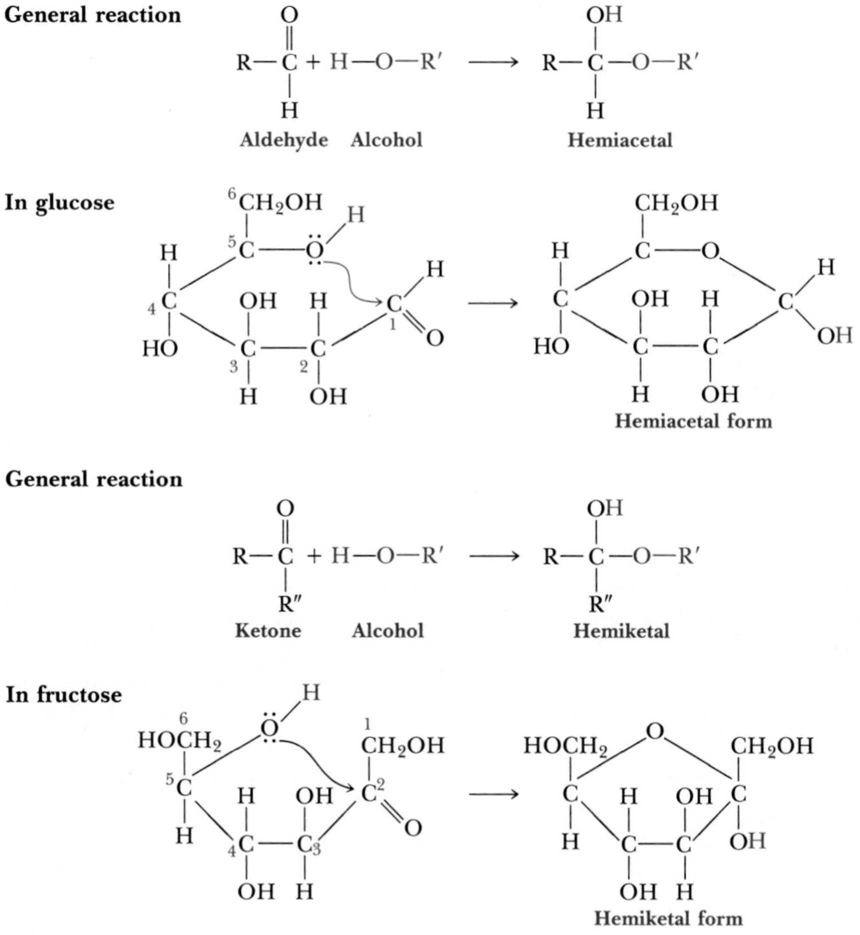

FIGURE 10.5

Formation of hemiacetals and hemiketals. An alcohol reacts with a carbonyl to give rise to a chiral center. In the six-carbon sugars, the hydroxyl group on carbon 5 reacts with the carbonyl group. In the cyclic form two possible configurations can occur at the carbon atom that was the carbonyl carbon in the open-chain form.

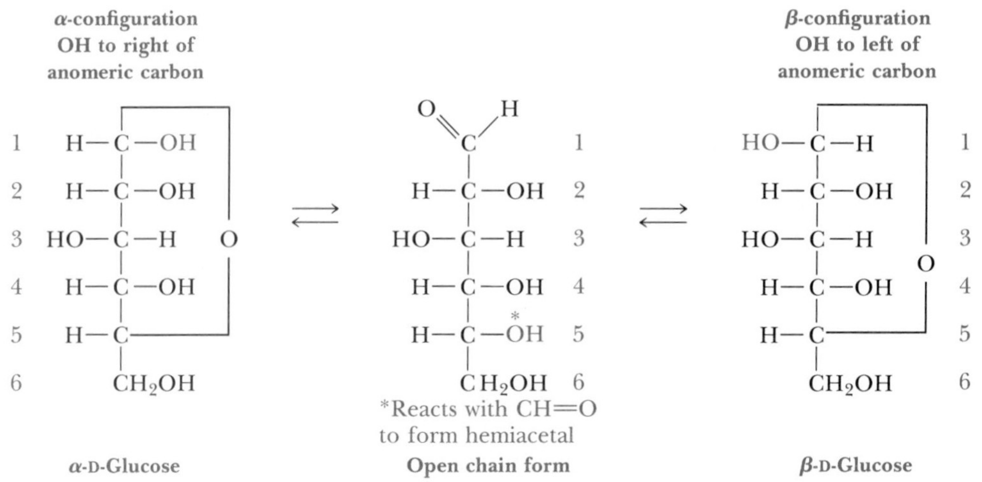

α-configuration
OH to right of
anomeric carbon

β-configuration
OH to left of
anomeric carbon

α-D-Glucose

*Reacts with CH=O
to form hemiacetal

Open chain form

β-D-Glucose

FIGURE 10.6

Fischer projection formulas of three forms of glucose. Note that the α and β forms can be converted to each other through the open-chain form. The configuration at carbon 5 determines the D designation.

carbonyl carbon becomes a new chiral center called the **anomeric carbon.** The cyclic sugar can take either of two different forms, designated α and β and called **anomers** of each other.

The Fischer projection of the α-anomer of a D sugar has the anomeric hydroxyl group to the right of the anomeric carbon (C—OH), and the β-anomer of a D sugar has the anomeric hydroxyl group to the left of the anomeric carbon (Figure 10.6). The free carbonyl species can readily form either the α- or β-anomer, and the anomers can be converted from one form to another through the free carbonyl species. In some biochemical reactions, any anomer of a given sugar can be used, but in other cases only one anomer occurs. For example, in living organisms only β-D-ribose and β-D-deoxyribose are found in RNA and DNA, respectively.

Fischer projection formulas are useful for describing the stereochemistry of sugars, but their long bonds and right-angle bends do not give a realistic picture of the bonding situation in the cyclic forms, nor do they accurately represent the overall shapes of the molecules. **Haworth projection formulas** are more useful for those purposes. In Haworth projections the cyclic structures of sugars are shown in perspective drawings as planar five- or six-membered rings viewed nearly edge on. A five-membered ring is called a **furanose** because of its resemblance to furan; a six-membered ring is called a **pyranose** because of its resemblance to pyran (Figure 10.7a, b). These cyclic formulas approximate the shapes of the actual molecules better for furanoses than for pyranoses. The five-membered rings of furanoses are in reality very nearly planar, but the six-membered rings of pyranoses actually exist in solution in the chair conformation (Figure 10.7c). Even though the Haworth formulas are approximations, they are very useful because they are easy to draw. Haworth projections represent the stereochemistry of sugars more realistically than do Fischer projections, and the

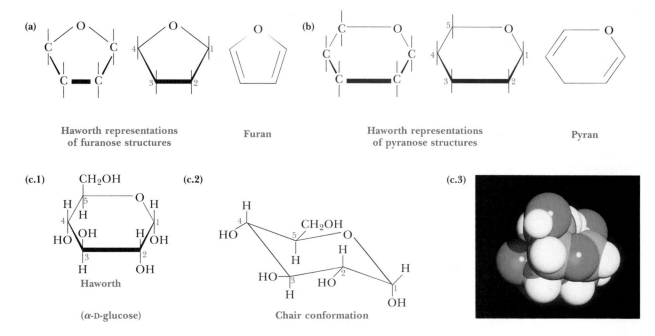

FIGURE 10.7

Haworth representations of sugar structures. (a) A comparison of the structure of furan with Haworth representations of furanoses. (b) A comparison of the structure of pyran with Haworth representations of pyranoses. (c) α-D-glucopyranose in the Haworth representation (c.1), in the chair conformation (c.2), and as a space-filling model (c.3).

Haworth scheme is adequate for our purposes. We shall continue to use Haworth projections in our discussion of sugars.

For a D sugar, any group that is written to the *right* of the carbon in a Fischer projection has a *downward* direction in a Haworth projection; any group that is written to the *left* in a Fischer projection has an *upward* direction in a Haworth projection. The terminal —CH_2OH group, which contains the carbon atom with the highest number in the numbering scheme, is shown in an upward direction. The structures of α- and β-D-glucose, which are both pyranoses, and of β-D-ribose, which is a furanose, illustrate this point (Figure 10.8).

10.2
REACTIONS OF MONOSACCHARIDES

Oxidation–Reduction Reactions

Oxidation and reduction reactions of sugars play key roles in biochemistry. Oxidation of sugars provides energy for organisms to carry out their life processes; the highest yield of energy from carbohydrates occurs when sugars are completely oxidized to CO_2 and H_2O in aerobic processes. The reverse of complete oxidation of sugars is the reduction of CO_2 and H_2O to form sugars, a process that takes place in photosynthesis.

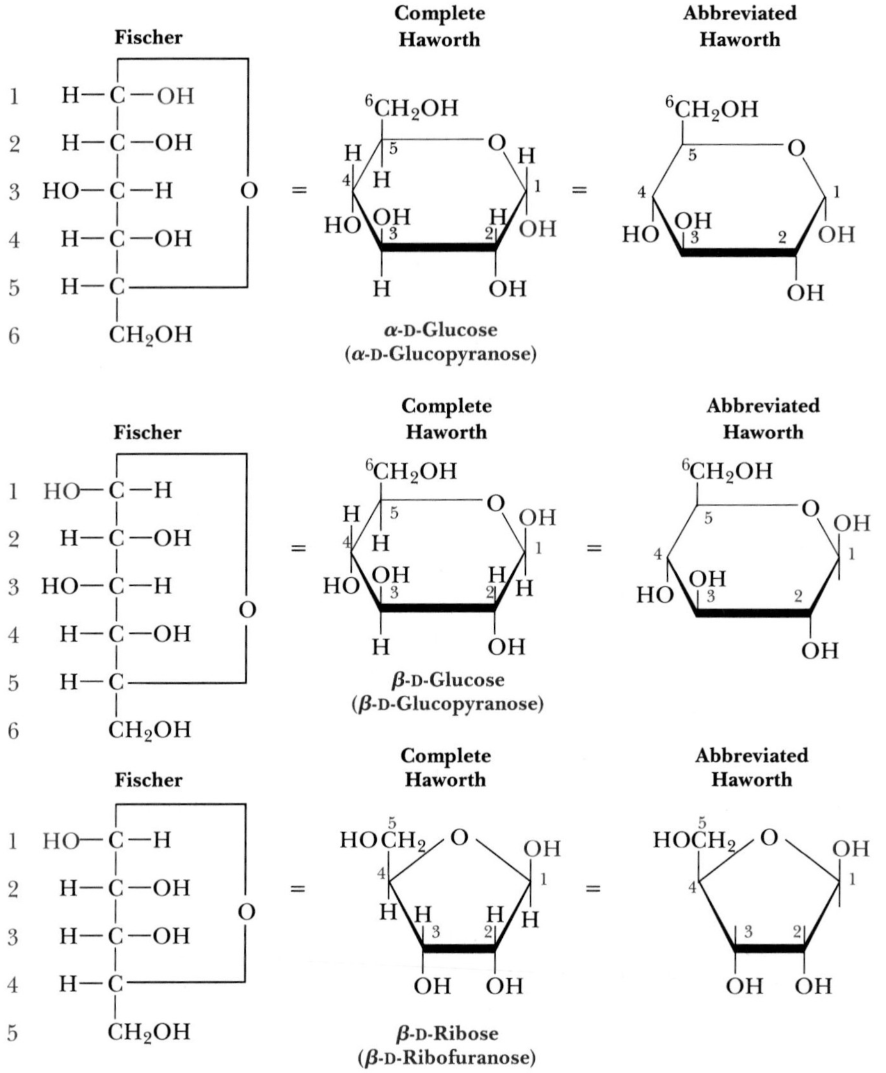

FIGURE 10.8

A comparison of the Fischer, complete Haworth, and abbreviated Haworth representations of α- and β-D-glucose (glucopyranose) and β-D-ribose (ribofuranose). In the Haworth representation, the α-anomer is represented with the OH group downward, and the β anomer is represented with the OH group upward (red).

Several oxidation reactions of sugars are of some importance in laboratory practice, because they can be used for identification of sugars. Aldehyde groups can be oxidized to give the carboxyl group that is characteristic of acids, and this reaction is the basis of a test for the presence of aldoses. When the aldehyde is oxidized, some oxidizing agent must be reduced. Aldoses are called **reducing sugars** because of this type of reaction; ketoses can also be reducing sugars because they isomerize to aldoses. In the cyclic form the compound produced by oxidation of an aldose is a **lactone** (a cyclic ester linking the carboxyl group and one of the sugar alcohols, as shown in Figure 10.9). A lactone of considerable importance to humans is discussed in Box 10.1.

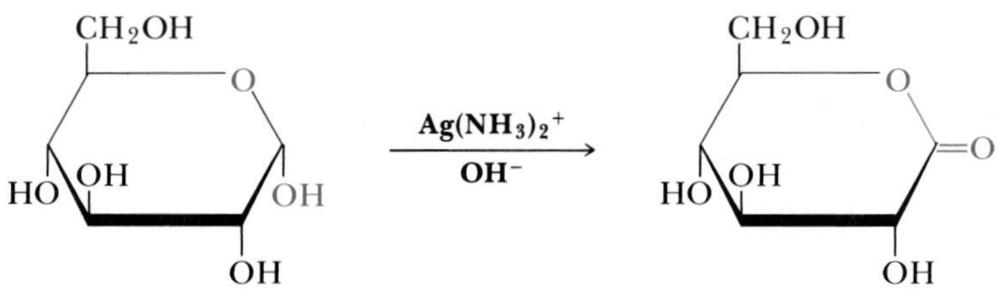

$$\text{α-D-glucose hemiacetal} \xrightarrow[\text{OH}^-]{\text{Ag(NH}_3)_2^+} \text{lactone}$$

FIGURE 10.9

An example of oxidation reaction of sugars: oxidation of α-D-glucose hemiacetal to give a lactone. Deposition of free silver as a silver mirror indicates that the reaction has taken place.

10.1

VITAMIN C IS RELATED TO SUGARS

Vitamin C (ascorbic acid) is an unsaturated lactone with a five-membered ring structure. Each carbon is bonded to a hydroxyl group except the carboxyl carbon involved in the cyclic ester bond. Most animals can synthesize vitamin C; the exceptions are guinea pigs and primates, including humans. As a result, guinea pigs and primates must acquire vitamin C in their diets. Air oxidation of ascorbic acid, followed by hydrolysis of the ester bond, leads to loss of activity as a vitamin. Consequently, lack of fresh food can cause vitamin C deficiencies, which in turn can lead to the disease scurvy (Chapter 4, Section 4.3). In this disease, defects in collagen structure cause skin lesions and fragile blood vessels. The presence of hydroxyproline is necessary for collagen stability, probably because of hydrogen-bonded cross-links between collagen strands. Ascorbic acid, in turn, is essential for the activity of prolyl hydroxylase, which converts proline residues in collagen to hydroxyproline.

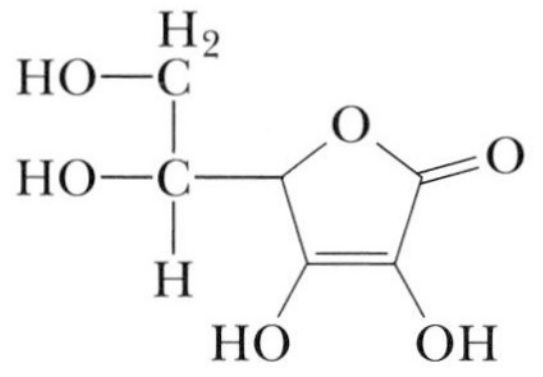

Ascorbic acid
(Vitamin C)

The British navy introduced citrus fruit into the diet of sailors in the 18th century to prevent scurvy during long sea voyages, and citrus fruit is still consumed by many for its vitamin C. Potatoes are another important source of vitamin C, not because potatoes contain a *high* concentration of ascorbic acid but because we eat so many potatoes.

Two types of reagents are used in the laboratory to detect the presence of reducing sugars. The first of these is Tollens' reagent, which uses the silver ammonia complex ion, $Ag(NH_3)_2^+$, as the oxidizing agent. A silver mirror is deposited on the wall of the test tube if a reducing sugar is present, as shown by the equation for the reaction:

$$\underset{\text{Sugar}}{RCHO} + 2\ Ag(NH_3)_2^+ + 2\ OH^- \longrightarrow RCOO^- + 2\ Ag + 3\ NH_3 + NH_4^+ + H_2O$$

The use of other types of oxidizing agents is based on the reaction of copper ion to produce a red precipitate of cuprous oxide, Cu_2O:

$$RCHO + 2\ Cu^{2+} + 5\ OH^- \longrightarrow RCOO^- + Cu_2O + 3\ H_2O$$

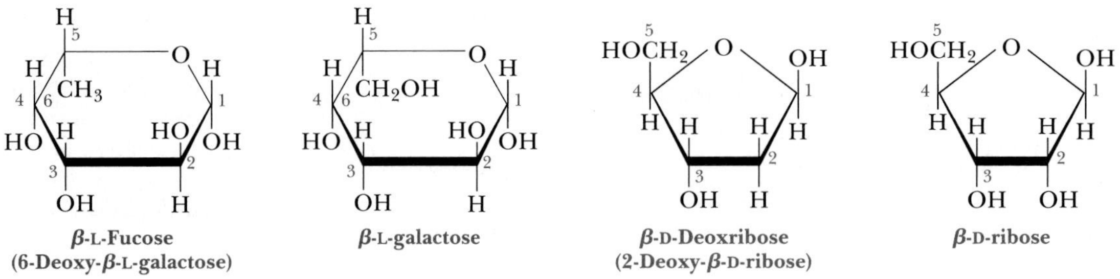

FIGURE 10.10

Structures of two deoxy sugars. The structures of the parent sugars are shown for comparison.

Two common reagents use copper ion as the oxidizing agent: in Fehling's solution the copper ion is dissolved in a buffer containing tartrate ion, and in Benedict's solution the copper ion is used with sodium carbonate and citrate buffer. In both cases the solution is quite basic. Ions such as tartrate and citrate can form complexes with the Cu^{2+}, which would otherwise form the insoluble hydroxide. Solubility considerations can make either Fehling's solution or Benedict's solution the reagent of choice for a particular test. A more recent method for detection of glucose, but not other reducing sugars, is based on the use of the enzyme glucose oxidase, which is specific for glucose.

In addition to oxidized sugars, there are some important reduced sugars. In **deoxy sugars** a hydrogen atom is substituted for one of the hydroxyl groups of the sugar. One of these deoxy sugars is L-fucose (L-6-deoxygalactose), which is found in the carbohydrate portions of some glycoproteins (Figure 10.10). The name "glycoprotein" indicates that these substances are conjugated proteins that contain some carbohydrate group (*glykos* is Greek for "sweet") in addition to the polypeptide chain. An even more important example of a deoxy sugar is D-2-deoxyribose, the sugar found in DNA (Figure 10.10).

When the carbonyl group of a sugar is reduced to a hydroxyl group, the resulting compound is one of the polyhydroxy alcohols known as **alditols.** Two compounds of this kind, xylitol and sorbitol, have commercial importance as sweeteners in sugarless chewing gum and candy.

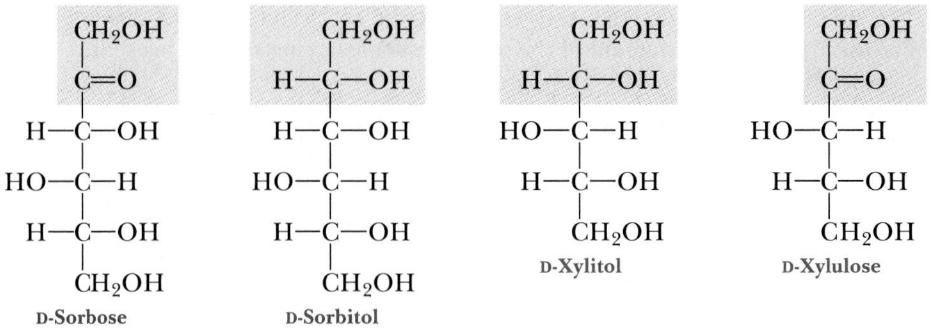

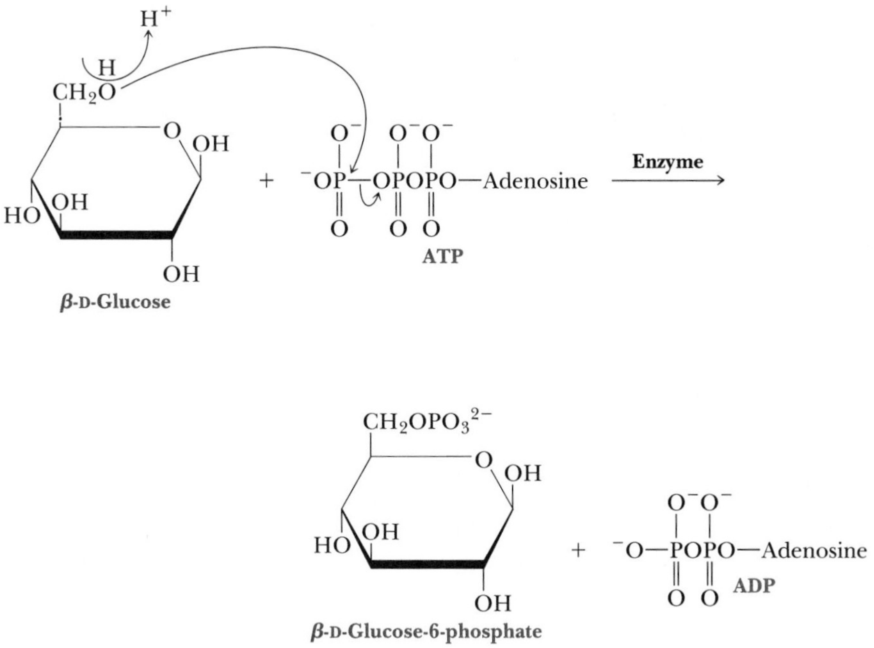

FIGURE 10.11

The formation of a phosphate ester of glucose. ATP is the phosphate group donor. The enzyme specifies the interaction with —C_6H_2OH.

Esterification Reactions

The hydroxyl groups of sugars behave exactly like all other alcohols in the sense that they can react with acids and derivatives of acids to form esters. The phosphate esters are particularly important ones because they are the usual intermediates in the breakdown of carbohydrates — not simply the breakdown of oligo- and polysaccharides to monosaccharides, but the further metabolism of sugars to provide energy. Phosphate esters are frequently formed by transfer of a phosphate group from ATP (adenosine triphosphate) to give the phosphorylated sugar and ADP (adenosine diphosphate), as shown in Figure 10.11. Such reactions play an important role in the metabolism of sugars (Chapter 11, Section 11.2).

The Formation of Glycosides

It is possible for a sugar hydroxyl group (ROH) to react with another hydroxyl (R'OH) to form a glycosidic linkage (R'—O—R). This type of reaction frequently involves the —OH group bonded to the anomeric carbon of a sugar in its cyclic form. (Recall that the anomeric carbon is the carbonyl carbon of the open-chain form of the sugar, and is the one that becomes a chiral center in the cyclic form.) Stated in a slightly different way, a hemiacetal can react with an alcohol such as methyl alcohol to give a **full**

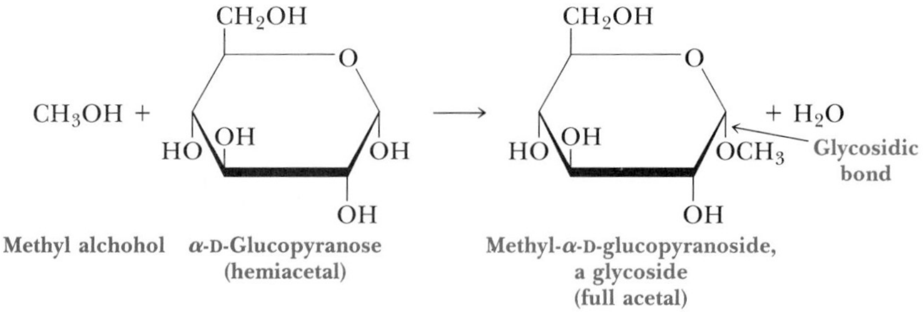

Methyl alchohol α-D-Glucopyranose
(hemiacetal)

Methyl-α-D-glucopyranoside,
a glycoside
(full acetal)

acetal or **glycoside** (Figure 10.12). The newly formed bond is called a **glycosidic bond.** The glycosidic bonds discussed in this chapter are *O*-glycosides, with each sugar bonded to an oxygen atom of another molecule. (We encountered *N*-glycosides in Chapter 6 when we discussed nucleosides and nucleotides, in which the sugar is bonded to a nitrogen atom of a base.) Glycosides derived from furanoses are called **furanosides,** and those derived from pyranoses are called **pyranosides.** (Recall from earlier in this chapter that a furanose is a cyclic sugar with a five-membered ring, whereas a pyranose is an analogous compound with a six-membered ring.)

Glycosidic bonds between monosaccharide units are the basis for the formation of oligosaccharides and polysaccharides. Glycosidic linkages can take various forms; either the α- or β-anomer of one sugar can be bonded to any one of the —OH groups on the other sugar. Many different combinations are found in nature. The —OH groups are numbered so that they can be distinguished, and the numbering scheme follows that of the carbon atoms. The notation for the glycosidic linkage between the two sugars specifies which anomeric form of the sugar is involved in the bond and also specifies which carbon atoms of the two sugars are linked together. Two ways in which two α-D-glucose molecules can be linked together are $\alpha(1 \longrightarrow 4)$ and $\alpha(1 \longrightarrow 6)$. In the first example, the α-anomeric carbon (C-1) of the first glucose molecule is joined in a glycosidic bond to the fourth carbon atom (C-4) of the second glucose molecule; the C-1 of the first glucose molecule is linked to the C-6 of the second glucose molecule in the second example (Figure 10.13). Another possibility of a glycosidic bond, this

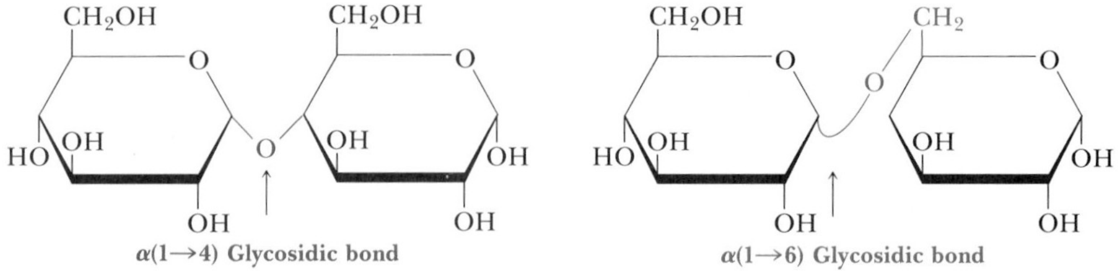

$\alpha(1{\rightarrow}4)$ Glycosidic bond

$\alpha(1{\rightarrow}6)$ Glycosidic bond

FIGURE 10.13

Two different disaccharides of α-D-glucose. These two chemical compounds have different properties because one has an $\alpha(1{\rightarrow}4)$ linkage and the other has an $\alpha(1{\rightarrow}6)$ linkage.

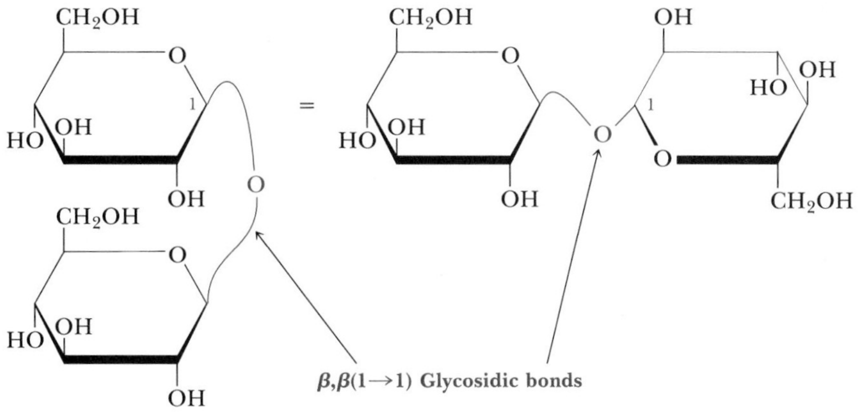

FIGURE 10.14

A disaccharide of β-D-glucose. Both anomeric carbons (C-1) are involved in the glycosidic linkage.

β,β(1→1) Glycosidic bonds

time between two β-D-glucose molecules, is a $\beta,\beta(1\longrightarrow 1)$ linkage. The anomeric forms at both C-1's must be specified because the linkage is between the two anomeric carbons, each of which is C-1 (Figure 10.14).

When oligosaccharides and polysaccharides form as a result of glycosidic bonding, their chemical natures depend on which monosaccharides are linked together and also on the particular glycosidic bond formed (i.e., which anomers and which carbon atoms are linked together). The difference between cellulose and starch depends on the glycosidic bond formed between glucose monomers. Because of the variation in glycosidic linkages, both linear and branched-chain polymers can be formed. If the internal monosaccharide residues that are incorporated in a polysaccharide form two glycoside bonds, the polymer will be linear. (Of course, the end residues will be involved in only one glycosidic linkage.) Some internal residues can form three glycosidic bonds, leading to the formation of branched-chain structures (Figure 10.15).

FIGURE 10.15

Linear and branched chain polymers of α-D-glucose. (a) The linear polyglucose chain occurs in amylose, and the branched-chain polymer occurs in amylopectin and glycogen (Section 10.4). All glycoside bonds are $\alpha(1\rightarrow 4)$.
(b) Branched-polyglucose-chain glycoside bonds are $\alpha(1\rightarrow 6)$ at branch points. Again, all glycoside bonds along the chain are $\alpha(1\rightarrow 4)$.

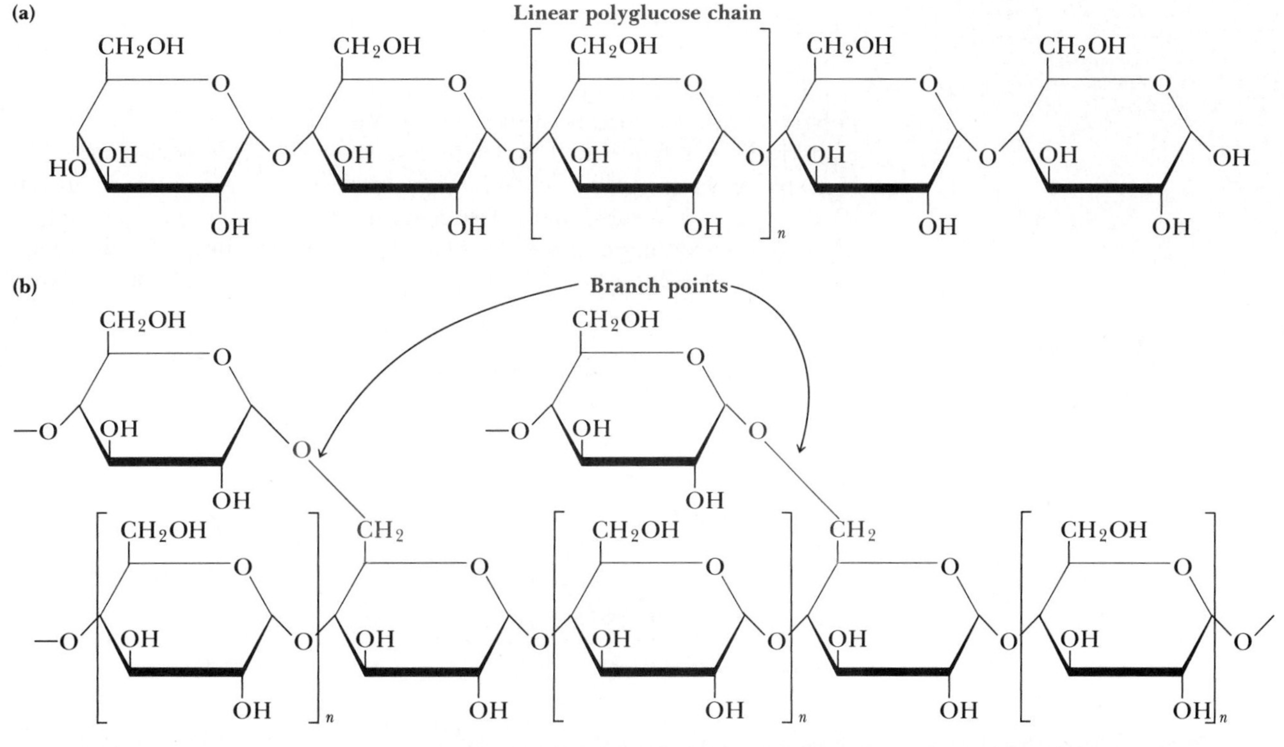

(a)   Linear polyglucose chain

(b)   Branch points

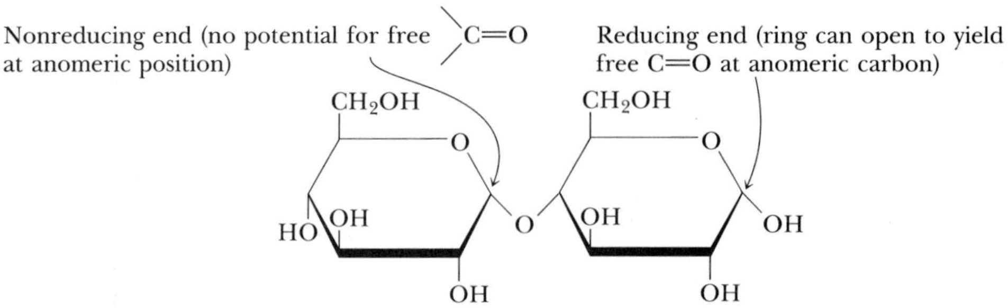

Dimer of α-D-glucose with α(1→4) linkage

FIGURE 10.16

A disaccharide with a free hemiacetal end is a reducing sugar because of the presence of a free anomeric aldehyde carbonyl or potential aldehyde group.

Another point about glycosides is worth mentioning. We have already seen that the anomeric carbon is frequently involved in the glycosidic linkage, and also that the test for the presence of sugars — specifically for reducing sugars — requires a reaction of the group at the anomeric carbon. The internal anomeric carbons in oligosaccharides are not free to give the test for reducing sugars. Only if the end residue is a free hemiacetal rather than a glycoside will there be a positive test for a reducing sugar (Figure 10.16). The level of detection can be important for such a test. A sample that contains only a few molecules of a large polysaccharide, each molecule with a single reducing end, might well produce a negative test because there are not enough reducing ends to detect.

Other Derivatives of Sugars

Amino sugars are an interesting class of compounds related to the monosaccharides. We shall not go into the chemistry of their formation, but it will be useful to have some acquaintance with them when we discuss polysaccharides. In sugars of this type, an amino group ($-NH_2$) or one of its derivatives is substituted for the hydroxyl group of the parent sugar. In **N-acetyl amino sugars** the amino group itself carries an acetyl group (CH_3-CO-) as a substituent. Two particularly important examples are N-acetyl-β-D-glucosamine and its derivative N-acetyl-β-muramic acid, which has an added carboxylic acid side chain (Figure 10.17). These two com-

FIGURE 10.17

The structures of N-acetyl-β-D-glucosamine and N-acetylmuramic acid.

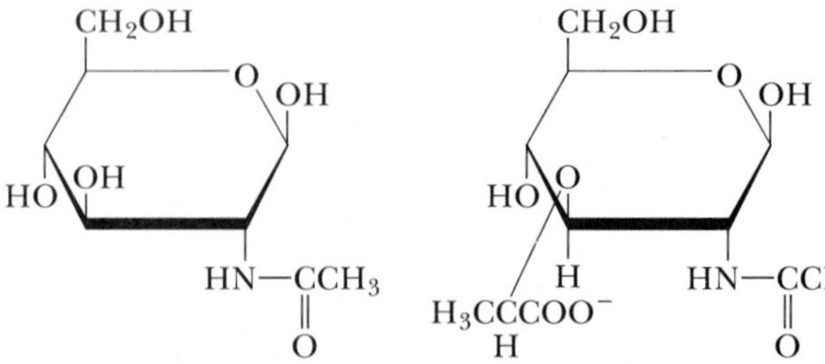

N-Acetyl-β-D-glucosamine

N-Acetylmuramic acid

pounds are components of bacterial cell walls. We did not specify whether N-acetylmuramic acid belongs to the L or the D series of configurations, and we did not specify the α- or β-anomer. This type of shorthand is the usual practice with β-D-glucose and its derivatives; the D configuration and the β-anomeric form are so common that we need not specify them all the time unless we want to make some specific point. The position of the amino group is also left unspecified, since discussion of amino sugars usually centers on a few compounds whose structures are well known.

10.3
OLIGOSACCHARIDES

Oligomers of sugars frequently occur as **disaccharides,** formed by the linking two monosaccharide units with glycosidic bonds. Three of the most important examples of oligosaccharides are disaccharides. They are sucrose, lactose, and maltose (Figure 10.18).

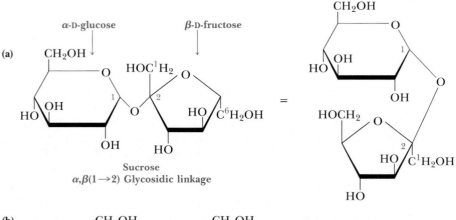

FIGURE 10.18

Structures of some disaccharides. (a) In sucrose, both anomeric carbons are involved in the glycosidic linkage; no free carbonyl (C=O) group can be obtained from ring opening. As a result, sucrose is not a reducing sugar. (b) In lactose, the anomeric carbon of glucose is not involved in the glycosidic bond ("α form" refers to this free anomeric end). A free carbonyl group can be obtained from ring opening. Lactose is a reducing sugar. (c) Cellobiose and maltose ("β form" and "α form" refer to the free anomeric end).

Sucrose is the common table sugar extracted from sugarcane and sugar beets. The monosaccharide units that make up sucrose are α-D-glucose and β-D-fructose. Glucose (an aldohexose) is a pyranose, and fructose (a ketohexose) is a furanose. The αC-1 carbon of the glucose is linked to the βC-2 carbon of the fructose (Figure 10.18a) in a glycosidic linkage that has the notation α, $\beta(1 \longrightarrow 2)$. Sucrose is not a reducing sugar because both anomeric groups are involved in the glycosidic linkage. Free glucose is a reducing sugar, and free fructose can also give a positive test even though it is a ketone rather than an aldehyde in the open-chain form. Fructose and ketoses in general can act as reducing sugars because they can isomerize to aldoses in a rather complex rearrangement reaction. (We need not concern ourselved with the details of this isomerization.)

When sucrose is consumed by animals, it is hydrolyzed to glucose and fructose, which are then degraded by metabolic processes to provide energy. Humans consume large quantities of sucrose, and excess consumption can contribute to health problems, which fact has led to a search for other sweetening agents. One that has been proposed is fructose itself. It is sweeter than sucrose, and therefore a smaller amount (by weight) of fructose than sucrose can produce the same sweetening effect with fewer "calories." Consequently, high-fructose corn syrup is frequently used in food processing. Also, one frequently sees the term "invert sugar" on food labels. That expression refers to the mixture of glucose and fructose produced by the hydrolysis of sucrose. The presence of fructose changes the texture of food, and the reaction to the change tends to depend on the preference of the consumer. Artificial sweeteners have been produced in the laboratory and have frequently been suspected of having harmful side effects; the ensuing controversies bear eloquent testimony to the human craving for sweets. Saccharin, for example, has been found to cause cancer in laboratory animals, as have cyclamates, but the applicability of these results to human carcinogenesis has been questioned by some. Aspartame (NutraSweet; Chapter 3, Section 3.5) has been suspected of causing neurological problems, especially in individuals whose metabolisms cannot tolerate phenylalanine (Chapter 17, Section 17.8). It is safe to predict that the search for nonfattening sweeteners will continue and that it will be accompanied by controversy.

Lactose (Box 10.2) is a disaccharide made up of β-D-galactose and D-glucose. Galactose is an epimer of glucose. In other words, the difference between glucose and galactose is inversion of configuration at C-4. The glycosidic linkage is $\beta(1 \longrightarrow 4)$, between the anomeric C-1 of the β form of galactose and the C-4 of glucose (Figure 10.18b). Since the anomeric carbon of glucose is not involved in the glycosidic linkage, it can be in either the α or β form. The two anomeric forms of lactose can be specified, and the designation refers to the glucose residue; galactose must be present as the β-anomer, since the β form of galactose is required by the structure of lactose. Lactose is a reducing sugar because the group at the anomeric carbon of the glucose portion is not involved in glycosidic linkage, so it is free to react with oxidizing agents.

Maltose is a disaccharide obtained from the hydrolysis of starch. It consists of two residues of D-glucose in an $\alpha(1 \longrightarrow 4)$ linkage. Maltose differs from **cellobiose,** a disaccharide that is obtained from the hydrolysis of

Box
10.2

LACTOSE INTOLERANCE

Lactose is sometimes referred to as milk sugar because it occurs in milk. Humans can be intolerant of milk and milk products for several reasons. In some adults a deficiency of the enzyme lactase in the intestinal villi causes a buildup of the disaccharide when milk products are ingested. This is because the enzyme is necessary to degrade lactose to galactose and glucose so that it can be absorbed into the bloodstream from the villi. Without the enzyme, built-up lactose in the intestine can be acted on by the lactase of intestinal bacteria (as opposed to the desirable lactase of the villi), producing hydrogen gas, carbon dioxide, and organic acids. These products of the bacterial lactase reaction lead to digestive problems such as bloating and diarrhea, as does the presence of undegraded lactose.

Lactase deficiency affects only about one tenth of the white population of the United States, but it is more common among African-Americans, Asians, Native Americans, and Hispanics. Even if the enzyme lactase is present so that lactose can be broken down by the body, other problems can occur. A different, but related, problem, can occur in the further metabolism of galactose, which must be isomerized to glucose if it is to enter the usual metabolic pathways. If the enzyme that catalyzes this reaction is missing and galactose builds up, a condition known as galactosemia can result. It is especially serious in infants and can lead to mental retardation.

cellulose, only in the glycosidic linkage. In cellobiose the two residues of D-glucose are bonded together in a $\beta(1 \longrightarrow 4)$ linkage (Figure 10.18c). Mammals can digest maltose but not cellobiose.

10.4
POLYSACCHARIDES

Polysaccharides that occur in organisms are usually composed of a very few types of monosaccharide components. A polymer that consists of only one type of monosaccharide is a **homopolysaccharide;** a polymer that consists of more than one type of monosaccharide is a **heteropolysaccharide.** Glucose is the most common monomer. When there is more than one type of monomer, frequently only two types of molecules occur in a repeating sequence. A complete characterization of a polysaccharide includes specification of which monomers are present and, if necessary, the sequence of monomers. It also requires that the type of glycosidic linkage be specified. We shall see the importance of the type of glycosidic linkage as we discuss different polysaccharides, since the nature of the linkage determines function. Cellulose and chitin are polysaccharides with β-glycosidic linkages and are structural materials. Starch and glycogen, also polysaccharides, have α-glycosidic linkages and serve as carbohydrate storage polymers in plant and animals, respectively.

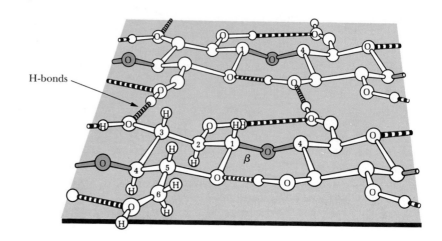

Repeating disaccharide
in cellulose
(β-cellobiose)

FIGURE 10.19

The polymeric structure of cellulose. β-Cellobiose is the repeating disaccharide. The monomer of cellulose is the β-anomer of glucose, which gives rise to long chains that can hydrogen bond to one another.

H-bonds

Cellulose and Starch

Cellulose is the major structural component of plants, especially of wood and plant fibers. It is a linear homopolysaccharide of β-D-glucose, and all residues are linked in β(1 → 4) glycosidic bonds (Figure 10.19). Individual

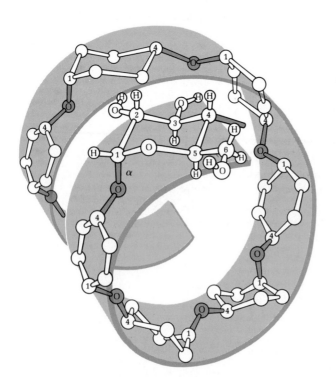

Glucose   Glucose

α(1→4)

FIGURE 10.20

The monomer of starch is the α-anomer of glucose, which gives rise to a chain that folds into a helical form. The repeating dimer has α(1→4) linkages throughout.

polysaccharide chains are hydrogen bonded together, giving plant fibers their mechanical strength. Animals lack enzymes, called *cellulases,* that hydrolyze cellulose to glucose. Such enzymes attack the β-linkage, which is common to structural polymers; the α-linkage, which animals can digest, is characteristic of energy-storage polymers such as starch (Figure 10.20). Cellulases are found in bacteria, including the bacteria that inhabit the digestive tracts of insects such as termites and grazing animals such as cattle and horses. The presence of these bacteria explains why cows and horses can live on grass and hay but humans cannot. The damage done by termites to the wooden parts of buildings arises from their ability — owing to the presence of suitable bacteria in their digestive tracts — to *use* the cellulose in wood as a nutrient.

Chitin

A polysaccharide that is similar to cellulose in both structure and function is **chitin,** which is also a linear homopolysaccharide with all the residues linked in $\beta(1 \rightarrow 4)$ glycosidic bonds. Chitin differs from cellulose in the nature of the monosaccharide unit; in cellulose the monomer is β-D-glucose, and in chitin the monomer is *N*-acetyl-β-D-glucosamine. The latter compound differs from glucose only in the substitution of the *N*-acetylamino group ($-NH-CO-CH_3$) for the hydroxyl group ($-OH$) on C-2 (Figure 10.21). Like cellulose, chitin plays a structural role and has a fair amount of mechanical strength because the individual strands are held together by

Termites and cattle can digest the cellulose in both grass and wood because their intestinal bacteria produce the enzyme cellulase, which hydrolyzes the β-glycosidic linkage in cellulose.

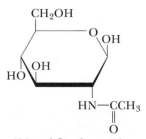

N-Acetyl-β-D-glucosamine

Repeating disaccharide in chitin

FIGURE 10.21

The polymeric structure of chitin. *N*-acetylglucosamine is the monomer, and a dimer of *N*-acetylglucosamine is the repeating disaccharide.

hydrogen bonds. It is a major structural component of the exoskeletons of invertebrates such as insects and crustaceans (a group that includes lobsters and shrimp), and it also occurs in cell walls of algae, fungi, and yeasts.

The Role of Polysaccharides in the Structures of Cell Walls

In organisms that have cell walls, such as bacteria and plants, the walls consist largely of polysaccharides. The cell walls of bacteria and plants have some biochemical differences, however, which serve to emphasize the differences between prokaryotes and eukaryotes.

Bacterial Cell Walls

Heteropolysaccharides are major components of bacterial cell walls. A distinguishing feature of prokaryotic cell walls is that the polysaccharides are cross-linked by peptides. The repeating unit of the polysaccharide consists of two residues held together by $\beta(1 \longrightarrow 4)$ glycosidic links, as was the case in cellulose and chitin. One of the two monomers is *N*-acetyl-D-glucosamine, which occurs in chitin, and the other monomer is *N*-acetylmuramic acid (Figure 10.22a). The structure of *N*-acetylmuramic acid differs from that of *N*-acetylglucosamine in the substitution of a lactic acid side chain [—O—CH(CH$_3$)—COOH] for the hydroxyl group (—OH) on carbon 3. *N*-acetylmuramic acid is found only in prokaryotic cell walls; it does not occur in eukaryotic cell walls.

The cross-links in bacterial cell walls consist of small peptides. We shall use one of the best-known examples as an illustration. In the cell wall of the bacterium *Staphylococcus aureus* an oligomer of four amino acids (a tetramer) is bonded to *N*-acetylmuramic acid, forming a side chain (Figure 10.22b). The tetrapeptides are themselves cross-linked by another small peptide, in this case consisting of five amino acids.

The carboxyl group of the lactic acid side chain of *N*-acetylmuramic acid forms a peptide bond with the N-terminal end of a tetrapeptide that has the sequence L-Ala-D-Gln-L-Lys-D-Ala. Recall that bacterial cell walls are one of the few places where D-amino acids occur in nature. The occurrence of D-amino acids and *N*-acetylmuramic acid in bacterial cell walls but not in plant cell walls indicates a biochemical as well as structural difference between prokaryotes and eukaryotes.

The tetrapeptide forms two cross-links, both of them to a pentapeptide that consists of five glycine residues, (Gly)$_5$. The glycine pentamers form peptide bonds to the C-terminal end and to the side-chain ϵ-amino group of the lysine in the tetrapeptide (Figure 10.22c). This extensive cross-linking produces a three-dimensional network of considerable mechanical strength, which is why bacterial cell walls are extremely difficult to disrupt. The material that results from the cross-linking of polysaccharides by peptides is a **peptidoglycan** (Figure 10.22d), so named because it has both peptide and carbohydrate components.

Plant Cell Walls

Plant cell walls consist largely of **cellulose.** The other important polysaccharide component found in plant cell walls is **pectin,** a polymer made up

(a)

NAM
(*N*-Acetylmuramic acid)

NAG
(*N*-Acetylglucosamine)

(b)

H—N

L-Ala

D-Gln

L-Lys—ϵ-NH$_3^+$

D-Ala

C=O

O$^-$

(c)

H—N

L-Ala

D-Gln

L-Lys—ϵ-NH—C—(Gly)$_5$—NH—   To tetrapeptide side chain

D-Ala

C=O

H—N

(Gly)$_5$

C=O

To tetrapeptide
side chain

(d)

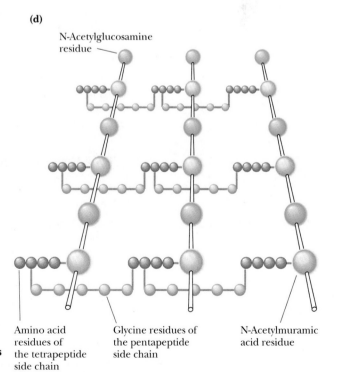

N-Acetylglucosamine
residue

Amino acid
residues of
the tetrapeptide
side chain

Glycine residues of
the pentapeptide
side chain

N-Acetylmuramic
acid residue

FIGURE 10.22

The structure of the peptidoglycan of the *Staphylococcus aureus* bacterial cell wall. (a) The repeating disaccharide. (b) The repeating disaccharide with the tetrapeptide side chain (shown in red). (c) Adding the pentaglycine cross-links (shown in red). (d) Schematic diagram of the peptidoglycan. The sugars are the larger circles. The red circles are the amino acid residues of the tetrapeptide, and the blue circles are the glycine residues of the pentapeptide.

FIGURE 10.23

The structure of lignin, a polymer of coniferyl alcohol.

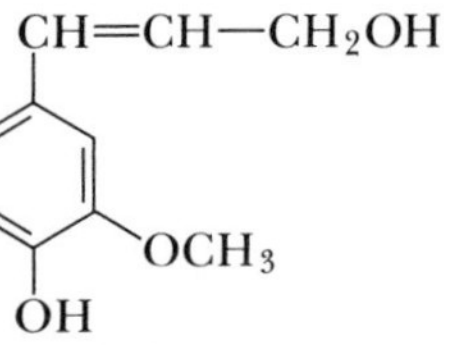

Coniferyl alcohol

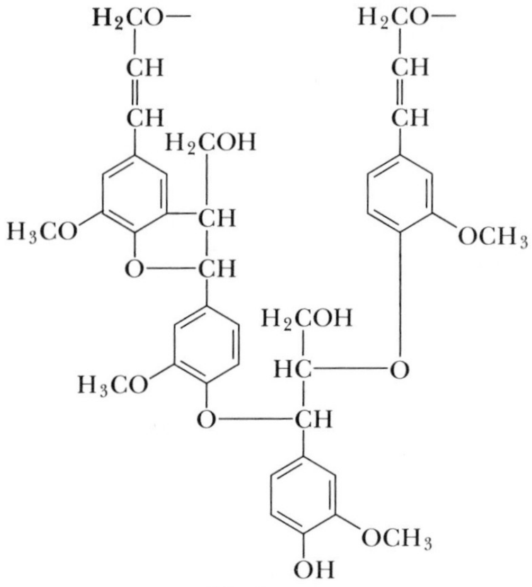

Lignin

mostly of D-galacturonic acid, a derivative of galactose in which the hydroxyl group on carbon 6 has been oxidized to a carboxyl group. Pectin is extracted from plants because it has commercial importance in the food

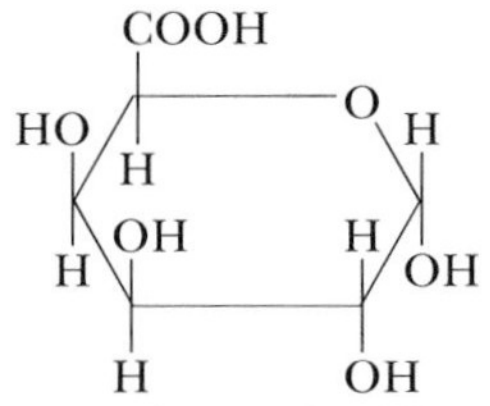

D-galacturonic acid

The structure of D-galacturonic acid.

processing industry as a gelling agent in fruit preserves, jams, and jellies. The major nonpolysaccharide component in plant cell walls, especially in woody plants, is **lignin** (Latin *lignum,* "wood"). Lignin is a polymer of coniferyl alcohol, and it is a very tough and durable material (Figure 10.23). Unlike bacterial cell walls, plant cell walls contain comparatively little peptide or protein.

The Forms of Starch

All the forms of polysaccharides (except starch) we have discussed so far have structural roles, but the importance of carbohydrates as energy sources suggests that there is also some use for polysaccharides in metabolism. We shall now discuss in more detail some polysaccharides, such as starches, that serve as vehicles for storage of glucose.

Starches are polymers of α-D-glucose that occur in plant cells, usually as starch granules in the cytosol. Note that there is an α-linkage in starch, in contrast with the β-linkage in cellulose. The types of starches can be distinguished from one another by their degrees of chain branching. **Amylose** is a linear polymer of glucose, with all the residues linked together by $\alpha(1 \longrightarrow 4)$ bonds. **Amylopectin** is a branched-chain polymer, with the branches starting at $\alpha(1 \longrightarrow 6)$ linkages along the chain of $\alpha(1 \longrightarrow 4)$ linkages. (Refer to Figure 10.15 for detailed structures involving these two kinds of glycosidic linkages.) The most usual conformation of amylose is a helix with six residues per turn. Iodine molecules can fit inside the helix to form a starch–iodine complex, which has a characteristic dark-blue color (Figure 10.24). The formation of this complex is a well-known test for the presence of starch. If there is a preferred conformation for amylopectin, it is not yet known. (It *is* known that the color of the product obtained when amylopectin and glycogen react with iodine is red-brown, not blue.)

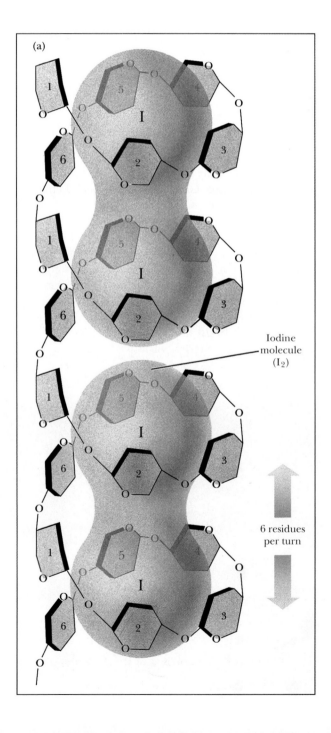

(a)

Iodine molecule (I_2)

6 residues per turn

FIGURE 10.24

The starch–iodine complex. Amylose occurs as a helix with six residues per turn. In the starch–iodine complex, the iodine molecules are parallel to the long axis of the helix. Four turns of the helix are shown here. Six turns of the helix, containing 36 glycosyl residues, are required to produce the characteristic blue color of the complex.

Since starches are storage molecules, there must be a mechanism for releasing glucose from starch when the organism needs energy. Both plants and animals contain enzymes that hydrolyze starches. Two of these enzymes, known as α- and β-**amylase** (the α and β do not signify anomeric forms in this case), attack $\alpha(1 \longrightarrow 4)$ linkages. β-Amylase is an **exoglycosidase** that cleaves from the nonreducing end of the polymer. Maltose, a dimer of glucose, is the product of reaction. The other enzyme, α-amylase, is an **endoglycosidase,** which can hydrolyze a glycosidic linkage anywhere along the chain to produce glucose and maltose. Amylose can be completely degraded to glucose and maltose by the two amylases, but amylopectin is not completely degraded because the branching linkages are not attacked. There are, however, **debranching enzymes** that occur in both plants and animals; they degrade the $\alpha(1 \longrightarrow 6)$ linkages. When these enzymes are combined with the amylases, they contribute to the complete degradation of both forms of starch.

Glycogen

Although starches occur only in plants, in animals there is a similar carbohydrate storage polymer. **Glycogen** is a branched-chain polymer of α-D-glucose, and in this respect it is similar to the amylopectin fraction of starch. Like amylopectin, glycogen consists of a chain of $\alpha(1 \longrightarrow 4)$ linkages with $\alpha(1 \longrightarrow 6)$ linkages at the branch points. The main difference between glycogen and amylopectin is that glycogen is more highly branched (Figure 10.25). Branch points occur about every 10 residues in glycogen and about every 25 residues in amylopectin. Glycogen is found in animal cells in granules similar to the starch granules in plant cells. Glycogen granules are observed in well-fed liver and muscle cells, but they are not seen in some other cell types, such as brain and heart cells under normal conditions. When the organism needs energy, various degradative enzymes remove

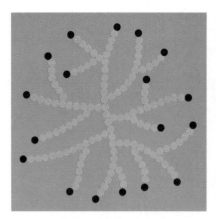
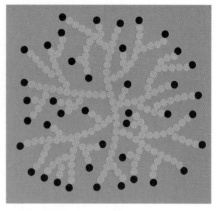

Starch amylopectin   Glycogen amylopectin

FIGURE 10.25

A comparison of the degrees of branching in starch amylopectin and glycogen amylopectin.

glucose units (Chapter 14, Section 14.1). **Glycogen phosphorylase** is one such enzyme; it cleaves one glucose at a time from the nonreducing end of a branch to produce glucose-1-phosphate, which then enters the metabolic pathways of carbohydrate breakdown. Debranching enzymes also play a role in the complete breakdown of glycogen. Some athletes, particularly long-distance runners, try to build up their glycogen reserves before a race by eating large amounts of carbohydrates.

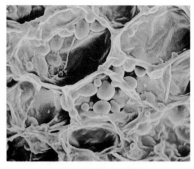

starch grains in plant

10.5
GLYCOPROTEINS

Glycoproteins contain carbohydrate residues in addition to the polypeptide chain. Some of the most important examples of glycoproteins are involved in the immune response; for example, **antibodies,** which bind to and immobilize antigens (the substances attacking the organism), are glycoproteins. Carbohydrates also play an important role as **antigenic determinants,** the portions of an antigenic molecule that antibodies recognize and to which they bind.

An example of the role of the oligosaccharide portion of glycoproteins as antigenic determinants is found in human blood groups. There are four human blood groups, A, B, AB, and O (Box 10.3). The distinctions between the groups depend on the oligosaccharide portions of the glycoproteins on

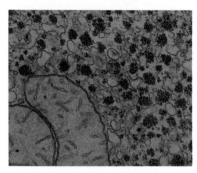

stored glycogen in liver

Electron micrographs of starch granules in a plant and glycogen granules in an animal.

Box 10.3

GLYCOPROTEINS AND BLOOD TRANSFUSIONS

If a transfusion is attempted with incompatible blood types, as when blood from a type A donor is given to a type B recipient, an antigen–antibody reaction takes place because the type B recipient has antibodies to the type A blood. The characteristic oligosaccharide residues of type A blood cells serve as the antigen. A cross-linking reaction occurs between antigens and antibodies, and the blood cells clump together. In the case of a transfusion of type B blood to a type A recipient, antibodies to type B blood produce the same result. Type O blood does not have either antigenic determinant, and so people with type O blood are considered universal donors. However, these people have antibodies to both type A and type B blood, and so they are not universal acceptors. Type AB persons have both antigenic determinants, and as a result they do not produce either type of antibody; they are universal acceptors.

Transfusion Relationships

PERSON HAS BLOOD TYPE	MAKES ANTIBODIES AGAINST	CAN RECEIVE FROM	CAN DONATE TO
O	A, B	O	O, A, B, AB
A	B	O, A	A, AB
B	A	O, B	B, AB
AB	None	O, A, B, AB	AB

β-*N*-Acetylgalactosamine   (1→3) β-Galactose (1→3) β-*N*-Acetylgalactosamine
↑2
|1
Nonreducing end   α-L-Fucose
A blood group antigen

α-Galactose   (1→3) β-Galactose (1→3) β-*N*-Acetylgalactosamine
↑2
|1
Nonreducing end   α-L-Fucose
B blood group antigen

FIGURE 10.26

The structures of the blood group antigenic determinants.

the surfaces of the blood cells called erythrocytes. In all blood types the oligosaccharide portion of the molecule contains the sugar L-fucose, mentioned earlier in this chapter as an example of a deoxy sugar. *N*-acetylgalactosamine is found at the nonreducing end of the oligosaccharide in the type A blood-group antigen. In type B blood, a D-galactrose takes the place of *N*-acetylgalactosamine. In type O blood neither of these terminal residues is present, and in type AB blood both kinds of oligosaccharide are present (Figure 10.26).

SUMMARY

The simplest examples of carbohydrates are monosaccharides, compounds that each contain a single carbonyl group and two or more hydroxyl groups. Monosaccharides frequently encountered in biochemistry are sugars that contain from three to seven carbon atoms. Sugars contain one or more chiral centers; the configurations of the possible stereoisomers can be represented by Fischer projection formulas. Sugars frequently exist as cyclic molecules rather than in open-chain form. Haworth projection formulas are more realistic representations of the cyclic forms of sugars than are Fischer projection formulas. Many stereoisomers are possible for five- and six-carbon sugars, but only a few of the possibilities are encountered frequently in nature. Monosaccharides can undergo various reactions, including oxidation and esterification, but the most important reaction by far is the formation of glycosidic linkages, which give rise to oligosaccharides and polysaccharides.

Three important examples of oligosaccharides are the disaccharides sucrose, lactose, and maltose. Sucrose is common table sugar, lactose occurs in milk, and maltose is obtained via the hydrolysis of starch. In polysaccharides the repeating unit of the polymer is frequently limited to one or two kinds of monomer. Cellulose and chitin are polymers based on single kinds of monomer units—glucose and *N*-acetylglucosamine, respectively. Both polymers play structural roles in organisms. Starch, found in plants, and glycogen, which occurs in animals, are energy-storage polymers of glucose. They differ from each other in the degree of branching in the polymer structure, and they differ from cellulose in the stereochemistry of the glycosidic linkage between monomers. In glycoproteins, carbohydrate residues are covalently linked to the polypeptide chain; glycoproteins play a role in the recognition sites of antigens.

E X E R C I S E S

1. Pectin, which occurs in plant cell walls, exists in nature as a polymer of D-galacturonic acid methylated at carbon 6 of the monomer. Draw a Haworth projection for a repeating disaccharide unit of pectin with one methylated and one unmethylated monomer unit in $\alpha(1 \longrightarrow 4)$ linkage. Is the methyl group attached by an ester or an ether linkage?

2. Draw a Haworth projection for the disaccharide gentibiose, given the following information:
   (a) It is a dimer of glucose.
   (b) The glycosidic linkage is $\beta(1 \longrightarrow 6)$.
   (c) The anomeric carbon not involved in the glycosidic linkage is in the α configuration.

3. An amylose chain is 5000 glucose units long. At how many places must it be cleaved to reduce the average length to 2500 units? To 1000? To 200? What percentage of the glycosidic links are hydrolyzed in each case? (Even partial hydrolysis can drastically alter the physical properties of polysaccharides and thus affect their structural role in organisms.)

4. Suppose that a polymer of glucose with alternating $\alpha(1 \longrightarrow 4)$ and $\beta(1 \longrightarrow 4)$ glycosidic linkages has just been discovered. Draw a Haworth projection for a repeating tetramer (two repeating dimers) of such a polysaccharide. Would you expect this polymer to have primarily a structural role or an energy storage role in organisms? What sort of organisms, if any, could use this polysaccharide as a food source?

5. Define the following terms: polysaccharide, furanose, pyranose, aldose, ketose, glycosidic bond, oligosaccharide, glycoprotein.

6. Name which, if any, of the following are epimers of D-glucose: D-mannose, D-galactose, D-ribose.

7. Name which, if any, of the following groups are *not* aldose–ketose pairs: D-ribose and D-ribulose, D-glucose and D-fructose, D-glyceraldehyde and dihydroxyacetone.

8. What is the metabolic basis for the observation that many adults cannot ingest large quantities of milk without developing gastric difficulties?

9. Following are Fischer projections for a group of five-carbon sugars, all of which are aldopentoses. Identify the pairs that are enantiomers and the pairs that are epimers. (The sugars shown here are not all the possible five-carbon sugars.)

(a)
```
      CHO
       |
  H—C—OH
       |
  H—C—OH
       |
  H—C—OH
       |
     CH2OH
```

(b)
```
      CHO
       |
  H—C—OH
       |
 HO—C—H
       |
 HO—C—H
       |
     CH2OH
```

(c)
```
      CHO
       |
  H—C—OH
       |
  H—C—OH
       |
 HO—C—H
       |
     CH2OH
```

(d)
```
      CHO
       |
 HO—C—H
       |
  H—C—OH
       |
  H—C—OH
       |
     CH2OH
```

(e)
```
      CHO
       |
  H—C—OH
       |
 HO—C—H
       |
  H—C—OH
       |
     CH2OH
```

(f)
```
      CHO
       |
 HO—C—H
       |
 HO—C—H
       |
 HO—C—H
       |
     CH2OH
```

10. Draw Haworth projection formulas for dimers of glucose with the following types of glycosidic linkages:
    (a) A $\beta(1 \longrightarrow 4)$ linkage (both molecules of glucose in the β form)
    (b) An $\alpha,\alpha(1 \longrightarrow 1)$ linkage
    (c) A $\beta(1 \longrightarrow 6)$ linkage (both molecules of glucose in the β form)

11. What are some of the main differences between the cell walls of plants and those of bacteria?

12. How does chitin differ from cellulose in structure and function?

13. How does glycogen differ from starch in structure and function?

14. How do the enzymes α-amylase and β-amylase differ from one another?

15. Briefly indicate the role of glycoproteins as antigenic determinants for blood groups.

A N N O T A T E D   B I B L I O G R A P H Y

Aspinall, G. O., ed. *The Polysaccharides.* 2 vols. New York: Academic Press, 1982. [Good coverage of topics of current interest in carbohydrate chemistry.]

Borman, S. Glycotechnology Drugs Begin To Emerge from the Lab. *Chem. Eng. News* **71** (26), 27–34 (1993). [How the pharmaceutical industry is making use of the important role of cell-surface carbohydrates in biological recognition.]

Kritchevsky, K., C. Bonfield, and J. Anderson, eds. *Dietary Fiber: Chemistry, Physiology, and Health Effects.* New York: Plenum Press, 1990. [A topic of considerable current interest, with explicit connections to the biochemistry of plant cell walls.]

Sharon, N. Carbohydrates. *Sci. Amer.* **243** (5), 90–102 (1980). [A good overview of structures.]

Sharon, N., and H. Lis. Carbohydrates in Cell Recognition. *Sci. Amer.* **268** (1), 82–89 (1993). [The development of drugs to stop infection and inflammation by targeting cell-surface carbohydrates.]

Takahashi, N., and T. Muramatsu. *Handbook of Endoglycosidases and Glyco-amidases.* Boca Raton, FL: CRC Press, 1992. [A source of practical information on how to manipulate biologically important carbohydrates.]

Most organic chemistry textbooks have one or more chapters on the structures and reactions of carbohydrates.

Glycolysis

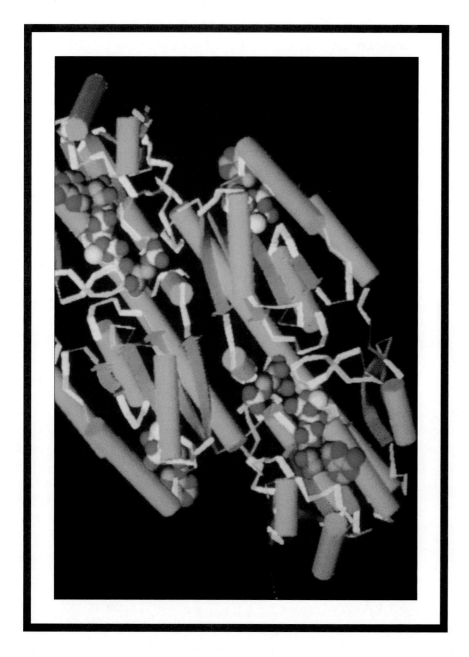

The structure of phosphofructokinase, a regulatory enzyme in glycolysis, showing bound substrate.

The complete oxidation of glucose to carbon dioxide and water (involving glycolysis, the citric acid cycle, and oxidative phosphorylation) yields the energy equivalent of 38 molecules of ATP. The first stage of glucose metabolism, *glycolysis,* an anaerobic process, yields only two molecules of ATP. Nevertheless, in sudden bursts of energy, such as the hundred-yard dash, the body temporarily prefers glycolysis. In sustained activity, glucose, stored as glycogen in muscle cells, may be quickly depleted. In such cases, pyruvate, the end-product of glycolysis, can be converted to lactate and exported from muscle to the liver. Under aerobic conditions glycolysis combines with further metabolic steps to give rise to considerably more ATP.

11.1
AN OVERVIEW OF THE GLYCOLYTIC PATHWAY

In **glycolysis,** one molecule of glucose (a six-carbon compound) is converted to fructose 1,6-*bis*phosphate (also a six-carbon compound), which eventually gives rise to two molecules of pyruvate ion (a three-carbon compound) (Figure 11.1). The glycolytic pathway (also called the Embden–Meyerhoff pathway) involves many steps, including the reactions in which metabolites of glucose are oxidized; there are other steps as well.

FIGURE 11.1

One molecule of glucose is converted to two molecules of pyruvate. Under aerobic conditions, pyruvate is oxidized to CO_2 and H_2O by the citric acid cycle (Chapter 12) and oxidative phosphorylation (Chapter 13). Under anaerobic conditions, lactate is produced, especially in muscle. Alcoholic fermentation occurs in yeast. The NADH produced in the conversion of glucose to pyruvate is reoxidized to NAD^+ in the subsequent reactions of pyruvate.

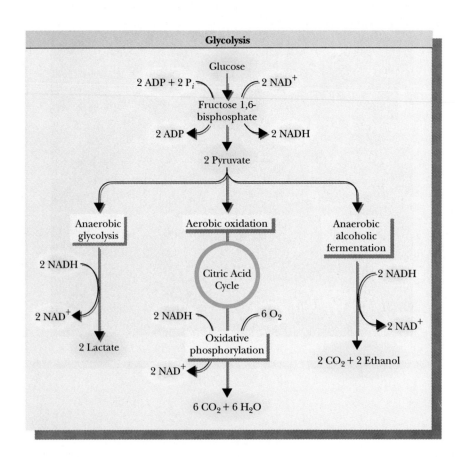

Each reaction in the pathway is catalyzed by an enzyme specific for that reaction. In two reactions in the pathway one molecule of ATP is hydrolyzed for each molecule of glucose metabolized; the energy released in the hydrolysis of these two ATP molecules makes coupled endergonic reactions possible. In each of two other reactions two molecules of ATP are produced by phosphorylation of ADP for each molecule of glucose, giving a total of four ATP molecules produced. A comparison of the number of ATP molecules used by hydrolysis (two) and the number produced (four) shows that there is a net gain of two ATP molecules for each molecule of glucose processed in glycolysis (Chapter 9, Section 9.10). Glycolysis plays a key role in the way organisms extract energy from nutrients.

When pyruvate is formed, it can have one of several fates (Figure 11.1). In aerobic metabolism (in the presence of oxygen), pyruvate loses carbon dioxide; the remaining two carbon atoms become linked to coenzyme A (Chapter 9, Section 9.11) as an acetyl group to form acetyl-CoA, which then enters the citric acid cycle (Chapter 12). There are two fates for pyruvate in anaerobic metabolism (in the absence of oxygen). In organisms capable of alcoholic fermentation, pyruvate loses carbon dioxide, this time producing acetaldehyde, which in turn is reduced to produce ethanol (Section 11.4). The more common fate of pyruvate in anaerobic metabolism is reduction to lactate, called **anaerobic glycolysis** to distinguish it from conversion of glucose to pyruvate, which is simply called glycolysis. Anaerobic metabolism is the only energy source in mammalian red blood cells, as well as in several species of bacteria, such as *Lactobacillus* in sour milk and *Botulina* in tainted canned foods. In all these reactions the conversion of glucose to product is an oxidation reaction, requiring an accompanying reduction reaction in which NAD^+ is converted to NADH, a point to which we shall return when we discuss the pathway in detail. The breakdown of glucose to pyruvate can be summarized as follows:

Glucose (six carbon atoms) $\longrightarrow$ 2 pyruvate (Three carbon atoms)

2 ATP + 4 ADP + 2 P_i $\longrightarrow$ 2 ADP + 4 ATP (Phosphorylation)

Glucose + 2 ADP + 2 P_i $\longrightarrow$ 2 pyruvate + 2 ATP (Net reaction)

Figure 11.2 shows the reaction sequence with the names of the compounds. All sugars in the pathway have the D configuration; we shall assume this point throughout this chapter.

A Summary of the Reactions of Glycolysis

Step 1. **Phosphorylation** of glucose to give glucose 6-phosphate (ATP is the source of the phosphate group). (See Equation 11.1, p. 346.)

Glucose + ATP $\longrightarrow$ glucose 6-phosphate + ADP

Step 2. **Isomerization** of glucose 6-phosphate to give fructose 6-phosphate. (See Equation 11.2, p. 347.)

Glucose 6-phosphate $\longrightarrow$ fructose 6-phosphate

Step 3. **Phosphorylation** of fructose 6-phosphate to give fructose 1,6-*bis*phosphate (ATP is the source of the phosphate group). (See Equation 11.3, p. 348.)

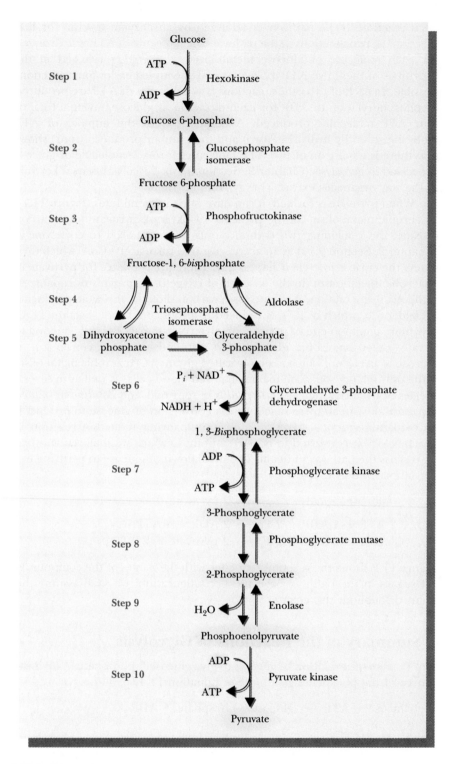

FIGURE 11.2

The pathway of glycolysis. Those reactions in which ATP or NADH appears are highlighted.

Fructose 6-phosphate + ATP $\longrightarrow$ fructose 1,6-*bis*phosphate + ADP

Step 4. **Cleavage** of fructose 1,6-*bis*phosphate to give two 3-carbon fragments, glyceraldehyde 3-phosphate and dihydroxyacetone phosphate. (See Equation 11.4, p. 349.)

Fructose 1,6-*bis*phosphate $\longrightarrow$
glyceraldehyde 3-phosphate + dihydroxyacetone phosphate

Step 5. **Isomerization** of dihydroxyacetone phosphate to give glyceraldehyde 3-phosphate. (See Equation 11.5, p. 349.)

Dihydroxyacetone phosphate $\longrightarrow$ glyceraldehyde 3-phosphate

Step 6. **Oxidation** (and phosphorylation) of glyceraldehyde 3-phosphate to give 1,3-*bis*phosphoglycerate. (See Equation 11.6, p. 351.)

Glyceraldehyde 3-phosphate + NAD^+ + P_i $\longrightarrow$
NADH + 1,3-*bis*phosphoglycerate + H^+

Step 7. **Transfer of a phosphate group** from 1,3-*bis*phosphoglycerate to ADP (phosphorylation of ADP to ATP) to give 3-phosphoglycerate. (See Equation 11.7, p. 354.)

1,3-*Bis*phosphoglycerate + ADP $\longrightarrow$ 3-phosphoglycerate + ATP

Step 8. **Isomerization** of 3-phosphoglycerate to give 2-phosphoglycerate. (See Equation 11.8, p. 356.)

3-Phosphoglycerate $\longrightarrow$ 2-phosphoglycerate

Step 9. **Dehydration** of 2-phosphoglycerate to give phosphoenolpyruvate. (See Equation 11.9, p. 356.)

2-Phosphoglycerate $\longrightarrow$ phosphoenolpyruvate + H_2O

Step 10. **Transfer of a phosphate group** from phosphoenolpyruvate to ADP (phosphorylation of ADP to ATP) to give pyruvate. (See Equation 11.10, p. 357.)

Phosphoenolpyruvate + ADP $\longrightarrow$ pyruvate + ATP

Note that only one of the ten steps in this pathway involves electron transfer reactions. We shall now look at each of these reactions in detail.

11.2
REACTIONS OF GLYCOLYSIS I: CONVERSION OF GLUCOSE TO GLYCERALDEHYDE 3-PHOSPHATE

The first steps of the glycolytic pathway prepare for the electron transfer and the eventual phosphorylation of ADP; these reactions make use of the free energy of hydrolysis of ATP.

Step 1. Glucose is phosphorylated to give glucose 6-phosphate. The phosphorylation of glucose is an endergonic reaction.

Glucose + P_i + $\longrightarrow$ glucose 6-phosphate + H_2O

$\Delta G^{\circ\prime} = 13.8$ kJ mol^{-1} = 3.3 kcal mol^{-1}

The hydrolysis of ATP is exergonic.

ATP + H_2O $\longrightarrow$ ADP + P_i

$\Delta G^{\circ\prime} = -30.5$ kJ mol^{-1} = -7.3 kcal mol^{-1}

These two reactions are coupled, so the overall reaction is the sum of the two and is exergonic.

Glucose + ATP $\longrightarrow$ glucose 6-phosphate + ADP

$\Delta G^{\circ\prime} = 13.8 + (-30.5) = -16.7$ kJ mol^{-1} = -4.0 kcal mol^{-1}

(11.1)

This reaction illustrates the use of chemical energy originally produced by the oxidation of nutrients and trapped by phosphorylation of ADP to ATP. Recall from Chapter 9, Section 9.10, that ATP does not represent stored energy just as an electric current does not represent stored energy. The chemical energy of nutrients is released by oxidation and is made available for immediate use on demand by being trapped as ATP.

The enzyme that catalyzes this reaction is **hexokinase.** The term "kinase" is applied to the class of ATP-dependent enzymes that transfer a phosphate group from ATP to a substrate. The substrate of hexokinase is not necessarily glucose, but can be any one of a number of hexoses such as glucose, fructose, and mannose. (In some organisms or tissues the enzyme **glucokinase** specifically phosphorylates glucose. The glucokinase in the human liver lowers blood glucose levels after one has eaten a meal.)

A large conformational change takes place in hexokinase when substrate is bound. It has been shown by x-ray crystallography that, in the absence of substrate, two lobes of the enzyme that surround the binding site are quite far apart. When glucose is bound, the two lobes move closer together and the glucose becomes almost completely surrounded by protein (Figure 11.3). This type of behavior is consistent with the induced-fit theory of enzyme action (Chapter 5, Section 5.4). In all kinases for which the structure is known, there is a cleft that closes when substrate is bound.

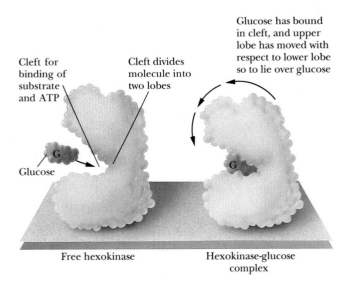

FIGURE 11.3

A comparison of the conformations of hexokinase and the hexokinase–glucose complex.

Cleft for binding of substrate and ATP

Cleft divides molecule into two lobes

Glucose has bound in cleft, and upper lobe has moved with respect to lower lobe so to lie over glucose

Glucose

Free hexokinase

Hexokinase-glucose complex

Step 2. Glucose 6-phosphate isomerizes to give fructose 6-phosphate. **Glucosephosphate isomerase** is the enzyme that catalyzes this reaction. The C-1 aldehyde group of glucose 6-phosphate is reduced to a hydroxyl group, and the C-2 hydroxyl group is oxidized to give the ketone group of fructose 6-phosphate, with no net oxidation or reduction. (Recall from Chapter 10, Section 10.1, that glucose is an aldose, a sugar whose open-chain, noncyclic structure contains an aldehyde group, while fructose is a ketose, a sugar whose corresponding structure contains a ketone group.) The phosphorylated forms, glucose 6-phosphate and fructose 6-phosphate, are an aldose and a ketose, respectively.

glucose 6-phosphate $\rightleftharpoons$ fructose 6-phosphate

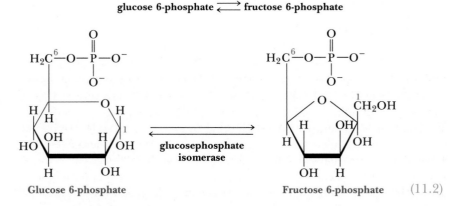

Glucose 6-phosphate

glucosephosphate isomerase

Fructose 6-phosphate   (11.2)

Step 3. Fructose 6-phosphate is further phosphorylated, producing fructose 1,6-*bis*phosphate. (See Equation 11.3 on page 348.)
The endergonic reaction of phosphorylation of fructose 6-phosphate is coupled to the exergonic reaction of hydrolysis of ATP, and the overall reaction is exergonic ($\Delta G^{\circ\prime} = -13.8$ kJ mol^{-1} = -3.3 kcal mol^{-1}), as was the case in Step 1. (See Table 11.1, p. 361.)

Fructose 6-phosphate + ATP $\longrightarrow$ fructose 1,6-*bis*phosphate + ADP

$$\text{Fructose 6-phosphate} + \text{ATP} \xrightarrow[\text{phosphofructokinase}]{\text{Mg}^{2+}} \text{Fructose 1,6-}bis\text{phosphate} + \text{ADP} \qquad (11.3)$$

The reaction in which fructose 6-phosphate is phosphorylated to give fructose 1,6-*bis*phosphate is the one in which the sugar is committed to glycolysis. Glucose 6-phosphate and fructose 6-phosphate can play roles in other pathways, but fructose 1,6-*bis*phosphate does not. Once fructose 1,6-*bis*phosphate is formed from the original sugar, no other pathways are available and the molecule must undergo the rest of the reactions of glycolysis. The phosphorylation of fructose 6-phosphate is highly exergonic and irreversible, and **phosphofructokinase,** the enzyme that catalyzes it, is the key regulatory enzyme in glycolysis.

Phosphofructokinase is a tetramer (in other words, it consists of four subunits) that is subject to allosteric regulation of the type we discussed in Chapter 5. This enzyme exists in slightly different forms in muscle and in liver; the two forms, designated M and L, respectively, are referred to as **isozymes.** The subunits differ slightly in amino acid composition, so the two isozymes can be separated from each other by electrophoresis (Interchapter A, Section A.1). The tetrameric form that occurs in muscle is designated M_4, while that in liver is designated L_4. In red blood cells, both isozymes, the M and the L forms, occur. All possible combinations of monomers combine to form the tetramer: M_4, M_3L, M_2L_2, ML_3, and L_4 (Figure 11.4). Individuals who lack the gene that directs the synthesis of the M form of the enzyme can carry on glycolysis in their livers but suffer muscle weakness because they lack the enzyme in muscle.

When the rate of the phosphofructokinase reaction is observed at varying concentrations of substrate (fructose 6-phosphate), the sigmoidal curve typical of allosteric enzymes is obtained. ATP is an allosteric effector in the reaction. High levels of ATP depress the rate of the reaction, and low levels of ATP stimulate the reaction. When there is a high level of ATP in the cell, a good deal of chemical energy is immediately available from hydrolysis

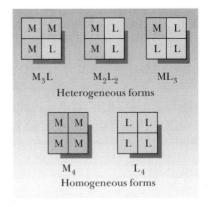

FIGURE 11.4

The possible isozymes of phosphofructokinase. The symbol M refers to the monomeric form that predominates in muscle, while the symbol L refers to the form that predominates in liver.

of ATP. The cell does not need to metabolize glucose for energy, so the presence of ATP inhibits the glycolytic pathway at this point. There is also another, more potent, allosteric effector of phosphofructokinase. This effector is fructose 2,6-*bis*phosphate; we shall discuss its mode of action in Chapter 14, Section 14.3, when we consider control mechanisms in carbohydrate metabolism.

Step 4. Fructose 1,6-*bis*phosphate is then split into two 3-carbon fragments.

fructose 1,6-*bis*phosphate $\rightleftharpoons$
D-glyceraldehyde 3-phosphate + dihydroxyacetone phosphate

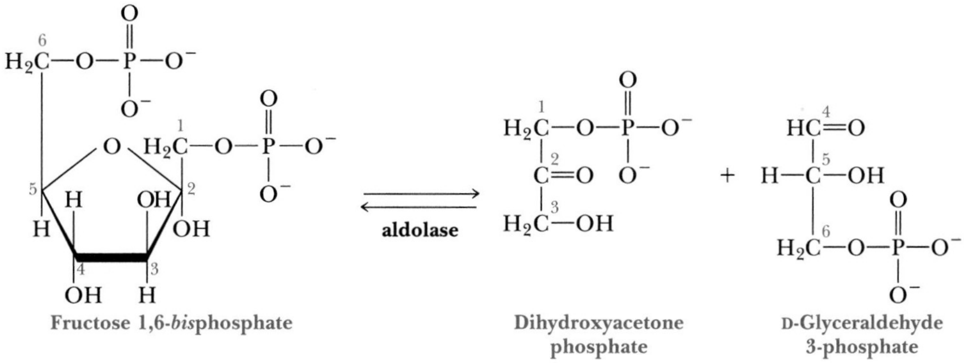

| Fructose 1,6-*bis*phosphate | Dihydroxyacetone phosphate | D-Glyceraldehyde 3-phosphate | (11.4) |

The cleavage reaction here is the reverse of an aldol condensation; the enzyme that catalyzes it is called **aldolase.** In the enzyme isolated from most animal sources (the one from muscle is the most extensively studied), the basic side chain of an essential lysine residue plays the key role in catalyzing this reaction. A compound known as a **Schiff base** is the key intermediate; it involves covalent bonding of the substrate to the ε-amino group of the essential lysine (Figure 11.5). The thiol group of a cysteine also acts as a base here.

Step 5. The dihydroxyacetone phosphate is converted to glyceraldehyde 3-phosphate.

dihydroxyacetone phosphate $\rightleftharpoons$ D-glyceraldehyde 3-phosphate

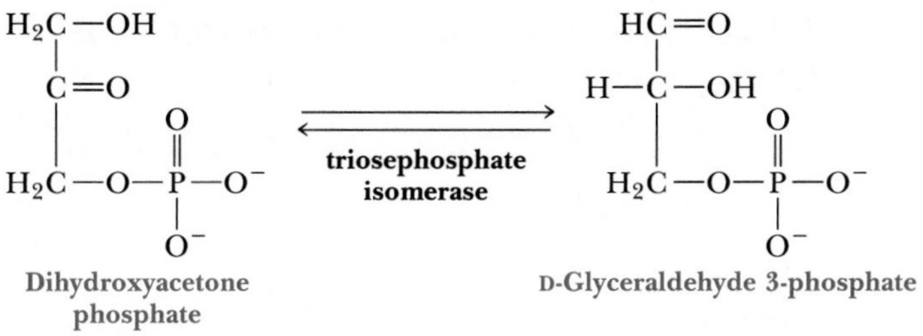

| Dihydroxyacetone phosphate | D-Glyceraldehyde 3-phosphate | (11.5) |

The enzyme that catalyzes this reaction is **triosephosphate isomerase.** (Both dihydroxyacetone and glyceraldehyde are trioses.) One molecule of glyceraldehyde 3-phosphate has already been produced by the aldolase reaction; we

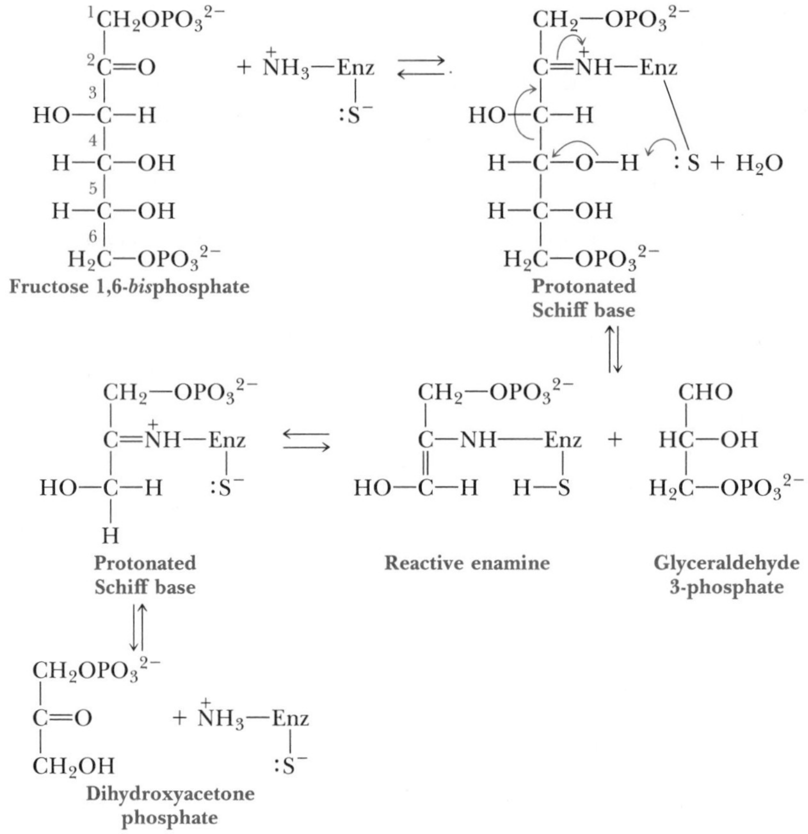

FIGURE 11.5

The aldol cleavage of fructose 1,6-*bis*phosphate. The ϵ-amino group of a lysine residue reacts with the substrate to form a Schiff base. (Enz stands for enzyme, and B for the basic group on the enzyme.)

Fructose 1,6-*bis*phosphate

Protonated Schiff base

Protonated Schiff base

Reactive enamine

Glyceraldehyde 3-phosphate

Dihydroxyacetone phosphate

now have a second molecule of glyceraldehyde 3-phosphate, produced by the triosephosphate isomerase reaction. The original molecule of glucose, which contains six carbon atoms, has now been converted to two molecules of glyceraldehyde 3-phosphate, each of which contains three carbon atoms.

11.3
REACTIONS OF GLYCOLYSIS II: CONVERSION OF GLYCERALDEHYDE 3-PHOSPHATE TO PYRUVATE

At this point a molecule of glucose (a six-carbon compound) that enters the pathway has been converted to two molecules of glyceraldehyde 3-phosphate. We have not seen any oxidation reactions yet, but now we shall encounter them. Keep in mind that in the rest of the pathway two molecules of each of the three-carbon compounds take part in every reaction for each original glucose molecule.

Step 6.   The next step is the oxidation of glyceraldehyde 3-phosphate to 1,3-*bis*phosphoglycerate.

$$\begin{array}{c} \text{glyceraldehyde 3-phosphate} + NAD^+ + P_i \rightleftharpoons \\ NADH + 1,3\text{-}\textit{bis}\text{phosphoglycerate} + H^+ \end{array}$$

$$\begin{array}{c}
\underset{\text{Glyceraldehyde 3-phosphate}}{
\begin{array}{l}
HC^1{=}O \\
| \\
HCOH \\
| \\
H_2C{-}O{-}\overset{\displaystyle O}{\overset{\|}{\underset{\displaystyle O^-}{P}}}{-}O^-
\end{array}}
\quad + NAD^+ + HO{-}\overset{\displaystyle O}{\overset{\|}{\underset{\displaystyle O^-}{P}}}{-}O^-
\quad \underset{\substack{\text{glyceraldehyde} \\ \text{3-phosphate} \\ \text{dehydrogenase}}}{\rightleftharpoons}
\end{array}$$

$$\underset{\text{1,3-\textit{Bis}phosphoglycerate}}{
\begin{array}{l}
\overset{\displaystyle O}{\overset{\|}{C^1}}{-}O{-}\overset{\displaystyle O}{\overset{\|}{\underset{\displaystyle O^-}{P}}}{-}O^- \\
| \\
HCOH \\
| \\
H_2C^3{-}O{-}\overset{\displaystyle O}{\overset{\|}{\underset{\displaystyle O^-}{P}}}{-}O^-
\end{array}}
\quad + NADH + H^+ \qquad (11.6)$$

This reaction, *the* characteristic reaction of glycolysis, should be looked at more closely. It involves the addition of a phosphate group to glyceraldehyde-3-phosphate as well as an electron-transfer reaction, from glyceraldehyde 3-phosphate to NAD^+. It will simplify discussion to consider the two parts separately.

The half reaction of oxidation is that of an aldehyde to a carboxylic acid group, where water can be considered to take part in the reaction.

$$RCHO + H_2O \longrightarrow RCOOH + 2\,H^+ + 2\,e^-$$

The half reaction of reduction is that of NAD^+ to NADH (Chapter 9, Section 9.9).

$$NAD^+ + 2\,H^+ + 2\,e^- \longrightarrow NADH + H^+$$

The overall redox reaction is thus

$$\underset{\substack{\text{Glyceraldehyde} \\ \text{3-phosphate}}}{RCHO} + H_2O + NAD^+ \longrightarrow \underset{\text{3-Phosphoglycerate}}{RCOOH} + H^+ + NADH$$

where R indicates the portions of the molecule other than the aldehyde and carboxylic acid groups, respectively. The oxidation reaction is exergonic under standard conditions ($\Delta G^{\circ\prime} = -43.1$ kJ mol$^{-1} = -10.3$ kcal mol^{-1}), but oxidation is only part of the overall reaction.

The phosphate group that is linked to the carboxyl group does not form an ester, since an ester linkage requires an alcohol and an acid. Instead, the carboxylic acid group and phosphoric acid form a mixed anhydride of two acids by loss of water (Chapter 2, Section 2.2),

$$3\text{-Phosphoglycerate} + P_i \longrightarrow 1,3\text{-}\textit{bis}\text{phosphoglycerate} + H_2O$$

where the substances involved in the reaction are in the ionized form appropriate at pH 7. Note that ATP and ADP do not appear in the equation. The source of the phosphate group is phosphate ion itself, rather than ATP. A reaction of this type in which P_i rather than ATP is the source of the phosphate group is called **substrate-level phosphorylation.** The phosphorylation reaction is endergonic under standard conditions ($\Delta G^{\circ\prime} = 49.3$ kJ mol^{-1} = 11.8 kcal mol^{-1}).

The overall reaction, including electron transfer and phosphorylation, is

$$RCHO + HPO_3^{2-} + NAD^+ \rightleftharpoons RC-OPO_3^{2-} + NADH + H^+$$

or

$$\text{Glyceraldehyde 3-phosphate} + P_i + NAD^+ \longrightarrow$$
$$\text{1,3-}bis\text{phosphoglycerate} + NADH + H^+$$

Let's show the two reactions that make up this reaction.

1. Oxidation of glyceraldehyde 3-phosphate ($\Delta G^{\circ\prime} = -43.1$ kJ $= -10.3$ kcal)

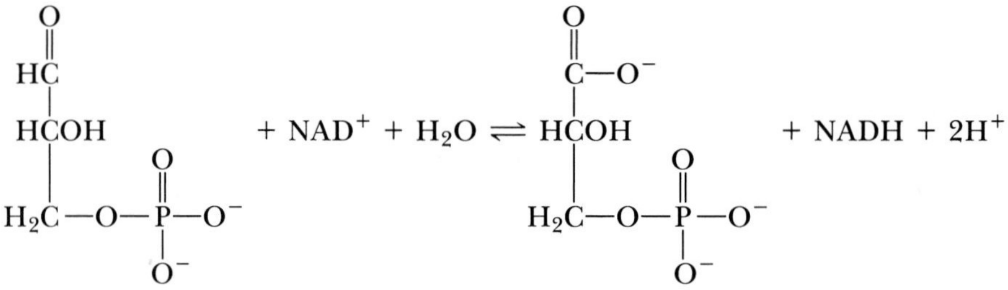

glyceraldehyde-
3-phosphate

3-phosphoglycerate

2. Phosphorylation of 3-phosphoglycerate ($\Delta G^{\circ\prime} = 49.3$ kJ $= 11.8$ kcal)

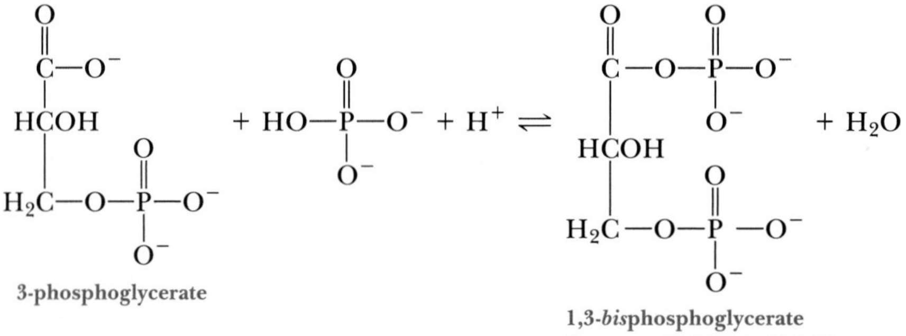

3-phosphoglycerate

1,3-*bis*phosphoglycerate
Sum = −43.1 kJ + 49.3 kJ
= 6.2 kJ = 1.5 kcal

The standard free energy change for the overall reaction is the sum of the values for the oxidation and phosphorylation reactions. The overall reaction is not far from equilibrium, being only slightly endergonic.

$$\Delta G^{\circ\prime} \text{ overall} = \Delta G^{\circ\prime} \text{ oxidation} + \Delta G^{\circ\prime} \text{ phosphorylation}$$
$$= (-43.1) + (49.3) \text{ kJ mol}^{-1}$$
$$= 6.2 \text{ kJ mol}^{-1} = 1.5 \text{ kcal mol}^{-1}$$

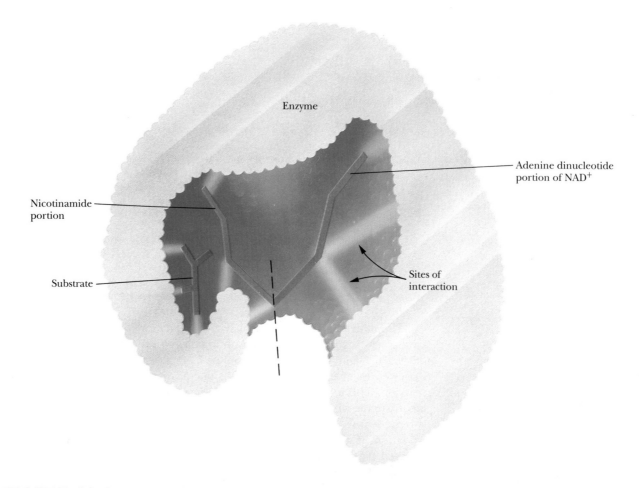

FIGURE 11.6

Schematic view of the binding site of an NADH-linked dehydrogenase. There are specific binding sites for the adenine nucleotide portion (shown in red to the right of the dashed line) and nicotinamide portion (shown in red to the left of the dashed line) of the coenzyme, in addition to the binding site for the substrate. Specific interactions with the enzyme hold the substrate and coenzyme in the proper positions. Sites of interaction are shown as a series of yellow lines.

This value of the standard free energy change is for the reaction of one mole of glyceraldehyde 3-phosphate; the value must be multiplied by two to get the value for each mole of glucose ($\Delta G^{\circ\prime} = 12.4$ kJ = 3.0 kcal).

The enzyme that catalyzes the conversion of glyceraldehyde 3-phosphate to 1,3-*bis*phosphoglycerate is **glyceraldehyde 3-phosphate dehydrogenase.** This enzyme is one of a class of similar enzymes, the NADH-linked dehydrogenases. The structures of a number of dehydrogenases of this type have been studied via x-ray crystallography. The overall structures are not strikingly similar, but the structure of the binding site for NADH is quite similar in all these enzymes (Figure 11.6). (The oxidizing agent is NAD^+; both oxidized and reduced forms of the coenzyme bind to the enzyme.) One portion of the binding site is specific for the nicotinamide ring, and one portion is specific for the adenine ring.

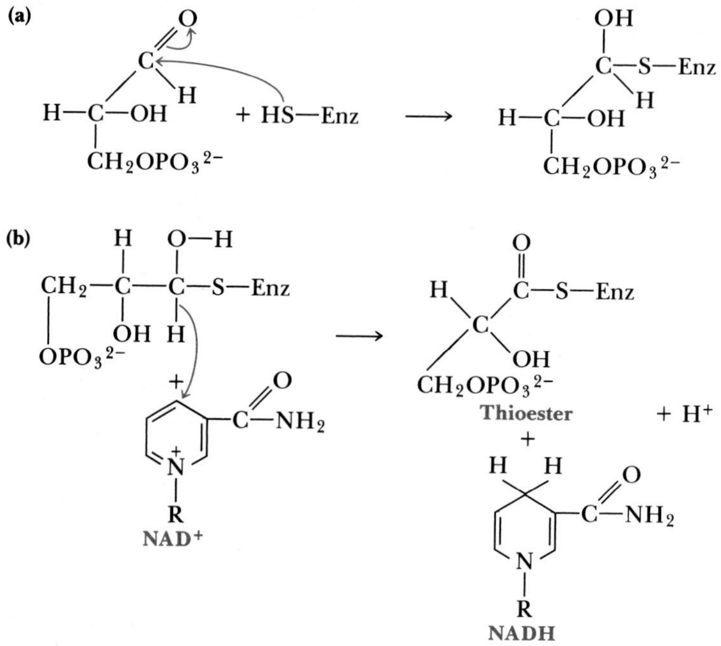

FIGURE 11.7

Hydride ion transfer from an enzyme-bound derivative of glyceraldehyde 3-phosphate to NAD^+ produces a thioester and NADH. (a) Formation of an enzyme-bound derivative of glyceraldehyde 3-phosphate. Enz is glyceraldehyde 3-phosphate dehydrogenase. (b) Hydride ion transfer. Note that H^- and H: are alternative ways of writing the formula for the hydride ion.

The molecule of glyceraldehyde 3-phosphate dehydrogenase is a tetramer, consisting of four identical subunits. Each subunit binds one molecule of NAD^+, and each subunit contains an essential cysteine residue. A thioester involving the cysteine residue is the key intermediate in this reaction. There is a **hydride ion (H^-) transfer** from the aldehyde to the nicotinamide ring of the NAD^+, forming NADH and the thioester. (An ester is a derivative of an acid; the first part of the reaction—namely, the oxidation of an aldehyde to an acid—has already taken place.) NADH is released by the enzyme, and NAD^+ binds in its place. Figure 11.7 shows some key features of the mechanism.

In the phosphorylation step, the thioester acts as a "high-energy" intermediate. Phosphate ion attacks the thioester, forming a mixed anhydride of the carboxylic and phosphoric acids, which is also a "high-energy" compound (Figure 11.8). This compound is 1,3-*bis*phosphoglycerate, the product of the reaction. Production of ATP requires a "high-energy" compound as starting material. The 1,3-*bis*phosphoglycerate fulfills this requirement and transfers a phosphate group to ADP in a highly exergonic reaction (i.e., it has a high phosphate-group transfer potential).

Step 7. The next step is one of the two reactions in which ATP is produced by phosphorylation of ADP:

1,3-*Bis*phosphoglycerate + ADP $\rightleftharpoons$ 3-phosphoglycerate + ATP

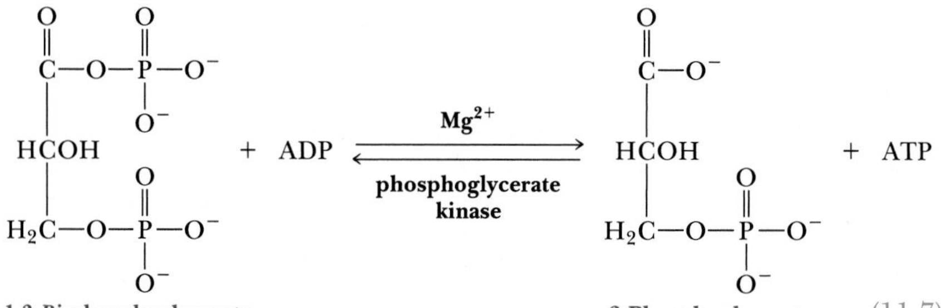

1,3-*Bis*phosphoglycerate   3-Phosphoglycerate   (11.7)

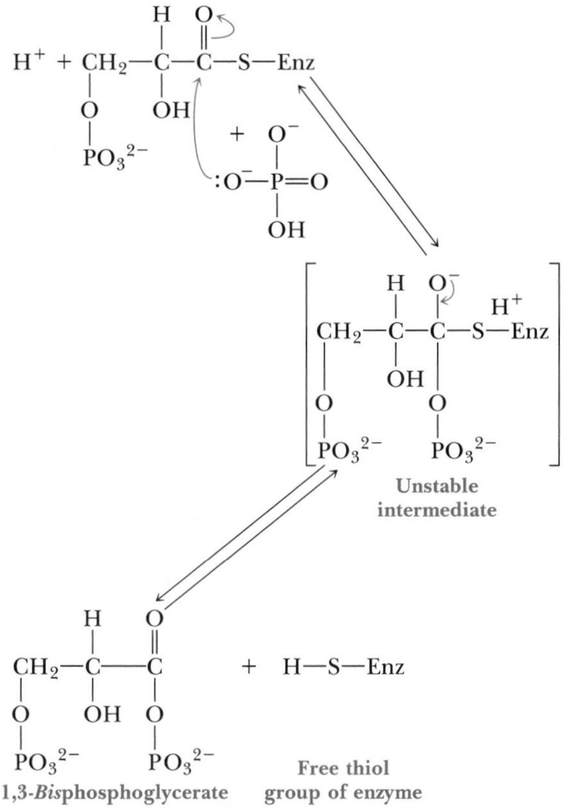

FIGURE 11.8

Phosphate ion attacks the thioester derivative of glyceraldehyde 3-phosphate dehydrogenase (Enz) to produce 1,3-*bis*phosphoglycerate and regenerate the thiol group of cysteine.

The enzyme that catalyzes this reaction is **phosphoglycerate kinase.** By now the term "kinase" should be familiar as the generic name for a class of ATP-dependent phosphate-group transfer enzymes. The most striking feature of the reaction has to do with energetics of phosphate-group transfer. In this step in glycolysis, a phosphate group is transferred from 1,3-*bis*phosphoglycerate to a molecule of ADP, producing ATP, the first of two such reactions in the glycolytic pathway. We already mentioned that 1,3-*bis*phosphoglycerate can easily transfer a phosphate group to other substances. The only requirement is that the standard free energy of the hydrolysis reaction is more negative than that for hydrolysis of the new phosphate compound being formed. Recall that the standard free energy of hydrolysis of 1,3-*bis*phosphoglycerate is -49.3 kJ mol^{-1}. We have already seen that the standard free energy of hydrolysis of ATP is -30.5 kJ mol^{-1}, and we must change the sign of the free energy change when the reverse reaction occurs:

$$ADP + P_i + H^+ \longrightarrow ATP + H_2O$$

$$\Delta G^{\circ\prime} = 30.5 \text{ kJ mol}^{-1} = 7.3 \text{ kcal mol}^{-1}$$

The net reaction is

$$\text{1,3-Bisphosphoglycerate} + \text{ADP} \longrightarrow \text{3-phosphoglycerate} + \text{ATP}$$

$$\Delta G^{\circ\prime} = -18.8 \text{ kJ mol}^{-1} = -4.5 \text{ kcal mol}^{-1}$$

Two molecules of ATP are produced by this reaction for each molecule of glucose that enters the glycolytic pathway. In the earlier stages of the pathway two molecules of ATP were invested to produce fructose 1,6-bisphosphate, and now they have been recovered. At this point the balance of ATP use and production is exactly even. The next few reactions will bring about the production of two more molecules of ATP for each original molecule of glucose, leading eventually to the net gain of two ATP molecules in glycolysis.

Step 8. In the next reaction the phosphate group is transferred from carbon 3 to carbon 2 of the glyceric acid backbone, setting the stage for the reactions that follow.

3-Phosphoglycerate $\rightleftharpoons$ 2-phosphoglycerate

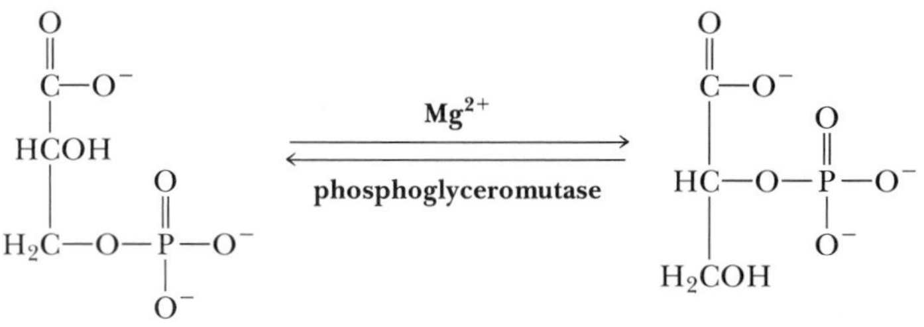

3-Phosphoglycerate   2-Phosphoglycerate   (11.8)

The enzyme that catalyzes this reaction is **phosphoglyceromutase.**

Step 9. The 2-phosphoglycerate molecule loses one molecule of water, producing phosphoenolpyruvate. This reaction does not involve electron transfer; it is a dehydration reaction. **Enolase,** the enzyme that catalyzes this reaction, requires Mg^{2+} as a cofactor. The water molecule that is eliminated binds to Mg^{2+} in the course of the reaction.

2-Phosphoglycerate $\rightleftharpoons$ phosphoenolpyruvate + H_2O

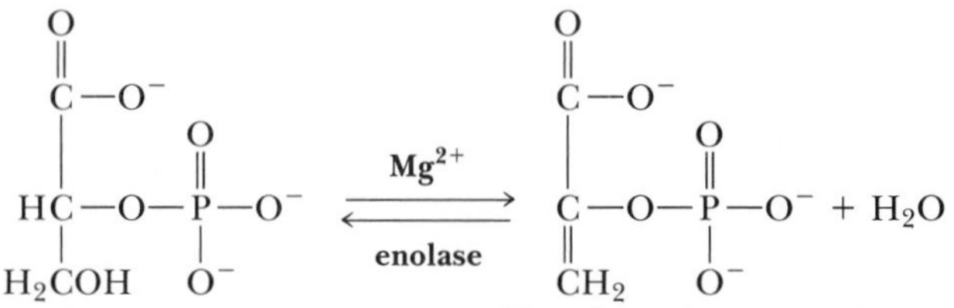

2-Phosphoglycerate   Phosphoenolpyruvate (PEP)   (11.9)

Step 10. Phosphoenolpyruvate then transfers its phosphate group to ADP, producing ATP and pyruvate.

$$\text{Phosphoenolpyruvate} + \text{ADP} \longrightarrow \text{pyruvate} + \text{ATP}$$

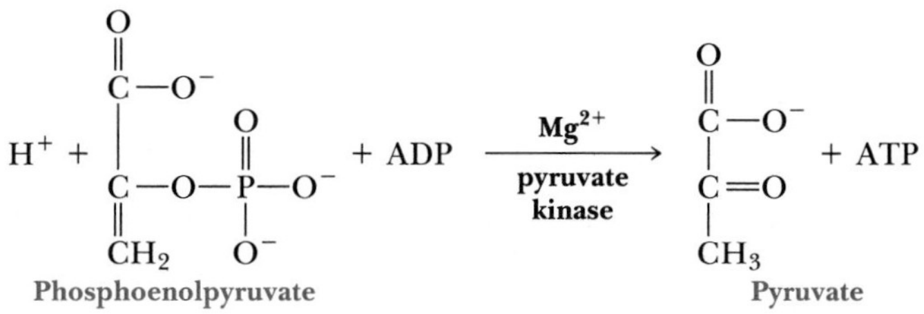

Phosphoenolpyruvate + ADP → Pyruvate + ATP (via pyruvate kinase, Mg²⁺)   (11.10)

The double bond shifts to the oxygen on carbon 2 and a hydrogen shifts to carbon 3. Phosphoenolpyruvate is a "high-energy" compound with a high phosphate-group transfer potential. The free energy of hydrolysis of this compound is more negative than that of ATP (-61.9 kJ mol^{-1} vs. -30.5 kJ mol^{-1}, or -14.8 kcal mol^{-1} vs. -7.3 kcal mol^{-1}). The reaction that occurs in this step can be considered to be the sum of the hydrolysis of phosphoenolpyruvate and the phosphorylation of ADP.

Phosphoenolpyruvate $\longrightarrow$ pyruvate + P$_i$

$\Delta G^{\circ\prime} = -61.9$ kJ mol^{-1} = -14.8 kcal mol^{-1}

ADP + P$_i$ $\longrightarrow$ ATP

$\Delta G^{\circ\prime} = 30.5$ kJ mol^{-1} = 7.3 kcal mol^{-1}

The net reaction is

Phosphoenolpyruvate + ADP $\longrightarrow$ pyruvate + ATP

$\Delta G^{\circ\prime} = -31.4$ kJ mol^{-1} = -7.5 kcal mol^{-1}

Since two moles of pyruvate are produced for each mole of glucose, twice as much energy is released for each mole of starting material.

Pyruvate kinase is the enzyme that catalyzes this reaction. Like phosphofructokinase, it is an allosteric enzyme consisting of four subunits. Pyruvate kinase is inhibited by ATP. The conversion of phosphoenolpyruvate to pyruvate slows down when the cell has a high concentration of ATP—that is to say, when the cell does not have a great need for energy in the form of ATP.

11.4
ANAEROBIC REACTIONS OF PYRUVATE

The Conversion of Pyruvate to Lactate in Muscle

The final reaction of anaerobic glycolysis is the reduction of pyruvate to lactate.

$$\text{Pyruvate} + \text{NADH} + \text{H}^+ \rightleftharpoons \text{lactate} + \text{NAD}^+$$

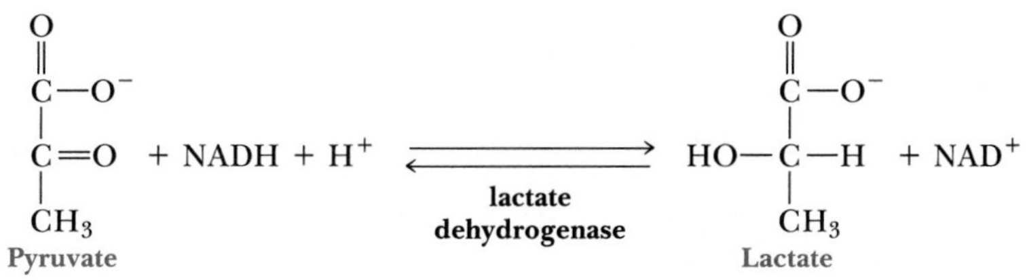

This reaction is also exergonic ($\Delta G^{\circ\prime} = -25.1$ kJ mol^{-1} = -6.0 kcal mol^{-1}); as before, we need to multiply this value by two to find the energy yield for each molecule of glucose that enters the pathway. Lactate is a dead end in metabolism, but it can be recycled to form pyruvate and even glucose by a pathway called gluconeogenesis ("new synthesis of glucose"), which we will discuss in Chapter 14, Section 14.2.

Lactate dehydrogenase (LDH) is the enzyme that catalyzes this reaction. Like glyceraldehyde 3-phosphate dehydrogenase, LDH is an NADH-linked dehydrogenase. LDH is an allosteric enzyme with four subunits. There are two kinds of subunits, designated M and H, which vary slightly in amino acid composition. The quaternary structure of the tetramer can vary according to the relative amounts of the two kinds of subunits. These variant forms of the enzyme are called isozymes (see Chapter 11, Section 11.2, for a discussion of the isozymes of phosphofructokinase). In human skeletal muscle the homogeneous tetramer of the M_4 type predominates, and in heart the other homogeneous possibility, the H_4 tetramer, is the predominant form. The heterogeneous forms, M_3H, M_2H_2, and MH_3, occur in blood serum. A very sensitive clinical test for heart disease is based on the existence of the various isozymic forms of this enzyme. The relative amounts of the H_4 and MH_3 isozymes in blood serum increase drastically after myocardial infarction (heart attack), compared with normal serum.

At this point one might ask why the reduction of pyruvate to lactate (a waste product in aerobic organisms) is the last step in anaerobic glycolysis, a pathway that provides energy for the organism by oxidation of nutrients. There is another point to consider about the reaction, one that involves the relative amounts of NAD$^+$ and NADH in a cell. The half reaction of reduction can be written

$$\text{Pyruvate} + 2\,H^+ + 2\,e^- \longrightarrow \text{lactate}$$

and the half reaction of oxidation is

$$\text{NADH} + H^+ \longrightarrow NAD^+ + 2\,e- + 2\,H^+$$

The overall reaction is, as we saw earlier,

$$\text{Pyruvate} + \text{NADH} + H^+ \longrightarrow \text{lactate} + NAD^+$$

The NADH produced from NAD$^+$ by the earlier oxidation of glyceraldehyde 3-phosphate is used up with no net change in the relative amounts of NADH and NAD$^+$ in the cell (Figure 11.9). This regeneration is needed under anaerobic conditions in the cell so that NAD$^+$ will be present for

FIGURE 11.9

The recycling of NAD$^+$ and NADH in anaerobic glycolysis.

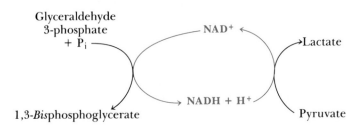

further glycolysis to take place. Without this regeneration, the oxidation reactions in anaerobic organisms would soon come to a halt because of the lack of NAD^+ to serve as an oxidizing agent in fermentative processes. On the other hand, NADH is a frequently encountered reducing agent in many reactions, and it is lost to the organism in lactate production. Aerobic metabolism makes more efficient use of reducing agents ("reducing power") such as NADH, because the conversion of pyruvate to lactate does not occur in aerobic metabolism. The NADH produced in the stages of glycolysis leading to the production of pyruvate is available for use in reactions in which a reducing agent is needed.

Alcoholic Fermentation

Two other reactions related to the glycolytic pathway lead to the production of ethanol by **alcoholic fermentation.** This process is one of the alternative fates of pyruvate (Section 11.1). In the first of the two reactions that lead to the production of ethanol, pyruvate is decarboxylated (loses carbon dioxide) to produce acetaldehyde. The enzyme that catalyzes this reaction is **pyruvate decarboxylase.**

This enzyme requires Mg^{2+} and a cofactor we have not met before, **thiamine pyrophosphate** (TPP). (Thiamine itself is vitamin B_1.) In TPP the carbon atom between the nitrogen and the sulfur in the thiazole ring (Figure 11.10) is highly reactive. It forms a carbanion (an ion with a negative charge on a carbon atom) quite easily, and the carbanion in turn attacks the carbonyl group of pyruvate to form an adduct. Carbon dioxide splits off, leaving a two-carbon fragment covalently bonded to TPP. There is a shift of electrons, and the two-carbon fragment splits off, producing acetaldehyde

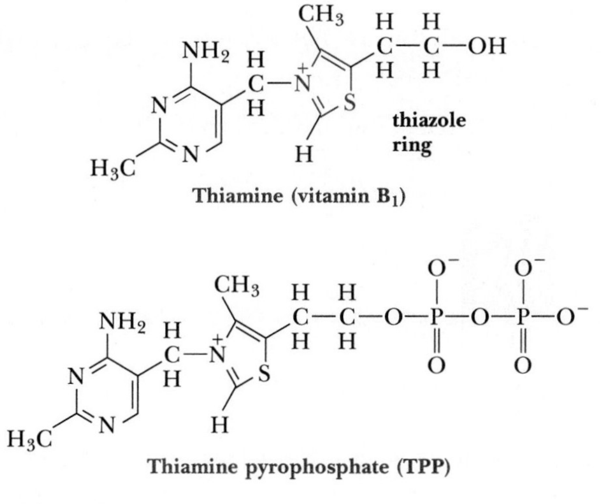

FIGURE 11.10

The structures of thiamine (vitamin B_1) and thiamine pyrophosphate (TPP), the active form of the coenzyme.

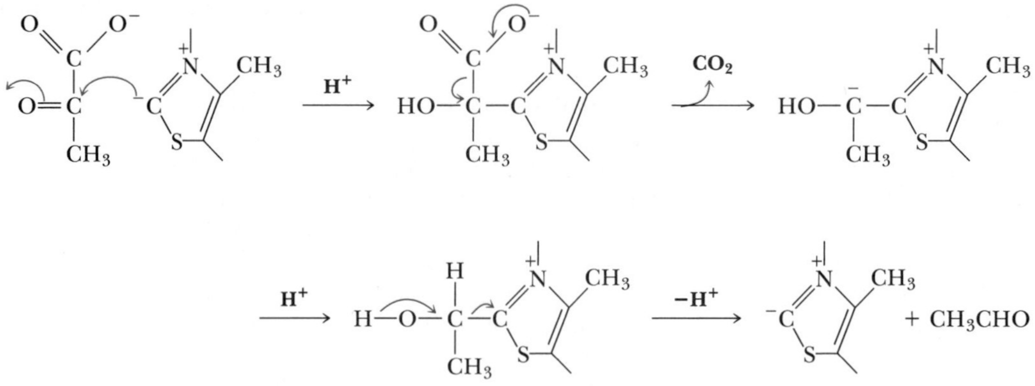

FIGURE 11.11

The mechanism of the pyruvate decarboxylase reaction. The carbanion form of the thiazole ring of TPP is strongly nucleophilic. The carbanion attacks the carbonyl carbon of pyruvate to form an adduct. Carbon dioxide splits out, leaving a two-carbon fragment (activated acetaldehyde) covalently bonded to the coenzyme. A shift of electrons releases acetaldehyde, regenerating the carbanion.

(Figure 11.11). (The two-carbon fragment bonded to TPP is sometimes called activated acetaldehyde.)

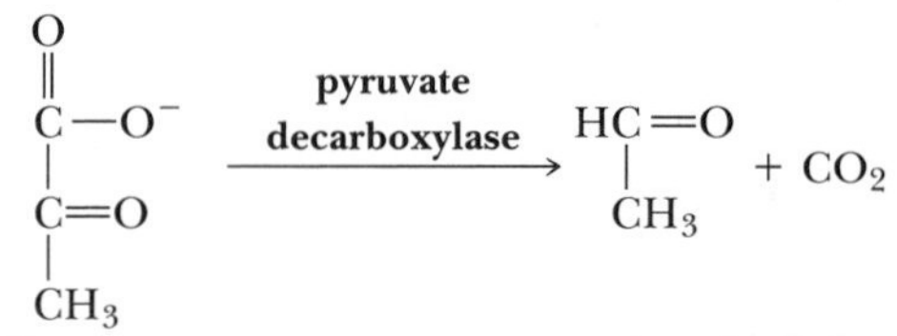

$$\text{pyruvate} \longrightarrow \text{acetaldehyde} + CO_2$$

The carbon dioxide produced is responsible for the bubbles in beer and sparkling wines. Acetaldehyde is then reduced to produce ethanol, and at the same time one molecule of NADH is oxidized to NAD^+ for each molecule of ethanol produced.

$$\text{Acetaldehyde} + NADH \longrightarrow \text{ethanol} + NAD^+$$

FETAL ALCOHOL SYNDROME

The complex of injuries to a fetus caused by maternal consumption of ethanol is called fetal alcohol syndrome. In catabolism of ethanol by the body, the first step is conversion to acetaldehyde—the reverse of the last reaction of alcoholic fermentation. The level of acetaldehyde in the blood of a pregnant woman is the key to detecting fetal alcohol syndrome. It has recently been shown that the acetaldehyde is transferred across the placenta and accumulates in the liver of the fetus. The acetaldehyde in turn is easily oxidized to breakdown products and is responsible for the harmful effects.

The reduction reaction of alcoholic fermentation is similar to the reduction of pyruvate to lactate, in the sense that it provides for recycling of NAD^+ and thus allows further anaerobic oxidation (fermentation) reactions. The net reaction for alcoholic fermentation is

Glucose + 2 ADP + 2 P_i + 2 H^+ $\longrightarrow$ 2 ethanol + 2 ATP + 2 CO_2 + 2 H_2O

NAD^+ and NADH do not appear explicitly in the net equation. It is essential that the recycling of NADH to NAD^+ takes place here, just as it does when lactate is produced, so that there can be further anaerobic oxidation. **Alcohol dehydrogenase,** the enzyme that catalyzes the conversion of acetaldehyde to ethanol, is similar to lactate dehydrogenase in many ways. The most striking similarity is that both are NADH-linked dehydrogenases. Both enzymes are allosteric enzymes, and both are tetramers.

11.5
ENERGY CONSIDERATIONS IN GLYCOLYSIS

Now that we have seen the reactions of the glycolytic pathway, we can do some bookkeeping and determine the standard free energy change for the entire pathway. This calculation is shown in Table 11.1. The conversion of pyruvate to lactate is also listed there.

The overall process of glycolysis is exergonic. The energy released in the exergonic phases of the process drives the endergonic reactions. The net reaction of glycolysis explicitly includes an important endergonic process, that of phosphorylation of two molecules of ADP.

2 ADP + 2 P_i $\longrightarrow$ 2 ATP

$\Delta G^{\circ\prime}$ reaction = 61 kJ = 14.6 kcal mol^{-1} glucose consumed

T A B L E   1 1 . 1   The Reactions of Glycolysis and Their Standard Free Energy Changes

STEP	REACTION	$\Delta G^{\circ\prime}$	
		kJ mol^{-1}	kcal mol^{-1}
1	Glucose + ATP $\longrightarrow$ glucose 6-phosphate + ADP	−16.7	−4.0
2	Glucose 6-phosphate $\longrightarrow$ fructose 6-phosphate	+1.7	+0.4
3	Fructose 6-phosphate + ATP $\longrightarrow$ fructose 1,6-*bis*phosphate + ADP	−13.8	−3.3
4	Fructose 1,6-*bis*phosphate $\longrightarrow$ dihydroxyacetone phosphate + glyceraldehyde 3-phosphate	+23.8	+5.7
5	Dihydroxyacetone phosphate $\longrightarrow$ glyceraldehyde 3-phosphate	+7.5	+1.8
6	2(Glyceraldehyde 3-phosphate + NAD^+ + P_i $\longrightarrow$ 1,3-*bis*phosphoglycerate + NADH + H^+)	+12.4	+3.0
7	2(1,3-*Bis*phosphoglycerate + ADP $\longrightarrow$ 3-phosphoglycerate + ATP)	−37.6	−9.0
8	2(3-Phosphoglycerate $\longrightarrow$ 2-phosphoglycerate)	+8.8	+2.1
9	2(2-Phosphoglycerate $\longrightarrow$ phosphoenolpyruvate + H_2O)	+3.4	+0.8
10	2(Phosphoenolpyruvate + ADP $\longrightarrow$ pyruvate + ATP)	−62.8	−15.0
Overall	Glucose + 2 ADP + 2 P_i + NAD^+ $\longrightarrow$ 2 pyruvate + 2 ATP + NADH + H^+	−73.3	−17.5
	2(Pyruvate + NADH + H^+ $\longrightarrow$ lactate + NAD^+)	−50.2	−12.0
	Glucose + 2 ADP + 2 P_i $\longrightarrow$ 2 lactate + 2 ATP	−123.5	−29.5

Without the production of ATP the reaction of one molecule of glucose to produce two molecules of pyruvate would be even more exergonic.

Glucose + 2 ADP + 2 P_i $\longrightarrow$ 2 pyruvate + 2 ATP

$\Delta G^{\circ\prime}$ = -73.3 kJ = -17.5 kcal mol^{-1} glucose consumed

$-$(2 ATP + 2 P_i $\longrightarrow$ 2 ATP)

$-$($\Delta G^{\circ\prime}$ reaction = 61 kJ = 14.6 kcal)

Overall: Glucose $\longrightarrow$ 2 pyruvate

$\Delta G^{\circ\prime}$ = -134.3 kJ = -32.1 kcal mol^{-1} glucose consumed

Without production of ATP, the energy released by the conversion of glucose to pyruvate would be lost to the organism and dissipated as heat. The energy required to produce the two molecules of ATP for each molecule of glucose can be recovered by the organism when the ATP is hydrolyzed in some metabolic process. We discussed this point briefly in Chapter 9 when we compared the thermodynamic efficiency of anaerobic and aerobic metabolism. The percentage of the energy released by the breakdown of glucose to lactate that is "captured" by the organism when ADP is phosphorylated to ATP is the efficiency of energy use in glycolysis; it is (61.0/184.5) × 100, or about 33%. The net release of energy in glycolysis, 123.5 kJ (29.5 kcal) for each mole of glucose, is not retained for use by the organism. Without the production of ATP to serve as a source of energy for other metabolic processes, the energy released by glycolysis would serve no purpose for the organism, except to help maintain body temperature in warm-blooded animals. A soft drink with ice can help keep you warm even on the coldest day of winter (if it is not a diet drink) because of its high sugar content.

The free energy changes we have listed in this chapter are the standard values, assuming the standard conditions such as 1 M concentrations of all solutes except hydrogen ion. Concentrations under physiological conditions can differ markedly from standard values. Fortunately, there are well-known methods (Chapter 9, Section 9.3) for calculating the difference in the free energy change. Also, large changes in concentrations frequently lead to relatively small differences in the free energy change, about a few kJ mol^{-1}. Some of the free energy changes may be different under physiological conditions from the values listed here for standard conditions, but the underlying principles and the conclusions drawn from them remain the same.

S U M M A R Y

In glycolysis, one molecule of glucose gives rise after a long series of reactions to two molecules of pyruvate. In two reactions in the pathway, one molecule of ATP is hydrolyzed for each molecule of glucose metabolized. In each of two other reactions, two molecules of ATP are produced by phosphorylation of ADP for each molecule of glucose, giving a total of four ATP molecules produced. There is a net gain of two ATP molecules for each molecule of glucose processed in glycolysis. A key intermediate is fructose 1,6-*bis*phosphate, and the enzyme that catalyzes its formation, phosphofructokinase, is an important controlling factor in the pathway.

Several metabolic fates are possible for pyruvate. In aerobic metabolism, pyruvate loses carbon dioxide; the remaining two carbon atoms become linked to coenzyme A as an acetyl group to form acetyl-CoA, which then enters the citric acid cycle. There are two fates for pyruvate in anaerobic metabolism. In organisms capable of alcoholic fermentation, pyruvate loses carbon dioxide, this time producing acetaldehyde, which in turn is reduced to produce ethanol. The common fate of pyruvate in anaerobic metabolism is reduction to lactate; in this chapter we concentrated on the conversion of glucose to lactate, called anaerobic glycolysis to distinguish it from conversion of glucose to pyruvate.

The anaerobic breakdown of glucose to lactate can be summarized as follows:

$$\text{Glucose} + 2\,\text{ADP} + 2\,\text{P}_i \longrightarrow 2\,\text{lactate} + 2\,\text{ATP}$$

The overall process of glycolysis is exergonic (see below). Without production of ATP, glycolysis would be still more exergonic, but the energy released would be lost to the organism and dissipated as heat.

	$\Delta G^{\circ\prime}$	
	kJ mol^{-1}	*kcal mol^{-1}*
Glucose + 2 ADP + 2 P$_i$ $\longrightarrow$ 2 pyruvate + 2 ATP	−73.3	−17.5
2(Pyruvate + NADH + H$^+$ $\longrightarrow$ lactate + NAD$^+$)	−50.2	−12.0
Glucose + 2 ADP + 2 P$_i$ $\longrightarrow$ 2 lactate + 2 ATP	−123.5	−29.5

E X E R C I S E S

1. What does the material of this chapter have to do with beer? With tired and aching muscles?
2. Which reaction or reactions we have met in this chapter require ATP? Which reaction or reactions produce ATP? List the enzymes that catalyze the reactions that require and that produce ATP.
3. Which reaction or reactions we have met in this chapter require NADH? Which reaction or reactions require NAD$^+$? List the enzymes that catalyze the reactions that require NADH and that require NAD$^+$.
4. Which of the enzymes discussed in this chapter are NADH-linked dehydrogenases?
5. Explain the origin of the name of the enzyme aldolase.
6. Show how the estimate of 33% efficiency of energy use in anaerobic glycolysis is derived.
7. Show that the reaction

   Glucose $\longrightarrow$ 2 glyceraldehyde 3-phosphate

   is slightly endergonic ($\Delta G^{\circ\prime}$ = 2.5 kJ mol^{-1} = 0.6 kcal mol^{-1}); that is, it is not too far from equilibrium. Use the data in Table 11.1.
8. Show by a series of equations the energetics of phosphorylation of ADP by phosphoenolpyruvate.
9. Using the Lewis electron-dot notation, show explicitly the transfer of electrons in the following redox reactions.
   (a) Pyruvate + NADH + H$^+$ $\longrightarrow$ lactate + NAD$^+$
   (b) Acetaldehyde + NADH + H$^+$ $\longrightarrow$ ethanol + NAD$^+$
   (c) Glyceraldehyde 3-phosphate + NAD$^+$ $\longrightarrow$ 3-phosphoglycerate + NADH + H$^+$ (redox reaction only)
10. What is the net gain of ATP molecules derived from the reactions of glycolysis?
11. What are the possible metabolic fates of pyruvate?

12. Is the reaction of 2-phosphoglycerate to phospho-enolpyruvate a redox reaction? Give the reason for your answer.

13. In what way is the observed mode of action of hexokinase consistent with the induced-fit theory of enzyme action?

14. How does ATP act as an allosteric effector in the mode of action of phosphofructokinase?

15. Define substrate-level phosphorylation and give an example from the reactions discussed in this chapter.

16. Define isozymes and give an example from the material discussed in this chapter.

17. Briefly discuss the role of thiamine pyrophosphate in enzymatic reactions, using material from this chapter to illustrate your points.

18. Beriberi is a disease caused by a deficiency of vitamin B_1 (thiamine) in the diet. Thiamine is the precursor of thiamine pyrophosphate. In view of what you have learned in this chapter, why is it not surprising that alcoholics tend to develop this disease?

19. It is well known among hunters that meat from animals that have been run to death tastes sour. Suggest a reason for this observation.

20. Why is the formation of fructose 1,6-*bis*phosphate a step in which control is likely to be exercised in the glycolytic pathway?

A N N O T A T E D   B I B L I O G R A P H Y

Bodner, G. M. Metabolism: Part I, Glycolysis, or the Embden–Meyerhoff Pathway. *J. Chem. Ed.* **63,** 566–570 (1986). [A clear, concise summary of the pathway. Part of a series on metabolism of carbohydrates and lipids.]

Boyer, P. D., ed. *The Enzymes.* Vols. 5–9. New York: Academic Press, 1972. [A standard reference with review articles on the glycolytic enzymes; lactate dehydrogenase and alcohol dehydrogenase appear in Volume 10.]

Florkin, M., and E. H. Stotz, eds. *Comprehensive Biochemistry.* New York: Elsevier, 1967. [Another standard reference. Vol. 17, *Carbohydrate Metabolism,* deals with glycolysis.]

Karl, P. I., B. H. J. Gordon, C. S. Lieber, and S. E. Fisher. Acetaldehyde Production and Transfer by the Perfused Human Placental Cotyledon. *Science* **242,** 273–275 (1988). [A report describing some of the processes involved in fetal alcohol syndrome.]

Lipmann, F. A Long Life in Times of Great Upheaval. *Ann. Rev. Biochem.* **53,** 1–33 (1984). [The reminiscences of a Nobel laureate whose research contributed greatly to the understanding of carbohydrate metabolism. Very interesting reading from the standpoints of autobiography and the author's contributions to biochemistry.]

The Citric Acid Cycle

Crystals of citric acid viewed under polarized light.

If the mitochondrion is the power plant of the cell, then the citric acid cycle operating inside the mitochondrion is its engine room. Here, metabolic fuels, especially glucose derived from carbohydrates, amino acids from proteins, and fatty acids from lipids, are all fed into the cycle ultimately to be oxidized to carbon dioxide and water. Their energy is transferred to electron carriers and finally to the terminal electron acceptor — oxygen. All metabolic fuels enter the citric acid cycle as acetyl-CoA. In the first stage of energy extraction from carbohydrates, glucose is catabolized to pyruvate in the ten anaerobic steps of glycolysis. In *aerobic* catabolism, pyruvate is converted to "high-energy" two-carbon acetyl-CoA, which then enters the citric acid cycle. As the first step of the cycle, acetyl-CoA combines with oxaloacetate to form citric acid. Each turn of the cycle releases two molecules of CO_2, regenerates the starting molecule oxaloacetate, and delivers reducing agents to the electron transport chain. In the final stage of glucose metabolism, proton flow across the inner mitochondrial membrane creates ATP, the energy currency of the cell. The citric acid cycle not only furnishes energy, it also provides intermediates for the biosynthesis of proteins, lipids, and the heme group.

12.1
THE ROLE OF THE CITRIC ACID CYCLE IN METABOLISM

The evolution of aerobic metabolism, by which nutrients are oxidized to carbon dioxide and water, was an important step in the history of life on earth. Organisms can obtain far more energy from nutrients by aerobic oxidation than by anaerobic oxidation. (Even yeast, usually thought of in terms of the anaerobic reaction of alcoholic fermentation, prefers the citric acid cycle and degrades glucose to carbon dioxide and water.) We saw in Chapter 11 that glycolysis produces only two molecules of ATP for each molecule of glucose metabolized; in this chapter and the next we shall see how 38 ATP can be produced from each molecule of glucose in complete aerobic oxidation to carbon dioxide and water. Three processes play roles in aerobic metabolism: the **citric acid cycle,** which we discuss in this chapter, and **electron transport** and **oxidative phosphorylation,** both of which we shall discuss in the next chapter. These three processes operate together in aerobic metabolism; separate discussion is a matter of convenience only.

Metabolism consists of catabolism, which is oxidative breakdown of nutrients, and anabolism, which is reductive synthesis of biomolecules. The citric acid cycle is **amphibolic,** meaning that it plays a role in both catabolism and anabolism. While the citric acid cycle is a part of the pathway of aerobic oxidation of nutrients (a catabolic pathway;) see Chapter 18, Section 18.2), some of the molecules that are included in this cycle are the starting points of biosynthetic (anabolic) pathways (see Section 18.3). Metabolic pathways operate simultaneously, even though we talk about them separately. We should always keep this point in mind.

There are two other common names for the citric acid cycle. One is the **Krebs cycle,** after Sir Hans Krebs, who first investigated the pathway (work for which he received a Nobel Prize in 1953). The other name is the

tricarboxylic acid cycle, from the fact that some of the molecules involved are acids with three carboxyl groups. We shall start our discussion with a general overview of the pathway and then go on to discussion of specific reactions.

12.2
OVERVIEW OF THE CITRIC ACID CYCLE

An important difference between glycolysis and the citric acid cycle is the part of the cell in which these pathways occur. In prokaryotes glycolysis takes place in the cytosol, while the citric acid cycle takes place on the plasma membrane. In eukaryotes glycolysis occurs in the cytosol, while the citric acid cycle takes place in mitochondria. Most of the enzymes of the citric acid cycle are present in the mitochondrial matrix.

A quick review of some aspects of mitochondrial structure is in order here, since we shall want to describe the exact location of each of the components of the citric acid cycle and the electron transport chain. Recall from Chapter 1 that a mitochondrion has an inner and an outer membrane (Figure 12.1). The region enclosed by the inner membrane is called the **matrix,** and there is an **intermembrane space** between the inner and outer membranes. The reactions of the citric acid cycle take place in the matrix, except for the one in which the intermediate electron acceptor is FAD. The enzyme that catalyzes the FAD-linked reaction is an integral part of the inner mitochondrial membrane.

FIGURE 12.1

The structure of a mitochondrion. (a) Scanning electron micrograph showing the internal structure of a mitochondrion (magnified 24,187×). (b) Interpretative drawing of the SEM. (c) Perspective drawing of a mitochondrion. (For an electron micrograph of mitochondrial structure, see Figure 1.4.)

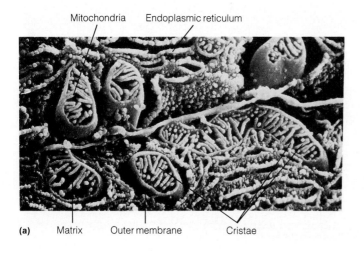

(a) Mitochondria Endoplasmic reticulum Matrix Outer membrane Cristae

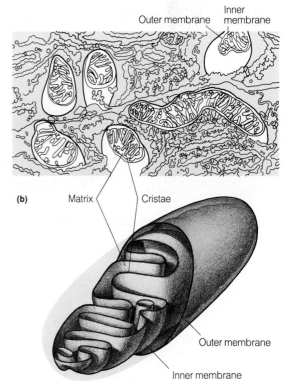

(b) Outer membrane Inner membrane Matrix Cristae

(c) Outer membrane Inner membrane

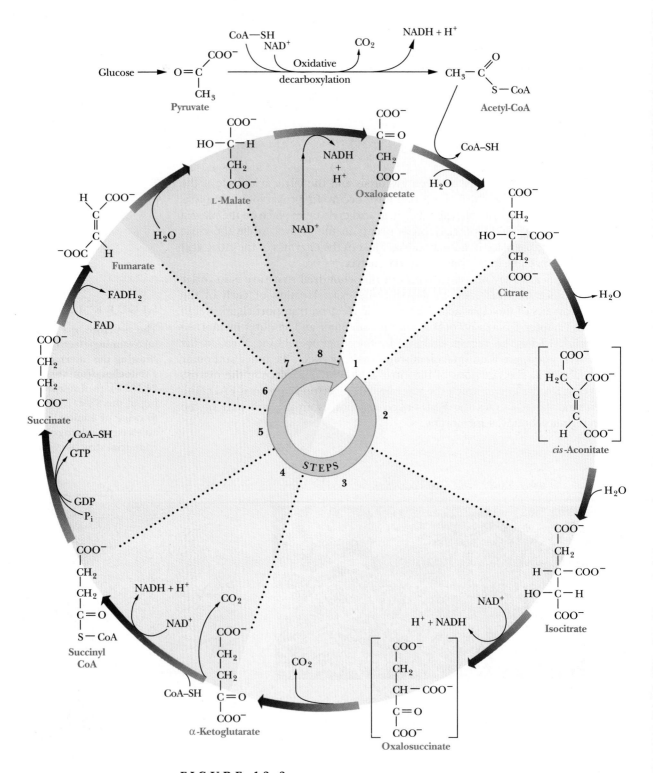

FIGURE 12.2

The citric acid cycle.

The citric acid cycle is shown in schematic form in Figure 12.2. Under aerobic conditions, pyruvate ion produced by glycolysis is oxidized further, with carbon dioxide and water as the final products. First the pyruvate is oxidized to one carbon dioxide molecule and to one acetyl group, which becomes linked to an intermediate, coenzyme A (CoA) (Chapter 9, Section 9.11; see p. 305). The acetyl-CoA enters the citric acid cycle. In the citric acid cycle two more molecules of carbon dioxide are produced for each molecule of acetyl-CoA that enters the cycle, and electrons are lost in the process. The immediate electron acceptor in all cases but one is NAD^+. In the one case where there is another intermediate electron acceptor, it is FAD (flavin adenine dinucleotide), which takes up electrons and hydrogen to produce $FADH_2$, the reduced form of flavin derived from riboflavin (vitamin B_2). The electrons are passed from NADH and $FADH_2$ through several stages of an electron transport chain with a different redox reaction at each step. The final electron acceptor is oxygen, with water as the product.

In the first reaction of the cycle the two-carbon acetyl group condenses with the four-carbon oxaloacetate ion to produce the six-carbon citrate ion. In the next few steps the citrate isomerizes, and then it both loses carbon dioxide and is oxidized. This process, called **oxidative decarboxylation,** produces the five-carbon compound α-ketoglutarate, which again is oxidatively decarboxylated to produce the four-carbon compound succinate. The cycle is completed by regeneration of oxaloacetate from succinate in several steps.

There are eight steps in the citric acid cycle, each catalyzed by a different enzyme. Four of the eight steps are oxidation reactions: those in Steps 3, 4, 6, and 8 (see Figure 12.2). The oxidizing agent is NAD^+ in all except Step 6, in which FAD plays the same role. In Step 5 a molecule of GDP (guanosine diphosphate) is phosphorylated to produce GTP (guanosine triphosphate). This reaction is equivalent to the production of ATP, since the phosphate group is easily transferred to ADP, producing GDP and ATP. GTP differs from ATP only in the substitution of guanine for adenine.

12.3
INDIVIDUAL REACTIONS OF THE CITRIC ACID CYCLE

Conversion of Pyruvate to Acetyl-CoA

An enzyme system called the **pyruvate dehydrogenase complex** is responsible for the conversion of pyruvate to carbon dioxide and the acetyl portion of acetyl-CoA. There is an —SH group at one end of the CoA molecule, which is the point at which the acetyl group is attached. As a result, CoA is frequently shown in equations as CoA-SH. Because CoA is a thiol (the sulfur (thio) analog of an alcohol), acetyl-CoA is a **thioester,** with a sulfur atom replacing an oxygen of the usual carboxylic ester. This difference is important, since thioesters are "high-energy" compounds. In other words, the hydrolysis of thioesters releases enough energy to drive other reactions. An oxidation reaction precedes the transfer of the acetyl group to the CoA.

The whole process involves several enzymes, all of which are part of the pyruvate dehydrogenase complex. The overall reaction

$$\text{Pyruvate} + \text{CoA-SH} + \text{NAD}^+ \longrightarrow \text{acetyl-CoA} + \text{CO}_2 + \text{H}^+ + \text{NADH}$$

is exergonic ($\Delta G^{\circ\prime} = -33.4$ kJ mol^{-1} = -8.0 kcal mol^{-1}), and NADH is recycled for further use.

The overall reaction of the pyruvate dehydrogenase complex

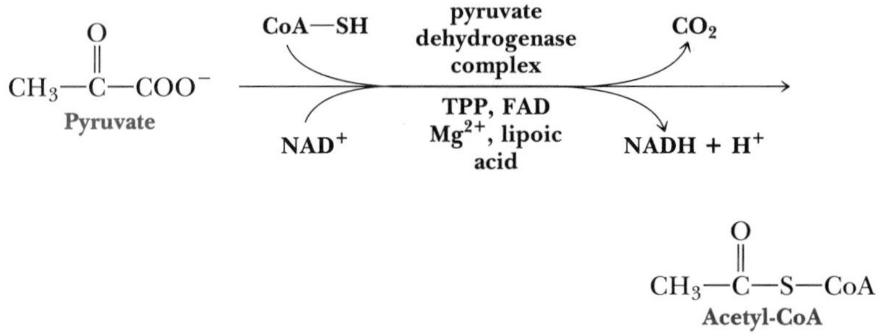

(12.1)

Three enzymes make up the pyruvate dehydrogenase complex. They are **pyruvate dehydrogenase, dihydrolipoyl transacetylase,** and **dihydrolipoyl dehydrogenase.** The reaction takes place in five steps. The last two enzymes catalyze reactions of **lipoic acid,** a compound that has a disulfide group in its oxidized form and two sulfhydryl groups in its reduced form. Lipoic acid differs in one respect from other coenzymes. It is a vitamin, rather than a metabolite of a vitamin, as is the case with many other coenzymes (Table 5.3).

Lipoic acid can act as an oxidizing agent; the reaction involves hydrogen transfer, which frequently accompanies biological oxidation reduction reactions (Chapter 9, Section 9.9). Another reaction of lipoic acid is the formation of a thioester linkage with the acetyl group before it is transferred

The dual role of lipoic acid as hydrogen acceptor (oxidizing agent) and acyl group transfer agent in the pyruvate dehydrogenase reaction

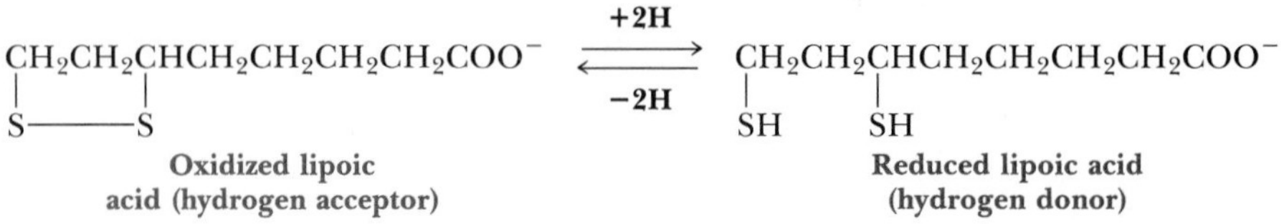

or

(Equation continues on next page)

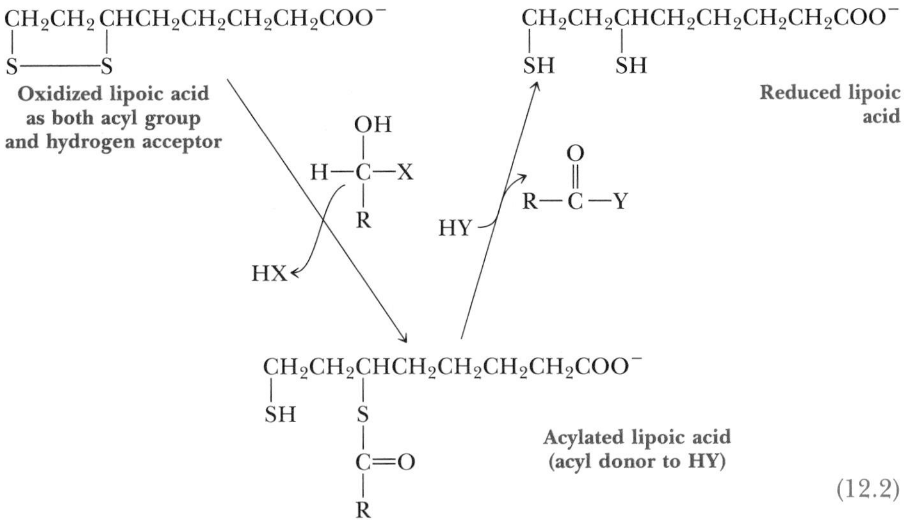

$$CH_2CH_2\;CHCH_2CH_2CH_2CH_2COO^-$$

| | |
| S——S |

**Oxidized lipoic acid
as both acyl group
and hydrogen acceptor**

$$CH_2CH_2CHCH_2CH_2CH_2CH_2COO^-$$

| |
| SH   SH |

**Reduced lipoic
acid**

$$H-\underset{R}{\overset{OH}{C}}-X$$

$$R-\overset{O}{\underset{\|}{C}}-Y$$

HY

HX

$$CH_2CH_2CHCH_2CH_2CH_2CH_2COO^-$$

| |
| SH   S |
	C=O
	R

**Acylated lipoic acid
(acyl donor to HY)**

(12.2)

to the acetyl-CoA. Lipoic acid can act simply as an oxidizing agent, or it can simultaneously take part in two reactions, a redox reaction and the shift of an acetyl group by transesterification.

The first step in the reaction sequence that converts pyruvate to carbon dioxide and acetyl-CoA is catalyzed by pyruvate dehydrogenase. This enzyme requires thiamine pyrophosphate (TPP) (a metabolite of vitamin B_1 [thiamine]) as a coenzyme. The coenzyme is not covalently bonded to the enzyme; they are held together by noncovalent interactions. Mg^{2+} is also required. We saw the action of TPP as a coenzyme in the conversion of pyruvate to acetaldehyde, catalyzed by pyruvate decarboxylase (Section 11.4). Since the mechanism is essentially the same here, we shall not discuss it in detail. In the pyruvate dehydrogenase reaction an α-keto acid, pyruvate, loses carbon dioxide; the remaining two-carbon unit becomes covalently bonded to TPP. The form in which the two-carbon unit is attached to the thiazole ring is a hydroxyethyl group.

**The five individual steps and the overall reaction
catalyzed by the pyruvate dehydrogenase complex**

E_{PDH} = pyruvate dehydrogenase E_{LDH} = dihydrolipoyl dehydrogenase
E_{TA} = dihydrolipoyl transacetylase   TPP = thiamine pyrophosphate

Step 1 $CH_3\overset{O}{\underset{\|}{C}}COO^- + E_{PDH} \sim TPP \xrightarrow{Mg^{2+}} E_{PDH} \sim TPP-\overset{OH}{\underset{|}{C}}HCH_3 + CO_2$

Pyruvate

**Noncovalent
linkage**

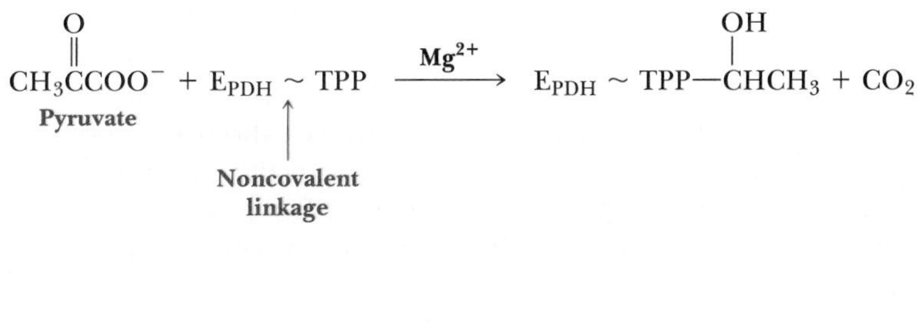

(Equation continues on next page)

Step 2 $E_{PDH} \sim TPP-\overset{\displaystyle OH}{\underset{\displaystyle |}{C}}HCH_3 + E_{TA}-NH-\overset{\displaystyle O}{\overset{\displaystyle \|}{C}}(CH_2)_4\,CH \begin{array}{c} CH_2 \\ \diagdown \\ S\!-\!-\!S \end{array} CH_2 \longrightarrow$

Covalent linkage between
lysine residue and lipoic acid

$E_{PDH} \sim TPP + E_{TA}-NH-\overset{\displaystyle O}{\overset{\displaystyle \|}{C}}(CH_2)_4\,CH \begin{array}{c} CH_2 \\ \\ \underset{\displaystyle \underset{S-\overset{O}{\overset{\|}{C}}CH_3}{|}}{CH_2SH} \end{array}$ Activated acyl
group as
thioester

Step 3 $E_{TA}-NH-\overset{\displaystyle O}{\overset{\displaystyle \|}{C}}(CH_2)_4\,CH \begin{array}{c} CH_2 \\ \\ \underset{\displaystyle \underset{S-\overset{O}{\overset{\|}{C}}CH_3}{|}}{CH_2SH} \end{array} + CoA-SH \longrightarrow$

$E_{TA}-NH-\overset{\displaystyle O}{\overset{\displaystyle \|}{C}}(CH_2)_4\,CH \begin{array}{c} CH_2 \\ \\ CH_2 \\ | \\ SH \end{array} + CH_3\overset{\displaystyle O}{\overset{\displaystyle \|}{C}}-S-CoA$
$\qquad\qquad\qquad\quad\underset{SH}{|}$ Acetyl-CoA

Step 4 $E_{TA}-NH-\overset{\displaystyle O}{\overset{\displaystyle \|}{C}}(CH_2)_4\,CH \begin{array}{c} CH_2 \\ \\ CH_2 \\ | \\ SH \end{array} + E_{LDH} \sim FAD \longrightarrow$
$\qquad\qquad\qquad\qquad\quad\underset{SH}{|}$

$E_{TA}-NH-\overset{\displaystyle O}{\overset{\displaystyle \|}{C}}(CH_2)_4\,CH \begin{array}{c} CH_2 \\ \diagdown \\ S\!-\!-\!S \end{array} CH_2 + E_{LDH} \sim FADH_2$

Step 5 $E_{LDH} \sim FADH_2 + NAD^+ \longrightarrow E_{LDH} \sim FAD + NADH$

Overall reaction $CH_3\overset{\displaystyle O}{\overset{\displaystyle \|}{C}}COO^- + NAD^+ + CoA-SH \xrightarrow[\substack{TPP,\ lipoic\ acid, \\ FAD,\ Mg^{2+}}]{\text{3 enzymes}}$
Pyruvate

$CH_3\overset{\displaystyle O}{\overset{\displaystyle \|}{C}}-S-CoA + CO_2 + NADH + H^+$
Acetyl-CoA
$\hfill (12.3)$

The second step of the reaction is catalyzed by dihydrolipoyl transacetylase. This enzyme requires lipoic acid as a coenzyme. The lipoic acid is covalently bonded to the enzyme by an amide bond to the ϵ-amino group of a lysine side chain. The two-carbon hydroxyethyl unit that originally came from pyruvate is transferred from the thiamine pyrophosphate to the lipoic acid. In the process the two-carbon unit is oxidized to produce an acetyl group. The disulfide group of the lipoic acid is the oxidizing agent, and the product of the reaction is a thioester. In other words, the acetyl group is now covalently bonded to the lipoic acid by a thioester linkage (see Equation 12.3).

The third step of the reaction is also catalyzed by dihydrolipoyl transacetylase. A molecule of CoA-SH attacks the thioester linkage, and the acetyl group is transferred to it. The acetyl group remains bound in a thioester linkage, this time as acetyl-CoA rather than esterified to lipoic acid. The reduced form of lipoic acid remains covalently bound to dihydrolipoyl transacetylase (see Equation 12.3). The reaction of pyruvate and CoA-SH has now reached the stage of the products, carbon dioxide and acetyl-CoA, but the lipoic acid coenzyme is in a reduced form. If the transacetylase is to catalyze further reactions, the lipoic acid must be regenerated.

In the fourth step of the overall reaction, the enzyme dihydrolipoyl dehydrogenase reoxidizes the reduced lipoic acid from the sulfhydryl to the disulfide form. The lipoic acid still remains covalently bonded to the transacetylase enzyme. The dehydrogenase also has a coenzyme, FAD (Chapter 9, Section 9.9), which is bound to the enzyme by noncovalent interactions. An electron acceptor is simply another name for an oxidizing agent, and the coenzyme serves this function here. As a result, FAD is reduced to $FADH_2$. In the fifth step of the reaction $FADH_2$ is reoxidized in turn. The oxidizing agent is NAD^+, and NADH is the product along with reoxidized FAD.

The reduction of NAD^+ to NADH accompanies the oxidation of pyruvate to the acetyl group, and the overall equation shows that there has been a transfer of two electrons from pyruvate to NAD^+ (Equation 12.3). The electrons gained by NAD^+ in generating NADH in this step are passed to the electron transport chain (the next step in aerobic metabolism). In the next chapter we shall see that the transfer of electrons from NADH ultimately to oxygen will give rise to three ATP. Two molecules of pyruvate are produced for each molecule of glucose, so that there will eventually be six ATP from each glucose from this step alone.

The reaction leading from pyruvate to acetyl-CoA is a complex one that requires three enzymes, each of which has its own coenzyme. The spatial orientation of the individual enzyme molecules with respect to one another is itself complex. In the enzyme isolated from *E. coli* the arrangement is quite compact so that the various steps of the reaction can be thoroughly coordinated. There is a core of 24 dihydrolipoyl transacetylase molecules. There is some evidence that the 24 polypeptide chains are arranged in eight trimers, with each trimer occupying the corner of a cube; this point is not definitely established, however. There are 12 dimers of pyruvate dehydrogenase, and they occupy the edges of the cube. Finally, six dimers of dihydrolipoyl dehydrogenase lie on the six faces of the cube (Figure 12.3). Note that many levels of structure combine to produce a suitable environment for the conversion of pyruvate to acetyl-CoA. Each of the enzyme molecules in this array has its own tertiary structure, and the array itself has the cubical structure we have just seen.

A compact arrangement such as the one in the pyruvate dehydrogenase multienzyme complex has two great advantages over an arrangement in which the various components are more widely dispersed. First, the various stages of the reaction can take place more efficiently because the reactants and the enzymes are so close to each other. The role of lipoic acid is particularly important here. Recall that the lipoic acid is covalently attached to the transacetylase enzyme that occupies a central position in the complex. The lipoic acid and the lysine side chain to which it is bonded are long enough to act as a "swinging arm," which can move to the site of each of the

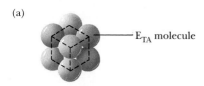

(a) — E_{TA} molecule

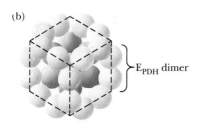

(b) — E_{PDH} dimer

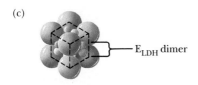

(c) — E_{LDH} dimer

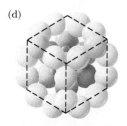

(d)

FIGURE 12.3

The molecular architecture of the pyruvate dehydrogenase complex from *Escherichia coli*. (a) Cubic cluster of 24 transacetylase (E_{TA}) molecules, possibly arranged as eight trimers. (b) Twelve pyruvate dehydrogenase (E_{PDH}) dimers on the edges of the transacetylase core. (c) Six dimers of dihydrolipoyl dehydrogenase (E_{LDH}) on each face of the transacetylase core; three dimers are shown here. (d) The complete pyruvate dehydrogenase aggregate.

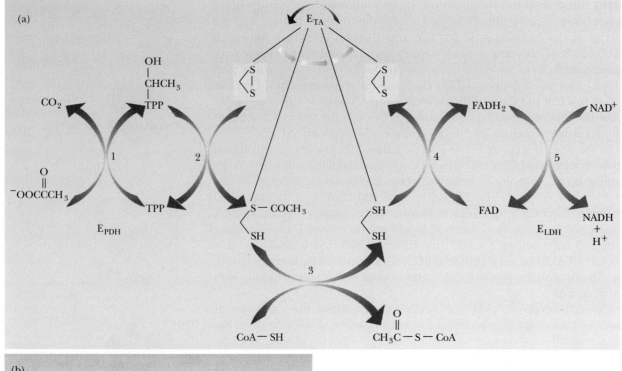

FIGURE 12.4

The role of the "swinging arm" in the reactions of the pyruvate dehydrogenase complex. (a) The numerals refer to the steps in Equation 12.3. The abbreviations for the enzymes are also those used in Equation 12.3. Steps 1 and 2 take place on the E_{PDH} subunit, Step 3 on the E_{TA}, and Steps 4 and 5 on the E_{LDH}. (b) The symbol —S—S refers to the lysyl–lipoamide "swinging arm," which can take different positions in the course of the reaction. Its structure is shown here.

steps of the reaction (Figure 12.4). As a result of the swinging arm action, the lipoic acid can move to the pyruvate dehydrogenase site to accept the two-carbon unit and then transfer it to the active site of the transacetylase. The acetyl group can then be transesterified to CoA-SH from the lipoic acid. Finally, the lipoic acid can swing to the active site of the dehydrogenase so that the sulfhydryl groups can be reoxidized to a disulfide.

A second advantage of a multienzyme complex is that regulatory controls can be applied more efficiently in such a system than with a single enzyme molecule. In the case of the pyruvate dehydrogenase complex, controlling factors are intimately associated with the multienzyme complex itself. The overall reaction is part of a pathway that releases energy. It is not surprising that the enzyme that catalyzes it is inhibited by ATP and NADH, since both compounds are abundant when a cell has a good deal of energy readily available. The end-products of a series of reactions inhibit the first reaction of the series, and the intermediate reactions do not take place when their products are not needed. It is consistent with this picture that the pyruvate dehydrogenase complex is activated by ADP, which is abundant when a cell needs energy. In mammals the actual mechanism by which the inhibition takes place is the phosphorylation of pyruvate dehydrogenase. A phosphate group is covalently bound to the enzyme in a reaction catalyzed

TABLE 12.1 **The Reactions of the Citric Acid Cycle**

STEP	REACTION	ENZYME
1	Acetyl-CoA + oxaloacetate + H_2O $\longrightarrow$ citrate + CoA-SH + H^+	Citrate synthase
2	Citrate $\longrightarrow$ isocitrate	Aconitase
3	Isocitrate + NAD^+ $\longrightarrow$ α-ketoglutarate + NADH + CO_2	Isocitrate dehydrogenase
4	α-Ketoglutarate + NAD^+ + CoA-SH $\longrightarrow$ succinyl-CoA + NADH + CO_2 + H^+	α-Ketoglutarate dehydrogenase
5	Succinyl-CoA + GDP + P_i $\longrightarrow$ succinate + GTP + CoA-SH	Succinyl-CoA synthetase
6	Succinate + FAD $\longrightarrow$ fumarate + $FADH_2$	Succinate dehydrogenase
7	Fumarate + H_2O $\longrightarrow$ L-malate	Fumarase
8	L-Malate + NAD^+ $\longrightarrow$ oxaloacetate + NADH + H^+	Malate dehydrogenase

by the enzyme **pyruvate dehydrogenase kinase.** When the need arises for pyruvate dehydrogenase to be activated, the hydrolysis of the phosphate ester linkage (dephosphorylation) is catalyzed by another enzyme, **phospho- protein phosphatase.** This latter enzyme is itself activated by Ca^{2+}. Both of these enzymes are associated with the intact pyruvate dehydrogenase complex, permitting effective control of the overall reaction from pyruvate to acetyl-CoA.

The Citric Acid Cycle Proper

The reactions of the citric acid cycle proper and the enzymes that catalyze them are listed in Table 12.1. We shall now discuss each of these reactions in turn.

Step 1. Formation of Citrate The first step of the citric acid cycle is the reaction of acetyl-CoA and oxaloacetate to form citrate and CoA-SH. This reaction is called a condensation because a new carbon–carbon bond is formed. The condensation reaction of acetyl-CoA and oxaloacetate to form citryl-CoA takes place in the first stage of the reaction; a basic group on the enzyme (Enz—B:) plays a major role here. The condensation is followed by the hydrolysis of citryl-CoA to give citrate and CoA-SH.

**The condensation of acetyl-CoA and oxaloacetate
to form citrate**

Overall reaction

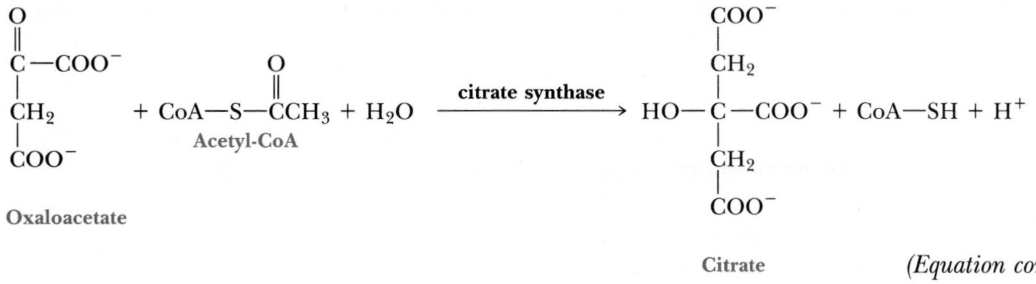

Oxaloacetate

Citrate

(Equation continues on next page)

Reaction mechanism

$$
\text{Enz—B:} \quad \underset{\overset{\displaystyle H}{\underset{\displaystyle H}{|}}}{H-C-}\overset{\displaystyle O}{\overset{\displaystyle \|}{C}}-S-CoA \qquad \text{Acetyl-CoA}
$$

$$\uparrow\downarrow$$

$$
\text{Enz—BH}^+ \qquad {}^-CH_2-\overset{\overset{\displaystyle O}{\displaystyle \|}}{C}-S-CoA
$$

$$
{}^-OOC-CH_2-\overset{\underset{\displaystyle O}{\displaystyle \|}}{C}-COO^-
$$

$$\longrightarrow \quad \underset{\overset{\displaystyle |}{\displaystyle CH_2COO^-}}{HO-\overset{\overset{\displaystyle CH_2\overset{\displaystyle O}{\overset{\displaystyle \|}{C}}-S-CoA}{\displaystyle |}}{C}-COO^-} \qquad \text{Citryl-CoA}$$

$$H^+$$

Oxaloacetate

$$\overset{\displaystyle H_2O}{\underset{\displaystyle CoA-SH}{\diagdown}}$$

$$
\underset{\overset{\displaystyle |}{\displaystyle CH_2COO^-}}{HO-\overset{\overset{\displaystyle CH_2COO^-}{\displaystyle |}}{C}-COO^-} \qquad \text{Citrate} \qquad (12.4)
$$

The reaction is catalyzed by the enzyme **citrate synthase,** originally called "condensing enzyme." It is an exergonic reaction ($\Delta G^{\circ\prime} = -32.8$ kJ mol^{-1} $= -7.8$ kcal mol^{-1}) because the hydrolysis of a thioester releases energy; thioesters are considered "high-energy" compounds.

Citrate synthase is an allosteric enzyme; thus, the inhibitor mechanisms we discussed in Chapter 5, Section 5.9, apply here. Both ATP and NADH inhibit the citrate synthase reaction, as we saw earlier with the pyruvate dehydrogenase reaction. Succinyl-CoA, a substance that appears later in the cycle, is an inhibitor as well. This is another example of the type of control mechanism in which products of a long series of reactions "turn off" the first reaction of the series.

Step 2. Isomerization of Citrate to Isocitrate   The second reaction of the citric acid cycle, the one catalyzed by aconitase, is the isomerization of citrate to isocitrate. The enzyme requires Fe^{2+}. One of the most interesting features of the reaction is that citrate, a symmetrical (achiral) compound, is converted to isocitrate, a chiral compound, a molecule that cannot be superimposed on its mirror image.

It is often possible for a chiral compound to have several different isomers. Isocitrate has four possible isomers, but only one of the four is produced by this reaction. (We shall not discuss nomenclature of the isomers of isocitrate here. See the exercises at the end of this chapter for a question about the other isomers.) Aconitase, the enzyme that catalyzes the conversion of citrate to isocitrate, is able to select one end of the citrate molecule in preference to the other.

PLANT POISONS AND THE CITRIC ACID CYCLE

A possible alternative substrate for citrate synthase is **fluoroacetyl-CoA.** The source of the fluoroacetyl-CoA is fluoroacetate, which is found in the leaves of various types of poisonous plants, including locoweeds. Animals that ingest these plants form fluoroacetyl-CoA, which in turn is converted to fluorocitrate by their citrate synthase. Fluorocitrate in turn is a potent inhibitor of **aconitase,** the enzyme that catalyzes the next reaction of the citric acid cycle. These plants are poisonous because they produce a potent inhibitor of life processes.

The poison "ten eighty" used by sheep ranchers is sodium fluoroacetate. Ranchers who want to protect their sheep from attacks by coyotes put the poison just outside the ranch fence. When the coyotes eat this poison they die. The mechanism of poisoning by "ten eighty" is the same as that by plant poisons.

The formation of fluorocitrate from fluoroacetate

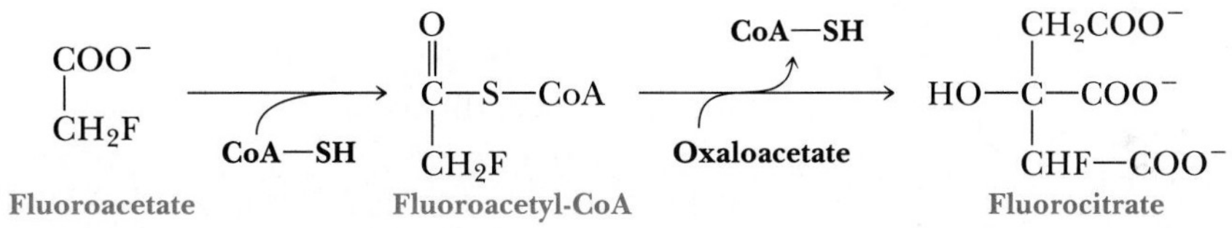

$$\underset{\text{Fluoroacetate}}{\overset{\displaystyle COO^-}{\underset{\displaystyle CH_2F}{|}}} \xrightarrow[\text{CoA—SH}]{} \underset{\text{Fluoroacetyl-CoA}}{\overset{\displaystyle O}{\underset{\displaystyle CH_2F}{\overset{\displaystyle \|}{\underset{|}{C-S-CoA}}}}} \xrightarrow[\text{Oxaloacetate}]{\text{CoA—SH}} \underset{\text{Fluorocitrate}}{\overset{\displaystyle CH_2COO^-}{\underset{\displaystyle CHF-COO^-}{HO-C-COO^-}}}$$

The formation of isocitrate (a chiral compound) from citrate (an achiral compound)

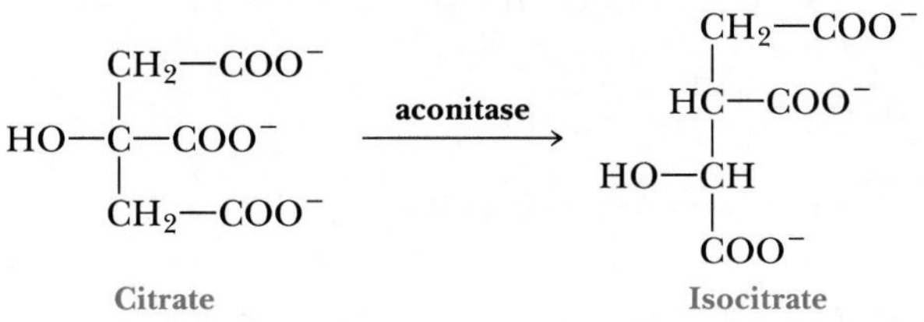

$$\underset{\text{Citrate}}{\overset{\displaystyle CH_2-COO^-}{\underset{\displaystyle CH_2-COO^-}{HO-C-COO^-}}} \xrightarrow{\text{aconitase}} \underset{\text{Isocitrate}}{\overset{\displaystyle CH_2-COO^-}{\underset{\displaystyle COO^-}{\overset{\displaystyle HC-COO^-}{HO-CH}}}} \qquad (12.5)$$

This type of behavior means that the enzyme can bind a symmetrical substrate in an unsymmetrical binding site. In Chapter 5, Section 5.13, we mentioned that this possiblity exists, and here we have an example of it. The enzyme forms an unsymmetrical three-point attachment to the citrate molecule. The reaction proceeds by removal of a water molecule from the citrate to produce *cis*-aconitate, and then water is added back to the *cis*-aconitate to give isocitrate.

**cis-aconitate as an intermediate in the conversion
of citrate to isocitrate
(B is a basic group on the enzyme)**

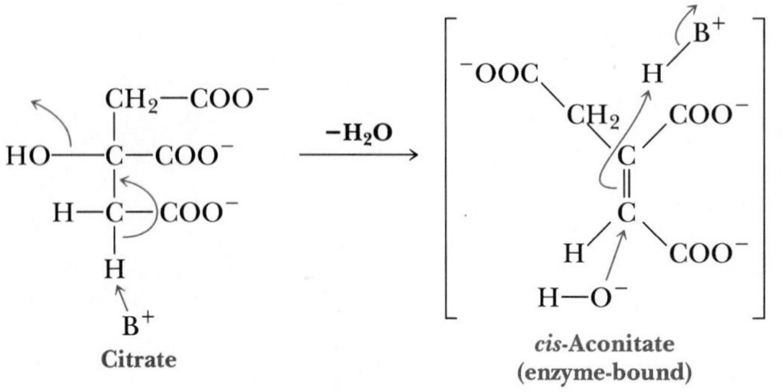

Citrate   *cis*-Aconitate
(enzyme-bound)   Isocitrate

(12.6)

The intermediate, *cis*-aconitate, remains bound to the enzyme during the course of the reaction. There is some evidence that the citrate is complexed to the Fe(II) in the active site of the enzyme in such a way that the citrate curls back on itself in a nearly circular conformation. Several authors have been unable to resist the temptation to call this situation the "ferrous wheel."

Step 3. Formation of α-Ketoglutarate and CO_2 The third step in the citric acid cycle is the oxidative decarboxylation of isocitrate to α-ketoglutarate and carbon dioxide. This reaction is the first of two oxidative decarboxylations of the citric acid cycle; the enzyme that catalyzes it is **isocitrate dehydrogenase.** The reaction takes place in two steps. First, isocitrate is oxidized to oxalosuccinate, which remains bound to the enzyme. Then oxalosuccinate is decarboxylated, and the carbon dioxide and α-ketoglutarate are released. This is the first of the reactions in which NADH is produced. One molecule of NADH is produced from NAD^+ at this state by the loss of two electrons in the oxidation. As we saw in our discussion of the pyruvate dehydrogenase complex, each NADH produced will lead to the production of three ATP in later stages of aerobic metabolism. Recall also that there will be two NADH, equivalent to six ATP for each original molecule of glucose.

The oxidative decarboxylation of isocitrate to α-ketoglutarate

Isocitrate   Oxalosuccinate
(enzyme bound) α-Ketoglutarate

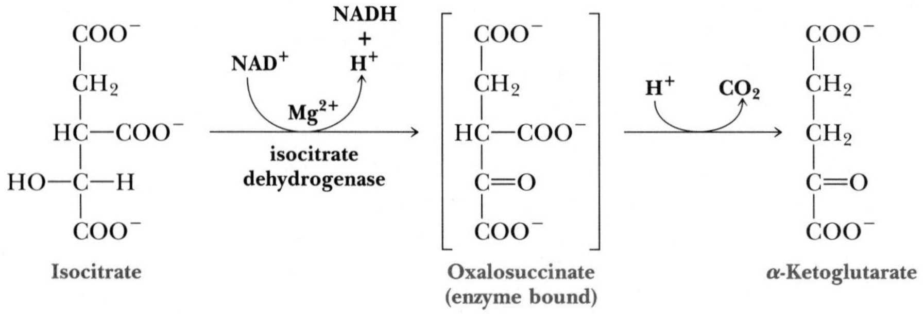

(12.7)

This reaction is a control site in the cycle. The enzyme is an oligomer, with a molecular weight on the order of 10^5. The reaction is controlled by the allosteric mechanism discussed in Chapter 5. The enzyme is inhibited by ATP and NADH and is activated by ADP and NAD^+. This pattern of inhibition by ATP and NADH and activation by ADP and NAD^+ is frequently encountered in catabolic reactions.

Step 4. Formation of Succinyl-CoA and CO_2 The second oxidative decarboxylation takes place in Step 4 of the citric acid cycle, in which carbon dioxide and succinyl-CoA are formed from α-ketoglutarate and CoA.

The conversion of α-ketoglutarate to succinyl-CoA

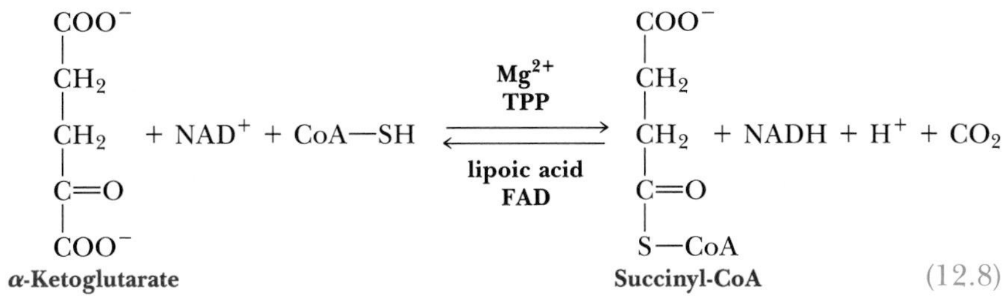

$$
\begin{array}{l}
\text{COO}^- \\
|\\
\text{CH}_2 \\
|\\
\text{CH}_2 \quad + \text{NAD}^+ + \text{CoA—SH} \\
|\\
\text{C}=\text{O} \\
|\\
\text{COO}^-
\end{array}
\quad
\xrightarrow[\substack{\text{lipoic acid}\\ \text{FAD}}]{\substack{\text{Mg}^{2+}\\ \text{TPP}}}
\quad
\begin{array}{l}
\text{COO}^- \\
|\\
\text{CH}_2 \\
|\\
\text{CH}_2 \quad + \text{NADH} + \text{H}^+ + \text{CO}_2 \\
|\\
\text{C}=\text{O} \\
|\\
\text{S—CoA}
\end{array}
$$

α-Ketoglutarate   Succinyl-CoA   (12.8)

This reaction is similar to the one in which acetyl-CoA is formed from pyruvate, with NADH produced from NAD^+. Once again each NADH will eventually give rise to three ATP, with six ATP from each original molecule of glucose.

The reaction occurs in several stages and is catalyzed by an enzyme system called the **α-ketoglutarate dehydrogenase complex,** which is very similar to the pyruvate dehydrogenase complex. Each of these multienzyme systems consists of three enzymes. The reaction takes place in several steps, and there is again a requirement for thiamine pyrophosphate (TPP), FAD, lipoic acid, and Mg^{2+}. This reaction is highly exergonic ($\Delta G^{\circ\prime} = -33.4$ kJ $mol^{-1} = -8.0$ kcal mol^{-1}), as is the one catalyzed by pyruvate dehydrogenase.

At this point, two molecules of CO_2 have been produced by the citric acid cycle. Removal of the CO_2 makes the citric acid cycle irreversible *in vivo,* although *in vitro* each separate reaction is reversible. One might suspect that the two molecules of CO_2 arise from the two carbon atoms of acetyl-CoA. Labeling studies have shown that this is not the case, but a full discussion of this point is beyond the scope of this text. We should also mention that the α-ketoglutarate dehydrogenase complex reaction is the third one in which we have encountered an enzyme that requires TPP. In all three cases — the pyruvate dehydrogenase complex, the α-ketoglutarate dehydrogenase complex, and pyruvate decarboxylase (Section 11.4) — the reaction involves transfer of an activated two-carbon group. (Refer to the end-of-chapter exercises for a treatment of the nature of the activated group in each reaction.)

Step 5. Formation of Succinate In the next step of the cycle, the thioester bond of succinyl-CoA is hydrolyzed to produce succinate and CoA-SH; an accompanying reaction is the phosphorylation of GDP to GTP. The whole reaction is catalyzed by the enzyme **succinyl-CoA synthetase.** In the reaction mechanism, a phosphate group covalently bonded to the enzyme is directly transferred to the GDP. The phosphorylation of GDP to GTP is endergonic, as is the corresponding ADP–ATP reaction ($\Delta G^{\circ\prime}$ = 30.5 kJ mol^{-1} = 7.3 kcal mol^{-1}).

The conversion of succinyl-CoA to succinate

$$
\begin{array}{c}
\text{COO}^- \\
| \\
\text{CH}_2 \\
| \\
\text{CH}_2 \\
| \\
\text{C}=\text{O} \\
| \\
\text{S}-\text{CoA} \\
\text{Succinyl-CoA}
\end{array}
\quad + \text{ GDP} + \text{P}_i
\quad \xrightarrow[\text{CoA synthetase}]{\text{succinyl-}}
\quad
\begin{array}{c}
\text{COO}^- \\
| \\
\text{CH}_2 \\
| \\
\text{CH}_2 \\
| \\
\text{COO}^- \\
\text{Succinate}
\end{array}
\quad + \text{ GTP} + \text{CoA}-\text{SH}
\qquad (12.9)
$$

The energy required for the phosphorylation of GDP to GTP is provided by the hydrolysis of succinyl-CoA to produce succinate and CoA. The free energy of hydrolysis ($\Delta G^{\circ\prime}$) of succinyl-CoA is -33.4 kJ mol^{-1} (-8.0 kcal mol^{-1}). The overall reaction is slightly exergonic ($\Delta G^{\circ\prime}$ = -3.3 kJ mol^{-1} = -0.8 kcal mol^{-1}) and, as a result, does not contribute greatly to the overall production of energy by the mitochondrion.

The enzyme **nucleoside diphosphosphate kinase** catalyzes the transfer of a phosphate group from GTP to ADP to give GDP and ATP.

$$\text{GTP} + \text{ADP} \longrightarrow \text{GDP} + \text{ATP}$$

This reaction step is called substrate-level phosphorylation to distinguish it from the type of reaction for production of ATP that is coupled to the electron transport chain. (We have not really been in a position to make this distinction until now, but we should keep it in mind henceforth.) The production of ATP in this reaction is the only place in the citric acid cycle in which chemical energy in the form of ATP is made available to the cell. Except for this reaction, the generation of ATP characteristic of aerobic metabolism is associated with the electron transport chain, the subject of the next chapter. As many as 38 molecules of ATP can be obtained from the oxidation of a single molecule of glucose by the combination of anaerobic and aerobic oxidation, compared to only two molecules of ATP produced by anaerobic glycolysis alone. The combined reactions that occur in mitochondria are of great importance to aerobic organisms.

Steps 6–8

In Steps 6 through 8 of the citric acid cycle the four-carbon succinate ion is converted to oxaloacetate ion to complete the cycle.

**The final stages of the citric acid cycle
succinate is converted to oxaloacetate
(conversion of a methylene group to a carbonyl group)**

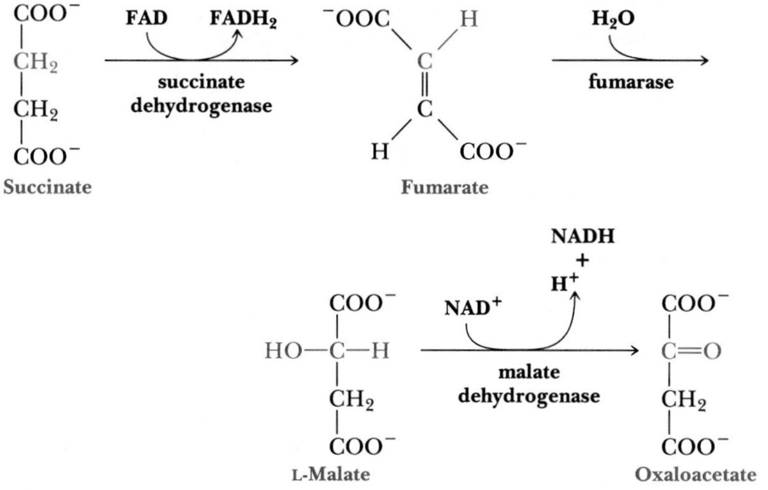

$$\text{(12.10)}$$

Step 6. Formation of Fumarate In Step 6, succinate is oxidized to fumarate, a reaction that is catalyzed by the enzyme **succinate dehydrogenase.** This enzyme is bound to the inner mitochondrial membrane. We shall have much more to say about the enzymes bound to the inner mitochondrial membrane in the next chapter. The other individual enzymes of the citric acid cycle are in the mitochondrial matrix. The electron acceptor, which is FAD rather than NAD^+, is covalently bonded to the enzyme; succinate dehydrogenase is called a flavoprotein because of the presence of FAD with its flavin moiety. In the succinate dehydrogenase reaction, FAD is reduced to $FADH_2$ and succinate is oxidized to fumarate.

The conversion of succinate to fumarate

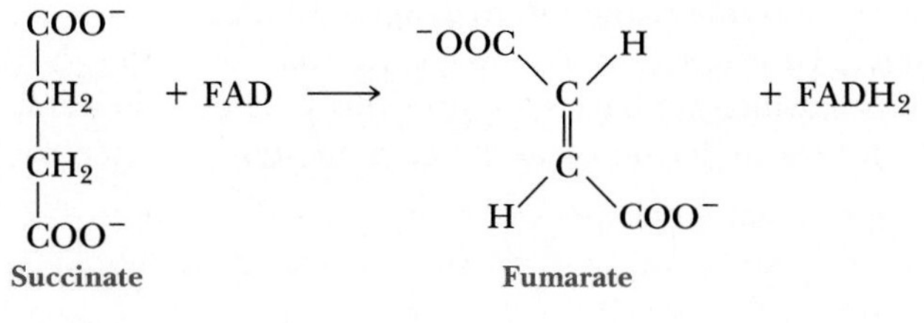

$$\text{(12.11)}$$

The overall reaction is

$$\text{Succinate} + \text{E—FAD} \longrightarrow \text{fumarate} + \text{E—FADH}_2$$

The E—FAD and E—$FADH_2$ in the equation indicate that the electron acceptor is covalently bonded to the enzyme. The $FADH_2$ group also passes electrons on to the electron transport chain and eventually to oxygen but will give rise to only two, rather than three, ATP, as is the case with NADH. As a result, each original glucose will give rise to four, not six, ATP from this step.

Succinate dehydrogenase contains four iron atoms, as well as sulfur atoms that are bonded to the iron atoms. The enzyme does not contain a heme group; it is also referred to as a **nonheme iron protein** or an **iron–sulfur protein.** This enzyme exhibits stereospecificity. Succinate itself is an achiral compound; when it is formed, the asymmetry of earlier compounds in the cycle is lost. The introduction of the double bond introduces another possibility for different isomers. The double bond produced by this reaction could have either the *trans* configuration (fumarate) or the *cis* configuration (maleate), but only the *trans* configuration is actually produced. This reaction is not a regulatory control point of the citric acid cycle.

Step 7. Formation of L-Malate   In reaction 7, which is catalyzed by the enzyme **fumarase,** water is added across the double bond of fumarate in a hydration reaction to give malate. Again there is stereospecificity in the reaction. Malate has two enantiomers, L- and D-malate, but only L-malate is produced. This reaction is another one in which no control of the cycle is apparent.

The conversion of fumarate to L-malate

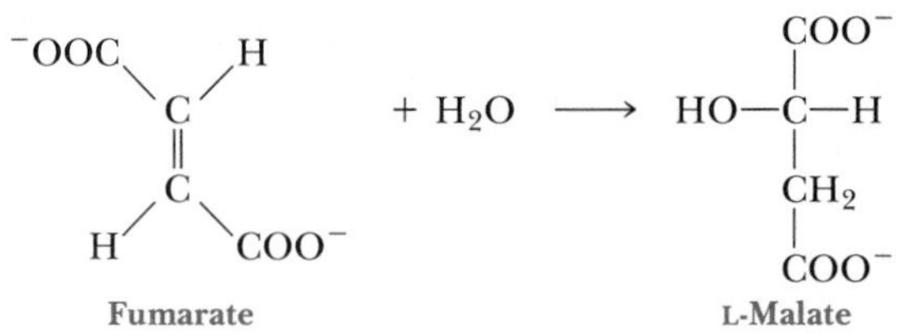

$$\text{Fumarate} + H_2O \longrightarrow \text{L-Malate} \qquad (12.12)$$

Step 8. Regeneration of Oxaloacetate   In Step 8, malate is oxidized to oxaloacetate, and another molecule of NAD^+ is reduced to NADH. This reaction is catalyzed by the enzyme **malate dehydrogenase** and is not a control point for the cycle. This reaction is similar to that of other NADH-linked dehydrogenases, which we have seen previously. The oxaloacetate can then react with another molecule of acetyl-CoA to start another round of the cycle.

The conversion of L-malate to oxaloacetate

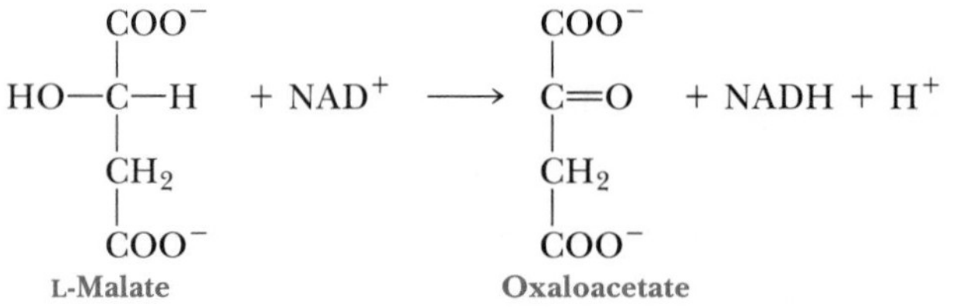

$$\text{L-Malate} + NAD^+ \longrightarrow \text{Oxaloacetate} + NADH + H^+ \qquad (12.13)$$

The oxidation of pyruvate by the pyruvate dehydrogenase complex and the citric acid cycle results in the production of three molecules of CO_2. As a result of these oxidation reactions, one molecule of GDP is phosphorylated to GTP, one molecule of FAD is reduced to $FADH_2$, and four molecules of NAD^+ are reduced to NADH. Of the four molecules of NADH produced, three come from the citric acid cycle and one from the reaction of the pyruvate dehydrogenase complex. The overall stoichiometry of the oxidation reactions is the sum of the pyruvate dehydrogenase reaction and the citric acid cycle. Note that there is no production of ATP from ADP *directly* from the citric acid cycle, but many more ATP will arise from reoxidation of NADH and $FADH_2$.

Pyruvate dehydrogenase complex:

$$\text{Pyruvate} + \text{CoA-SH} + NAD^+ \longrightarrow \text{acetyl-CoA} + \text{NADH} + CO_2 + H^+$$

Citric acid cycle:

$$\text{Acetyl-CoA} + 3\ NAD^+ + FAD + GDP + P_i + 2\ H_2O \longrightarrow$$
$$2\ CO_2 + \text{CoA-SH} + 3\ \text{NADH} + 3\ H^+ + FADH_2 + GTP$$

Overall reaction:

$$\text{Pyruvate} + 4\ NAD^+ + FAD + GDP + P_i + 2\ H_2O \longrightarrow$$
$$3\ CO_2 + 4\ \text{NADH} + FADH_2 + GTP + 4\ H^+$$

Eventual ATP production per pyruvate:

$$4\ \text{NADH} \longrightarrow 12\ \text{ATP}$$
$$1\ FADH_2 \longrightarrow 2\ \text{ATP}$$
$$1\ \text{GTP} \longrightarrow 1\ \text{ATP}$$

Total: 15 ATP per pyruvate or 30 ATP per glucose

There were also 2 ATP produced per glucose in glycolysis and 2 NADH, which will give rise to another 6 ATP (8 more ATP). In the next chapter we shall say more about the subject of ATP production from the complete oxidation of glucose.

At this point we would do well to recapitulate what we have said about the citric acid cycle. This is done in Figure 12.5. Note that Figure 12.5 differs from our earlier representation of the citric acid cycle, Figure 12.2, in several important respects. We have added the names of the enzymes that catalyze the steps, we have pointed out the three places where carbon dioxide is evolved, we have indicated the phosphorylation of GDP to GTP, and we have pointed out the involvement of the coenzymes of electron transfer. The reoxidation of these coenzymes, the intermediate electron carriers NADH and $FADH_2$, will occupy our attention in the next chapter. There we shall see how they eventually pass electrons to oxygen, the ultimate electron acceptor.

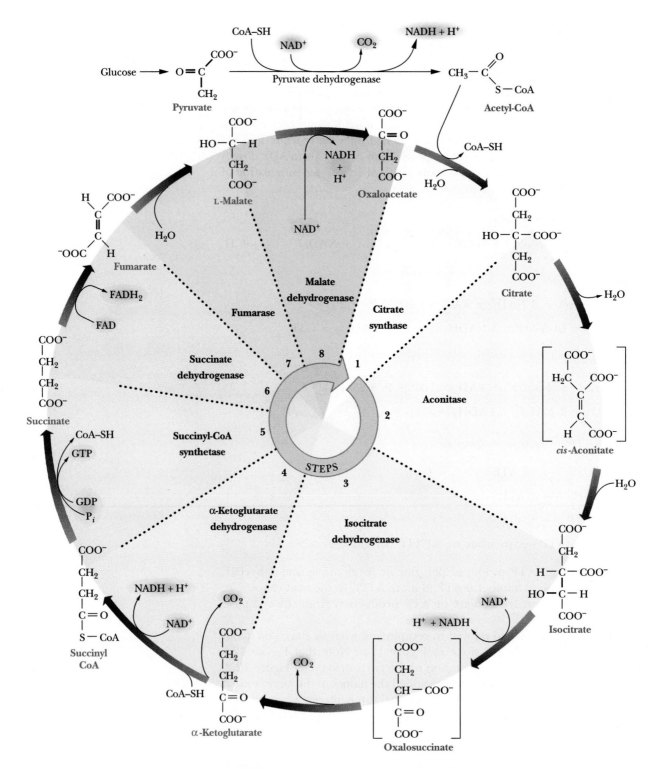

FIGURE 12.5

A recapitulation of the citric acid cycle. Note the names of the enzymes. The loss of CO_2 is indicated, as is the phosphorylation of GDP to GTP. The production of NADH and $FADH_2$ is also indicated.

12.4
ENERGETICS AND CONTROL
OF THE CITRIC ACID CYCLE

The reaction of pyruvate to acetyl-CoA is exergonic, as we have seen ($\Delta G^{\circ\prime} = -33.4$ kJ mol^{-1} = -8.0 kcal mol^{-1}). The citric acid cycle itself is also exergonic ($\Delta G^{\circ\prime} = -44.3$ kJ mol^{-1} = -10.6 kcal mol^{-1}), and you will be asked in Exercise 17 to confirm this point. The standard free energy changes for the individual reactions are listed in Table 12.2. Of the individual reactions of the cycle, only one is strongly endergonic, the oxidation of malate to oxaloacetate ($\Delta G^{\circ\prime} = +29.2$ kJ mol^{-1} = $+7.0$ kcal mol^{-1}). This endergonic reaction is, however, coupled to one of the strongly exergonic reactions of the cycle, the condensation of acetyl-CoA and oxaloacetate to produce citrate and acetyl-CoA ($\Delta G^{\circ\prime} = -33.4$ kJ mol^{-1} = -8.0 kcal mol^{-1}). In addition to the energy released by the oxidation reactions, there is more release of energy to come in the electron transport chain. When the four NADH and single FADH$_2$ produced by the pyruvate dehydrogenase complex and citric acid cycle are reoxidized by the electron transport chain, considerable quantities of ATP are produced.

Control of the citric acid cycle is exercised at three points; that is, there are three enzymes within the citric acid cycle that play a regulatory role (Figure 12.6). There is also a control point outside the cycle itself, namely the reaction in which pyruvate is oxidatively decarboxylated to produce the acetyl-CoA needed for the first reaction of the citric acid cycle proper. As we have already seen, the pyruvate dehydrogenase complex is inhibited by ATP and NADH. (There is also product inhibition by acetyl-CoA in this reaction.)

Within the citric acid cycle itself the three control points are the reactions catalyzed by citrate synthase, isocitrate dehydrogenase, and the α-ketoglutarate dehydrogenase complex. We have already mentioned that the first reaction of the cycle is one in which regulatory control appears, as is to be expected in the first reaction of any pathway. Citrate synthase is an allosteric enzyme inhibited by ATP, NADH, and succinyl-CoA.

T A B L E 1 2 . 2 The Energetics of Conversion of Pyruvate to CO$_2$

STEP	REACTION	$\Delta G^{\circ\prime}$ kJ mol^{-1}	$\Delta G^{\circ\prime}$ kcal mol^{-1}
	Pyruvate + CoA-SH + NAD$^+$ $\longrightarrow$ acetyl-CoA + NADH + CO$_2$ + H$^+$	-33.4	-8.0
1	Acetyl-CoA + oxaloacetate + H$_2$O $\longrightarrow$ citrate + CoA-SH + H$^+$	-32.2	-7.7
2	Citrate $\longrightarrow$ isocitrate	$+6.3$	$+1.5$
3	Isocitrate + NAD$^+$ $\longrightarrow$ α-ketoglutarate + NADH + CO$_2$	-7.1	-1.7
4	α-Ketoglutarate + NAD$^+$ + CoA-SH $\longrightarrow$ succinyl-CoA + NADH + CO$_2$ + H$^+$	-33.4	-8.0
5	Succinyl-CoA + GDP + P$_i$ $\longrightarrow$ succinate + GTP + CoA-SH	-3.3	-0.8
6	Succinate + FAD $\longrightarrow$ fumarate + FADH$_2$	≈ 0	≈ 0
7	Fumarate + H$_2$O $\longrightarrow$ L-malate	-3.8	-0.9
8	L-Malate + NAD$^+$ $\longrightarrow$ oxaloacetate + NADH + H$^+$	$+29.2$	$+7.0$
Overall	Pyruvate + 4 NAD$^+$ + FAD + GDP + P$_i$ + 2 H$_2$O $\longrightarrow$ 3 CO$_2$ + 4 NADH + FADH$_2$ + GTP + 4 H$^+$	-77.7	-18.6

FIGURE 12.6

Control points in the conversion of pyruvate to acetyl-CoA and in the citric acid cycle.

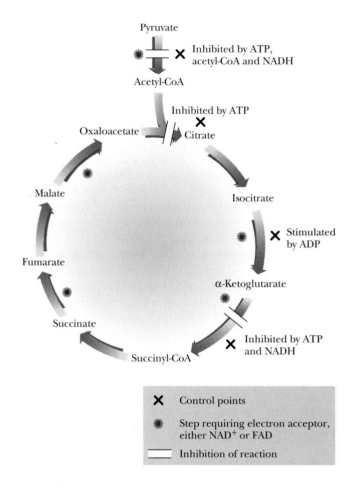

The second regulatory site is the isocitrate dehydrogenase reaction. In this case, ADP is an allosteric activator of the enzyme. We have called attention to the recurring pattern in which ATP and NADH inhibit enzymes of the pathway, and ADP and NAD^+ activate these enzymes.

The α-ketoglutarate dehydrogenase complex is the third regulatory site. As before, ATP and NADH are inhibitors and ADP and NAD^+ are activators. This recurring theme in metabolism reflects the way in which a cell can adjust to an active state or to a resting state.

When a cell is metabolically active it uses ATP and NADH at a great rate, producing large amounts of ADP and NAD^+ (Table 12.3). In other words, when the ratio ATP/ADP is low, the cell is using energy and needs to release more energy from stored nutrients. A low $NADH/NAD^+$ ratio is also characteristic of an active metabolic state. On the other hand, a resting cell has fairly high levels of ATP and NADH. The ratios of ATP/ADP and $NADH/NAD^+$ are also high in resting cells, which do not need to maintain a high level of oxidation to produce energy.

When cells have low energy requirements (they have a high "energy charge") with high ATP/ADP and $NADH/NAD^+$ ratios, the presence of so much ATP and NADH serves as a signal to "shut down" the enzymes responsible for oxidative reactions. When cells have a low energy charge, characterized by low ATP/ADP and $NADH/NAD^+$ ratios, the need to release more energy and to generate more ATP serves as a signal to "turn

TABLE 12.3 **Relationship Between the Metabolic State of a Cell and the ATP/ADP and NADH/NAD$^+$ Ratios**

Cells in a resting metabolic state
Need and use comparatively little energy
High ATP, low ADP levels imply high (ATP/ADP)
High NADH, low NAD$^+$ levels imply high (NADH/NAD$^+$)

Cells in a highly active metabolic state
Need and use more energy than resting cells
Low ATP, high ADP levels imply low (ATP/ADP)
Low NADH, high NAD$^+$ levels imply low (NADH/NAD$^+$)

on" the oxidative enzymes. This relationship of energy requirements to enzyme activity is the basis for the overall regulatory mechanism exerted at a few key control points in metabolic pathways.

12.5
THE GLYOXYLATE CYCLE: A RELATED PATHWAY

In plants and in some bacteria, but not in animals, acetyl-CoA can serve as the starting material for the biosynthesis of carbohydrates. Animals can convert carbohydrates to fats, but not fats to carbohydrates. (Acetyl-CoA is produced in the catabolism of fatty acids.) Two enzymes are responsible for the ability of plants and bacteria to produce glucose from fatty acids. **Isocitrate lyase** cleaves isocitrate, producing glyoxylate and succinate. **Malate synthase** catalyzes the reaction of glyoxylate with acetyl-CoA to produce malate.

The unique reactions of the glyoxylate cycle

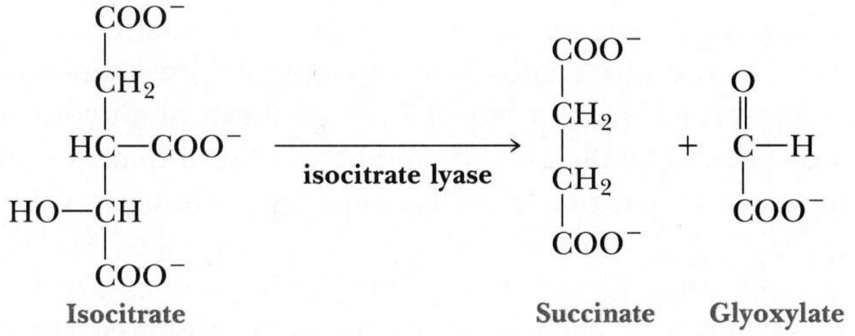

The conversion of isocitrate to glyoxylate and succinate

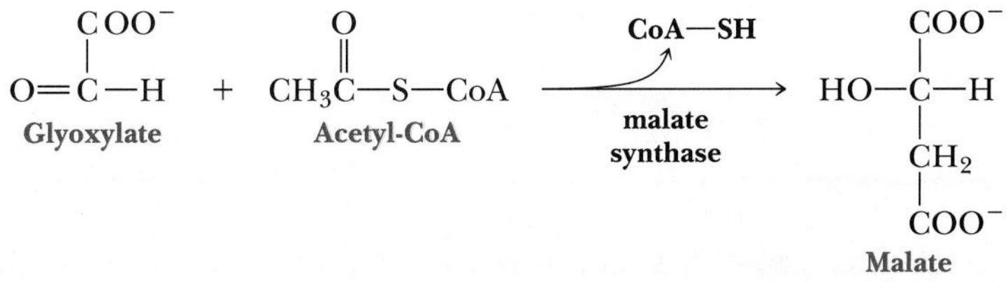

(12.14)

The reaction of glyoxylate with acetyl-CoA to produce malate

FIGURE 12.7

The glyoxylate cycle. This pathway results in the net conversion of two acetyl-CoA to oxaloacetate. The unique reactions of the glyoxylate cycle are shown in green, and the main features of the citric acid cycle are shown in blue.

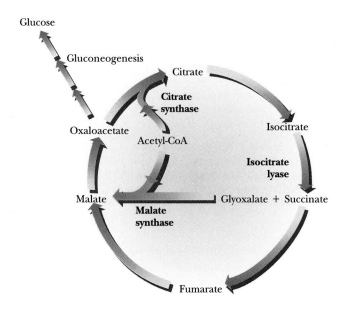

These two reactions in succession bypass the two oxidative decarboxylation steps of the citric acid cycle. The net result is an alternative pathway, the **glyoxylate cycle** (Figure 12.7). Two molecules of acetyl-CoA enter the glyoxylate cycle; they give rise to one molecule of malate and eventually to one molecule of oxaloacetate. Two 2-carbon units (the acetyl groups of acetyl-CoA) give rise to a 4-carbon unit (malate), which is then converted to oxaloacetate (also a 4-carbon compound). Glucose can then be produced from oxaloacetate by gluconeogenesis.

Specialized organelles in plants, called **glyoxysomes,** are the sites of the glyoxylate cycle. This pathway is particularly important in germinating seeds. The fatty acids stored in the seeds are broken down for energy during germination. First the fatty acids give rise to acetyl-CoA, which can enter the citric acid cycle and go on to release energy in the ways we have already seen. The citric acid cycle and the glyoxylate cycle can operate simultaneously. Acetyl-CoA also serves as the starting point for the synthesis of glucose and any other compounds needed by the growing seedling. (Recall that carbohydrates play an important structural, as well as energy-producing, role in plants.)

The glyoxylate cycle also occurs in bacteria. This point is far from surprising, since many types of bacteria can live on very limited carbon sources. They have metabolic pathways that can produce all the biomolecules they need from quite simple molecules. The glyoxylate cycle is one example of how bacteria manage this feat.

12.6
A FINAL NOTE

The citric acid cycle is considered part of aerobic metabolism, but we have not encountered any reactions in this chapter in which oxygen takes part.

The reactions of the citric acid cycle are intimately related to those of electron transport and oxidative phosphorylation, which do eventually lead to oxygen. The citric acid cycle provides a vital link between the chemical energy of nutrients and the chemical energy of ATP. Many molecules of ATP can be generated as a result of coupling to oxygen, and we shall see that the number depends on the NADH and $FADH_2$ generated in the citric acid cycle.

S U M M A R Y

The citric acid cycle plays a central role in metabolism. It is the first part of aerobic metabolism; it is also amphibolic (both catabolic and anabolic). Unlike glycolysis, which takes place in the cytosol, the citric acid cycle occurs in mitochondria. Most of the enzymes of the citric acid cycle are in the mitochondrial matrix. (Succinate dehydrogenase is localized in the inner mitochondrial membrane.)

Pyruvate produced by glycolysis is transformed by oxidative decarboxylation into acetyl-CoA in the presence of coenzyme A. Acetyl-CoA then enters the citric acid cycle by reacting with oxaloacetate to produce citrate. The reactions of the citric acid cycle include two other oxidative decarboxylations, which transform the six-carbon compound citrate into the four-carbon compound succinate. The cycle is completed by regeneration of oxaloacetate from succinate in a multistep process that includes two other oxidation reactions. The overall reaction, starting with pyruvate, is

$$\text{Pyruvate} + 4\ NAD^+ + FAD + GDP + P_i + 2\ H_2O \longrightarrow$$
$$3\ CO_2 + 4\ NADH + FADH_2 + GTP + 4\ H^+$$

NAD^+ and FAD are the electron acceptors in the oxidation reactions. The cycle is strongly exergonic. Control of the citric acid cycle is exercised at three points. There is also a control point outside the cycle, the reaction in which pyruvate is oxidatively decarboxylated to produce acetyl-CoA. Within the citric acid cycle the three control points are the reactions catalyzed by citrate synthase, isocitrate dehydrogenase, and the α-ketoglutarate dehydrogenase complex. In general, ATP and NADH are inhibitors and ADP and NAD^+ are activators of the enzymes at the control points.

In plants and bacteria there is a pathway related to the citric acid cycle, the glyoxylate cycle. The two oxidative decarboxylations of the citric acid cycle are bypassed. This pathway plays a role in the ability of plants to convert acetyl-CoA to carbohydrates, a process that does not occur in animals.

E X E R C I S E S

1. We have seen one of the four possible isomers of isocitrate, the one produced in the aconitase reaction. Draw the configurations of the other three.

2. Draw the structures of the activated two-carbon groups bound to thiamine pyrophosphate in three enzymes that contain this coenzyme. (*Hint:* Keto-enol tautomerism may enter into the picture.)

3. Why is the citric acid cycle considered part of aerobic metabolism even though molecular oxygen does not appear in any reaction?

4. ATP is a competitive inhibitor of NADH binding to malate dehydrogenase, as are ADP and AMP. Suggest a structural basis for this inhibition.

5. How does an increase in the ADP/ATP ratio affect the activity of isocitrate dehydrogenase?

6. How does an increase in the $NADH/NAD^+$ ratio affect the activity of pyruvate dehydrogenase?

7. Would you expect the citric acid cycle to be more or less active when a cell has a high ATP/ADP ratio and a high $NADH/NAD^+$ ratio? Give the reason for your answer.

8. Show by Lewis electron-dot structures of the appropriate portions of the molecule where electrons are lost in the following conversions:
   (a) Pyruvate to acetyl-CoA

(b) Isocitrate to α-ketoglutarate

(c) α-Ketoglutarate to succinyl-CoA

(d) Succinate to fumarate

(e) Malate to oxaloacetate

9. Is the conversion of fumarate to malate a redox (electron transfer) reaction or not? Give the reason for your answer.

10. Prepare a sketch showing how the individual reactions of the three enzymes of the pyruvate dehydrogenase complex give rise to the overall reaction.

11. Why is the reaction catalyzed by citrate synthase considered a condensation reaction?

12. Would you expect the $\Delta G^{\circ\prime}$ for the hydrolysis of a thioester to be (a) large and negative, (b) large and positive, (c) small and negative, or (d) small and positive? Give the reason for your answer.

13. In what part of the cell does the citric acid cycle take place? Does this differ from the part of the cell where glycolysis occurs?

14. What electron acceptors play a role in the citric acid cycle?

15. Briefly describe the dual role of lipoic acid in the pyruvate dehydrogenase complex.

16. Discuss oxidative decarboxylation, using a reaction from this chapter to illustrate your points.

17. Some reactions of the citric acid cycle are endergonic; show how the overall cycle is exergonic. (See Table 12.2.)

18. Describe the conversion of acetyl-CoA to oxaloacetate in the glyoxylate cycle.

A N N O T A T E D   B I B L I O G R A P H Y

Bodner, C. M. The Tricarboxylic Acid (TCA), Citric Acid, or Krebs Cycle. *J. Chem. Ed.* **63,** 673–677 (1986). [A concise and well-written summary of the citric acid cycle. Part of a series on metabolism.]

Boyer, P. D., ed. *The Enzymes,* 3rd ed. New York: Academic Press, 1975. [There are reviews on aconitase in Volume 5 and on dehydrogenases in Volume 11.]

Krebs, H. A. *Reminiscences and Reflections.* New York: Oxford University Press, 1981. [A review of the citric acid cycle along with the autobiography.]

Popjak, G. Stereospecificity of Enzyme Reactions. In Boyer, P. D., ed., *The Enzymes,* 3rd ed. Vol. 2. New York: Academic Press, 1970. [A review of stereochemical aspects of the citric acid cycle.]

See also the bibliographies for Chapters 10 and 11.

Chapter 13

Electron Transport and Oxidative Phosphorylation

Outline

Mitochondria, shown here, are the sites of the citric acid cycle, electron transport, and oxidative phosphorylation.

391

Energy derived from the oxidation of metabolic fuels is ultimately converted to ATP, the quick energy currency of the cell. In eukaryotic cells, under aerobic conditions, ATP is generated by the power of electron transport *along* the inner membrane of the mitochondrion coupled with proton transport *across* the inner membrane. The electron transport chain is actually four closely related enzyme complexes embedded in the inner mitochondrial membrane. In a series of oxidation–reduction transfers, they conduct electrons along the membrane from one complex to another until the electrons reach their final destination, where they combine with molecular oxygen to reduce O_2 to $2 H_2O$. The energy of electron transport can then be used by these same enzyme complexes to pump protons across the membrane out into the intermembrane space. The reverse flow of protons back through the membrane into the inner matrix can be used to generate ATP. Also embedded in the inner membrane is an ATP synthase complex that binds ADP and phosphate ion to synthesize ATP. The flow of protons through the ATP synthase from the intermembrane space to the inner matrix releases the new ATP that has been synthesized. This process is very similar to the production of ATP by photosynthesis in the thylakoid membrane of the chloroplast in green plants.

13.1
THE ROLE OF ELECTRON TRANSPORT IN METABOLISM

Aerobic metabolism is a highly efficient way for an organism to extract energy from nutrients. In eukaryotic cells the aerobic processes (including conversion of pyruvate to acetyl-CoA, the citric acid cycle, and electron transport) all occur in the mitochondria, while the anaerobic process, glycolysis, takes place outside the mitochondria in the cytosol. We have not yet seen any reactions in which oxygen plays a part, but in this chapter we shall discuss the role of oxygen in metabolism as the final acceptor of electrons in the **electron transport chain.** The reactions of the electron transport chain take place in the inner mitochondrial membrane.

The energy released by the oxidation of nutrients is used by organisms in the form of chemical energy of ATP. Production of ATP in the cell requires the process of **oxidative phosphorylation,** in which ADP is phosphorylated to give ATP. The production of ATP by oxidative phosphorylation is a process separate from electron transport, but the reactions of the electron transport chain are strongly linked to one another and are tightly coupled to synthesis of ATP by phosphorylation of ADP. The operation of the electron transport chain leads to pumping of hydrogen ions (protons) across the inner mitochondrial membrane, creating a pH gradient (also called a **proton gradient**); this proton gradient represents stored energy and provides the basis of the coupling mechanism (Figure 13.1). Oxidative phosphorylation gives rise to most of the ATP production associated with the complete oxidation of glucose.

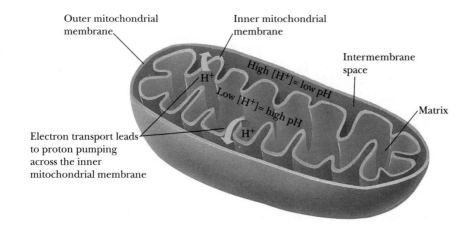

Outer mitochondrial membrane

Inner mitochondrial membrane

Intermembrane space

High [H$^+$] = low pH

Low [H$^+$] = high pH

Matrix

Electron transport leads to proton pumping across the inner mitochondrial membrane

FIGURE 13.1

A proton gradient is established across the inner mitochondrial membrane as a result of electron transport. Transfer of electrons through the electron transport chain leads to the pumping of protons from the matrix to the intermembrane space. The proton gradient (also called the pH gradient), together with the membrane potential (a voltage across the membrane), provides the basis of the coupling mechanism that drives ATP synthesis.

The NADH and FADH$_2$ molecules generated in glycolysis and the citric acid cycle transfer electrons to oxygen in the series of reactions known collectively as the electron transport chain. The NADH and FADH$_2$ are oxidized to NAD$^+$ and FAD, and can be used again in various metabolic pathways. Oxygen, the ultimate electron acceptor, is reduced to water; this completes the process by which glucose is completely oxidized to carbon dioxide and water. We have already seen how carbon dioxide is produced from pyruvate, which in turn is produced from glucose, by the pyruvate dehydrogenase complex and the citric acid cycle. In this chapter we shall see how water is produced.

The complete series of oxidation–reduction reactions of the electron transport chain is presented in schematic form in Figure 13.2. A particularly noteworthy point about electron transport is that three molecules of ATP are generated for each molecule of NADH that enters the electron

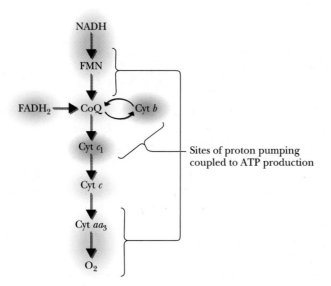

Sites of proton pumping coupled to ATP production

FIGURE 13.2

Schematic representation of the electron transport chain, showing sites of proton pumping coupled to oxidative phosphorylation. FMN is the flavin coenzyme *f*lavin *mono*nucleotide, which differs from FAD in not having an adenine nucleotide. CoQ is coenzyme Q (see Figure 13.3). Cyt *b*, cyt *c*$_1$, cyt *c*, and cyt *aa*$_3$ are the heme-containing proteins cytochrome *b*, cytochrome *c*$_1$, cytochrome *c*, and cytochrome *aa*$_3$, respectively.

transport chain, and two molecules of ATP are produced for each molecule of $FADH_2$. The general outline of the process is that NADH passes electrons to coenzyme Q, as does $FADH_2$, providing an alternative mode of entry into the electron transport chain. Electrons are then passed from coenzyme Q to a series of proteins called cytochromes, designated by lowercase letters, and eventually to oxygen.

13.2
ELECTRON TRANSPORT FROM NADH TO O_2 REQUIRES FOUR MEMBRANE-BOUND COMPLEXES

Intact mitochondria isolated from cells can carry out all the reactions of the electron transport chain; the electron transport apparatus can also be resolved into its component parts by a process called fractionation. Four separate **respiratory complexes** can be isolated from the inner mitochondrial membrane. These complexes are multienzyme systems; in the last chapter we encountered other examples of such multienzyme complexes, such as the pyruvate dehydrogenase complex and the α-ketoglutarate dehydrogenase complex. Each of the respiratory complexes can carry out the reactions of a portion of the electron transport chain.

Complex I The first complex, **NADH-CoQ oxidoreductase,** catalyzes the first steps of electron transport, namely the transfer of electrons from NADH to coenzyme Q. This complex is an integral part of the inner mitochondrial membrane and includes, among other subunits, several proteins that contain an iron–sulfur cluster and the flavoprotein that oxidizes NADH. (The total number of subunits is over 20. This complex is a subject of active research, which has proven to be a challenging task because of its complexity. It is particularly difficult to generalize about the nature of the iron–sulfur clusters because they vary from species to species.) The flavoprotein has a flavin coenzyme, FMN, which differs from FAD in not having an adenine nucleotide.

The reaction occurs in several steps with successive oxidation and reduction of the flavoprotein and the iron–sulfur moiety. The first step is the transfer of electrons from NADH to the flavin portion of the flavoprotein:

$$NADH + H^+ + E\text{—}FMN \longrightarrow NAD^+ + E\text{—}FMNH_2$$

where the notation E—FMN indicates that the flavin is covalently bonded to the enzyme. In the second step the reduced flavoprotein is reoxidized and coenzyme Q (represented simply as CoQ) is reduced to $CoQH_2$ as a pair of electrons is passed on (Figure 13.3). Coenzyme Q is also called ubiquinone. The electrons are passed first to the iron–sulfur clusters and then to coenzyme Q.

$$E\text{—}FMNH_2 + Fe\text{—}S_{oxidized} \longrightarrow E\text{—}FMN + Fe\text{—}S_{reduced} + 2\ H^+$$

$$Fe\text{—}S_{reduced} + CoQ + 2H^+ \longrightarrow Fe\text{—}S_{oxidized} + CoQH_2$$

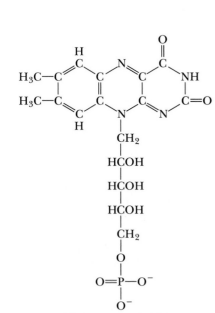

The structure of FMN
(*Flavin mononucleotide*)

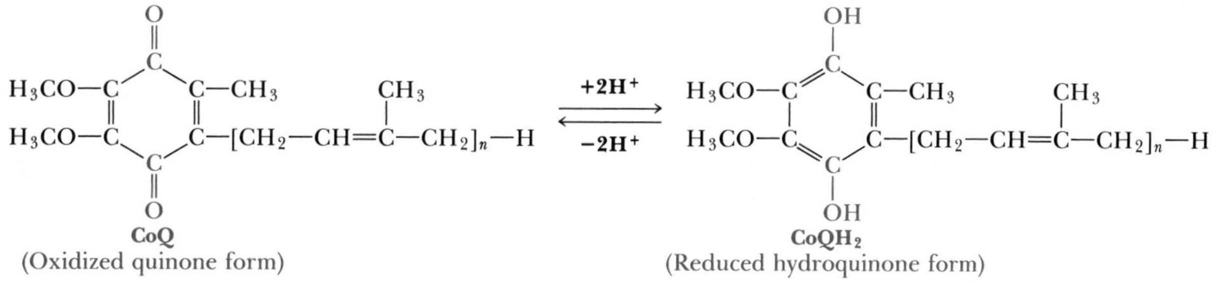

FIGURE 13.3

The oxidized and reduced forms of coenzyme Q. Coenzyme Q is also called ubiquinone.

The notation Fe—S indicates the iron–sulfur clusters. The overall equation for the reaction is

$$NADH + H^+ + CoQ \longrightarrow NAD^+ + CoQH_2$$

This reaction is one of the three responsible for the proton pumping (Figure 13.4) that creates the pH (proton) gradient. The standard free energy change ($\Delta G^{\circ\prime} = -81$ kJ mol^{-1} $= -19.4$ kcal mol^{-1}) indicates that the reaction is strongly exergonic, releasing enough energy to drive the phosphorylation of ADP to ATP (Figure 13.5).

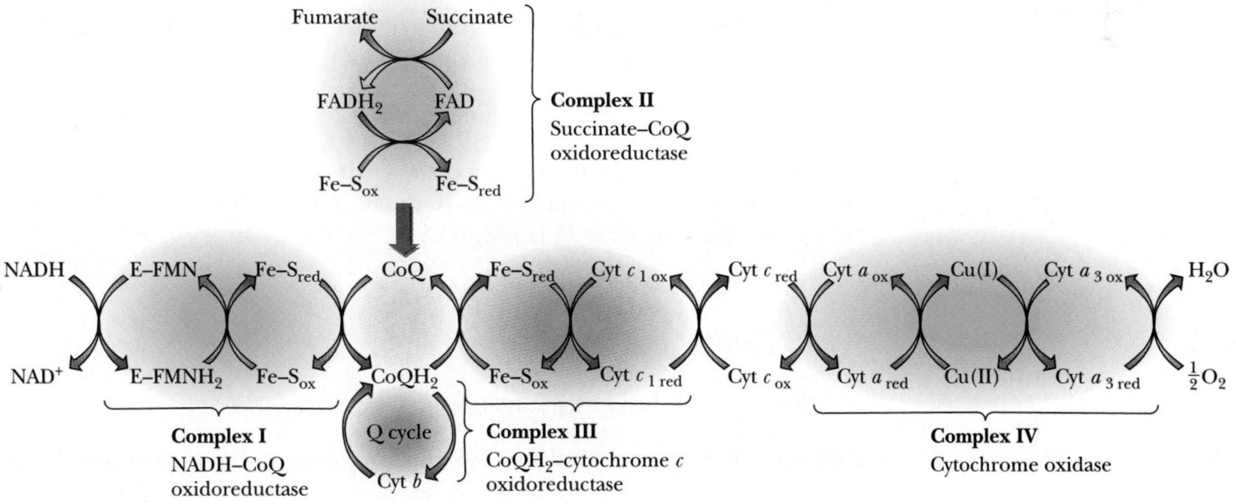

FIGURE 13.4

The electron transport chain, showing the respiratory complexes. In the reduced cytochromes the iron is in the Fe(II) oxidation state, while in the oxidized cytochromes the oxygen is in the Fe(III) oxidation state.

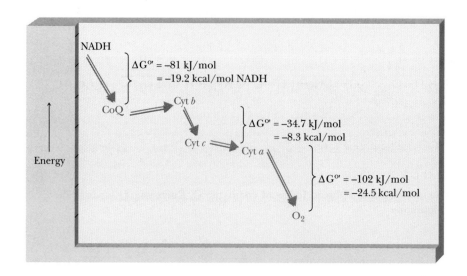

Coenzyme Q itself is mobile; that is to say, it is free to move in the membrane and to pass the electrons it has gained to the third complex for further transport to oxygen. We shall now see that the second complex also transfers electrons from an oxidizable substrate to coenzyme Q.

Complex II   The second of the four membrane-bound complexes, **succinate-CoQ oxidoreductase,** also catalyzes the transfer of electrons to coenzyme CoQ. However, its source of electrons (in other words, the substance being oxidized) differs from the oxidizable substrate (NADH) acted on by NADH-CoQ oxidoreductase, the first of the four respiratory complexes. In this case the substrate is succinate from the citric acid cycle, which is oxidized to fumarate by a flavin enzyme (see Figure 13.4).

$$\text{Succinate} + \text{E—FAD} \longrightarrow \text{fumarate} + \text{E—FADH}_2$$

The notation E—FAD indicates that the flavin portion is covalently bonded to the enzyme. The flavin group is reoxidized in the next stage of the reaction, and coenzyme Q is reduced.

$$\text{E—FADH}_2 + \text{CoQ} \longrightarrow \text{E—FAD} + \text{CoQH}_2$$

The overall reaction is

$$\text{Succinate} + \text{CoQ} \longrightarrow \text{fumarate} + \text{CoQH}_2$$

We already saw the first step of this reaction when we discussed the oxidation of succinate to fumarate as part of the citric acid cycle. The enzyme traditionally called succinate dehydrogenase, which catalyzes the oxidation of succinate to fumarate (Chapter 12, Section 12.3), has been shown by later work to be a part of this enzyme complex. Recall that the succinate dehydrogenase portion consists of a flavoprotein and an iron–sulfur protein. The other components of Complex II are a *b*-type cytochrome and two iron–sulfur proteins. The whole complex is an integral part of the inner mitochondrial membrane. The standard free energy change ($\Delta G^{\circ\prime}$) is

$- 13.5$ kJ mol^{-1} $= -3.2$ kcal mol^{-1}. The overall reaction is exergonic, but there is not enough energy from this reaction to drive ATP production.

In further steps of the electron transport chain, electrons are passed from coenzyme Q, which is then reoxidized, to the first of a series of very similar proteins called cytochromes. Each of these proteins contains a heme group, and in each heme group the iron ion is successively reduced to Fe(II) and reoxidized to Fe(III). This situation differs from that of the iron in the heme group of hemoglobin, which remains in the reduced form as Fe(II) through the entire process of oxygen transport in the bloodstream. There are also some structural differences between the heme group in hemoglobin and the heme groups in the various types of cytochromes.

To be strictly accurate, the successive oxidation–reduction reactions of the cytochromes are not simply

$$Fe(III) + e^- \longrightarrow Fe(II) \text{(reduction)}$$

and

$$Fe(II) \longrightarrow Fe(III) + e^- \text{(oxidation)}$$

The free energy of each reaction, $\Delta G^{\circ\prime}$, differs from the others because of the influences of the various types of hemes and protein structures. Each of the proteins is slightly different in structure and thus each protein has slightly different properties, including the tendency to participate in oxidation–reduction reactions. The different types of cytochromes are distinguished by lowercase letters (a, b, c); further distinctions are possible with subscripts, as in c_1.

Complex III The third complex, **$CoQH_2$–cytochrome c oxidoreductase** (also called cytochrome reductase), catalyzes the oxidation of coenzyme Q. The electrons produced by this oxidation reaction are passed along to cytochrome c in a multistep process. The overall reaction is

$$CoQH_2 + 2 \text{ cyt } c[Fe(III)] \longrightarrow CoQ + 2 \text{ cyt } c[(Fe(II)] + 2 \text{ H}^+$$

Recall that the oxidation of coenzyme Q involves two electrons, whereas the reduction of Fe(III) to Fe(II) requires only one electron. Therefore, two molecules of cytochrome c are required for every molecule of coenzyme Q. The components of this complex include cytochrome b (actually two b-type cytochromes), cytochrome c_1, and several iron–sulfur proteins (Figure 13.4).

The third complex is an integral part of the inner mitochondrial membrane. Coenzyme Q is soluble in the lipid component of the mitochondrial membrane and is separated from the complex in the fractionation process that resolves the electron transport apparatus into its component parts, but the coenzyme is probably close to respiratory complexes in the intact membrane (Figure 13.6). Cytochrome c itself is not part of the complex but is loosely bound to the outer surface of the inner mitochondrial membrane, facing the intermembrane space. It is noteworthy that these two important electron carriers, coenzyme Q and cytochrome c, are not part of the respiratory complexes but can move freely in the membrane.

The flow of electrons from reduced coenzyme Q to the other components of the complex does not take a simple direct path. It is becoming clear that a cyclic flow of electrons involves coenzyme Q twice. This behavior

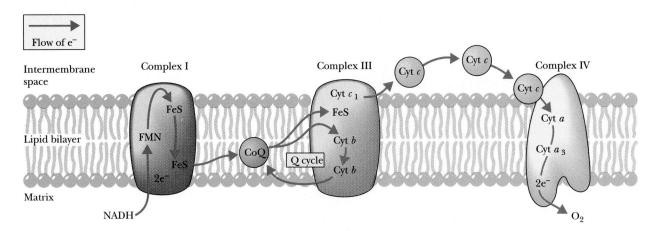

FIGURE 13.6

**The compositions and locations of respiratory complexes in the inner mito-
chondrial membrane, showing the flow of electrons from NADH to O_2. Complex II is not involved and not shown. NADH has accepted electrons from
substrates such as pyruvate, isocitrate, α-ketoglutarate, and malate. Note that
the binding site for NADH is on the matrix side of the membrane. Coenzyme
Q is soluble in the lipid bilayer. Complex III contains two b-type cytochromes,
which are involved in the Q cycle (see text). Cytochrome c is loosely bound to
the membrane, facing the intermembrane space. In Complex IV the binding
site for oxygen lies on the side toward the matrix.**

depends on the fact that, as a quinone, coenzyme Q can exist in three forms
(Figure 13.7). The semiquinone form, which is intermediate between the
oxidized and reduced forms, is of crucial importance here. Because of the
crucial involvement of coenzyme Q, this portion of the pathway is called the
Q cycle.

In part of the Q cycle *one* electron is passed from reduced coenzyme Q
to the iron–sulfur clusters to cytochrome c_1, leaving coenzyme Q in the
semiquinone form.

$$CoQH_2 \longrightarrow Fe\!-\!S \longrightarrow cyt\ c_1$$

The notation Fe—S indicates the iron–sulfur clusters. The series of
reactions involving coenzyme Q and cytochrome c_1, but omitting the
iron–sulfur proteins, can be written as follows:

$$CoQH_2 + cytochrome\ c_1(oxidized) \longrightarrow$$

$$cytochrome\ c_1(reduced) + CoQ^{\,-}(semiquinone\ anion) + 2\ H^+$$

The semiquinone, along with the oxidized and reduced forms of coenzyme
Q, participates in a cyclic process in which the two b cytochromes are
reduced and oxidized in turn. A second molecule of coenzyme Q is involved,
transferring a second electron to cytochrome c_1, and from there to the
mobile carrier cytochrome c. We are going to omit a number of details of the
process in the interest of simplicity. Each of the two molecules of coenzyme
Q involved in the Q cycle loses one electron. The net result is the same as

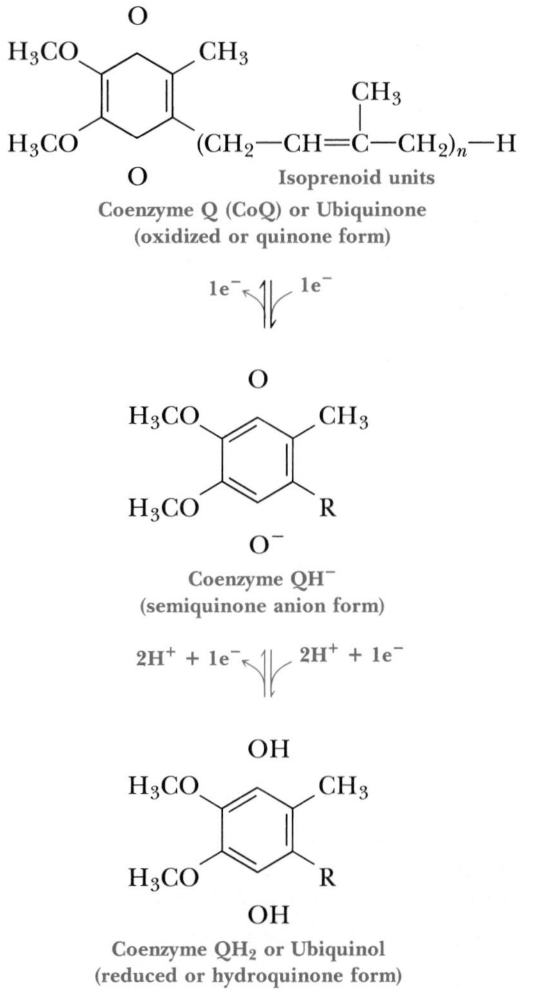

FIGURE 13.7

The oxidized and reduced forms of coenzyme Q, showing the intermediate semiquinone anion form involved in the Q cycle.

if one molecule of CoQ had lost two electrons. It is known that one molecule of $CoQH_2$ is regenerated and one is oxidized to CoQ, which is consistent with this picture. Most importantly, the Q cycle provides a mechanism for electrons to be transferred one at a time from coenzyme Q to cytochrome c_1.

Proton pumping, to which ATP production is coupled, occurs as a result of the reactions of this complex. The Q cycle is implicated in the process, and the whole topic is under active investigation. The standard free energy change ($\Delta G^{\circ\prime}$) is -34.2 kJ $= -8.2$ kcal for each mole of NADH that enters the electron transport chain (see Figure 13.5). The phosphorylation of ADP requires 30.5 kJ $mol^{-1} = 7.3$ kcal mol^{-1}, and the reaction catalyzed by the third complex supplies enough energy to drive the production of ATP.

Complex IV The fourth complex, **cytochrome oxidase,** catalyzes the final steps of electron transport, the transfer of electrons from cytochrome c to oxygen. The overall reaction is

$$2 \text{ Cyt } c[Fe(II)] + 2H^+ + \tfrac{1}{2} O_2 \longrightarrow 2 \text{ cyt } c[Fe(III)] + H_2O$$

Proton pumping also takes place as a result of this reaction. Like the other respiratory complexes, cytochrome oxidase is an integral part of the inner mitochondrial membrane and contains cytochromes a and a_3, as well as two Cu^{2+} ions that are involved in the electron transport process. Taken as a whole, this complex contains about ten subunits. In the flow of electrons, the copper ions are intermediate electron acceptors that lie between the two a-type cytochromes in the sequence

$$\text{Cyt } c \longrightarrow \text{cyt } a \longrightarrow Cu^{2+} \longrightarrow \text{cyt } a_3 \longrightarrow O_2$$

To show the reactions of the cytochromes more explicitly,

Cytochrome c [reduced, Fe(II)] + cytochrome aa_3[oxidized, Fe(III)] $\longrightarrow$

cytochrome aa_3[reduced, Fe(II)] + cytochrome c [oxidized, Fe(III)]

Cytochromes a and a_3 taken together form the complex known as cytochrome oxidase. The reduced cytochrome oxidase is then oxidized by oxygen, which is itself reduced to water. The half reaction for the reduction of oxygen (oxygen acts as an oxidizing agent) is

$$\tfrac{1}{2}O_2 + 2\,H^+ + 2\,e^- \longrightarrow H_2O$$

The overall reaction is

2 Cytochrome aa_3[reduced, Fe(II)] + $\tfrac{1}{2}O_2$ + 2 H$^+$ $\longrightarrow$

2 cytochrome aa_3[oxidized, Fe(III)] + H$_2$O

Note that in this final reaction we have finally seen the link to molecular oxygen in aerobic metabolism.

The standard free energy change ($\Delta G^{\circ\prime}$) is -110 kJ $= -26.3$ kcal for each mole of NADH that enters the electron transport chain (see Figure 13.5). We have now seen the three places in the respiratory chain where electron transport is coupled to ATP production by proton pumping. These three places are the NADH dehydrogenase reaction, the oxidation of cytochrome b, and the reaction of cytochrome oxidase with oxygen. Table 13.1 summarizes the energetics of electron transport reactions.

T A B L E 1 3 . 1   The Energetics of Electron Transport Reactions

REACTION	$\Delta G^{\circ\prime}$	
	kJ (mol NADH^{-1})	*kcal (mol NADH^{-1})*
NADH + H$^+$ + E—FMN $\longrightarrow$ NAD$^+$ + E—FMNH$_2$	-38.6	-9.2
E—FMNH$_2$ + CoQ $\longrightarrow$ E—FMN + CoQH$_2$	-42.5	-10.2
CoQH$_2$ + 2 cyt b[Fe(III)] $\longrightarrow$ CoQ + 2H$^+$ + 2 cyt b[Fe(II)]	$+11.6$	$+2.8$
2 cyt b[Fe(II)] + 2 cyt c_1 [Fe(III)] $\longrightarrow$ 2 cyt c_1[Fe(II)] + 2 cyt b[Fe(III)]	-34.7	-8.3
2 cyt c_1[Fe(II)] + 2 cyt c [Fe(III)] $\longrightarrow$ 2 cyt c[Fe(II)] + 2 cyt c_1[Fe(III)]	-5.8	-1.4
2 cyt c [Fe(II)] + 2 cyt (a/a_3) [Fe(III)] $\longrightarrow$ 2 cyt (a/a_3) [Fe(II)] + 2 cyt c[Fe(III)]	-7.7	-1.8
2 cyt (a/a_3) [Fe(II)] + $\tfrac{1}{2}O_2$ + 2 H$^+$ $\longrightarrow$ 2 cyt (a/a_3) [Fe(III)] + H$_2$O	-102.3	-24.5
Overall reaction:		
NADH + H$^+$ + $\tfrac{1}{2}O_2$ $\longrightarrow$ NAD$^+$ + H$_2$O	-220	-52.6

13.3
THE COUPLING OF OXIDATION TO PHOSPHORYLATION

Some of the energy released by the oxidation reactions in the electron transport chain is used to drive the phosphorylation of three molecules of ADP. The phosphorylation of each mole of ADP requires 30.5 kJ = 7.3 kcal, and we have seen how each of the reactions catalyzed by three of the four respiratory complexes provides more than enough energy to drive this reaction. It is a common theme in metabolism that energy to be used by cells is converted to the chemical energy of ATP as needed. The energy-releasing oxidation reactions give rise to proton pumping and thus to the pH gradient across the inner mitochondrial membrane. The energy of the electrochemical potential (voltage drop) across the membrane is converted to the chemical energy of ATP by the coupling process. The process of phosphorylation is not the same as that of electron transport, even though the two are intimately associated.

A coupling factor is needed to link oxidation and phosphorylation. A complex protein oligomer, separate from the electron transport complexes, serves this function; the complete protein spans the inner mitochondrial membrane and projects into the matrix as well. The portion of the protein that spans the membrane is called F_0. It consists of four different kinds of polypeptide chains, and research is in progress to characterize it further. The portion that projects into the matrix is called F_1; it consists of five different kinds of polypeptide chains in the ratio $\alpha_3\beta_3\gamma\delta\epsilon$. Electron micrographs of mitochondria show the projections into the matrix from the inner mitochondrial membrane (Figure 13.8). The F_1 sphere is the site of ATP synthesis. The whole protein complex is called **ATP synthase.** It is also known as mitochondrial ATPase, because the reverse reaction of ATP hydrolysis, as well as phosphorylation, can be catalyzed by the enzyme. The

FIGURE 13.8

The F_0–F_1 complex (ATP synthase), the site of ATP synthesis. (a) A model for the F_0–F_1 complex subunits, shown schematically. (b) Electron micrograph of projections into the matrix space of a mitochondrion.

(a)

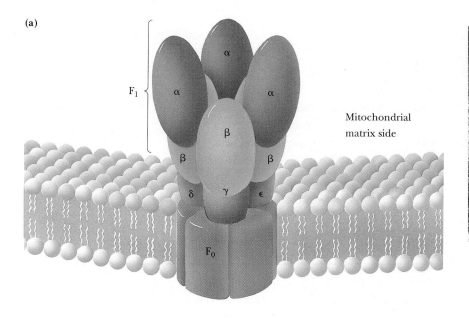

Mitochondrial matrix side

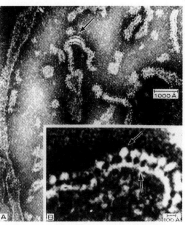

hydrolytic reaction was discovered before the reaction of the synthesis of ATP, hence the name.

Compounds known as **uncouplers** inhibit the phosphorylation of ADP without affecting electron transport. A well-known example of an uncoupler is **2,4-dinitrophenol.** Various antibiotics such as **valinomycin** and **gramicidin A** are also uncouplers (Figure 13.9). When mitochondrial oxidation processes are operating normally, electron transport from NADH or $FADH_2$ to oxygen results in the production of ATP. When an uncoupler is present, oxygen is still reduced to H_2O, but ATP is not produced. If the uncoupler is removed, ATP synthesis linked to electron transport resumes.

A term called the **P/O ratio** is used to indicate the coupling of ATP production to electron transport. The P/O ratio gives the number of moles of P_i consumed in the reaction $ADP + P_i \longrightarrow ATP$ for each mole of oxygen atoms consumed in the reaction $\frac{1}{2} O_2 + 2 H^+ + 2 e^- \longrightarrow H_2O$. As we have already seen, three moles of ATP are produced when one mole of NADH is oxidized to NAD^+. Recall that oxygen is the ultimate acceptor of the electrons from NADH and that $\frac{1}{2}$ mole of O_2 molecules (one mole of oxygen atoms) is reduced for each mole of NADH oxidized. The experimentally determined P/O ratio is 3 when NADH is the substrate oxidized. The P/O ratio is 2 when $FADH_2$ is the substrate oxidized (also an experimentally determined value); we have seen that only two moles of ATP are produced for each mole of $FADH_2$ oxidized.

FIGURE 13.9

Some uncouplers of oxidative phosphorylation: 2,4-dinitrophenol, valinomycin, and gramicidin A.

2,4-Dinitrophenol (DNP)

L-Lactate   L-Valine   D-Hydroisovalerate   D-Valine

Repeating unit of valinomycin

(Valinomycin is a cyclic trimer of four repeating units.)

HN—L-Val–Gly–L-Ala–D-Leu–L-Ala–D-Val–L-Val–D-Val–L-Trp–D-Leu–L-Trp–D-Leu–L-Trp–D-Leu–L-Trp—C

1   5   6   10   11   15

Gramicidin A

(Note alternating L- and D- amino acids)

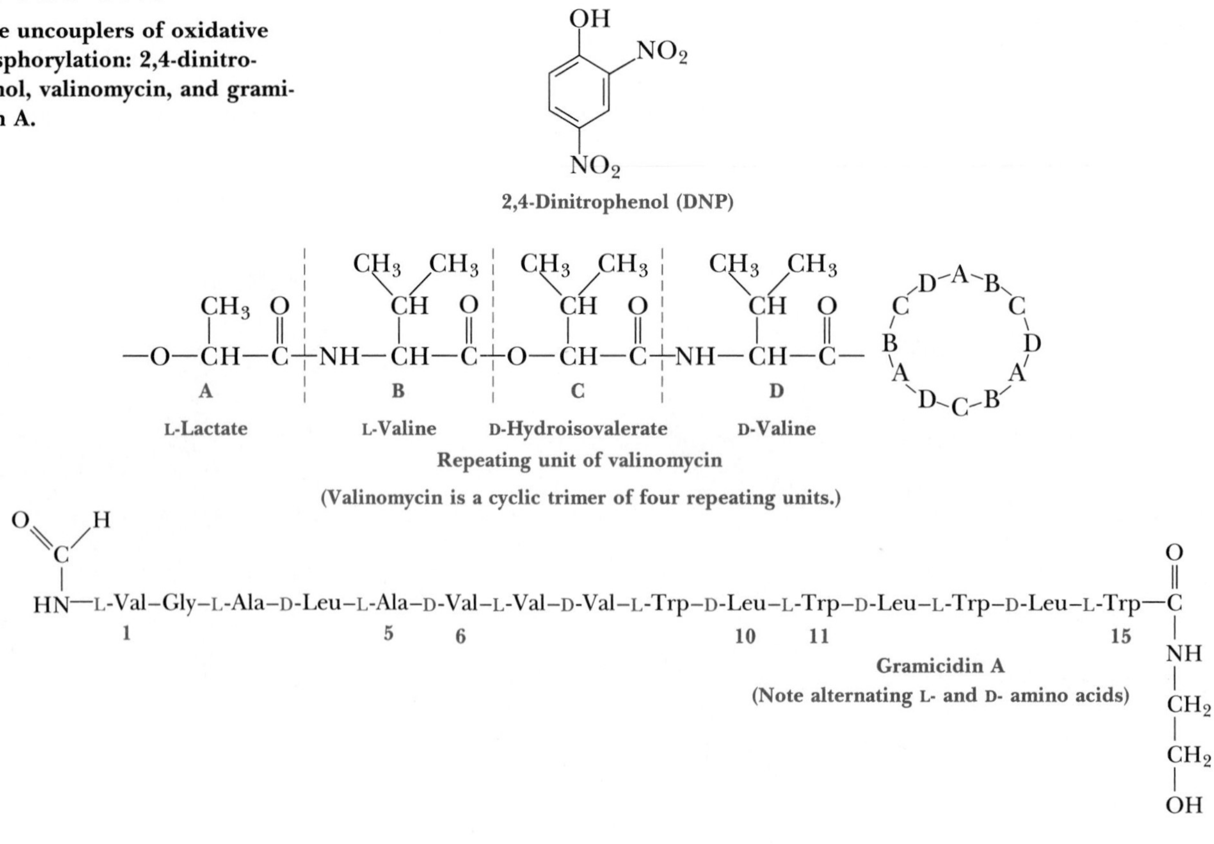

13.4
THE MECHANISM OF COUPLING IN OXIDATIVE PHOSPHORYLATION

Several mechanisms have been proposed to account for the coupling of electron transport and ATP production. The mechanism that currently has the most support is chemiosmotic coupling, which may or may not include a consideration of conformational coupling.

Chemiosmotic Coupling

As originally proposed, the **chemiosmotic coupling** mechanism was based entirely on the difference in hydrogen ion concentration between the intermembrane space and the matrix of an actively respiring mitochondrion. In other words, the proton (hydrogen ion, H^+) gradient across the inner mitochondrial membrane is the crux of the matter. The proton gradient exists because the various proteins that serve as electron carriers in the respiratory chain are not symmetrically oriented with respect to the two sides of the inner mitochondrial membrane, nor do they react in the same way with respect to the matrix and the intermembrane space (Figure 13.10). In the process of electron transport these proteins take up hydrogen ions from the matrix to transfer them in redox reactions; these electron carriers subsequently release hydrogen ions into the intermembrane space when they are reoxidized, creating the proton gradient. The reactions of NADH, coenzyme Q, and molecular oxygen (O_2) all require hydrogen ions, which come from the matrix side of the membrane. The respiratory complexes transfer the protons to the intermembrane space. As a result, there is a

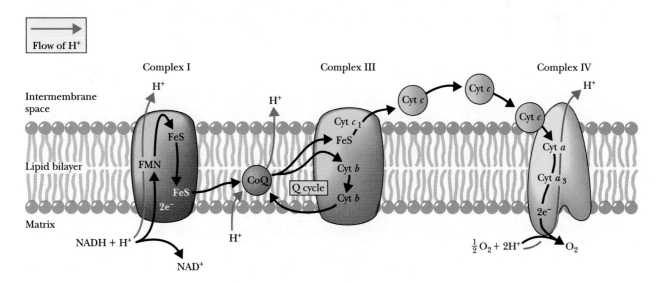

FIGURE 13.10

The creation of a proton gradient in chemiosmotic coupling. The overall effect of the electron transport reaction series is to move protons (H^+) out of the matrix into the intermembrane space, creating a difference in pH across the membrane.

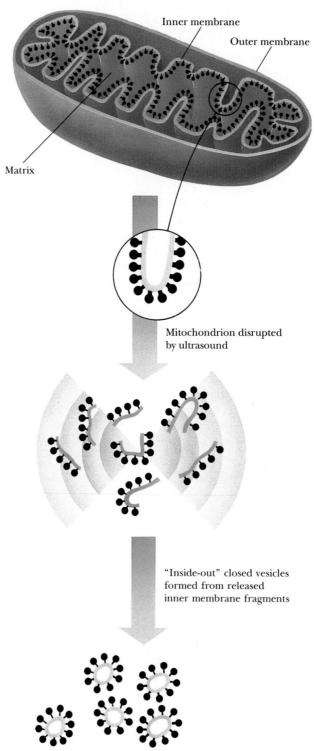

Inner membrane

Outer membrane

Matrix

Mitochondrion disrupted
by ultrasound

"Inside-out" closed vesicles
formed from released
inner membrane fragments

Purified submitochondrial
particles

FIGURE 13.11

Closed vesicles prepared from mitochondria can pump protons and produce
ATP.

higher concentration of hydrogen ions in the intermembrane space than in the matrix; this condition is precisely what we mean by a proton gradient. It is known that the intermembrane space has a lower pH than the matrix, which is another way of saying that there is a higher concentration of hydrogen ions in the intermembrane space than in the matrix. The proton gradient in turn can drive the production of ATP that occurs when the protons flow back into the matrix.

Since chemiosmotic coupling was first suggested in 1961, a considerable body of experimental evidence has accumulated to support it.

1. A system with definite inside and outside compartments (closed vesicles) is essential for oxidative phosphorylation. The process does not occur in soluble preparations or in membrane fragments without compartmentalization.
2. Submitochondrial preparations that contain closed vesicles can be prepared; such vesicles can carry out oxidative phosphorylation, and the asymmetrical orientation of the respiratory complexes with respect to the membrane can be demonstrated (Figure 13.11).
3. A model system for oxidative phosphorylation can be constructed with proton pumping in the absence of electron transport. The model system consists of reconstituted membrane vesicles, mitochondrial ATP synthase, and a proton pump. The pump is bacteriorhodopsin, a protein found in the purple membrane of halobacteria. The proton pumping takes place when the protein is illuminated (Figure 13.12).
4. The existence of the pH gradient has been demonstrated and confirmed experimentally.

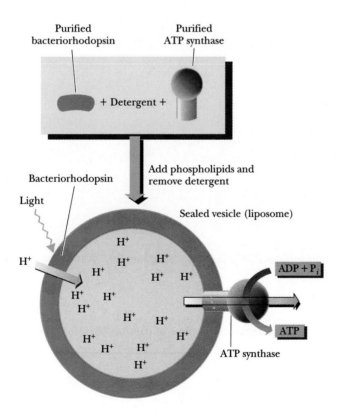

FIGURE 13.12

ATP can be produced by closed vesicles with bacteriorhodopsin as a proton pump.

The way in which the proton gradient leads to the production of ATP depends on ion channels through the inner mitochondrial membrane; these channels are a feature of the structure of ATP synthase. Protons flow back into the matrix through ion channels in the ATP synthase; the F_0 part of the protein is the proton channel. The flow of protons is accompanied by formation of ATP, which takes place in the F_1 unit (Figure 13.13). The unique feature of chemiosmotic coupling is the direct linkage of the proton gradient to the phosphorylation reaction. The details of the way in which phosphorylation takes place as a result of the linkage to the proton gradient are not explicitly specified in this mechanism.

A reasonable mode of action for uncouplers can be proposed in light of the existence of a proton gradient. Dinitrophenol is an acid; its conjugate base, dinitrophenolate anion, is the actual uncoupler, since it can react with hydrogen ions in the intermembrane space, reducing the difference in hydrogen ion concentration between the two sides of the inner mitochondrial membrane. The antibiotic uncouplers such as gramicidin A and valinomycin are **ionophores,** creating a channel through which ions such as H^+, K^+, and Na^+ can pass through the membrane. The proton gradient is overcome, resulting in the uncoupling of oxidation and phosphorylation. Box 13.1 discusses a natural uncoupler.

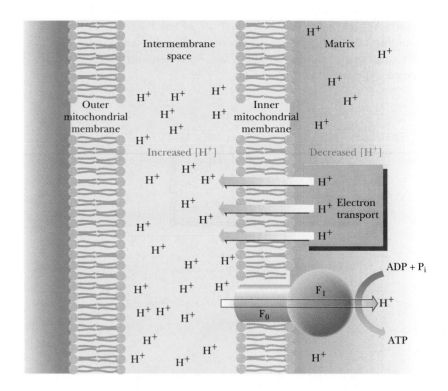

FIGURE 13.13

Formation of ATP accompanies the flow of protons back into the mitochondrial matrix.

BROWN ADIPOSE TISSUE: A CASE OF USEFUL INEFFICIENCY

When electron transport generates a proton gradient, some of the energy produced takes the form of heat. There are two situations in which dissipation of energy as heat is useful to organisms: cold-induced nonshivering thermogenesis (production of heat) and diet-induced thermogenesis. Cold-induced nonshivering thermogenesis enables animals to survive in the cold once they have become adapted to such conditions, and diet-induced thermogenesis prevents the development of obesity in spite of prolonged overeating. These two processes may be the same biochemically; it is firmly established that they occur principally, if not exclusively, in brown adipose tissue (BAT), which is rich in mitochondria. (Brown fat takes its color from the large number of mitochondria present in it, unlike the usual white fat cells.) The key to this "inefficient" use of energy in brown adipose tissue appears to be a mitochondrial protein called thermogenin, also referred to as the "uncoupling protein." When this membrane-bound protein is activated in thermogenesis, it serves as a proton channel through the inner mitochondrial membrane. Like all other uncouplers it "punches a hole" in the mitochondrial membrane and decreases the effect of the proton gradient. Protons flow back into the matrix through thermogenin, bypassing the ATP synthase complex.

Very little research on the biochemistry or physiology of brown adipose tissue has been done in humans. Most of the work on both obesity and adaptation to cold stress has been done on small mammals such as rats, mice, and hamsters. What role, if any, brown fat deposits play in the development or prevention of obesity in humans is an open question for researchers.

Conformational Aspects of Coupling

In conformational coupling the proton gradient is indirectly related to ATP production. The proton gradient leads to conformational changes in a number of proteins, particularly in the ATP synthase itself. It appears from recent evidence that the effect of the proton gradient is not the actual formation of ATP but the release of tightly bound ATP from the synthase as a result of the conformational change (Figure 13.14). There are three sites for substrate on the synthase and three possible conformational states: open (O), with low affinity for substrate; loose-binding (L), which is not catalytical-

FIGURE 13.14

The role of conformational change in releasing ATP from ATP synthase. According to the binding-change mechanism, the effect of the proton flux is to cause a conformational change that leads to the release of already formed ATP from ATP synthase. [From R. L. Cross, D. Cunningham, and J. K. Tamura, 1984, *Curr. Top. Cell. Regul.* 24:336.]

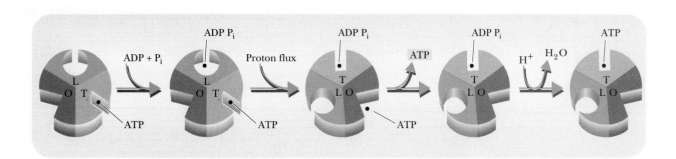

ly active; and tight-binding (T), which is catalytically active. At any given time, each of the sites is in one of three different conformational states. These states interconvert as a result of the proton flux through the synthase. ATP already formed by the synthase is bound at a site in the T conformation, while ADP and P_i bind at a site in the L conformation. A proton flux converts the site in the T conformation to the O conformation, releasing the ATP. The site at which ADP and P_i are bound assumes the T conformation, which can then give rise to ATP.

Electron micrographs have shown that the conformation of the inner mitochondrial membrane and of the cristae is distinctly different in the resting and active states. This evidence supports the idea that conformational changes play a role in the coupling of oxidation and phosphorylation.

13.5
CYTOCHROMES AND OTHER IRON-CONTAINING PROTEINS OF ELECTRON TRANSPORT

In contrast to the electron carriers in the early stages of electron transport, such as NADH, FMN, and CoQ, the cytochromes are macromolecules. These proteins are found in all types of organisms and are typically located in membranes. In eukaryotes the usual site is the inner mitochondrial membrane, but cytochromes can also occur in the endoplasmic reticulum.

The amino acid sequences of cytochromes from many different types of organisms, particularly cytochrome c, have been determined. The evolutionary implications of sequence similarities and differences among organisms constitute a topic of considerable interest to biochemists. There is a discussion of the topic in the article by Dickerson listed in the bibliography at the end of this chapter.

FIGURE 13.15

The heme group of cytochromes. (a) Structure of the heme of all b cytochromes and of hemoglobin and myoglobin. The wedge bonds show the fifth and sixth coordination sites of the iron atom. (b) A comparison of the side chains of a and c cytochromes to those of b cytochromes.

(a)

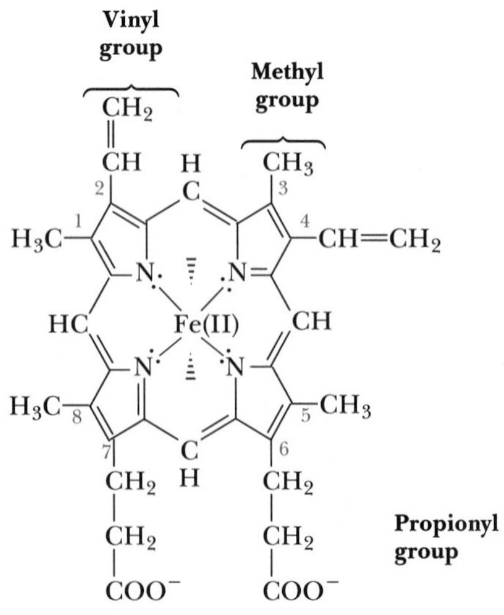

(b)

POSITION	a CYTOCHROMES	c CYTOCHROMES
1	Same	Same
2 (in a)	$-CH-CH_2-(CH_2-CH=C-CH_2)_3H$ $\quad\ \ \|$ $\quad OH \qquad\qquad\qquad\qquad\ CH_3$	
2 (in c)		$-CHCH_3$ $\quad\ \|$ $\quad S-$protein (Covalent attachment)
3	Same	Same
4	Same	$-CHCH_3$ $\quad\ \|$ $\quad S-$protein
5	Same	Same
6	Same	Same
7	Same	Same
8	$-C=O$ (Formyl group) $\quad\ \|$ $\quad H$	Same

All cytochromes contain the heme group, a part of the structure of hemoglobin and myoglobin (Chapter 4, Section 4.4). In the cytochromes the iron of the heme group does not bind to oxygen; instead, the iron is involved in the series of redox reactions, which we have already seen. There are differences in the side chains of the heme group of the cytochromes involved in the various stages of electron transport (Figure 13.15). These structural differences, combined with the variations in the polypeptide chain and in the way in which the polypeptide chain is attached to the heme, account for the differences in properties among the cytochromes in the electron transport chain.

The absorption spectrum of a reduced cytochrome is not the same as that of the same cytochrome in the oxidized form. The spectrum of a reduced cytochrome usually contains three peaks, designated α, β, and γ. (The γ-peak is also called the Soret band.) The α- and β-peaks are not found in the spectrum of oxidized cytochromes (Figure 13.16). The presence or absence of the α- and β-peaks enables us to distinguish the oxidized and reduced forms, and the relative intensities of all the peaks make it possible to determine the relative amounts of the two forms.

(a)

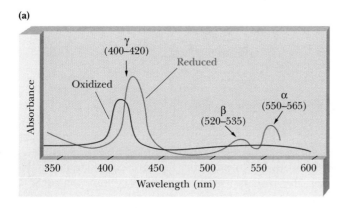

(b)

CYTOCHROME (SOURCE)	Absorption Bands (nm)		
	α	β	γ
b	563	532	429
c	550	521	415
c_1 Mitochondria of animals, plants,	554	524	418
a yeasts, and fungi	600	Absent	439
a_3	604	Absent	443
b_1 (E. coli)	558	528	425
b_2 (yeast)	557	528	424

FIGURE 13.16

Characteristic absorption peaks of cytochromes. (a) A typical absorption spectrum. The wavelength ranges of absorption of light in the α-, β-, and γ-peaks of reduced cytochromes are indicated. (The γ-peak is also called the Soret band.) (b) Table of wavelengths of maximum absorption of light by selected cytochromes.

Nonheme iron proteins do not contain a heme group, as their name indicates. Many of the most important proteins in this category contain sulfur, as is the case with the nonheme iron proteins involved in electron

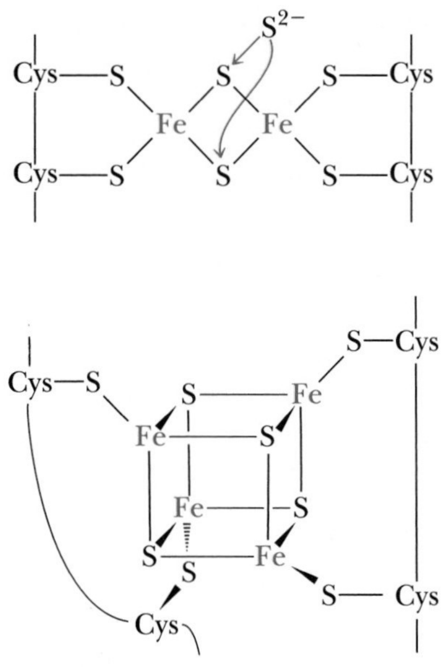

Iron–sulfur bonding in nonheme iron proteins.

transport; these are the iron–sulfur proteins that are components of the respiratory complexes. The iron is usually bound to cysteine or to S^{2-}. There are still many questions about the location and mode of action of iron–sulfur proteins in mitochondria.

13.6
RESPIRATORY INHIBITORS BLOCK THE FLOW OF ELECTRONS IN ELECTRON TRANSPORT

If a pipeline is blocked, there will be a backup. Liquid will accumulate upstream of the blockage point, but there will be less liquid downstream. In electron transport the flow of electrons is from one compound to another rather than along a pipe, but the analogy of a blocked pipeline can be useful for understanding the workings of the pathway. When a flow of electrons is blocked in a series of redox reactions, there will be an accumulation of reduced compounds before the blockage point in the pathway. Recall that reduction is a gain of electrons, and oxidation represents a loss of electrons. The compounds that come after the blockage point will be lacking electrons and will tend to be found in the oxidized form (Figure 13.17). By using respiratory inhibitors we can gather additional evidence to establish the order of components in the electron transport pathway.

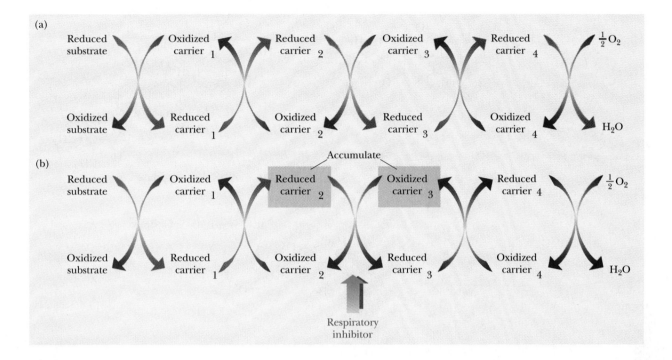

FIGURE 13.17

The effect of respiratory inhibitors. (a) No inhibitor present. Schematic view of electron transport. The orange arrows indicate the flow of electrons. (b) Inhibitor present. The flow of electrons from carrier 2 to carrier 3 is blocked by the respiratory inhibitor. Reduced carrier 2 accumulates, as does oxidized carrier 3, since they cannot react with one another.

The use of respiratory inhibitors to determine the order of the electron transport chain depends on determining the relative amounts of oxidized and reduced forms of the various electron carriers in intact mitochondria. The logic of the experiment can be seen from the analogy of the blocked pipe. In this case the reduced form of the carrier upstream (e.g., the iron–sulfur cluster of complex III) will accumulate because it cannot pass electrons farther in the chain. Likewise, the oxidized form of the carrier downstream (cytochrome c_1 in our example) will also accumulate because the supply of electrons that it could accept has been cut off (Figure 13.17). By use of careful techniques, intact mitochondria can be isolated from cells and can carry out electron transport if an oxidizable substrate is available. If electron transport in mitochondria occurs in the presence and absence of a respiratory inhibitor, there will be different relative amounts of oxidized and reduced forms of the electron carriers.

The type of experiment done to determine the relative amounts of oxidized and reduced forms of electron carriers depends on the spectroscopic properties of these substances. The oxidized and reduced forms of cytochromes can be distinguished from one another by the presence of the α- and β-peaks characteristic of the reduced form. Specialized spectroscopic techniques exist to detect the presence of electron carriers in intact

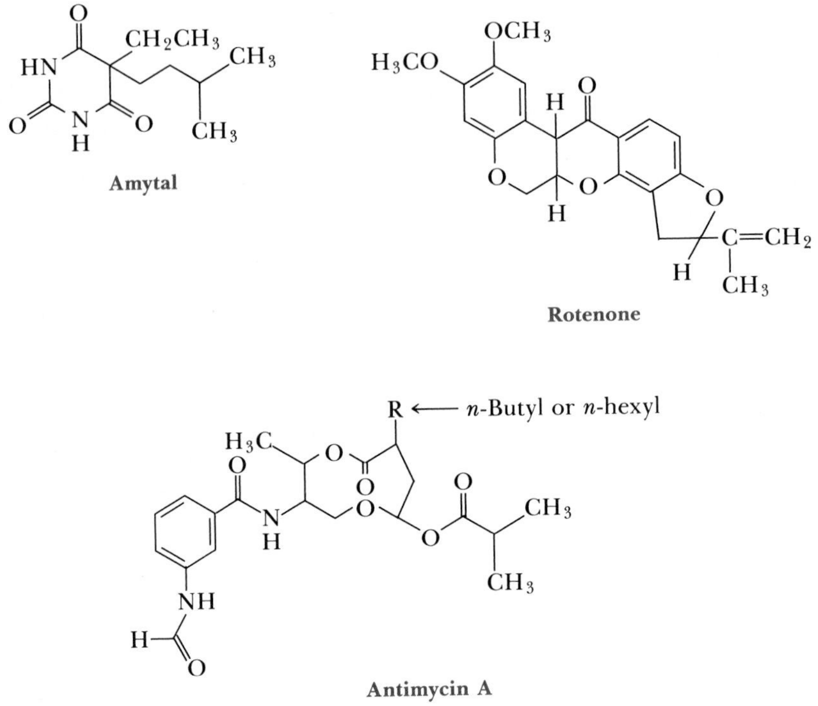

FIGURE 13.18

Structures of some respiratory inhibitors.

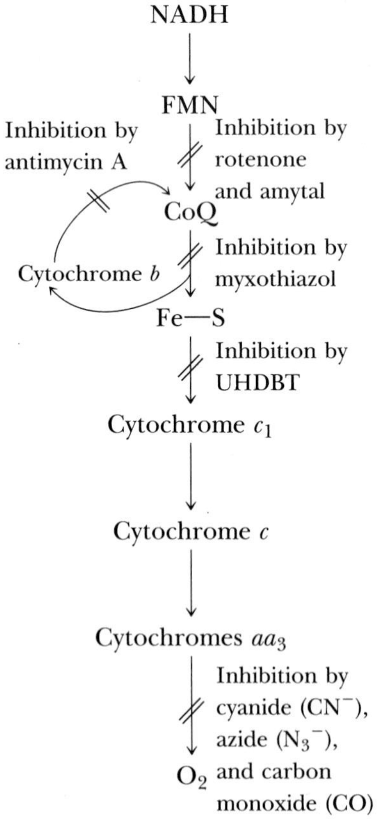

FIGURE 13.19

Sites of action of some respiratory inhibitors.

mitochondria. The individual types of cytochromes can be identified by the wavelength at which the peak appears, and the relative amounts can be determined from the intensities of the peaks.

There are three sites in the electron transport chain at which inhibitors have an effect, and we shall look at some classic examples. At the first site, barbiturates (of which amytal is an example) block the transfer of electrons from the flavoprotein NADH reductase to coenzyme Q. Rotenone is another inhibitor that is active at this site. (This compound is used as an insecticide; it is highly toxic to fish but not to humans.) The second site at which blockage can occur is that of electron transfer involving the b cytochromes, coenzyme Q, and cytochrome c_1; the classic inhibitor associated with this blockage is the antibiotic antimycin A (Figure 13.18). More recently developed inhibitors that are active in this part of the electron transport chain include myxothiazol and 5-n-undecyl-6-hydroxy-4,7-dioxobenzothiazol (UHDBT). These compounds played a role in establishing the existence of the Q cycle. The third site subject to blockage is the transfer of electrons from the cytochrome a/a_3 complex to oxygen. Several potent inhibitors operate at this site (Figure 13.19), such as cyanide (CN^-), azide (N_3^-), and carbon monoxide (CO). Note that each of the three sites of action of respiratory inhibitors corresponds to one of the respiratory complexes. Research is continuing with some of the more recently developed inhibitors; the goal of additional work is to elucidate more of the details of the electron transport process.

13.7
SHUTTLE MECHANISMS MEDIATE TRANSPORT OF METABOLITES BETWEEN MITOCHONDRIA AND CYTOSOL

NADH is produced by glycolysis, which occurs in the cytosol, but NADH in the cytosol cannot cross the mitochondrial membrane to enter the electron transport chain. However, the electrons can be transferred to a carrier that can cross the membrane. The number of ATP molecules generated depends on the nature of the carrier, which varies according to the type of cell in which it occurs.

One carrier system that has been extensively studied in insect flight muscle is the **glycerol phosphate shuttle.** This mechanism makes use of the fact that glycerol phosphate can cross the mitochondrial membrane. The glycerol phosphate is produced by the reduction of dihydroxyacetone phosphate; in the course of the reaction NADH is oxidized to NAD^+. Inside the mitochondrion the glycerol phosphate is reoxidized to dihydroxyacetone phosphate, which can pass back out into the cytosol. In this reaction the oxidizing agent (which is itself reduced) is FAD, and the product is $FADH_2$ (Figure 13.20). The $FADH_2$ then passes electrons through the electron transport chain, leading to the production of two molecules of ATP for each molecule of cytosolic NADH. This mechanism has also been observed in mammalian muscle and brain.

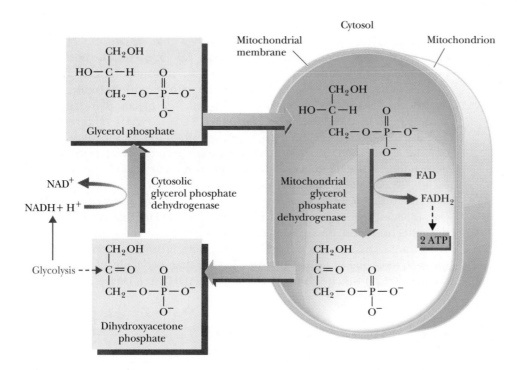

FIGURE 13.20

The glycerol phosphate shuttle.

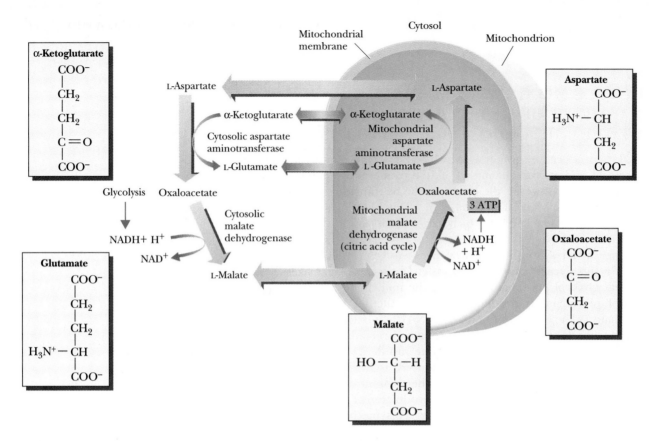

FIGURE 13.21

The malate–aspartate shuttle.

A more complex and more efficient shuttle mechanism is the **malate–aspartate shuttle,** which has been found in mammalian kidney, liver, and heart. This shuttle makes use of the fact that malate can cross the mitochondrial membrane, while oxaloacetate cannot do so. The noteworthy point about this shuttle mechanism is that the transfer of electrons from NADH in the cytosol produces NADH in the mitochondrion. In the cytosol, oxaloacetate is reduced to malate by the cytosolic malate dehydrogenase, accompanied by the oxidation of cytosolic NADH to NAD^+ (Figure 13.21). The malate then crosses the mitochondrial membrane. In the mitochondrion, the conversion of malate back to oxaloacetate is catalyzed by the mitochondrial malate dehydrogenase (one of the enzymes of the citric acid cycle). Oxaloacetate is converted to aspartate, which can also cross the mitochondrial membrane. Aspartate is converted to oxaloacetate in the cytosol, completing the cycle of reactions. The NADH that is produced in the mitochondrion thus passes electrons to the electron transport chain. With the malate–aspartate shuttle, three molecules of ATP are produced for each molecule of cytosolic NADH rather than two molecules of ATP in the glycerol phosphate shuttle, which uses $FADH_2$ as a carrier.

13.8

THE ATP YIELD FROM COMPLETE OXIDATION OF GLUCOSE

In Chapters 11 through 13 we have discussed many aspects of the complete oxidation of glucose to carbon dioxide and water. At this point it is useful to

TABLE 13.2 **The Balance Sheet for Oxidation of One Molecule of Glucose**

	NADH MOLECULES	FADH$_2$ MOLECULES	ATP MOLECULES	
Cytoplasmic Reactions				
1. Glucose $\longrightarrow$ glucose 6-phosphate			-1	
2. Fructose 6-phosphate $\longrightarrow$ fructose 1,6-*bis*phosphate			-1	
3. 1,3-*bis*phosphoglycerate $\longrightarrow$ 3-phosphoglycerate (2 molecules)			$+2$	
4. Phosphoenolpyruvate $\longrightarrow$ pyruvate (2 molecules)			$+2$ $+2$	
Oxidation of glyceraldehyde 3-phosphate (2 molecules)	$+2$			
Mitochondrial Reactions				
Pyruvate $\longrightarrow$ acetyl-CoA (2 molecules)	$+2$			
Citric Acid Cycle				
1. Succinyl-CoA formed GDP $\longrightarrow$ GTP (2 molecules)			$+2$	
2. Oxidation of succinate (2 molecules)		$+2$		
3. Oxidation of 2 molecules each of isocitrate, α-ketoglutarate, and malate	$+6$			
Oxidative Phosphorylation–Electron Transport				
1. Reoxidation of NADH produced in glycolysis	-2		$+6$ (liver)	$+4$ (muscle, brain)
2. Reoxidation of NADH from the pyruvate $\longrightarrow$ acetyl-CoA step	-2		$+6$	$+6$
3. Reoxidation of FADH$_2$ produced in citric acid cycle		-2	$+4$	$+4$
4. Reoxidation of NADH from citric acid cycle	-6		$+18$	$+18$
	0	0	38	36

Note that there is no net change in the number of molecules of NADH or FADH$_2$.

do some bookkeeping to see how many molecules of ATP are produced for each molecule of glucose oxidized. Recall that some ATP is produced in glycolysis, but that far more ATP is produced by aerobic metabolism. Table 13.2 summarizes ATP production, and also follows the recycling of NADH and FADH$_2$.

S U M M A R Y

In the final stages of aerobic metabolism, electrons are transferred from NADH to oxygen (the ultimate electron acceptor) in a series of oxidation–reduction reactions known as the electron transport chain. This series of reactions creates a pH gradient across the inner mitochondrial membrane. The stored energy of the pH gradient drives the process of oxidative phosphorylation, which is separate from electron transport. The two processes, electron transport and oxidative phosphorylation, are coupled by the mechanism of chemiosmotic coupling, which ultimately owes its existence to the pH gradient.

Three molecules of ATP are generated for each molecule of NADH that enters the electron transport chain and two molecules of ATP for each molecule of FADH$_2$. The general outline of the process is that

NADH passes electrons to coenzyme Q, as does FADH$_2$, providing an alternative mode of entry into the electron transport chain. Electrons are then passed from coenzyme Q to the cytochromes and eventually to oxygen. The energy involved in each reaction can be characterized by a standard free energy change ($\Delta G^{\circ\prime}$).

Four separate respiratory complexes can be isolated from the inner mitochondrial membrane. Each of the respiratory complexes can carry out the reactions of a portion of the electron transport chain. In addition to the respiratory complexes, two electron carriers, coenzyme Q and cytochrome c, are not bound to the complexes but are free to move within the membrane. Many of the workings of the electron transport chain have been elucidated by experiments using respiratory inhibitors. The respiratory complexes play an important role in proton pumping across the inner mitochondrial membrane, creating the pH gradient.

A complex protein oligomer is the coupling factor that links oxidation and phosphorylation; the complete protein spans the inner mitochondrial membrane and projects into the matrix as well. The portion of the protein that spans the membrane is called F$_0$; it consists of four different kinds of polypeptide chains. The portion that projects into the matrix is called F$_1$; it consists of five different kinds of polypeptide chains in the ratio $\alpha_3\beta_3\gamma\delta\epsilon$. The F$_1$ sphere is the site of ATP synthesis. The whole protein complex is called ATP synthase. It is also known as mitochondrial ATPase.

Chemiosmotic coupling is the mechanism most widely used to explain the manner in which electron transport and oxidative phosphorylation are coupled to one another. In this mechanism the proton gradient is directly linked to the phosphorylation process. The way in which the proton gradient leads to the production of ATP depends on ion channels through the inner mitochondrial membrane; these channels are a feature of the structure of ATP synthase. Protons flow back into the matrix through ion channels in ATP synthase; the F$_0$ part of the protein is the proton channel. The flow of protons is accompanied by formation of ATP, which occurs in the F$_1$ unit. In the conformational coupling mechanism the proton gradient is indirectly related to ATP production. It appears from recent evidence that the effect of the proton gradient is not the formation of ATP but the release of tightly bound ATP from the synthase as a result of the conformational change.

Two shuttle mechanisms — the glycerol phosphate shuttle and the malate–aspartate shuttle — transfer the electrons, but not the NADH, produced in cytosolic reactions into the mitochondrion. In the first of the two shuttles, which is found in muscle and brain, the electrons are transferred to FAD; in the second, which is found in kidney, liver, and heart, the electrons are transferred to NAD$^+$. With the malate–aspartate shuttle, three molecules of ATP are produced for each molecule of cytosolic NADH rather than two ATP in the glycerol phosphate shuttle, a point that affects the overall yield of ATP in these tissues.

E X E R C I S E S

1. Briefly summarize the steps in the electron transport chain from NADH to oxygen.
2. What yield of ATP can be expected from complete oxidation of each of the following substrates by the reactions of glycolysis, the citric acid cycle, and oxidative phosphorylation?
   (a) Fructose 1,6-*bis*phosphate
   (b) Glucose
   (c) Phosphoenolpyruvate
   (d) Glyceraldehyde 3-phosphate
   (e) NADH
   (f) Pyruvate
3. Comment on the fact that the reduction of pyruvate to lactate, catalyzed by lactate dehydrogenase, is strongly

exergonic (recall this from Chapter 11) even though the standard free energy change for the half reaction

Pyruvate + 2 H$^+$ + 2e^- $\longrightarrow$ lactate

is positive ($\Delta G^{\circ\prime}$ = 36.2 kJ/mol = 8.8 kcal/mol), indicating an endergonic reaction.

4. The free energy change ($\Delta G^{\circ\prime}$) for the oxidation of the cytochrome a/a_3 complex by molecular oxygen is -102.3 kJ = -24.5 kcal for each mole of electron pairs transferred. What is the maximum number of moles of ATP that could be produced in the process? How many moles of ATP are actually produced? What is the efficiency of the process, expressed as a percentage?
5. What is the effect of each of the following substances on

electron transport and production of ATP? Be specific about which reaction is affected.

(a) Azide
(b) Antimycin A
(c) Amytal
(d) Rotenone
(e) Dinitrophenol
(f) Gramicidin A
(g) Carbon monoxide

6. What is the approximate P/O ratio that can be expected if intact mitochondria are incubated in the presence of oxygen, along with added succinate?

7. Cytochrome oxidase and succinate-CoQ oxidoreductase are isolated from mitochondria and are incubated in the presence of oxygen, along with cytochrome c, coenzyme Q, and succinate. What is the overall oxidation–reduction reaction that can be expected to take place?

8. Two biochemistry students are about to use mitochondria isolated from rat liver for an experiment on oxidative phosphorylation. The directions for the experiment specify addition of purified cytochrome c from any source to the reaction mixture. Why is the added cytochrome c needed, and why does the source not have to be the same as that of the mitochondria?

9. Briefly summarize the main arguments of the chemiosmotic coupling hypothesis.

10. Describe the role of the F_1 portion of ATP synthase in oxidative phosphorylation.

11. How does the yield of ATP from complete oxidation of one molecule of glucose in muscle and brain differ from that in liver, heart, and kidney? What is the underlying reason for this difference?

12. Using the information in Table 13.1, calculate $\Delta G^{\circ\prime}$ for the following reaction.

$$2 \text{ Cytochrome } a/a_{3\text{oxid}}[\text{Fe(III)}] + 2 \text{ cytochrome } b_{\text{red}}[\text{Fe(II)}] \longrightarrow$$
$$2 \text{ cytochrome } a/a_{3\text{red}}[\text{Fe(II)}] + 2 \text{ cytochrome } b_{\text{oxid}}[\text{Fe(III)}]$$

13. List the reactions of electron transport from NADH to oxygen. Show how a P/O ratio of 3 is obtained when NADH is the starting point of the electron transport chain.

14. Show how the reactions of the electron transport chain differ from those in Exercise 13 when $FADH_2$ is the starting point for electron transport. Show how the P/O ratio is 2 for electron transport starting from $FADH_2$.

A N N O T A T E D   B I B L I O G R A P H Y

Cannon, B., and J. Nedergaard. The Biochemistry of an Inefficient Tissue: Brown Adipose Tissue. *Essays in Biochemistry* **20**, 110–164 (1985). [A review describing the usefulness to mammals of the "inefficient" production of heat in brown fat.]

Dickerson, R. E. Cytochrome c and the Evolution of Energy Metabolism. *Sci. Amer.* **242** (3), 136–152 (1980). [An account of the evolutionary implications of cytochrome c structure.]

Fillingame, R. The Proton-Translocating Pumps of Oxidative Phosphorylation. *Ann. Rev. Biochem.* **49**, 1079–1114 (1980). [A review of chemiosmotic coupling.]

Hatefi, Y. The Mitochondrial Electron Transport and Oxidative Phosphorylation System. *Ann. Rev. Biochem.* **54**, 1015–1069 (1985). [A review that emphasizes the coupling between oxidation and phosphorylation.]

Hinkle, P. C., and R. E. McCarty. How Cells Make ATP. *Sci. Amer.* **238** (3), 104–123 (1978). [Chemiosmotic coupling and the mode of action of uncouplers. Getting old, but very good.]

Lane, M. D., P. L. Pedersen, and A. S. Mildvan. The Mitochondrion Updated. *Science* **234**, 526–527 (1986). [A report on an international conference on bioenergetics and energy coupling.]

Mitchell, P. Keilin's Respiratory Chain Concept and Its Chemiosmotic Consequences. *Science* **206**, 1148–1159 (1979). [A Nobel Prize lecture by the scientist who first proposed the chemiosmotic coupling hypothesis.]

Moser, C. C., et al. Nature of Biological Electron Transfer. *Nature* **355**, 796–802 (1992). [An advanced treatment of electron transfer in biological systems.]

Trumpower, B. The Protonmotive Q Cycle: Energy Transduction by Coupling of Proton Translocation to Electron Transfer by the Cytochrome bc_1 Complex. *J. Biol. Chem.* **265**, 11409–11412 (1990). [An advanced article that goes into detail about the Q cycle.]

Vignais, P. V., and J. Lunardi. Chemical Probes of the Mitochondrial ATP Synthesis and Translocation. *Ann. Rev. Biochem.* **54**, 977–1014 (1985). [A review about the synthesis and use of ATP.]

Chapter 14

Further Aspects of Carbohydrate Metabolism

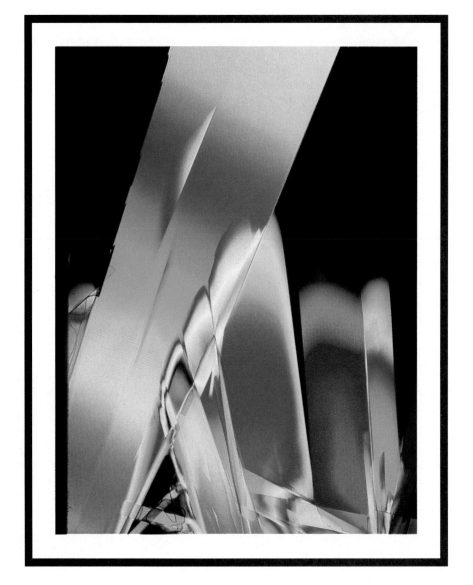

Crystals of thiamine (vitamin B_1). Thiamine pyrophosphate is a cofactor in a number of enzymes.

We usually think of carbohydrates as "quick energy." It is certainly true that the body can mobilize carbohydrates more easily than fat, even though fats contain more energy on a gram-for-gram basis. Free glucose in the bloodstream is quickly depleted, but we have stored glucose in readily accessible form as the polymer glycogen. In sustained activity glycogen in muscle cells is broken down to glucose in a quick response to the need for energy. Before long the supply of glycogen is depleted in turn. In such cases, pyruvate, the end-product of glycolysis, can be converted to lactate and exported from muscle to the liver. There, with the help of its energy reserves of ATP, the liver can convert lactate to "new glucose" and return the glucose to muscle for another round of glycolysis. This process, called *gluconeogenesis,* is essentially (with some exceptions) glycolysis in reverse. New glucose can also be made from pyruvate combined with molecules from the citric acid cycle. This "homemade" glucose produced by gluconeogenesis is available to supply glucose to the brain, which has little reserve of its own. The brain requires a steady stream of energy in the form of glucose (usually its only fuel), no matter what the level of mental activity is, or whether the person is awake or asleep.

14.1
GLYCOGEN METABOLISM

When we digest a meal high in carbohydrates, we have a supply of glucose that exceeds our immediate needs. We store glucose as a polymer, glycogen (Chapter 10, Section 10.4), that is similar to the starches found in plants; glycogen differs from the amylopectin form of starch only in the degree of chain branching. In fact, glycogen is sometimes called "animal starch" because of this similarity. A look at the metabolism of glycogen will give us some insights into how glucose can be stored in this form and made available on demand. In the degradation of glycogen, several glucose residues can be released at one time, one from each end of a branch, rather than one at a time as would be the case in a linear polymer. This feature is useful to an organism in meeting short-term demands for energy by increasing the glucose supply as quickly as possible (Figure 14.1). It has been shown by

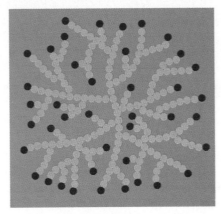

Glycogen

FIGURE 14.1

The highly branched structure of glycogen makes it possible for several glucose residues to be released at once to meet energy needs. This would not be possible with a linear polymer.

mathematical modeling that the structure of glycogen is *optimized* for its ability to store and deliver energy quickly and for the longest amount of time possible. The key to this optimization is the average chain length of the branches (13 residues). If the average chain length were much greater or much shorter, glycogen would not be as efficient a vehicle for energy storage and release on demand. Experimental results support the conclusions reached from the mathematical modeling.

Breakdown of Glycogen

Glycogen is primarily found in liver and muscle. Release of glycogen in liver is triggered by low levels of glucose in blood. Liver glycogen breaks down to glucose 6-phosphate, which is hydrolyzed to give glucose. The release of glucose from the liver by this breakdown of glycogen replenishes the supply of glucose in the blood. In muscle, glucose 6-phosphate obtained from glycogen breakdown enters the glycolytic pathway rather than being exported to the bloodstream.

Three reactions play roles in the conversion of glycogen to glucose 6-phosphate. In the first reaction, each glucose residue cleaved from glycogen reacts with phosphate ion to give glucose 1-phosphate.

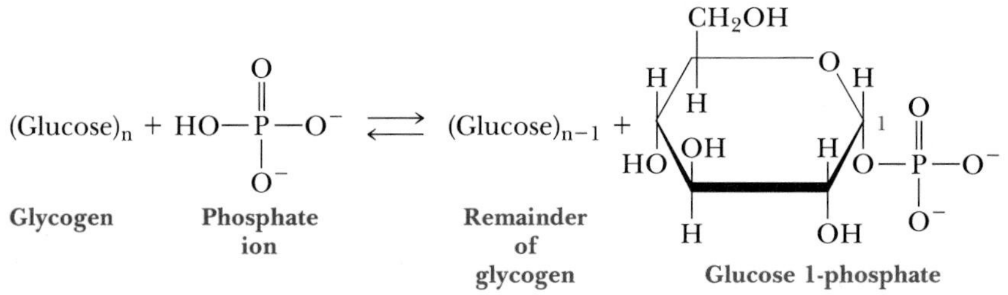

In a second reaction, glucose 1-phosphate isomerizes to give glucose 6-phosphate.

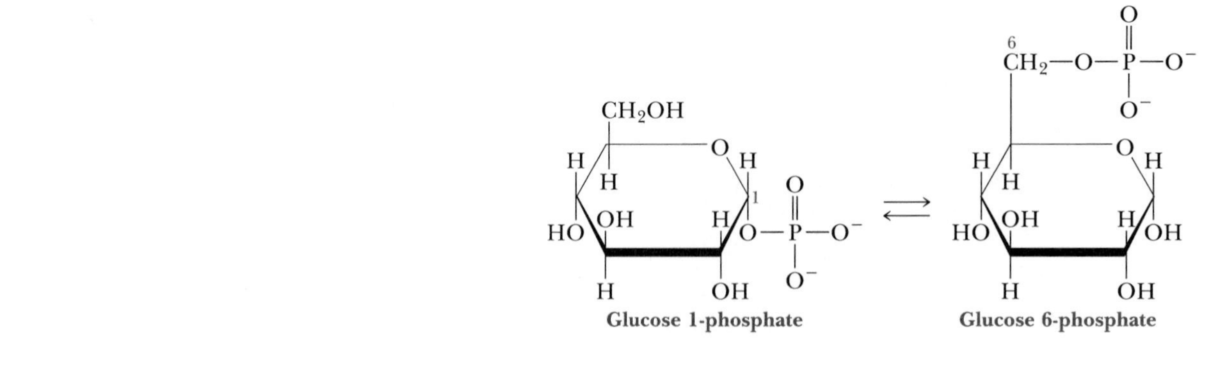

Complete breakdown of glycogen also requires a debranching reaction to hydrolyze the glycosidic bonds of the glucose residues at branch points in the glycogen structure.

The enzyme that catalyzes the first of these reactions is *glycogen phosphorylase;* the second reaction is catalyzed by *phosphoglucomutase.*

$$\text{Glycogen} + P_i \xrightleftharpoons[]{\substack{\textbf{Glycogen}\\ \textbf{phosphorylase}}} \text{glucose 1-phosphate} + \text{remainder of glycogen}$$

$$\text{Glucose 1-phosphate} \xrightleftharpoons[]{\textbf{Phosphoglucomutase}} \text{glycogen 6-phosphate}$$

Glycogen phosphorylase cleaves the $(1 \longrightarrow 4)$ linkages in glycogen. Complete breakdown requires **debranching enzymes** that degrade the $\alpha(1 \longrightarrow 6)$ linkages. Note that no ATP is hydrolyzed in the first reaction. In the glycolytic pathway we saw another example of phosphorylation of a substrate directly by phosphate ion without involvement of ATP: the phosphorylation of glyceraldehyde 3-phosphate to 1,3-*bis*phosphoglycerate. As we saw earlier, a phosphorylation reaction that does not involve the "high-energy" phosphate of ATP is called substrate-level phosphorylation. This is an alternative mode of entry to the glycolytic pathway that "saves" one molecule of ATP for each molecule of glucose because it bypasses the first step in glycolysis. When glycogen rather than glucose is the starting material for glycolysis, there is a net gain of 3 ATP molecules for each glucose monomer, rather than 2 ATP as when glucose itself is the starting point.

The debranching of glycogen involves the transfer of a "limit branch" of three glucose residues to the end of another branch, where they are subsequently removed by glycogen phosphorylase. The same glycogen debranching enzyme then hydrolyzes the $\alpha(1 \longrightarrow 6)$ glycosidic bond of the last glucose residue remaining at the branch point (Figure 14.2).

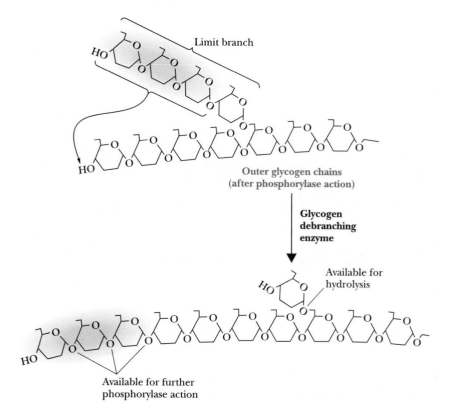

Limit branch

Outer glycogen chains
(after phosphorylase action)

Glycogen
debranching
enzyme

Available for
hydrolysis

Available for further
phosphorylase action

FIGURE 14.2

The mode of action of the de-branching enzyme in glycogen breakdown. The enzyme transfers three $\alpha(1\rightarrow4)$-linked glucose residues from a limit branch to the end of another branch. The same enzyme also catalyzes the hydrolysis of the $\alpha(1\rightarrow6)$-linked residue at the branch point.

When an organism needs energy quickly, glycogen breakdown is important. Muscle tissue can mobilize glycogen more easily than fat and can do so anaerobically. It is only after glycogen supplies are depleted that fat burning becomes important. Some athletes, particularly long-distance runners, try to build up their glycogen reserves before a race by eating large amounts of carbohydrates.

Formation of Glycogen from Glucose

The formation of glycogen from glucose is not the exact reversal of the breakdown of glycogen to glucose. The synthesis of glycogen requires energy, which is provided by the hydrolysis of a nucleoside triphosphate, UTP.

In the first stage of glycogen synthesis, glucose 1-phosphate reacts with UTP to produce uridine diphosphate glucose (also called UDP-glucose or UDPG) and pyrophosphate (PP_i).

Athletes break down their existing glycogen supplies in the course of strenuous activity. Especially during activities that require short bursts of speed, gluconeogenesis plays an important role.

Uridine disphosphate glucose (UDPG)

The enzyme that catalyzes this reaction is *UDP-glucose pyrophosphorylase*. The exchange of one phosphoric anhydride bond for another has a free energy change close to zero. The release of energy comes about when the enzyme inorganic pyrophosphatase catalyzes the hydrolysis of pyrophosphate to two phosphate ions, a strongly exergonic reaction.

	$\Delta G^{\circ\prime}$	
	kJ mol^{-1}	*kcal mol^{-1}*
Glucose 1-phosphate + UTP $\rightleftharpoons$ UDPG + PP_i	~0	~0
H_2O + PP_i $\longrightarrow$ 2 P_i	−30.5	−7.3
Overall Glucose 1-phosphate + UTP $\longrightarrow$ UDPG + 2 P_i	−30.5	−7.3

It is common in biochemistry to see the energy released by the hydrolysis of pyrophosphate combined with the free energy of hydrolysis of a nucleo-

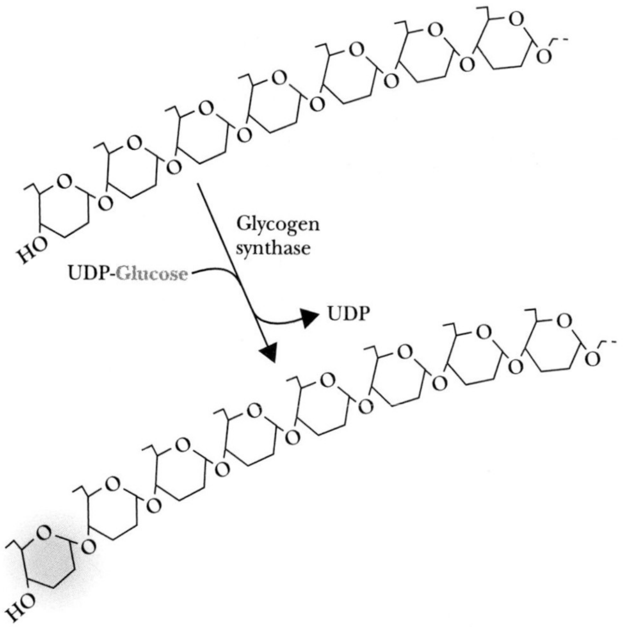

FIGURE 14.3

The reaction catalyzed by glycogen synthase. A glucose residue is transferred from UDPG to the growing end of a glycogen chain in an $\alpha(1\rightarrow4)$ linkage.

side triphosphate. The coupling of these two exergonic reactions to a reaction that is not energetically favorable allows an otherwise endergonic reaction to take place. The supply of UTP is replenished by an exchange reaction with ATP, catalyzed by nucleoside phosphate kinase:

$$UDP + ATP \rightleftharpoons UTP + ADP$$

This exchange reaction makes the hydrolysis of any nucleoside triphosphate energetically equivalent to the hydrolysis of ATP.

The addition of UDPG to a growing chain of glycogen is the next step in glycogen synthesis. Each step involves formation of a new $\alpha(1\longrightarrow4)$ glycosidic bond in a reaction catalyzed by the enzyme *glycogen synthase* (Figure 14.3). This enzyme cannot simply form a bond between two isolated glucose molecules; it must add to an existing chain with $\alpha(1\longrightarrow4)$ glycosidic linkages. The initiation of glycogen synthesis requires a primer for this reason. The hydroxyl group of a specific tyrosine of the protein *glycogenin* serves this purpose; a glucose residue is linked to this tyrosine hydroxyl in the first stage of glycogen synthesis.

Synthesis of glycogen requires the formation of $\alpha(1\longrightarrow6)$ as well as $\alpha(1\longrightarrow4)$ glycosidic linkages. A **branching enzyme** accomplishes this task. It does so by transferring a segment about seven residues in length from the end of a growing chain to a branch point where it catalyzes the formation of the required $\alpha(1\longrightarrow6)$ glycosidic linkage (Figure 14.4). Note that this enzyme has already catalyzed the breaking of an $\alpha(1\longrightarrow4)$ glycosidic linkage in the process of transferring the oligosaccharide segment. Each transferred

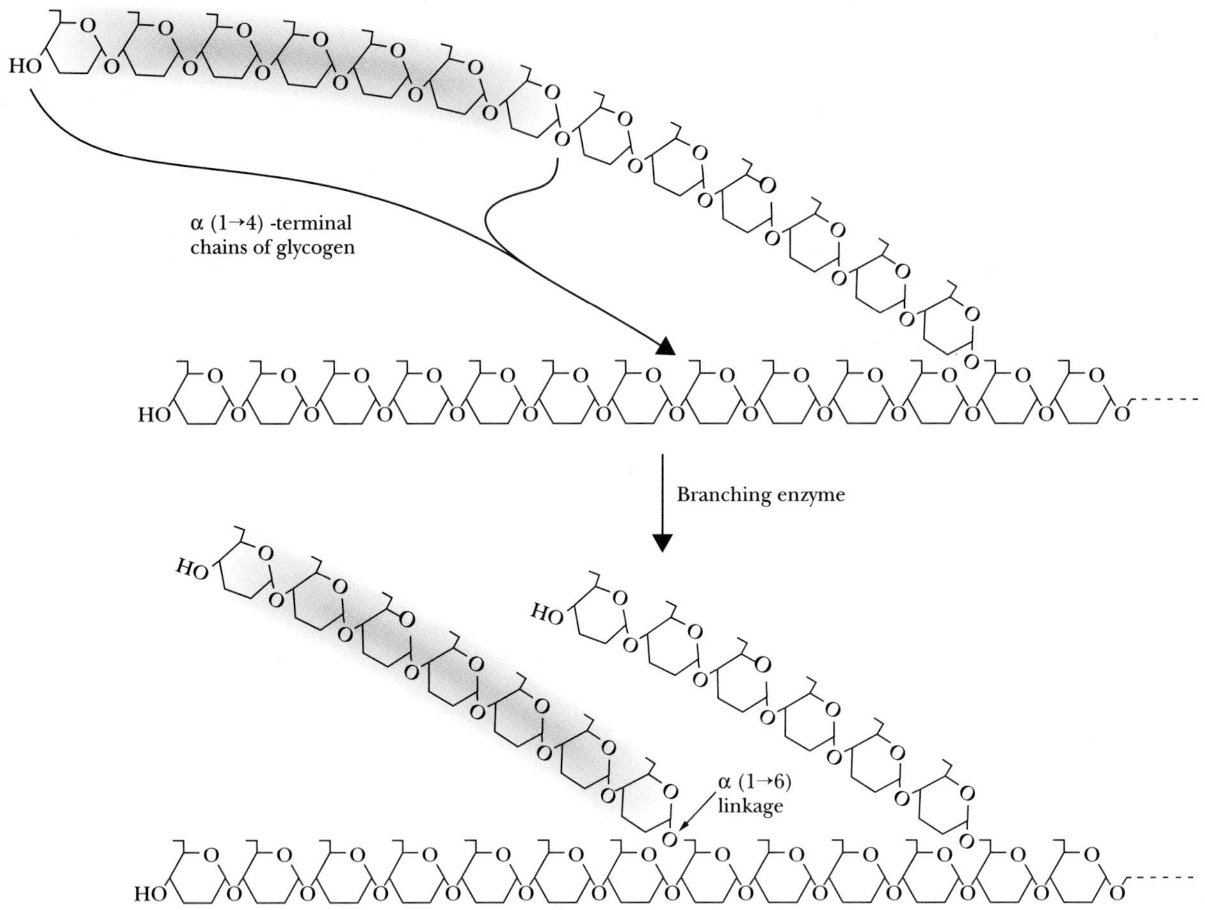

α (1→4) -terminal
chains of glycogen

Branching enzyme

α (1→6)
linkage

FIGURE 14.4

The mode of action of the branching enzyme in glycogen synthesis. A segment seven residues long is transferred from a growing branch to a new branch point, where an $\alpha(1\rightarrow 6)$ linkage is formed.

segment must come from a chain at least 11 residues long; each new branch point must be at least four residues away from the nearest existing branch point.

Control of Glycogen Metabolism: A Case Study in Control Mechanisms

How does an organism ensure that glycogen synthesis and glycogen breakdown do not operate simultaneously? If this were to occur, the main result would be the hydrolysis of UTP, which would waste chemical energy stored in the phosphoric anhydride bonds. A major controlling factor lies in the behavior of glycogen phosphorylase. This enzyme is not only subject to allosteric control, it represents another control feature as well: covalent modification. We saw an earlier example of this kind of control in the sodium–potassium pump in Chapter 8, Section 8.6. In that example phosphorylation and dephosphorylation of an enzyme determined whether or not it was active, and a similar effect takes place here.

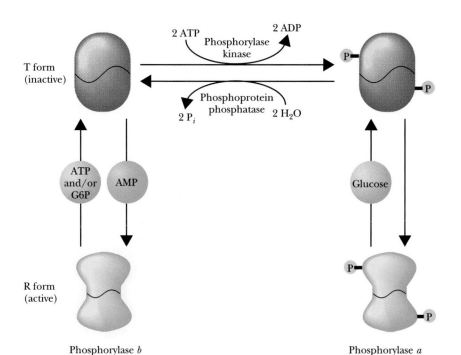

T form
(inactive)

2 ATP   2 ADP
Phosphorylase
kinase

Phosphoprotein
phosphatase
2 P$_i$ 2 H$_2$O

ATP
and/or
G6P   AMP

Glucose

R form
(active)

Phosphorylase *b*

Phosphorylase *a*

FIGURE 14.5

Glycogen phosphorylase activity is subject to allosteric control and covalent modification. Phosphorylation of the *a* form of the enzyme converts it to the *b* form. Only the T form is subject to modification and demodification. The *a* and *b* forms respond to different allosteric effectors (see text).

Figure 14.5 summarizes some of the salient control features that affect glycogen phosphorylase activity. The enzyme is a dimer that exists in two forms, the inactive T form and the active R form. In the T form (and *only* in the T form) it can be modified by phosphorylation of a specific serine residue on each of the two subunits. The esterification of the serines to phosphoric acid is catalyzed by the enzyme *phosphorylase kinase;* the dephosphorylation is catalyzed by *phosphoprotein phosphatase.* The phosphorylated form of glycogen phosphorylase is called **phosphorylase *a*,** and the dephosphorylated form is called **phosphorylase *b*.**

Phosphorylase *a* does not respond to the allosteric effectors that control the behavior of phosphorylase *b,* and *vice versa* (Figure 14.5). Glucose is an allosteric inhibitor of phosphorylase *a*, while ATP and glucose 6-phosphate (G6P) are allosteric inhibitors of phosphorylase *b;* AMP is an activator of phosphorylase *b*. These differences ensure that glycogen will be degraded when there is a need for energy and for ATP, as is the case with high [AMP], low [G6P], and low [ATP]. When the reverse is true (low [AMP], high [G6P], and high [ATP]), the need for energy and, consequently, the need for glycogen breakdown are less. "Shutting down" glycogen phosphorylase activity is the appropriate response. The combination of covalent modification and allosteric control of the process allows for a degree of fine tuning that would not be possible with either mechanism alone.

The activity of glycogen synthase is subject to the same combination of allosteric control and covalent modification. The only difference is that the response to effectors is the opposite of that of glycogen phosphorylase. The same enzymes (phosphorylase kinase and phosphoprotein phosphatase) that modify glycogen phosphorylase perform the same functions with glycogen synthase. The fact that two target enzymes, glycogen phosphorylase and glycogen synthase, are modified in the same way by the same enzymes links the opposing processes of synthesis and breakdown of glycogen even more intimately.

Finally, the modifying enzymes are themselves subject to covalent modification and allosteric control. This feature complicates the process considerably but adds the possibility of an amplified response to small changes in conditions. A small change in the concentration of an allosteric effector of a modifying enzyme can cause a large change in the concentra-

FIGURE 14.6

The pathway of gluconeogenesis. The enzymes in red are unique to this pathway. The enzymes that catalyze reversible reactions are shared with the glycolytic pathway.

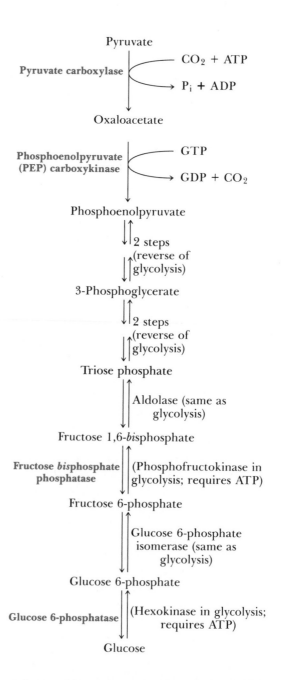

tion of an active, modified target enzyme; this amplification response is due to the fact that the substrate for the modifying enzyme is itself an enzyme. At this point the situation has become very complex indeed, but it is a good example of how opposing processes of breakdown and synthesis can be controlled to the advantage of an organism. When we see in the next section how glucose is synthesized from lactate, we shall have another example, one that we can contrast with glycolysis to explore in more detail how carbohydrate metabolism is controlled.

14.2
GLUCONEOGENESIS

The conversion of pyruvate to glucose occurs by a process called **gluconeogenesis.** Gluconeogenesis is not the exact reversal of glycolysis. Some of the reactions of glycolysis are essentially irreversible; these reactions are bypassed in gluconeogenesis. An analogy is a hiker who goes directly down a steep slope but who climbs back up the hill by an alternative, easier route. We shall see that the biosynthesis and the degradation of many important biomolecules follow different pathways.

There are three irreversible steps in glycolysis, and it is in these three reactions that the differences between glycolysis and gluconeogenesis are found. The first is the production of pyruvate (and ATP) from phosphoenolpyruvate. The second is the production of fructose 1,6-*bis*phosphate from fructose 6-phosphate, and the third is the production of glucose 6-phosphate from glucose. The first of these reactions is exergonic, and the reverse reaction is endergonic. Reversing the second and third reactions requires the production of ATP from ADP, which is also an endergonic reaction. The net result of gluconeogenesis is the reversal of these three glycolytic reactions, but the pathway is different, with different reactions and different enzymes (Figure 14.6).

Oxaloacetate Is an Intermediate in the Production of Phosphoenolpyruvate in Gluconeogenesis

The conversion of pyruvate to phosphoenolpyruvate in gluconeogenesis takes place in two steps. The first step is the reaction of pyruvate and carbon dioxide to give oxaloacetate. This step requires energy, which is available from the hydrolysis of ATP.

$$\text{Pyruvate} + \text{ATP} + CO_2 + H_2O \longrightarrow \text{oxaloacetate} + \text{ADP} + P_i + 2\,H^+$$

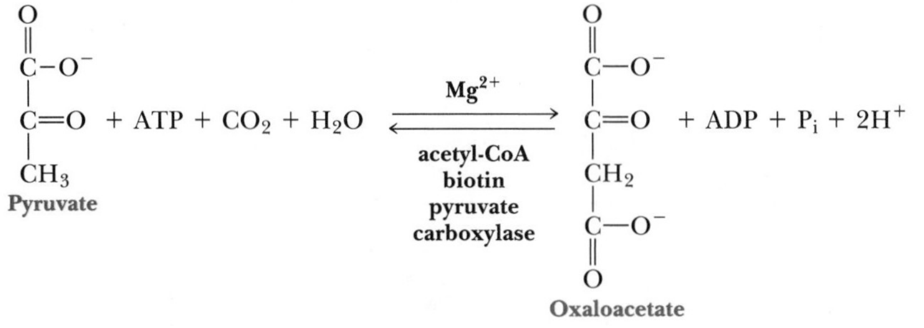

Oxaloacetate

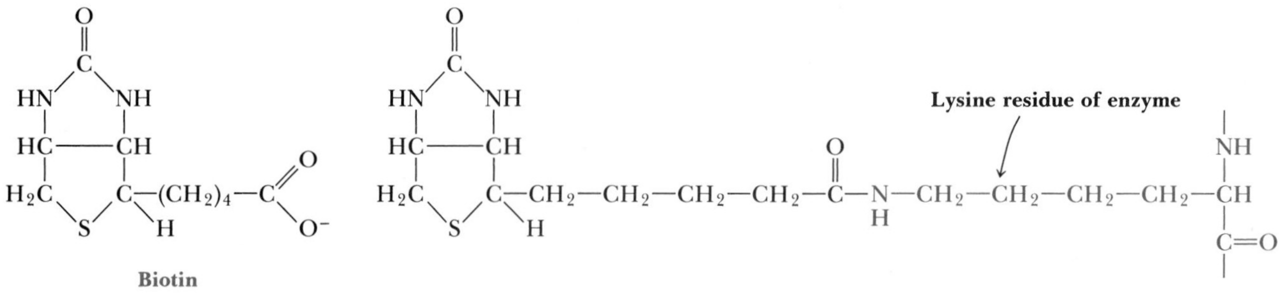

FIGURE 14.7

The structure of biotin and its mode of attachment to pyruvate carboxylase.

The enzyme that catalyzes this reaction is *pyruvate carboxylase,* an allosteric enzyme. Acetyl-CoA is an allosteric effector that activates pyruvate carboxylase. This reaction is a control point in the pathway because the enzyme that catalyzes it is subject to allosteric regulation. If high levels of acetyl-CoA are present (in other words, if there is more acetyl-CoA than is needed to supply the citric acid cycle), pyruvate (a precursor of acetyl-CoA) can be diverted to gluconeogenesis. Magnesium ion (Mg^{2+}) and biotin are also required for effective catalysis. We have seen Mg^{2+} as a cofactor before, but we have not seen biotin, and it requires some discussion.

Biotin is a carrier of carbon dioxide; it has a specific site for covalent attachment of CO_2 (Figure 14.7). The carboxyl group of the biotin forms an amide bond with the ϵ-amino group of a specific lysine side chain of pyruvate carboxylase. The CO_2 is attached to the biotin, which in turn is covalently bonded to the enzyme, and then the CO_2 is shifted to pyruvate to form oxaloacetate (Figure 14.8).

FIGURE 14.8

The two stages of the pyruvate carboxylase reaction. (a) CO_2 is attached to the biotinylated enzyme. (b) CO_2 is transferred from the biotinylated enzyme to pyruvate, forming oxaloacetate.

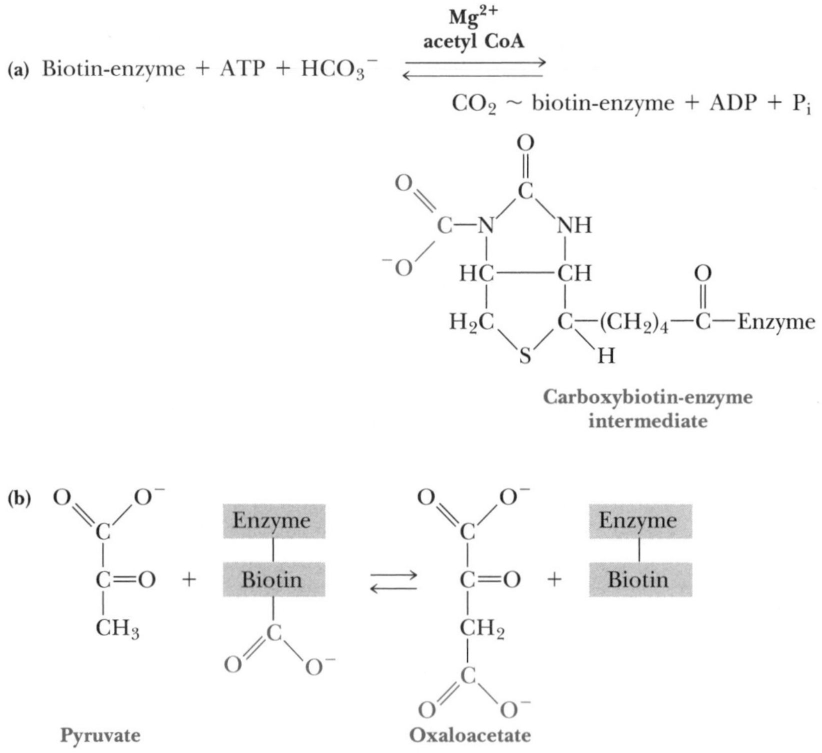

Carboxybiotin-enzyme intermediate

Pyruvate   Oxaloacetate

The conversion of oxaloacetate to phosphoenolpyruvate is catalyzed by the enzyme *phosphoenolpyruvate carboxykinase.* This reaction also involves hydrolysis of a nucleoside triphosphate, GTP in this case rather than ATP.

$$\text{Oxaloacetate} + \text{GTP} \xrightarrow{\text{Mg}^{2+}} \text{phosphoenolpyruvate} + CO_2 + \text{GDP}$$

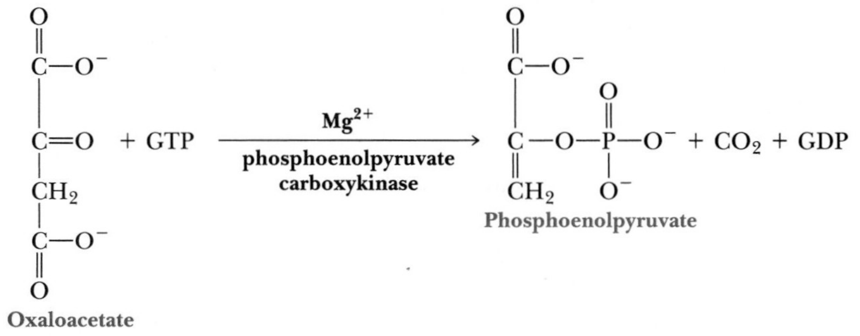

The successive carboxylation and decarboxylation reactions are both close to equilibrium and, as a result, the conversion of pyruvate to phosphoenolpyruvate is also close to equilibrium ($\Delta G^{\circ\prime} = 2.1$ kJ mol^{-1} = 0.5 kcal mol^{-1}). A small increase in the level of oxaloacetate can drive the equilibrium to the right, and a small increase in the level of phosphoenolpyruvate can drive it to the left. A concept well known in general chemistry, the **law of mass action,** relates the concentrations of reactants and products in a system at equilibrium to the equilibrium constant. The concept of equilibrium also implies that a reaction will proceed to the right on addition of reactants and to the left on addition of products.

$$\text{Pyruvate} + \text{ATP} + \text{GTP} \rightleftharpoons \text{phosphoenolpyruvate} + \text{ADP} + \text{GDP} + P_i$$

The Role of Sugar Phosphates in Gluconeogenesis

The other two reactions in which gluconeogenesis differs from glycolysis are ones in which a phosphate-ester bond to a sugar–hydroxyl group is hydrolyzed. Both reactions are catalyzed by phosphatases, and both reactions are exergonic. The first reaction is the hydrolysis of fructose 1,6-*bis*phosphate to produce fructose 6-phosphate and phosphate ion ($\Delta G^{\circ\prime} = -16.7$ kJ mol^{-1} = -4.0 kcal mol^{-1}).

$$\text{Fructose 1,6-}bis\text{phosphate} \longrightarrow \text{fructose 6-phosphate} + P_i$$

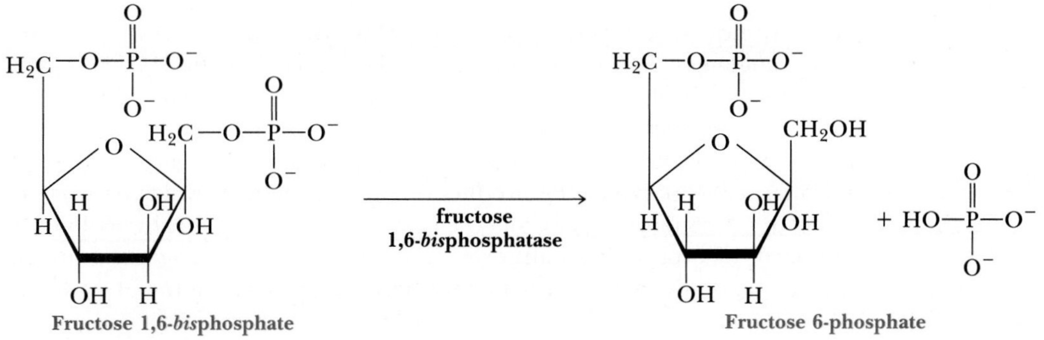

This reaction is catalyzed by the enzyme *fructose 1,6-bisphosphatase,* an allosteric enzyme strongly inhibited by adenosine monophosphate (AMP) but stimulated by ATP. Because of allosteric regulation, this reaction is also a control point in the pathway. When the cell has an ample supply of ATP, the formation rather than the breakdown of glucose is favored. This enzyme is inhibited by fructose 2,6-*bis*phosphate, a compound we met in Chapter 11, Section 11.2, as an extremely potent activator of phosphofructokinase. We shall return to this point in the next section.

The second reaction is the hydrolysis of glucose 6-phosphate to glucose and phosphate ion ($\Delta G^{\circ\prime} = -13.8$ kJ mol^{-1} = -3.3 kcal mol^{-1}). The enzyme that catalyzes this reaction is *glucose 6-phosphatase.*

$$\text{Glucose 6-phosphate} \longrightarrow \text{glucose} + P_i$$

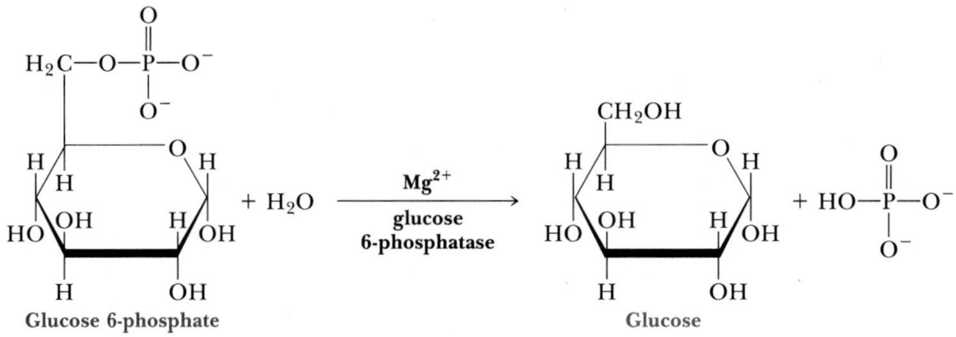

When we discussed glycolysis in Chapter 11, we saw that both of the phosphorylation reactions that are the reverses of the two phosphatase-catalyzed reactions are endergonic. In glycolysis the phosphorylation reactions must be coupled to the hydrolysis of ATP to make them exergonic and thus energetically allowed. In gluconeogenesis the organism can make direct use of the fact that the hydrolysis reactions of the sugar phosphates are exergonic. The corresponding reactions are not the reverse of each other in the two pathways. They differ from each other in whether they require ATP and in the enzymes involved.

14.3
CONTROL MECHANISMS IN CARBOHYDRATE METABOLISM

We have now seen several aspects of carbohydrate metabolism: glycolysis, gluconeogenesis, and the reciprocal breakdown and synthesis of glycogen. Glucose has a central role in all these processes. It is the starting point for glycolysis, in which it is broken down to pyruvate, and for the synthesis of glycogen, in which many glucose residues combine to give the glycogen polymer. Glucose is also the product of gluconeogenesis, which has the net effect of reversing glycolysis; glucose is also obtained from the breakdown of glycogen. The opposing pathways, glycolysis and gluconeogenesis on the one hand and the breakdown and synthesis of glycogen on the other hand,

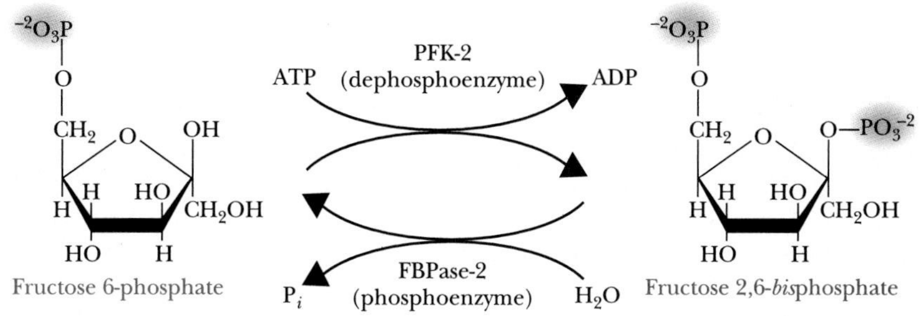

Fructose 6-phosphate

PFK-2
(dephosphoenzyme)

ATP   ADP

FBPase-2
(phosphoenzyme)

P_i H_2O

Fructose 2,6-*bis*phosphate

FIGURE 14.9

The formation and breakdown of fructose 2,6-*bis*phosphate (F2,6P) are catalyzed by two enzyme activities on the same protein. These two enzyme activities are controlled by a phosphorylation/ dephosphorylation mechanism. Phosphorylation activates the enzyme that degrades F2,6P while dephosphorylation activates the enzyme that produces it.

are not the exact reversal of each other, even though the net results are simply those of reversal. It is time to see how all these related pathways are controlled.

An important element in the control process involves fructose 2,6-*bis*phosphate (F2,6P). We mentioned in Chapter 11, Section 11.2 that this compound is an important allosteric activator of phosphofructokinase, the key enzyme of glycolysis; it is also an inhibitor of fructose *bis*phosphate phosphatase, which plays a role in gluconeogenesis. A high concentration of F2,6P stimulates glycolysis, whereas a low concentration stimulates gluconeogenesis. The concentration of F2,6P in a cell depends on the balance between its synthesis (catalyzed by *phosphofructokinase-2* [PFK-2]) and its breakdown (catalyzed by *fructose bisphosphatase-2* [FBPase-2]). The enzymes that control the formation and breakdown of F2,6P are themselves controlled by a phosphorylation/dephosphorylation mechanism similar to that we have already seen in the case of glycogen phosphorylase and glycogen synthase (Figure 14.9). Both enzyme activities are located on the same protein (a dimer of about 100 kD molecular weight). Phosphorylation of the dimeric protein leads to an increase in activity of FBPase-2 and a decrease in the concentration of F2,6P, ultimately stimulating gluconeogenesis. Dephosphorylation of the dimeric protein leads to an increase in PFK-2 activity and an increase in the concentration of F2,6P, ultimately stimulating glycolysis. The net result is similar to the control of glycogen synthesis and breakdown that we saw in Section 14.1.

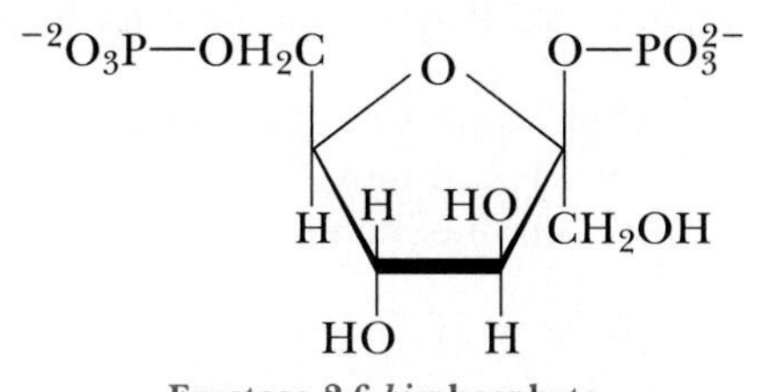

Fructose 2,6-*bis*phosphate
(F2,6P)

Table 14.1 summarizes important mechanisms of metabolic control; even though we discuss them here in the context of carbohydrate metabolism, they apply to all aspects of metabolism. Of the four kinds of control mechanisms listed in Table 14.1 — allosteric control, covalent modification, substrate cycling, and genetic control — we have already seen examples of the first two, allosteric control and covalent modification. We need more

T A B L E 1 4 . 1   Mechanisms of Metabolic Control

TYPE OF CONTROL	MODE OF OPERATION	EXAMPLES
Allosteric	Effectors (substrates, products, or coenzymes) of a pathway inhibit or activate an enzyme. (Responds rapidly to external stimuli.)	ATCase (Section 5.5); phosphofructokinase (Section 11.2)
Covalent modification	Inhibition or activation of enzyme depends on formation or breaking of a bond, frequently by phosphorylation or dephosphorylation. (Responds rapidly to external stimuli.)	Sodium–potassium pump (Section 8.6); glycogen phosphorylase, glycogen synthase (Section 14.1)
Substrate cycles	Two opposing reactions, such as formation and breakdown of a given substance, are catalyzed by different enzymes, which can be activated or inhibited separately. (Responds rapidly to external stimuli.)	Formation and breakdown of fructose 1,6-*bis*phosphate in glycolysis (Chapter 11), and gluconeogenesis (Section 14.2)
Genetic control	The amount of enzyme present is increased by protein synthesis. (Longer-term control than the other mechanisms listed here.)	Induction of β-galactosidase (Section 20.9)

information about the synthesis of proteins and nucleic acids to discuss genetic control, so we shall defer the topic to Chapter 20. Substrate cycling is a mechanism that we can profitably discuss here.

The term **substrate cycling** refers to the fact that opposing reactions can be catalyzed by different enzymes. Consequently the opposing reactions can be independently regulated and have different rates. It would not be possible to have different rates with the same enzyme, because a catalyst speeds up a reaction and the reverse of the reaction at the same rate (Chapter 5, Section 5.2). We shall use the conversion of fructose 6-phosphate to fructose 1,6-*bis*phosphate and then back to fructose 6-phosphate as an example of a substrate cycle. In glycolysis, the reaction catalyzed by phosphofructokinase is highly exergonic under physiological conditions ($\Delta G = -25.9$ kJ·mol^{-1} = -6.2 kcal·mol^{-1}).

$$\text{Fructose 6-phosphate} + \text{ATP} \longrightarrow \text{fructose 1,6-}bis\text{phosphate} + \text{ADP}$$

The opposing reaction, which is part of gluconeogenesis, is also exergonic ($\Delta G = -8.6$ kJ·mol^{-1} = -2.1 kcal·mol^{-1} under physiological conditions) and is catalyzed by another enzyme, namely fructose 1,6-*bis*phosphatase.

$$\text{Fructose 1,6-}bis\text{phosphate} + \text{H}_2\text{O} \longrightarrow \text{fructose 6-phosphate} + \text{P}_i$$

Note that the opposing reactions are not the exact reverse of one another. If we add the two opposing reactions together, we obtain the net reaction

$$\text{ATP} + \text{H}_2\text{O} \rightleftharpoons \text{ADP} + \text{P}_i$$

Hydrolysis of ATP is the energetic price that is paid for independent control of the opposing reactions.

Using combinations of these control mechanisms, an organism can set up a division of labor among tissues and organs to maintain control of

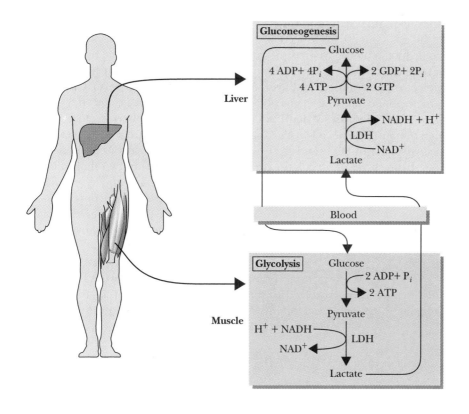

FIGURE 14.10

The Cori cycle. Lactate produced in muscles by glycolysis is transported by the blood to the liver. Gluconeogenesis in the liver converts the lactate back to glucose, which can be carried back to the muscles by the blood. Glucose can be stored as glycogen until it is degraded by glycogenolysis.

glucose metabolism. A particularly clear example is found in the Cori cycle, and we shall discuss it here. Shown in Figure 14.10, the Cori cycle is named for Gerty and Carl Cori, who first described it. There is cycling of glucose due to glycolysis in muscle and gluconeogenesis in liver. Glycolysis in skeletal muscle produces lactate under conditions of oxygen debt such as a sprint. Skeletal muscle has comparatively few mitochondria, so metabolism is largely anaerobic in this tissue. The buildup of lactate is responsible for the muscular aches that follow strenuous exercise. Gluconeogenesis recycles the lactate that is produced (lactate is first oxidized to pyruvate). The process

Gerty and Carl Cori, codiscoverers of the Cori cycle.

occurs to a great extent in the liver after the lactate is transported there by the blood. Glucose produced in the liver is transported back to skeletal muscle by the blood, where it becomes an energy store for the next burst of exercise. Note that we have here a division of labor between two different types of organs, muscle and liver. In the same cell (of whatever type) these two metabolic pathways, glycolysis and gluconeogenesis, are not highly active simultaneously. When the cell needs ATP, glycolysis is more active; when there is little need for ATP, gluconeogenesis is more active. Because of the hydrolysis of ATP and GTP in the reactions of gluconeogenesis that differ from those of glycolysis, the overall pathway from two molecules of pyruvate back to one molecule of glucose is exergonic ($\Delta G^{\circ\prime} = -37.6$ kJ $mol^{-1} = -9.0$ kcal mol^{-1}, for one mole of glucose). The conversion of pyruvate to lactate is exergonic, which means that the reverse reaction is endergonic. The energy released by the exergonic conversion of pyruvate to glucose by gluconeogenesis facilitates the conversion of lactate to pyruvate.

Note that the Cori cycle requires the net hydrolysis of two ATP and two GTP. While ATP is produced by the glycolytic part of the cycle, the portion involving gluconeogenesis requires yet more ATP in addition to GTP.

Glycolysis:

$$\text{Glucose} + 2\ NAD^+ + 2\ ADP + 2\ P_i \longrightarrow$$
$$2\ \text{pyruvate} + 2\ NADH + 4\ H^+ + 2\ ATP + 2\ H_2O$$

Gluconeogenesis:

$$2\ \text{Pyruvate} + 2\ NADH + 4\ H^+ + \textbf{4 ATP} + \textbf{2 GTP} + 6\ H_2O \longrightarrow$$
$$\text{glucose} + 2\ NAD^+ + 4\ ADP + 2\ GDP + 6\ P_i$$

Overall:

$$2\ ATP + 2\ GTP + 4\ H_2O \longrightarrow 2\ ADP + 2\ GDP + 4\ P_i$$

The hydrolysis of both ATP and GTP is the price of increased simultaneous control of the two opposing pathways.

14.4
THE PENTOSE PHOSPHATE PATHWAY – AN ALTERNATIVE PATHWAY FOR GLUCOSE METABOLISM

The **pentose phosphate pathway** is an alternative to glycolysis and differs from it in several important ways. In glycolysis, one of our most important concerns was the production of ATP. In the pentose phosphate pathway, the production of ATP is not the crux of the matter. As the name of the pathway indicates, five-carbon sugars, including ribose, are produced from glucose. Ribose plays an important role in the structure of nucleic acids. Another important facet of the pentose phosphate pathway is the production of NADPH, a compound that differs from NADH by having one extra phosphate group esterified to the ribose ring of the adenine nucleotide portion of the molecule (Figure 14.11). A more important difference is the

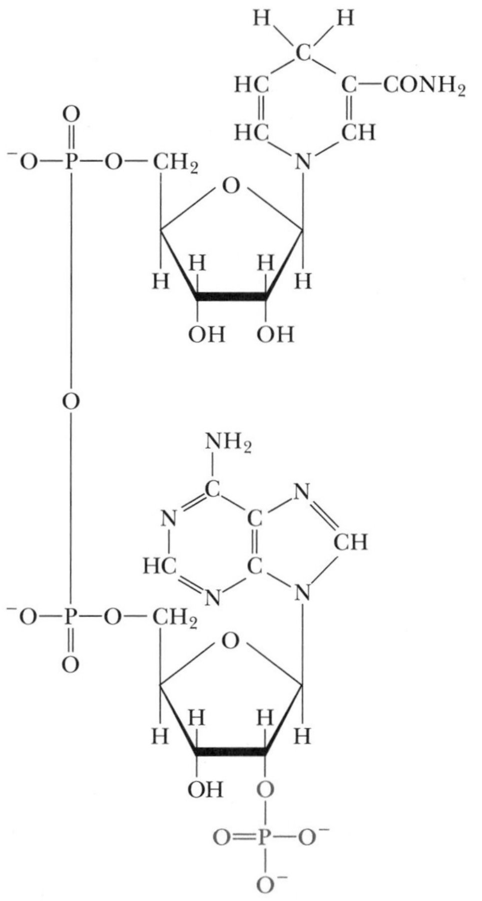

FIGURE 14.11

The structure of reduced nicotinamide adenine dinucleotide phosphate (NADPH).

way these two coenzymes function. NADH is produced in the oxidative reactions that give rise to ATP. NADPH is a reducing agent in biosynthesis, which by its very nature is a reductive process.

The pentose phosphate pathway begins with a series of oxidation reactions that produce NADPH and five-carbon sugars. The remainder of the pathway involves nonoxidative reshuing of the carbon skeletons of the sugars involved. The products of these nonoxidative reactions include substances such as fructose 6-phosphate and glyceraldehyde 3-phosphate, which play a role in glycolysis.

Oxidative Reactions of the Pentose Phosphate Pathway

In the first reaction of the pathway, glucose 6-phosphate is oxidized to 6-phosphoglucono-δ-lactone (Figure 14.12). The enzyme that catalyzes this reaction is *glucose 6-phosphate dehydrogenase*. Note that NADPH is produced by the reaction. The same sort of reaction takes place in the oxidation of sugars by metal ions (Chapter 10, Section 10.2). The hemiacetal form of the sugar is a cyclic structure resulting from the addition of one of the hydroxyl groups to the aldehyde group. It is oxidized to a lactone, which is a cyclic ester formed between the resulting carboxylic acid group and an alcohol group elsewhere in the molecule.

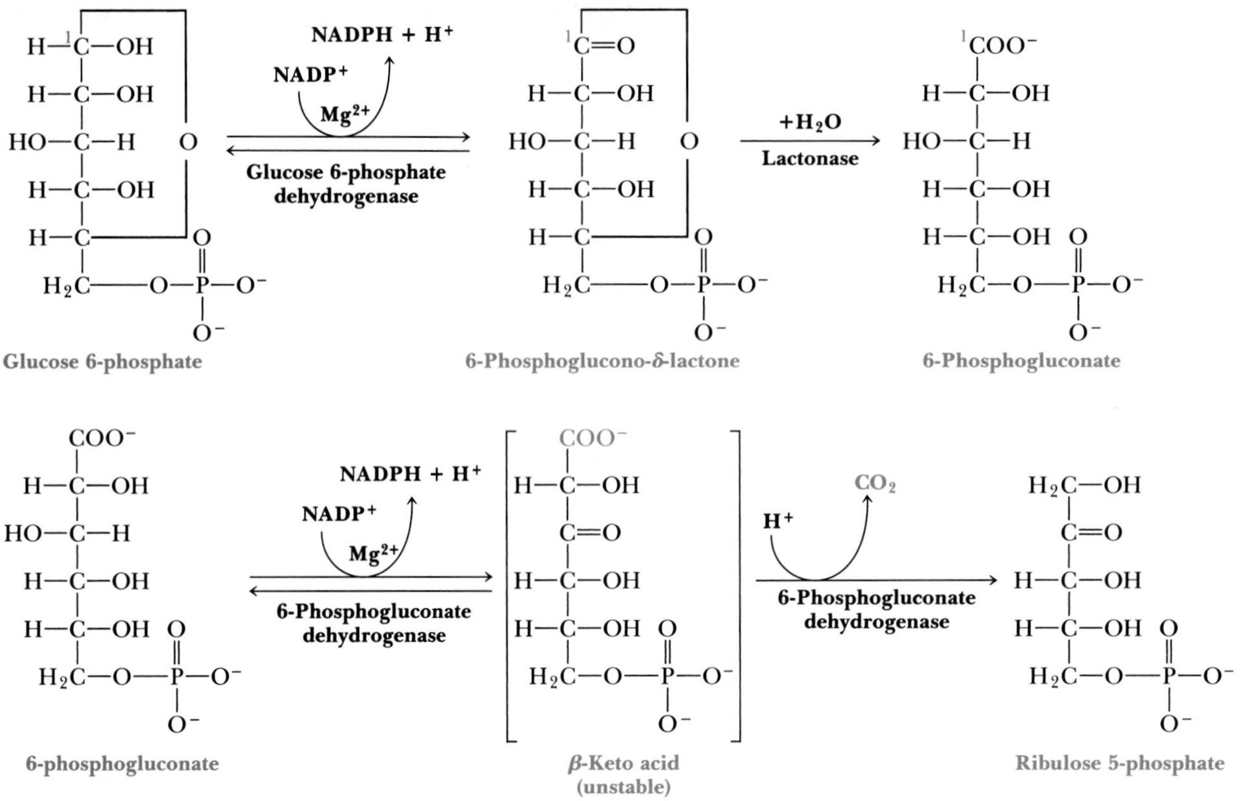

FIGURE 14.12

The oxidative reactions of the pentose phosphate pathway.

The cyclic ester bond of 6-phosphoglucono-δ-lactone is hydrolyzed in the next reaction. The open-chain compound 6-phosphogluconate (the gluconic acid is present in its ionized form) is the product of this reaction, which is catalyzed by the enzyme *lactonase* (Figure 14.12).

The next reaction is an oxidative decarboxylation, and NADPH is produced once again. The 6-phosphogluconate molecule loses its carboxyl group, which is released as carbon dioxide, and the five-carbon keto-sugar (ketose) ribulose 5-phosphate is the other product. The enzyme that catalyzes this reaction is *6-phosphogluconate dehydrogenase*. Note that in the process the C-3 hydroxyl group of the 6-phosphogluconate is oxidized to form a β-keto acid, which is unstable and readily decarboxylates to form ribulose 5-phosphate.

Nonoxidative Reactions of the Pentose Phosphate Pathway

In the remaining steps of the pentose phosphate pathway, several reactions involve transfer of two- and three-carbon units. In order to keep track of the carbon backbone of the sugars and their aldehyde and ketone functional groups, we shall write the formulas in the open-chain form.

There are two different reactions in which ribulose 5-phosphate isomerizes. In one of these reactions, catalyzed by *phosphopentose 3-epimerase*, there is an inversion of configuration around carbon atom 3, producing xylulose 5-phosphate, which is also a ketose (Figure 14.13). The other isomerization reaction, catalyzed by *phosphopentose isomerase,* produces a sugar with an aldehyde group (an aldose) rather than a ketone. In this second reaction, ribulose 5-phosphate isomerizes to ribose 5-phosphate (Figure 14.13).

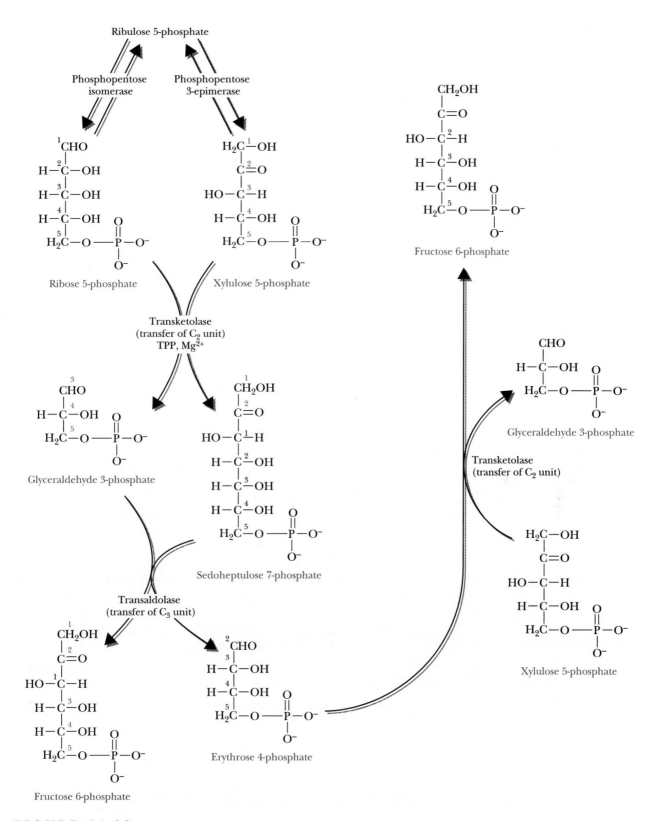

FIGURE 14.13

The nonoxidative reactions of the pentose phosphate pathway. Carbons from xylulose are numbered in color.

Ribose 5-phosphate is a necessary building block for the synthesis of nucleic acids and coenzymes such as NADH.

The group-transfer reactions that link the pentose phosphate pathway with glycolysis require the two 5-carbon sugars produced by the isomerization of ribulose 5-phosphate. Two molecules of xylulose 5-phosphate and one molecule of ribose 5-phosphate rearrange to give two molecules of fructose 6-phosphate and one molecule of glyceraldehyde 3-phosphate. In other words, three molecules of pentose (with five carbon atoms each) give two molecules of hexose (with six carbon atoms each) and one molecule of a triose (with three carbon atoms). The total number of carbon atoms (15) does not change, but there is considerable rearrangement as a result of group transfer.

Two enzymes, *transketolase* and *transaldolase,* are responsible for the reshuffling of the carbon atoms of sugars such as ribose 5-phosphate and xylulose 5-phosphate in the remainder of the pathway, which consists of three reactions. Transketolase transfers a two-carbon unit, and transaldolase transfers a three-carbon unit. Transketolase catalyzes the first and third reactions in the rearrangement process, and transaldolase catalyzes the second reaction. The results of these transfers can be summarized in Table 14.2. In the first of these reactions a two-carbon unit from xylulose 5-phosphate (five carbons) is transferred to ribose 5-phosphate (five carbons) to give sedoheptulose 7-phosphate (seven carbons) and glyceraldehyde 3-phosphate (three carbons), as shown in Figure 14.13. This reaction is catalyzed by transketolase.

In the reaction catalyzed by transaldolase, a three-carbon unit is transferred from the seven-carbon sedoheptulose 7-phosphate to the three-carbon glyceraldehyde 3-phosphate (Figure 14.13). The products of the reaction are fructose 6-phosphate (six carbons) and erythrose 4-phosphate (four carbons).

In the final reaction of this type in the pathway, xylulose 5-phosphate reacts with erythrose 4-phosphate. This reaction is catalyzed by transketolase. The products of the reaction are fructose 6-phosphate and glyceraldehyde 3-phosphate (Figure 14.13).

In the pentose phosphate pathway, glucose 6-phosphate can be converted to fructose 6-phosphate and glyceraldehyde 3-phosphate by a means

TABLE 14.2   Group Transfer Reactions in the Pentose Phosphate Pathway

	REACTANT	ENZYME	PRODUCTS
	$C_5 + C_5$	Transketolase $\rightleftharpoons$ Two-carbon shift	$C_7 + C_3$
	$C_7 + C_3$	Transaldolase $\rightleftharpoons$ Three-carbon shift	$C_6 + C_4$
	$C_5 + C_4$	Transketolase $\rightleftharpoons$ Two-carbon shift	$C_6 + C_3$
Net reaction	$3\ C_5$	$\rightleftharpoons$	$2\ C_6 + C_3$

other than the glycolytic pathway. For this reason the pentose phosphate pathway is also called the **hexose monophosphate shunt,** and this name is used in some texts. A major feature of the pentose phosphate pathway is the production of ribose 5-phosphate and NADPH. The control mechanisms of the pentose phosphate pathway can respond to the varying needs of organisms for either or both of these compounds.

Control of the Pentose Phosphate Pathway

As we have seen, the reactions catalyzed by transketolase and transaldolase are reversible, which allows the pentose phosphate pathway to respond to the needs of an organism. The starting material, glucose 6-phosphate, will undergo different reactions depending on whether there is a greater need for ribose 5-phosphate or for NADPH. The operation of the oxidative portion of the pathway depends strongly on the organism's requirement for NADPH. The need for ribose 5-phosphate can be met in other ways, since ribose 5-phosphate can be obtained from glycolytic intermediates without the oxidative reactions of the pentose phosphate pathway (Figure 14.14).

If the organism needs more NADPH than ribose 5-phosphate, the reaction series goes through the complete pathway just discussed. The

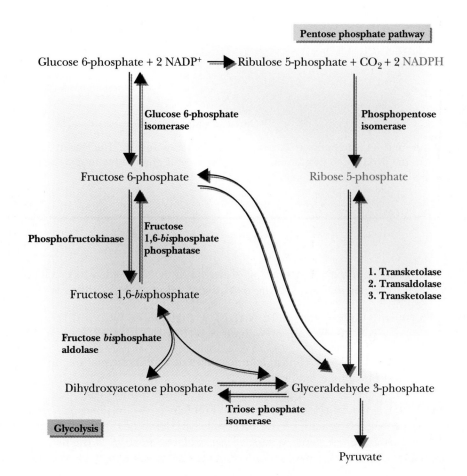

FIGURE 14.14

Relationships between the pentose phosphate pathway and glycolysis. If the organism needs NADPH more than ribose 5-phosphate, the entire pentose phosphate pathway is operative. If the organism needs ribose 5-phosphate more than NADPH, the nonoxidative reactions of the pentose phosphate pathway, operating in reverse, produce ribose 5-phosphate (see text).

THE PENTOSE PHOSPHATE PATHWAY AND HEMOLYTIC ANEMIA

The pentose phosphate pathway is the only source of NADPH in red blood cells, which as a result are highly dependent on the proper functioning of the enzymes involved. A glucose 6-phosphate dehydrogenase deficiency leads to an NADPH deficiency, which can in turn lead to **hemolytic anemia** because of wholesale destruction of red blood cells.

The relationship between NADPH deficiency and anemia is an indirect one. NADPH is required to reduce the peptide glutathione from the disulfide to the free thiol form. The presence of the reduced form of glutathione is neces-

sary for the maintenance of the sulfhydryl groups of hemoglobin and other proteins in their reduced forms, as well as for keeping the Fe(II) of hemoglobin in its reduced form. Glutathione also maintains the integrity of red cells by reacting with peroxides that would otherwise degrade fatty acid side chains in the cell membrane. About 11% of African Americans are affected by glucose 6-phosphate dehydrogenase deficiency. This condition, like the sickle-cell trait, leads to increased resistance to malaria, accounting for some of its persistence in the gene pool in spite of its otherwise deleterious consequences.

(a)

$$
\begin{array}{c}
\text{COO}^- \\
| \\
{}^+\text{H}_3\text{N}-\text{CH} \qquad \gamma\text{-Glu} \\
| \\
(\text{CH}_2)_2 \\
| \\
\text{C}=\text{O} \\
| \\
\text{NH} \qquad\qquad | \\
| \\
\text{HS}-\text{CH}_2-\text{CH} \qquad \text{Cys} \\
| \\
\text{C}=\text{O} \\
| \\
\text{NH} \qquad\qquad | \\
| \\
\text{CH}_2 \qquad \text{Gly} \\
| \\
\text{COO}^-
\end{array}
$$

Reduced glutathione
(γ-glutamylcysteinylglycine)

(b)

$$
\begin{array}{c}
\gamma\text{-Glu}-\text{Cys}-\text{Gly} \\
| \\
\text{S} \\
| \\
\text{S} \\
| \\
\gamma\text{-Glu}-\text{Cys}-\text{Gly}
\end{array}
+ \text{NADPH} + \text{H}^+
\xrightarrow{\text{Glutathione reductase}}
2\ \gamma\text{-Glu}-\text{Cys}-\text{Gly} + \text{NADP}^+ \\
\qquad\qquad\qquad\qquad\qquad\qquad\qquad\qquad |\\
\qquad\qquad\qquad\qquad\qquad\qquad\qquad\qquad \text{SH}
$$

(c)

$$
2\ \gamma\text{-Glu}-\text{Cys}-\text{Gly} + \text{R-S-S-R} \longrightarrow \gamma\text{-Glu}-\text{Cys}-\text{Gly} + 2\ \text{RSH} \\
\qquad\qquad | \qquad\qquad\qquad\qquad\qquad\qquad\qquad\qquad | \\
\qquad\qquad \text{SH} \qquad\qquad\qquad\qquad\qquad\qquad\qquad \text{S} \\
\qquad\qquad\qquad\qquad\qquad\qquad\qquad\qquad\qquad\qquad | \\
\qquad\qquad\qquad\qquad\qquad\qquad\qquad\qquad\qquad\qquad \text{S} \\
\qquad\qquad\qquad\qquad\qquad\qquad\qquad\qquad\qquad\qquad | \\
\qquad\qquad\qquad\qquad\qquad\qquad\qquad\qquad \gamma\text{-Glu}-\text{Cys}-\text{Gly}
$$

Glutathione and its reactions. (a) The structure of glutathione. (b) The role of NADPH in the production of glutathione. (c) The role of glutathione in maintaining the reduced form of protein sulfhydryl groups.

oxidative reactions at the beginning of the pathway are needed to produce NADPH. The net reaction for the oxidative portion of the pathway is

$$
\text{6 Glucose 6-phosphate} + 12\ \text{NADP}^+ + 6\ \text{H}_2\text{O} \longrightarrow
$$
$$
\text{6 ribose 5-phosphate} + 6\ \text{CO}_2 + 12\ \text{NADPH} + 12\ \text{H}^+
$$

Box 14.1 discusses a clinical manifestation of an enzyme malfunction in the pentose phosphate pathway.

If the organism has a greater need for ribose 5-phosphate than for NADPH, fructose 6-phosphate and glyceraldehyde 3-phosphate can give rise to ribose 5-phosphate by the successive operation of the transketolase and transaldolase reactions, bypassing the oxidative portion of the pentose phosphate pathway (Figure 14.14). The reactions catalyzed by transketolase and transaldolase are reversible, and this fact plays an important role in the ability of the organism to adjust its metabolism to changes in conditions. We shall now take a look at the mode of action of these two enzymes.

Transaldolase has many features in common with the enzyme aldolase, which we met in the glycolytic pathway. Both an aldol cleavage and an aldol condensation occur at different stages of the reaction. We already saw the mechanism of aldol cleavage, involving the formation of a Schiff base, when we discussed the aldolase reaction in glycolysis, and we need not discuss this point further.

Transketolase resembles pyruvate decarboxylase, the enzyme that converts pyruvate to acetaldehyde (Chapter 11, Section 11.4), in that it also requires Mg^{2+} and thiamine pyrophosphate (TPP). As in the pyruvate decarboxylase reaction, a carbanion plays a crucial role in the reaction mechanism, which is similar to that of the conversion of pyruvate to acetaldehyde.

S U M M A R Y

When an organism has available a supply of extra glucose, more than is immediately needed as a source of energy extracted in glycolysis, it forms glycogen, a polymer of glucose. Glycogen can be readily broken down to glucose in response to energy needs. Control mechanisms ensure that both formation and breakdown of glycogen are not active simultaneously, a situation that would waste energy.

The conversion of pyruvate (the product of glycolysis) to glucose takes place by a process called gluconeogenesis. Gluconeogenesis is not the exact reversal of glycolysis. There are three irreversible steps in glycolysis, and it is in these three reactions that gluconeogenesis differs from glycolysis. The net result of gluconeogenesis is the reversal of these three glycolytic reactions, but the pathway is different, with different reactions and different enzymes. In the same cell,

glycolysis and gluconeogenesis are not highly active simultaneously. When the cell needs ATP, glycolysis is more active; when there is little need for ATP, gluconeogenesis is more active. Glycolysis and gluconeogenesis play roles in the Cori cycle. The division of labor between liver and muscle allows glycolysis and gluconeogenesis to take place in different organs to serve the needs of an organism.

The pentose phosphate pathway is an alternative pathway for glucose metabolism. In this pathway five-carbon sugars, including ribose, are produced from glucose. In the oxidative reactions of the pathway, NADPH is produced as well. Control of the pathway allows the organism to adjust the relative levels of production of five-carbon sugars and of NADPH according to its needs.

E X E R C I S E S

1. Which reaction or reactions discussed in this chapter require ATP? Which reaction or reactions produce ATP? List the enzymes that catalyze the reactions that require and that produce ATP.
2. List three differences in structure or function between NADH and NADPH.
3. What reactions in this chapter require acetyl-CoA or biotin?
4. What is the connection between material in this chapter and hemolytic anemia?
5. Which steps of glycolysis are irreversible? What bearing does this observation have on the reactions in which

gluconeogenesis differs from glycolysis?

6. Using the Lewis electron-dot notation, show explicitly the transfer of electrons in the following redox reaction.

$$\text{Glucose 6-phosphate} + \text{NADP}^+ \longrightarrow$$
$$\text{6-phosphoglucono-}\delta\text{-lactone} + \text{NADPH} + \text{H}^+$$

7. Does the net gain of ATP in glycolysis differ when glycogen, rather than glucose, is the starting material? If so, what is the change?

8. What is a major difference between transketolase and transaldolase?

9. You are planning to go on a strenuous hike and are advised to eat plenty of high-carbohydrate foods, such as bread and pasta, for several days beforehand. Suggest a reason for the advice.

ANNOTATED BIBLIOGRAPHY

Florkin, M., and E. H. Stotz, eds. *Comprehensive Biochemistry.* New York: Elsevier, 1967. [A standard reference. Volume 17, *Carbohydrate Metabolism,* deals with glycolysis and related topics.]

Hers, H. G., and L. Hue. Gluconeogenesis and Related Aspects of Glycolysis. *Ann. Rev. Biochem.* **52,** 617–653 (1983). [A review that concentrates on the relationship between glycolysis and gluconeogenesis.]

Horecker, B. L. Transaldolase and Transketolase. *Comprehensive Biochem.,* Vol. 15 (1973). [A review of these two enzymes and their mechanism of action.]

Lipmann, F. A Long Life in Times of Great Upheaval. *Ann. Rev. Biochem.* **53,** 1–33 (1984). [The reminiscences of a Nobel laureate whose research contributed greatly to the understanding of carbohydrate metabolism. Very interesting reading from the standpoint of autobiography and the author's contributions to biochemistry.]

Meister, A., and M. E. Anderson. Glutathione. *Ann. Rev. Biochem.* **52,** 711–760 (1983). [A review of several aspects of the action of this peptide.]

Lipid Metabolism

A human heart artery blocked by cholesterol (artherosclerosis), viewed under polarized light.

In the energy economy of the cell, glucose reserves are like ready cash, whereas lipid reserves are like a fat savings account. The potential energy of lipids resides in the fatty acid chains of triacylglycerols. When there are excess calories, fatty acids are *synthesized* and stored in fat cells. When energy demands are great, fatty acids are *catabolized* to liberate energy. The synthesis of fatty acids begins with acetyl-CoA, after which carbon atoms are added to the growing hydrocarbon chain, usually two at a time. Catabolism proceeds in the opposite direction — beginning with the carboxyl group and ending with acetyl-CoA. The fragmented hydrocarbon chain and acetyl-CoA are both oxidized in the citric acid cycle to provide energy that is temporarily stored as ATP. Pathways of lipid synthesis and catabolism of lipids occur simultaneously, but in different parts of the cell. If these pathways were to operate at the same time in the same place, the lipids would be taken apart as soon as they were synthesized. To avoid such a futile cycle, the pathways of lipid synthesis and catabolism differ in important ways: (1) the reactions proceed in different directions; (2) synthesis takes place in the cytosol while catabolism takes place in the mitochondrion; and (3) NADPH is the donor of "high-energy" electrons in lipid synthesis, whereas FAD and NAD^+ are electron acceptors in lipid catabolism.

15.1
THE METABOLISM OF LIPIDS PROVIDES PATHWAYS FOR THE GENERATION AND STORAGE OF ENERGY

In the past few chapters we have seen how energy can be released by the catabolic breakdown of carbohydrates in aerobic and anaerobic processes. Earlier, in Chapter 10, we saw that there are carbohydrate polymers, such as starch in plants and glycogen in animals, which represent stored energy in the sense that these carbohydrates can be hydrolyzed to monomers and then oxidized to provide energy in response to the needs of an organism. In this chapter we shall see how the metabolic oxidation of lipids releases large quantities of energy and how lipids represent an even more efficient way of storing chemical energy.

15.2
CATABOLISM OF LIPIDS

The oxidation of fatty acids is the chief source of energy in the catabolism of lipids; in fact, lipids that are sterols (steroids that have a hydroxyl group as part of their structure; Chapter 8, Section 8.2) are not catabolized as a source of energy, but are excreted. Both triacylglycerols, which are the main storage form of the chemical energy of lipids, and phosphoacylglycerols, which are important components of biological membranes, have fatty acids as part of their covalently bonded structures. In both types of compounds, the bond between the fatty acid and the rest of the molecule can be hydrolyzed (Figure 15.1), with the reaction catalyzed by suitable groups of

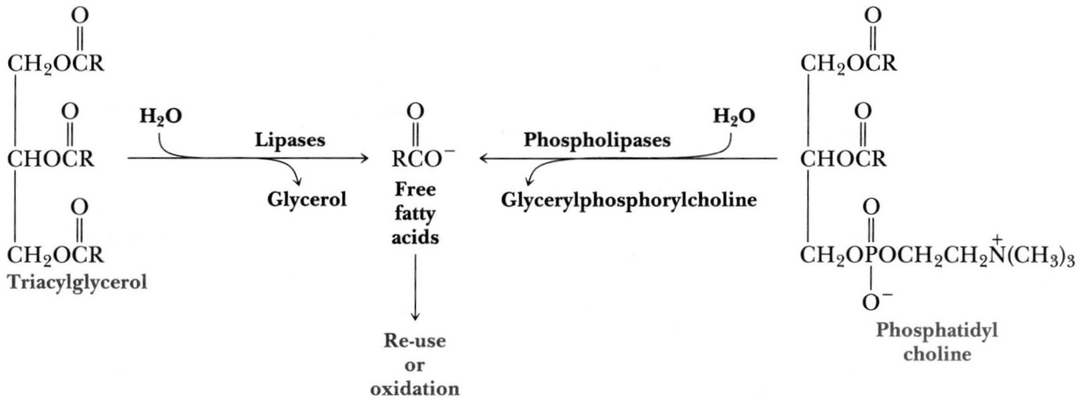

$$\begin{array}{ccc}
\underset{\text{Triacylglycerol}}{\begin{array}{c}
\overset{\overset{O}{\parallel}}{CH_2OCR} \\[2pt]
\overset{\overset{O}{\parallel}}{CHOCR} \\[2pt]
\overset{\overset{O}{\parallel}}{CH_2OCR}
\end{array}}
& \xrightarrow[\text{Glycerol}]{\substack{H_2O \\ \text{Lipases}}}
& \underset{\substack{\text{Free} \\ \text{fatty} \\ \text{acids}}}{\overset{\overset{O}{\parallel}}{RCO^-}}
& \xleftarrow[\text{Glycerylphosphorylcholine}]{\substack{H_2O \\ \text{Phospholipases}}}
& \underset{\substack{\text{Phosphatidyl} \\ \text{choline}}}{\begin{array}{c}
\overset{\overset{O}{\parallel}}{CH_2OCR} \\[2pt]
\overset{\overset{O}{\parallel}}{CHOCR} \\[2pt]
\overset{\overset{O}{\parallel}}{CH_2OPOCH_2CH_2\overset{+}{N}(CH_3)_3} \\[2pt]
\underset{O^-}{|}
\end{array}}
\end{array}$$

Re-use
or
oxidation

FIGURE 15.1

The release of fatty acids for future use. The source of fatty acids can be a triacylglycerol *(left)* or a phospholipid such as phosphatidylcholine *(right)*.

enzymes, **lipases** in the case of triacylglycerols (Chapter 8, Section 8.2) and **phospholipases** in the case of phosphoacylglycerols.

Several different phospholipases can be distinguished on the basis of sites at which they hydrolyze phospholipids (Figure 15.2). Phospholipase A_2 is widely distributed in nature; it is also being actively studied by biochemists interested in its structure and mode of action, which involves hydrolysis of phospholipids at the surfaces of micelles (Section 2.1). The reason for the use of the name phospholipase A_2 rather than phospholipase B is found in a bit of biochemical history. At one time it was thought that phospholipase A_2 hydrolyzed the fatty acid at the A position. Later biochemists found that the point of attack was at the B position, but they kept the name phospholipase A with the added subscript. Phospholipase D occurs in spider venom and is responsible for the tissue damage that accompanies spider bites. Snake venoms also contain phospholipases; the concentration of phospholipases is particularly high in venoms with comparatively low concentrations of the toxins (usually small peptides) that are characteristic of some kinds of venom. The lipid products of hydrolysis lyse red blood cells, preventing clot

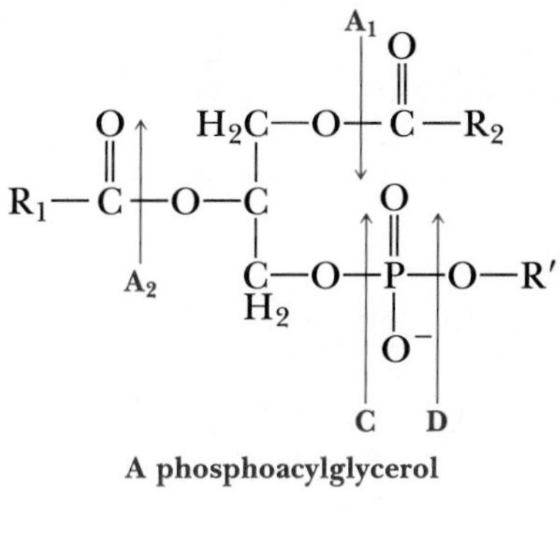

$$\begin{array}{c}
\overset{A_1}{}\quad \overset{O}{\parallel} \\
\overset{O}{\parallel}\quad H_2C-O\!+\!C-R_2 \\
R_1-C-O-C\quad\quad \overset{O}{\parallel} \\
\underset{A_2}{}\quad\quad \underset{H_2}{C}-O-\overset{\displaystyle}{\underset{\displaystyle O^-}{\overset{\displaystyle O}{\overset{\parallel}{P}}}}\!-O-R' \\
\quad C\quad D
\end{array}$$

A phosphoacylglycerol

FIGURE 15.2

Several phospholipases hydrolyze phosphoacyglycertols. They are designated A_1, A_2, C, and D. Their sites of action are shown. The site of action of phospholipase A_2 is the B site, and the name phospholipase A_2 is the result of historical accident (see text).

formation. Snakebite victims bleed to death in this situation. The hydrolytic reactions of lipids take place in the cytosol, as does the activation of the fatty acid, the step that prepares for the series of oxidation reactions.

Fatty acid oxidation begins with **activation** of the molecule. In this reaction a thioester bond is formed between the carboxyl group of the fatty acid and the thiol group of coenzyme A (CoA-SH). The activated form of the fatty acid is an acyl-CoA, the exact nature of which depends on the nature of the fatty acid itself. Keep in mind throughout this discussion that all acyl-CoA molecules are thioesters, since the fatty acid is esterified to the thiol group of CoA. The enzyme that catalyzes formation of the ester bond, an *acyl-CoA synthetase*, requires ATP for its action. In the course of the reaction, an acyl adenylate intermediate is formed. The acyl group is then transferred to CoA-SH. ATP is converted to AMP and PP_i, rather than to ADP and P_i. The PP_i is hydrolyzed to 2 P_i; the hydrolysis of two "high-energy" phosphate bonds provides energy for the activation of the fatty acid and is equivalent to the use of 2 ATP. Note also that the hydrolysis of ATP to AMP and 2 P_i represents an increase in entropy (Figure 15.3). There are several enzymes of this type, some specific for longer- and some for shorter-chain fatty acids. Both saturated and unsaturated fatty acids can serve as substrates for these enzymes. The esterification takes place in the cytosol, but the rest of the reactions of fatty acid oxidation occur in the mitochondrial matrix. The activated fatty acid must be transported into the mitochondrion so that the rest of the oxidation process can proceed.

The acyl-CoA can cross the outer mitochondrial membrane but not the inner membrane (Figure 15.4). In the intermembrane space the acyl group is transferred to carnitine by transesterification; this reaction is catalyzed by the enzyme carnitine acyltransferase, which is located in the inner membrane. Acyl carnitine, a compound that can cross the inner mitochondrial membrane, is formed. In the matrix the acyl group is transferred from carnitine to mitochondrial CoA-SH by another transesterification reaction.

In the matrix a repeated sequence of reactions successively cleaves two-carbon units from the fatty acid, starting from the carboxyl end. This process is called **β-oxidation,** since the oxidative cleavage takes place at the

Step 1. $RCOO^- + ATP \xrightarrow{\quad PP_i \quad} [\overset{\overset{\displaystyle O}{\displaystyle \|}}{R}C{-}AMP]$

Acyl adenylate
intermediate

Step 2. $[\overset{\overset{\displaystyle O}{\displaystyle \|}}{R}C{-}AMP] + CoA{-}SH \longrightarrow \overset{\overset{\displaystyle O}{\displaystyle \|}}{R}C{-}SCoA + AMP$

Thioester
(activated acyl group)

Overall Reaction:
$RCOO^- + ATP + CoA{-}SH \xrightarrow[\text{synthetase}]{\text{Acyl-CoA}} \overset{\overset{\displaystyle O}{\displaystyle \|}}{R}C{-}SCoA + AMP + PP_i$

FIGURE 15.3

The formation of an acyl-CoA.

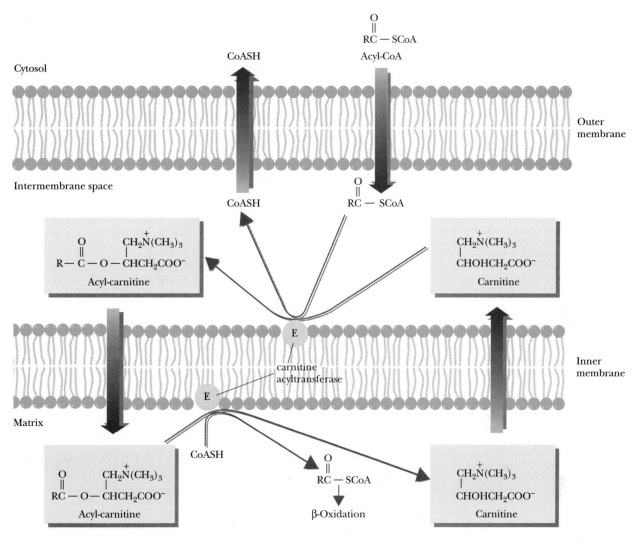

FIGURE 15.4

The role of carnitine in the transfer of acyl groups to the mitochondrial matrix.

β-carbon of the acyl group esterified to CoA. The β-carbon of the original fatty acid becomes the carboxyl carbon in the next stage of degradation. The whole cycle requires four reactions (Figure 15.5).

1. The acyl-CoA is *oxidized* to an α, β unsaturated acyl-CoA (also called a β-enoyl-CoA). The product has the *trans* arrangement at the double bond. This reaction is catalyzed by an FAD-dependent acyl-CoA dehydrogenase.
2. The unsaturated acyl-CoA is *hydrated* to produce a β-hydroxyacyl-CoA. This reaction is catalyzed by the enzyme enoyl-CoA hydratase.
3. A second *oxidation* reaction is catalyzed by β-hydroxyacyl-CoA dehydrogenase, an NADH-dependent enzyme. The product is a β-ketoacyl-CoA.
4. The enzyme thiolase catalyzes the *cleavage* of the β-ketoacyl-CoA; a molecule of CoA is required for the reaction. The products are acetyl-CoA and an acyl-CoA that is two carbons shorter than the original

FIGURE 15.5

The β-oxidation cycle for fatty acids.

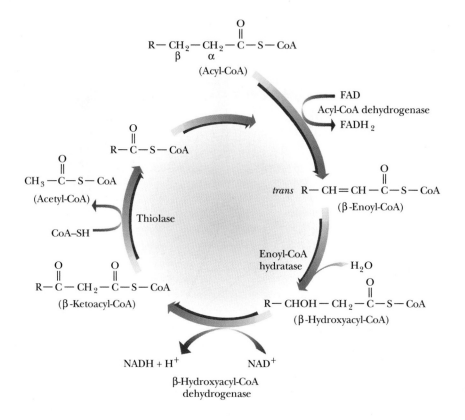

molecule that entered the β-oxidation cycle. The CoA is needed in this reaction to form the new thioester bond in the smaller acyl-CoA molecule. This smaller molecule then undergoes another round of the β-oxidation cycle.

When a fatty acid with an even number of carbon atoms undergoes successive rounds of the β-oxidation cycle, the product is acetyl-CoA. (Fatty acids with even numbers of carbon atoms are the ones normally found in nature, so acetyl-CoA is the usual product of fatty acid catabolism.) The number of molecules of acetyl-CoA produced is equal to half the number of carbon atoms in the original fatty acid. For example, stearic acid contains 18 carbon atoms and gives rise to 9 molecules of acetyl-CoA. Note that the conversion of one 18-carbon stearic acid molecule to nine 2-carbon acetyl units requires eight, not nine, cycles of β-oxidation (Figure 15.6). The acetyl-CoA enters the citric acid cycle, with the rest of the oxidation of fatty

FIGURE 15.6

Stearic acid (18 carbons) gives rise to nine 2-carbon units after eight cycles of β-oxidation. The ninth 2-carbon unit remains esterified to CoA after eight cycles of β-oxidation have removed eight successive two-carbon units, starting at the carboxyl end on the right.

acids to carbon dioxide and water taking place through the citric acid cycle and electron transport. Recall that most of the enzymes of the citric acid cycle are located in the mitochondrial matrix, and we have just seen that the β-oxidation cycle takes place in the matrix as well.

15.3
THE ENERGY YIELD FROM THE OXIDATION OF FATTY ACIDS

In carbohydrate metabolism the energy released by oxidation reactions is used to drive the production of ATP, with most of the ATP produced in aerobic processes. In the same aerobic processes—namely, the citric acid cycle and oxidative phosphorylation—the energy released by the oxidation of fatty acids can also be used to produce ATP. There are two sources of ATP to keep in mind when calculating the overall yield of ATP. The first source is the reoxidation of the NADH and $FADH_2$ produced by the β-oxidation of the fatty acid to acetyl-CoA. The second source is ATP production from the processing of the acetyl-CoA through the citric acid cycle and oxidative phosphorylation. We shall use the oxidation of stearic acid, which contains 18 carbon atoms, as our example.

Eight cycles of β-oxidation are required to convert 1 mole of stearic acid to 9 moles of acetyl-CoA; in the process 8 moles of FAD are reduced to $FADH_2$, and 8 moles of NAD^+ are reduced to NADH.

$$CH_3(CH_2)_{16}CO—S—CoA + 8\ FAD + 8\ NAD^+ + 8\ H_2O + 8\ CoA-SH \longrightarrow$$
$$9\ CH_3CO—S—CoA + 8\ FADH_2 + 8\ NADH + 8\ H^+$$

The 9 moles of acetyl-CoA produced from each mole of stearic acid enter the citric acid cycle. One mole of $FADH_2$ and 3 moles of NADH are produced for each mole of acetyl-CoA that enters the citric acid cycle. At the same time, 1 mole of GDP is phosphorylated to produce GTP for each turn of the citric acid cycle.

$$9\ CH_3CO—S—CoA + 9\ FAD + 27\ NAD^+ + 9\ GDP + 9\ P_i + 27\ H_2O \longrightarrow$$
$$18\ CO_2 + 9\ CoA-SH + 9\ FADH_2 + 27\ NADH + 9\ GTP + 27\ H^+$$

The $FADH_2$ and NADH produced by β-oxidation and by the citric acid cycle enter the electron transport chain, and ATP is produced by oxidative phosphorylation. In our example there are 17 moles of $FADH_2$, 8 from β-oxidation and 9 from the citric acid cycle; there are also 35 moles of NADH, 8 from β-oxidation and 27 from the citric acid cycle. Three moles of ATP are produced for each mole of NADH that enters the electron transport chain, and 2 moles of ATP result from each mole of $FADH_2$.

$$17\ FADH_2 + 8\tfrac{1}{2}\ O_2 + 34\ ADP + 34\ P_i \longrightarrow 17\ FAD + 34\ ATP + 17\ H_2O$$
$$35\ NADH + 35\ H^+ + 17\tfrac{1}{2}\ O_2 + 105\ ADP + 105\ P_i \longrightarrow 35\ NAD^+ + 105\ ATP + 35\ H_2O$$

The overall yield of ATP from the oxidation of stearic acid can be obtained by adding the equations for β-oxidation, for the citric acid cycle, and for oxidative phosphorylation. In this calculation we take GDP as equivalent to ADP and GTP as equivalent to ATP, which means that the equivalent of 9 ATP must be added to those produced in the reoxidation of FADH$_2$ and NADH. There are 9 ATP equivalent to the 9 GTP from the citric acid cycle, 34 ATP from the reoxidation of FADH$_2$, and 105 ATP from the reoxidation of NADH, for a grand total of 148 ATP.

$$CH_3(CH_2)_{16}CO—S—CoA + 26\ O_2 + 148\ ADP + 148\ P_i \longrightarrow$$
$$18\ CO_2 + 17\ H_2O + 148\ ATP + CoA\text{-}SH$$

The activation step in which stearyl-CoA was formed is not included in this calculation, and we must subtract the ATP that was required for that step. Even though only 1 ATP was required, two "high-energy" phosphate bonds are lost because of the production of AMP and PP$_i$. The pyrophosphate must be hydrolyzed to phosphate (P$_i$) before it can be recycled in metabolic intermediates. As a result we must subtract the equivalent of 2 ATP for the activation step. The net yield of ATP becomes 146 moles of ATP for each mole of stearic acid that is completely oxidized. See Table 15.1 for a balance sheet.

As a comparison, note that 38 moles of ATP can be obtained from the complete oxidation of 1 mole of glucose; but glucose contains 6, rather than 18, carbon atoms. Three glucose molecules contain 18 carbon atoms, and a more interesting comparison is the ATP yield from the oxidation of three glucose molecules, which is $3 \times 38 = 114$ ATP for the same number of carbon atoms. The yield of ATP from the oxidation of the lipid is still higher than that from the carbohydrate, even for the same number of carbon atoms. The reason is that a fatty acid is all hydrocarbon except for the carboxyl group; that is, it exists in a highly reduced state. A sugar is already partly oxidized because of the presence of its oxygen-containing groups.

Another point of interest is that water is produced in the oxidation of fatty acids. We have already seen that water is also produced in the complete oxidation of carbohydrates. The production of **metabolic water** is a common feature of aerobic metabolism. This process can be a source of water for organisms that live in desert environments. Camels are a well-known example; the stored lipids in their humps are a source of both energy and water during long trips through the desert. The kangaroo rat is an even

T A B L E   1 5 . 1   **The Balance Sheet for Oxidation of One Molecule of Stearic Acid**

	NADH MOLECULES	FADH$_2$ MOLECULES	ATP MOLECULES
REACTION			
1. Stearic acid $\longrightarrow$ stearyl-CoA (activation step)			−2
2. Stearyl-CoA $\longrightarrow$ 9 acetyl-CoA (8 cycles of β-oxidation)	+8	+8	
3. 9 Acetyl-CoA $\longrightarrow$ 9 CO$_2$ (citric acid cycle)	+27	+9	+9
GDP $\longrightarrow$ GTP (9 molecules)			
4. Reoxidation of NADH from β-oxidation cycle	−8		+24
5. Reoxidation of NADH from citric acid cycle	−27		+81
6. Reoxidation of FADH$_2$ from β-oxidation cycle		−8	+16
7. Reoxidation of FADH$_2$ from citric acid cycle		−9	+18
	0	0	+146

Note that there is no net change in the number of molecules of NADH or FADH$_2$.

more striking example of adaptation to an arid environment. This animal has been observed to live indefinitely without having to drink water. It lives on a diet of seeds, which are rich in lipids but contain little water. The metabolic water that the kangaroo rat produces is adequate for all its water needs. This metabolic response to arid conditions is usually accompanied by a reduced output of urine.

15.4
SOME ADDITIONAL REACTIONS IN THE OXIDATION OF FATTY ACIDS

Fatty acids with odd numbers of carbon atoms are not as frequently encountered in nature as are the ones with even numbers of carbon atoms. Odd-numbered fatty acids also undergo the β-oxidation process (Figure 15.7). The last cycle of β-oxidation produces one molecule of propionyl-CoA. An enzymatic pathway exists to convert propionyl-CoA to succinyl-CoA, which then enters the citric acid cycle. In this pathway, propionyl-CoA is first carboxylated to methyl malonyl-CoA in a reaction catalyzed by propionyl-CoA carboxylase, which then undergoes rearrangement to form succinyl-CoA; since propionyl-CoA is also a product of the catabolism of several amino acids, the conversion of propionyl-CoA to succinyl-CoA also figures in amino acid metabolism (Chapter 17, Section 17.7).

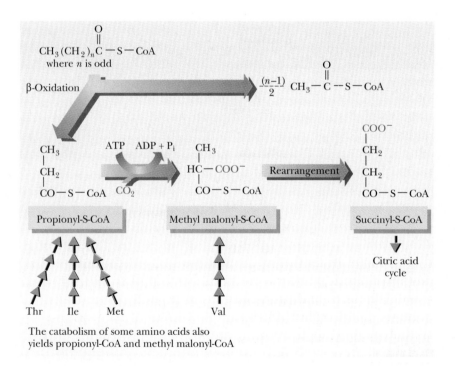

The catabolism of some amino acids also yields propionyl-CoA and methyl malonyl-CoA

FIGURE 15.7

The oxidation of a fatty acid containing an odd number of carbon atoms.

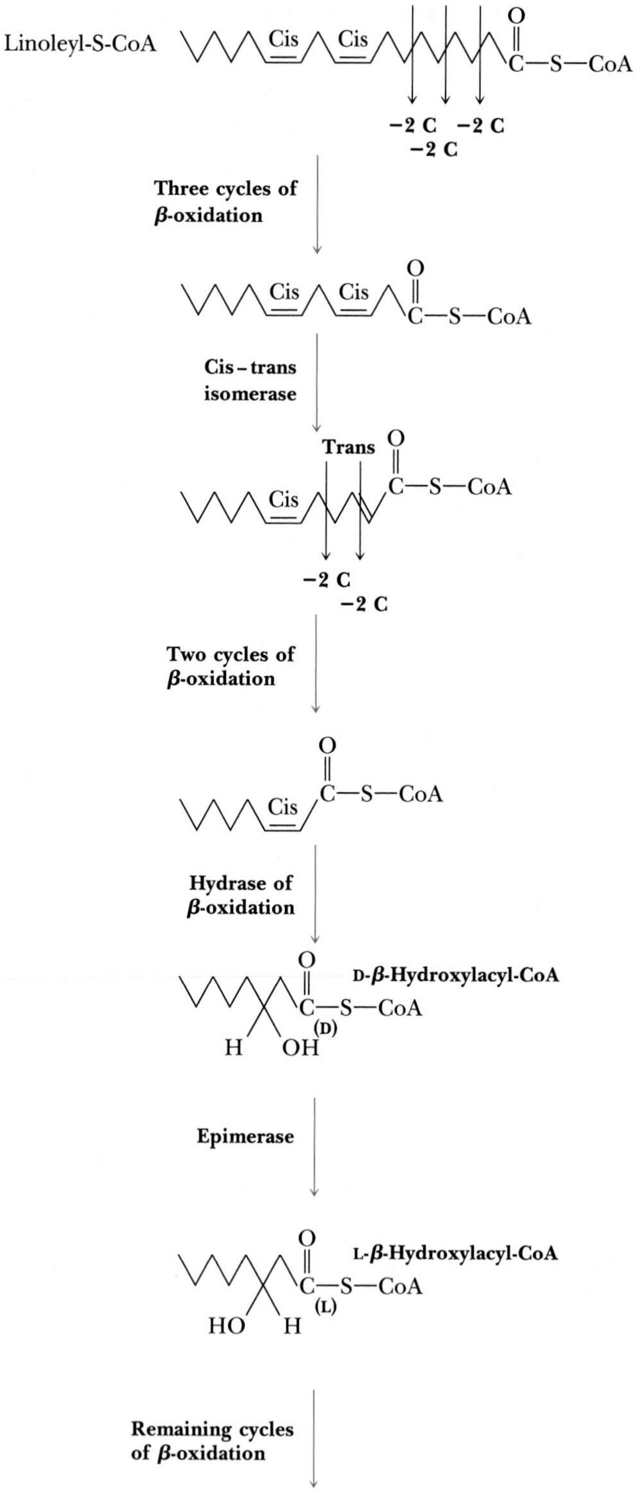

FIGURE 15.8

The oxidation of unsaturated fatty acids.

The conversion of unsaturated fatty acids to acetyl-CoA requires two reactions that are not encountered in the oxidation of saturated acids, a *cis-trans* isomerization and an epimerization (Figure 15.8). Successive rounds of β-oxidation of linoleic acid, which has two double bonds, provides an example of these reactions. The process of β-oxidation gives rise to unsaturated fatty acids in which the double bond is in the *trans* arrangement, whereas the double bonds in most naturally occurring fatty acids are in the *cis* arrangement; in the case of linoleic acid there are two *cis* double bonds, between carbons 9 and 10 and between carbons 12 and 13. Three rounds of β-oxidation produce a 12-carbon unsaturated fatty acid with the two *cis* double bonds between carbons 3 and 4 and between carbons 6 and 7. The hydratase of the β-oxidation cycle requires a *trans* double bond between carbon atoms 2 and 3 as a substrate. A **cis-trans isomerase** produces a *trans* double bond between carbons 2 and 3 from the *cis* double bond between carbons 3 and 4.

Two more cycles of β-oxidation give rise to an eight-carbon fatty acid with a *cis* double bond between carbons 2 and 3. The position of the double bond makes the eight-carbon intermediate a possible substrate for the hydratase, but the stereochemistry of the product presents a problem in the next step of the cycle. The β-hydroxy product has the D configuration, and the dehydrogenase that catalyzes the next stage of β-oxidation requires the L configuration. An *epimerase* catalyzes the required change of configuration, and the oxidation of the fatty acid can proceed. Unsaturated fatty acids make up a large enough portion of the fatty acids in storage fat (40% for oleic acid alone) to make the reactions of the *cis-trans* isomerase and the epimerase of some importance.

15.5
THE FORMATION OF "KETONE BODIES"

Substances related to acetone ("ketone bodies") are produced when an excess of acetyl-CoA arises from β-oxidation. This condition occurs when not enough oxaloacetate is available to react with the large amounts of acetyl-CoA that could enter the citric acid cycle. Oxaloacetate in turn arises from glycolysis, since it is formed from pyruvate in a reaction catalyzed by pyruvate carboxylase. A situation like this can come about when an organism has a high intake of lipids and a low intake of carbohydrates, but there are other possible causes as well, such as starvation and diabetes. Starvation conditions cause an organism to break down fats for energy, leading to the production of large amounts of acetyl-CoA by β-oxidation. The amount of acetyl-CoA is excessive by comparison with the amount of oxaloacetate available to react with it. In the case of diabetics, the cause of the imbalance is not inadequate intake of carbohydrates, but rather the inability to metabolize them.

The reactions that result in "ketone bodies" start with the condensation of two molecules of acetyl-CoA to produce acetoacetyl-CoA. **Acetoacetate** arises from acetoacetyl-CoA and can have two possible fates. A reduction reaction can produce **β-hydroxybutyrate** from acetoacetate. The other possible reaction is the decarboxylation of acetoacetate to give **acetone**

FIGURE 15.9

The formation of acetoacetate and related compounds from acetyl-CoA.

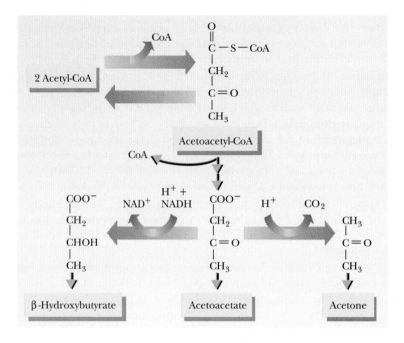

FIGURE 15.9

The formation of acetoacetate and related compounds from acetyl-CoA.

(Figure 15.9). The odor of acetone can frequently be detected on the breath of diabetics.

Even though glucose is the usual fuel in most tissues and organs, acetoacetate can be and is used as a fuel. In heart muscle and the renal cortex, acetoacetate is the preferred source of energy. Even in organs such as the brain, in which glucose is the preferred fuel, starvation conditions can lead to the use of acetoacetate for energy. In this situation acetoacetate is converted to two molecules of acetyl-CoA, which can then enter the citric acid cycle.

15.6
THE ANABOLISM OF FATTY ACIDS

The anabolism of fatty acids is not simply a reversal of the reactions of β-oxidation. Anabolism and catabolism are not, in general, the exact reverse of each other; for instance, gluconeogenesis (Chapter 14, Section 14.2) is not simply a reversal of the reactions of glycolysis. A first example of the differences between the degradation and the biosynthesis of fatty acids is that the anabolic reactions take place in the cytosol. We have just seen that the degradative reactions of β-oxidation take place in the mitochondrial matrix. The first step in fatty acid biosynthesis is transport of acetyl-CoA to the cytosol.

Acetyl-CoA can be formed by β-oxidation of fatty acids or by decarboxylation of pyruvate. (Degradation of certain amino acids also produces acetyl-CoA; see Chapter 17, Section 17.7.) Most of these reactions take place in the mitochondria, requiring a transport mechanism to export acetyl-CoA to the cytosol for fatty acid biosynthesis. The transport mecha-

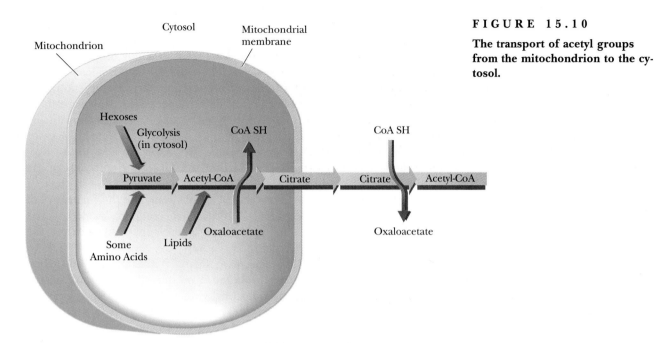

FIGURE 15.10

The transport of acetyl groups from the mitochondrion to the cytosol.

nism is based on the fact that citrate can cross the mitochondrial membrane. Acetyl-CoA condenses with oxaloacetate (which cannot cross the mitochondrial membrane) to form citrate (recall that this is the first reaction of the citric acid cycle). The citrate that is exported to the cytosol can undergo the reverse reaction, producing oxaloacetate and acetyl-CoA (Figure 15.10). Acetyl-CoA enters the pathway for fatty acid biosynthesis, while oxaloacetate undergoes a series of reactions in which there is a substitution of NADPH for NADH (see Lipid Anabolism in Chapter 18, Section 18.4). This substitution exercises control over the pathway, since NADPH is required for fatty acid anabolism.

In the cytosol, acetyl-CoA is carboxylated, producing **malonyl-CoA,** a key intermediate in fatty acid biosynthesis (Figure 15.11). This reaction is catalyzed by the *acetyl-CoA carboxylase* complex, which consists of three enzymes and which requires Mn^{2+} and biotin for activity, in addition to ATP. We have already seen that enzymes catalyzing reactions that take place in several steps frequently consist of several separate protein molecules, and this enzyme follows that pattern. In this case acetyl-CoA carboxylase consists of the three proteins *biotin carboxylase,* the *biotin carrier protein,* and *carboxyl transferase.* Biotin carboxylase catalyzes the transfer of the carboxyl group to

FIGURE 15.11

The formation of malonyl-CoA, catalyzed by acetyl-CoA carboxylase.

$$H_3C-\overset{\overset{\displaystyle O}{\|}}{C}-S-CoA + ATP + HCO_3^- \xrightarrow[\text{Mn}^{2+}]{\text{Biotin}}$$

Acetyl-CoA

$$^-OOC-CH_2-\overset{\overset{\displaystyle O}{\|}}{C}-S-CoA + ADP + P_i + H^+$$

Malonyl-CoA

biotin. The "activated CO_2" (the carboxyl group derived from the bicarbonate ion HCO_3^-) is covalently bound to biotin. Biotin (whether carboxylated or not) is bound to the biotin carrier protein by an amide linkage to the ϵ-amino group of a lysine side chain. The amide linkage to the side chain that bonds biotin to the carrier protein is long enough and flexible enough to move the carboxylated biotin into position to transfer the carboxyl group to acetyl-CoA in the reaction catalyzed by carboxyl transferase, producing malonyl-CoA (Figure 15.12).

The biosynthesis of fatty acids involves the successive addition of two-carbon units to the growing chain. Two of the three carbon atoms of the malonyl group of malonyl-CoA are added to the growing fatty acid chain at each stage of the biosynthetic reaction. This reaction, like the formation of the malonyl-CoA itself, requires a multienzyme complex located in the cytosol and not attached to any membrane. The complex, made up of the individual enzymes, is called fatty acid synthase.

The usual product of fatty acid anabolism is **palmitate,** the 16-carbon saturated fatty acid. All 16 carbons come from the acetyl group of acetyl-CoA; we have already seen how malonyl-CoA, the immediate precursor, arises from acetyl-CoA. But first there is a priming step in which one molecule of acetyl-CoA is required for each molecule of palmitate produced. In this priming step the acetyl group from acetyl-CoA is transferred to an acyl carrier protein (ACP), which is considered a part of the fatty acid synthase complex (Figure 15.13, Step 1a). The acetyl group is bound to the protein as a thioester. The group on the protein to which the acetyl group is bonded is the 4'-phosphopantetheine group, which in turn is bonded to a

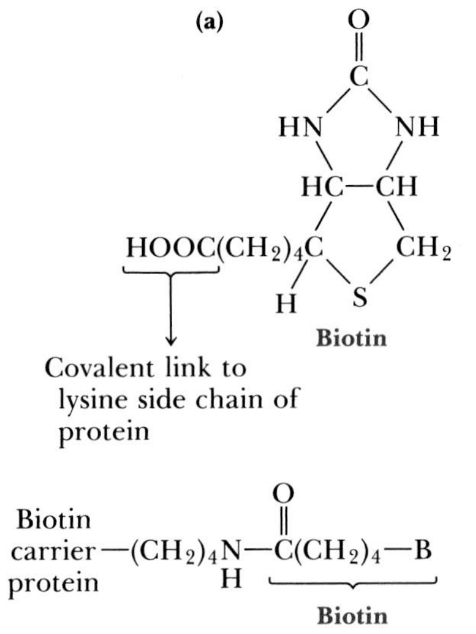

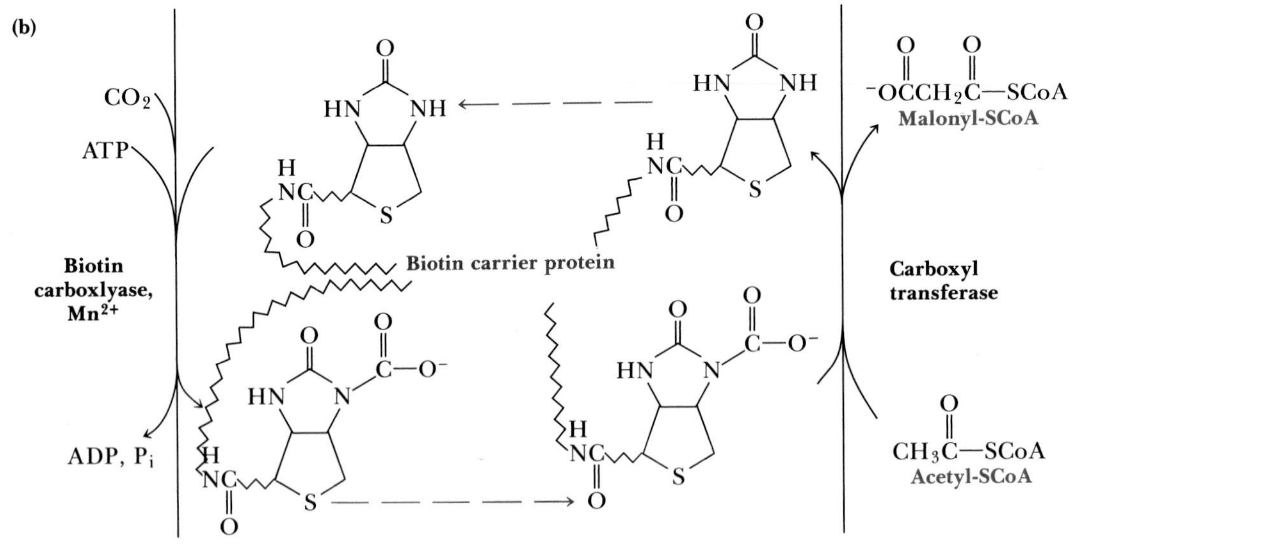

(a)

Covalent link to
lysine side chain of
protein

Biotin

Biotin
carrier —$(CH_2)_4N-\overset{\overset{\displaystyle O}{\|}}{C}(CH_2)_4$—B
protein   H

Biotin

(b)

CO_2

ATP

**Biotin
carboxylase,
Mn^{2+}**

ADP, P_i

Biotin carrier protein

Malonyl-SCoA

$^-OCCH_2C-SCoA$

**Carboxyl
transferase**

$CH_3C-SCoA$
Acetyl-SCoA

FIGURE 15.12

The mode of action of the three proteins that make up acetyl-CoA carboxylase. (a) The structure of biotin, showing the site of its covalent link to the carrier protein. (b) In the actual reaction, biotin moves from one side to another and remains covalently linked to the biotin carrier protein.

Step 1. Priming of the system by acetyl-CoA

a. ACP-Acyltransferase reaction

$$H_3C-\overset{\overset{\displaystyle O}{\|}}{C}-S-CoA + ACP-SH \rightleftharpoons H_3C-\overset{\overset{\displaystyle O}{\|}}{C}-S-ACP + CoA-SH$$

Acetyl-CoA   **Acetyl-ACP**

b. Transfer to β-ketoacyl-ACP synthase

$$H_3C-\overset{\overset{\displaystyle O}{\|}}{C}-S-ACP + Synthase-SH \rightleftharpoons H_3C-\overset{\overset{\displaystyle O}{\|}}{C}-S-Synthase + ACP-SH$$

Acetyl-ACP   **Acetyl-synthase**

Step 2. ACP-malonyltransferase reaction (malonyl transfer to system)

$$^-OOC-CH_2-\overset{\overset{\displaystyle O}{\|}}{C}-S-CoA + ACP-SH \rightleftharpoons {}^-OOC-CH_2-\overset{\overset{\displaystyle O}{\|}}{C}-S-ACP + CoA-SH$$

Malonyl-CoA   **Malonyl-ACP**

Step 3. β-Ketoacyl-ACP synthase reaction (condensation)

$$H_3C-\overset{\overset{\displaystyle O}{\|}}{C}-S-Synthase + {}^-OOC-CH_2-\overset{\overset{\displaystyle O}{\|}}{C}-S-ACP \rightleftharpoons H_3C-\overset{\overset{\displaystyle O}{\|}}{C}-CH_2-\overset{\overset{\displaystyle O}{\|}}{C}-S-ACP + CO_2 + Synthase-SH$$

Acetoacetyl-synthase   **Malonyl-ACP**   **Acetoacetyl-ACP**

Step 4. β-Ketoacyl-ACP reductase reaction (first reduction)

$$H_3C-\overset{\overset{\displaystyle O}{\|}}{C}-CH_2-\overset{\overset{\displaystyle O}{\|}}{C}-S-ACP + NADPH + H^+ \rightleftharpoons H_3C-\overset{\overset{\displaystyle OH}{|}}{CH}-CH_2-\overset{\overset{\displaystyle O}{\|}}{C}-S-ACP + NADP^+$$

Acetoacetyl-ACP   D-**β-Hydroxybutyryl-ACP**

Step 5. β-Hydroxyacyl-ACP dehydratase (dehydration)

$$H_3C-\overset{\overset{\displaystyle OH}{|}}{CH}-CH_2-\overset{\overset{\displaystyle O}{\|}}{C}-S-ACP \rightleftharpoons H_3C-\overset{trans}{CH}=CH-\overset{\overset{\displaystyle O}{\|}}{C}-S-ACP + H_2O$$

D-**β-Hydroxybutyryl-ACP**   **Crotonyl-ACP**

Step 6. β-Enoyl-ACP reductase (second reduction)

$$H_3C-\overset{trans}{CH}=CH-\overset{\overset{\displaystyle O}{\|}}{C}-S-ACP + NADPH + H^+ \rightleftharpoons H_3C-CH_2-CH_2-\overset{\overset{\displaystyle O}{\|}}{C}-S-ACP + NADP^+$$

Crotonyl-ACP   **Butyryl-ACP**

F I G U R E   1 5 . 1 3

The first cycle of palmitate synthesis. ACP is the acyl carrier protein.

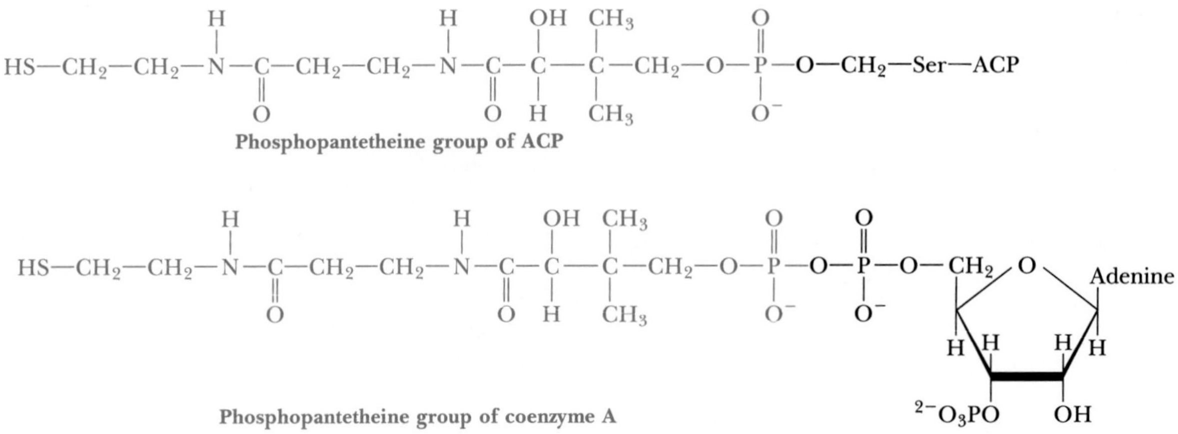

Phosphopantetheine group of ACP

Phosphopantetheine group of coenzyme A

FIGURE 15.14

Structural similarities between coenzyme A and the phosphopantetheine group of ACP.

serine side chain; note in Figure 15.14 that this group is structurally similar to CoA-SH itself. The acetyl group is transferred from CoA-SH, to which it is bound by a thioester linkage, to the ACP; the acetyl group is bound to the ACP by a thioester linkage.

The acetyl group is transferred in turn from the ACP to another protein, to which it is bound by a thioester linkage to a cysteine-SH; the other protein is β-ketoacyl-S-ACP-synthase (Figure 15.13, Step 1b). The first of the successive additions of two of the three malonyl carbons to the fatty acid starts at this point. The malonyl group itself is transferred from a thioester linkage with CoA-SH to another thioester bond to the ACP (Figure 15.13, Step 2).

The next step is a condensation reaction that produces acetoacetyl-S-ACP (Figure 15.13, Step 3). In other words, the principal product of this reaction is an acetoacetyl group bound to the ACP by a thioester linkage. Two of the four carbons of acetoacetate come from the priming acetyl group, and the other two come from the malonyl group. The carbon atoms that arise from the malonyl group are the one directly bonded to the sulfur and the one in the $-CH_2-$ group next to it. The CH_3CO- group comes from the priming acetyl group. The other carbon of the malonyl group is released as CO_2; the CO_2 that is lost is the original CO_2 that was used to carboxylate the acetyl-CoA to produce malonyl-CoA. The synthase is no longer involved in a thioester linkage.

Acetoacetyl-ACP is converted to butyryl-ACP by a series of reactions involving two reductions and a dehydration (Figure 15.13, Steps 4-6). In the first reduction the β-keto group is reduced to an alcohol, giving rise to D-β-hydroxybutyryl-ACP. In the process NADPH is oxidized to $NADP^+$; the enzyme that catalyzes this reaction is β-ketoacyl-ACP reductase (Figure 15.13, Step 4). The dehydration step, catalyzed by β-hydroxyacyl-ACP dehydratase, produces crotonyl-ACP (Figure 15.13, Step 5). Note that the double bond is in the *trans* configuration. A second reduction reaction, catalyzed by β-enoyl-ACP reductase, produces butyryl-ACP (Figure 15.13,

Step 6). In this reaction NADPH is the coenzyme, as it was in the first reduction reaction in this series.

In the second round of fatty acid biosynthesis, butyryl-ACP plays the same role as acetyl-ACP in the first round. The butyryl group is transferred to the synthase, and a malonyl group is transferred to the ACP. Once again there is a condensation reaction with malonyl-ACP (Figure 15.15). In this second round the condensation produces a six-carbon β-ketoacyl-ACP. The two added carbon atoms come from the malonyl group, as they did in the first round. The reduction and dehydration reactions take place as before, giving rise to hexanoyl-ACP. The same series of reactions is repeated until palmitoyl-ACP is produced. In mammalian systems the process stops at C_{16}, since the fatty acid synthase does not produce longer chains. Mammals produce longer-chain fatty acids by modification of the fatty acids formed by the synthase reaction.

Fatty acid synthases from different types of organisms have markedly different characteristics. In *Escherichia coli* the multienzyme system consists of an aggregate of separate enzymes, including a separate ACP (Figure

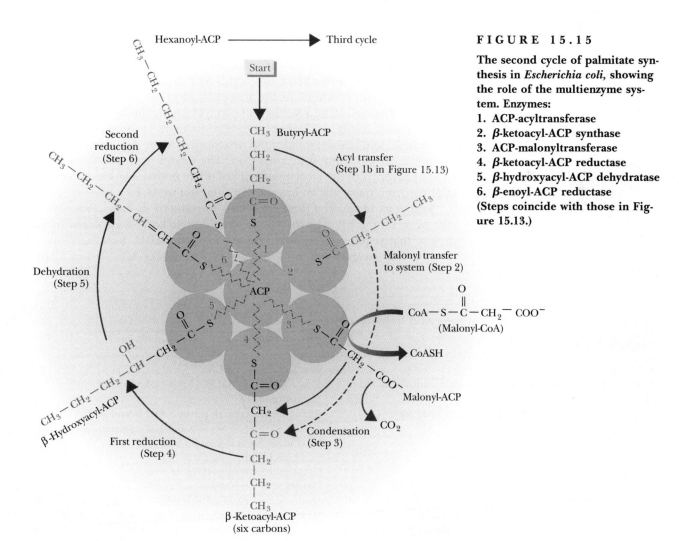

FIGURE 15.15

The second cycle of palmitate synthesis in *Escherichia coli*, showing the role of the multienzyme system. Enzymes:
1. ACP-acyltransferase
2. β-ketoacyl-ACP synthase
3. ACP-malonyltransferase
4. β-ketoacyl-ACP reductase
5. β-hydroxyacyl-ACP dehydratase
6. β-enoyl-ACP reductase
(Steps coincide with those in Figure 15.13.)

15.15). The ACP is of primary importance to the complex and is considered to occupy a central position in it. The phosphopantetheine group plays the role of a "swinging arm," much like that of biotin, which was discussed earlier in this chapter. This bacterial system has been extensively studied and has been considered a typical example of a fatty acid synthase. In mammals and in yeast, however, the fatty acid synthase consists of only two subunits (two different kinds of polypeptide chain), in contrast to the larger number in the bacterial system. Each of the subunits is a **multifunctional enzyme** that catalyzes reactions requiring several different proteins in the *E. coli* system. The ACP and the enzyme for the condensation reaction appear to be the two subunits in the mammalian and yeast systems. The growing fatty acid chain swings back and forth between the two on the "swinging arm," and different parts of these two polypeptide chains catalyze the other stages of the addition of each two-carbon unit. Like the bacterial system, the eukaryotic system keeps all the components of the reaction in proximity to one another.

Several additional reactions are required for the elongation of fatty acid chains and the introduction of double bonds. When mammals produce fatty acids with longer chains than that of palmitate, the reaction does not involve cytosolic fatty acid synthase. There are two sites for the chain-lengthening reactions: the endoplasmic reticulum (ER) and the mitochondrion. In the chain-lengthening reactions in the mitochondrion, the intermediates are of the acyl-CoA type rather than the acyl-S-ACP type. In other words, the chain-lengthening reactions in the mitochondrion are the reverse of the catabolic reactions of fatty acids, with acetyl-CoA as the source of added carbon atoms; this is a difference between the main pathway of fatty acid biosynthesis and these modification reactions. In the ER the source of additional carbon atoms is malonyl-CoA. The modification reactions in the ER also differ from the biosynthesis of palmitate in that, like the mitochondrial reaction, there are no intermediates bound to ACP.

Reactions in which a double bond is introduced in fatty acids mainly take place on the ER. The insertion of the double bond is catalyzed by an oxidase that requires molecular oxygen (O_2) and NADH. Reactions linked to molecular oxygen are comparatively rare (Chapter 12, Section 12.5).

T A B L E 1 5 . 2   A Comparison of Fatty Acid Degradation and Biosynthesis

DEGRADATION	BIOSYNTHESIS
1. Product is acetyl-CoA	Precursor is acetyl-CoA
2. Malonyl-CoA is not involved; no requirement for biotin	Malonyl-CoA is source of two-carbon units; biotin required
3. Oxidative process; requires NAD^+ and FAD and produces ATP	Reductive process; requires NADPH and ATP
4. Fatty acids form thioesters with CoA-SH	Fatty acids form thioesters with acyl carrier proteins (ACP-SH)
5. Starts at carboxyl end (CH_3CO_2—)	Starts at methyl end (CH_3CH_2—)
6. Occurs in the cytosol, catalyzed by an ordered multienzyme complex	Occurs in the mitochondrial matrix, with no ordered aggregate of enzymes
7. β-Hydroxyacyl intermediates have the L configuration	β-Hydroxyacyl intermediates have the D configuration

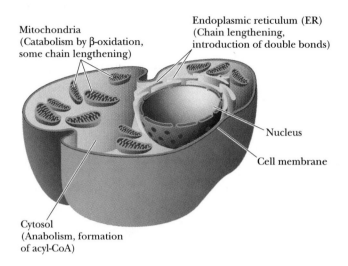

Mitochondria
(Catabolism by β-oxidation,
some chain lengthening)

Endoplasmic reticulum (ER)
(Chain lengthening,
introduction of double bonds)

Nucleus

Cell membrane

Cytosol
(Anabolism, formation
of acyl-CoA)

FIGURE 15.16

A portion of an animal cell, showing the sites of various aspects of
fatty acid metabolism. The cytosol
is the site of fatty anabolism. It is
also the site of formation of acyl-
CoA, which is transported to the
mitochondria for catabolism by
the β-oxidation process. Some
chain-lengthening reactions (beyond C_{16}) take place in the mitochondria. Other chain-lengthening
reactions take place in the endoplasmic reticulum (ER), as do reactions that introduce double
bonds.

Mammals cannot introduce a double bond beyond carbon atom 9 (counting
from the carboxyl end) of the fatty acid chain. As a result, linoleate
[CH_3—$(CH_2)_4$—CH=CH—CH_2—CH=CH—$(CH_2)_7$—COO^-], with two
double bonds, and linolenate [CH_3—$(CH_2)_4$—CH=CH—CH_2—CH=
CH—CH_2—CH=CH—$(CH_2)_4$—COO^-], with three double bonds, must
be included in the diets of mammals. They are **essential fatty acids** because
they   are   precursors   of   other   lipids,   including   prostaglandins.
Even though both the anabolism and the catabolism of fatty acids
require successive reactions of two-carbon units, the two pathways are not
the exact reversal of each other. The differences between the two pathways
can be summarized in Table 15.2. The sites in the cell in which various
anabolic and catabolic reactions take place are shown in Figure 15.16.

15.7
THE ANABOLISM OF ACYLGLYCEROLS
AND COMPOUND LIPIDS

Other lipids, including triacylglycerols, phosphoacylglycerols, and steroids,
are derived from fatty acids and metabolites of fatty acids such as
acetoacetyl-CoA. Free fatty acids do not occur in the cell to any great extent;
they are normally found incorporated in triacylglycerols and phosphoacylglycerols. The biosynthesis of these two types of compounds takes place
principally on the ER of liver cells or fat cells (adipocytes).

Triacylglycerols

The glycerol portion of lipids is derived from glycerol 3-phosphate, a
compound available from glycolysis. Another source is glycerol released by
degradation of acylglycerols. An acyl group of fatty acid is transferred from
an acyl-CoA in a reaction catalyzed by the enzyme *glycerol phosphate*

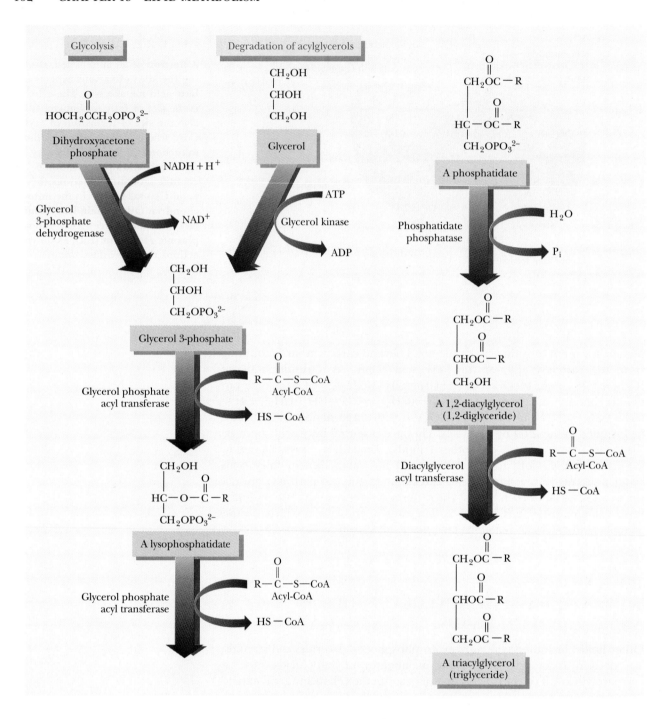

FIGURE 15.17

Pathways for the biosynthesis of triacylglycerols.

acyltransferase. The products of this reaction are CoA-SH and a **lysophosphatidate** (a monoacylglycerol phosphate) (Figure 15.17). The acyl group is shown as esterified at carbon atom 2 (C-2) in this series of equations, but it is equally likely that it is esterified at C-1. A second acylation reaction takes place, catalyzed by the same enzyme, producing a **phosphatidate** (a diacylglyceryl phosphate). Phosphatidates occur in membranes and are precursors of other phospholipids. The phosphate group of the phosphatidate is removed by hydrolysis in a reaction catalyzed by *phosphatidate phosphatase,* producing a **diacylglycerol.** A third acyl group is added in a reaction catalyzed by *diacylglycerol acyltransferase.* As before, the source of the acyl group is an acyl-CoA rather than the free fatty acid.

Phosphoacylglycerols

Phosphoacylglycerols (phosphoglycerides) are based on phosphatidates, with the phosphate group esterified to another alcohol, frequently a nitrogen-containing alcohol such as ethanolamine (see Phosphoacylglycerols [Phosphoglycerides] in Chapter 8, Section 8.2). The conversion of phosphatidates to other phospholipids frequently requires the presence of nucleoside triphosphates, particularly **cytidine triphosphate** (CTP). The role of CTP depends on the type of organism, since the details of the biosynthetic pathway are not the same in mammals and bacteria. We shall use a comparison of the synthesis of phosphatidylethanolamine in mammals and in bacteria (Figure 15.18) as a case study of the kinds of reactions commonly encountered in phosphoglyceride biosynthesis.

In mammals the synthesis of **phosphatidylethanolamine** requires two preceding steps in which the component parts are processed. The first of these two steps is the removal by hydrolysis of the phosphate group of the phosphatidate, producing a diacylglycerol; the second step is the reaction of ethanolamine phosphate with CTP to produce pyrophosphate (PP_i) and cytidine diphosphate ethanolamine (CDP-ethanolamine). The CDP-ethanolamine and diacylglycerol react to form phosphatidylethanolamine. In bacteria, CTP reacts instead with phosphatidate itself to produce cytidine diphosphodiacylglycerol (a CDP diglyceride). The CDP diglyceride reacts with ethanolamine phosphate to form phosphatidylethanolamine.

Sphingolipids

The structural basis of sphingolipids is not glycerol but **sphingosine,** a long-chain amine (see Sphingolipids in Chapter 8, Section 8.2). The precursors of sphingosine are palmitoyl-CoA and the amino acid serine, which react to produce dihydrosphingosine. The carboxyl group of the serine is lost as CO_2 in the course of this reaction (Figure 15.19). An oxidation reaction introduces a double bond, with sphingosine as the resulting compound. Reaction of the amino group of sphingosine with another acyl-CoA to form an amide bond results in an **N-acylsphingosine,** also called a **ceramide.** Ceramides in turn are the parent compounds of sphingomyelins, cerebrosides, and gangliosides. Attachment of phosphorylcholine to the primary alcohol group of a ceramide produces a **sphingomyelin,** whereas attachment of sugars such as glucose at the same site produces **cerebrosides. Gangliosides** are formed from ceramides by attachment of oligosaccharides

(Text continues on p. 466.)

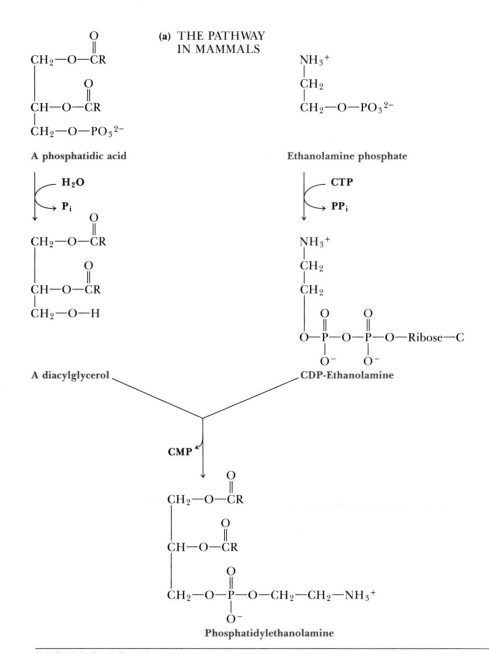

FIGURE 15.18

A comparison of the biosynthesis of phosphatidylethanolamine in mammals and in bacteria.

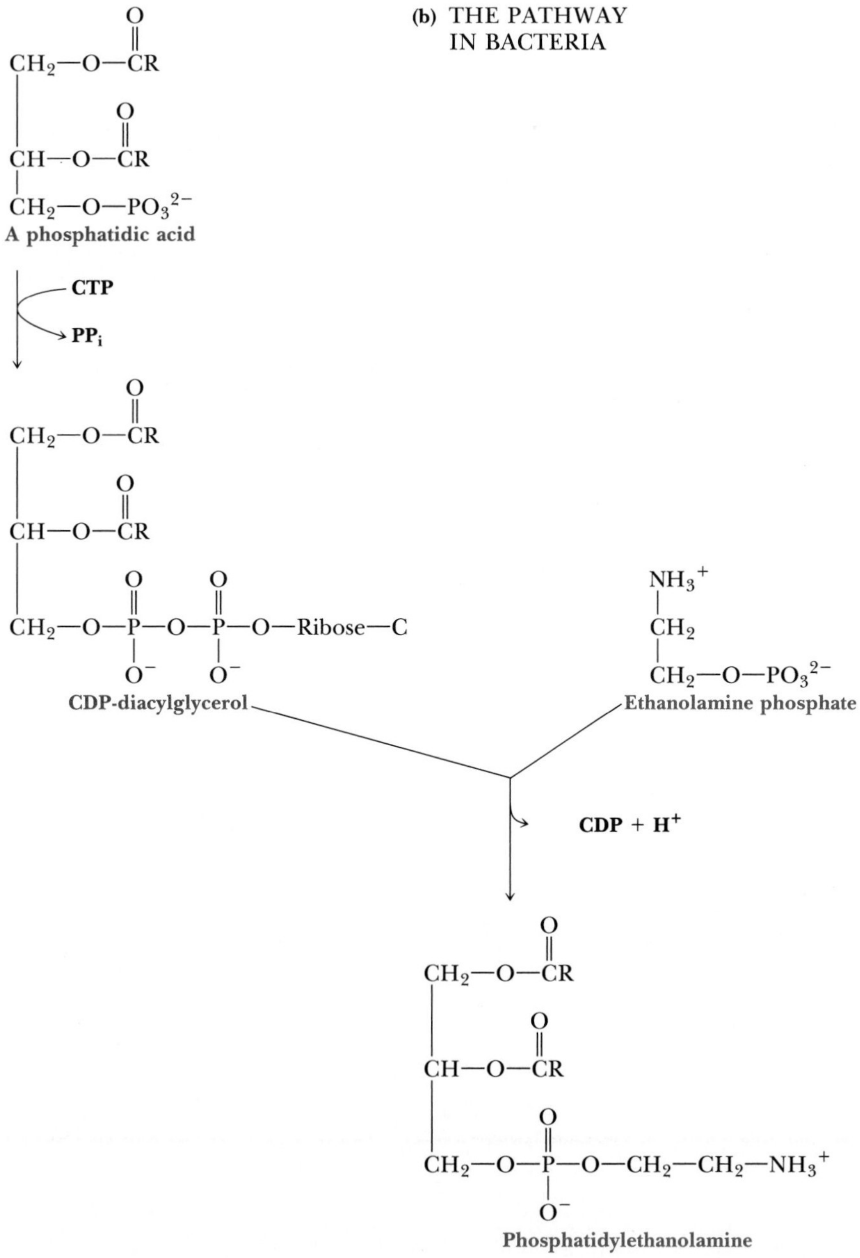

(b) THE PATHWAY IN BACTERIA

FIGURE 15.18 (continued)

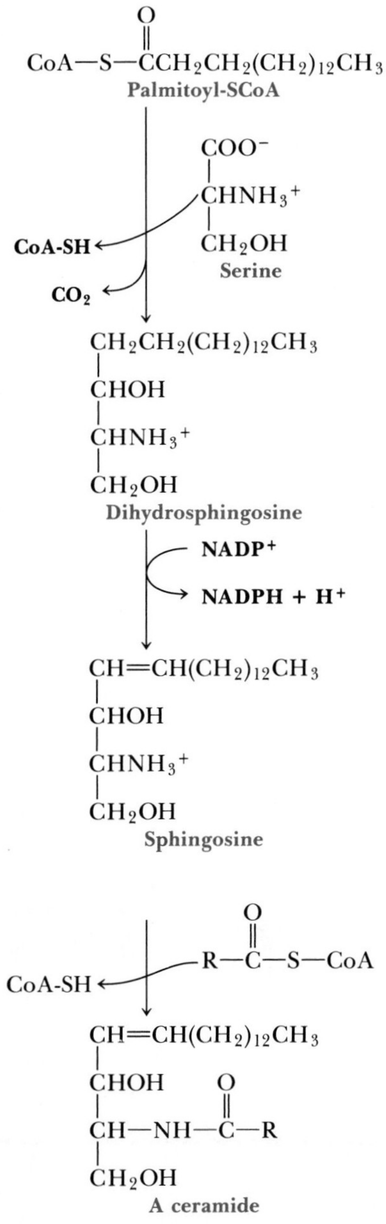

FIGURE 15.19

The biosynthesis of sphingolipids. When ceramides are formed, they can react (a) with choline to yield sphingomyelins, (b) with sugars to yield cerebrosides, or (c) with sugars and sialic acid to yield gangliosides.

Box 15.1

TAY–SACHS DISEASE

Tay–Sachs disease is an inborn error of lipid metabolism with particularly tragic consequences. In this disease, there is a blockage in the catabolism of gangliosides (see Sphingolipids in Chapter 8, Section 8.2). The enzyme hexosaminidase A, responsible for the hydrolysis of N-acetylgalactosamine from ganglioside GM_2, is missing. Inclusions of accumulated GM_2 appear in the neurons of affected individuals. Those affected with Tay-Sachs disease appear normal as newborns, but by the age of one year they exhibit the characteristic symptoms of weakness, retardation, and blindness. This disease is fatal by age three or four. It is possible to detect the disease during fetal development via amniocentesis, a technique based on assay of amniotic fluid or amniotic cells for various enzyme activities.

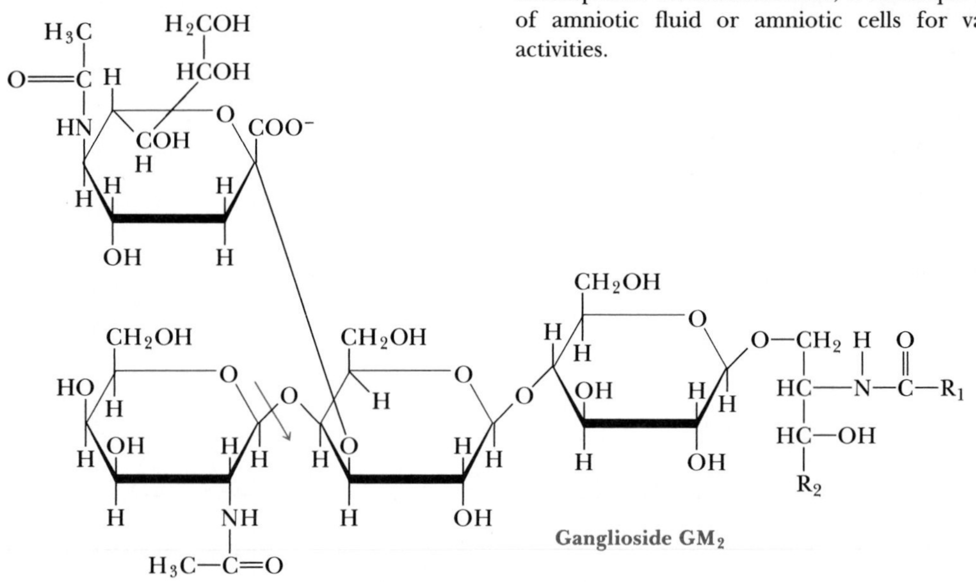

Ganglioside GM_2

The enzyme that catalyzes the hydrolysis of the bond indicated by a red arrow is missing in Tay–Sachs disease.

---

that contain a sialic acid residue, also at the primary alcohol group. See Sphingolipids in Chapter 8, Section 8.2, for the structures of these compounds.

Gangliosides play a part in **Tay–Sachs disease** (Box 15.1). Because the enzyme hexosaminidase A is missing, the catabolism of gangliosides is blocked, creating conditions that invariably prove fatal to affected individuals.

15.8
THE ANABOLISM OF CHOLESTEROL

The ultimate precursor of all the carbon atoms in cholesterol and in the other steroids that are derived from cholesterol is the acetyl group of acetyl-CoA. There are many steps in the biosynthesis of steroids. The condensation of three acetyl groups produces mevalonate, which contains six carbons. Decarboxylation of mevalonate produces the five-carbon isoprene unit frequently encountered in the structure of lipids. The involvement of isoprene units is a key point in the biosynthesis of steroids and of many other compounds that have the generic name *terpenes*. Vitamins A, E, and K come from these reactions, which humans cannot carry out. That is why these compounds are vitamins that we require; vitamin D, the remaining lipid-soluble vitamin, is derived from cholesterol (Chapter 8, Section 8.8). Isoprene units are involved in the biosynthesis of ubiquinone (coenzyme Q) and of derivatives of proteins and tRNA with specific five-carbon units attached.

Six isoprene units condense to form squalene, which contains 30 carbon atoms. Finally squalene is converted to cholesterol, which contains 27 carbon atoms (Figure 15.20); squalene can also be converted to other sterols.

$$\text{Acetate} \longrightarrow \text{mevalonate} \longrightarrow [\text{isoprene}] \longrightarrow \text{squalene} \longrightarrow \text{cholesterol}$$
$$C_2 \qquad\qquad C_6 \qquad\qquad\quad C_5 \qquad\qquad\quad C_{30} \qquad\qquad C_{27}$$

It is well established that 12 of the carbon atoms of cholesterol arise from the carboxyl carbon of the acetyl group; these are the carbon atoms labeled

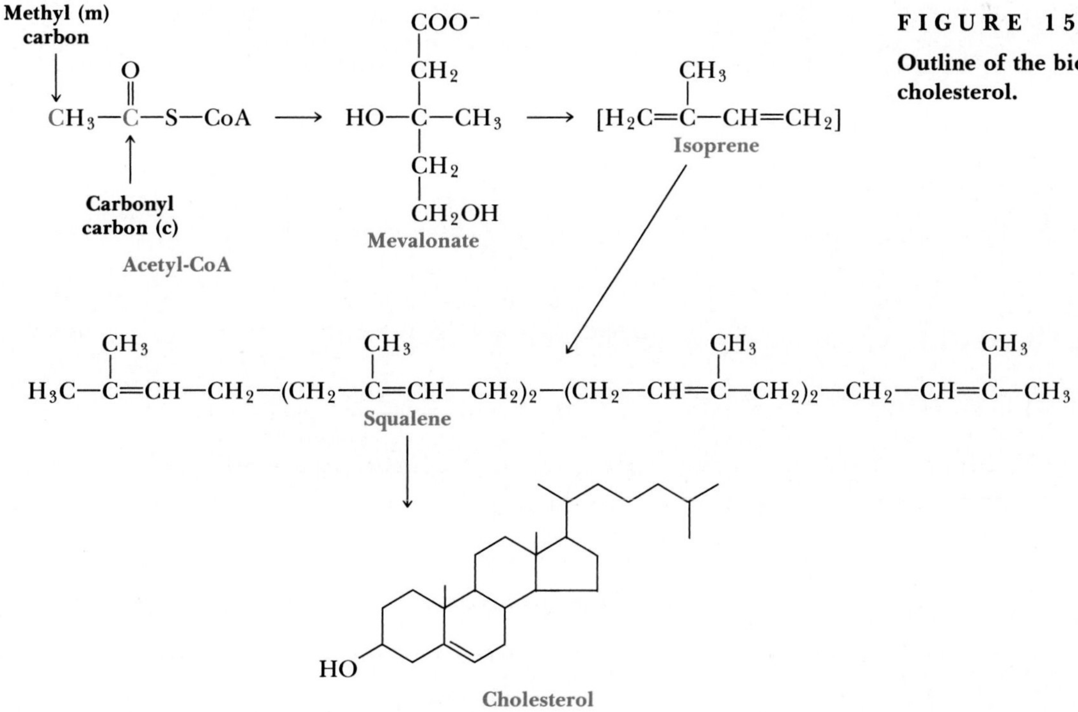

FIGURE 15.20

Outline of the biosynthesis of cholesterol.

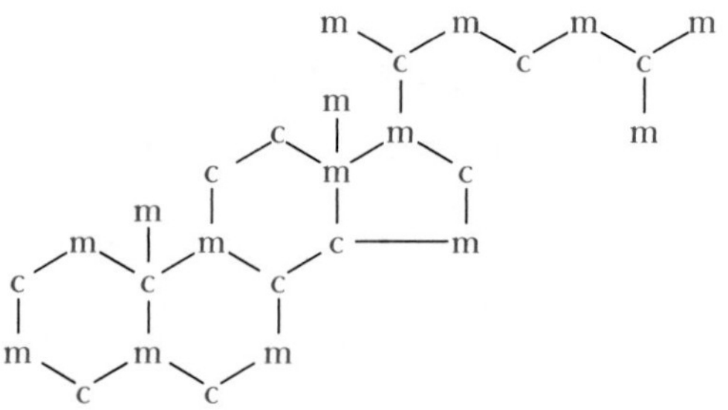

FIGURE 15.21

The labeling pattern of cholesterol. The m's are methyl carbons and the c's are carbonyl carbons.

"c" in Figure 15.21. The other 15 carbon atoms arise from the methyl carbon of the acetyl group; these are the carbon atoms labeled "m." We shall now look at the individual steps of the process in more detail.

The conversion of three acetyl groups of acetyl-CoA to **mevalonate** takes place in several steps (Figure 15.22). We already saw the first of these steps, the production of acetoacetyl-CoA from two molecules of acetyl-CoA,

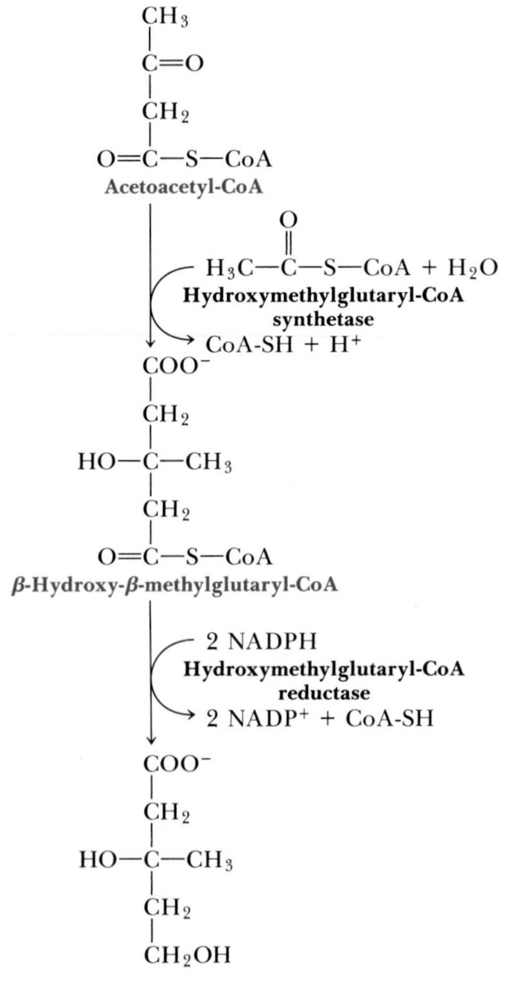

FIGURE 15.22

The biosynthesis of mevalonate.

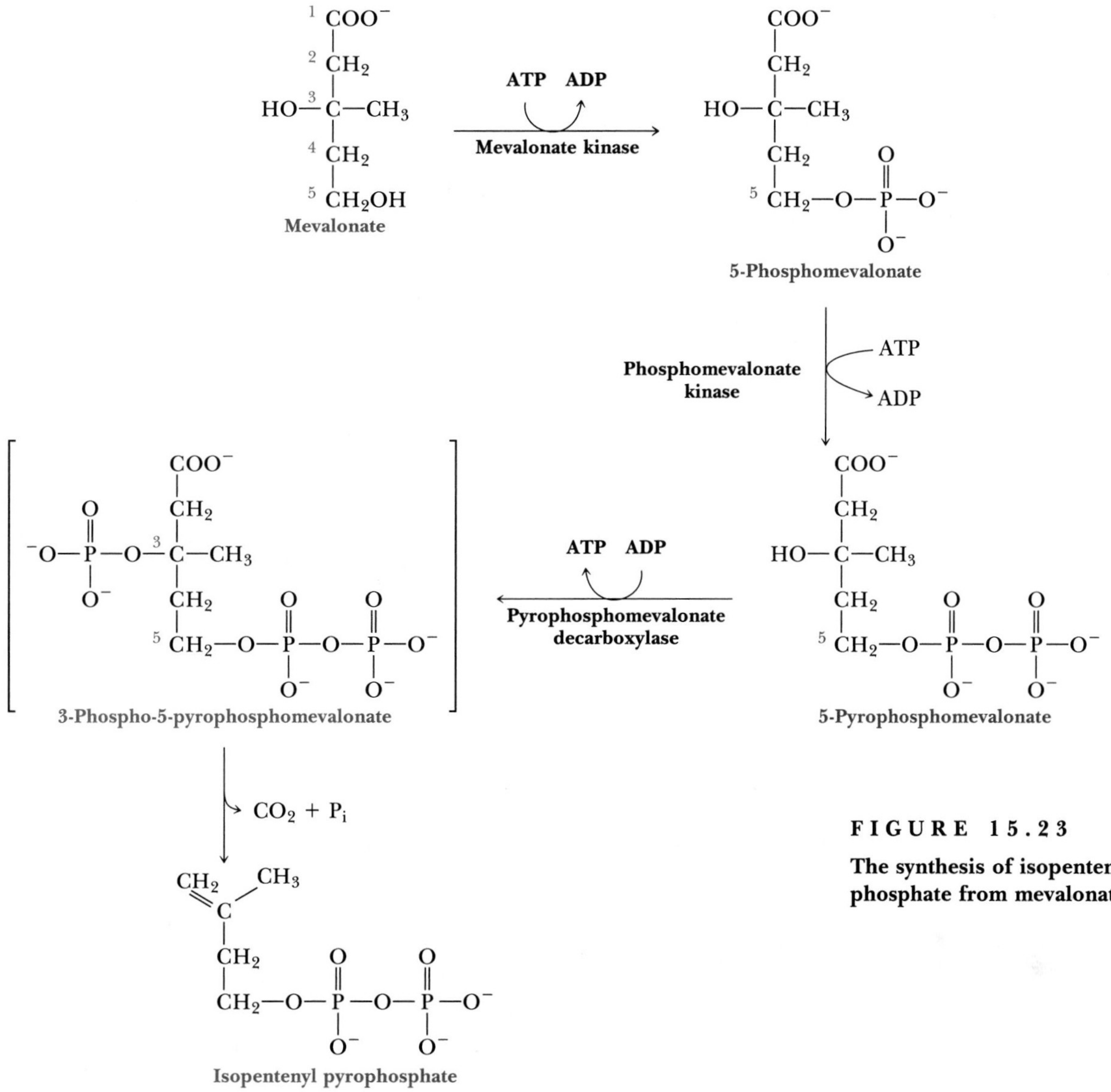

FIGURE 15.23

The synthesis of isopentenyl pyrophosphate from mevalonate.

when we discussed the formation of ketone bodies and the anabolism of fatty acids. A third molecule of acetyl-CoA condenses with acetoacetyl-CoA to produce **β-hydroxy-β-methylglutaryl-CoA** (also called HMG-CoA and 3-hydroxy-3-methylglutaryl-CoA). This reaction is catalyzed by the enzyme hydroxymethylglutaryl-CoA synthase; one molecule of CoA-SH is released in the process. In the next reaction the production of mevalonate from hydroxymethylglutaryl-CoA is catalyzed by the enzyme hydroxymethylglutaryl-CoA reductase. A carboxyl group, the one esterified to CoA-SH, is reduced to a hydroxyl group, and the CoA-SH is released. This step is inhibited by high levels of cholesterol; it is also a possible target for design of drugs to lower cholesterol levels in the body.

Mevalonate is then converted to an isoprenoid unit by a combination of phosphorylation, decarboxylation, and dephosphorylation reactions (Figure 15.23). Three successive phosphorylation reactions, each of which is catalyzed by an enzyme that requires ATP, give rise to **3-phospho-5-**

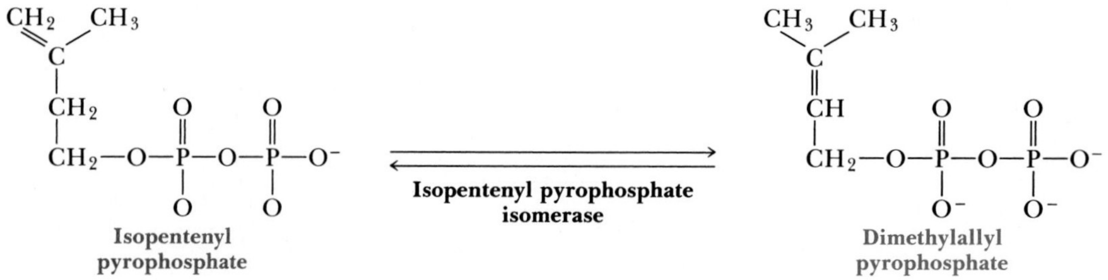

FIGURE 15.24

Isomerization of isopentenyl pyrophosphate.

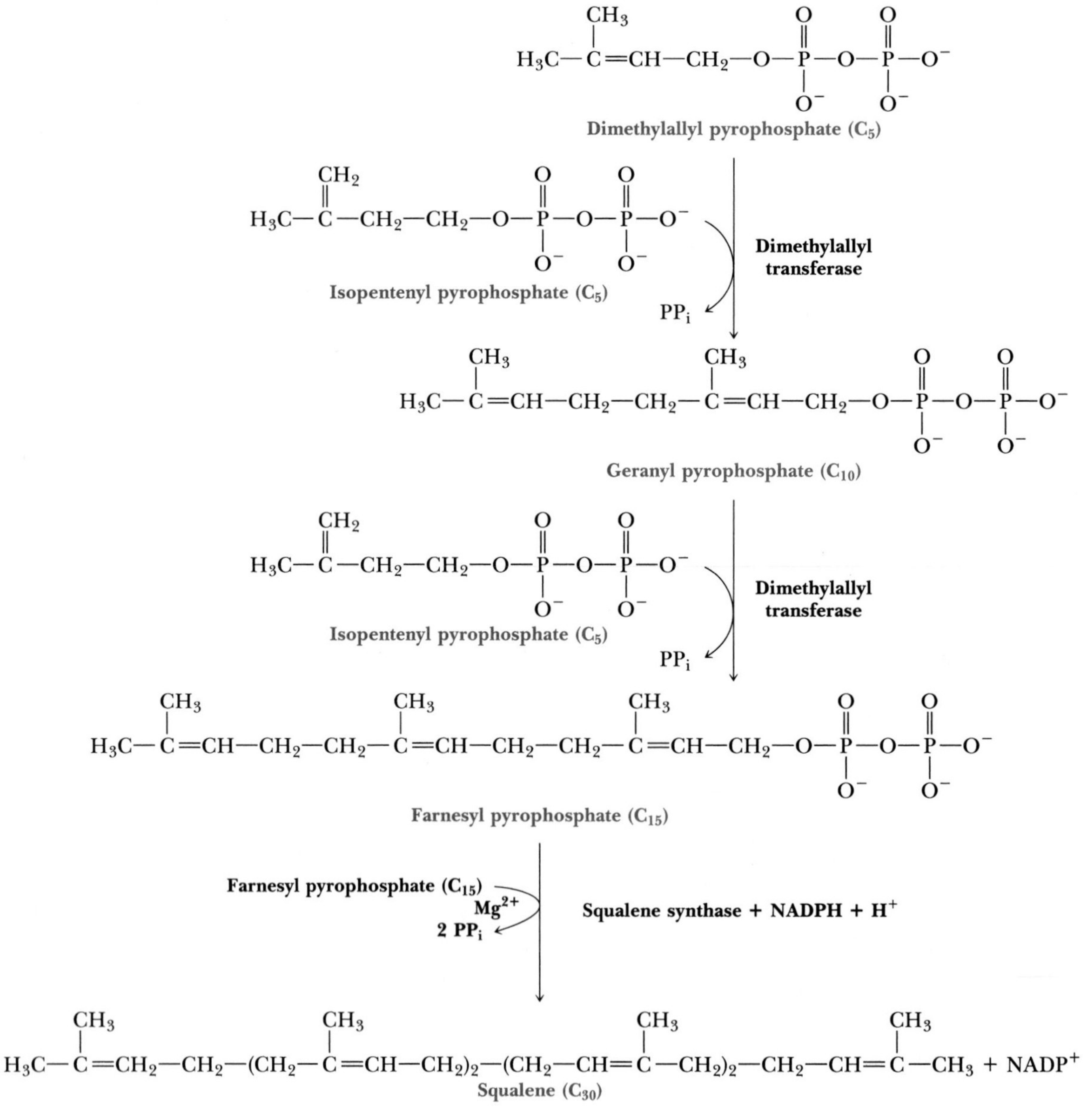

FIGURE 15.25

The synthesis of squalene.

pyrophosphomevalonate. This last compound is unstable, and it undergoes a decarboxylation and a dephosphorylation reaction to produce **isopentenyl pyrophosphate,** a five-carbon isoprenoid derivative. Isopentenyl pyrophosphate and **dimethylallyl pyrophosphate,** another isoprenoid derivative, can be interconverted in a rearrangement reaction catalyzed by the enzyme isopentenyl pyrophosphate isomerase (Figure 15.24).

Condensation of isoprenoid units then leads to the production of squalene (Figure 15.25) and ultimately of cholesterol. Both of the isoprenoid derivatives we have met so far are required; isopentenyl pyrophosphate and dimethylallyl pyrophosphate condense with each other in a reaction catalyzed by dimethylallyl transferase (also called prenyl transferase) to produce **geranyl pyrophosphate,** a ten-carbon compound. Another condensation reaction takes place, this time between geranyl pyrophosphate and isopentenyl pyrophosphate. This reaction is again catalyzed by dimethylallyl transferase; **farnesyl pyrophosphate,** a 15-carbon compound, is produced. Two molecules of farnesyl pyrophosphate condense to form **squalene,** a 30-carbon compound. The reaction is catalyzed by squalene synthase, and NADPH is required for the reaction.

Figure 15.26 shows the conversion of squalene to cholesterol. The details of this conversion are far from simple. Squalene is converted to **squalene epoxide** in a reaction that requires both NADPH and molecular oxygen (O_2). This reaction is catalyzed by squalene monooxygenase. Squalene epoxide then undergoes a complex cyclization reaction to form **lanosterol.** This remarkable reaction is catalyzed by squalene epoxide cyclase. The mechanism of the reaction is a concerted reaction, that is, one in which each part is essential for any other part to take place. No portion of a concerted reaction can be left out or changed, because it all takes place simultaneously rather than in a sequence of steps. The conversion of lanosterol to cholesterol is a complex process. It is known that 19 steps are required to remove three methyl groups and to move a double bond, but we shall not discuss the details of the process.

Cholesterol Is a Precursor of Other Steroids

Once cholesterol is formed, it can be converted to other steroids of widely varying physiological function. The smooth ER is an important site for both the synthesis of cholesterol and its conversion to other steroids. Most of the cholesterol formed in the liver, which is the principal site of cholesterol synthesis in mammals, is converted to **bile acids** such as cholate and glycocholate (Figure 15.27). These compounds aid in the digestion of lipid droplets by emulsifying them and rendering them more accessible to enzyme attack.

Cholesterol is the precursor of important steroid hormones (Figure 15.28) in addition to the bile acids. Like all hormones, whatever their chemical nature (Chapter 18, Section 18.5), steroid hormones serve as signals from outside a cell that regulate metabolic processes within a cell. Steroids are best known as sex hormones (they are a component of birth control pills), but they play other roles as well. **Pregnenolone** is formed from cholesterol, and **progesterone** is formed from pregnenolone. Progesterone is a sex hormone and is a precursor for other sex hormones such as **testosterone** and **estradiol** (an estrogen). Other types of steroid hormones also arise from progesterone. The role of sex hormones in sexual matura-

(Text continues on p. 474.)

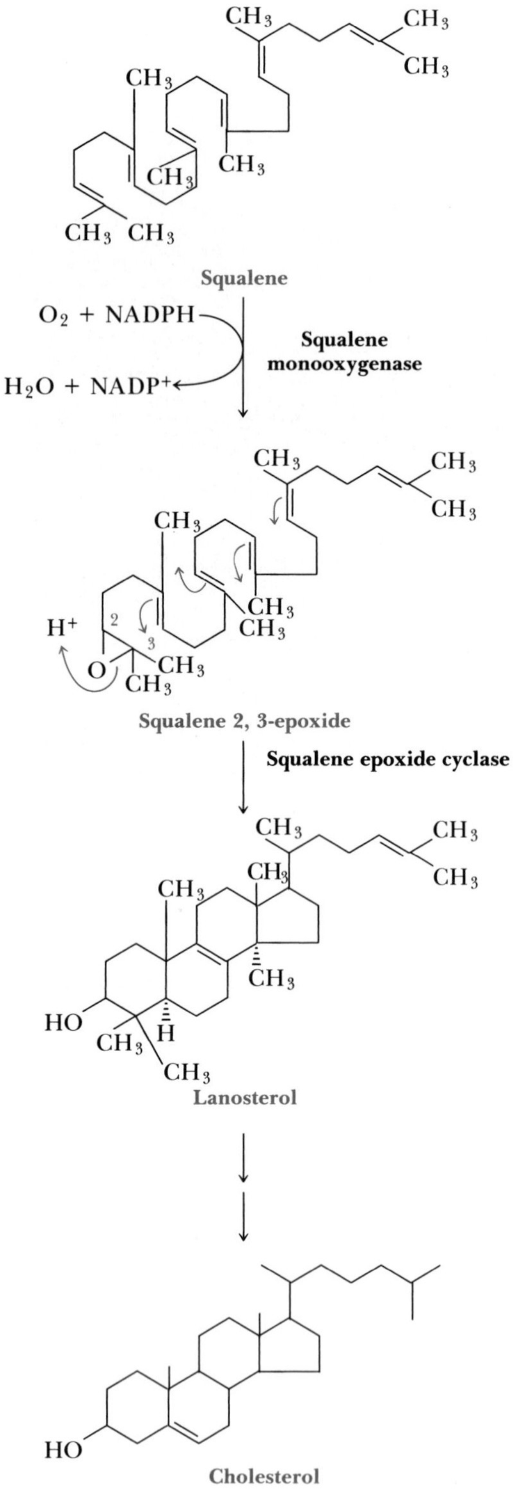

FIGURE 15.26

The conversion of squalene to cholesterol.

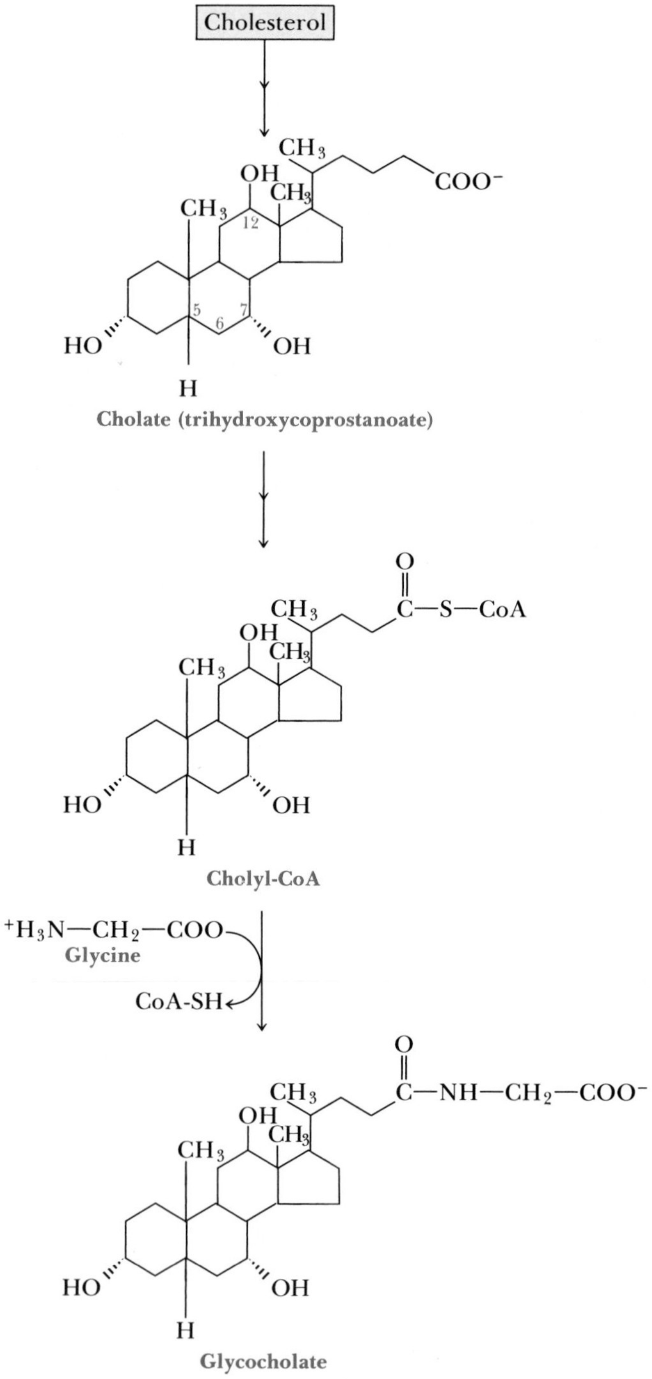

FIGURE 15.27

The synthesis of bile acids from cholesterol.

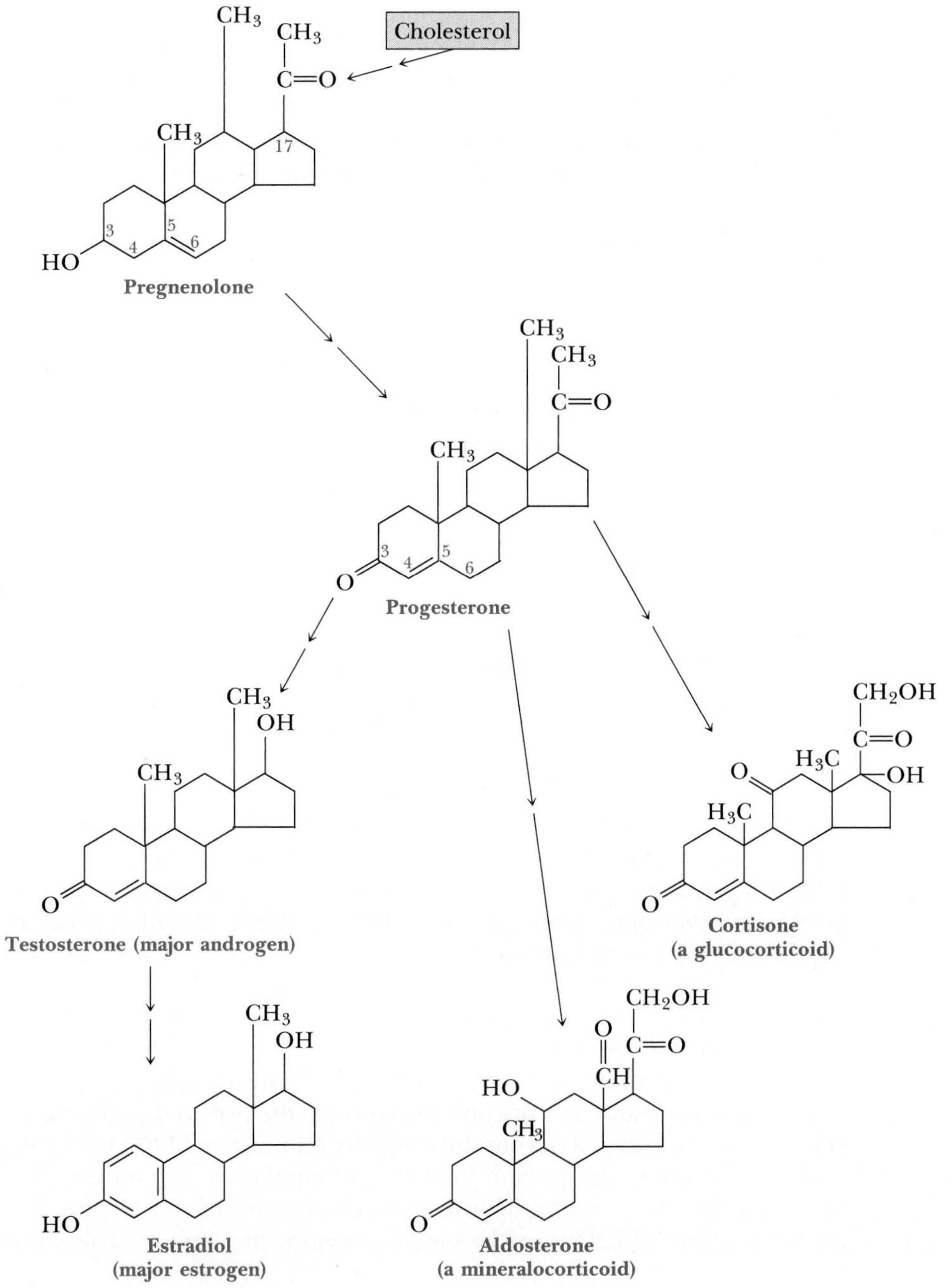

FIGURE 15.28

The synthesis of steroid hormones from cholesterol.

tion is discussed in Chapter 18, Section 18.5. **Cortisone** is an example of **glucocorticoids,** a group of hormones that play a role in carbohydrate metabolism, as the name implies, as well as in the metabolism of proteins and fatty acids. **Mineralocorticoids** constitute another class of hormones that are involved in the metabolism of electrolytes, including metal ions ("minerals") and water. **Aldosterone** is an example of a mineralocorticoid. In cells in which cholesterol is converted to steroid hormones, an enlarged smooth ER is frequently observed, providing a site for the process to take place.

The Role of Cholesterol in Heart Disease

Atherosclerosis is a condition in which arteries are blocked to a greater or lesser extent by the deposition of cholesterol plaques, leading to heart attacks. The process by which the clogging of arteries occurs is a complex one. Both diet and genetics are instrumental in the development of atherosclerosis. A diet high in cholesterol and fats, particularly saturated fats, will lead to a high level of cholesterol in the bloodstream. The body also makes its own cholesterol because this steroid is a necessary component of cell membranes. It is possible for more cholesterol to come from endogenous sources (synthesized within the body) than from the diet.

Cholesterol must be packaged for transport in the bloodstream; several classes of lipoproteins (summarized in Table 15.3) are involved in the transport of lipids in blood. These lipoprotein aggregates are usually classified by their densities. Besides chylomicrons, they include very-low-density lipoproteins (VLDL), intermediate-density lipoproteins (IDL), low-density lipoproteins (LDL), and high-density lipoproteins (HDL). LDL and HDL will play the major role in our discussion of heart disease. The protein portions of these aggregates can vary widely. The major lipids are generally cholesterol and its esters, in which the hydroxyl group is esterified to a fatty acid; triacylglycerols are also found in these aggregates. Chylomicrons are involved in the transport of dietary lipids, whereas the other lipoproteins primarily deal with endogenous lipids.

Figure 15.29 shows the architecture of an LDL particle. The interior consists of many molecules of cholesteryl esters (the hydroxyl group of the cholesterol is esterified to an unsaturated fatty acid such as linoleate); on the surface, a protein (apoprotein B-100), phospholipids, and unesterified cholesterol are in contact with the aqueous medium of the plasma. The protein portions of LDL particles bind to receptor sites on the surface of a

T A B L E   1 5 . 3   **Major Classes of Lipoproteins in Human Plasma**

LIPOPROTEIN CLASS	DENSITY $(g\ mL^{-1})$
Chylomicrons	<0.95
VLDL	0.95–1.006
IDL	1.006–1.019
LDL	1.019–1.063
HDL	1.063–1.210

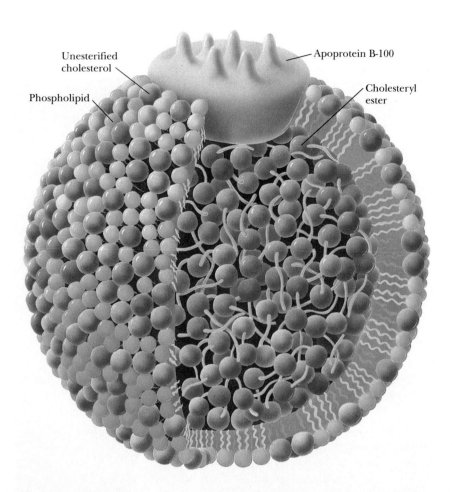

Unesterified
cholesterol

Phospholipid

Apoprotein B-100

Cholesteryl
ester

FIGURE 15.29

Schematic diagram of an LDL particle. (From M. S. Brown and J. L. Goldstein, 1984, How LDL Receptors Influence Cholesterol and Atherosclerosis, *Sci. Amer.* **251 (5), 58–66.)**

typical cell. Refer to Membrane Receptors in Chapter 8, Section 8.6, for a discussion of the process by which LDL particles are taken into the cell as one aspect of receptor action. This process is typical of the mechanism of uptake of lipids by cells, and we shall use the processing of LDL as a case study. LDL is the major player in the development of atherosclerosis.

LDL particles are degraded in the cell. The protein portion is hydrolyzed to the component amino acids, while the cholesterol esters are hydrolyzed to cholesterol and fatty acids. Free cholesterol can then be used directly as a component of membranes; the fatty acids can have any of the catabolic or anabolic fates reviewed earlier in this chapter (Figure 15.30). Cholesterol not needed for membrane synthesis can be stored as oleate or palmitoleate esters in which the fatty acid is esterified to the hydroxyl group of cholesterol. The production of these esters is catalyzed by acyl-CoA cholesterol acyltransferase (ACAT), and the presence of free cholesterol increases the enzymatic activity of ACAT. In addition, cholesterol inhibits

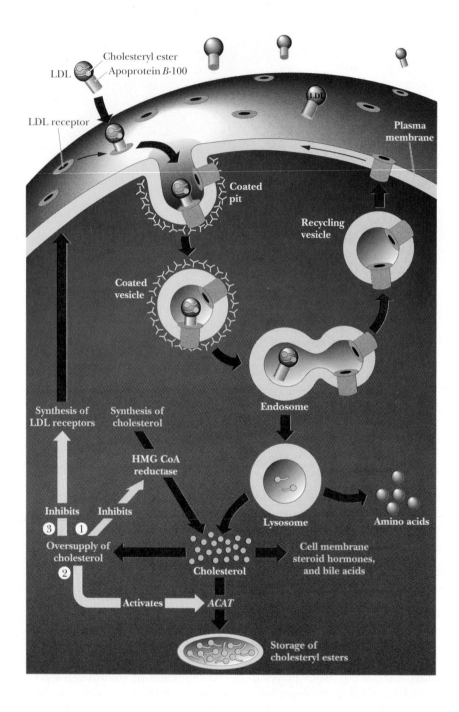

both the synthesis and the activity of the enzyme hydroxymethylglutaryl-
CoA reductase (HMG-CoA reductase). This enzyme catalyzes the production of mevalonate, the reaction that is the committed step in cholesterol
biosynthesis. This point has important implications. Dietary cholesterol
suppresses the synthesis of cholesterol by the body, especially in tissues
other than the liver. A third effect of the presence of free cholesterol in the
cell is inhibition of synthesis of LDL receptors. As a result of reduction in

the number of receptors, *intra*cellular synthesis of cholesterol is inhibited, and the level of LDL in the blood increases, leading to the deposition of atherosclerotic plaques.

The crucial role of LDL receptors in maintaining the level of cholesterol in the bloodstream is especially clear in the case of **familial hypercholester-olemia,** which results from a defect in the gene that codes for active receptors. An individual who has one gene that codes for active receptor and one defective gene is heterozygous for this trait (a heterozygote). Heterozygotes have blood cholesterol levels that are above average and therefore are at higher risk for heart disease than the general population. An individual with two defective genes and thus no active LDL receptor is homozygous (a homozygote). Homozygotes have very high blood cholesterol levels from birth, and there are recorded cases of heart attacks in two-year-olds with this condition. Patients who are homozygous for familial hyper-cholesterolemia usually die before age 20. Another genetic abnormality involved in hypercholesterolemia is the one that gives rise to a faulty protein component of lipoprotein E, involved in the uptake of lipids by the cell. The unfortunate result is the same.

Before we leave this discussion, we should mention the "good" choles-terol, high-density lipoprotein (HDL). Unlike LDL, which transports choles-terol from the liver to the rest of the body, HDL transports it back to the liver for degradation to bile acids. It is desirable to have low levels of cholesterol and LDL in the bloodstream, but it is also desirable to have as high a proportion of total cholesterol as possible in the form of HDL. It is well known that high levels of LDL and low levels of HDL are correlated with the development of heart disease. Factors that are known to increase HDL levels decrease the probability of heart disease; regular, strenuous exercise increases HDL, whereas smoking reduces it.

SUMMARY

The catabolic oxidation of lipids releases large quan-tities of energy, whereas the anabolic formation of lipids represents an efficient way of storing chemical energy.

The oxidation of fatty acids is the chief source of energy in the catabolism of lipids. After an initial activation step in the cytosol, the breakdown of fatty acids takes place in the mitochondrial matrix by the process of β-oxidation. In this process, two-carbon units are successively removed from the carboxyl end of the fatty acid to produce acetyl-CoA, which subse-quently enters the citric acid cycle. There is a net yield of 146 ATP molecules for each molecule of stearic acid (an 18-carbon compound) that is completely oxidized to carbon dioxide and water. The pathway of catabolism of fatty acids includes reactions in which

unsaturated, as well as saturated, fatty acids can be metabolized. Odd-numbered fatty acids can also be metabolized by converting their unique breakdown product, propionyl-CoA, to succinyl-CoA, an interme-diate of the citric acid cycle. "Ketone bodies" are substances related to acetone that are produced when an excess of acetyl-CoA results from β-oxidation. This situation can arise from a large intake of lipids and a low intake of carbohydrates or can occur in diabetes, in which the inability to metabolize carbohydrates causes an imbalance in the breakdown products of carbohydrates and lipids.

The anabolism of fatty acids proceeds by a differ-ent pathway from β-oxidation. Some of the most important differences between the two processes are the requirement for biotin in anabolism and not in catabolism, and the requirement for NADPH in anabolism rather than the NAD^+ required in catabolism. Fatty acid biosynthesis occurs in the cytosol, catalyzed by an ordered multienzyme complex; fatty acid catabolism occurs in the mitochondrial matrix, with no ordered aggregate of enzymes.

Most compound lipids, such as triacylglycerols, phosphoacylglycerols, and sphingolipids, have fatty acids as precursors. In the case of steroids, the starting material is acetyl-CoA. Isoprene units are formed from acetyl-CoA in the early stages of a lengthy process that leads ultimately to cholesterol. Cholesterol in turn is the precursor of the other steroids. Both dietary cholesterol and genetic factors influence the role of cholesterol in heart disease.

E X E R C I S E S

1. Compare and contrast the pathways of fatty acid breakdown and biosynthesis. What features do these two pathways have in common? How do they differ?
2. Calculate the ATP yield for the complete oxidation of one molecule of palmitic acid (16 carbons).
3. Why does the degradation of palmitic acid (see Exercise 2) to eight molecules of acetyl-CoA require seven, rather than eight, rounds of the β-oxidation process?
4. Outline the role of carnitine in the transport of acyl-CoA molecules into the mitochondrion.
5. You hear a fellow student say that the oxidation of unsaturated fatty acids requires exactly the same group of enzymes as the oxidation of saturated fatty acids. Is the statement true or false, and why?
6. It is frequently said that camels store water in their humps for long desert journeys. How would you modify this statement on the basis of information in this chapter?
7. You meet someone who has a pronounced odor of acetone on his breath. Why is it not surprising to you to discover that he is diabetic?

8. Outline the steps involved in the production of malonyl-CoA from acetyl-CoA.
9. Why are linoleate and linolenate considered essential fatty acids?
10. Is it possible to convert fatty acids to other lipids without acyl-CoA intermediates?
11. Discuss the role of isoprenoid units in the biosynthesis of cholesterol.
12. A cholesterol sample is prepared using acetyl-CoA labeled with ^{14}C at the carboxyl group as precursor. Which carbon atoms of cholesterol are labeled?
13. What is the role of citrate in the transport of acetyl groups from the mitochondrion to the cytosol?
14. What structural feature do all steroids have in common? What are the biosynthetic implications of this common feature?
15. Why must cholesterol be packaged for transport rather than occurring freely in the bloodstream?

A N N O T A T E D   B I B L I O G R A P H Y

Bodner, C. M. Lipids. *J. Chem. Ed.* **63**, 772–775 (1986). [Part of a series of concise and clearly written articles on metabolism.]

Brown, M. S., and J. L. Goldstein. How LDL Receptors Influence Cholesterol and Atherosclerosis. *Sci. Amer.* **251** (5), 58–66 (1984). [A description of the role of cholesterol in heart disease by the winners of the 1985 Nobel Prize in medicine.]

Krutch, J. W. *The Voice of the Desert.* New York: Morrow, 1975. [Chapter 7, "The Mouse That Never Drinks," is a description, primarily from a naturalist's point of view, of the kangaroo rat, but it does make the point that metabolic water is this animal's only source of water.]

Lawn, R. Lipoprotein(a) in Heart Disease. *Sci. Amer.* **266** (6), 54–60 (1992). [Relates the properties of lipids and protein structure to the blockage of arteries characteristic of heart disease.]

McCarry, J. D., and D. W. Foster. Regulation of Hepatic Fatty Acid Oxidation and Ketone Body Production. *Ann. Rev. Biochem.* **49**, 395–420 (1980). [A review.]

Wakil, S. J., and E. M. Barnes. Fatty Acid Metabolism. *Compr. Biochem.* **18**, 57–104 (1971). [Extensive coverage of the topic.]

Chapter 16

Photosynthesis

Photosynthesis linked to oxygen plays an essential role in all life, plant and animal.

The drama of photosynthesis, converting sunlight to energy-rich carbohydrates, is played out in the chloroplast "theater" of the green plant. In each chloroplast there are stacks of thylakoid discs. The thylakoid membrane inside each disc is the lighted stage where the drama of Act I is performed. Here the energy of light is captured by electrons of chlorophyll molecules. The excited electrons are passed along a series of acceptors in an electron transport chain. In the process, a molecule of water is split and oxygen is released into the atmosphere. At the same time, protons pumped out of the thylakoid membrane drive the production of ATP. Excited electrons reduce $NADP^+$ to NADPH, and the stored energy is used in Act II for the biosynthesis of glucose, which takes place in the dark of the stroma outside the thylakoid membrane. Carbon dioxide from the atmosphere is combined with a five-carbon sugar to produce, through an intermediate, two 3-carbon sugars and eventually the energetic six-carbon molecule of glucose. The energy to drive this biosynthesis comes from ATP and the reducing power of NADPH, *n*icotinamide *a*denine *d*inucleotide *p*hosphate. Plants at the bottom of the food chain toil in the sun to store energy and generate oxygen for the benefit of all us animals.

16.1
THE REACTIONS OF PHOTOSYNTHESIS

It is well known that the photosynthetic organisms, such as green plants, convert carbon dioxide (CO_2) and water to carbohydrates such as glucose (written here as $C_6H_{12}O_6$) and molecular oxygen (O_2).

$$6\ CO_2 + 6\ H_2O \longrightarrow C_6H_{12}O_6 + 6\ O_2$$

The equation actually represents two processes. One, the splitting of water to produce oxygen, is a reaction that requires light energy from the sun, and the other, the fixation of CO_2 to give sugars, can and does take place in the dark but only because it uses solar energy indirectly. The actual storage form of the carbohydrates produced is not glucose but oligosaccharides (e.g., sucrose in sugarcane and sugar beets) and polysaccharides (starch and cellulose). However, it is customary and convenient to write the carbohydrate product as glucose, and we shall follow this time-honored practice.

In the light reaction, water is converted to oxygen by oxidation, and $NADP^+$ is reduced to NADPH. The light reaction is coupled to the phosphorylation of ADP to ATP in a process called **photophosphorylation.**

$$H_2O + NADP^+ + \longrightarrow NADPH + H^+ + \tfrac{1}{2} O_2$$

$$ADP + P_i \longrightarrow ATP$$

The light reaction of photosynthesis in turn consists of two parts, accomplished by two distinct but related photosystems. One part of the reaction is the reduction of $NADP^+$ to NADPH, carried out by **Photosystem I.** The second part of the reaction is the splitting of water to produce oxygen, carried out by **Photosystem II.** Both photosystems carry out redox (electron transfer) reactions. Photosystem I generates a mild oxidizing agent, and

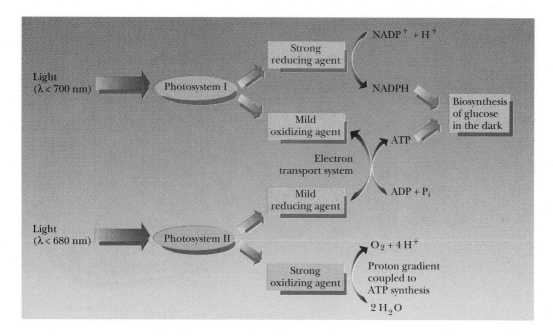

FIGURE 16.1

Photosystems I and II. NADPH is generated by Photosystem I for subsequent use in the dark reaction of photosynthesis. ATP is generated by photophosphorylation linked to electron transport between Photosystems I and II. Photosystem II splits water, giving rise to a proton gradient coupled to ATP synthesis.

Photosystem II generates a mild reducing agent (Figure 16.1). The mild oxidizing agent and mild reducing agent interact with each other indirectly through an electron transport chain that links the two photosystems. The production of ATP is linked to electron transport in a process similar to that seen in the production of ATP by mitochondrial electron transport.

In the dark reaction, the ATP and NADPH produced in the light reaction provide the energy and reducing power for the fixation of CO_2. The dark reaction is also a redox process, since the carbon in carbohydrates is in a more reduced state than the highly oxidized carbon in CO_2.

16.2
CHLOROPLASTS AND CHLOROPHYLLS

In prokaryotes such as cyanobacteria, photosynthesis takes place in granules bound to the plasma membrane. The site of photosynthesis in eukaryotes such as green plants and green algae is the **chloroplast** (Figure 16.2), a membrane-bounded organelle that we discussed in Chapter 1, Section 1.5.

Like the mitochondrion, the chloroplast has inner and outer membranes and an intermembrane space. Within the chloroplast are bodies called **grana,** which consist of stacks of flattened membranes called **thylakoid disks.** The thylakoid disks are formed by the folding of the inner

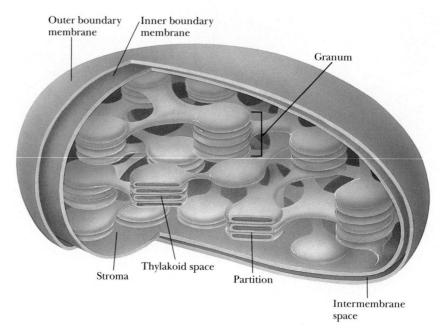

Outer boundary membrane

Inner boundary membrane

Granum

Stroma

Thylakoid space

Partition

Intermembrane space

FIGURE 16.2

Membrane structures in chloroplasts.

membrane of the chloroplast; in this respect they resemble the cristae of mitochondria. The grana are connected by membranes called intergranal lamellae. The trapping of light and the production of oxygen take place in the thylakoid disks. The dark reaction, in which CO_2 is fixed to carbohydrates, takes place in the soluble portion of the chloroplast, called the **stroma.** The stroma plays the same role in the structure of the chloroplast as does the matrix in the mitochondrion. In addition to the stroma, there is a **thylakoid space** within the thylakoid disks themselves.

It is well established that the primary event in photosynthesis is the absorption of light by **chlorophyll.** The higher-energy states (excited states) of chlorophyll are useful in photosynthesis because the light energy can be passed along and converted to chemical energy in the light reaction. There are two principal types of chlorophyll, **chlorophyll *a*** and **chlorophyll *b*.** Eukaryotes such as green plants and green algae contain both chlorophyll *a* and chlorophyll *b*. Prokaryotes such as cyanobacteria (formerly called blue-green algae) contain only chlorophyll *a*. Photosynthetic bacteria other than cyanobacteria have bacteriochlorophylls, with **bacteriochlorophyll *a*** being the most common. Organisms such as green and purple sulfur bacteria, which contain bacteriochlorophylls, do not use water as the ultimate source of electrons for the redox reactions of photosynthesis, nor do they produce oxygen. Instead, they use other electron sources such as H_2S, which produces elemental sulfur instead of oxygen. Organisms that contain bacteriochlorophyll are anaerobic and have only one photosystem.

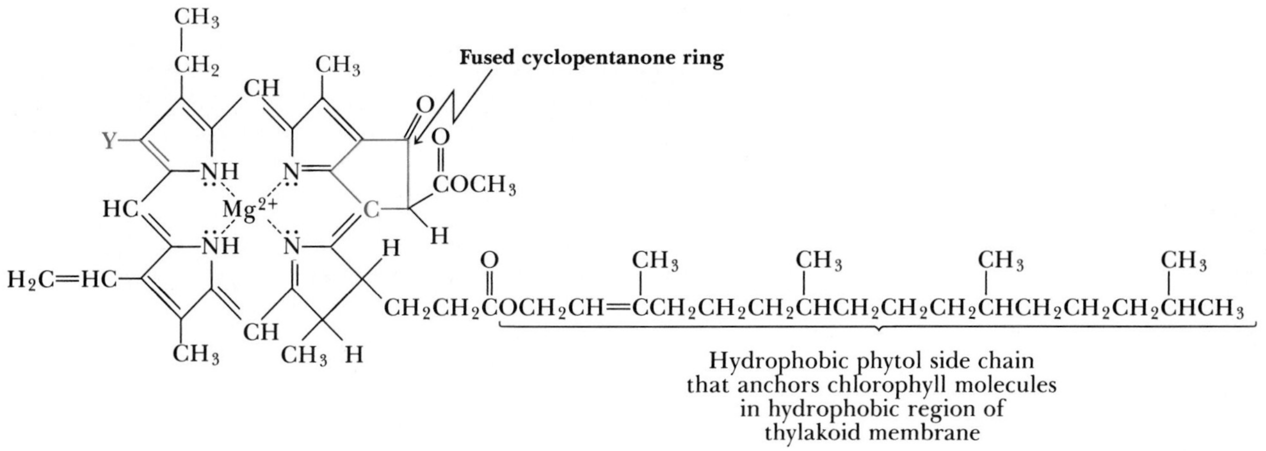

CH_3
CH_2
CH
CH_3

Fused cyclopentanone ring

Y—

—NH

N—

Mg^{2+}

HC

O
O
$COCH_3$

—NH

N—

H

$H_2C=HC$—

C

H

CH_3
CH_3 H
CH

O
$\|$
$CH_2CH_2COCH_2CH=CCH_2CH_2CH_2CHCH_2CH_2CH_2CHCH_2CH_2CH_2CHCH_3$
  CH_3 CH_3 CH_3 CH_3

Hydrophobic phytol side chain
that anchors chlorophyll molecules
in hydrophobic region of
thylakoid membrane

Y is —CH₃ in chlorophyll *a*
Y is —CHO in chlorophyll *b*
Y is —CH₃ in bacteriochlorophyll *a*
   (and highlighted bond is saturated)

FIGURE 16.3

Molecular structures of chlorophyll *a*, chlorophyll *b*, and bacteriochlorophyll *a*.

The structure of chlorophyll is based on the tetrapyrrole ring of porphyrins, which occurs in the heme group of myoglobin, hemoglobin, and the cytochromes (Figure 16.3) (see Chapter 4, Section 4.4). The metal ion bound to the tetrapyrrole ring is magnesium, Mg(II), rather than iron, which occurs in heme. Another difference between chlorophyll and heme is the presence of a cyclopentanone ring fused to the tetrapyrrole ring. There is a long hydrophobic side chain, the phytol group, which contains four isoprenoid units (five-carbon units that are basic building blocks in many lipids; (Chapter 15, Section 15.8) and which binds to the thylakoid membrane by hydrophobic interactions. The phytol group is bound to the rest of the chlorophyll molecule by an ester linkage between the alcohol group of the phytol and a propionic acid side chain on the porphyrin ring. The difference between chlorophyll *a* and chlorophyll *b* lies in the substitution of an aldehyde group for a methyl group on the porphyrin ring. The difference between bacteriochlorophyll *a* and chlorophyll *a* is that a double bond in the porphyrin ring of chlorophyll *a* is saturated in bacteriochlorophyll *a*. The lack of a conjugated system (alternating double and single bonds) in the porphyrin ring of bacteriochlorophylls causes a significant difference in the absorption of light by bacteriochlorophyll *a* compared with chlorophyll *a* and *b*.

The absorption spectra of chlorophyll *a* and chlorophyll *b* differ slightly (Figure 16.4). Both absorb light in the red and blue portions of the visible spectrum (600–700 nm and 400–500 nm, respectively), and the presence of both types of chlorophyll guarantees that more wavelengths of the visible

(a)

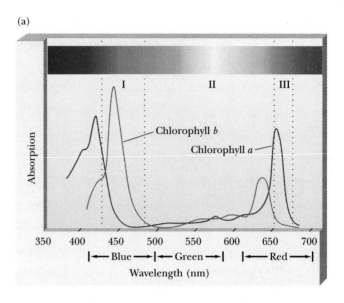

(b)

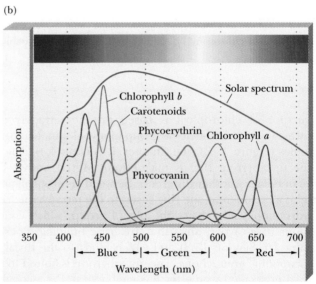

FIGURE 16.4

(a) The absorption of visible light by chlorophylls *a* and *b*. The areas marked I, II, and III are regions of the spectrum that give rise to chloroplast activity. There is greater activity in regions I and III, which are close to major absorption peaks. There are high levels of O_2 production when light from regions I and III is absorbed by chloroplasts. Lower (but measurable) activity is seen in region II, where some of the accessory pigments absorb. (b) The absorption of light by accessory pigments (superimposed on the absorption of chlorophylls *a* and *b*). The accessory pigments absorb light and transfer their energy to chlorophyll.

spectrum are absorbed than would be the case with either one individually. Recall that chlorophyll *a* is found in all photosynthetic organisms that produce oxygen. Chlorophyll *b* is found in eukaryotes such as green plants and green algae, but it occurs in smaller amounts than chlorophyll *a.* The presence of chlorophyll *b,* however, increases the portion of the visible spectrum that is absorbed and thus enhances the efficiency of photosynthesis in green plants compared with cyanobacteria. In addition to chlorophyll, various **accessory pigments** absorb light and transfer energy to chlorophylls (Figure 16.4b). Bacteriochlorophylls, the molecular form characteristic of photosynthetic organisms that do not produce oxygen, absorb light at longer wavelengths. The wavelength of maximum absorption of bacteriochlorophyll *a* is 780 nm; other bacteriochlorophylls have absorption maxima at still longer wavelengths, such as 870 or 1050 nm. Light of wavelength longer than 800 nm is part of the infrared, rather than the visible, region of the spectrum. The wavelength of light absorbed plays a critical role in the light reaction of photosynthesis because the energy of light is inversely related to wavelength (see Box 16.1).

Chlorophyll molecules are arranged in **photosynthetic units.** Most of the chlorophyll molecules in the unit simply gather light (antennae chlorophylls). The light-harvesting molecules then pass their excitation energy along to a specialized pair of chlorophyll molecules at a **reaction center.** When the light energy reaches the reaction center, the chemical reactions of photosynthesis begin. The different environments of the antennae chlorophylls and the reaction-center chlorophylls give different properties to the

16.1

THE RELATIONSHIP BETWEEN WAVELENGTH AND ENERGY OF LIGHT

A well-known equation relates the wavelength and energy of light, a point of crucial importance for our purposes. Max Planck established in the early 20th century that the energy of light is directly proportional to its frequency:

$$E = h\nu$$

where E is energy, h is a constant (Planck's constant), and ν is the frequency of the light. The wavelength of light is related to the frequency:

$$\nu = \frac{c}{\lambda}$$

where λ is wavelength, ν is frequency, and c is the velocity of light. We can rewrite the expression for the energy of light in terms of wavelength rather than frequency.

$$E = h\nu = \frac{hc}{\lambda}$$

Light of shorter wavelength (higher frequency) is higher in energy than light of longer wavelength (lower frequency).

In the visible spectrum, blue light has a shorter wavelength (λ), higher frequency (ν), and higher energy (E) than red light. Intermediate values of all these quantities are observed for other colors of the visible spectrum.

Highest frequency ——————————→ Lowest frequency (ν)
Highest energy ——————————→ Lowest energy (E)
Shortest wavelength ——————————→ Longest wavelength (λ)

two different kinds of molecules. In a typical photosynthetic unit, there are several hundred light-harvesting antennae chlorophylls for each unique chlorophyll at a reaction center.

The precise nature of reaction centers is the subject of active research. The most extensively studied system is that from bacteria of the genus *Rhodopseudomonas*. These bacteria do not produce molecular oxygen as a result of their photosynthetic activities, but enough similarities exist between the photosynthetic reactions of *Rhodopseudomonas* and photosynthesis linked to oxygen to lead scientists to draw conclusions about the nature of reaction centers in all organisms. The detailed process that goes on at the reaction center of *Rhodopseudomonas* is important enough to warrant further discussion.

It is well established that there is a pair of bacteriochlorophyll molecules in the reaction center of *Rhodopseudomonas viridis*; the critical pair of chlorophylls is embedded in a protein complex that is in turn an integral part of the photosynthetic membrane. (We shall refer to the bacteriochlorophylls simply as chlorophylls in the interest of simplifying the discussion.) Accessory pigments, which also play a role in the light-trapping process, have specific positions close to the special pair of chlorophylls. The absorption of light by the special pair of chlorophylls raises them to a higher energy level (Figure 16.5a). An electron is passed to a series of accessory pigments (Figure 16.5b). The first of these accessory pigments is pheophytin, which is structurally similar to chlorophyll, differing only by having two hydrogens in place of the magnesium. The electron is passed along to the pheophytin, raising it in turn to an excited energy level. (Note that the electron travels on only one of two possible paths. Research is in progress to determine why this is so.) The next electron acceptor is menaquinone (Q_A); it is structurally similar to coenzyme Q, which plays a role in the mitochondrial electron transport chain. The final electron acceptor, which is also raised to an excited state, is coenzyme Q itself (ubiquinone, called Q_B here). The electron that was passed to Q_B is replaced by an electron donated by a cytochrome, which acquires a positive charge in the process (Figure 16.5c). The cytochrome is not bound to the membrane and diffuses away, carrying its positive charge with it. The whole process takes place in less than 10^{-3} second. The positive and negative charges have traveled in opposite

The structures of menaquinone and ubiquinone.

Menaquinone

Ubiquinone

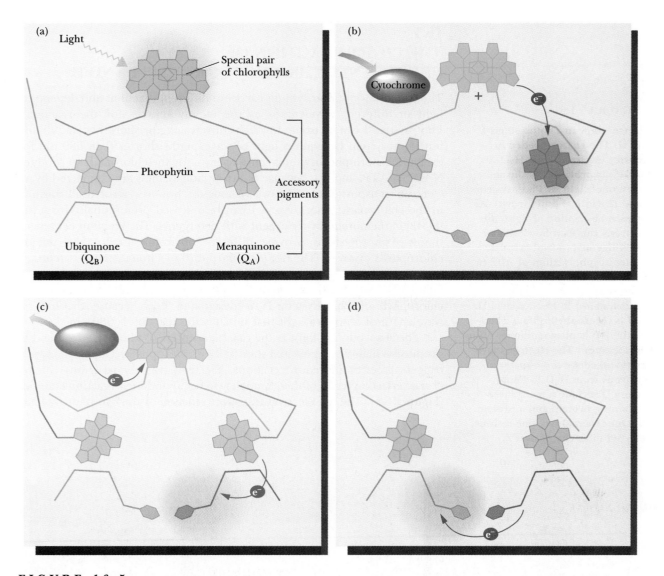

FIGURE 16.5

Molecular events that take place at the photosynthetic reaction center of *Rhodopseudomonas*. (a) The special pair of chlorophylls is raised to a higher energy level by absorption of light. The chlorophylls in the reaction center are bacteriochlorophyll, rather than chlorophyll *a* or *b*. (b) An excited electron is passed to pheophytin on one side of the reaction center. A cytochrome not bound to the membrane diffuses into the region of the special pair of chlorophylls. (c) The cytochrome donates an electron to the reaction center. The cytochrome, which is now positively charged, diffuses away. The excited electron is passed to menaquinone. (d) The electron is then passed to ubiquinone. The resulting separation of charge represents stored energy.

directions from the chlorophyll pair and are separated from each other. This situation is similar to the proton gradient in mitochondria, where the existence of the proton gradient is ultimately responsible for oxidative phosphorylation. The separation of charge is equivalent to a battery, a form of stored energy. The reaction center has acted as a transducer, converting light energy to a form usable by the cell to carry out the energy-requiring reactions of photosynthesis. The processes that take place in *Rhodopseudomonas* serve as a model for reaction centers in photosynthesis linked to oxygen.

16.3
THE LIGHT REACTION OF PHOTOSYNTHESIS: PHOTOSYSTEMS I AND II

The two different photosystems carry out different reactions and depend on light of different wavelengths as the energy sources for the reactions. Photosystem I can be excited by light of wavelengths shorter than 700 nm, but Photosystem II requires light of wavelengths shorter than 680 nm for excitation. Both photosystems must operate for the chloroplast to produce NADPH, ATP, and O_2, because the two photosystems are connected by an electron transport chain. The two systems are, however, structurally distinct in the chloroplast; Photosystem I can be released preferentially from the thylakoid membrane by treatment with detergents. The reaction centers of the two photosystems provide different environments for the unique chlorophylls involved. The unique chlorophyll of Photosystem I is referred to as P_{700}, P for pigment and 700 for the longest wavelength of absorbed light (700 nm) that initiates the reaction. Similarly, the reaction-center chlorophyll of Photosystem II is designated P_{680} because the longest wavelength of absorbed light that initiates the reaction is 680 nm.

The absorption of light at the reaction centers of Photosystems I and II produces a high-energy excited state in their chlorophylls. This event in turn triggers a series of redox reactions. Electron transfer reactions can be characterized in terms of their tendency to occur and their reaction energies (Figure 16.6). The net electron transport reaction of the two photosystems

FIGURE 16.6

Electron flow in Photosystems I and II. The energy needed to transfer electrons from H_2O to $NADP^+$ is provided by the absorption of light by Photosystems I and II (vertical [up] arrows). After each absorption of light, the electrons can then flow "downhill" (diagonal [down] arrows). Photophosphorylation of ADP to yield ATP is coupled to the electron transport chain that links Photosystem I to Photosystem II. (Chl is chlorophyll; Phe is pheophytin; PQ is plastoquinone; PC is plastocyanin.) The electron carriers that mediate the transfer of electrons from H_2O to Photosystem II include a manganese-containing protein and a protein with an essential tyrosine residue, referred to as component Z.

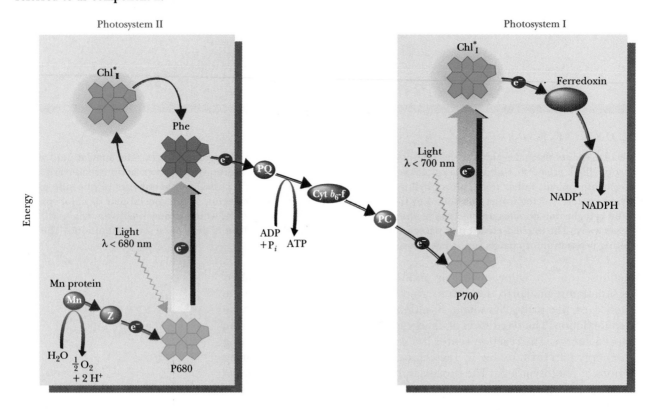

taken together is, except for the substitution of NADPH for NADH, the reverse of mitochondrial electron transport. The half reaction of reduction is that of $NADP^+$ to NADPH, whereas the half reaction of oxidation is that of water to oxygen.

$$NADP^+ + 2\,H^+ + 2\,e^- \longrightarrow NADPH + H^+$$
$$H_2O \longrightarrow \tfrac{1}{2}\,O_2 + 2\,H^+ + 2\,e^-$$
$$\overline{NADP^+ + H_2O \longrightarrow NADPH + H^+ + \tfrac{1}{2}\,O_2}$$

This is an endergonic reaction with a positive $\Delta G^{\circ\prime} = +220$ kJ $= +52.6$ kcal mol^{-1}. The light energy absorbed by the chlorophylls in both photosystems provides the energy that allows this endergonic reaction to take place.

The absorption of light by the chlorophyll of Photosystem I supplies the energy that ultimately allows the photoreduction of $NADP^+$ to take place. The absorption process can be written

$$Chl_I + h\nu\ (700\ nm) \longrightarrow Chl_I^*$$

where Chl_I and Chl_I^* are the P_{700} chlorophyll in the ground (unexcited) and excited states, respectively, and $h\nu$ is the light absorbed. Chl_I^* becomes a strong reducing agent; it can easily give electrons to another substance. After a series of electron transfers, the final substance reduced is $NADP^+$.

Since electrons have been transferred to $NADP^+$, producing NADPH, Chl_I^+ now lacks an electron. The needed electron is supplied by Photosystem II. The absorption of light by the chlorophyll of Photosystem II provides the energy that ultimately allows the photooxidation of water, the final electron source. The absorption process in Photosystem II can be written

$$Chl_{II} + h\nu\ (680\ nm) \longrightarrow Chl_{II}^*$$

where Chl_{II} is the P_{680} chlorophyll. Chl_{II}^* becomes a strong enough reducing agent to pass electrons to Chl_I^+ through an electron transport chain coupled to the phosphorylation of ADP. The energy for the reactions that allow electrons to flow from Photosystem II to Photosystem I is supplied by the absorption of light by Chl_{II}, the reaction-center chlorophyll of Photosystem II. Chl_{II}^+, now lacking an electron, is a strong enough oxidizing agent to oxidize water to oxygen. The water, as it is oxidized to oxygen, is the ultimate source of the electrons flowing through the system.

There are two places in the reaction scheme of the two photosystems where the absorption of light supplies energy to make endergonic reactions take place (Figure 16.6). Neither reaction-center chlorophyll is a strong enough reducing agent to pass electrons to the next substance in the reaction sequence, but the absorption of light by the chlorophylls of both photosystems provides enough energy for such reactions to take place. When Chl_I absorbs light, enough energy is provided to allow the ultimate reduction of $NADP^+$ to take place. (Note that the energy difference is shown on the vertical axis of Figure 16.6. This type of diagram is also called a "Z scheme.") Similarly, the absorption of light by Chl_{II} allows electrons to be passed to the electron transport chain that links Photosystem II and

Photosystem I and generates a strong enough oxidizing agent to split water, producing oxygen. In both photosystems, the result of supplying energy is analogous to pumping water uphill.

Photosystem I: Reduction of NADP$^+$

As we just saw, the absorption of light by Chl_I leads to the series of electron transfer reactions of Photosystem I. The substance to which the excited-state chlorophyll, Chl_I^*, gives an electron is apparently a molecule of chlorophyll *a;* this transfer of electrons is mediated by processes that take place in the reaction center. The next electron acceptor in the series is bound ferredoxin, an iron-sulfur protein occurring in the membrane in Photosystem I. The bound ferredoxin passes its electron to a molecule of soluble ferredoxin. Soluble ferredoxin in turn reduces an FAD enzyme called ferredoxin-NADP reductase. The FAD portion of the enzyme reduces NADP$^+$ to NADPH (Figure 16.6). We can summarize the main features of the process in two equations, in which the notation ferredoxin refers to the soluble form of the protein.

$$Chl_I^* + ferredoxin_{oxidized} \longrightarrow Chl_I^+ + ferredoxin_{reduced}$$

$$2\ Ferredoxin_{red} + H^+ + NADP^+ \xrightarrow{\text{Ferredoxin-NADP reductase}} 2\ ferredoxin_{ox} + NADPH$$

Two points can be noted about this series of equations. The first is a matter of bookkeeping. Chl_I^* donates one electron to ferredoxin, but the electron transfer reactions of FAD and NADP$^+$ involve two electrons. Thus, an electron from each of two ferredoxins is required for the production of NADPH. The second point is the fate of the oxidized chlorophyll (Chl_I^+), which has lost an electron. This pigment is colorless in its oxidized form, and it must be reduced back to the light-absorbing form to continue photosynthesis. The electrons needed to reduce the oxidized chlorophyll are supplied by Photosystem II.

Photosystem II: Water Is Split To Produce Oxygen

Photosystem II is more complex than Photosystem I, and the components of Photosystem II are more difficult to isolate from the thylakoid membranes than are those of Photosystem I. The oxidation of water by Photosystem II to produce oxygen is the ultimate source of electrons in photosynthesis. These electrons are passed from Photosystem II to Photosystem I by the electron transport chain. The oxidized Chl_I^+ produced by Photosystem I is reduced by electrons donated by Photosystem II.

In Photosystem II, as in Photosystem I, the absorption of light by chlorophyll in the reaction center produces an excited state of chlorophyll. In this case the wavelength of light is 680 nm rather than 700 nm; the reaction-center chlorophyll of Photosystem II is also referred to as P_{680}.

$$Chl_{III} + h\nu \ (680 \text{ nm}) \longrightarrow Chl_{II}^*$$

As in Photosystem I, the excited chlorophyll is a reducing agent and passes an electron to a primary acceptor. In Photosystem II the primary electron

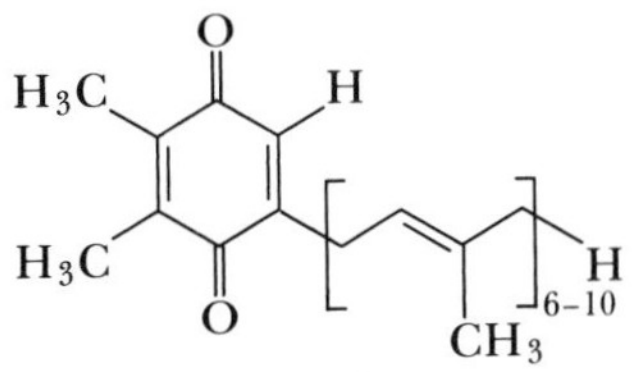

Plastoquinone

FIGURE 16.7

The structure of plastoquinone. The length of the aliphatic side chain varies in different organisms.

acceptor is a molecule of **pheophytin** (Phe), one of the accessory pigments of the photosynthetic apparatus. The structure of pheophytin differs from that of chlorophyll *a* in the substitution of two hydrogens for the magnesium in the ring system. As in Photosystem I, the transfer of electrons is mediated by events that take place at the reaction center. The next electron acceptor is **plastoquinone** (PQ). The structure of plastoquinone (Figure 16.7) is similar to that of coenzyme Q (ubiquinone), a part of the respiratory electron transport chain (Chapter 13, Section 13.4).

The electron transport chain that links the two photosystems consists of pheophytin, plastoquinone, a complex of plant cytochromes (the b_6–f complex), a copper-containing protein called **plastocyanin** (PC), and the oxidized form of P_{700} (Chl_I^+) (see Figure 16.6). The b_6–f complex of plant cytochromes consists of two *b*-type cytochromes (cytochrome b_6) and a *c*-type cytochrome (cytochrome *f*). This complex is similar in structure to the bc_1 complex in mitochondria and occupies a similar central position in an electron transport chain. This part of the photosynthetic apparatus is the subject of active research. There is a possibility that a Q-cycle (recall this from Chapter 13, Section 13.2) may operate here as well, and the object of some of this research is to establish definitely whether this is so. In plastocyanin the copper ion is the actual electron carrier; the copper ion exists as Cu(II) and Cu(I) in the oxidized and reduced forms, respectively. ATP is generated in a process coupled to this electron transport chain, as is the case in respiration.

When the oxidized chlorophyll of P_{700} accepts electrons from the electron transport chain, it is reduced. The reaction-center chlorophyll of Photosystem II (P_{680}) is now in an oxidized state, but it gains electrons and is reduced as a result of the oxidation of water.

$$2\ H_2O + 4\ Chl_{II}^+ \longrightarrow O_2 + 4\ H^+ + 4\ Chl_{II}$$

There are intermediate steps in this reaction, since four electrons are required for the oxidation of water, and Chl_{II}^+ (P_{680}^+) can accept only one electron at a time. A manganese-containing protein and several other protein components are required. The immediate electron donor, designated Z, to the P_{680} chlorophyll is a tyrosine residue of one of the protein components that does not contain manganese. Several quinones serve as intermediate electron transfer agents to accommodate four electrons donated by one water molecule. Redox reactions of manganese also play a role here. (See the article by Govindjee and Coleman listed in the bibliography at the end of this chapter for a discussion of the workings of this complex.)

FIGURE 16.8

Cyclic electron flow coupled to photophosphorylation in Photosystem I. Note that water is not split and that no NADPH is produced. (Chl is chlorophyll; Phe is pheophytin; PQ is plastoquinone; PC is plastocyanin.)

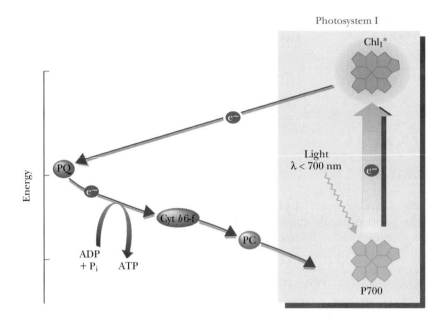

The net reaction for the two photosystems together is the flow of electrons from H_2O to $NADP^+$ (see Figure 16.6).

$$2 \; H_2O + 2 \; NADP^+ \longrightarrow O_2 + 2 \; NADPH + 2 \; H^+$$

Cyclic Electron Transport in Photosystem I

In addition to the electron transfer reactions just described, it is possible for cyclic electron transport in Photosystem I to be coupled to the production of ATP (Figure 16.8). No NADPH is produced in this process. Photosystem II is not involved, and no O_2 is generated. Cyclic phosphorylation takes place when there is a high $NADPH/NADP^+$ ratio in the cell: there is not enough $NADP^+$ present in the cell to accept all the electrons generated by the excitation of Chl_{II}.

16.4
A PROTON GRADIENT DRIVES THE PRODUCTION OF ATP IN PHOTOSYNTHESIS

In Chapter 13 we saw that a proton gradient across the inner mitochondrial membrane drives the phosphorylation of ADP in respiration. The mechanism of photophosphorylation is essentially the same as that of the production of ATP in the respiratory electron transport chain. In fact, some of the strongest evidence for the chemiosmotic coupling of phosphorylation to electron transport has been obtained from experiments on chloroplasts rather than mitochondria. Chloroplasts can synthesize ATP from ADP and P_i *in the dark* if they are provided with a pH gradient.

16.2

SOME HERBICIDES INHIBIT PHOTOSYNTHESIS

The main purpose of herbicides is to kill weeds so that they do not choke out desirable plants. One way of doing this is by selectively inhibiting photosynthesis in the weeds and not in the desired plants. A prime example is the use of 2,4-D and 2,4,5-T to kill broad-leaved weeds such as dandelions without affecting the growth of grass. In terms of acreage, lawn grasses are the most widely grown crop in the United States.

A number of other herbicides interfere with photosynthesis in some way. Amitrole inhibits biosynthesis of chlorophyll and carotenoids. The affected plants present a bleached appearance before they die because of the loss of their characteristic pigments. Another herbicide, atrazine, inhibits the splitting of water to hydrogen ion and oxygen. Still other herbicides interfere with electron transfer in the two photosystems. In Photosystem II, diuron inhibits electron transfer to plastoquinone, whereas bigyridylium herbicides accept electrons by competing with the electron acceptors in Photosystem I. The inhibitors active in Photosystem I include diquat and paraquat. The latter substance attained

Broad-leaved roadside weeds killed by herbicides which selectively inhibit photosynthesis. Note the unaffected corn in the background.

some notoriety when it was sprayed on marijuana fields to destroy the growing plants.

If isolated chloroplasts are allowed to equilibrate in a pH 4 buffer for several hours, their internal pH will be equal to 4. If the pH of the buffer is raised rapidly to 8 and if ADP and P_i are added simultaneously, ATP will be produced (Figure 16.9). The production of ATP does not require the presence of light; the proton gradient produced by the pH difference supplies the driving force for phosphorylation. This experiment provides solid evidence for the chemiosmotic coupling mechanism.

Several reactions contribute to the generation of a proton gradient in chloroplasts in an actively photosynthesizing cell. The splitting of water releases H^+ into the thylakoid space. Electron transport from Photosystem II and Photosystem I also helps create the proton gradient by involving plastoquinone and cytochromes in the process. Then Photosystem I reduces $NADP^+$ by using H^+ in the stroma to produce NADPH. As a result, the pH

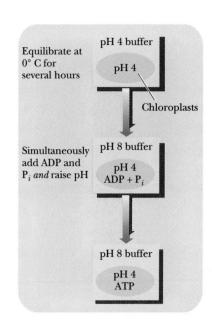

FIGURE 16.9

ATP is synthesized by chloroplasts in the dark in the presence of a proton gradient, ADP, and P_i.

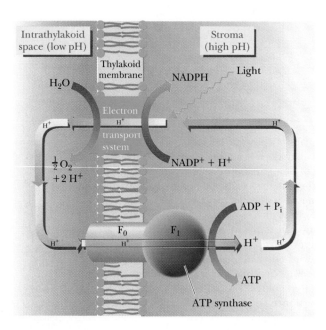

FIGURE 16.10

The relationship between photophosphorylation and the proton gradient in chloroplasts. Photosynthetic electron transport pumps H^+ out of the stroma to the intrathylakoid space to form the proton gradient (high pH in the stroma, low pH in the intrathylakoid space). The flow of H^+ back to the stroma through the ATP synthase provides the energy for synthesis of ATP from ADP and P_i.

of the thylakoid space is lower than that of the stroma (Figure 16.10). We saw a similar situation in Chapter 13 when we discussed the pumping of protons from the mitochondrial matrix into the intermembrane space. The coupling factor in chloroplasts is similar to the mitochondrial coupling factor, F_1. The chloroplast coupling factor is designated CF_1, where the C serves to distinguish it from its mitochondrial counterpart. Some evidence exists that the components of the electron chain in chloroplasts are arranged asymmetrically in the thylakoid membrane, as is the case in mitochondria. An important consequence of this asymmetrical arrangement is the release of the ATP and NADPH produced by the light reaction into the stroma, where they provide energy and reducing power for the dark reaction of photosynthesis.

In mitochondrial electron transport there are four respiratory complexes connected by soluble electron carriers. The electron transport apparatus of the thylakoid membrane is similar in that it consists of several large membrane-bound complexes. They are PSII (the Photosystem II complex), the cytochrome b_6–f complex, and PSI (the Photosystem I complex). As in mitochondrial electron transport, several soluble electron carriers form the connection between the protein complexes. In the thylakoid membrane the soluble carriers are plastoquinone and plastocyanin, which have a role similar to that of coenzyme Q and cytochrome c in mitochondria (Figure 16.11). The proton gradient created by electron transport drives the synthesis of ATP in chloroplasts, as in mitochondria.

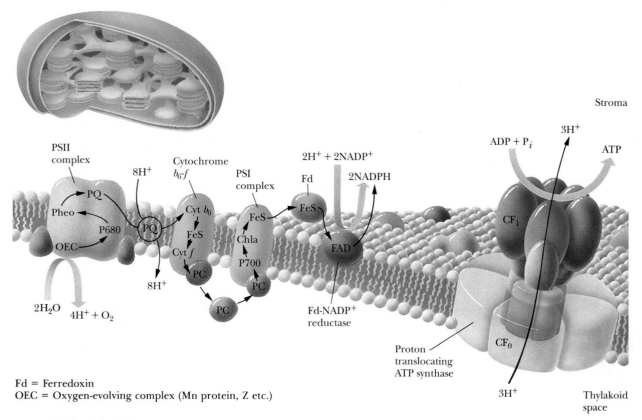

Fd = Ferredoxin
OEC = Oxygen-evolving complex (Mn protein, Z etc.)

FIGURE 16.11

The components of the electron transport chain of the thylakoid membrane. This schematic representation shows Photosystem II (PSII), the cytochrome b_6–f complex, and Photosystem I (PSI), along with the soluble electron carriers plastoquinone (PQ) and plastocyanin (PC). The action of the electron transport chain sets up a proton gradient across the thylakoid membrane, coupled to synthesis of ATP by the CF_0–CF_1 ATP synthase. (After D. R. Ort and N. E. Good, 1988, *Trends Biochem. Sci.* 13, 469.)

16.5
A COMPARISON OF PHOTOSYNTHESIS WITH AND WITHOUT OXYGEN: EVOLUTIONARY IMPLICATIONS

Photosynthetic prokaryotes other than cyanobacteria have only one photosystem and do not produce oxygen. The chlorophyll in these organisms is different from that found in photosystems linked to oxygen (Figure 16.12). Anaerobic photosynthesis is not as efficient as photosynthesis linked to oxygen, but the anaerobic version of the process appears to be an evolutionary way station. Anaerobic photosynthesis is a means for organisms to use solar energy to satisfy their needs for food and energy. Although it is efficient in the production of ATP, its efficiency is less than that of aerobic photosynthesis for carbon fixation.

A possible scenario for the development of photosynthesis starts with heterotrophic (*heterotrophs* are organisms that depend on their environment

FIGURE 16.12

The two possible electron transfer pathways in a photosynthetic anaerobe. Both cyclic and noncyclic forms of photophosphorylation are shown. HX is any compound (such as H_2S) that can be a hydrogen donor. (From L. Margulis, 1985, *Early Life*, Science Books International, Boston, p. 45.)

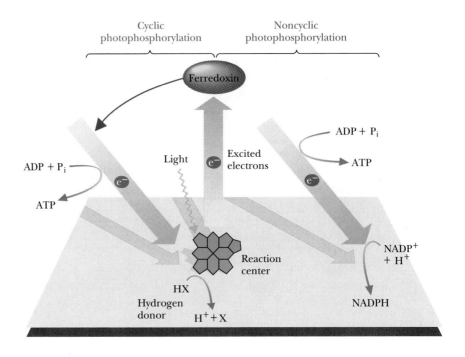

for organic nutrients and for energy) bacteria that contain some form of chlorophyll, probably bacteriochlorophyll. In such organisms the light energy absorbed by chlorophyll can be trapped in the forms of ATP and NADPH. The important point about such a series of reactions is that photophosphorylation takes place, ensuring an independent supply of ATP for the organism. In addition, the supply of NADPH facilitates synthesis of biomolecules from simple sources such as CO_2. Under conditions of limited food supply, organisms that can synthesize their own nutrients have a selective advantage. Organisms of this sort are *autotrophs* (not dependent on an external source of biomolecules) but are also anaerobes. The ultimate electron source that they use is not water but some more easily oxidized substance, such as H_2S, as is the case with present-day green sulfur bacteria (and purple sulfur bacteria), or various organic compounds, as is the case with present-day purple nonsulfur bacteria. These organisms do not possess an oxidizing agent powerful enough to split water, which is a far more abundant electron source than H_2S or organic compounds. The ability to use water as an electron source confers a further evolutionary advantage.

As is frequently the case in biological oxidation–reduction reactions, hydrogens as well as electrons are transferred from a donor to an acceptor. In green plants, green algae, and cyanobacteria, the hydrogen donor and acceptor are H_2O and CO_2, respectively, with oxygen as a product. Other organisms such as bacteria and fungi carry out photosynthesis in which there is a hydrogen donor other than water. Some possible donors include H_2S, $H_2S_2O_3$, and succinic acid. As an example, if H_2S is the source of hydrogens and electrons, a schematic equation for photosynthesis can be written with sulfur, rather than oxygen, as a product.

$$CO_2 \quad + \ 2\,H_2S \ \longrightarrow \ (CH_2O) \quad + 2\,S + H_2O$$
H-acceptor   H-donor   Carbohydrate

(It is possible for the hydrogen acceptor to be NO_2^- or NO_3^-, in which case NH_3 is a product.) Photosynthesis linked to oxygen with carbon dioxide as the ultimate hydrogen acceptor is a special case of a far more general process, widely distributed among many classes of organisms.

Cyanobacteria were apparently the first organisms that developed the ability to use water as the ultimate reducing agent in photosynthesis. As we have seen, this feat required the development of a second photosystem as well as a new variety of chlorophyll, chlorophyll *a* rather than bacteriochlorophyll in this case. Chlorophyll *b* had not yet appeared on the scene, since it occurs only in eukaryotes, but with cyanobacteria the basic system of aerobic photosynthesis was in place. As a result of aerobic photosynthesis by cyanobacteria, the earth acquired its present atmosphere with its high levels of oxygen. The existence of all other aerobic organisms depends ultimately on the activities of cyanobacteria.

16.6
THE DARK REACTION OF PHOTOSYNTHESIS: PATH OF CARBON

Carbon dioxide fixation takes place in the stroma. The equation for the overall reaction, like all equations for photosynthetic processes, is deceptively simple.

$$6\ CO_2 + 12\ NADPH \xrightarrow[\text{Enzymes}]{\text{ATP}} C_6H_{12}O_6 + 12\ NADP^+$$

The actual reaction pathway has some features in common with glycolysis and some in common with the pentose phosphate pathway.

The *net reaction* of six molecules of carbon dioxide to produce one molecule of glucose requires the carboxylation of six molecules of a five-carbon key intermediate, **ribulose 1,5-*bis*phosphate,** to form six molecules of an unstable six-carbon intermediate, which then splits to give 12

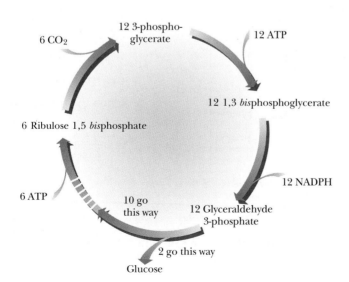

FIGURE 16.13

The main features of the Calvin cycle. Glucose is produced, and ribulose 1,5-*bis*phosphate is regenerated.

molecules of **3-phosphoglycerate.** Of these, two molecules of 3-phosphogly-
cerate react in turn, ultimately producing glucose. The remaining ten
molecules of 3-phosphoglycerate are used to regenerate the six molecules of
ribulose 1,5-*bis*phosphate. The overall reaction pathway is cyclic and is
called the **Calvin cycle** (Figure 16.13) after the scientist who first investigated it, Melvin Calvin, winner of the 1961 Nobel Prize in chemistry.

The first reaction of the Calvin cycle is the condensation of ribulose
1,5-*bis*phosphate with carbon dioxide to form a six-carbon intermediate,
2-carboxy-3-ketoribitol 1,5-*bis*phosphate, which quickly hydrolyzes to give
two molecules of 3-phosphoglycerate (Figure 16.14). The reaction is
catalyzed by the enzyme *ribulose 1,5-bisphosphate carboxylase* (also called
ribulose 1,5-*bis*phosphate carboxylase:oxygenase). This enzyme is located on
the stromal side of the thylakoid membrane and is probably one of the most
abundant proteins in nature, since it accounts for about 15% of the total
protein in chloroplasts. The molecular weight of ribulose 1,5-*bis*phosphate
carboxylase is about 560,000, and it consists of eight large subunits
(molecular weight, 55,000) and eight small subunits (molecular weight,

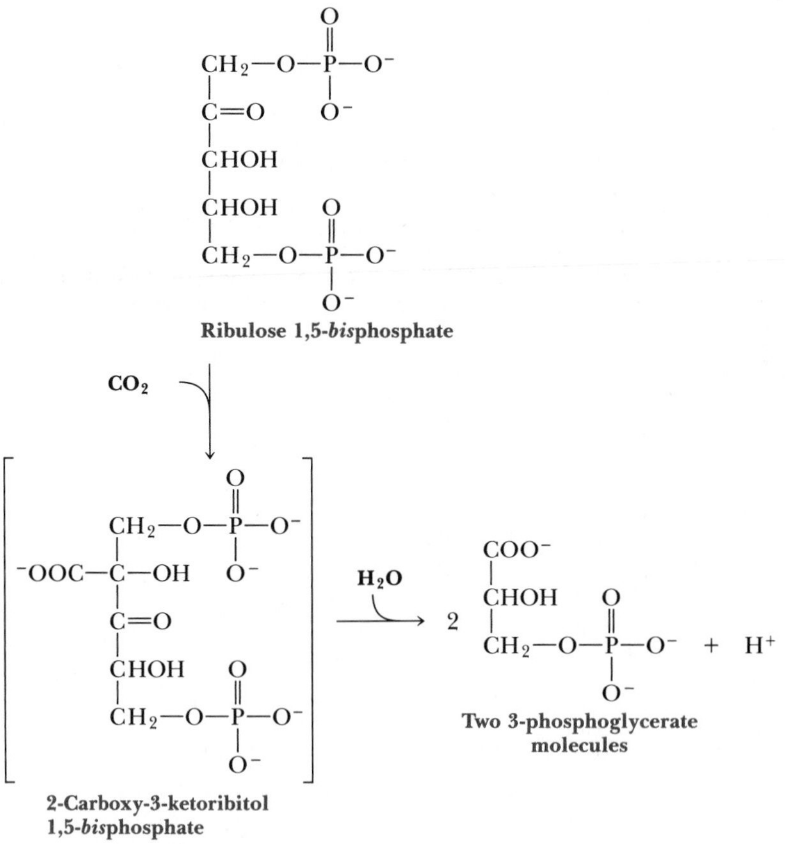

FIGURE 16.14

The reaction of ribulose 1,5-*bis*phosphate with CO_2 ultimately produces two
molecules of 3-phosphoglycerate.

15,000) (Figure 16.15). The sequence of the large subunit is encoded by a chloroplast gene, and that of the small subunit is encoded by a nuclear gene. The endosymbiotic theory for the development of eukaryotes (Chapter 1, Section 1.8) favors the idea of independent genetic material in organelles. The large subunit (chloroplast gene) is catalytic, whereas the small subunit (nuclear gene) plays a regulatory role, an observation that is consistent with an endosymbiotic origin for organelles such as chloroplasts.

The incorporation of CO_2 into 3-phosphoglycerate represents the actual fixation process; the remaining reactions are those of carbohydrates. The next two reactions lead to the reduction of 3-phosphoglycerate to form glyceraldehyde 3-phosphate. The reduction takes place in the same fashion as in gluconeogenesis, except for one unique feature (Figure 16.16a): the reactions in chloroplasts require NADPH rather than NADH for the

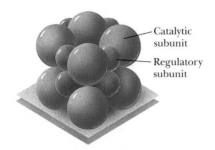

FIGURE 16.15

The subunit structure of ribulose 1,5-*bis*phosphate carboxylase.

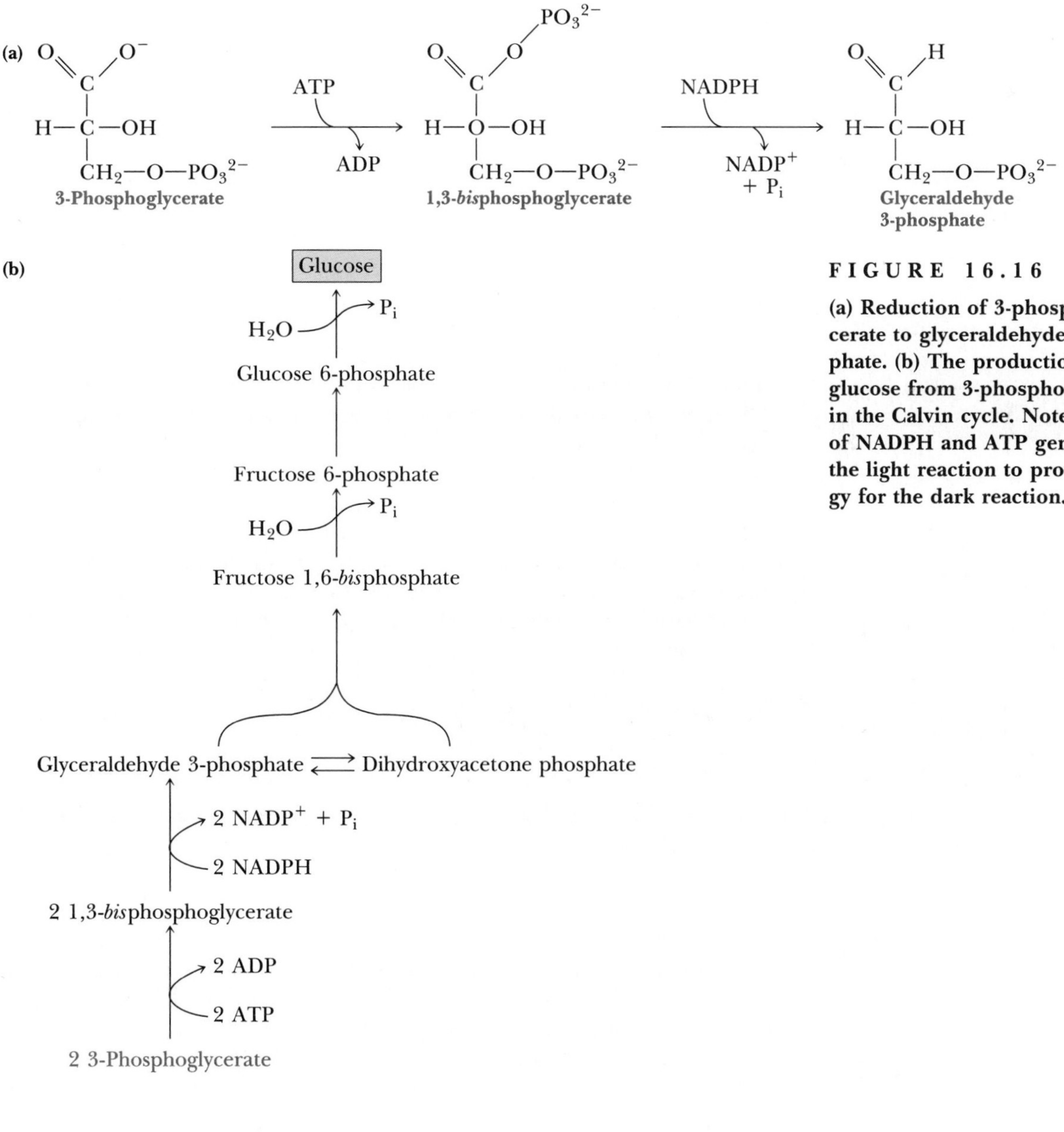

(a) Reduction of 3-phosphoglycerate to glyceraldehyde 3-phosphate.

3-Phosphoglycerate →(ATP → ADP)→ 1,3-*bis*phosphoglycerate →(NADPH → NADP⁺ + Pᵢ)→ Glyceraldehyde 3-phosphate

FIGURE 16.16

(a) Reduction of 3-phosphogly-cerate to glyceraldehyde 3-phosphate. (b) The production of glucose from 3-phosphoglycerate in the Calvin cycle. Note the use of NADPH and ATP generated in the light reaction to provide energy for the dark reaction.

(b)

Glucose

↑ (H₂O → Pᵢ)

Glucose 6-phosphate

↑

Fructose 6-phosphate

↑ (H₂O → Pᵢ)

Fructose 1,6-*bis*phosphate

↑

Glyceraldehyde 3-phosphate ⇌ Dihydroxyacetone phosphate

↑ (2 NADP⁺ + Pᵢ ← 2 NADPH)

2 1,3-*bis*phosphoglycerate

↑ (2 ADP ← 2 ATP)

2 3-Phosphoglycerate

reduction of 1,3-*bis*phosphoglycerate to glyceraldehyde 3-phosphate. When glyceraldehyde 3-phosphate is formed, it can have two alternative fates: one is the production of six-carbon sugars, and the other is the regeneration of ribulose 1,5-*bis*phosphate.

Production of Six-Carbon Sugars

The formation of glucose from glyceraldehyde 3-phosphate takes place in the same manner as in gluconeogenesis (Figure 16.16b). The conversion of glyceraldehyde 3-phosphate to dihydroxyacetone phosphate takes place easily (Chapter 11, Section 11.2). Dihydroxyacetone phosphate in turn reacts with glyceraldehyde 3-phosphate, in a series of reactions we have already seen, to give rise to fructose 6-phosphate and ultimately to glucose. Since we have already seen these reactions, we shall not discuss them again.

Regeneration of Ribulose 1,5-*Bis*phosphate

This process is readily divided into four steps: *preparation, reshuffling, isomerization,* and *phosphorylation.* The preparation begins with conversion of some of the glyceraldehyde 3-phosphate to dihydroxyacetone phosphate (catalyzed by triosephosphate isomerase). (This reaction also functions in the production of six-carbon sugars.) Portions of both the glyceraldehyde 3-phosphate and the dihydroxyacetone phosphate are then condensed to form fructose 1,6-*bis*phosphate (catalyzed by aldolase). Fructose 1,6-*bis*-phosphate is hydrolyzed to fructose 6-phosphate (catalyzed by diphosphatase). With a supply of glyceraldehyde 3-phosphate, dihydroxyacetone phosphate, and fructose 6-phosphate now available, the reshuing can begin.

Most of the reactions of the reshuing process are the same as ones we have already seen as part of the pentose phosphate pathway (Chapter 14, Section 14.4). Consequently, we shall concentrate just on the main outline of the process, and summarize the results later in Figures 16.18 and 16.19. Three reactions catalyzed in turn by *transketolase, aldolase,* and again transketolase are the reactions of rearrangement of carbon skeletons in the reshuing phase of the Calvin cycle. The results can be summarized as

$$
\begin{aligned}
C_6 + C_3 &\longrightarrow C_4 + C_5 \\
C_4 + C_3 &\longrightarrow C_7 \\
C_7 + C_3 &\longrightarrow C_5 + C_5 \\
\hline
\text{Net } C_6 + 3\,C_3 &\longrightarrow 3\,C_5
\end{aligned}
$$

The C_6 is fructose 6-phosphate, and the 3 C_3 are two molecules of glyceraldehyde 3-phosphate and one molecule of dihydroxyacetone phosphate. The 3 C_5 are two molecules of xylulose 5-phosphate and one molecule of ribose 5-phosphate.

The isomerization step involves the conversion of both ribose 5-phosphate and xylulose 5-phosphate to ribulose 5-phosphate. *Ribose 5-phosphate*

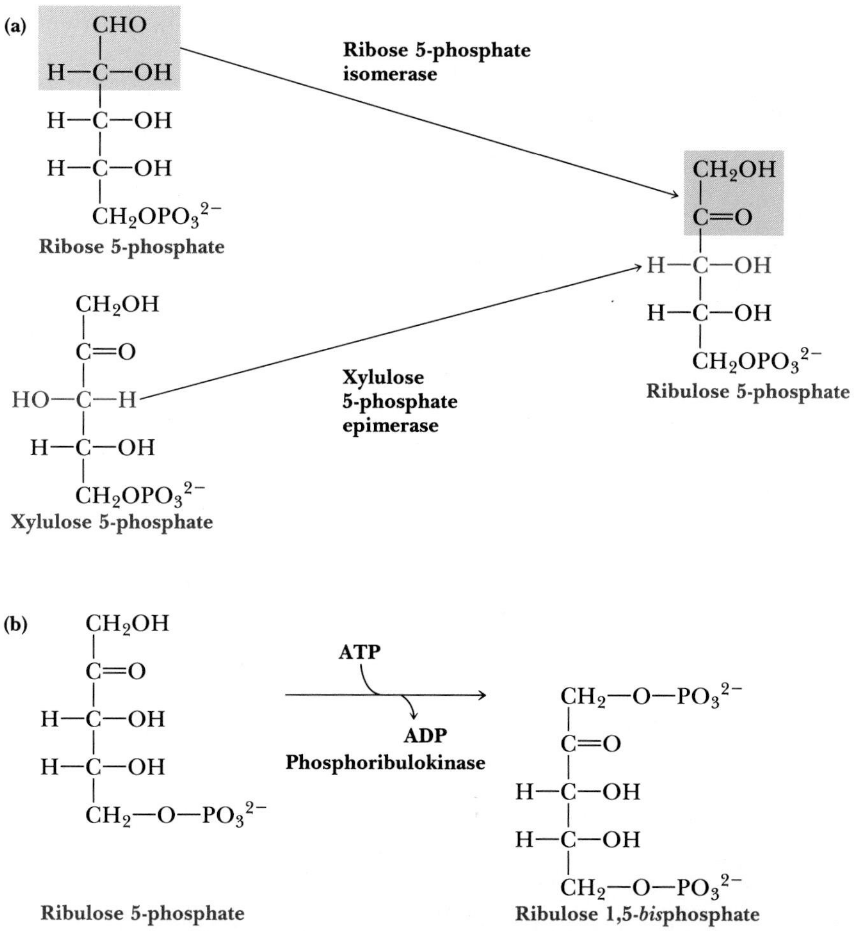

FIGURE 16.17

(a) The production of ribulose 5-phosphate from ribose 5-phosphate and from xylulose 5-phosphate. (b) The reaction catalyzed by phosphoribulokinase.

isomerase catalyzes the conversion of ribose 5-phosphate to ribulose 5-phosphate, and *xylulose 5-phosphate epimerase* catalyzes the conversion of xylulose 5-phosphate to ribulose 5-phosphate (Figure 16.17a). The reverses of both these reactions take place in the pentose phosphate pathway, catalyzed by the same enzymes.

In the final step, ribulose 1,5-*bis*phosphate is regenerated by the phosphorylation of ribulose 5-phosphate (Figure 16.17b). This reaction requires ATP and is catalyzed by the enzyme *phosphoribulokinase*. The reactions leading to the regeneration of ribulose 1,5-*bis*phosphate are summarized in Figure 16.18, in which a net equation is obtained by adding all the reactions. Now we are in a position to examine the stoichiometry of the dark reaction of photosynthesis.

FIGURE 16.18

The reactions that regenerate ribulose 1,5-*bis*phosphate in the Calvin cycle. In one turn of the cycle, the net process occurs twice.

1. 2 Glyceraldehyde 3-PO_3^{2-} ⟶ 2 Dihydroxyacetone PO_3^{2-}
    2 × 3 carbons   2 × 3 carbons

2. Glyceraldehyde 3-PO_3^{2-} + Dihydroxyacetone PO_3^{2-} ⟶
    3 carbons   3 carbons

Fructose 1,6-*bis*-PO_3^{4-}
6 carbons

3. Fructose 1,6-*bis*-PO_3^{4-} ⟶ Fructose 6-PO_3^{2-} + P_i
    6 carbons   6 carbons

4. Fructose 6-PO_3^{2-} + Glyceraldehyde 3-PO_3^{2-} ⟶
    6 carbons   3 carbons

Erythrose 4-PO_3^{2-} + Xylulose 5-PO_3^{2-}
4 carbons   5 carbons

5. Erythrose 4-PO_3^{2-} + Dihydroxyacetone PO_3^{2-} ⟶
    4 carbons   3 carbons

Sedoheptulose 1,7-*bis*-PO_3^{4-}
7 carbons

6. Sedoheptulose 1,7-*bis*-PO_3^{4-} ⟶ Sedoheptulose 7-PO_3^{2-} + P_i
    7 carbons   7 carbons

7. Sedoheptulose 7-PO_3^{2-} + Glyceraldehyde 3-PO_3^{2-} ⟶
    7 carbons   3 carbons

Ribose 5-PO_3^{2-} + Xylulose 5-PO_3^{2-}
5 carbons   5 carbons

8. Ribose 5-PO_3^{2-} ⟶ Ribulose 5-PO_3^{2-}
    5 carbons   5 carbons

9. 2 Xylulose 5-PO_3^{2-} ⟶ 2 Ribulose 5-PO_3^{2-}
    2 × 5 carbons   2 × 5 carbons

10. 3 Ribulose 5-PO_3^{2-} + 3 ATP ⟶ 3 Ribulose 1,5-*bis*-PO_3^{4-} + 3 ADP
     3 × 5 carbons   3 × 5 carbons

NET: 5 Glyceraldehyde 3-PO_3^{2-} + 3 ATP ⟶
     5 × 3 carbons

3 Ribulose 1,5-*bis*-PO_3^{4-} + 3 ADP + 2 P_i
3 × 5 carbons

Stoichiometry of the Calvin Cycle

It will be convenient to refer to Figures 16.13 and 16.19 during our discussion. We shall follow what happens to six molecules of CO_2 in the course of one turn of the Calvin cycle.

For each CO_2 that reacts with one molecule of ribulose 1,5-*bis*phosphate, two molecules of 3-phosphoglycerate are produced. Conversion of each molecule of 3-phosphoglycerate to glyceraldehyde 3-phosphate

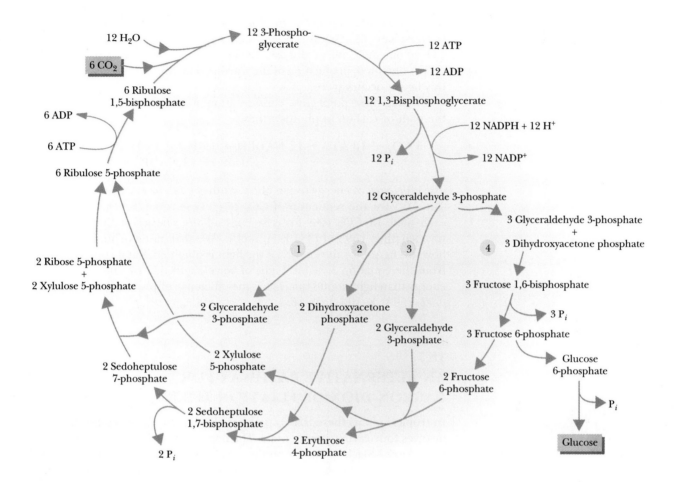

FIGURE 16.19

The complete Calvin cycle, showing the regeneration of ribulose 1,5-*bis*phosphate. Note that when glyceraldehyde 3-phosphate is formed, it (or dihydroxyacetone phosphate, to which it is easily converted) can have all four possible fates. The possible pathways are numbered. (See Figure 16.18 for the balanced equation.)

requires 1 ADP and 1 NADPH. For six molecules of CO_2 we can write the equation

$6\ CO_2$ + 6 ribulose 1,5-*bis*phosphate + 12 ATP + 12 NADPH + 12 H$^+$ + 12 H$_2$O $\longrightarrow$
(30 carbons)

12 glyceraldehyde 3-phosphate + 12 ADP + 12 P$_i$ + 12 NADP$^+$
(36 carbons)

The important point here is the requirement for 12 ATP and 12 NADPH for each molecule of glucose. Ten of the 12 glyceraldehyde 3-phosphate

(C$_3$) molecules are regenerated to ribulose 1,5-*bis*phosphate (Figure 16.19), accounting for 30 of the 36 carbon atoms in 12 molecules of glyceraldehyde 3-phosphate. The remaining six carbon atoms (two glyceraldehyde 3-phosphates) are converted to glucose. The regeneration of ribulose 1,5-*bis*phosphate also requires 6 ATP in the process of the net conversion of 6 CO$_2$ to one molecule of glucose. See Figure 16.18 to see how this figure of 6 ATP is obtained; in one turn of the Calvin cycle, the overall process shown in this figure occurs twice.

Taking these points into consideration, we arrive at the *net* equation for the path of carbon in photosynthesis.

$$6 \ CO_2 + 18 \ ATP + 12 \ NADPH + 12 \ H^+ + 12 \ H_2O \longrightarrow$$
$$glucose + 12 \ NADP^+ + 18 \ ADP + 18 \ P_i$$

The efficiency of energy use in photosynthesis can be calculated fairly easily. The $\Delta G^{\circ\prime}$ for the reduction of CO$_2$ to glucose is +478 kJ (+114 kcal) for each mole of CO$_2$ (see Exercise 3), and the energy of light of 600-nm wavelength is 1593 kJ (381 kcal) mol^{-1}. We shall not explain in detail here how this figure for the energy of the light is obtained, but it comes ultimately from the equation $E = h\nu$. Light of wavelength 680 or 700 nm has lower energy than light at 600 nm. Thus, the efficiency of photosynthesis is at least (477/1593) $\times$ 100, or 30%.

16.7
AN ALTERNATIVE PATHWAY FOR CARBON DIOXIDE FIXATION (OPTIONAL)

In tropical plants there is a C$_4$ pathway (Figure 16.20), so named because it involves four-carbon compounds. The operation of this pathway (also called the Hatch–Slack pathway) ultimately leads to the C$_3$ (based on 3-phosphoglycerate) pathway of the Calvin cycle. (There are other C$_4$ pathways as well, but this one is most widely studied.)

When CO$_2$ enters the outermost cells of the plant (the mesophyll cells) through the leaf pores, it reacts with phosphoenolpyruvate to produce oxaloacetate and P$_i$. Oxaloacetate is reduced to malate, with the concomitant oxidation of NADPH. Malate then enters the bundle-sheath cells (the next layer) through channels that connect two kinds of cells.

In the bundle-sheath cells, malate is decarboxylated to give pyruvate and CO$_2$. In the process, NADP$^+$ is reduced to NADPH (Figure 16.21). The CO$_2$ reacts with ribulose 1,5-*bis*phosphate to enter the Calvin cycle. Pyruvate is transported back to the mesophyll cells, where it is phosphorylated to phosphoenolpyruvate, which can react with CO$_2$ to start another round of the C$_4$ pathway. When pyruvate is phosphorylated, ATP is hydrolyzed to AMP and PP$_i$. This situation represents a loss of two "high-energy" phosphate bonds, equivalent to the use of 2 ATP. Consequently, the C$_4$ pathway requires two more ATP than the Calvin cycle alone for each CO$_2$ incorporated into glucose. Even though more ATP is required for the C$_4$ pathway than for the Calvin cycle, there is abundant light to produce the extra ATP by the light reaction of photosynthesis.

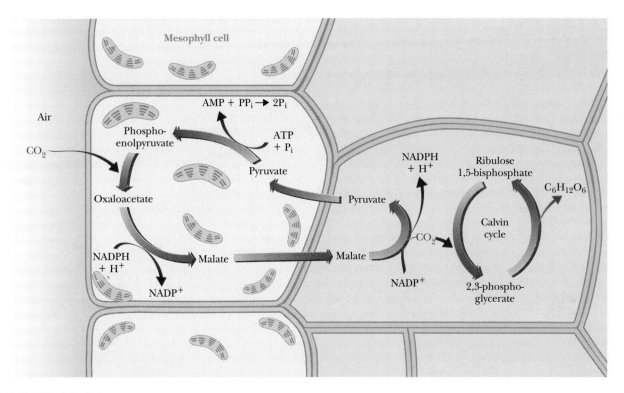

FIGURE 16.20

The C$_4$ pathway.

FIGURE 16.21

The characteristic reactions of the C$_4$ pathway.

$$CH_2{=}C{-}COO^- \overset{OPO_3^{2-}}{\underset{}{}} + CO_2 \xrightarrow{\textbf{PEP carboxylase}} {}^-OOC{-}CH_2{-}\overset{O}{\overset{\|}{C}}{-}COO^- + P_i$$

Phosphoenolpyruvate
(PEP)

Oxaloacetate

$$^-OOC{-}CH_2{-}\overset{O}{\overset{\|}{C}}{-}COO^- + NADPH + H^+ \xrightarrow{\substack{\textbf{Malate} \\ \textbf{dehydrogenase}}} {}^-OOC{-}CH_2{-}\overset{H}{\underset{OH}{\overset{|}{C}}}{-}COO^- + NADP^+$$

Oxaloacetate

L-Malate

$$^-OOC{-}CH_2{-}\overset{H}{\underset{OH}{\overset{|}{C}}}{-}COO^- + NADP^+ \xrightarrow{\substack{\textbf{Malic} \\ \textbf{enzyme}}} CH_3{-}\overset{O}{\overset{\|}{C}}{-}COO^- + CO_2 + NADPH + H^+$$

L-Malate

Pyruvate

$$CH_3{-}\overset{O}{\overset{\|}{C}}{-}COO^- + ATP + P_i \xrightarrow{\substack{\textbf{Pyruvate} \\ \textbf{phosphate} \\ \textbf{dikinase}}} CH_2{=}C{-}COO^- \overset{}{\underset{OPO_3^{2-}}{}} + AMP + PP_i$$

Pyruvate

Phosphoenolpyruvate

FIGURE 16.22

The characteristic reactions of photorespiration.

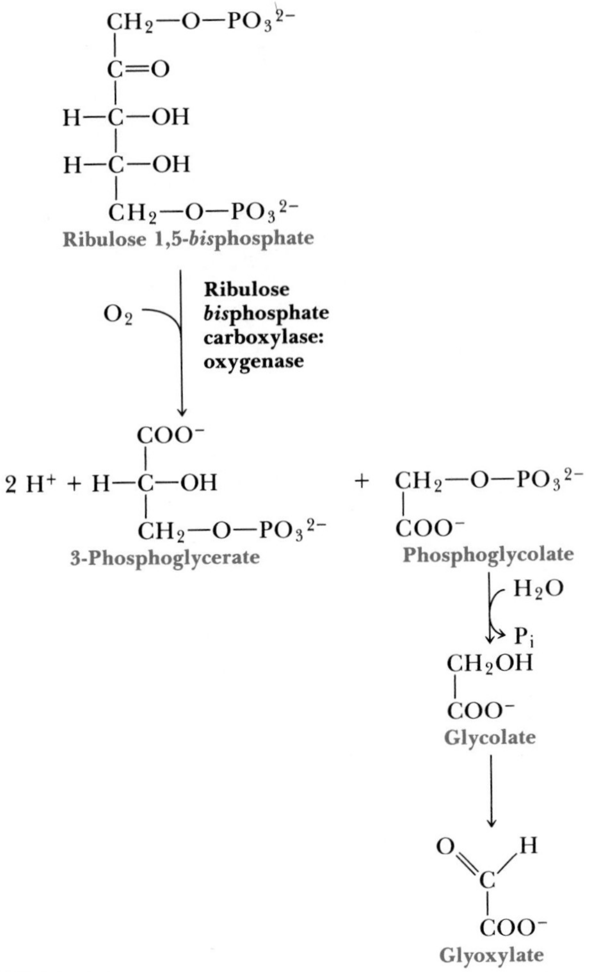

$$CH_2\text{—}O\text{—}PO_3{}^{2-}$$
$$|$$
$$C\text{=}O$$
$$|$$
$$H\text{—}C\text{—}OH$$
$$|$$
$$H\text{—}C\text{—}OH$$
$$|$$
$$CH_2\text{—}O\text{—}PO_3{}^{2-}$$

Ribulose 1,5-bisphosphate

O_2 — | **Ribulose bisphosphate carboxylase: oxygenase**

$$COO^-$$
$$|$$
$$2\,H^+ + H\text{—}C\text{—}OH$$
$$|$$
$$CH_2\text{—}O\text{—}PO_3{}^{2-}$$

3-Phosphoglycerate

$+$

$$CH_2\text{—}O\text{—}PO_3{}^{2-}$$
$$|$$
$$COO^-$$

Phosphoglycolate

$\curvearrowright H_2O$
$\searrow P_i$

$$CH_2OH$$
$$|$$
$$COO^-$$

Glycolate

$$O\diagdown\,\diagup H$$
$$C$$
$$|$$
$$COO^-$$

Glyoxylate

Note that the C_4 pathway fixes CO_2 in the mesophyll cells only to unfix it in the bundle-sheath cells, where CO_2 then enters the C_3 pathway. This observation raises the question of the advantage to tropical plants of using the C_4 pathway. The conventional wisdom on the subject focuses on the role of CO_2, but there is more to the question than that. According to the conventional view, the point of the C_4 pathway is that it concentrates CO_2 and, as a result, accelerates the process of photosynthesis. Leaves of tropical plants have small pores to minimize water loss, and these small pores decrease CO_2 entry into the plant. In tropical areas, where there is abundant light, the amount of CO_2 available to plants controls the rate of photosynthesis. The C_4 pathway deals with the situation, allowing tropical plants to grow more quickly and to produce more biomass per unit of leaf area than plants that use the C_3 pathway. A more comprehensive view of the subject includes a consideration of the role of oxygen and the process of **photorespiration.**

In the dark, green leaf cells of both C_3 and C_4 plants carry out respiration and phosphorylation in their mitochondria. The energy source is the store of compounds produced by photosynthesis in the light. In the light, C_3 plants also carry on respiration. This respiration is not entirely a mitochondrial process because it continues at a lower rate in the presence of cyanide, an inhibitor of the mitochondrial electron transport chain (Chapter 13, Section 13.6). Photorespiration is the cyanide-insensitive respiration of C_3 plants. The process is almost completely absent in C_4 plants.

Although the actual biological role of photorespiration is not known, several points are well established. The process in many ways is wasteful of reducing power and ATP. Oxidative phosphorylation does not accompany photorespiration, and reducing agents such as NADH are needed for the series of redox reactions linked to oxygen. The principal substrate oxidized in photorespiration is **glycolate** (Figure 16.22). The product of the oxidation reaction, which takes place in peroxisomes of leaf cells (Chapter 1, Section 1.5), is **glyoxylate.** Glycolate arises ultimately from the oxidative breakdown of ribulose 1,5-*bis*phosphate. The enzyme that catalyzes this reaction is ribulose 1,5-*bis*phosphate carboxylase:oxygenase, acting as an oxygenase (linked to O_2) rather than as the carboxylase (linked to CO_2) that fixes CO_2 into 3-phosphoglycerate.

When levels of O_2 are high compared with those of CO_2, ribulose 1,5-*bis*phosphate is oxygenated to produce phosphoglycolate (which gives rise to glycolate) and 3-phosphoglycerate by photorespiration, rather than the two molecules of 3-phosphoglycerate that arise from the carboxylation reaction. This situation occurs in C_3 plants. In C_4 plants, the small pores decrease the entry not only of CO_2 but also of O_2 into the leaves. The ratio of CO_2 to O_2 in the bundle-sheath cells is relatively high as a result of the operation of the C_4 pathway, favoring the carboxylation reaction. The C_4 pathway is an advantageous one for tropical plants because it allows such plants to dispense with photorespiration.

S U M M A R Y

The equation for photosynthesis

$$6\ CO_2 + 6\ H_2O \longrightarrow C_6H_{12}O_6 + 6\ O_2$$

actually represents two processes. One, the splitting of water to produce oxygen, is a reaction that requires light energy from the sun, and the other, the fixation of CO_2 to give sugars, can and does take place in the dark, but only because it uses solar energy indirectly. In the light reaction, water is oxidized to produce oxygen, accompanied by the reduction of $NADP^+$ to NADPH. The light reaction is coupled to the phosphorylation of ADP to ATP. In the dark reaction, the ATP and NADPH produced in the light reaction provide the energy and reducing power for the fixation of CO_2.

The light reaction of photosynthesis takes place in eukaryotes in the thylakoid membranes of chloro-plasts. The trapping of light takes place at a reaction center within the chloroplast; the process requires a pair of chlorophylls in a unique environment.

The light reaction consists of two parts, each carried out by a separate photosystem. The reduction of $NADP^+$ to NADPH is accomplished by Photosystem I, whereas the splitting of water to produce oxygen is done by Photosystem II. The two photosystems are linked by an electron transport chain coupled to the production of ATP. A proton gradient drives the production of ATP in photosynthesis, as it does in mitochondrial respiration.

The dark reaction of photosynthesis involves the net synthesis of one molecule of glucose from six molecules of CO_2. The net reaction of six molecules of CO_2 to produce one molecule of glucose requires the carboxylation of six molecules of a five-carbon key intermediate, ribulose 1,5-*bis*phosphate, ultimately forming 12 molecules of 3-phosphoglycerate. Of these, two molecules of 3-phosphoglycerate react to give rise to glucose. The remaining ten molecules of 3-phosphoglycerate are used to regenerate the six molecules of ribulose 1,5-*bis*phosphate. The overall reaction pathway is cyclic and is called the Calvin cycle.

In addition to the Calvin cycle, there is an alternative pathway for CO_2 fixation in tropical plants, called the C_4 pathway because it involves four carbon compounds. In this pathway, CO_2 reacts in the outer (mesophyll) cells with phosphoenolpyruvate to pro-duce oxaloacetate and P_i. Oxaloacetate in turn is reduced to malate. Malate is transported from mesophyll cells, where it is produced, to inner (bundle-sheath) cells, where it is ultimately passed to the Calvin cycle. Plants in which the C_4 pathway operates grow more quickly and produce more biomass per unit of leaf area than C_3 plants, in which only the Calvin cycle operates. In C_3 plants the process of photorespiration, which wastes energy and reducing power, is operative. In C_4 plants photorespiration is almost totally absent.

E X E R C I S E S

1. In cyclic photophosphorylation in Photosystem I, ATP is produced, even though water is not split. Explain how the process takes place.

2. Uncouplers of oxidative phosphorylation in mitochondria also uncouple photoelectron transport and ATP synthesis in chloroplasts. Give an explanation for this observation.

3. Using information from Chapter 9, Section 9.10, show how the $\Delta G^{\circ\prime}$ of 477 kJ (114 kcal) is obtained for each mole of CO_2 fixed in photosynthesis. The reaction in question is $6 \ CO_2 + 6 \ H_2O \longrightarrow$ glucose $+ \ 6 \ O_2$.

4. Chlorophyll is green because it absorbs green light less than it absorbs light of other wavelengths. The accessory pigments in the leaves of deciduous trees tend to be red and yellow, but their color is masked by that of the chlorophyll. Suggest a connection between these points and the appearance of fall foliage colors in many sections of the country.

5. How can a proton gradient be created in cyclic photo-phosphorylation in Photosystem I?

6. Suppose that a prokaryotic organism that contains both chlorophyll *a* and chlorophyll *b* has been discovered. Comment on the evolutionary implications of such a discovery.

7. Outline the events that take place at the photosynthetic reaction center in *Rhodopseudomonas*.

8. What are the two places where light energy is required in the light reaction of photosynthesis? Why must energy be supplied at precisely these points?

9. If photosynthesizing plants are grown in the presence of $^{14}CO_2$, is every carbon atom of the glucose that is produced labeled with the radioactive carbon? Why or why not?

10. How does the production of sugars by tropical plants differ from the same reactions in the Calvin cycle?

A N N O T A T E D   B I B L I O G R A P H Y

Barber, J. Has the Mangano-protein of the Water Splitting Reaction of Photosynthesis Been Isolated? *Trends Biochem. Sci.* **9,** 79–80 (1984). [A report on an important topic in research on Photosystem II.]

Bering, C. L. Energy Interconversions in Photosynthesis. *J. Chem. Ed.* **62,** 659–664 (1985). [A short discussion of basic concepts of photosynthesis, concentrating on the light reaction and photosystems.]

Bishop, M. B., and C. B. Bishop. Photosynthesis and Carbon Dioxide Fixation. *J. Chem. Ed.* **64,** 302–305 (1987). [Concentrates on the Calvin cycle.]

Danks, S. M., E. H. Evans, and P. A. Whittaker. *Photosynthetic Systems: Structure, Function and Assembly.* New York: Wiley, 1983. [A short book with excellent electron micrographs of chloroplasts and related structures in Chapter 1.]

Deisenhofer, J., and H. Michel. The Photosynthetic Reaction Center from the Purple Bacterium *Rhodopseudomonas viridis. Science* **245,** 1463–1473 (1989). [The authors' Nobel Prize address, describing their work on the structure of the reaction center.]

Deisenhofer, J., H. Michel, and R. Huber. The Structural Basis of Photosynthetic Light Reactions in Bacteria. *Trends*

Biochem. Sci. **10**, 243–248 (1985). [A discussion of the photosynthetic reaction center in bacteria.]

Dennis, D. T. *The Biochemistry of Energy Utilization in Plants.* New York: Chapman and Hall, 1987. [A short book on plant biochemistry.]

Fox, M., W. Jones, and D. Watkins. Light-Harvesting Polymer Systems. *Chem. Eng. News* **71** (11), 38–48 (1993). [Photosynthetic reaction centers serve as a model for artificial systems that convert solar light energy to chemical energy.]

Govindjee, ed. *Photosynthesis.* Vol. 1, *Energy Conversion in Plants and Bacteria;* Vol. 11, *Development, Carbon Metabolism and Plant Productivity.* New York: Academic Press, 1982. [A fairly advanced discussion of photosynthesis.]

Govindjee, and W. J. Coleman. How Plants Make Oxygen. *Sci. Amer.* **262** (2), 50–58 (1990). [Focuses on the water-splitting apparatus of Photosystem II.]

Halliwell, B. *Chloroplast Metabolism: The Structure and Function of Chloroplasts in Green Leaf Cells.* New York: Oxford University Press, 1981. [A detailed description of chloroplast activity.]

Hipkins, M. F., and N. R. Baker, eds. *Photosynthesis: Energy Transduction: A Practical Approach.* Oxford, England: IRL Press, 1986. [A collection of articles about research methods used to study photosynthesis.]

Karplus, P., M. Daniels, and J. Herriott. Atomic Structure of Ferredoxin-NADP$^+$ Reductase: Prototype for a Structurally Novel Flavoenzyme Family. *Science* **251**, 60–66 (1991). [The structure of a key enzyme involved in nitrogen and sulfur metabolism as well as in photosynthesis.]

Margulis, L. *Early Life.* Boston: Science Books International, 1982. [Chapters 2 and 3 discuss the evolutionary development of photosynthesis.]

Norris, J., and M. Schiffer. Photosynthetic Reaction Centers in Bacteria. *Chem. Eng. News* **68** (31), 22–37 (1993). [A review that covers both structures and reactions.]

Youvan, D. C., and B. L. Marrs. Molecular Mechanisms of Photosynthesis. *Sci. Amer.* **256** (6), 42–48 (1987). [A detailed description of a bacterial photosynthetic reaction center and the molecular events that take place there.]

Zuber, H. Structure of Light-Harvesting Antenna Complexes of Photosynthetic Bacteria, Cyanobacteria and Red Algae. *Trends Biochem. Sci.* **11**, 414–419 (1986). [Concentrates on the protein portion of the photosynthetic reaction center.]

Chapter 17

The Metabolism of Nitrogen

Outline

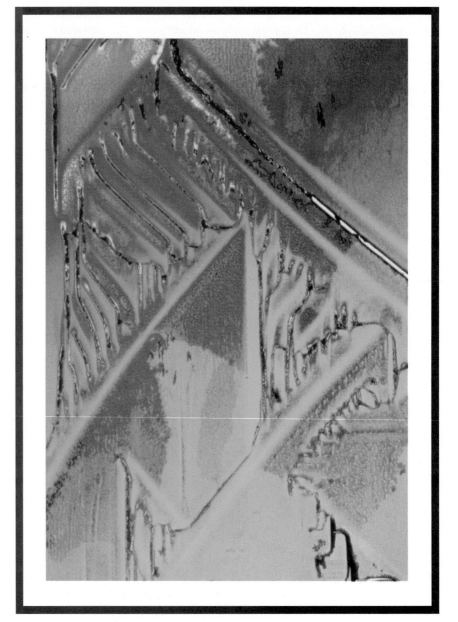

Crystals of urea viewed under polarized light.

Surprisingly, nitrogen found in living organisms does not come directly from the atmosphere, where it composes 80% of atmospheric gas. Instead, organic nitrogen enters the biological realm through bacteria living in the root nodules of leguminous plants such as peas and beans. They are able to convert the triple-bonded nitrogen molecule N_2 to ammonia, NH_3, which is incorporated into the amino acid glutamate, which in turn is a precursor of proline and arginine. Thus, the 20 amino acids may be formed by the transformation from one amino acid into another, often by long and complex metabolic pathways. Although amino acids act principally as the building blocks of proteins, they also contribute to the synthesis of a variety of biologically important molecules. These molecules include the nitrogenous bases of DNA and RNA, the nucleotide coenzyme electron carriers NAD^+ and $NADP^+$, and the oxygen-binding pyrrole ring of hemoglobin, as well as small hormones and neurotransmitters. When amino acids are deaminated (deamination is the removal of the α-amino group, NH_3^+), their carbon skeletons can be fed into the citric acid cycle to be oxidized to carbon dioxide and water. Alternatively, they may be used as precursors of other biomolecules, including glucose and fatty acids. In many ways, then, amino acids can be seen as connecting links between metabolic pathways.

17.1
AN OVERVIEW OF NITROGEN METABOLISM

We have seen the structures of many types of compounds that contain nitrogen, including amino acids, porphyrins, and nucleotides, but we have not discussed their metabolism. The metabolic pathways we have dealt with up to now have mainly involved compounds of carbon, hydrogen, and oxygen, such as sugars and fatty acids. Several important topics can be included in our discussion of the metabolism of nitrogen. The first of these is **nitrogen fixation,** the process by which inorganic molecular nitrogen from the atmosphere (N_2) is incorporated first into ammonia and then into organic compounds that are of use to organisms. Nitrate ion (NO_3^-), another kind of inorganic nitrogen, is the form in which nitrogen is found in the soil, and many fertilizers contain nitrates, frequently potassium nitrate. The process of **nitrification** (nitrate reduction to ammonia) provides another way for organisms to obtain nitrogen. Nitrate ion and nitrite ion (NO_2^-) are also involved in **denitrification** reactions, which return nitrogen to the atmosphere (Figure 17.1).

Ammonia formed by either pathway, nitrogen fixation or nitrification, enters the biosphere. Ammonia is converted to organic nitrogen by plants, and organic nitrogen is passed to animals through food chains. Finally, waste products such as urea are excreted and degraded to ammonia by microorganisms. The word "ammonia" comes from *sal ammoniac* (ammonium chloride), which was first prepared from the dung of camels at the temple of Jupiter Ammon in North Africa. The process of death and decay releases ammonia in both plants and animals. Denitrifying bacteria reverse the conversion of ammonia to nitrate and then recycle the NO_3^- as free N_2 (Figure 17.1).

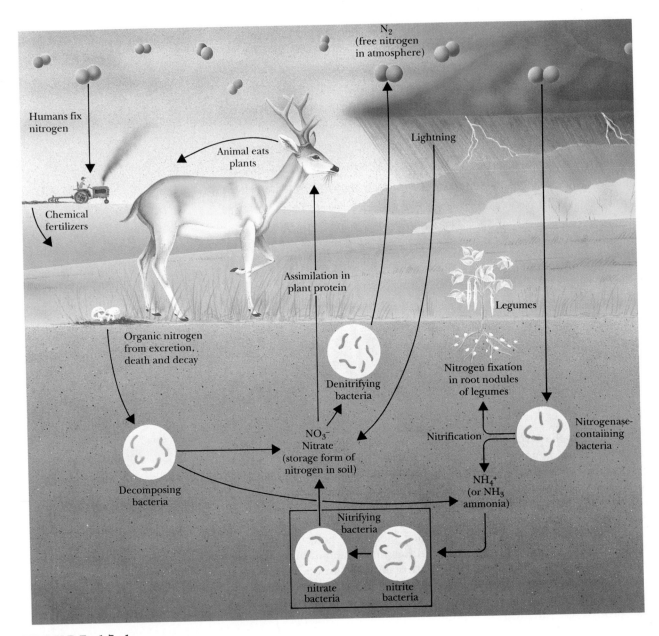

FIGURE 17.1

The flow of nitrogen to the biosphere.

The topic of nitrogen metabolism includes the biosynthesis and break-down of *amino acids, purines,* and *pyrimidines;* also, the metabolism of *porphyrins* is related to that of amino acids. Many of these pathways, particularly anabolic ones, are long and complex. Where the amount of material is large and highly detailed, we shall concentrate on the most important points. Specifically, we shall concentrate on overall patterns and on interesting reactions of wide applicability. We shall also be interested in health-related aspects of this material. Other reactions will be found in Interchapter B, which follows. It can be considered a repository of supplementary material for this chapter.

17.2
NITROGEN FIXATION

Bacteria are responsible for the reduction of N_2 to ammonia (NH_3). Typical nitrogen-fixing bacteria are symbiotic organisms that form nodules on the roots of leguminous plants such as beans and alfalfa. Many free-living microbes and some cyanobacteria also fix nitrogen. Plants and animals cannot carry out nitrogen fixation. This conversion of molecular nitrogen to ammonia is the only source of nitrogen in the biosphere except for that provided by nitrates. The conjugate acid form of NH_3, ammonium ion (NH_4^+), is the form of nitrogen that is used in the first stages of the synthesis of organic compounds. Parenthetically, NH_3 obtained by chemical synthesis from nitrogen and hydrogen is the starting point for the production of many synthetic fertilizers, which frequently contain nitrates.

The **nitrogenase** enzyme complex found in nitrogen-fixing bacteria catalyzes the production of ammonia from molecular nitrogen. The half reaction of reduction (Figure 17.2a) is

$$N_2 + 6\,e^- + 12\ \text{ATP} + 12\ H_2O + 8\ H^+ \longrightarrow$$
$$2\ NH_4^+ + 12\ \text{ADP} + 12\ P_i$$

Several proteins are included in the nitrogenase complex. Ferredoxin is one of them (this protein also plays an important role in electron transfer in photosynthesis; Chapter 16, Section 16.3). There are also two proteins specific to the nitrogenase reaction. One is an iron–molybdenum (Fe-Mo) protein, and the other is a nonheme (iron–sulfur) protein, usually referred to as the Fe protein. The flow of electrons is from ferredoxin to the Fe

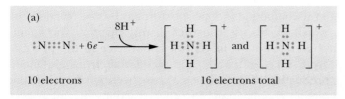

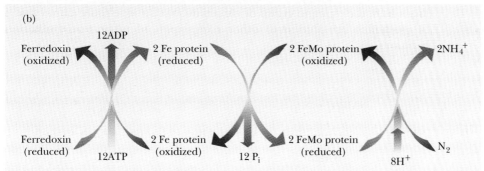

FIGURE 17.2

Some aspects of the nitrogenase reaction. (a) The reduction of N_2 to 2 NH_4^+. (b) The path of electrons from ferredoxin to N_2.

protein to the Fe-Mo protein to nitrogen (Figure 17.2b). The nature of the nitrogenase complex is a subject of active research.

17.3
FEEDBACK CONTROL: A UNIFYING THEME IN NITROGEN METABOLISM

The biosynthetic pathways that produce amino acids and nucleobases (purines and pyrimidines) are long and complex, requiring a large investment of energy by the organism. If there is a high level of some end-product, such as an amino acid or a nucleotide, the cell saves energy by not making that compound. However, the cell needs some signal not to produce more of that particular compound. The signal is frequently a **feedback inhibition** mechanism, in which the end-product of a metabolic pathway inhibits an enzyme at the beginning of the pathway. We saw an example of such a control mechanism when we discussed the allosteric enzyme aspartate transcarbamoylase in Chapter 5, Section 5.9. This enzyme catalyzes one of the early stages of pyrimidine nucleotide biosynthesis, and it is inhibited by the end-product of that pathway, namely cytidine triphosphate (CTP). Feedback inhibition is frequently encountered in the biosynthesis of amino acids and nucleotides.

17.4
THE METABOLISM OF AMINO ACIDS: ANABOLISM

General Features

Ammonia is toxic in high concentrations, and so it must be incorporated into biologically useful compounds when it is formed by the reactions of nitrogen fixation, which we discussed earlier in this chapter. The amino acids glutamate and glutamine are of central importance in the process. **Glutamate** arises from α-ketoglutarate, and **glutamine** from glutamate (Figure 17.3). The production of glutamate is a reductive amination, and the production of glutamine is amidation. In other reactions of amino acid anabolism, the α-amino group of glutamate and the side-chain amino group of glutamine are shifted to other compounds in other **transamination** reactions.

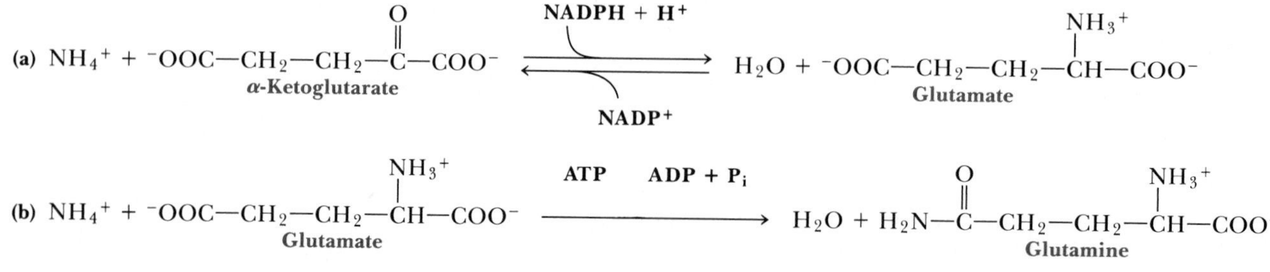

FIGURE 17.3

(a) The production of glutamate from α-ketoglutarate. (b) The production of glutamine from glutamate.

(a) $NH_4^+ + {}^-OOC-CH_2-CH_2-\overset{\overset{\displaystyle O}{\|}}{C}-COO^- \underset{NADP^+}{\overset{NADPH + H^+}{\rightleftharpoons}} H_2O + {}^-OOC-CH_2-CH_2-\overset{\overset{\displaystyle NH_3^+}{|}}{CH}-COO^-$

α-Ketoglutarate   Glutamate

(b) $NH_4^+ + {}^-OOC-CH_2-CH_2-\overset{\overset{\displaystyle NH_3^+}{|}}{CH}-COO^- \xrightarrow{\;\; ATP \quad ADP + P_i \;\;} H_2O + H_2N-\overset{\overset{\displaystyle O}{\|}}{C}-CH_2-CH_2-\overset{\overset{\displaystyle NH_3^+}{|}}{CH}-COO^-$

Glutamate   Glutamine

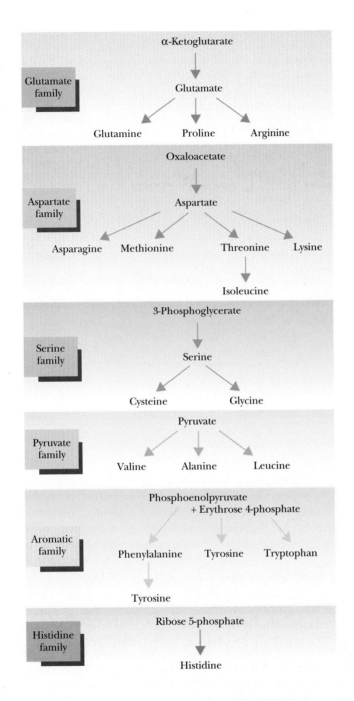

FIGURE 17.4

Families of amino acids based on biosynthetic pathways. Each family has a common precursor.

The biosynthesis of amino acids involves a common set of reactions. In addition to transamination reactions, *transfer of one-carbon units,* such as formyl or methyl groups, occurs frequently. We are not going to discuss all the details of the reactions that give rise to amino acids. We can, however, organize this material by grouping amino acids into families based on common precursors (Figure 17.4). The reactions of some of the individual families of amino acids provide good examples of those reactions that are of general importance, transamination and one-carbon transfer. Other reactions for the biosynthesis of amino acids can be found in Interchapter B.

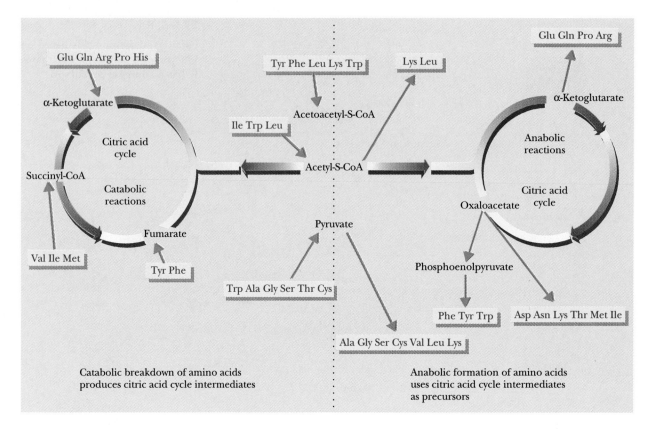

Catabolic breakdown of amino acids produces citric acid cycle intermediates

Anabolic formation of amino acids uses citric acid cycle intermediates as precursors

FIGURE 17.5

The relationship between amino acid metabolism and the citric acid cycle.

We can also make some generalizations about amino acid metabolism in terms of the relationship of the carbon skeleton to the citric acid cycle and the related reactions of pyruvate and acetyl-CoA (Figure 17.5). The citric acid cycle is amphibolic; it has a part in both catabolism and anabolism. The anabolic aspect of the citric acid cycle is of interest in amino acid biosynthesis. The catabolic aspect is apparent in the breakdown of amino acids, leading to their eventual excretion, which takes place in reactions related to the citric acid cycle.

Transamination Reactions: The Role of Glutamate and of Pyridoxal Phosphate

Glutamate is formed from NH_4^+ and α-ketoglutarate in a reductive amination and that requires NADPH. This reaction is reversible and is catalyzed by *glutamate dehydrogenase.*

$$NH_4^+ + \alpha\text{-ketoglutarate} + NADPH + H^+ \longrightarrow$$
$$\text{glutamate} + NADP^+ + H_2O$$

Glutamate is a major donor of amino groups in reactions of amino acid anabolism, and α-ketoglutarate is a major acceptor of amino groups (see Figure 17.3a).

The conversion of glutamate to glutamine is catalyzed by **glutamine synthetase** in a reaction that requires ATP (see Figure 17.3b).

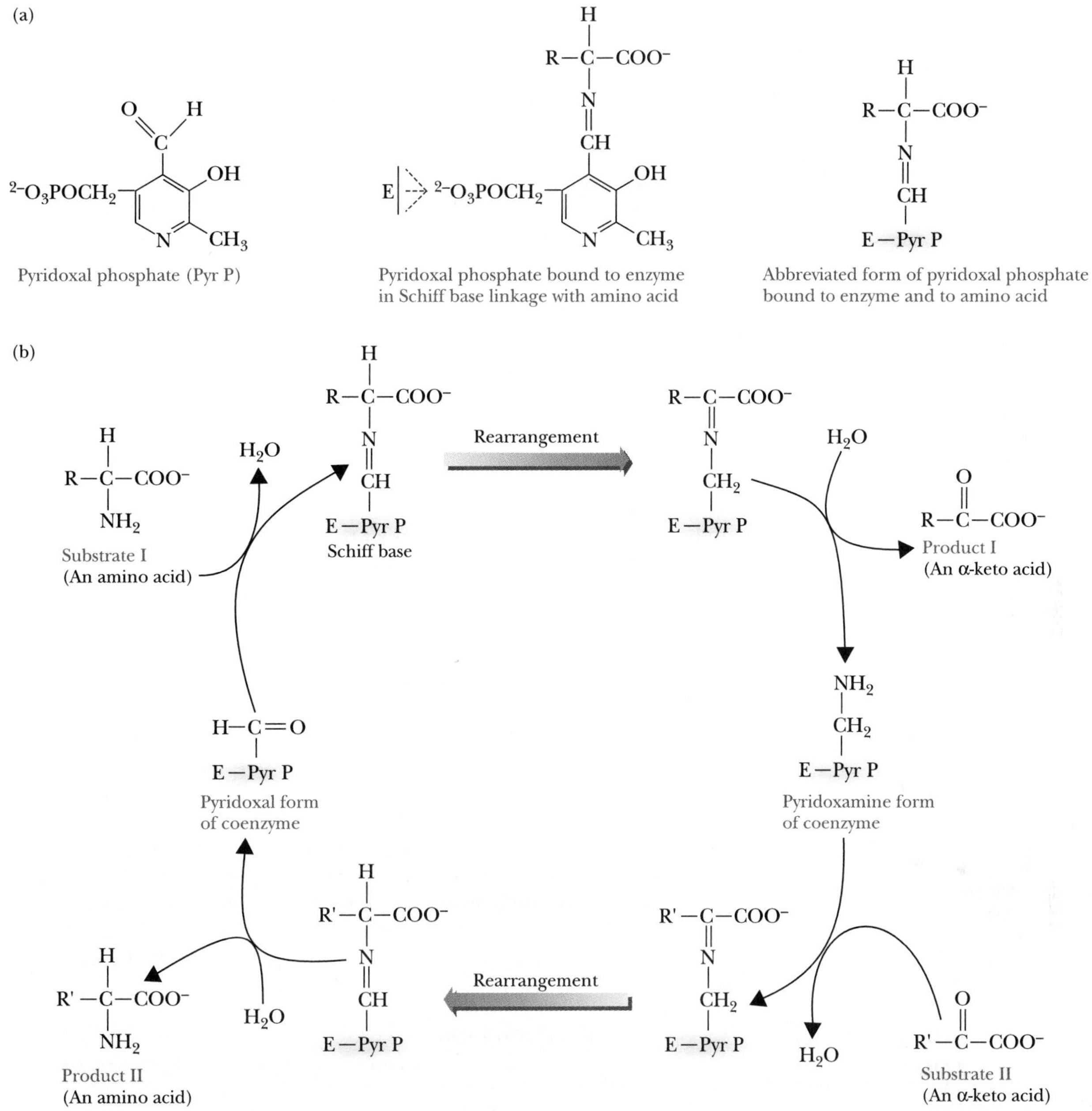

(a)

Pyridoxal phosphate (Pyr P)

Pyridoxal phosphate bound to enzyme in Schiff base linkage with amino acid

Abbreviated form of pyridoxal phosphate bound to enzyme and to amino acid

(b)

$$NH_4^+ + \text{glutamate} + ATP \longrightarrow \text{glutamine} + ADP + P_i + H_2O$$

These reactions fix inorganic nitrogen (NH_3), forming organic (carbon-containing) nitrogen compounds, such as amino acids.

Quite frequently, enzymes that catalyze transamination reactions require pyridoxal phosphate as a coenzyme (Figure 17.6). We discussed this compound in Chapter 5, Section 5.13, as a typical example of a coenzyme, and here we can see its mode of action in context.

Pyridoxal phosphate forms a Schiff base with the amino group of Substrate I (the amino group donor) (Figure 17.6a). The next stage is a rearrangement followed by hydrolysis, which removes Product I (the α-keto acid corresponding to Substrate I). The coenzyme now carries the amino

FIGURE 17.6

The role of pyridoxal phosphate in transamination reactions. (a) The mode of binding of pyridoxal phosphate (PyrP) to the enzyme (E) and to the substrate amino acid. (b) The reaction itself. The original substrate, an amino acid, is deaminated, while an α-keto acid is aminated to form an amino acid. The net reaction is one of transamination. Note that the coenzyme is regenerated and that the original substrate and final product are both amino acids.

517

group (pyridoxamine). Substrate II (another α-keto acid) then forms a Schiff base with pyridoxamine. Again there is a rearrangement followed by a hydrolysis, which gives rise to Product II (an amino acid) and regenerates pyridoxal phosphate. The net reaction (Figure 17.6b) is that an amino acid (Substrate I) reacts with an α-keto acid (Substrate II) to form an α-keto acid (Product I) and an amino acid (Product II). The amino group has been transferred from Substrate I to Substrate II, forming the amino acid, Product II.

One-Carbon Transfers and the Serine Family

In addition to transamination reactions, one-carbon transfer reactions occur frequently in amino acid biosynthesis. A good example of a one-carbon transfer can be found in the reactions that produce the amino acids of the serine family. This family also includes glycine and cysteine. Serine and glycine themselves are frequently precursors in other biosynthetic pathways. A discussion of the synthesis of cysteine will give us some insight into the metabolism of sulfur as well as nitrogen.

The ultimate precursor of serine is 3-phosphoglycerate, which is obtainable from the glycolytic pathway. The hydroxyl group on carbon 2 is oxidized to a keto group, giving an α-keto acid. A transamination reaction in which glutamate is the nitrogen donor produces 3-phosphoserine. This reaction is an example of a transamination involving glutamate and α-ketoglutarate. Hydrolysis of the phosphate group then gives rise to serine (Figure 17.7).

The conversion of serine to glycine involves the transfer of a one-carbon unit from serine to an acceptor. This reaction is catalyzed by *serine hydroxymethylase,* with pyridoxal phosphate as a coenzyme. The acceptor in this reaction is **tetrahydrofolate,** a derivative of folic acid and a frequently encountered carrier of one-carbon units in metabolic pathways. Its structure has three parts: a substituted pteridine ring, *p*-aminobenzoic acid, and glutamic acid (Figure 17.8).

Serine + tetrahydrofolate $\longrightarrow$

glycine + methylenetetrahydrofolate + H_2O

The one-carbon unit transferred in this reaction is bound to tetrahydrofolate, forming N^5, N^{10}-methylenetetrahydrofolate, in which the methylene (one-carbon) unit is bound to two of the nitrogens of the carrier (Figure 17.9). Tetrahydrofolate is not the only carrier of one-carbon units. We have already encountered biotin, a carrier of CO_2, and we have discussed the role that biotin plays in gluconeogenesis (Chapter 14, Section 14.2) and in the anabolism of fatty acids (Chapter 15, Section 15.6).

The conversion of serine to cysteine involves some interesting reactions. The source of the sulfur in animals differs from that in plants and bacteria. In plants and bacteria, serine is acetylated to form O-acetylserine. This

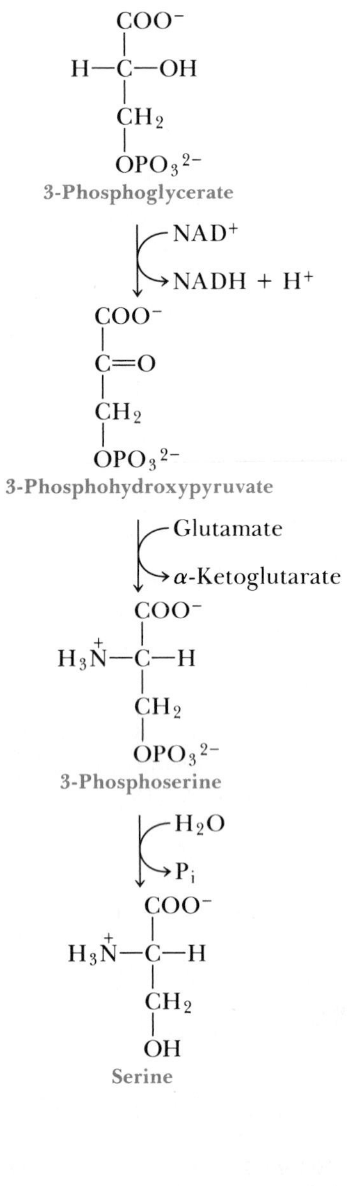

FIGURE 17.7

The biosynthesis of serine.

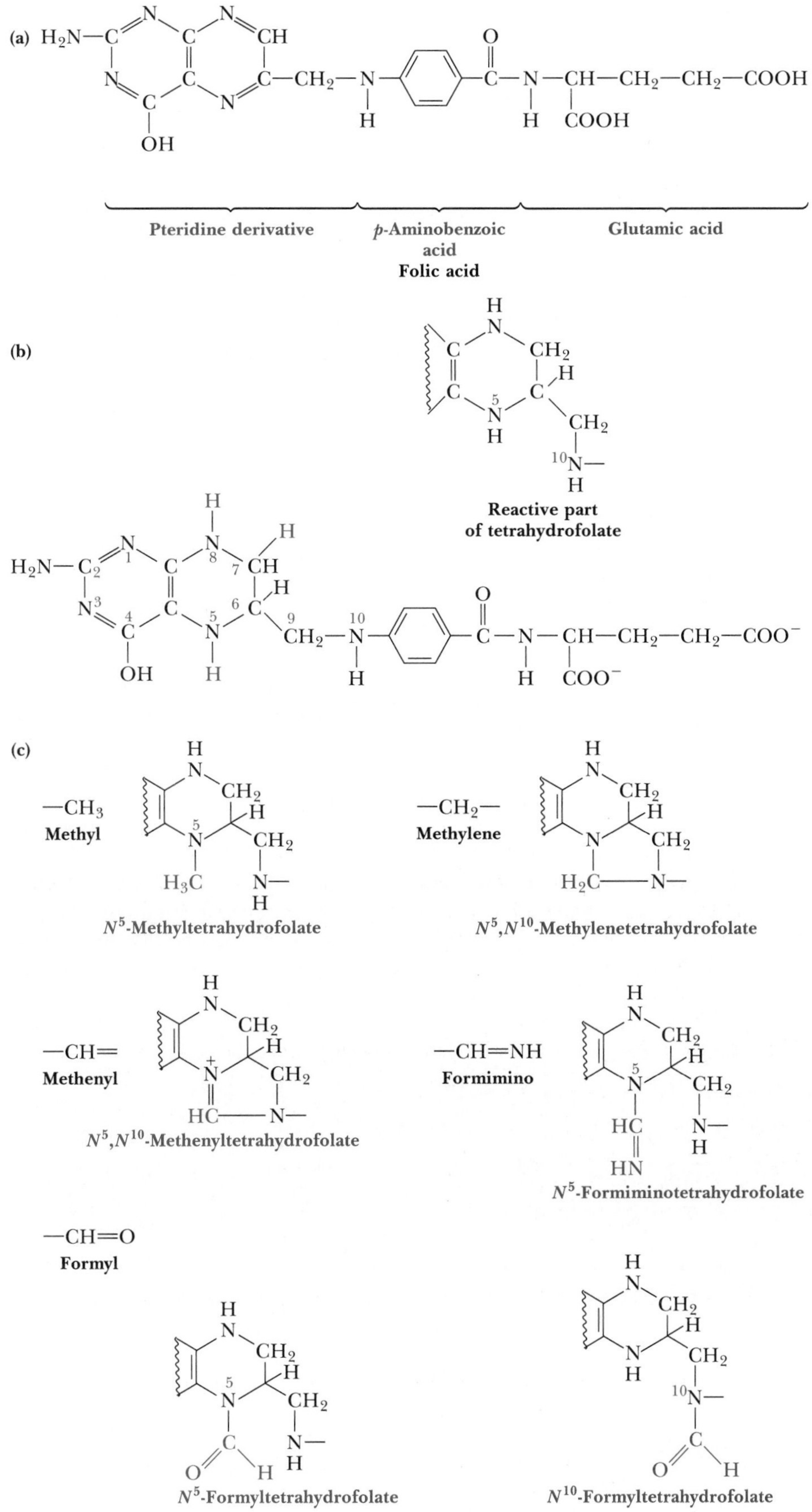

FIGURE 17.8

(a) The structure of folic acid, shown in nonionized form. (b) The structure of tetrahydrofolate, shown in ionized form. Hydrogens are added at positions 5, 6, 7, and 8. The nitrogens at positions 5 and 10 are involved in the reactions of tetrahydrofolate. (c) The one-carbon groups transferred by tetrahydrofolate, showing the forms in which they are bound. Note that there are two positions, at nitrogen 5 and nitrogen 10, at which a formyl group is bound to tetrahydrofolate. There is only one binding site for the other one-carbon groups.

(a) Folic acid

Pteridine derivative · p-Aminobenzoic acid · Glutamic acid

(b) Reactive part of tetrahydrofolate

(c)
Methyl — N^5-Methyltetrahydrofolate

Methylene — N^5,N^{10}-Methylenetetrahydrofolate

Methenyl — N^5,N^{10}-Methenyltetrahydrofolate

Formimino — N^5-Formiminotetrahydrofolate

Formyl — N^5-Formyltetrahydrofolate — N^{10}-Formyltetrahydrofolate

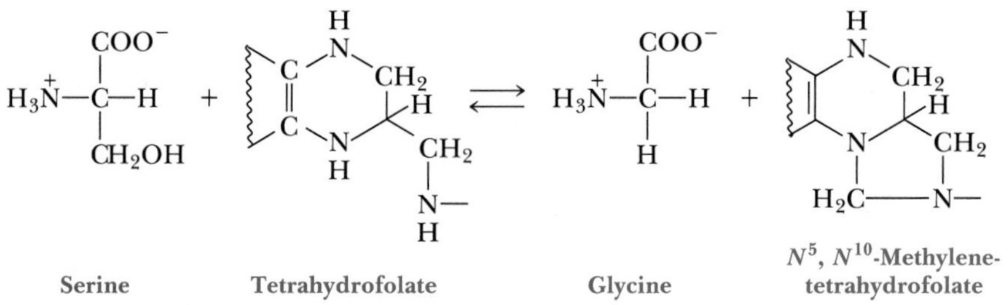

FIGURE 17.9

The conversion of serine to glycine, showing the role of tetrahydrofolate.

reaction is catalyzed by *serine acyltransferase,* with acetyl-CoA as the acyl donor (Figure 17.10). Conversion of O-acetylserine to cysteine requires production of sulfide by a sulfur donor. The sulfur donor for plants and bacteria is 3'-phospho-5'-adenyl sulfate. The sulfate group is reduced first to sulfite and then to sulfide (Figure 17.11). The sulfide, in the conjugate acid form HS⁻, displaces the acetyl group of the O-acetylserine to produce cysteine. Animals form cysteine from serine by a different pathway, since they do not have the enzymes to carry out the sulfate-to-sulfide conversion that we have just seen. The reaction sequence in animals involves the amino acid methionine.

Methionine, which is produced by reactions of the aspartate family (Interchapter B, Section B.2) in bacteria and plants, cannot be produced by animals. It must be obtained from dietary sources. It is an **essential amino acid** because it cannot be synthesized by the body. The ingested methionine reacts with ATP to form **S-adenosylmethionine,** which has a highly reactive methyl group (Figure 17.12). This compound is a carrier of methyl groups in many reactions. The methyl group from S-adenosylmethionine can be transferred to any one of a number of acceptors, producing S-adenosylhomocysteine. Hydrolysis of S-adenosylhomocysteine in turn produces homocysteine. Cysteine can be synthesized from serine and homocysteine, and this pathway for cysteine biosynthesis is the only one available to

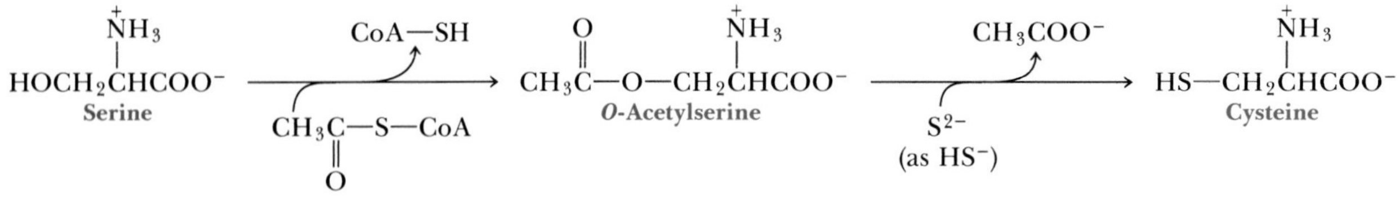

FIGURE 17.10

The biosynthesis of cysteine in plants and bacteria.

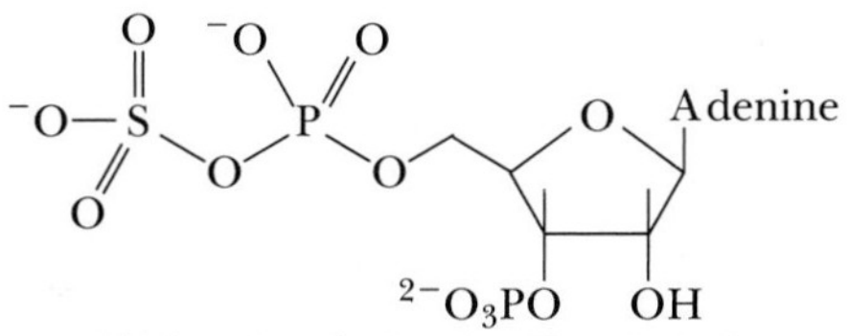

3′-Phospho-5′-adenyl sulfate (PAPS)

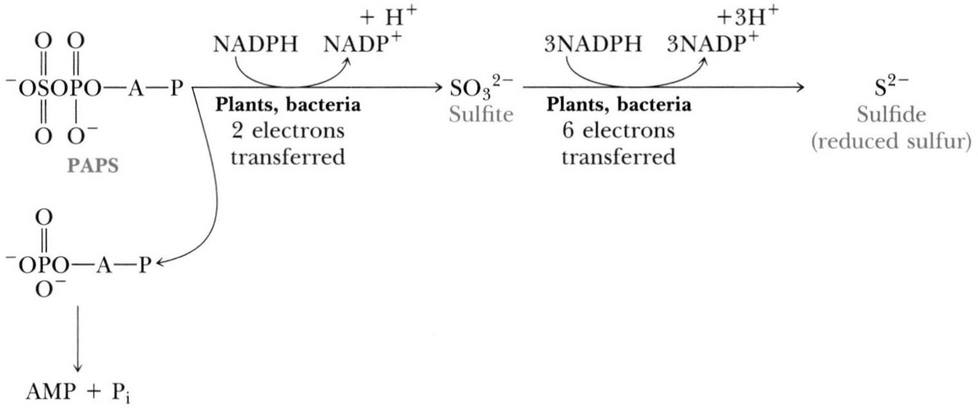

FIGURE 17.11

Electron transfer reactions of sulfur in plants and bacteria.

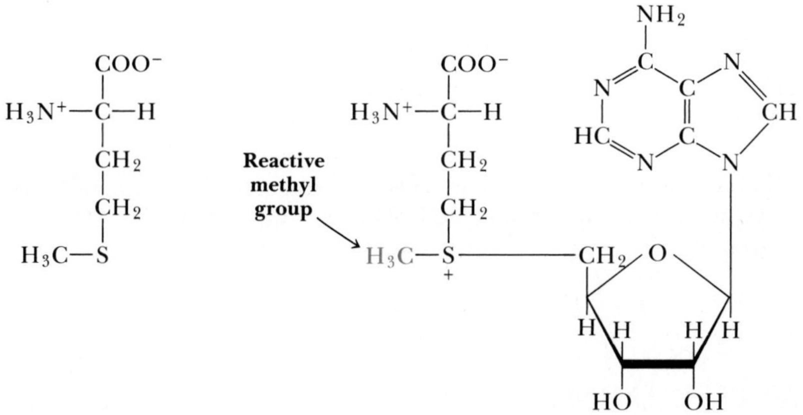

Methionine

S-Adenosylmethionine

FIGURE 17.12

The structure of *S*-adenosylmethionine, with the structure of methionine shown for comparison.

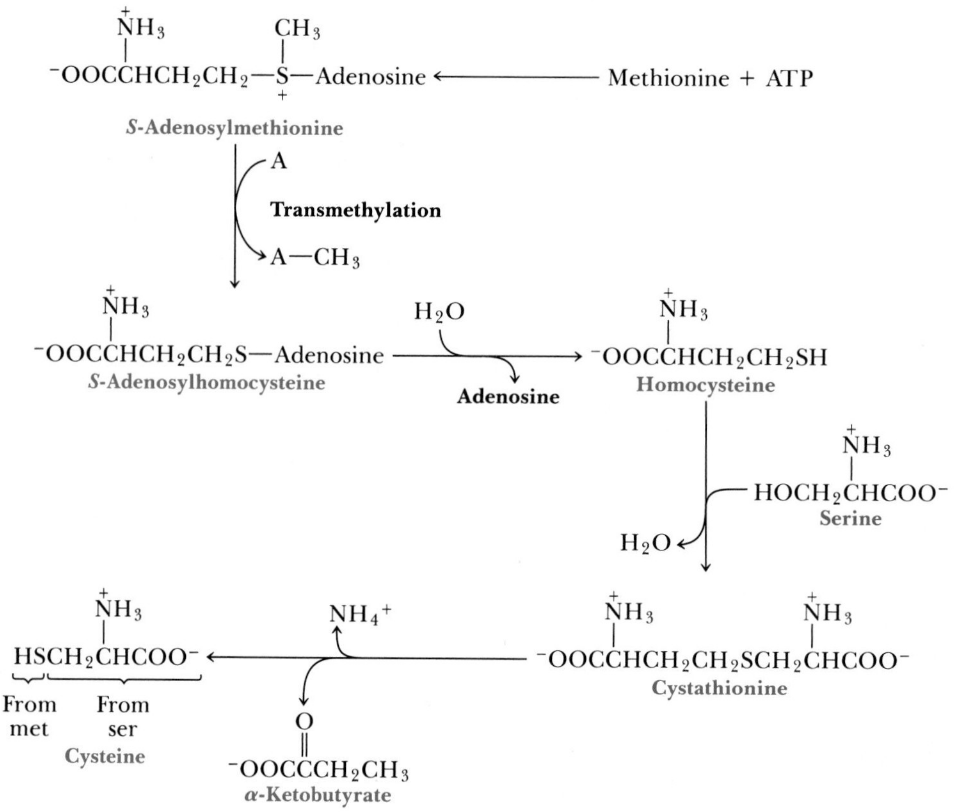

FIGURE 17.13

The biosynthesis of cysteine in animals. (A is acceptor.)

animals (Figure 17.13). Serine and homocysteine react to produce cystathionine, which hydrolyzes to form cysteine, NH_4^+, and α-ketobutyrate.

It is worth noting that we have now seen three important carriers of one-carbon units: biotin, a carrier of CO_2; tetrahydrofolate (FH_4), a carrier of methylene and formyl groups; and S-adenosylmethionine, a carrier of methyl groups.

17.5
ESSENTIAL AMINO ACIDS

The biosynthesis of proteins requires the presence of all the constituent amino acids. If 1 of the 20 amino acids is missing or in short supply, protein biosynthesis is inhibited. Some organisms, such as *Escherichia coli*, can synthesize all the amino acids that they need. Other species, including humans, must obtain some amino acids from dietary sources. The essential amino acids in human nutrition are listed in Table 17.1. The body can synthesize some of these amino acids but not in sufficient quantities for its needs, especially in the case of growing children. This last point applies particularly to children's requirement for arginine and histidine. Amino

TABLE 17.1   Amino Acid Requirements in Humans

ESSENTIAL	NONESSENTIAL
Arginine	Alanine
Histidine	Asparagine
Isoleucine	Aspartate
Leucine	Cysteine
Lysine	Glutamate
Methionine	Glutamine
Phenylalanine	Glycine
Threonine	Proline
Tryptophan	Serine
Valine	Tyrosine

acids are not stored (except in proteins), and dietary sources of essential amino acids are needed at regular intervals. Protein deficiency, especially a prolonged deficiency in sources that contain essential amino acids, leads to the disease **kwashiorkor.** The problem in this disease, particularly severe in growing children, is not simply starvation but the breakdown of the body's own proteins.

17.6
CATABOLISM OF AMINO ACIDS: THE UREA CYCLE

The first step to consider in the catabolism of amino acids is the removal of nitrogen by transamination. Transamination reactions are also important in the anabolism of amino acids, so it is important to remind ourselves that anabolic and catabolic pathways are not the exact reverse of each other, nor do they involve exactly the same group of enzymes. In catabolism, the amino nitrogen of the original amino acid is transferred to α-ketoglutarate to produce glutamate, leaving behind the carbon skeleton. The fates of the carbon skeleton and of the nitrogen can be considered separately.

Breakdown of the carbon skeleton of amino acids follows two general pathways, the difference between the two pathways depending on the type of end-product. A **glucogenic** amino acid is one that yields pyruvate or oxaloacetate on degradation. Oxaloacetate is the starting point for the production of glucose by gluconeogenesis. A **ketogenic** amino acid is one that breaks down to acetyl-CoA or acetoacetyl-CoA, leading to the formation of ketone bodies (Table 17.2, see Chapter 15, Section 15.5). The carbon skeletons of the amino acids give rise to metabolic intermediates such as pyruvate, acetyl-CoA, acetoacetyl-CoA, α-ketoglutarate, succinyl-CoA, fumarate, and oxaloacetate (see Figure 17.5). Oxaloacetate is a key intermediate in the breakdown of the carbon skeleton of amino acids because of its dual role in the citric acid cycle and in gluconeogenesis. The amino acids degraded to acetyl-CoA and acetoacetyl-CoA are used in the

T A B L E   1 7 . 2   **Glucogenic and Ketogenic Amino Acids**

GLUCOGENIC	KETOGENIC	GLUCOGENIC AND KETOGENIC
Aspartate	Leucine	Isoleucine
Asparagine		Lysine
Alanine		Phenylalanine
Glycine		Tryptophan
Serine		Tyrosine
Threonine		
Cysteine		
Glutamate		
Glutamine		
Arginine		
Proline		
Histidine		
Valine		
Methionine		

FIGURE 17.14

Nitrogen-containing products of amino acid catabolism.

$$NH_3$$
Ammonia
as
$$NH_4^+$$
Ammonium
ion

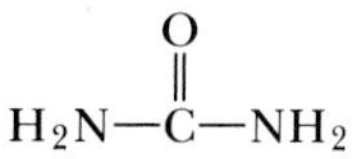

Urea

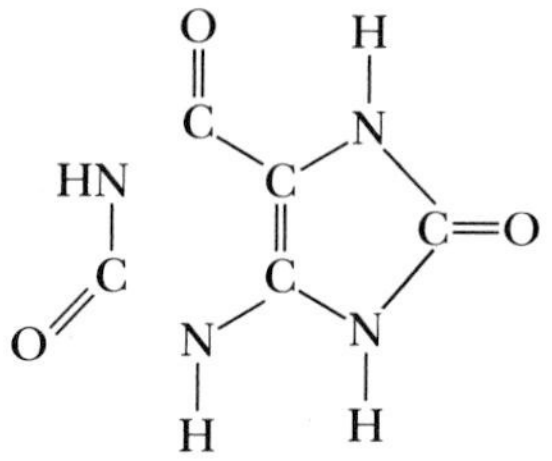

Uric acid

citric acid cycle, but mammals cannot synthesize glucose from acetyl-CoA. This fact is the source of the distinction between glucogenic and ketogenic amino acids. Glucogenic amino acids can be converted to glucose, with oxaloacetate as an intermediate, but ketogenic amino acids cannot.

The nitrogen portion of amino acids is involved in transamination reactions in breakdown as well as in biosynthesis. Excess nitrogen is excreted in one of three forms: *ammonia* (as ammonium ion), *urea,* and *uric acid* (Figure 17.14). Animals, such as fish, that live in an aquatic environment excrete nitrogen as ammonia; they are protected from the toxic effects of high concentrations of ammonia not only by the removal of ammonia from their bodies but also by rapid dilution of the excreted ammonia by the water in the environment. The principal waste product of nitrogen metabolism in terrestrial animals is urea (a water-soluble compound); its reactions provide some interesting comparisons with the citric acid cycle. Birds excrete nitrogen in the form of uric acid, which is insoluble in water. They do not have to carry the excess weight of water, which could hamper flight, to rid themselves of waste products.

The fate of nitrogen on which we shall concentrate is the production of urea in the **urea cycle** (Figure 17.15). The nitrogens that enter the urea cycle do so first as ammonia in the form of ammonium ion. The immediate precursor is glutamate, but the ammonia nitrogens of glutamate have ultimately come from many sources as a result of transamination reactions. A condensation reaction between the ammonium ion and carbon dioxide produces **carbamoyl phosphate** in a reaction that requires two molecules of ATP for each molecule of carbamoyl phosphate. Carbamoyl phosphate reacts with **ornithine** to form **citrulline.** A second nitrogen enters the urea cycle when aspartate reacts with citrulline to form **argininosuccinate** in another reaction that requires ATP (AMP and PP$_i$ are produced in this reaction). The amino group of the aspartate is the source of the second nitrogen in the urea that will be formed in this series of reactions.

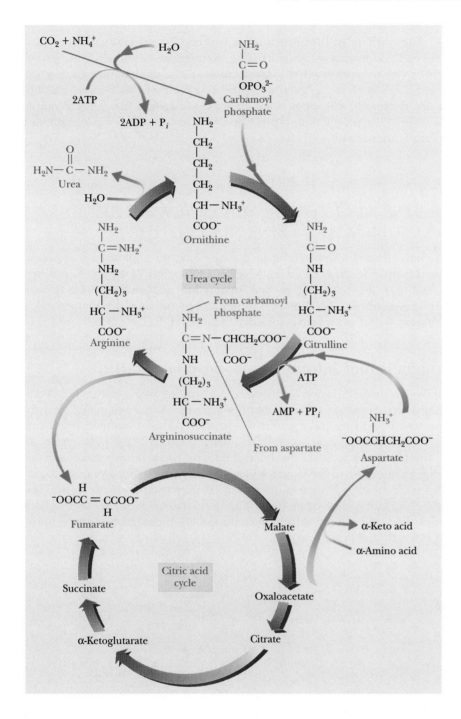

FIGURE   17.15

The urea cycle and some of its links to the citric acid cycle.

Argininosuccinate is split to produce **arginine** and **fumarate**. Finally, arginine is hydrolyzed to give urea and to regenerate ornithine. The biosynthesis of arginine from ornithine is discussed in Interchapter B. Another way of looking at the urea cycle is to consider arginine the immediate precursor of urea and to see it as producing ornithine in the process. According to this point of view, the rest of the cycle is the regeneration of arginine from ornithine.

The synthesis of fumarate is a link between the urea cycle and the citric acid cycle. Fumarate is, of course, an intermediate of the citric acid cycle, and it can be converted to oxaloacetate. A transamination reaction can convert oxaloacetate to aspartate, providing another link between the two cycles (Figure 17.15). Four "high-energy" phosphate bonds are required because of the production of pyrophosphate in the conversion of aspartate to argininosuccinate.

17.7
PURINE NUCLEOTIDE METABOLISM: THE ANABOLISM OF PURINE NUCLEOTIDES

We have already discussed the formation of ribose 5-phosphate as part of the pentose phosphate pathway (Chapter 14, Section 14.4). The biosynthetic pathway for both purine and pyrimidine nucleotides makes use of preformed ribose 5-phosphate. Purines and pyrimidines are synthesized in different ways, and we shall consider them separately.

Anabolism of Inosine Monophosphate (IMP)

In the synthesis of purine nucleotides, the growing ring system is bonded to the ribose phosphate while the purine skeleton is being assembled, first the five-membered ring and then the six-membered ring, eventually producing inosine 5'-monophosphate. All four nitrogen atoms of the purine ring are derived from amino acids: two from glutamine, one from aspartate, and one from glycine. Two of the five carbon atoms (adjacent to the glycine nitrogen) also come from glycine, two more come from tetrahydrofolate derivatives, and the fifth comes from CO_2 (Figure 17.16). The series of reactions producing IMP is long and complex; the details of the process can be found in Interchapter B, Section B.7.

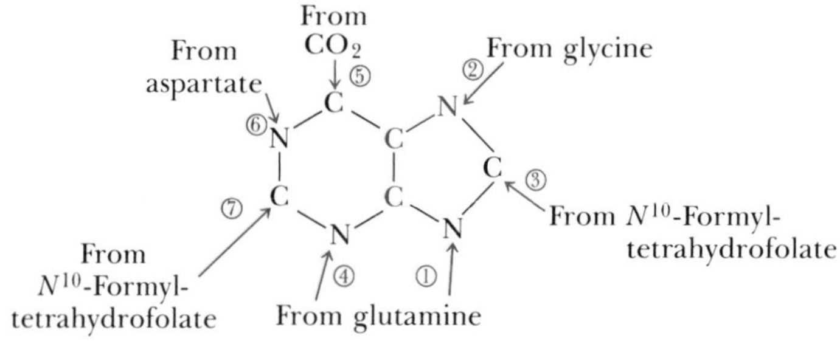

FIGURE 17.16

Sources of the atoms in the purine ring in purine nucleotide biosynthesis. The numbering system indicates the order in which each atom or group of atoms is added.

The Conversion of IMP to AMP and GMP

IMP is the precursor of both AMP and GMP. The conversion of IMP to AMP takes place in two stages (Figure 17.17). The first step is the reaction of aspartate with IMP to form adenylosuccinate. This reaction is catalyzed by adenylosuccinate synthetase and requires GTP, not ATP. The cleavage of fumarate from adenylosuccinate to produce AMP is catalyzed by adenylosuccinase. This enzyme also functions in the synthesis of the six-membered ring of IMP.

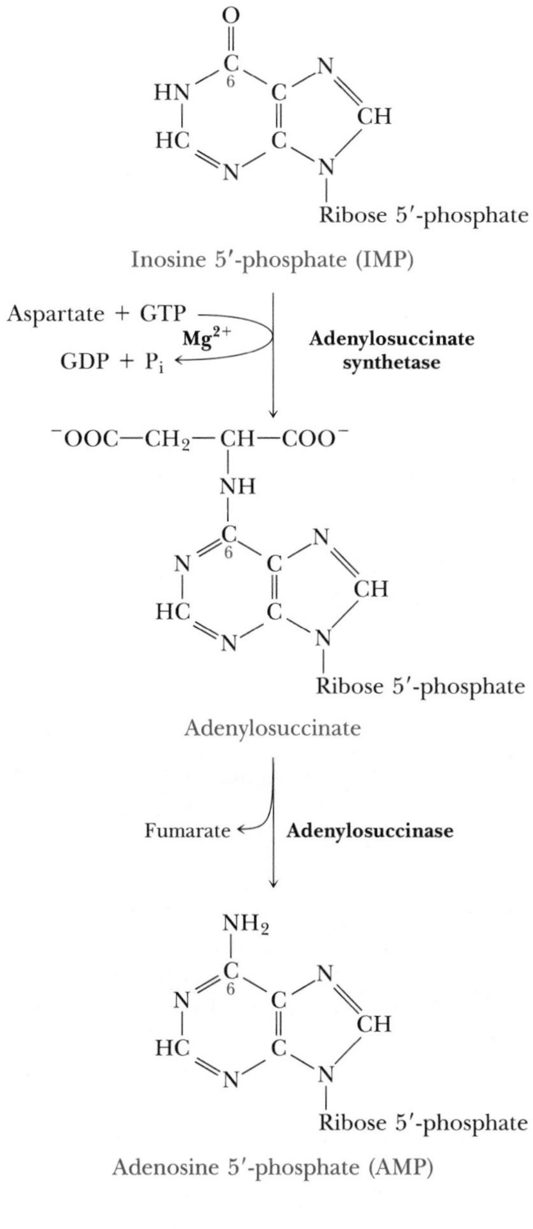

FIGURE 17.17

The synthesis of AMP from IMP.

FIGURE 17.18

The synthesis of GMP from IMP.

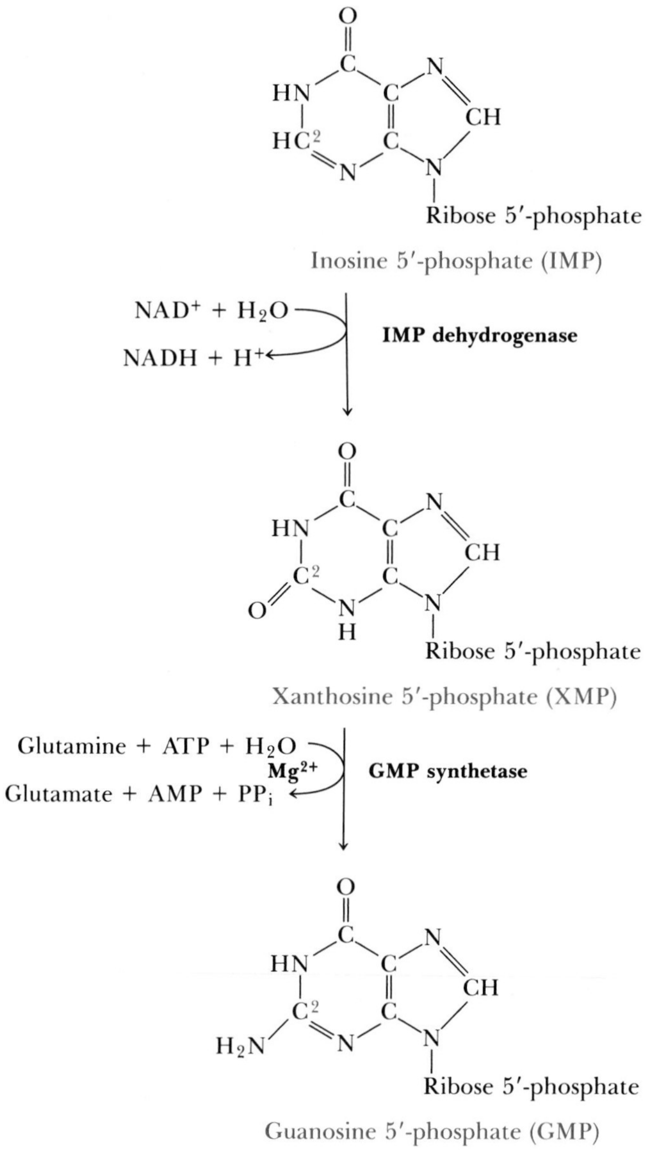

Inosine 5'-phosphate (IMP)

Xanthosine 5'-phosphate (XMP)

Guanosine 5'-phosphate (GMP)

The conversion of IMP to GMP also takes place in two stages (Figure 17.18). The first of the two steps is an oxidation in which the C—H group at the C-2 position is converted to a keto group. The oxidizing agent in the reaction is NAD^+, and the enzyme involved is IMP dehydrogenase. The nucleotide formed by the oxidation reaction is xanthosine 5'-phosphate (XMP). An amino group from the side chain of glutamine replaces the C-2 keto group of XMP to produce GMP. This reaction is catalyzed by GMP synthetase; ATP is hydrolyzed to AMP and PP_i in the process. Note that there is some control over the relative levels of purine nucleotides; GTP is needed for the synthesis of adenine nucleotides, whereas ATP is required for the synthesis of guanine nucleotides. Each of the purine nucleotides must occur at a reasonably high level for the other to be synthesized.

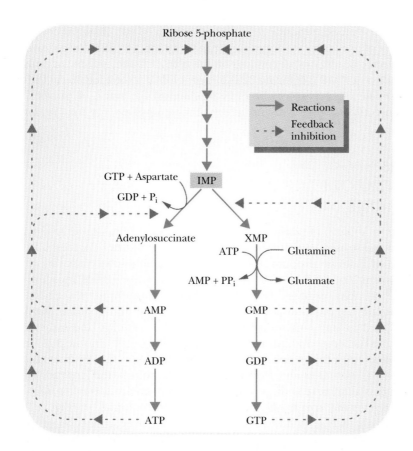

FIGURE 17.19

The role of feedback inhibition in regulation of purine nucleotide biosynthesis.

Subsequent phosphorylation reactions produce purine nucleoside diphosphates (ADP and GDP) and triphosphates (ATP and GTP). The purine nucleoside monophosphates, diphosphates, and triphosphates are all feedback inhibitors of the first stages of their own biosynthesis. Also, AMP, ADP, and ATP inhibit the conversion of IMP to adenine nucleotides, and GMP, GDP, and GTP inhibit the conversion of IMP to xanthylate and to guanine nucleotides (Figure 17.19).

Energy Requirements for Production of AMP and GMP

The production of IMP starting with ribose 5-phosphate requires the equivalent of 7 ATP (see Interchapter B, Section B.7, and the article by Meyer *et al.* in the bibliography at the end of this chapter). The conversion of IMP to AMP requires hydrolysis of an additional phosphate ester bond — in this case, that of GTP. In the formation of AMP from ribose 5-phosphate, the equivalent of 8 ATP is needed. The conversion of IMP to GMP requires two "high-energy" bonds, since a reaction occurs in which ATP is hydrolyzed to AMP and PP_i. For the production of GMP from ribose 5-phosphate, the equivalent of 9 ATP is necessary. The anaerobic oxidation of glucose produces only 2 ATP for each molecule of glucose (Chapter 11, Section 11.1). Anaerobic organisms require four molecules of glucose (which produce 8 ATP) for each AMP they form, or five molecules of glucose (which produce 10 ATP) for each GMP. The process is more efficient for aerobic organisms. Since 36 or 38 ATP result from each molecule of glucose, depending on the type of tissue, aerobic organisms can produce 4 AMP (requiring 32 ATP) or 4 GMP (requiring 36 ATP) for each molecule of glucose oxidized. A mechanism for reuse of purines, rather than complete turnover and new synthesis, saves energy for organisms.

17.8
CATABOLIC REACTIONS OF PURINE NUCLEOTIDES

The catabolism of purine nucleotides proceeds by hydrolysis to the nucleoside and subsequently to the free base, which is further degraded. Deamination of guanine produces xanthine, and deamination of adenine produces hypoxanthine (the base corresponding to the nucleoside inosine) (Figure 17.20a). Hypoxanthine can be oxidized to xanthine, so this base is a common degradation product of both adenine and guanine. Xanthine is oxidized in turn to **uric acid.** In birds, some reptiles, and insects, and in Dalmatian dogs and primates (including humans), uric acid is the end-product of purine metabolism and is excreted. In all other terrestrial animals, including all other mammals, allantoin is the product excreted, whereas allantoate is the product in fish. Allantoate is further degraded to glyoxylate and urea by microorganisms and some amphibia (Figure 17.20b). **Gout** is a disease in humans that is caused by overproduction of uric acid. Deposits of uric acid (which is not soluble in water) accumulate in the joints of the hands and feet. Benjamin Franklin suffered from gout, and he left eloquent testimony to the painful nature of this disease. **Allopurinol** is a compound used to treat gout; it inhibits the degradation of hypoxanthine to

Allopurinol, a substance used in the treatment of gout.

xanthine and of xanthine to uric acid, preventing the buildup of uric acid deposits.

Salvage reactions are important in the metabolism of purine nucleotides because of the amount of energy required for the synthesis of the purine bases. A free purine base that has been cleaved from a nucleotide can produce the corresponding nucleotide by reacting with the compound phosphoribosylpyrophosphate (PRPP), formed by a transfer of a pyrophosphate group from ATP to ribose 5-phosphate (Figure 17.21, p. 532).

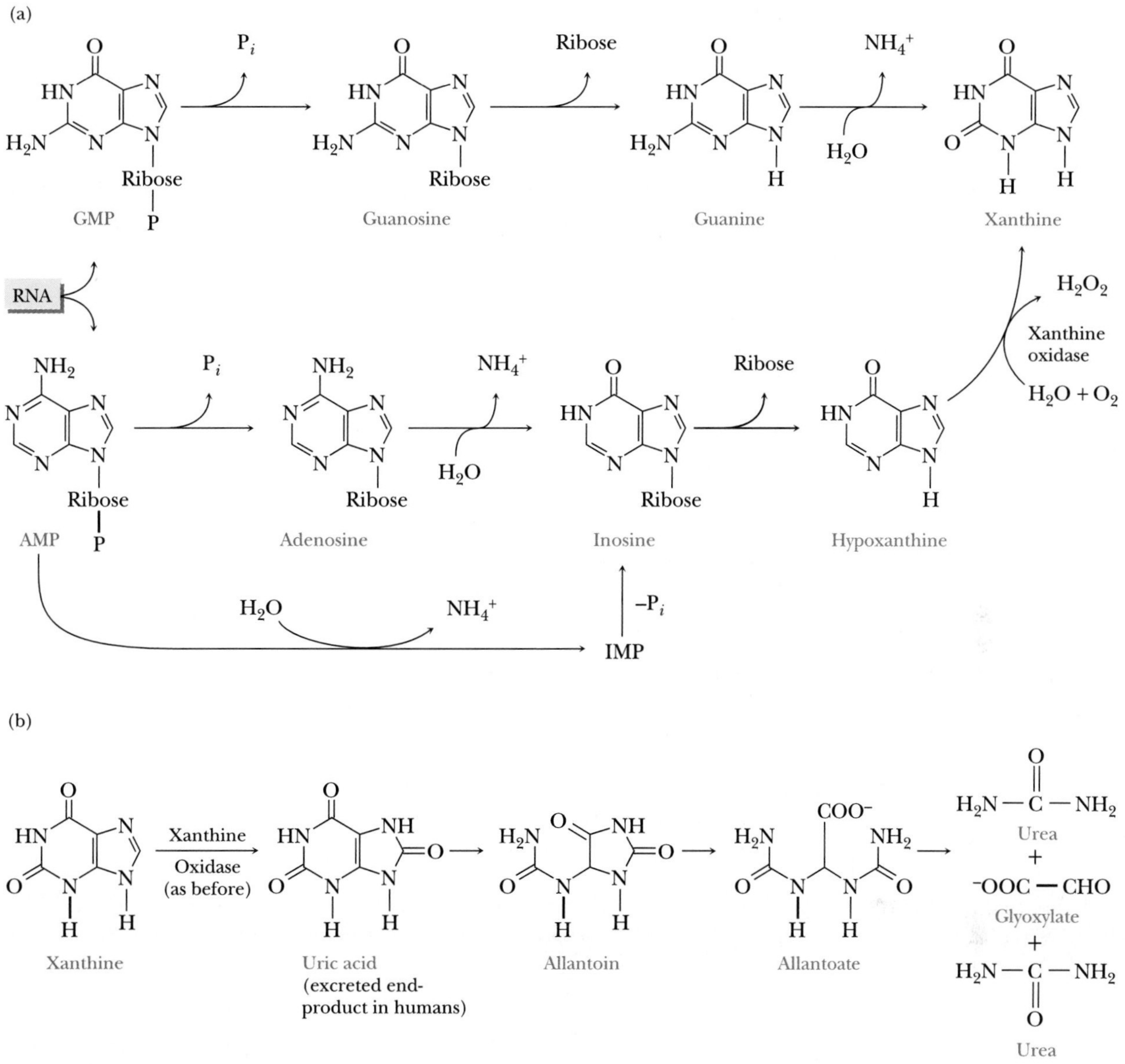

(a)

GMP → Guanosine → Guanine → Xanthine

AMP → Adenosine → Inosine → Hypoxanthine

(b)

Xanthine → Uric acid (excreted end-product in humans) → Allantoin → Allantoate → Urea + Glyoxylate + Urea

FIGURE 17.20

The reactions of purine catabolism. (a) Purine nucleotides are converted to the free base and then to xanthine. (b) Catabolic reactions of xanthine.

Two different enzymes with different specificities with respect to the purine base catalyze salvage reactions. The reaction

$$\text{Adenine} + \text{PRPP} \longrightarrow \text{AMP} + \text{PP}_i$$

is catalyzed by adenine phosphoribosyltransferase. The corresponding reactions of guanine and hypoxanthine

$$\text{Hypoxanthine} + \text{PRPP} \xrightarrow{\text{HPRT}} \text{IMP} + \text{PP}_i$$

$$\text{Guanine} + \text{PRPP} \xrightarrow{\text{HPRT}} \text{GMP} + \text{PP}_i$$

are catalyzed by hypoxanthine–guanine phosphoribosyltransferase (HPRT). A deficiency in HPRT can result in a serious disorder, the **Lesch–Nyhan syndrome** (Box 17.1).

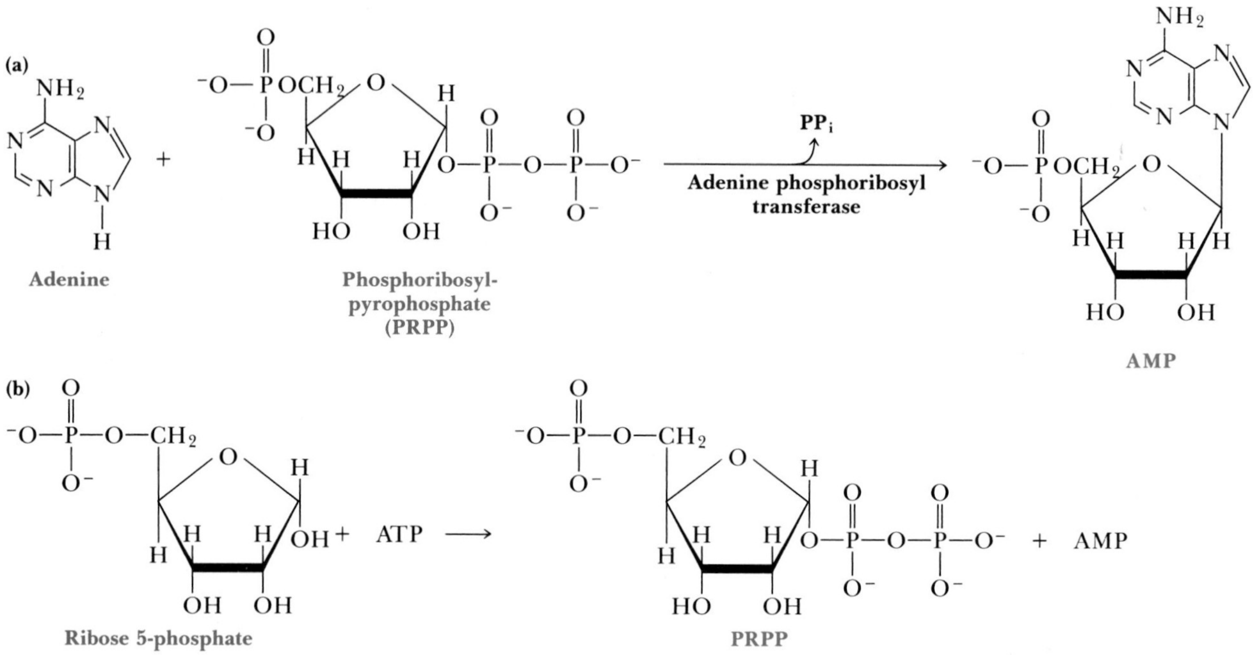

FIGURE 17.21

Purine salvage. (a) Adenine is the purine in this example. There are analogous reactions for salvage of guanine and hypoxanthine (see text). (b) The formation of PRPP.

LESCH–NYHAN SYNDROME

A deficiency of the HPRT enzyme is the cause of the Lesch–Nyhan syndrome, a genetic disease. The biochemical consequences include an elevated concentration of PRPP and increased production of purines and uric acid. The accumulation of uric acid leads to kidney stones and gout, but the most striking clinical manifestations are neurological. There is a compulsive tendency toward self-mutilation among patients with the Lesch–Nyhan syndrome; they tend to bite off fingertips and parts of the lips. The development of kidney stones and gouty symptoms can be prevented by the administration of allopurinol, but there is no real treatment for the self-destructive behavior and the mental retardation and spasticity that accompany it. The diverse manifestations of this disease show clearly that metabolism is extremely complex and that the failure of one enzyme to function can have consequences that reach far beyond the reaction that it catalyzes.

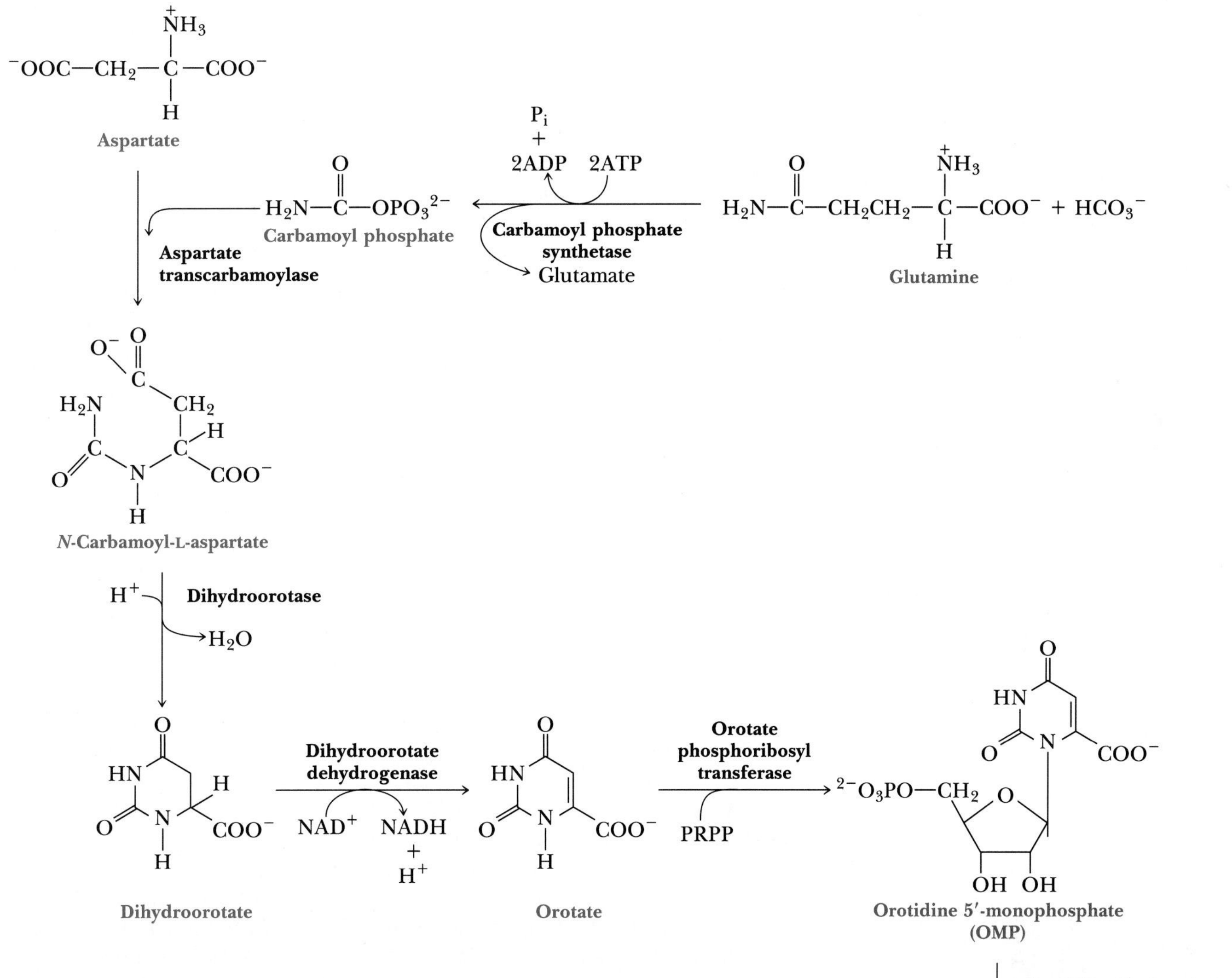

The biosynthesis of the pyrimidine ring, with the enzymes and intermediates shown: Aspartate, Carbamoyl phosphate, Glutamine, N-Carbamoyl-L-aspartate, Dihydroorotate, Orotate, Orotidine 5′-monophosphate (OMP), and Uridine 5′-monophosphate (UMP).

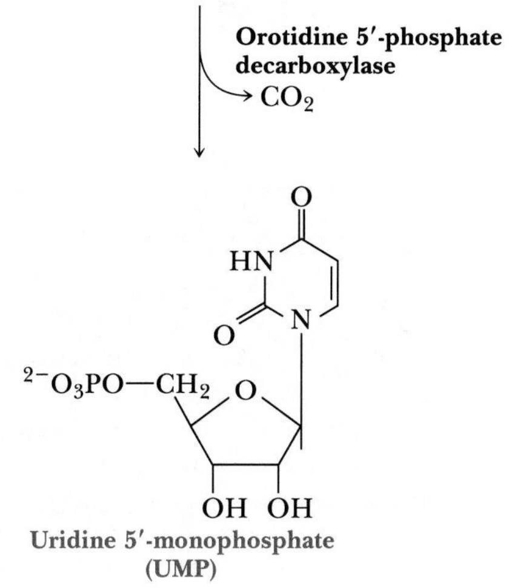

FIGURE 17.22

The biosynthesis of UMP.

17.9
PYRIMIDINE NUCLEOTIDE METABOLISM: ANABOLISM AND CATABOLISM

The Anabolism of Pyrimidine Nucleotides

The overall scheme of pyrimidine nucleotide biosynthesis differs from that of purine nucleotides in that the pyrimidine ring is assembled before it is attached to ribose 5-phosphate. The carbon and nitrogen atoms of the pyrimidine ring come from carbamoyl phosphate and aspartate. The production of carbamoyl phosphate for pyrimidine biosynthesis takes place in the cytosol, and the nitrogen donor is glutamine. (We already saw a reaction for the production of carbamoyl phosphate when we discussed the urea cycle in Section 17.6. That reaction differs from this one because it takes place in mitochondria and the nitrogen donor is NH_4^+.)

$$HCO_3^- + glutamine + 2\ ATP + H_2O \longrightarrow$$
$$carbamoyl\ phosphate + glutamate + 2\ ADP + P_i$$

The reaction of carbamoyl phosphate with aspartate to produce *N*-carbamoylaspartate is the committed step in pyrimidine biosynthesis. The

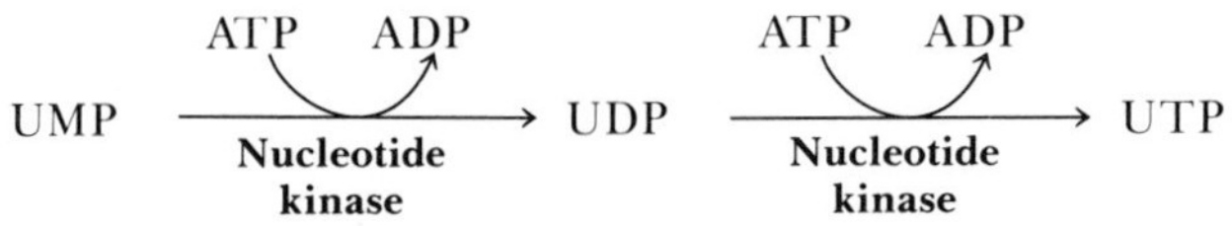

FIGURE 17.23

The conversion of UMP to UTP.

compounds involved in reactions up to this point in the pathway can play other roles in metabolism; after this point, *N*-carbamoylaspartate can be used only to produce pyrimidines — thus the term "committed step."

This reaction is catalyzed by aspartate transcarbamoylase, which we discussed in detail in Chapter 5 as a prime example of an allosteric enzyme subject to feedback regulation. The next step, the conversion of *N*-carbamoylaspartate to dihydroorotate, takes place in a reaction that involves an intramolecular dehydration (loss of water) as well as cyclization. This reaction is catalyzed by dihydroorotase. Dihydroorotate is converted to orotate by dihydroorotate dehydrogenase, with the concomitant conversion of NAD^+ to NADH. A pyrimidine nucleotide is now formed by the reaction of orotate with PRPP to give orotidine 5′-phosphate (OMP), which is a reaction similar to the one that takes place in purine salvage (Section 17.8). Orotate phosphoribosyl transferase catalyzes this reaction. Finally, orotidine 5′-phosphate decarboxylase catalyzes the conversion of OMP to UMP (uridine 5′-phosphate), which is the precursor of the remaining pyrimidine nucleotides (Figure 17.22, page 533).

Two successive phosphorylation reactions convert UMP to UTP (Figure 17.23). The conversion of uracil to cytosine takes place in the triphosphate form, catalyzed by CTP synthetase (Figure 17.24). Glutamine is the nitrogen donor, and ATP is required, as we saw earlier in similar reactions.

$$\text{UTP} + \text{glutamine} + \text{ATP} \longrightarrow \text{CTP} + \text{glutamate} + \text{ADP} + P_i$$

Feedback inhibition in pyrimidine nucleotide biosynthesis takes place in several ways. CTP is an inhibitor of aspartate transcarbamoylase and of CTP synthetase. UMP is an inhibitor of an even earlier step, the one catalyzed by carbamoyl phosphate synthetase (Figure 17.25).

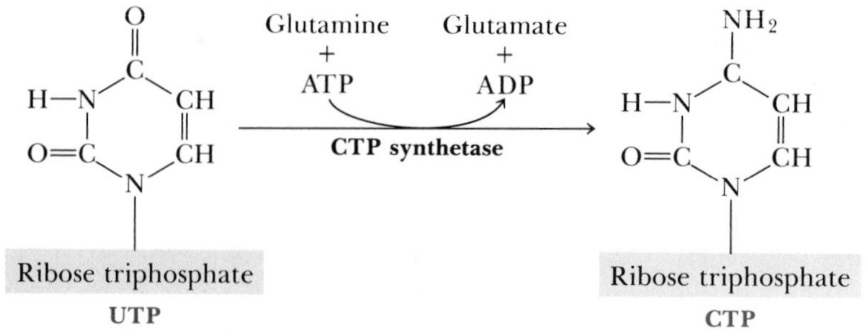

FIGURE 17.24

The conversion of UTP to CTP.

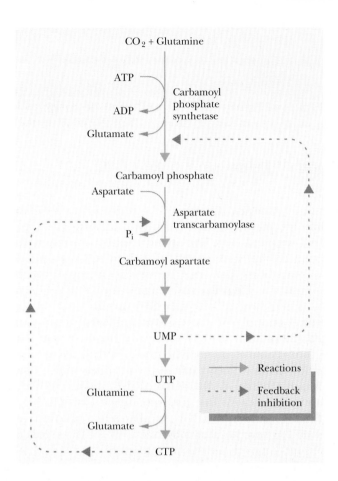

FIGURE 17.25

The role of feedback inhibition in the regulation of pyrimidine nucleotide biosynthesis.

Pyrimidine Catabolism

Pyrimidine nucleotides are broken down first to the nucleoside and then to the base, as purine nucleotides are. Cytosine can be deaminated to uracil, and the double bond of the uracil ring is reduced to produce dihydrouracil. The ring opens to produce N-carbamoylpropionate, which in turn is broken down to NH_4^+, CO_2, and β-alanine (Figure 17.26).

FIGURE 17.26

The catabolism of pyrimidines.

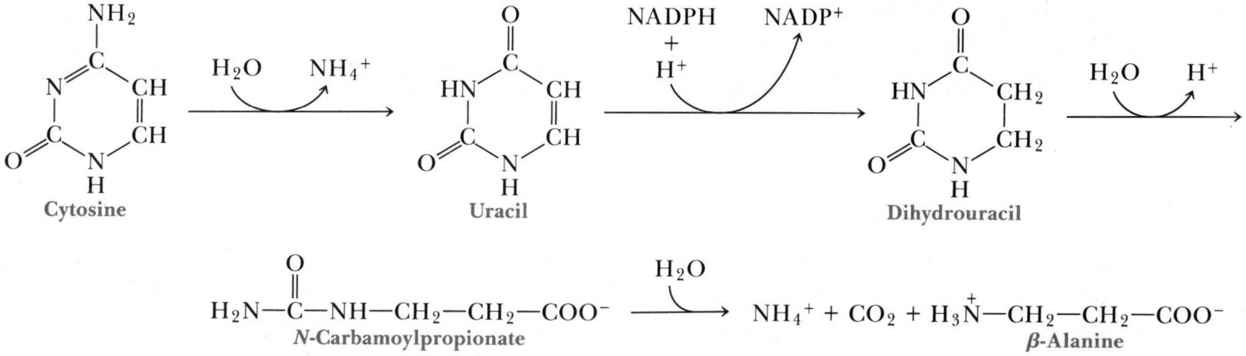

$$H_2N-\overset{\overset{\displaystyle O}{\|}}{C}-NH-CH_2-CH_2-COO^- \quad \xrightarrow{H_2O} \quad NH_4^+ + CO_2 + H_3\overset{+}{N}-CH_2-CH_2-COO^-$$

N-Carbamoylpropionate
β-Alanine

17.10
THE REDUCTION OF RIBONUCLEOTIDES TO DEOXYRIBONUCLEOTIDES

Ribonucleoside diphosphates are reduced to 2'-deoxyribonucleoside diphosphates in all organisms (Figure 17.27a); NADPH is the reducing agent.

$$\text{Ribonucleoside diphosphate} + \text{NADPH} + \text{H}^+ \longrightarrow$$
$$\text{deoxyribonucleoside diphosphate} + \text{NADP}^+ + \text{H}_2\text{O}$$

The actual process, which is catalyzed by *ribonucleotide reductase,* is more complex than the preceding equation would indicate and involves some intermediate electron carriers. The ribonucleotide reductase system from *E. coli* has been extensively studied, and its mode of action gives some clues to the nature of the process. Two other proteins are required, thioredoxin and thioredoxin reductase. **Thioredoxin** contains a disulfide (S—S) group in its oxidized form and two sulfhydryl (—SH) groups in its reduced form. NADPH reduces thioredoxin in a reaction catalyzed by *thioredoxin reductase.* The reduced thioredoxin in turn reduces a ribonucleoside diphosphate (NDP) to a deoxyribonucleoside diphosphate (dNDP) (Figure 17.27b), and it is this reaction that is actually catalyzed by ribonucleotide reductase. Note that this reaction produces dADP, dGDP, dCDP, and dUDP. The first three are phosphorylated to give the corresponding triphosphates, which are substrates for the synthesis of DNA. Another required substrate for DNA synthesis is dTTP, and we shall now see how dTTP is produced from dUDP.

FIGURE 17.27

(a) The conversion of ribonucleotides to deoxyribonucleotides. TR (—S—S—) and TR (—SH)$_2$ refer to the oxidized (disulfide) and reduced (sulfhydryl) forms of thioredoxidin. (b) The structures of NDP and dNDP.

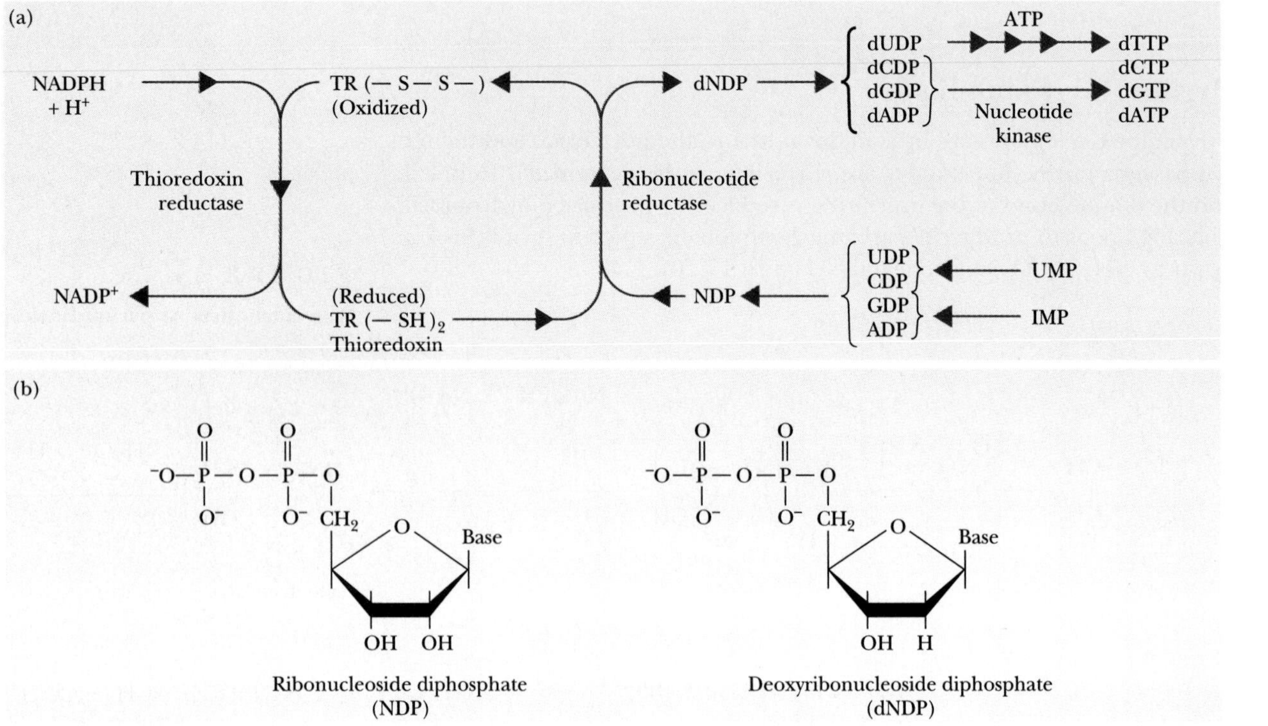

17.11
THE CONVERSION OF dUDP TO dTTP

A one-carbon transfer is required for the conversion of uracil to thymine by attachment of the methyl group. The most important reaction in this conversion is that catalyzed by *thymidylate synthetase* (Figure 17.28). The source of the one-carbon unit is N^5, N^{10}-methylenetetrahydrofolate, which is converted to dihydrofolate in the process. The metabolically active form of the one-carbon carrier is tetrahydrofolate. Dihydrofolate must be reduced to tetrahydrofolate for this series of reactions to continue, and this process requires NADPH and *folate reductase.*

Since a supply of dTTP is necessary for DNA synthesis, inhibition of enzymes that catalyze the production of dTTP will inhibit the growth of rapidly dividing cells. Cancer cells, like all fast-growing cells, depend on continued DNA synthesis for growth. Inhibitors of thymidylate synthetase, such as fluorouracil (see Exercise 5), and inhibitors of folate reductase, such as aminopterin and methotrexate (structural analogues of folate), have been used in cancer chemotherapy (Figure 17.29). The intent of such therapy is to inhibit the formation of dTTP and thus of DNA in cancer cells, causing the death of the cancer cells with minimal effect on normal cells, which grow more slowly. Chemotherapy has adverse side effects because of the highly toxic nature of most of the drugs involved, and enormous amounts of research are focused on finding safe and effective forms of treatment.

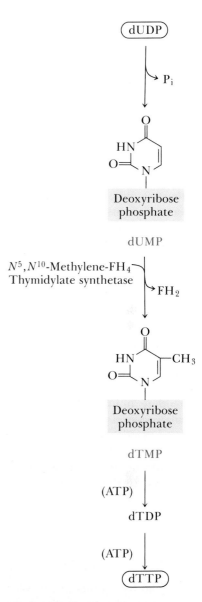

FIGURE 17.28

The conversion of dUDP and dTTP. (FH$_4$ is tetrahydrofolate; FH$_2$ is dihydrofolate.)

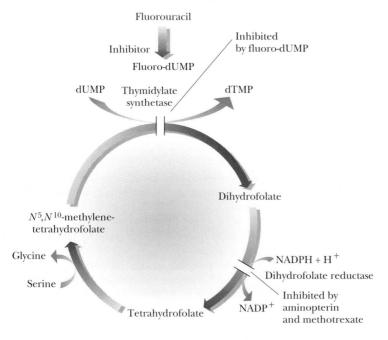

FIGURE 17.29

Targeting of thymidylate synthetase and dihydrofolate reductase in cancer chemotherapy. The actions of both enzymes are blocked by inhibitors. Fluoro-dUMP blocks the methylation of dUMP. Aminopterin and methotrexate block the reduction of dihydrofolate to tetrahydrofolate.

S U M M A R Y

The metabolism of nitrogen encompasses a number of topics, including the anabolism and catabolism of amino acids, porphyrins, and nucleotides. Atmospheric nitrogen is the ultimate source of this element in biomolecules. Nitrogen fixation is the process by which molecular nitrogen from the atmosphere is made available to organisms, in the form of ammonia. Nitrification reactions convert NO_3^- to NH_3 and provide another source of nitrogen. Feedback control mechanisms are a unifying factor in biosynthetic pathways involving nitrogen compounds.

In the anabolism of amino acids, transamination reactions play an important role. Glutamate and glutamine are frequently the amino group donors. The enzymes that catalyze transamination reactions frequently require pyridoxal phosphate as a coenzyme. One-carbon transfers also operate in the anabolism of amino acids. Carriers are required for the one-carbon groups transferred. Tetrahydrofolate is a carrier of methylene and formyl groups, and S-adenosylmethionine is a carrier of methyl groups. Some species, including humans, cannot synthesize all the amino acids required for protein synthesis and must obtain these essential amino acids from dietary sources.

The catabolism of amino acids has two parts: the fate of the nitrogen and the fate of the carbon skeleton. In the urea cycle, nitrogen released by the catabolism of amino acids is converted to urea. The carbon skeleton is converted to pyruvate or oxaloacetate in the case of glucogenic amino acids, or to acetyl-CoA or acetoacetyl-CoA in the case of ketogenic amino acids.

The anabolic pathway of nucleotide synthesis involving purines differs from that involving pyrimidines. Both pathways make use of preformed ribose 5-phosphate but differ with regard to the point in the pathway at which the sugar phosphate is attached to the base. In the case of purine nucleotides, the growing base is attached to the sugar phosphate during the synthesis. In pyrimidine biosynthesis, the base is first formed and then attached to the sugar phosphate. In catabolism, purine bases are frequently salvaged and reattached to sugar phosphates. Otherwise, purines are broken down to uric acid. Pyrimidines are degraded to β-alanine.

Deoxyribonucleotides for DNA synthesis are produced by the reduction of ribonucleoside diphosphates to deoxyribonucleoside diphosphates. Another reaction specifically needed to produce substrates for DNA synthesis is the conversion of uracil to thymine. This pathway, which requires a tetrahydrofolate derivative as the carrier for one-carbon transfer, is a target for cancer chemotherapy.

E X E R C I S E S

1. Lysine is frequently added to cereal products as a means of enhancing the quality of the product as a protein source. Suggest a reason for the choice of lysine.
2. Sketch the structure of folic acid; also show by a sketch how it serves as a carrier of one-carbon groups.
3. How many α-amino acids participate directly in the urea cycle? Of these, how many can be used for protein synthesis?
4. Write an equation for the net reaction of the urea cycle; show how the urea cycle is linked to the citric acid cycle.
5. Suggest a mode of action for fluorouracil in cancer chemotherapy.

5-Fluorouracil.

6. Why is there no net gain if homocysteine is converted to methionine with S-adenosylmethionine as the methyl donor?
7. Show, by the equation for a typical reaction, why glutamate plays a central role in the biosynthesis of amino acids.
8. List the essential amino acids for a phenylketonuric adult and compare them with the requirements for a normal adult.
9. What is an important difference between the biosynthesis of purine nucleotides and that of pyrimidine nucleotides?
10. List the intermediates in the flow of nitrogen from N_2 to the N-7 nitrogen of purines.
11. Sulfanilamide and related sulfa drugs were widely used to treat diseases of bacterial origin before penicillin and more advanced drugs were readily available. The inhibi-

tory effect of sulfanilamide on bacterial growth can be reversed by *p*-aminobenzoate. Suggest a mode of action for sulfanilamide.

$$H_2N-\!\!\!\bigcirc\!\!\!-SO_2NH_2$$

Sulfanilamide.

12. By means of a structural formula, show how *S*-adenosylmethionine is a carrier of methyl groups.
13. Comment briefly on the usefulness to organisms of feedback control mechanisms in long biosynthetic pathways.

14. People on high-protein diets are advised to drink lots of water. Why?
15. Chemotherapy patients receiving cytotoxic (cell-killing) agents such as FdUMP (the UMP analogue that contains fluorouracil) and methotrexate temporarily go bald. Why?
16. How many "high-energy" phosphate bonds must be hydrolyzed in the pathway that produces GMP from guanine and PRPP by the PRPP salvage reaction, compared with the number of such bonds hydrolyzed in the pathway leading to IMP and then to GMP?

A N N O T A T E D   B I B L I O G R A P H Y

Bender, D. A. *Amino Acid Metabolism.* 2nd ed. New York: John Wiley, 1985. [A general treatment of the topic, with a particularly good section on tryptophan metabolism.]

Benkovic, S. On the Mechanism of Action of Folate- and Biopterin-Requiring Enzymes. *Ann. Rev. Biochem.* **49,** 227–254 (1980). [A review of one-carbon transfers.]

Braunstein, A. E. Amino Group Transfer. In Boyer, P. D., ed. *The Enzymes.* 3rd ed. Vol. 9. New York: Academic Press, 1973. [Getting old, but a standard reference.]

Karplus, P., M. Daniels, and J. Herriott. Atomic Structure of Ferredoxin–NADP$^+$ Reductase: Prototype for a Structurally Novel Flavoenzyme Family. *Science* **251,** 60–66 (1991). [The structure of a key enzyme involved in nitrogen and sulfur metabolism, as well as in photosynthesis.]

Meyer, E., N. Leonard, B. Bhat, J. Stubbe, and J. Smith. Purification and Characterization of the *pur*E, *pur*K, and *pur*C Gene Products: Identification of a Previously Unrecognized Energy Requirement in the Purine Biosynthetic Pathway. *Biochemistry* **31,** 5022–5032 (1992). [The discovery of a hitherto unsuspected requirement for additional ATP in the biosynthesis of purines.]

Orme-Johnson, W., Nitrogenase Structure: Where To Now? *Science* **257,** 1639–1640 (1992). [Thoughts about nitrogen fixation based on the determination of the structure of nitrogenase by x-ray crystallography.]

Stadtman, E. R. Mechanisms of Enzyme Regulation in Metabolism. In Boyer, P. D., ed. *The Enzymes.* 3rd ed. Vol. 1. New York: Academic Press, 1970. [A review dealing with the importance of feedback control mechanisms.]

The Anabolism of Nitrogen-Containing Compounds

Crystals of L-tryptophan viewed under polarized light.

Amino acids are more than the building blocks of protein molecules. They are the ports of entry for organic nitrogen into a variety of nitrogen-containing compounds that include the following: (1) the nitrogenous bases of DNA and RNA; (2) the nucleotide coenzymes, such as NAD^+ and $NADP^+$, that serve as electron carriers in energy metabolism; (3) heme, an oxygen-binding component of the proteins myoglobin and hemoglobin; and (4) small, physiologically active molecules such as the hormone epinephrine (adrenalin), and the neurotransmitter serotonin. Of the 20 amino acids, 9 are synthesized by plants and microorganisms but not by humans. These amino acids must be supplied from the diet and are called *essential amino acids.* Although each of the amino acids is synthesized by a different pathway, they may be grouped into families. For example, α-ketoglutarate, a citric acid intermediate, is the precursor of the nonessential amino acid, glutamate; and glutamate is the precursor of glutamine, proline, and arginine. The pathways of the essential amino acids are more complex. In bacteria, for example, aspartate is the precursor of lysine, methionine, and threonine. Surplus amino acids are deaminated and nitrogen is excreted as urea, while their carbon skeletons are fed into the citric acid cycle to be used as fuel or recycled as components of other biomolecules.

B.1
GLUTAMATE AS A PRECURSOR OF PROLINE AND ARGININE

Two amino acids, proline and arginine, can be classified as members of the glutamate family in addition to the ones we saw in Chapter 17. Proline is synthesized in several steps, involving the production of glutamate-γ-semialdehyde, which then cyclizes, ultimately forming proline (Figure B.1). The conversion of glutamate to the semialdehyde is inhibited by proline, providing an example of feedback inhibition in nitrogen metabolism. In addition to being a precursor of proline, glutamate-γ-semialdehyde undergoes other reactions, eventually forming ornithine, which gives rise to arginine in the urea cycle (Figure B.2) (see Chapter 17, Section 17.6). Arginine is an inhibitor of an early stage of this pathway.

B.2
THE ASPARTATE FAMILY

Aspartate, the precursor of the other amino acids in this family, is formed from oxaloacetate by a transamination reaction. Aspartate in turn undergoes further reactions to produce isoleucine, threonine, lysine, methionine,

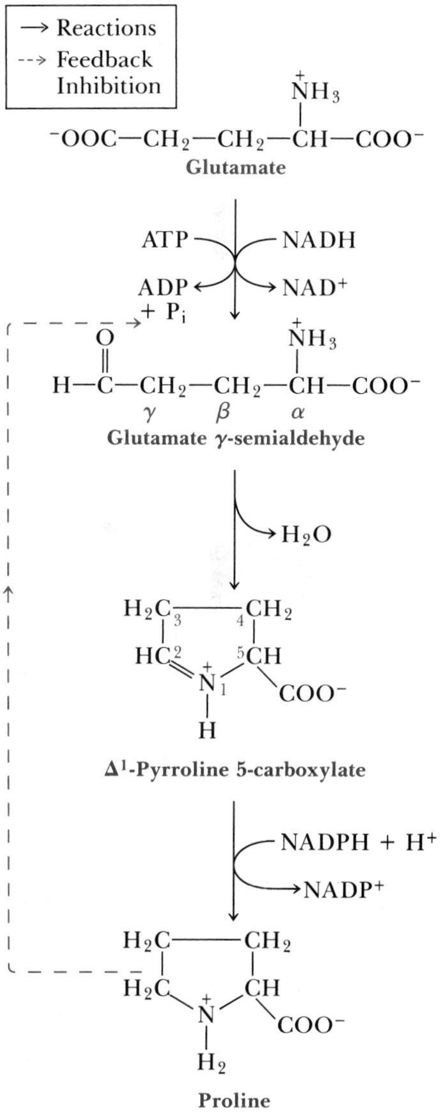

FIGURE B.1

The biosynthesis of proline. (Δ^1 refers to the double bond at position 1 in the ring.)

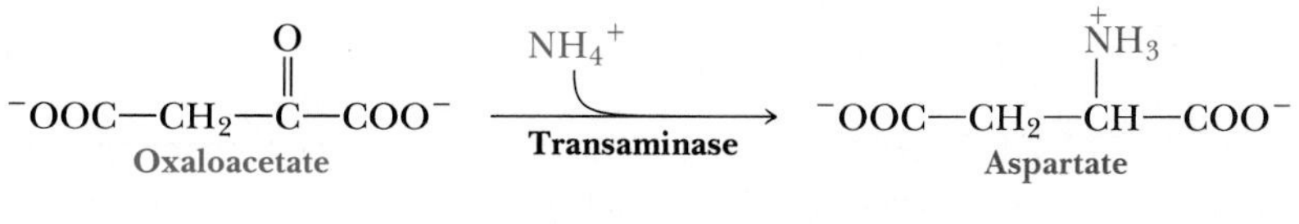

Aspartate is produced by transamination of oxaloacetate.

FIGURE B.2

The biosynthesis of arginine.

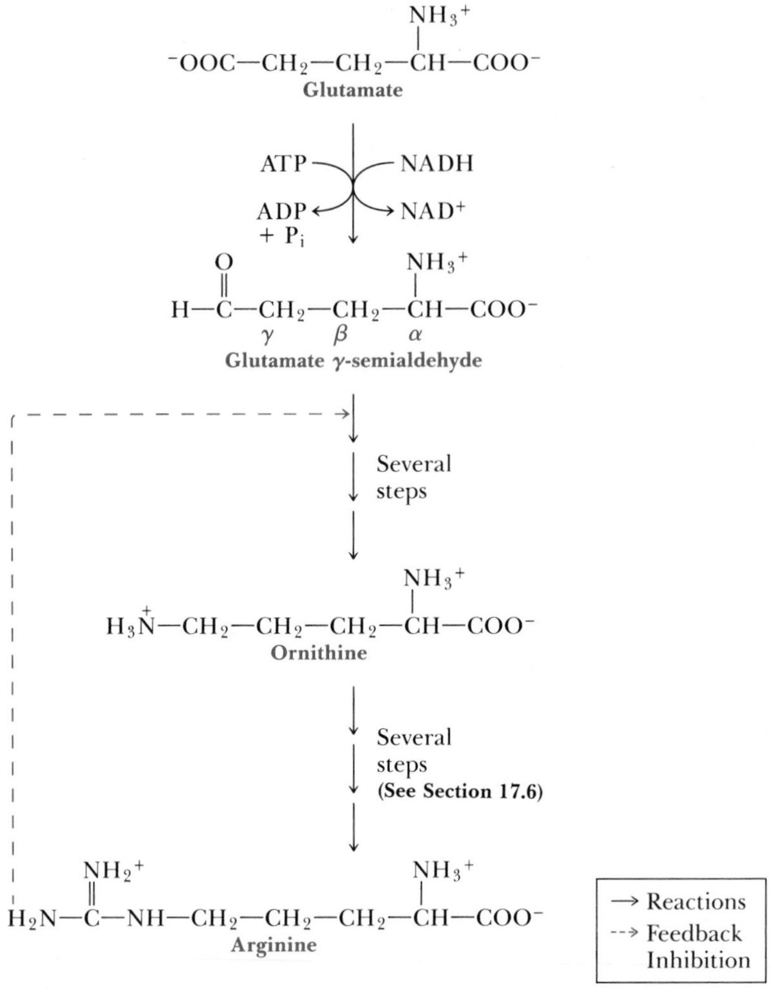

and asparagine. The pathway that produces this group of amino acids comprises several branches. In methionine biosynthesis, the side-chain carboxylate of aspartate is phosphorylated in a reaction requiring ATP. The carboxyl group that has been phosphorylated is then reduced in two stages, first to the aldehyde, then to the alcohol, producing homoserine (Figure B.3). In bacteria the next step in methionine biosynthesis is the reaction of succinyl-CoA with homoserine to form *O*-succinylhomoserine. (There is an analogous reaction in plants, involving the reaction of homoserine with acetyl-CoA to form *O*-acetylhomoserine.) The next step in bacterial methionine synthesis is the transfer of sulfur from cysteine to *O*-succinylhomoserine to form succinate and cystathionine. A hydrolysis reaction removes pyruvate and ammonia, producing homocysteine. Note that homocysteine differs from homoserine only in the substitution of a thiol for a hydroxyl group in the side chain; homocysteine is also produced by plants in analogous reactions. The final step in methionine biosynthesis in both plants and bacteria is the transfer of a one-carbon unit from N^5-methyltetrahydro-

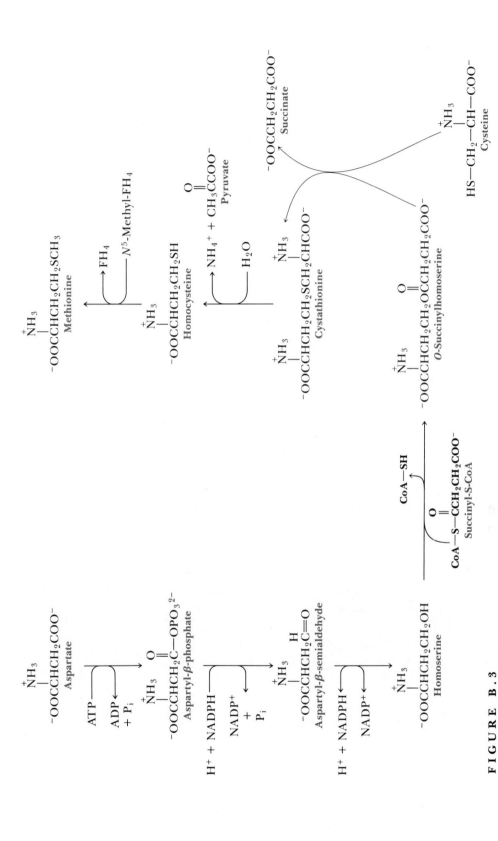

FIGURE B.3

The biosynthesis of methionine in bacteria. (FH_4 is tetrahydrofolate.)

folate to produce methionine from homocysteine. In animals no biosynthetic pathway for methionine exists, and thus it is an essential amino acid (see One-Carbon Transfers and the Serine Family in Chapter 17, Section 17.4).

We shall not discuss in detail the biosynthesis of the other members of the aspartate family. Aspartyl-β-semialdehyde, which plays a role in the biosynthesis of methionine, is also converted to lysine in a multistep process. Homoserine can be converted to threonine as well as to methionine. Threonine in turn is converted to isoleucine in a multistep process. We see in Figure B.4 that feedback inhibition occurs at nearly every step in this biosynthetic pathway.

FIGURE B.4

Feedback control in biosynthesis of amino acids of the aspartate family.

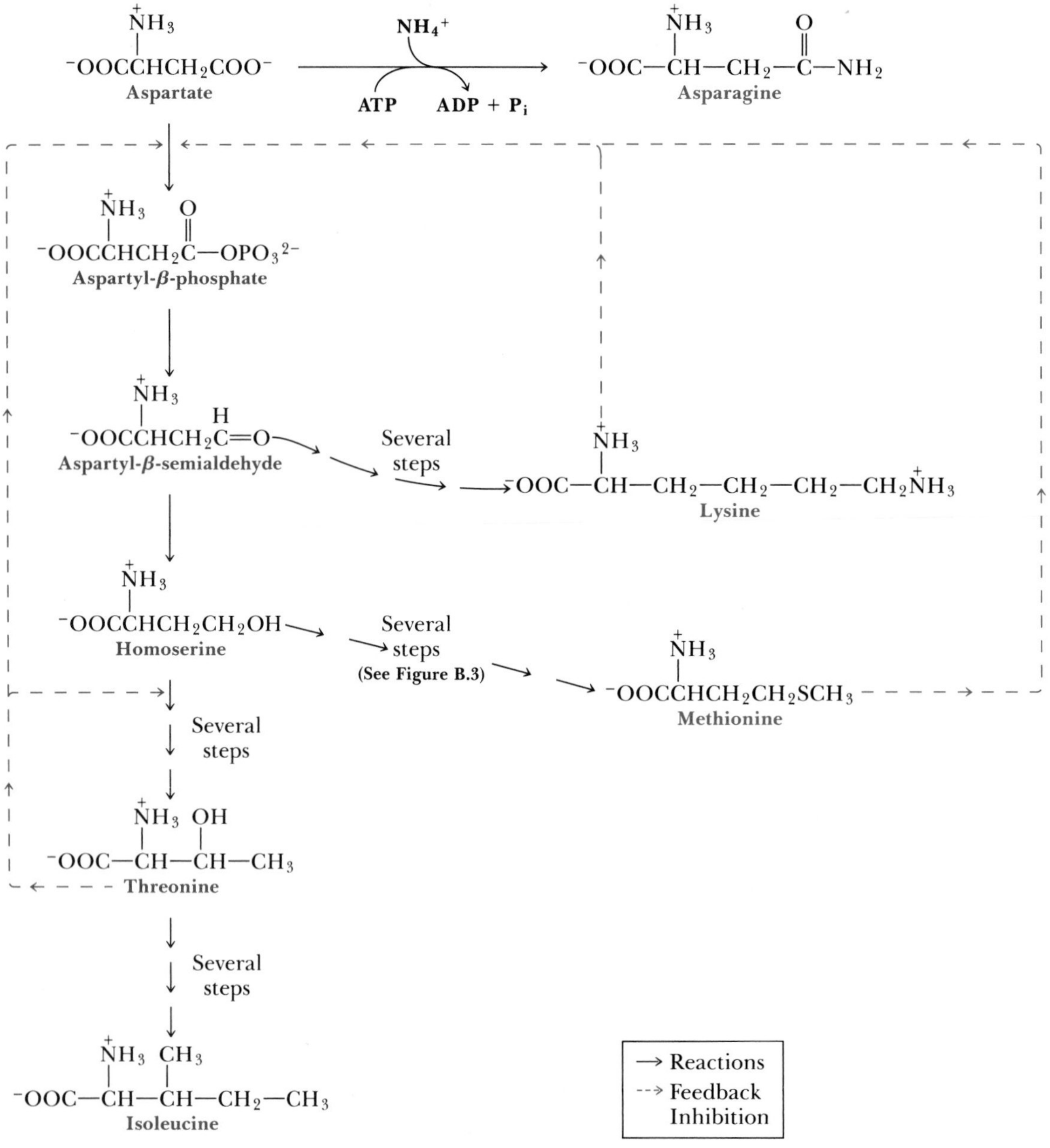

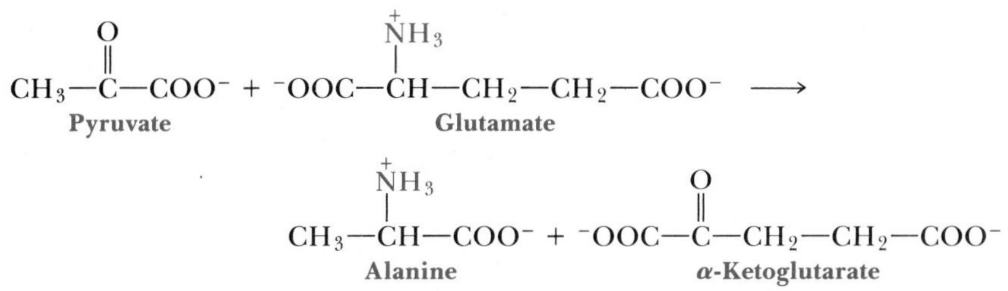

Alanine is produced from pyruvate by transamination.

B.3
THE PYRUVATE FAMILY

The carbon skeletons of alanine, valine, and leucine are derived from pyruvate, the key metabolic intermediate that links glycolysis and the citric acid cycle. Alanine is obtained from pyruvate directly by a transamination reaction involving glutamate (Figure B.5). The formation of valine and leucine is a more complicated process. Several steps are necessary for the conversion of pyruvate to α-ketoisovalerate (Figure B.6). This intermediate can be converted to valine by transamination, or it can undergo several

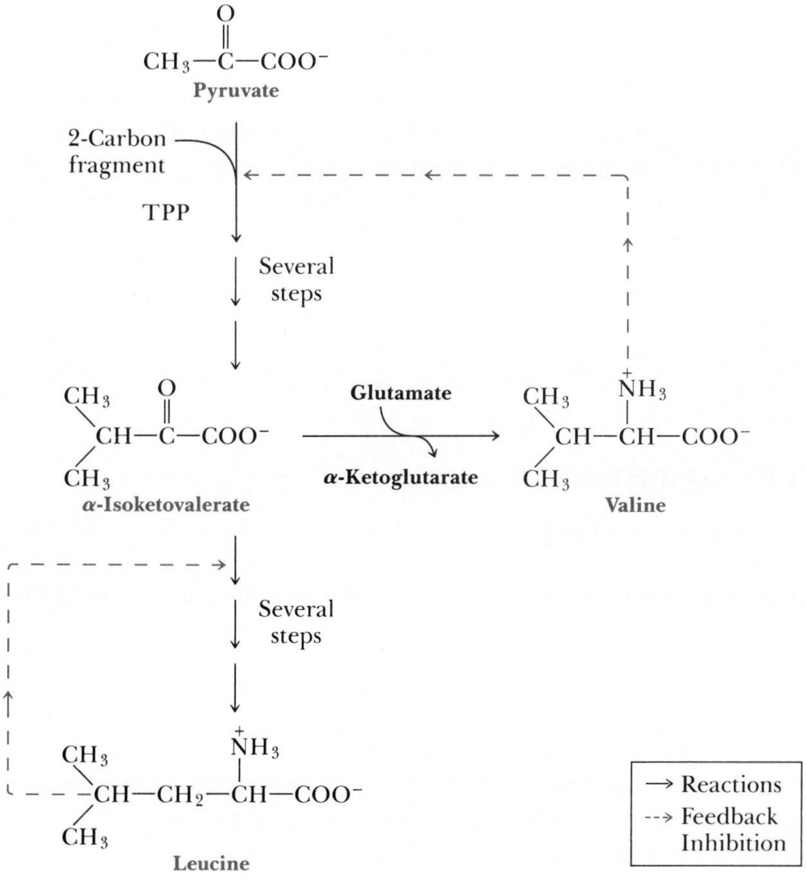

The biosynthesis of leucine and valine from pyruvate, showing feedback inhibition. (TPP is thiamine pyrophosphate.)

other changes to produce α-ketoisocaproate, which is converted to leucine by transamination. Both valine and leucine are inhibitors of early steps in their own biosynthesis.

B.4
THE AROMATIC AMINO ACIDS

The ultimate precursors of the carbon skeletons of the aromatic amino acids are phosphoenolpyruvate, which is a glycolytic intermediate, and erythrose 4-phosphate, which is an intermediate in the pentose phosphate pathway. The reaction of these two compounds eventually leads to the formation of shikimate, which in turn is converted to chorismate. The pathway for the synthesis of aromatic amino acids branches at chorismate; one branch leads to phenylalanine and tyrosine (Figure B.7), and the other branch leads to tryptophan. (Chorismate is also a precursor of folic acid, coenzyme Q [CoQ], and plastoquinone, but we shall not discuss those pathways.) Chorismate can be converted to phenylpyruvate (which leads to phenylalanine in a transamination reaction) or to p-hydroxyphenylpyruvate (which similarly leads to tyrosine by transamination). Phenylalanine can also be converted to tyrosine by a hydroxylation reaction.

The conversion of chorismate to tryptophan (Figure B.8) starts with the substitution of an amino group (which comes from the amide side chain of glutamine) on the ring system of chorismate, accompanied by the elimination of pyruvate, giving rise to anthranilate. Phosphoribosylpyrophosphate (PRPP) (see Chapter 17, Section 17.9 and Figure 17.21) reacts with anthranilate, eventually giving rise to indole-3-glycerol phosphate. Finally, serine reacts with indole-3-glycerol phosphate to produce tryptophan and glyceraldehyde 3-phosphate. This reaction is catalyzed by tryptophan synthase, an enzyme that requires pyridoxal phosphate. The most extensive studies of tryptophan synthase have been done in *Neurospora crassa* and in bacterial systems such as *Escherichia coli* and *Salmonella typhimurium*. Tryptophan is another example of an essential amino acid, since humans cannot synthesize it.

B.5
HISTIDINE BIOSYNTHESIS

Histidine is an essential amino acid for children, and it may be essential for adults as well; this last point has not been definitely established. The essential amino acids tend to have long and complex biosynthetic pathways, and histidine is no exception (Figure B.9). Most of the carbon skeleton of histidine comes from PRPP (Sections 17.9 and B.4). One carbon and one nitrogen of the imidazole ring come from ATP, and the second nitrogen of the imidazole ring comes from the side-chain amide of glutamine. The amino group comes from glutamate by the familiar transamination reaction. This pathway was established by studies in *E. coli* and *Salmonella*. Histidine is a feedback inhibitor of the first stage of its own biosynthesis.
(Text continues on page 550.)

The biosynthesis of phenylalanine and tyrosine.

FIGURE B.8

The conversion of chorismate to tryptophan in *E. coli*. (P) is a phosphate group.

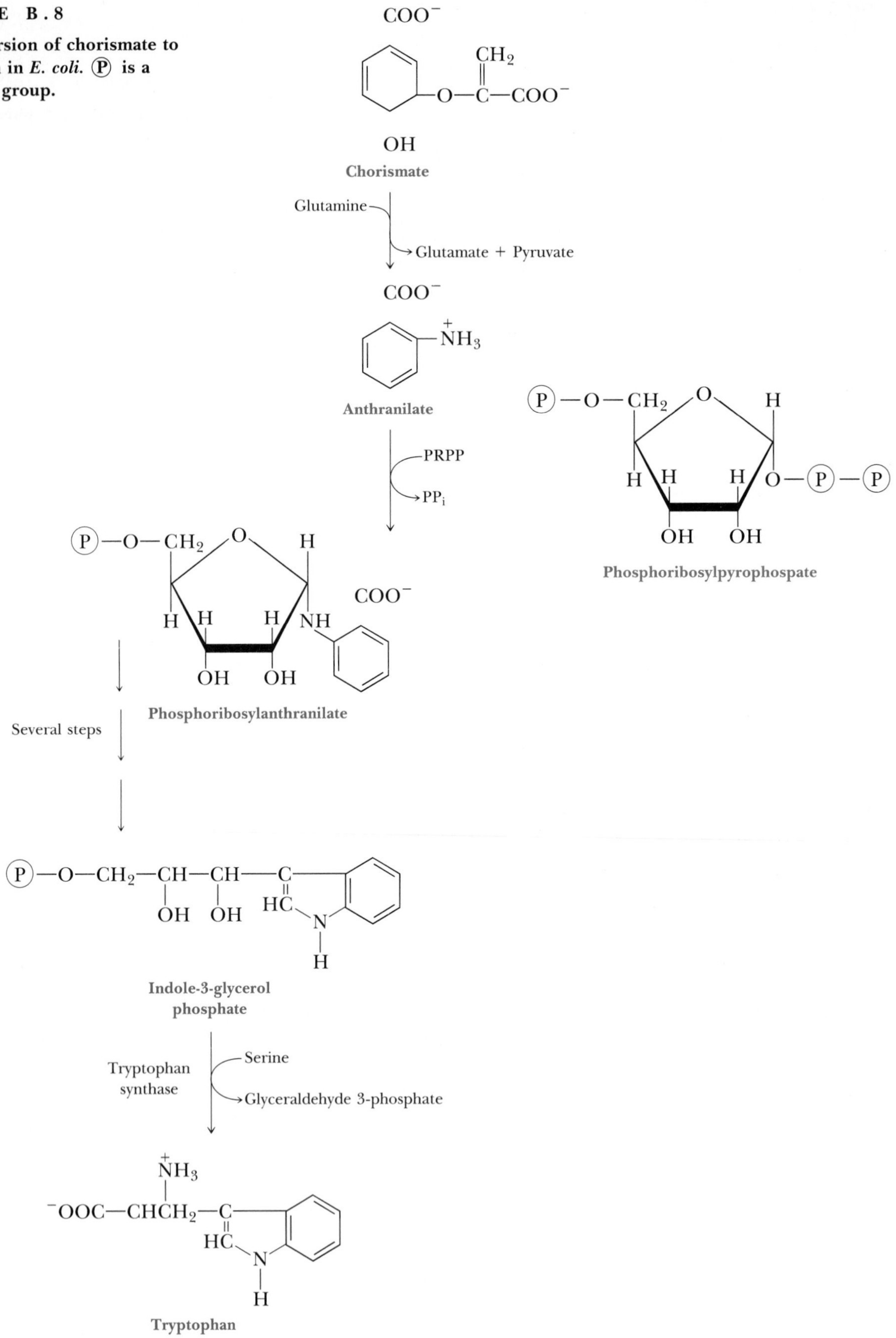

Chorismate

Glutamine

Glutamate + Pyruvate

Anthranilate

PRPP

PP$_i$

Phosphoribosylpyrophospate

Phosphoribosylanthranilate

Several steps

Indole-3-glycerol phosphate

Tryptophan synthase

Serine

Glyceraldehyde 3-phosphate

Tryptophan

548

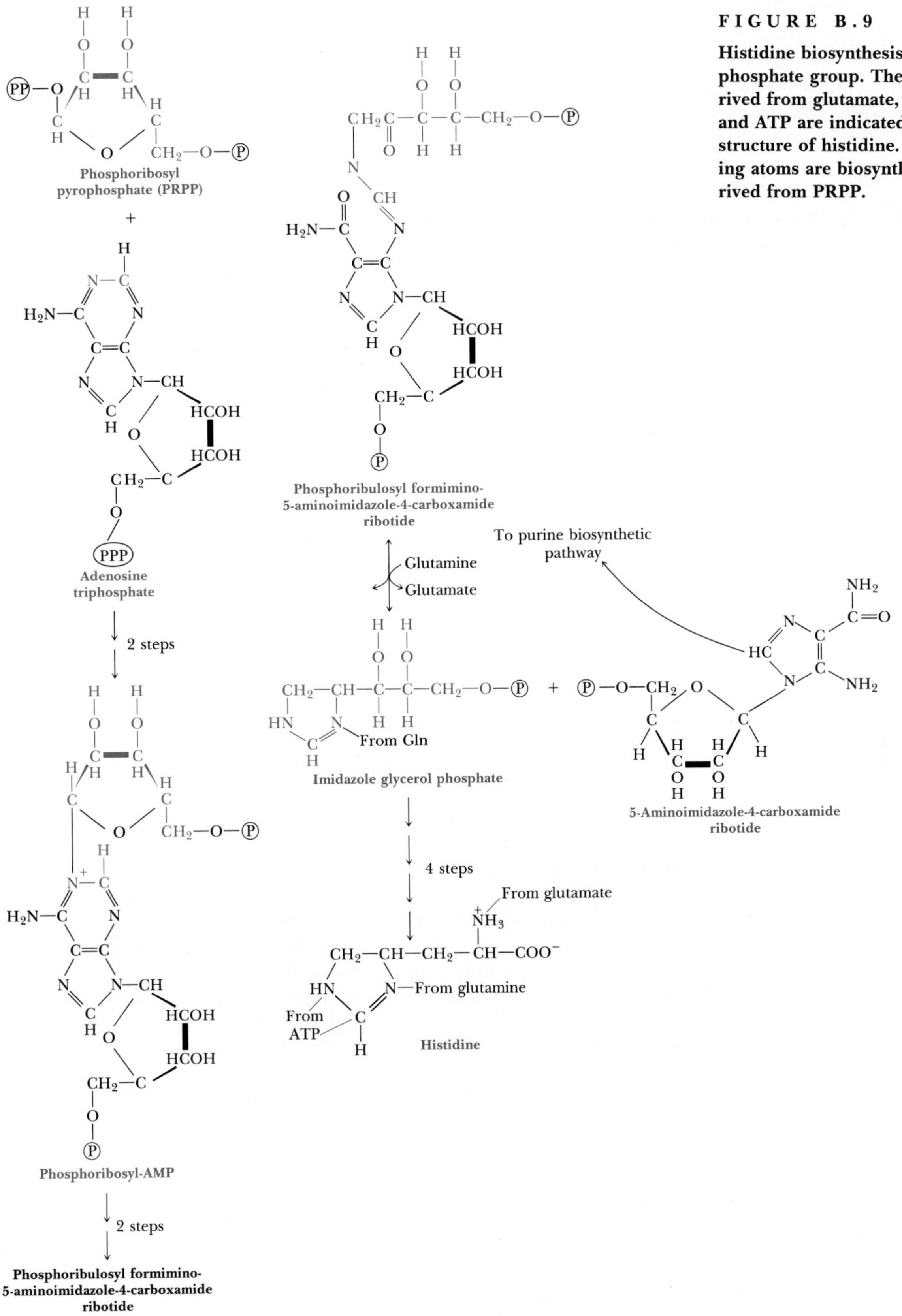

FIGURE B.9

Histidine biosynthesis. (P) is a phosphate group. The atoms derived from glutamate, glutamine, and ATP are indicated in the structure of histidine. The remaining atoms are biosynthetically derived from PRPP.

B.6
THE ANABOLISM OF PORPHYRINS

The biosynthesis of porphyrins and heme is shown in Figure B.10. Glycine and succinyl-CoA (an intermediate in the citric acid cycle) are the precursors of the porphyrin rings of hemes and chlorophylls. The first step of this pathway is a condensation reaction followed by the loss of CO_2, leading to the formation of δ-aminolevulinate. This step is inhibited by the presence of lead; the decreased synthesis of hemoglobin accounts for the anemia that can accompany lead poisoning. Two molecules of δ-aminolevulinate condense in turn, producing porphobilinogen, a molecule that contains a *pyrrole ring.* Four molecules of porphobilinogen condense further to form a **linear tetrapyrrole.** Three methylene bridges have been formed, and three ammonium ions have been released in the process. Finally, the linear tetrapyrrole cyclizes to form cyclic tetrapyrrole, **uroporphyrinogen III.** Once again, an ammonium ion is lost in the process. The remaining steps are those of modifying the side chains of the porphyrin ring to produce the form that occurs in hemoglobin. In uroporphyrinogen III, the side chains are acetate or propionate groups. The conversion from uroporphyrinogen III to protoporphyrin IX requires the decarboxylation of the acetate side chains to methyl groups and the modification of two of the four propionate side chains to vinyl groups. The addition of Fe^{2+} then produces heme.

Box B.1 explains how the deficient metabolism of amino acids, as well as other inborn errors of metabolism, can have serious consequences in affected persons.

B.7
THE ANABOLIC PATHWAY FOR IMP

The biosynthetic pathway for purine nucleotides starts with ribose 5-phosphate; the reactions of the series attach the growing purine ring to the sugar phosphate. The form in which ribose 5-phosphate is used in the synthesis of purine nucleotides is PRPP. We shall use this pathway as a case study showing all the steps in a long and complex biosynthesis (Figure B.11).

PRPP reacts with glutamine as the nitrogen source to produce 5-phospho-β-D-ribosylamine. Glycine is then attached to the phosphoribosylamine by forming an amide linkage between the carboxyl group of the glycine and the amino group that came from glutamine. The product of this reaction is 5′-phosphoribosylglycinamide. This reaction also requires ATP. The next step is the addition of a one-carbon formyl group to the free amino group of the glycine. The source of the formyl group is a tetrahydrofolate derivative, N^{10}-formyltetrahydrofolate. The compound produced by the formylation reaction is 5′-phosphoribosyl-N-formylglycinamide. The keto group of an amide (the linkage between the carboxyl group contributed by glycine and the amino group contributed by the first glutamine) is next converted to an imino group, making use of nitrogen from a second glutamine. In the terminology of organic chemistry, the amide is converted to an amidine. The amidine produced is 5′-phosphoribosyl-N-formylamidine. Both ATP and Mg^{2+} are required in this reaction. Finally, a

(Text continues on page 554.)

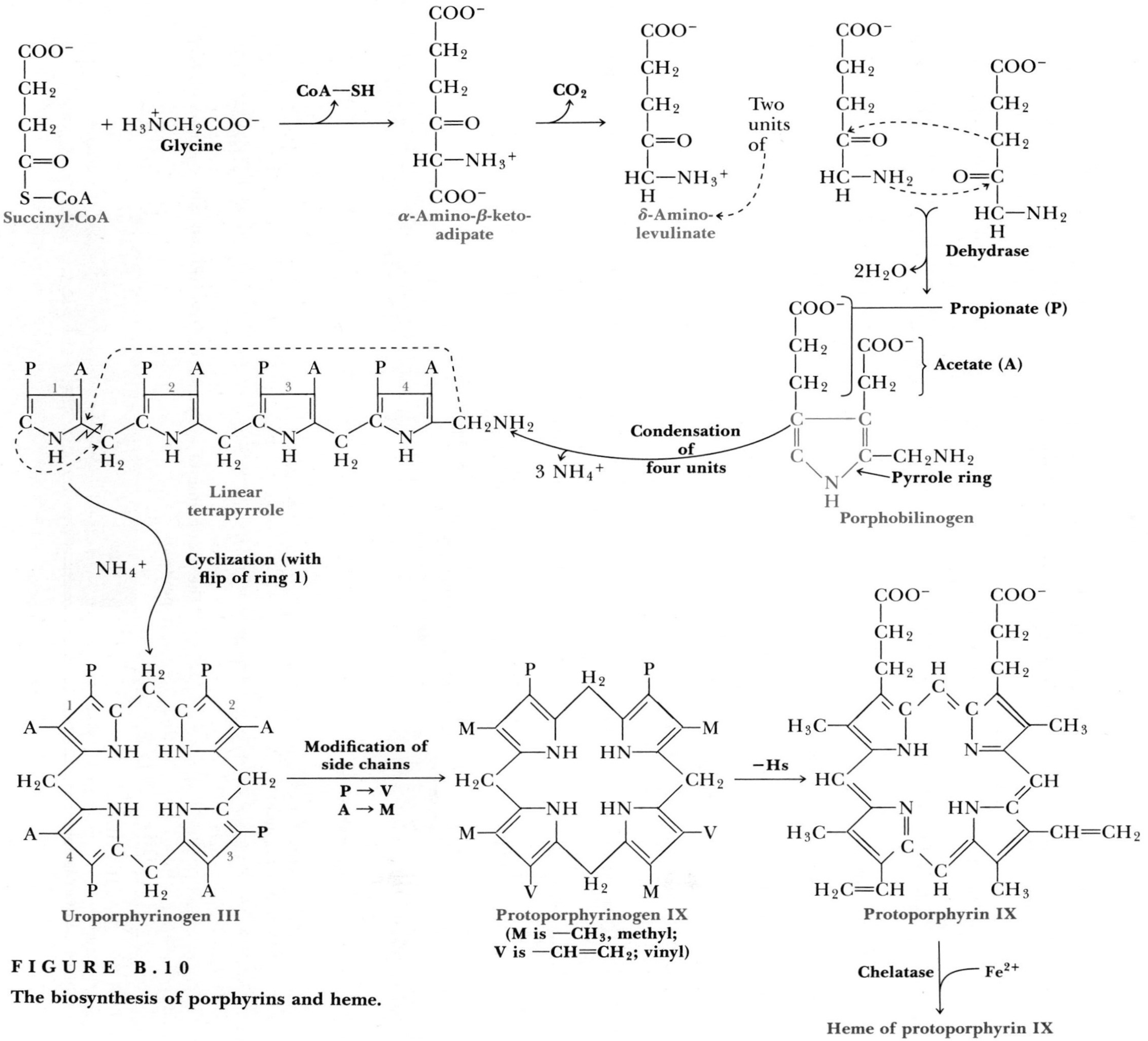

FIGURE B.10

The biosynthesis of porphyrins and heme.

Box

B.1

INBORN ERRORS OF METABOLISM INVOLVING
AMINO ACIDS AND THEIR DERIVATIVES

Mutations leading to deficiencies in enzymes that catalyze reactions of amino acid metabolism frequently have drastic consequences, many of them leading to severe forms of mental retardation. **Phenylketonuria** (PKU) is a well-known example. The enzyme that is missing in this case is phenylalanine hydroxylase, which is responsible for converting phenylalanine to tyrosine. Phenylalanine, phenylpyruvate, phenyllactate, and phenylacetate accumulate in the blood and urine. Available evidence suggests that phenylpyruvate, which is a phenyl ketone, causes mental retardation by interfering with the conversion of pyruvate to acetyl-CoA in the brain. Metabolic activity in the brain is reduced as a result. Fortunately, this condition can be detected easily in newborns, and the consequences can be avoided by keeping the child on a diet that is restricted in phenylalanine.

Albinism also results from the lack of an enzyme required for the metabolism of an aromatic amino acid. In this case the enzyme is **tyrosinase,** which is involved in the pathway by which tyrosine is converted to melanin, the material responsible for pigmentation of skin and hair. A temperature-sensitive form of tyrosinase is responsible for another easily visible effect on the pigmentation of an organism: the characteristic pattern of light and dark fur seen in Siamese cats. The tyrosinase found in these cats is inactive at their normal body temperature but is more active in parts of the body that have lower temperatures, such as the paws, tail, and ears. The more active enzyme leads to the production of melanin in these areas, and the fur is darker there.

Deficiencies in porphyrin metabolism can have severe neurological consequences. There are several classes of diseases, collectively known as **porphyrias,** associated with the lack of enzymes that catalyze reactions of porphyrins. In all cases the urine of the affected person is red; the neurological consequences can include a form of mania. King George III of England probably suffered from one type of porphyria, and his irrational behavior, which grew worse toward the end of his life, affected the course of history at the time of the American Revolution.

Tyrosine

Phenylalanine hydroxylase   **Enzyme deficiency in PKU**

Phenylalanine

Transaminase

Reactions involved in the development of phenylketonuria (PKU). A deficiency in the enzyme that catalyzes the conversion of phenylalanine to tyrosine leads to the accumulation of phenylpyruvate, a phenyl ketone.

—CH₂CCOO⁻ (O)

Phenylpyruvate
(a phenyl ketone)

| 2H⁺

OH
—CH₂CHCOO⁻

Phenyllactate

↳CO₂

—CH₂COO⁻

Phenylacetate

$$\text{Phenylpyruvate} \xrightarrow{2H^+}$$

Tyrosinase

HO———CH₂—C—COO⁻ (with H on top, NH₃⁺ below)

Tyrosine

→ → →

Melanin polymer

The production of melanin from tyrosine takes place in a multistep pathway.

Siamese cat.

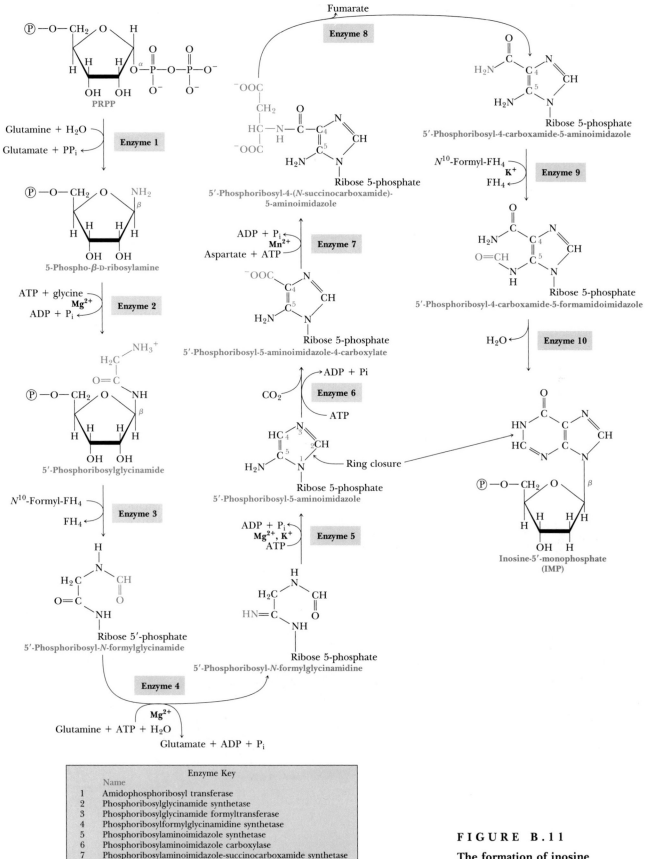

FIGURE B.11

The formation of inosine 5'-monophosphate starting with PRPP.

553

ring closure takes place to produce the five-membered imidazole ring of the growing purine skeleton. The ring closure takes place between the carbon that came from the formyl group and the glycosidic nitrogen (the one bonded to the ribose). Once again ATP is required, and the product formed is 5'-phosphoribosyl-5-aminoimidazole. (The numbering used on the five-membered ring at this point is that for an imidazole rather than that for a purine.)

Three of the six atoms of the six-membered ring, two carbons and a nitrogen, are already in place. Two carbon atoms and one nitrogen atom have yet to be added. The first of the two carbons comes from CO_2 in a carboxylation reaction. This reaction requires ATP. The product of the carboxylation reaction is 5'-phosphoribosyl-5-aminoimidazole-4-carboxylate (Figure B.11). The nitrogen atom to be added to the six-membered ring comes from aspartate in a two-step process. The carboxyl group just added to the growing purine reacts with the amino group of aspartate to give 5'-phosphoribosyl-4-(N-succinocarboxamide)-5-aminoimidazole. This reaction again requires ATP. Fumarate is then split off from the succinocarboxamide. The amino group is the only portion of the aspartate that will appear in the purine that is eventually formed. Addition of the final carbon atom is another one-carbon transfer involving N^{10}-formyltetrahydrofolate. The product of this reaction is 5'-phosphoribosyl-4-carboxamide-5-formamido-imidazole.

The last reaction is a ring closure that produces the six-membered ring of purines. The compound formed is inosine 5'-monophosphate (IMP), which is a purine nucleotide. Note that 5 ATP were required for the synthesis of the purine ring, in addition to the two "high-energy" bonds (in the form of pyrophosphate) needed for the formation of PRPP, giving a total of 7 ATP required for the production of IMP starting with ribose 5-phosphate.

Metabolism in Perspective

Crystals of ascorbic acid (vitamin C) viewed under polarized light.

In the process of *catabolism,* large biopolymers are broken down to smaller molecules, whereas in *anabolism,* small precursors are built up into larger molecules. The citric acid cycle, as the hub of metabolic pathways, serves to connect the breakdown and synthesis of proteins, carbohydrates, and lipids. Most of the metabolites of major nutrients can be fed into the citric acid cycle as acetyl-CoA and oxidized to produce energy. Alternatively, intermediates of the citric acid cycle can be converted to other biomolecules. Thus, the citric acid cycle is *amphibolic,* meaning that its action is both catabolic and anabolic. Overproduction of molecules in the metabolic pathways can be controlled by regulatory enzymes that act by feedback inhibition to limit the creation of new molecules. On the other hand, hormones can send signals to speed up their production. Under stress, the hypothalamus in the brain sends hormone-releasing factors to the pituitary gland, which initiates a hormone cascade that increases the production of glucose to provide the extra energy needed to deal with a stressful situation. The body is subjected to even greater stress when infections arise from invasion by disease-causing agents. The immune system deals with this situation on both the cellular and molecular levels. All aspects of biochemistry operate in concert, determining at the molecular level the responses that cells and whole organisms will make to the outside world.

18.1
ALL METABOLIC PATHWAYS ARE RELATED

In the preceding chapters we learned about a number of individual metabolic pathways. Some metabolites, such as pyruvate, oxaloacetate, and acetyl-CoA, appear in more than one pathway. Furthermore, reactions of metabolism can take place simultaneously, and it is important to consider control mechanisms by which some reactions and pathways are turned on and off. We shall now focus on some of the relationships among pathways by considering the related pathways themselves and some of the physiological responses to biochemical events.

The **citric acid cycle** plays a central role in metabolism. Three main points can be considered in assigning a central role to the citric acid cycle. The first of these is its part in the *catabolism* of nutrients of the main types: carbohydrates, lipids, and proteins. The second is the function of the citric acid cycle in the *anabolism* of sugars, lipids, and amino acids. The third and final point is the relationship between individual metabolic pathways and the citric acid cycle. When we discuss these broader considerations, we can and should address questions that involve more than individual cells and the reactions that go on in them, such as questions of what goes on in tissues and in whole organs. In this chapter we shall look at three such topics — nutrition, hormonal control, and the immune system.

18.2
THE CITRIC ACID CYCLE IN CATABOLISM

The nutrients taken in by an organism can include large molecules. This observation is especially true in the case of animals, which ingest polysaccharides and proteins, which are polymers, as well as lipids. Nucleic acids constitute a very small percentage of the nutrients present in foodstuffs, and we shall not consider their catabolism.

The first step in the breakdown of nutrients is the degradation of large molecules to smaller ones. Polysaccharides are hydrolyzed by specific enzymes to produce sugar monomers; an example is the breakdown of starch by amylases. Lipases hydrolyze triacylglycerols to give fatty acids and glycerol. Proteins are digested by proteases, with amino acids as the end-products. Sugars, fatty acids, and amino acids then enter their specific catabolic pathways.

In Chapter 11 we discussed the glycolytic pathway by which sugars are converted to pyruvate, which then enters the citric acid cycle. In Chapter 15 we saw how fatty acids are converted to acetyl-CoA; we learned about the fate of acetyl-CoA in the citric acid cycle in Chapter 12. Amino acids enter the cycle by various paths. We discussed catabolic reactions of amino acids in Chapter 17.

Figure 18.1 shows schematically the various catabolic pathways that feed into the citric acid cycle. The catabolic reactions occur in the cytosol; the citric acid cycle takes place in mitochondria. Many of the end-products of catabolism cross the mitochondrial membrane and then participate in the

FIGURE 18.1

A summary of catabolism, showing the central role of the citric acid cycle. Note that the end-products of the catabolism of carbohydrates, lipids, and amino acids all appear. (PEP is phosphoenolpyruvate; α-KG is α-ketoglutarate; TA is transamination; ⟶⟶⟶ is a multistep pathway.)

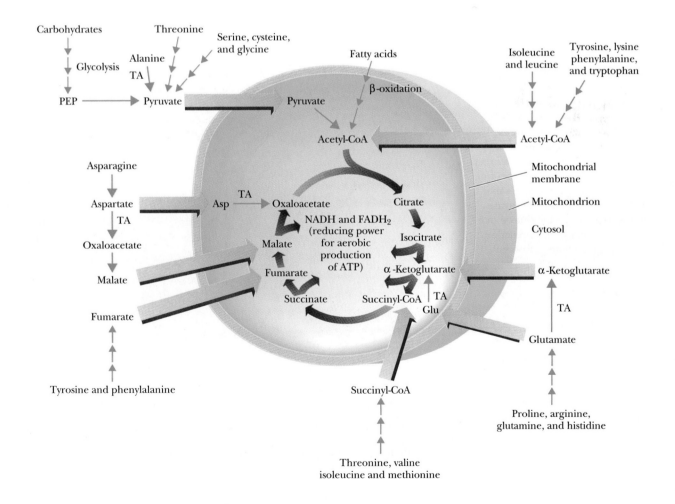

citric acid cycle. This figure also shows the outline of pathways by which amino acids are converted to components of the citric acid cycle. Be sure to notice that sugars, fatty acids, and amino acids are all included in this overall catabolic scheme.

18.3
THE CITRIC ACID CYCLE IN ANABOLISM

The citric acid cycle is a source of starting materials for the biosynthesis of many important biomolecules, but the supply of the starting materials that are components of the cycle must be replenished if the cycle is to continue operating. In particular, the oxaloacetate in an organism must be maintained at a level sufficient to allow acetyl-CoA to enter the cycle. In some organisms acetyl-CoA can be converted to oxaloacetate and other citric acid cycle intermediates, but mammals cannot do this. In mammals oxaloacetate is produced from pyruvate by the enzyme *pyruvate carboxylase* (Figure 18.2). We already encountered this enzyme and this reaction in the context of gluconeogenesis (see Oxaloacetate Is an Intermediate in the Production of Phosphoenolpyruvate in Gluconeogenesis, Chapter 14, Section 14.2), and here we have another highly important role for this enzyme and the reaction it catalyzes. This type of reaction is called **anaplerotic,** a word derived from

FIGURE 18.2

The necessity of anaplerotic reactions in mammals. An anabolic reaction uses a citric acid cycle intermediate (α-ketoglutarate is transaminated to glutamate in our example), competing with the rest of the cycle. The concentration of acetyl-CoA rises and signals the allosteric activation of pyruvate carboxylase to produce more oxaloacetate. (*Anaplerotic reaction. **Part of glyoxylate pathway.)

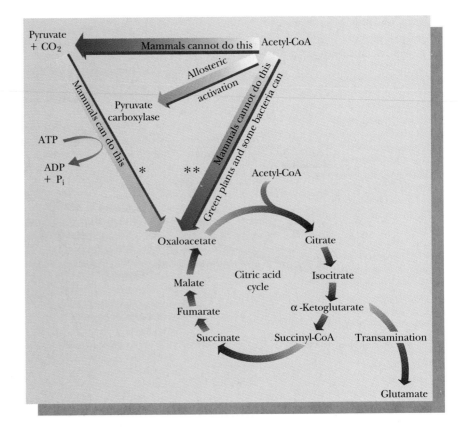

a Greek expression meaning "to fill up," since an anaplerotic reaction maintains an adequate supply of a metabolic intermediate. The supply of oxaloacetate would soon be depleted if there were no means of producing it from a readily available precursor.

This reaction, which produces oxaloacetate from pyruvate, provides a connection between the amphibolic citric acid cycle and the anabolism of sugars by gluconeogenesis. On this same topic of carbohydrate anabolism, we should note again that pyruvate cannot be produced from acetyl-CoA in mammals. Since acetyl-CoA is the end-product of catabolism of fatty acids, we can see that mammals could not exist with fats or acetate as the sole carbon source. The intermediates of carbohydrate metabolism would soon be depleted. Carbohydrates are the principal energy and carbon source in animals (Figure 18.2). Plants can carry out the conversion of acetyl-CoA to pyruvate and oxaloacetate, so they can exist without carbohydrates as a carbon source. The conversion of pyruvate to acetyl-CoA does take place in both plants and animals (see Conversion of Pyruvate to Acetyl-CoA, Chapter 12, Section 12.3).

The anabolic reactions of gluconeogenesis take place in the cytosol. Two mechanisms exist for the transfer of molecules needed for gluconeogenesis from mitochondria to the cytosol. One mechanism takes advantage of the fact that phosphoenolpyruvate can be formed from oxaloacetate in the mitochondrial matrix (this reaction is the next step in gluconeogenesis); phosphoenolpyruvate is then transferred to the cytosol, where the remaining reactions take place (Figure 18.3). Oxaloacetate is not transported across the mitochondrial membrane. The other mechanism relies on the fact

FIGURE 18.3

Transfer of the starting materials of gluconeogenesis from the mitochondrion to the cytosol. Note that phosphoenolpyruvate (PEP) can be transferred from the mitochondrion to the cytosol, as can malate. Oxaloacetate is not transported across the mitochondrial membrane. ((1) is PEP carboxykinase in mitochondria; (2) is PEP carboxykinase in cytosol; other symbols are as in Figure 18.1.)

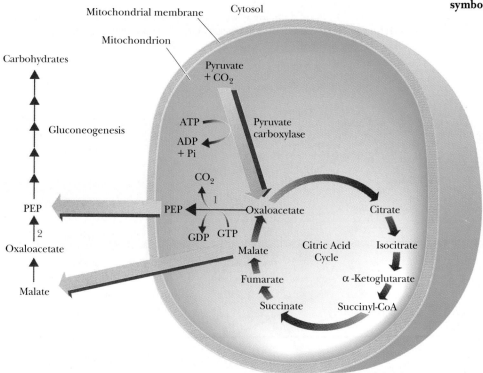

that malate, which is another intermediate of the citric acid cycle, can be transferred to the cytosol. There is a *malate dehydrogenase* enzyme in the cytosol as well as in mitochondria, and malate can be converted to oxaloacetate in the cytosol.

$$\text{Malate} + \text{NAD}^+ + \text{H}^+ \longrightarrow \text{oxaloacetate} + \text{NADH}$$

Oxaloacetate is then converted to phosphoenolpyruvate, leading to the rest of the steps of gluconeogenesis (Figure 18.3).

Gluconeogenesis has many steps in common with the production of glucose in photosynthesis, but photosynthesis also has many reactions in common with the pentose phosphate pathway. Thus, nature has evolved common strategies to deal with carbohydrate metabolism in all its aspects.

Lipid Anabolism

The starting point of lipid anabolism is acetyl-CoA. The anabolic reactions of lipid metabolism, like those of carbohydrate metabolism, take place in the cytosol; these reactions are catalyzed by soluble enzymes that are not bound to membranes. Acetyl-CoA is mainly produced in mitochondria, whether from pyruvate or from breakdown of fatty acids. It is not clear whether acetyl-CoA is directly transferred to the cytosol (Figure 18.4), but an indirect transfer mechanism does exist in which citrate is transferred to the

FIGURE 18.4

Transfer of the starting materials of lipid anabolism from the mitochrondrion to the cytosol. ((1) is ATP-citrate lyase; other symbols are as in Figure 18.1.) It is not definitely established whether acetyl-CoA is transported from the mito-chondrion to the cytosol.

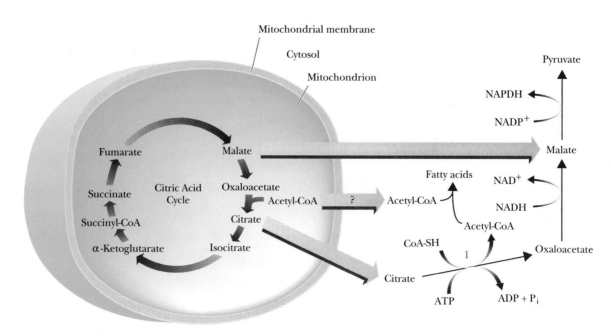

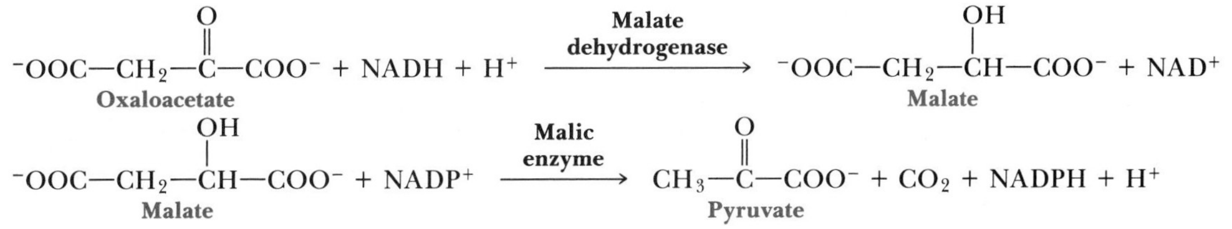

$$^-OOC-CH_2-\overset{\overset{\displaystyle O}{\|}}{C}-COO^- + NADH + H^+ \xrightarrow[\text{dehydrogenase}]{\text{Malate}} {}^-OOC-CH_2-\overset{\overset{\displaystyle OH}{|}}{CH}-COO^- + NAD^+$$

Oxaloacetate   Malate

$$^-OOC-CH_2-\overset{\overset{\displaystyle OH}{|}}{CH}-COO^- + NADP^+ \xrightarrow[\text{enzyme}]{\text{Malic}} CH_3-\overset{\overset{\displaystyle O}{\|}}{C}-COO^- + CO_2 + NADPH + H^+$$

Malate   Pyruvate

FIGURE 18.5

Reactions involving citric acid cycle intermediates that produce NADPH for fatty acid anabolism. Note that these reactions take place in the cytosol.

cytosol. Citrate reacts with CoA-SH to produce citryl-CoA, which is then cleaved to yield oxaloacetate and acetyl-CoA. The enzyme that catalyzes this reaction requires ATP and is called ATP-citrate lyase. The overall reaction is

Citrate + CoA-SH + ATP $\longrightarrow$ acetyl-CoA + oxaloacetate + ADP + P$_i$

Acetyl-CoA is the starting point for lipid anabolism in both plants and animals. (An important source of acetyl-CoA is the catabolism of carbohydrates. We have just seen that animals cannot convert lipids to carbohydrates, but they can convert carbohydrates to lipids. The efficiency of the conversion of carbohydrates to lipids in animals is a source of considerable chagrin to many humans.)

Oxaloacetate can be reduced to malate by the reverse of a reaction we saw in the last section in the context of carbohydrate anabolism.

Oxaloacetate + NADH + H$^+$ $\longrightarrow$ malate + NAD$^+$

Malate can move into and out of mitochondria by active transport processes, and the malate produced in this reaction can be used again in the citric acid cycle. However, malate need not be transported back into mitochondria but can be oxidatively decarboxylated to pyruvate by *malic enzyme*, which requires NADP$^+$.

Malate + NADP$^+$ $\longrightarrow$ pyruvate + CO$_2$ + NADPH + H$^+$

These last two reactions are a reduction reaction followed by an oxidation; there is *no net oxidation*. There is, however, a *substitution of NADPH for NADH*. This last point is an important one, since many of the enzymes of fatty acid synthesis require NADPH. The pentose phosphate pathway (Chapter 14, Section 14.4) is the principal source of NADPH in most organisms, but here we have another source as well (Figure 18.5).

The two ways of producing NADPH clearly indicate that all metabolic pathways are related. The malate-citryl-CoA shuttle is a control mechanism in lipid anabolism, whereas the pentose phosphate pathway is part of carbohydrate metabolism. Both carbohydrates and lipids are important energy sources in many organisms, particularly animals.

Anabolism of Amino Acids and Other Metabolites

The anabolic reactions that produce amino acids have, as a starting point, those intermediates of the citric acid cycle that can cross the mitochondrial membrane into the cytosol. We have already seen that malate can cross the

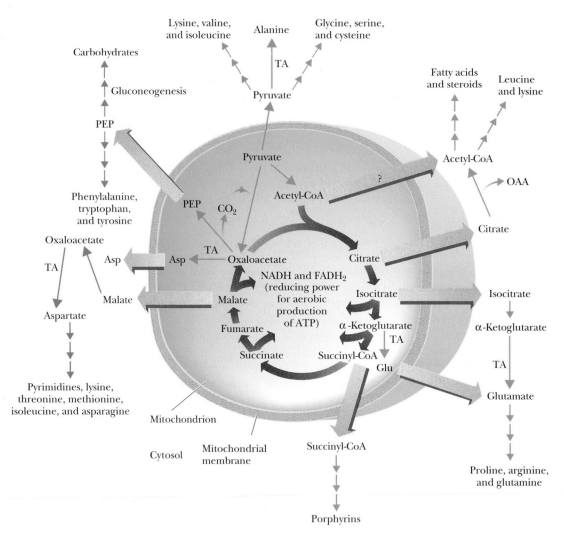

FIGURE 18.6

A summary of anabolism, showing the central role of the citric acid cycle. Note that there are pathways for the biosynthesis of carbohydrates, lipids, and amino acids. It is not definitely established whether acetyl-CoA is transported from the mitochondrion to the cytosol. OAA is oxaloacetate. Symbols are as in Figure 18.1.

mitochondrial membrane and give rise to oxaloacetate in the cytosol. Oxaloacetate can undergo a transamination reaction to produce aspartate, and aspartate in turn can undergo further reactions to form not only amino acids but also other nitrogen-containing metabolites, such as pyrimidines. Similarly, isocitrate can cross the mitochondrial membrane and produce α-ketoglutarate in the cytosol. Glutamate arises from α-ketoglutarate as a result of another transamination reaction, and glutamate undergoes further reactions to form still more amino acids. Succinyl-CoA is another citric acid cycle intermediate that can cross the mitochondrial membrane. It gives rise not to amino acids but to the porphyrin ring of the heme group.

The overall outline of anabolic reactions is shown in Figure 18.6. We used the same type of diagram in Figure 18.1 to show the overall outline of catabolism. The similarity of the two schematic diagrams points out that catabolism and anabolism, while not exactly the same, are closely related. The operation of any metabolic pathway, anabolic or catabolic, can be "speeded up" or "slowed down" in response to the needs of an organism by control mechanisms such as feedback control. Regulation of metabolism takes place in similar ways in many different pathways.

18.4
BIOCHEMISTRY AND NUTRITION

The molecules that we process by catabolic reactions ultimately come from outside the body because we are heterotrophic organisms (dependent on external food sources). We shall devote this section to a brief look at how the foods we eat are sources of substrates for catabolic reactions. We should also bear in mind that nutrition is related to physiology as well as to biochemistry. This last point is certainly appropriate in view of the fact that many early biochemists were physiologists by training.

Required Nutrients

In humans, the catabolism of **macronutrients** (carbohydrates, fats, and proteins) to supply energy is an important aspect of nutrition. In the United States, most diets provide more than an adequate number of nutritional calories. The typical American diet is high enough in fat that essential fatty acids (Chapter 15, Section 15.6) are seldom, if ever, deficient. The only concern is that the diet contain an adequate supply of protein. If the intake of protein is sufficient, the supply of essential amino acids (Chapter 17, Section 17.5) is normally sufficient as well. Packaging on food items frequently lists the protein content in terms of both the number of grams of protein and the percentage of the daily value (DV) suggested by the Food and Nutrition Board under the auspices of the National Research Council of the National Academy of Sciences (see Table 18.1). Daily values have replaced the recommended daily allowances (RDAs) formerly seen on food packaging.

 Micronutrients (vitamins and minerals) are also listed on food packaging. The vitamins we require are compounds that are necessary for metabolic processes; our bodies cannot synthesize them or cannot synthesize them in amounts sufficient for our needs. As a result, we must obtain vitamins from dietary sources. DVs are listed for the fat-soluble vitamins—vitamins A, D, and E (Chapter 8, Section 8.3)—but care must be taken to avoid overdoses of these vitamins. Overdoses can be toxic when excess amounts of fat-soluble vitamins accumulate in adipose tissue; excess vitamin A is especially toxic. With water-soluble vitamins, turnover is frequent enough that the danger of excess is not normally a problem. The water-soluble vitamins with listed DVs are vitamin C, necessary for the prevention of scurvy (Chapter 4, Section 4.3); and the B vitamins—niacin, pantothenic acid, vitamin B_6, riboflavin, thiamine, folic acid, biotin, and vitamin B_{12}. The B vitamins are the precursors of the metabolically important coenzymes listed in Table 5.2 in Chapter 5. In that table, references to the reactions in which the coenzymes play a role are given.

TABLE 18.1   **Daily Values for the Average Man and Woman, Aged 19 to 22**

NUTRIENT	MAN	WOMAN
Protein	56 g	44 g
Lipid-soluble vitamins		
Vitamin A	1 mg RE*	0.8 mg RE*
Vitamin D	7.5 μ†	7.5 μg†
Vitamin E	10 mg α-TE††	8 mg α-TE††
Water-soluble vitamins		
Vitamin C	60 mg	60 mg
Thiamine (vitamin B$_1$)	1.5 mg	1.1 mg
Riboflavin (vitamin B$_2$)	1.7 mg	1.3 mg
Vitamin B$_6$	3 μg	3 μg
Vitamin B$_{12}$	19 mg	14 mg
Niacin	3 μg	3 μg
Folic acid	19 mg	14 mg
Pantothenic acid (estimate)	10 mg	10 mg
Biotin (estimate)	0.3 mg	0.3 mg
Minerals		
Calcium	800 mg	800 mg
Phosphorus	800 mg	800 mg
Magnesium	350 mg	300 mg
Zinc	15 mg	15 mg
Iron	10 mg	18 mg
Copper (estimate)	3 mg	3 mg
Iodine	150 μg	150 μg

*RE = retinol equivalent, where 1 retinol equivalent = 1 μg retinol or 6 μg β-carotene. See Section 8.5.
†As cholecalciferol. See Section 8.5.
††α-TE = α-tocopherol equivalent, where 1 α-TE = 1 mg D-α-tocopherol. See Section 8.5. Data from the Food and Nutrition Board, National Academy of Sciences–National Research Council, Washington, D.C., 1988.

Minerals in the nutritional sense are inorganic substances required in the ionic or free element form for life processes. The macrominerals (those needed in the largest amounts) are sodium, potassium, chloride, magnesium, phosphorus, and calcium. The required amounts of all these minerals, except calcium, can easily be satisfied by a normal diet. Deficiencies of calcium can, and frequently do, occur. Such deficits lead to bone fragility, with concomitant risk of fracture, which is a problem especially for elderly women. Calcium supplements are indicated in such cases. Requirements for some microminerals (trace minerals) are not always clear. It is known, for example, from biochemical evidence that chromium is necessary for glucose metabolism and manganese for bone formation, but no deficiencies of these elements have been recorded. Requirements have been established for iron, copper, zinc, iodide, and fluoride; there are DVs for all these minerals except fluoride. In the case of copper and zinc, needs are easily met by dietary sources, and overdoses can be toxic. A deficiency of iodide, leading to an enlarged thyroid gland (Chapter 18, Section 18.5), has been a problem in some parts of the United States for many years. Fluoride is necessary to prevent tooth decay in children and, with that end in mind, has been added to water supplies, sometimes causing considerable controversy. Iron is important because it is part of the structure of the ubiquitous heme proteins. Women of childbearing age are more susceptible to iron deficiencies than are other segments of the population, and in some cases supplements are advised. These recommended levels vary with the age of the individual and are subject to adjustment for level of activity.

A more recent approach to publicizing healthful food selection has been the development of the Food Guide Pyramid, a graphic display that focuses on a diet sufficient in nutrients but without excesses. The goal was to use a well-chosen diet to promote good health. In order to avoid confusion, the development of this scheme had to take into account the fact that many people are somewhat familiar with the older recommendations about food groups. The newer recommendations pay particular attention to increasing the amount of fiber and decreasing the amount of fat in the typical diet. Variety and moderation were key concepts of the graphic presentation. From the biochemical point of view, these recommendations translate into a diet based primarily on carbohydrates, with enough protein to meet needs for essential amino acids (Chapter 17, Section 17.5). Lipids should not contribute more than 30% of daily calories, but the typical American diet currently is about 45% fat. High-fat diets have been linked to heart disease and to some kinds of cancer, so the recommendation about lipid intake is of considerable importance. (See the article by Willett in the bibliography at the end of this chapter for more on this topic.)

Food Guide Pyramid
A Guide to Daily Food Choices

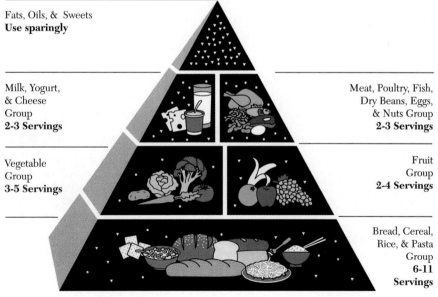

Fats, Oils, & Sweets
Use sparingly

Milk, Yogurt,
& Cheese
Group
2-3 Servings

Meat, Poultry, Fish,
Dry Beans, Eggs,
& Nuts Group
2-3 Servings

Vegetable
Group
3-5 Servings

Fruit
Group
2-4 Servings

Bread, Cereal,
Rice, & Pasta
Group
**6-11
Servings**

Key
•Fat (naturally occurring and added)   ▼ Sugars (added)
These symbols show fats, oils, and added sugars in foods.

USDA, 1992

The Food Guide Pyramid. The recommended choices reflect a diet primarily based on carbohydrates. Smaller amounts of proteins and lipids are sufficient to meet the body's needs.

18.5
HORMONES AND SECOND MESSENGERS

The metabolic processes within a given cell can be, and frequently are, regulated by signals from outside the cell. A usual means of intercellular communication takes place through the workings of the **endocrine system,** in which the ductless glands produce **hormones** as intercellular messengers. Hormones are transported from the sites of their synthesis to the sites of action by the bloodstream (Figure 18.7). In terms of their chemical nature, some typical hormones are steroids, such as estrogens, androgens, and mineralocorticoids (Chapter 15, Section 15.8); polypeptides, such as insulin and endorphins (Chapter 3, Section 3.5); and amino acid derivatives, such as epinephrine and norepinephrine (Table 18.2).

Hormones have several important functions in the body. They help to maintain **homeostasis,** the balance of biological activities in the body. The effect of insulin in keeping the blood glucose level within narrow limits is an example of this function. The operation of epinephrine and norepinephrine in the "fight or flight" syndrome is an example of the way in which hormones mediate response to external stimuli. Finally, hormones play roles in growth and development, as seen in the roles of growth hormone and the sex hormones. The methods and insights of both biochemistry and physiology help illuminate the workings of the endocrine system.

The release of hormones exerts control on the cells of target organs; other control mechanisms in turn determine the workings of the endocrine gland that releases the hormone in question. Simple feedback mechanisms can be postulated, in which the action of the hormone leads to feedback

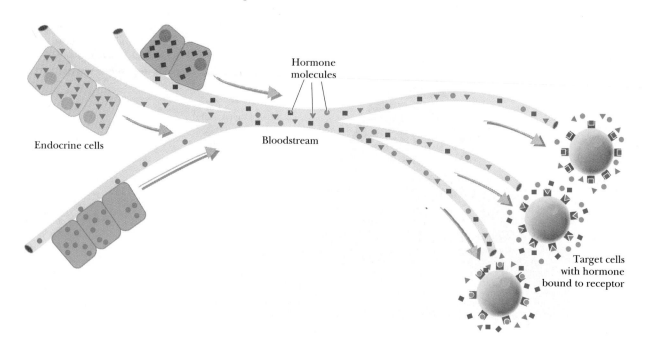

FIGURE 18.7

Endocrine cells secrete hormone into the bloodstream, which transports them to target cells.

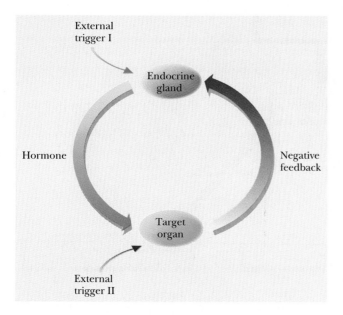

External
trigger I

Endocrine
gland

Hormone

Negative
feedback

Target
organ

External
trigger II

FIGURE 18.8

A simple feedback control system involving an endocrine gland and a target organ.

inhibition of the release of hormone (Figure 18.8). The workings of the endocrine system are in fact much less simple, with the added complexity allowing for a greater degree of control. To take a rather restricted example, insulin is released in response to a rapid rise in the level of blood glucose. In the absence of control mechanisms, an excess of insulin can produce **hypoglycemia,** the condition of low blood glucose. In addition to negative feedback control on the release of insulin, the action of the hormone glucagon tends to increase the level of glucose in the bloodstream. The two hormones together regulate blood glucose.

A more sophisticated control system involves the action of the *hypothalamus,* the *pituitary,* and specific *endocrine* glands (Figure 18.9). The central nervous system sends a signal to the hypothalamus. The **hypothalamus** secretes a hormone-releasing factor, which in turn stimulates release of a trophic hormone by the anterior pituitary (Table 18.2). (The action of the hypothalamus on the posterior pituitary is mediated by nerve impulses.) **Trophic hormones** act on specific **endocrine glands,** which release the hormones to be transported to target organs. Note that feedback control is exerted at every stage of the process. Even more fine-tuning is possible with zymogen activation mechanisms (Chapter 5, Section 5.10), which exist for many well-known hormones.

The releasing factors and trophic hormones listed in Table 18.2 tend to be polypeptides, but the chemical natures of the hormones released by specific endocrine glands show greater variation. Thyroxine, for example, produced by the thyroid, is an iodinated derivative of the amino acid tyrosine (Chapter 3, Section 3.2). Abnormally low levels of thyroxine lead to **hypothyroidism,** characterized by lethargy and obesity, whereas increased levels produce the opposite effect (**hyperthyroidism**). Low levels of iodine

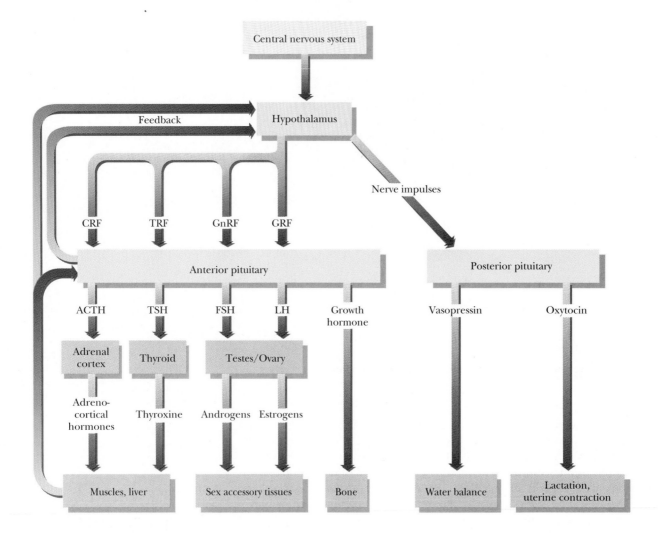

FIGURE 18.9

Hormonal control systems showing the role of the hypothalamus, pituitary, and target tissues. See Table 18.2 for the names of the hormones.

in the diet often lead to hypothyroidism and an enlarged thyroid gland (**goiter**). This condition has largely been eliminated by the addition of sodium iodide to commercial table salt ("iodized" salt). (It is virtually impossible to find table salt that is not iodized.)

Steroid hormones (Chapter 15, Section 15.8) are produced by the adrenal cortex and the gonads (testes in males, ovaries in females). The **adrenocortical hormones** include **glucocorticoids,** which affect carbohydrate metabolism, modulate inflammatory reactions, and are involved in reactions to stress. The **mineralocorticoids** control the level of excretion of water and salt by the kidney. If the adrenal cortex does not function adequately, one result is **Addison's disease,** characterized by hypoglycemia, weakness, and increased susceptibility to stress. This disease is eventually fatal unless treated by administration of mineralocorticoids and glucocorticoids to make up for what is missing. The opposite condition, *hyperfunction,* is frequently caused by a tumor of the adrenal cortex or of the pituitary. The characteristic clinical manifestation is **Cushing's syndrome,** marked by hyperglycemia, water retention, and the easily recognized "moon face."

TABLE 18.2 **Selected Human Hormones**

	SOURCE	MAJOR EFFECTS
Polypeptides		
Corticotropin-releasing factor (CRF)	Hypothalamus	Stimulates release of ACTH
Gonadotropin-releasing factor (GnRF)	Hypothalamus	Stimulates FSH and LH release
Thyrotropin-releasing factor (TRF)	Hypothalamus	Stimulates TSH release
Growth hormone-releasing factor (GRF)	Hypothalamus	Stimulates growth hormone release
Adrenocorticotropic hormone (ACTH)	Anterior pituitary	Stimulates release of adrenocorticosteroids
Thyrotropin (TSH)	Anterior pituitary	Stimulates thyroxine release
Follicle-stimulating hormone (FSH)	Anterior pituitary	In ovaries, stimulates ovulation and estrogen synthesis; in testes, stimulates spermatogenesis
Luteinizing hormone (LH)	Anterior pituitary	In ovaries, stimulates estrogen and progesterone synthesis; in testes, stimulates androgen synthesis
Met-enkephalin	Anterior pituitary	Has opioid effects on central nervous system
Leu-enkephalin	Anterior pituitary	Has opioid effects on central nervous system
β-Endorphin	Anterior pituitary	Has opioid effects on central nervous system
Vasopressin	Posterior pituitary	Stimulates water resorption by kidney and raises blood pressure
Oxytocin	Posterior pituitary	Stimulates uterine contractions and flow of milk
Insulin	Pancreas (β-cells of islets of Langerhans)	Stimulates uptake of glucose from bloodstream
Glucagon	Pancreas (α-cells of islets of Langerhans)	Stimulates release of glucose to bloodstream
Steroids		
Glucocorticoids	Adrenal cortex	Decrease inflammation, increase resistance to stress
Mineralocorticoids	Adrenal cortex	Maintain salt and water balance
Estrogens	Gonads and adrenal cortex	Stimulate development of secondary sex characteristics, particulary in females
Androgens	Gonads and adrenal cortex	Stimulate development of secondary sex characteristics, particularly in males
Amino acid derivatives		
Epinephrine	Adrenal medulla	Increases heart rate and blood pressure
Norepinephrine	Adrenal medulla	Decreases peripheral circulation, stimulates lipolysis in adipose tissue
Thyroxine	Thyroid	Stimulates metabolism generally

The adrenal cortex produces some steroid sex hormones, the *androgens* and *estrogens,* but the main site of production is the gonads. Estrogens are required for female sexual maturation and function, but not for embryonic sexual development of female mammals. Information on sexual differentiation related to the actions of steroid hormones has been obtained from studies of chromosome abnormalities in human patients of infertility clinics. While normal males have the XY genotype (have X and Y chromosomes) and normal females exhibit the XX genotype (have two X chromosomes), there are XXY and XO (missing chromosome) individuals who are phenotypic (external-appearing) males and phenotypic females, respectively, but who are sterile. Extremely rare individuals are XX males and XY females. An XX male has a small segment of a normal Y chromosome attached to one of the

X chromosomes; an XY female has a Y chromosome that is missing this segment. A protein encoded by this part of the Y chromosome, the **testis-determining factor** (TDF), controls development of the undifferentiated embryonic gonads. In the presence of this factor, the gonads develop as testes; in its absence, they develop as ovaries. If a genotypic male animal has his gonads surgically removed, he develops as a phenotypic female. Embryonic mammals develop as phenotypic females in the absence of male sex hormones.

As a final example, we shall discuss growth hormone (GH), which is a polypeptide. When overproduction of GH occurs, it is usually because of a pituitary tumor. If this condition occurs while the skeleton is still growing, the result is **giantism.** If the skeleton has stopped growing before the onset of GH overproduction, the result is **acromegaly,** characterized by enlarged hands, feet, and facial features. Underproduction of GH leads to **dwarfism,** but this condition can be treated by the injection of human GH before the skeleton reaches maturity. Animal GH is ineffective in treating dwarfism in humans. Supplies of human GH were very limited when it could be obtained only from cadavers, but it can now be synthesized by recombinant DNA techniques. (Another discussion of peptide hormones can be found in Box 5.1 in Chapter 5, which treats oxytocin and vasopressin.)

Second Messengers

When a hormone binds to the specific receptor for it on a target cell, it sets off a chain of events in which the actual response within the cell is elicited. Several kinds of receptors are known. The receptors for steroid hormones tend to occur within the cell rather than as part of the membrane (steroids can pass the plasma membrane); steroid–receptor complexes affect the transcription of specific proteins. More frequently, the receptor proteins are a part of the plasma membrane. Binding of hormone to the receptor triggers a change in concentration of a second messenger. The **second messenger** brings about the changes within the cell as a result of a series of reactions.

Cyclic AMP (adenosine-3', 5'-cyclic phosphate, cAMP) is one example of a second messenger. The mode of action starts with binding of a hormone

Cyclic AMP

Cyclic AMP (adenosine-3',5'-cyclic phosphate, cAMP).

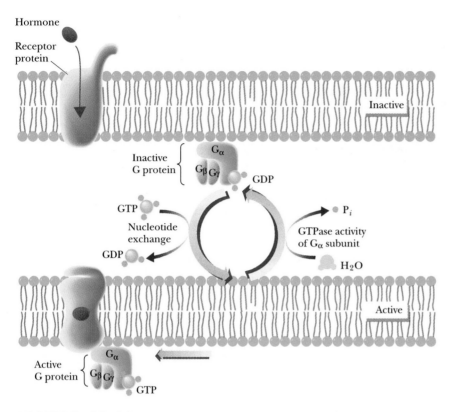

FIGURE 18.10

Activation of the G protein. The inactive G protein is a trimer, consisting of α, β, and γ subunits, with GDP bound to the α subunit. In the process of activation, GTP binds to the α subunit, replacing the GDP that is already there. The system returns to its inactive state when the GTPase activity of the α subunit hydrolyzes GTP to GDP.

to a specific receptor, which triggers the production of cAMP from ATP, catalyzed by *adenylate cyclase*. This reaction is mediated by a stimulatory G protein, a trimer consisting of three subunits — α, β, and γ. Binding of the hormone to the receptor activates the G protein, the α subunit binds GTP, giving rise to the name of the protein. The active protein has GTPase activity and hydrolyzes GTP, returning the G protein to the inactive state. GDP remains bound to the α subunit and must be exchanged for GTP when the protein is activated the next time (Figure 18.10). The G protein and adenylate cyclase are bound to the plasma membrane, while cAMP is released into the interior of the cell to act as a second messenger. Some examples are known in which the binding of hormone to receptor inhibits rather than stimulates adenylate cyclase. A G protein with a different kind of α subunit mediates the process. The modified G protein is referred to as an inhibitory G protein to distinguish it from the kind that stimulates response to hormone binding.

In eukaryotic cells the usual mode of action of cAMP is to stimulate a cAMP-dependent protein kinase, a tetramer consisting of two regulatory subunits and two catalytic subunits. When cAMP binds to the dimer of

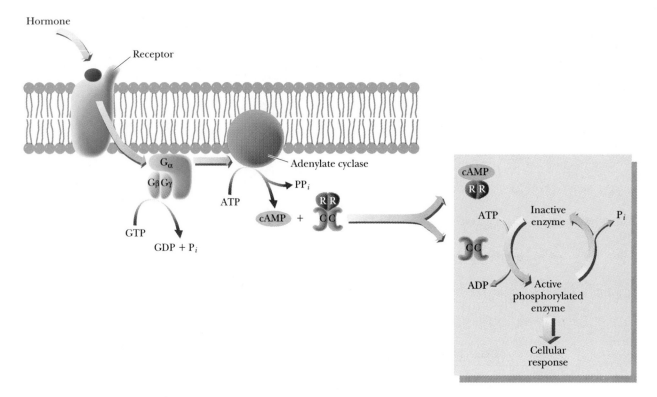

FIGURE 18.11

The activation of adenylate cyclase by the binding of hormone to the receptor and the mode of action of cAMP. The binding of hormone to the receptor leads to the production of cAMP from ATP, catalyzed by adenylate cyclase; this reaction is mediated by a G protein. Once cAMP is formed, it stimulates a protein kinase by binding to the regulatory subunits. The active catalytic subunits are released and catalyze the phosphorylation of a target enzyme. The target enzyme elicits the response of the cell to the hormonal signal. This scheme applies in situations in which phosphorylation activates the target enzyme.

regulatory subunits, the two active catalytic subunits are released. The active kinase catalyzes the phosphorylation of some target enzyme (Figure 18.11). (In the scheme shown in Figure 18.11, phosphorylation activates the enzyme. Cases are also known in which phosphorylation inactivates a target enzyme, e.g., glycogen synthase [Chapter 14, Section 14.1].) The usual site of phosphorylation is the hydroxyl group of a serine or a threonine. ATP is the source of the phosphate group transferred to the enzyme. The target enzyme then elicits the cellular response.

The G protein is permanently activated by cholera toxin, leading to excessive stimulation of adenylate cyclase and chronic elevation of cAMP levels. The main danger in **cholera,** caused by the bacterium *Vibrio cholerae,* is severe dehydration as a result of diarrhea. The unregulated activity of adenylate cyclase in epithelial cells leads to diarrhea, since cAMP in epithelial cells stimulates active transport of Na^+. Excessive cAMP in epithelial cells produces a large flow of Na^+ and water from the epithelial cells to the intestines. If the lost fluid and salts can be replaced in cholera victims, the immune system can deal with the actual infection within a few days.

Calcium ion (Ca^{2+}) is involved in another ubiquitous second-messenger scheme. Much of the calcium-mediated response depends on release of Ca^{2+} from intracellular reservoirs, similar to the release of Ca^{2+} from the sarcoplasmic reticulum in the action of the neuromuscular junction (Chapter 8, Section 8.7). A component of the inner layer of the phospholipid bilayer, *phosphatidylinositol 4,5-bisphosphate* (PIP$_2$), is also required in this scheme (Figure 18.12).

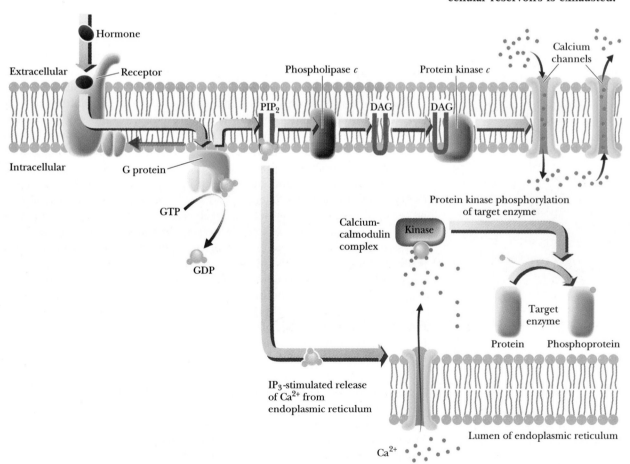

R₁ and R₂ = fatty acid residues

P = phosphate moiety

Phosphatidylinositol 4,5-bisphosphate (PIP₂)

Diacylglycerol (DAG) moiety

Inositol 1,4,5-triphosphate (IP₃) moiety

FIGURE 18.12

The PIP₂ second messenger scheme. When a hormone binds to a receptor, it activates phospholipase C, in a process mediated by a G protein. Phospholipase C hydrolyzes PIP₂ to IP₃ and DAG. IP₃ stimulates the release of Ca²⁺ from intracellular reservoirs in the ER. A complex formed between Ca²⁺ and the calcium-binding protein calmodulin activates a cytosolic protein kinase for phosphorylation of a target enzyme. DAG remains bound to the plasma membrane, where it activates the membrane-bound protein kinase C (PKC). PKC is involved in the phosphorylation of a number of target enzymes. PKC also phosphorylates channel proteins that control the flow of Ca²⁺ in and out of the cell. Ca²⁺ from extracellular sources can produce sustained responses even when the supply of Ca²⁺ in intracellular reservoirs is exhausted.

When the external trigger binds to its receptor on the cell membrane, it activates *phospholipase C* (Chapter 15, Section 15.2), which hydrolyzes PIP$_2$ to *inositol 1,4,5-triphosphate* (IP$_3$) and a *diacylglycerol* (DAG) in a process mediated by a different member of the family of G proteins. The IP$_3$ is the actual second messenger. It diffuses through the cytosol to the endoplasmic reticulum (ER), where it stimulates the release of Ca^{2+}. A complex is formed between the calcium-binding protein calmodulin and Ca^{2+}. This calcium–calmodulin complex activates a cytosolic protein kinase, which phosphorylates target enzymes in the same fashion as in the cAMP second-messenger scheme. DAG also plays a role in this scheme; it is nonpolar and diffuses through the plasma membrane. When DAG encounters the membrane-bound protein kinase C, it too acts as a second messenger by activating this enzyme (actually a family of enzymes). **Protein kinase C** also phosphorylates target enzymes, including channel proteins that control the flow of Ca^{2+} into and out of the cell. By controlling the flow of Ca^{2+}, this second-messenger system can produce sustained responses even when the supply of Ca^{2+} in the intracellular reservoirs becomes exhausted. (For more information on this point, see the article by Rasmussen listed in the bibliography at the end of the chapter.)

18.6
HORMONAL CONTROL IN METABOLISM

Now that we know something about the effects of hormones in triggering responses within the cell, we can return to and expand on some earlier points about metabolic control. In Chapter 14, Section 14.3, we discussed some points about control mechanisms in carbohydrate metabolism. We saw at that time how glycolysis and gluconeogenesis can be regulated and how glycogen synthesis and breakdown can respond to the body's needs. Phosphorylation and dephosphorylation of the appropriate enzymes played a large role there, and that whole scheme is subject to hormonal action.

Three hormones play a part in the regulation of carbohydrate metabolism: epinephrine, glucagon, and insulin. Epinephrine acts on muscle tissue to raise levels of glucose on demand, while glucagon acts on the liver, also to increase availability of glucose. Feedback control plays a role in the process and ensures that the amount of glucose made available does not reach an excessive level (Section 18.5). The role of insulin is to trigger the feedback response that achieves this further control.

Epinephrine (also called adrenalin) is structurally related to the amino acid tyrosine. This compound is released from the adrenal glands in response to stress (the "fight or flight" response). When it binds to specific receptors, it sets off a chain of events that leads to increased levels of glucose in the blood. Glucagon (a peptide that contains 29 amino acid residues) is released by the α-cells of the islets of Langerhans in the pancreas, and it too binds to specific receptors to set off a chain of events to make glucose available to the organism. Each time a single hormone molecule, whether epinephrine or glucagon, binds to its specific receptor, it activates a number of stimulatory G proteins. This effect starts an amplification of the hormonal signal. Each active G protein in turn stimulates adenylate cyclase several times before the G protein is inactivated by its own GTPase activity, leading

NH$_3$
|
CH—COO
|
CH$_2$

OH
Tyrosine

CH$_3$
|
NH$_2$
|
CH$_2$
|
CHOH

HO

OH
Epinephrine

Tyrosine and epinephrine. The hormone epinephrine is metabolically derived from the amino acid tyrosine.

to still more amplification. The cAMP produced by the increased adenylate cyclase activity allows for increased activity of the cAMP-dependent protein kinase, phosphorylating target enzymes that lead to increased glucose levels. In particular, this means an increase in the activity of the enzymes involved in gluconeogenesis and glycogen breakdown as well as a decrease in the activity of enzymes involved in glycolysis and glycogen synthesis. The series of amplifying steps is called a **cascade,** and the cumulative effect is the underlying reason why small amounts of hormones can have such strongly marked effects.

Figure 18.13 shows how the binding of epinephrine to specific receptors leads to increased glycogen breakdown in muscle and suppression of glycogen synthesis. The hormonal stimulation leads to activation of adenylate cyclase, which in turn activates the cAMP-dependent protein kinase responsible for activating glycogen phosphorylase and inactivating glycogen synthase.

The effect of glucagon binding to receptors in stimulating gluconeogenesis in the liver and suppressing glycolysis depends on changes in the

FIGURE 18.13

When epinephrine binds to its receptor, the binding activates a stimulatory G protein, which in turn activates adenylate cyclase. The cAMP thus produced activates a cAMP-dependent protein kinase. The phosphorylation reactions catalyzed by the cAMP-dependent kinase suppress the activity of glycogen synthase and enhance that of phosphorylase kinase. Glycogen phosphorylase is activated by phosphorylase kinase, leading to glycogen breakdown.

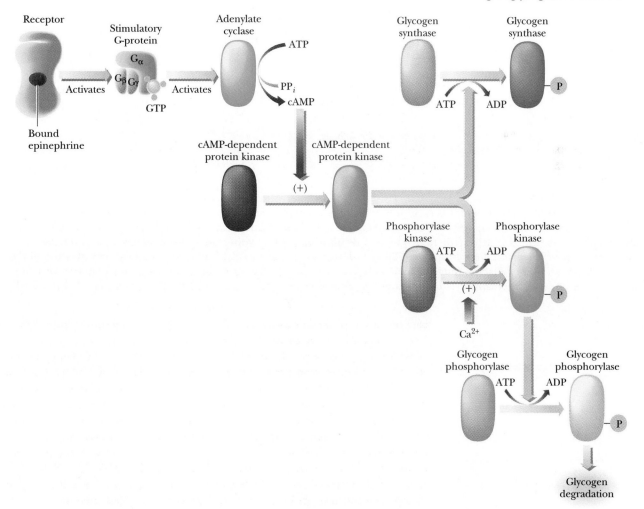

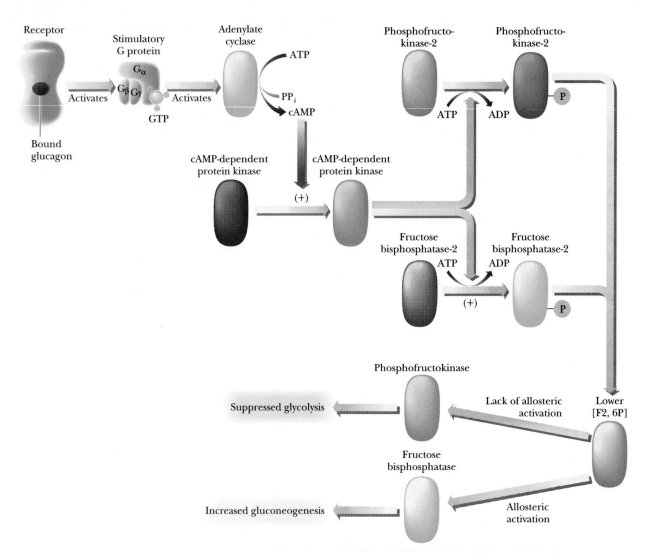

FIGURE 18.14

Binding of glucagon to its receptor sets off the chain of events that leads to the activation of a cAMP-dependent protein kinase. The enzymes phosphorylated in this case are phosphofructokinase-2, which is inactivated, and fructose *bis*phosphatase-2, which is activated. The combined result of phosphorylating these two enzymes is to lower the concentration of fructose 2,6-*bis*phosphate (F2,6P). A lower concentration of F2,6P leads to allosteric activation of the enzyme fructose *bis*phosphatase, thus enhancing gluconeogenesis. At the same time, the lower concentration of F2,6P implies that phosphofructokinase is lacking a potent allosteric activator, with the result that glycolysis is suppressed.

concentration of the key allosteric effector fructose 2,6-*bis*phosphate. Recall from Chapter 14, Section 14.3, that this compound is an important allosteric activator of phosphofructokinase, the key enzyme of glycolysis; it is also an inhibitor of fructose *bis*phosphate phosphatase, which plays a role in gluconeogenesis. A high concentration of F2,6P stimulates glycolysis, whereas a low concentration stimulates gluconeogenesis. The concentration of F2,6P in a cell depends on the balance between its synthesis (catalyzed by phosphofructokinase-2 [PFK-2]) and its breakdown (catalyzed by fructose *bis*phosphatase-2 [FBPase-2]). The enzymes that control the formation and breakdown of F2,6P are themselves controlled by a phosphorylation/de-phosphorylation mechanism, which in turn is subject to the same sort of hormonal control we just discussed for the enzymes of glycogen metabolism. Figure 18.14 summarizes the chain of events that leads to increased gluconeogenesis in the liver as a result of the binding of glucagon to its specific receptor.

18.7
THE IMMUNE SYSTEM: THE BODY'S DEFENSES

The distinctive characteristic of the immune system is the ability to *distinguish self from nonself.* It is this ability that enables the cells and molecules responsible for immunity to recognize and destroy pathogens (disease-causing agents) such as viruses and bacteria when they invade the body. Since infectious diseases can be fatal, the operation of the immune system can be a matter of life or death. Striking confirmation of this last point is apparent in the lives of those who have **AIDS** (acquired immune deficiency syndrome). This disease so weakens the immune system that those who suffer from it become prey to infections that proceed unchecked, with ultimately fatal consequences. Suppression of the immune system can save lives as well as take them. The development of drugs that suppress the immune system has made *organ transplants* possible. Recipients of hearts, lungs, kidneys, or livers tolerate the transplanted organs without rejecting them because these drugs have thwarted the way in which the immune system tries to attack the grafts; however, the immune suppression also makes them more susceptible to infections than they would be otherwise.

It is also possible for the immune system to go awry in distinguishing self from nonself. The result is **autoimmune diseases,** in which the immune system attacks the body's own tissues. Examples include rheumatoid arthritis, insulin-dependent diabetes, and multiple sclerosis (Chapter 8, Box 8.1). A significant portion of research on the immune system is directed toward developing approaches for treating these diseases. **Allergies** are another example of improper functioning of the immune system. Millions suffer from asthma as a result of allergies to plant pollens and to other allergens (substances that trigger allergic attacks). Food allergies can evoke violent reactions to the point of being life-threatening.

Over the years researchers have unraveled some of the mysteries of the immune system and have used its properties as a therapeutic aid. The first **vaccine,** that against smallpox, was developed about 200 years ago. Since that time it has been used so effectively as a preventive measure that smallpox has been eradicated. The action of vaccines of this sort depends on exposure to the infectious agent in a weakened form. The immune system mounts an attack, and *the immune system retains "memory" of the exposure.* In subsequent encounters with the same pathogen, the immune system can mount a quick and effective attack. It is hoped that current research can be carried to the point of developing vaccines that can treat AIDS in persons already infected. Other strategies are directed at finding treatments for autoimmune diseases. Still others are attempting to use the immune system to attack and destroy cancer cells.

We need an understanding of how the immune system operates to go into more detail about how some of these goals might be achieved. There are two important aspects to the process: those that operate on the cellular level and those that operate on the molecular level. We shall discuss them in turn.

Allergic reactions arise when the immune system attacks innocuous substances. Allergies to plant pollens are common, producing well-known symptoms such as sneezing.

Edward Jenner developed the world's first vaccine in 1796. It was a safe and effective way to prevent smallpox and has led to eradication of this disease.

The Immune System: Cellular Aspects

A major component of the immune system is the class of white blood cells called **lymphocytes.** Like all blood cells, they arise from common precursor cells (stem cells) in the bone marrow. Unlike other blood cells, however, they can leave the blood vessels and circulate in the lymphatic system. Lymphoid tissues such as lymph nodes, the spleen, and, above all, the thymus gland play important roles in the workings of the immune system.

Two kinds of lymphocytes can be distinguished: *T* cells and *B* cells. ***T* cells** develop primarily in the *t*hymus gland and ***B* cells** primarily in the *b*one marrow, accounting for their names (Figure 18.15). Much of the cellular aspect of immunity is the province of the *T* cells, whereas much of the molecular aspect depends on the activities of the *B* cells.

T cells can have a number of functions. As *T* cells differentiate, each becomes specialized for one of the possible functions. The first of these possibilities, that of **killer *T* cells,** involves surface receptors that recognize and bind to **antigens,** the foreign substances that trigger the immune response. The antigens are presented to the *T* cell by other white blood cells called macrophages. The macrophages ingest and process antigens, and then present them to *T* cells. The processed antigen frequently takes the form of a short peptide bound to a protein of the *major histocompatability complex (MHC I)* on the surface of the macrophage. The macrophage also presents another molecule, a protein of a family known as B7, which binds to another *T* cell surface protein called CD28; the exact nature of the B7 protein involved is a subject of active research (see the article by Cohen listed in the bibliography at the end of this chapter). Figure 18.16 shows the interactions that take place in the binding process. The combination of the two signals leads to *T* cell growth and differentiation, producing killer *T* cells. Proliferation of killer *T* cells is also triggered when macrophages bound to *T* cells produce small proteins called **interleukins.** The *T* cells make an interleukin receptor protein as long as they are bound to the macrophage but do not do so when they are no longer bound. (Interleukins

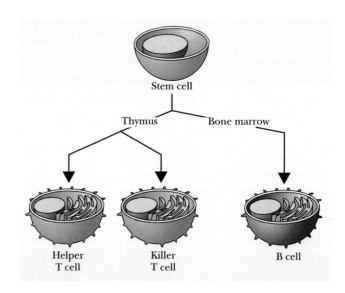

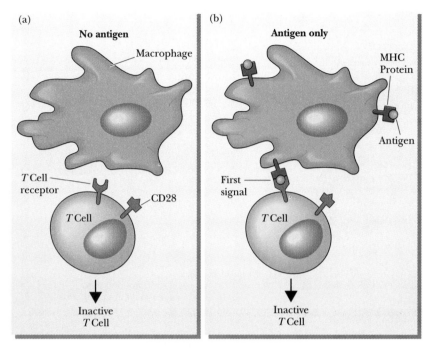

FIGURE 18.16

A two-stage process leads to the growth and differentiation of *T* cells.
(a) In the absence of antigen, proliferation of *T* cells does not take place.
(b) In the presence of antigen alone, the *T* cell receptor binds to antigen
presented on the surface of a macrophage cell by the MHC protein. There
is still no proliferation of *T* cells because the second signal is missing. In
this way the body can avoid an inappropriate response to its own antigens.
(c) When an infection takes place, a B7 protein is produced in response to
the infection. The B7 protein on the surface of the infected cell binds to a
CD28 protein on the surface of the immature *T* cell, giving the second
signal that allows it to grow and proliferate.

are part of a class of substances called **cytokines;** this term refers to soluble protein factors produced by one cell that specifically affect another cell. Cytokines that play a role in the operation of lymphocytes are also called **lymphokines.**) In this way, *T* cells do not proliferate in uncontrolled fashion.

T cells that bind to a given antigen and *only to that antigen* grow when these conditions are fulfilled. Note the specificity of which the immune system is capable. Many substances, including ones that do not exist in nature, can be antigens. The remarkable adaptability of the immune system in dealing with so many possible challenges is another of its main features. The process by which only those cells that respond to a given antigen grow in preference to other *T* cells is called **clonal selection** (Figure 18.17). The immune system can thus be versatile in its responses to the challenges it meets.

As their name implies, killer *T* cells destroy antigen-infected cells. They do so by binding to them and by releasing a protein that perforates the plasma membrane of the infected cell. This aspect of the immune system is

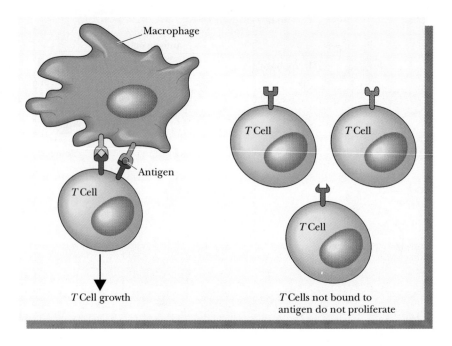

particularly effective in preventing the spread of viral infection by killing virus-infected host cells. In a situation such as this, the antigen can be considered to be all or part of the coat protein of the virus. When the infection subsides, some memory cells remain, conferring immunity against later attacks from the same source.

T cells play a second role in the immune system. Another class of T cells develops receptors for another class of antigen-presenting MHC proteins, in this case MHC II. These will become **helper T cells,** which develop in much the same way as killer T cells. *Helper T cells are the specific targets of the human immunodeficiency virus (HIV), which causes AIDS.* The function of helper T cells is not to kill infected cells but to aid in the development of the B cells. Maturing B cells display the MHC II protein, with processed antigen, on their surfaces. Note particularly that the MHC proteins play a key role in the immune system. This property has led to a considerable amount of research to determine their structure, including structure determination by x-ray crystallography. The MHC II of the B cells is the binding site for helper T cells. The binding of helper T cells to B cells releases interleukins and triggers the development of B cells to plasma cells (Figure 18.18). Both B cells and plasma cells produce **antibodies** (also known as **immunoglobulins**), the proteins that will occupy most of our time as we discuss the molecular aspects of the immune response. B cells display antibodies on their surfaces in addition to the MHC II proteins. The antibodies recognize and bind to antigens. This property allows B cells to absorb antigens for processing. Plasma cells release circulating antibodies into the bloodstream,

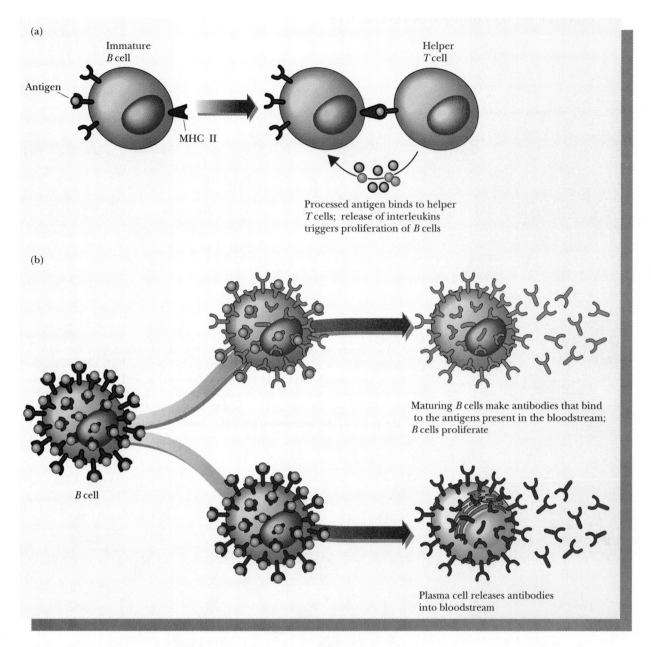

FIGURE 18.18

Helper *T* cells aid in the development of *B* cells. (a) A helper *T* cell has a receptor for the MHC II protein on the surfaces of immature *B* cells. When helper *T* cells bind to the processed antigen presented by the MHC II protein, they release interleukins and trigger the maturation and proliferation of *B* cells. (b) *B* cells have antibodies on their surfaces, which allow them to bind to antigens. The *B* cells with antibodies for the antigens present grow and develop. When *B* cells develop into plasma cells, they release circulating antibodies into the bloodstream.

where they bind to antigen, marking it for destruction by the immune system. We now need to take a detailed look at these highly important proteins.

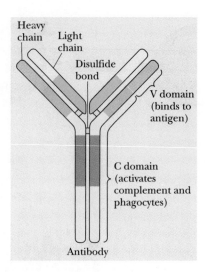

FIGURE 18.19

A typical antibody molecule is a Y-shaped molecule consisting of two identical light chains and two identical heavy chains linked by disulfide bonds. Each light chain and each heavy chain has a variable region and a constant region. The variable region, which is at the prongs of the Y, binds to antigen. The constant region, toward the stem of the Y, activates phagocytes and complement, the parts of the immune system that destroy antibody-bound antigen.

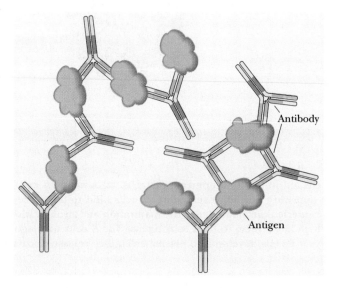

FIGURE 18.20

An antigen–antibody reaction forms a precipitate. An antigen, such as a bacterium or virus, typically has several binding sites for antibodies. Each variable region of an antibody (each prong of the Y) can bind to a different antigen. The aggregate thus formed precipitates and is attacked by phagocytes and the complement system.

The Immune System: Molecular Aspects

Antibodies are Y-shaped molecules, consisting of two identical heavy chains and two identical light chains, held together by disulfide bonds (Figure 18.19). They are glycoproteins, with oligosaccharides linked to their heavy chains. (There are different classes of antibodies, based on differences in the heavy chains. In some of these classes, heavy chains are linked to form dimers, trimers, or pentamers.) Each light chain and each heavy chain has a constant region and a variable region. The variable region (also called the V domain) is found at the prongs of the Y and is the part of the antibody that binds to the antigen (Figure 18.20). The binding sites for the antibody on the antigen are called **epitopes.** Most antigens have several such binding sites, so that the immune system will have several possible avenues of attack for naturally occurring antigens. Each antibody can bind to two antigens, and each antigen usually has several binding sites for antibody, giving rise to a precipitate that is the basis of experimental methods for immunological research. The constant region (the C domain) is located at the hinge and the stem of the Y; it is this part of the antibody that is recognized by phagocytes and by the complement system (the portion of the immune system that destroys antibody-bound antigen). Box 18.1 describes a clinical application of antigen–antibody binding as a test for disease.

How does the body produce so many highly diverse antibodies to respond to essentially any possible antigen? The number of possible

18.1

AN IMMUNOLOGICAL TEST FOR SYPHILIS

A clinical test for syphilis depends on the presence on the cell's surface of large quantities of an antigenic determinant characteristic of the bacteria that cause this disease. What is moderately unusual about the nature of the antigenic determinant is that it is neither a protein nor an oligosaccharide (the usual kinds of antigens) but a lipid. Cardiolipin, the substance in question, is the only lipid known to be antigenic. Those who are infected with syphilis have high levels of antibody to cardiolipin in their blood serum, providing the basis of the test. The method involves determining the levels of the anti-cardiolipin antibody in the blood by the precipitation reaction based on cross-linking of antigen and antibody.

$$R_2-\overset{\overset{\displaystyle O}{\|}}{C}-O-\overset{\overset{\displaystyle CH_2-O-\overset{\overset{\displaystyle O}{\|}}{C}-R_1}{|}}{\underset{\underset{\displaystyle CH_2-O-\underset{\underset{\displaystyle O^-}{|}}{\overset{\overset{\displaystyle O}{\|}}{P}}-OCH_2\underset{\underset{\displaystyle OH}{|}}{CH}CH_2-O-\underset{\underset{\displaystyle O^-}{|}}{\overset{\overset{\displaystyle O}{\|}}{P}}-O-CH_2}{|}}{C}$$

Cardiolipin

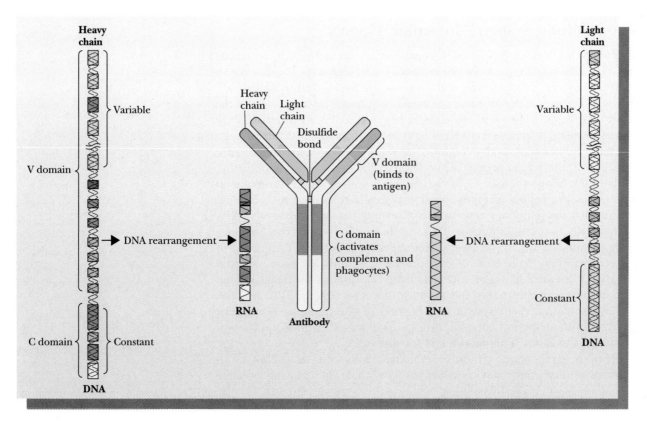

FIGURE 18.21

The heavy and light chains of antibodies are encoded by genes that consist of a number of DNA segments. These segments rearrange and in the process give rise to genes for different chains in each *B* cell. Since the joining is highly variable, comparatively few gene segments give rise to millions of distinct antibodies.

antibodies is virtually unlimited, as is the number of words in the English language. In a language, the letters of the alphabet can be arranged in countless ways to give a variety of words, and the same possibility for enormous numbers of rearrangements exists with the gene segments that code for portions of antibody chains. Antibody genes are inherited as small fragments that join together to form a complete gene in individual *B* cells as they develop (Figure 18.21). When gene segments are joined, the enzymes that catalyze the process add random DNA bases to the ends of segments being spliced, allowing for the wide variety observed experimentally. This rearrangement process takes place in the genes for both the light and heavy chains. In addition, it is well known that *B* lymphocytes have a particularly high rate of somatic mutation, in which changes in the base sequence of DNA occur as the cell develops. As we saw in Chapter 7, Section 7.7, changes outside the germ cells apply only to the organism in which they take place and are not passed on to succeeding generations.

Each *B* cell (and each progeny plasma cell) produces only one kind of antibody. In principle each such cell should be a source of a supply of homogeneous antibody by cloning. This is not possible in practice because

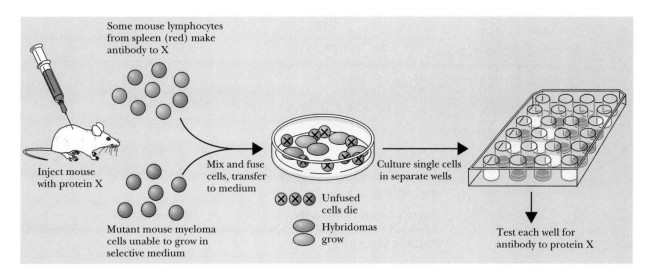

FIGURE 18.22

A procedure for producing monoclonal antibodies against a protein antigen X. A mouse is immunized against the antigen X, and some of its spleen lymphocytes produce antibody. The lymphocytes are fused with mutant myeloma cells that cannot grow in a given medium because they lack an enzyme found in the lymphocytes. Unfused cells die because lymphocytes cannot grow in culture and the mutant myeloma cells cannot survive in this medium. The individual cells are grown in culture in separate wells and tested for antibody to protein X.

lymphocytes do not grow continuously in culture. In the late 1970s Georges Köhler and César Milstein developed a method to circumvent this problem, a feat for which they received the Nobel Prize in physiology or medicine in 1984. The technique requires fusing lymphocytes that make the desired antibody with mouse myeloma cells. The resulting **hybridoma** (hybrid myeloma), like all cancer cells, can be cloned in culture (Figure 18.22) and produces the desired antibody. Since the clones are the progeny of a single cell, they produce homogeneous **monoclonal antibodies.** In this way it is possible to produce antibodies to almost any antigen in quantity. Monoclonal antibodies can be used to assay for biological substances that can act as antigens. A striking example of their usefulness is in testing blood for the presence of HIV; this procedure has become routine to protect the public blood supply.

18.8
NITRIC OXIDE IN BIOCHEMISTRY:
A CASE STUDY IN CONNECTIONS

A particularly striking example of the fact that all biochemical reactions are intimately related to one another and to physiological processes has recently come to light. The simple molecule in question, nitric oxide (NO), has a wide variety of effects on cells. It has been found to affect about a dozen different

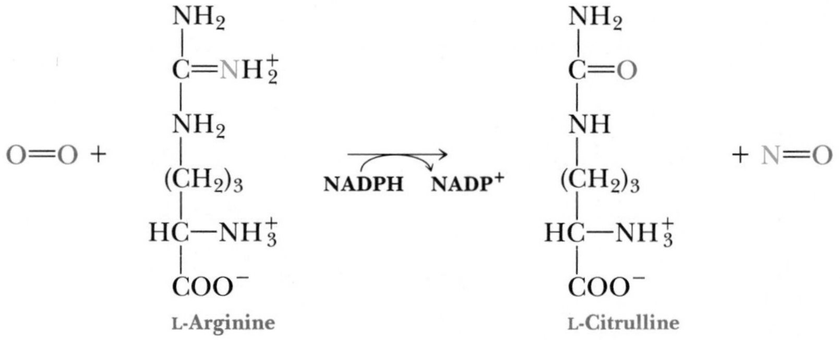

FIGURE 18.23

Arginine reacts with oxygen to produce citrulline and NO. The reaction is catalyzed by the enzyme nitric oxide synthase, with concomitant oxidation of NADPH to NADP$^+$.

kinds of cells, with particularly striking effects in the immune system, in the transmission of nerve impulses, and in the control of blood pressure. Nitric oxide is not only a small molecule, it is electrically neutral, and thus it can pass through cell membranes by simple diffusion (Chapter 8, Section 8.6). NO does not require a specific receptor or channel protein, so it is free to move to a number of sites and to act on them. In all the sites where NO is active it is produced by the reaction of arginine with molecular oxygen to give citrulline as well as NO (Figure 18.23). Note that NADPH is oxidized to NADP$^+$ in the process. The reaction is catalyzed by the enzyme **nitric oxide synthase** (NOS).

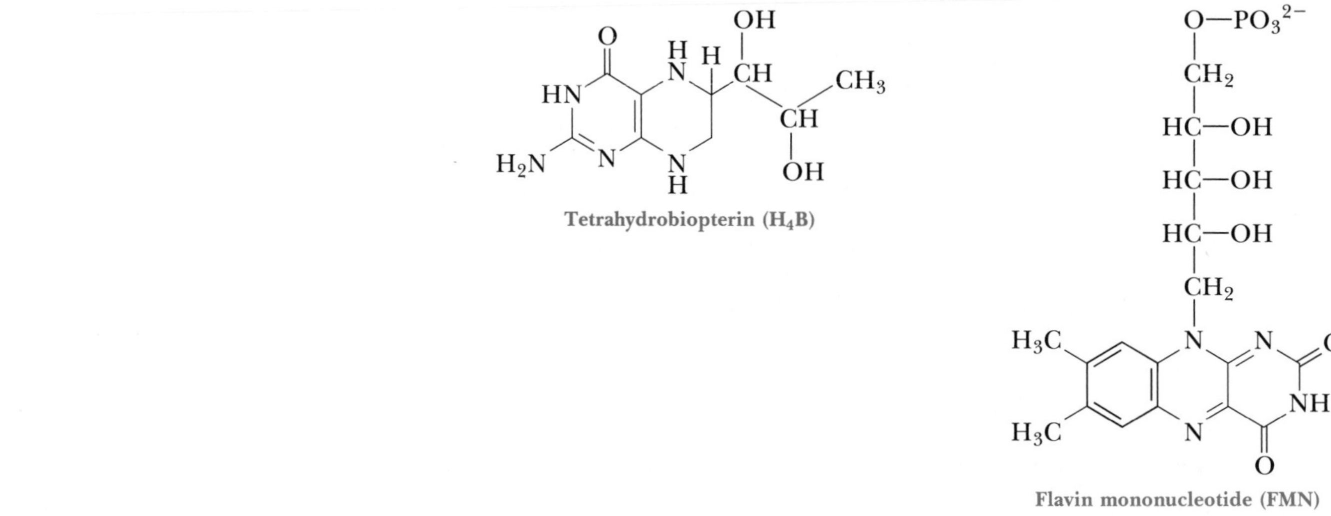

Tetrahydrobiopterin (H$_4$B)

Flavin mononucleotide (FMN)

The structures of tetrahydrobiopterin and FMN. Note that FMN differs from FAD only in the fact that it is missing the adenine nucleotide moiety found in FAD.

NOS exists in several slightly different forms, depending on the kind of cell in which it is found. It is known that the forms of NOS found in both the nervous system and the immune system are dimers with two identical subunits with molecular weights in the range of 130,000 to 150,000 daltons. All forms of the enzyme currently known contain four prosthetic groups: FAD, FMN, heme, and tetrahydrobiopterin (H_4B). The structure of NOS does not differ greatly in any of these different systems. The enzyme must be activated in order to catalyze the production of NO, and the way it is activated is different in different tissues.

In the nervous system NOS is activated when a calcium–calmodulin complex binds to it. (This mode of activation also takes place in vascular endothelial cells, which play a role in the control of blood pressure.) The formation of the calcium–calmodulin complex requires inflow of calcium ion into the cell. Figure 18.24 shows the process that takes place in brain cells (neurons) as a result of a chemical signal traversing the **synapse** (the space between the neurons). Note that there are several similarities to the action of calcium as a second messenger (Section 18.5) and to the action of the neuromuscular junction (Chapter 8, Section 8.7). The neuron activated by a nerve impulse (the presynaptic neuron), at left, releases glutamate into

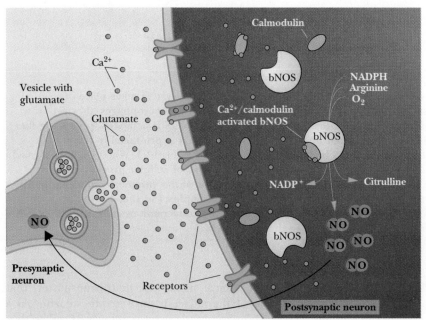

FIGURE 18.24

The activation of nitric oxide synthase in brain (bNOS). A neuron activated by a nerve impulse (the presynaptic neuron), at left, releases glutamate into the synapse. Glutamate binds to a membrane receptor on the adjacent postsynaptic neuron. As a result of the binding, a channel opens in the receptor, allowing Ca^{2+} to flow into the cell and to bind to calmodulin. The calcium–calmodulin complex binds to bNOS, activating it so that it can catalyze the formation of NO. The NO thus formed can diffuse to other neurons to reinforce neural connections.

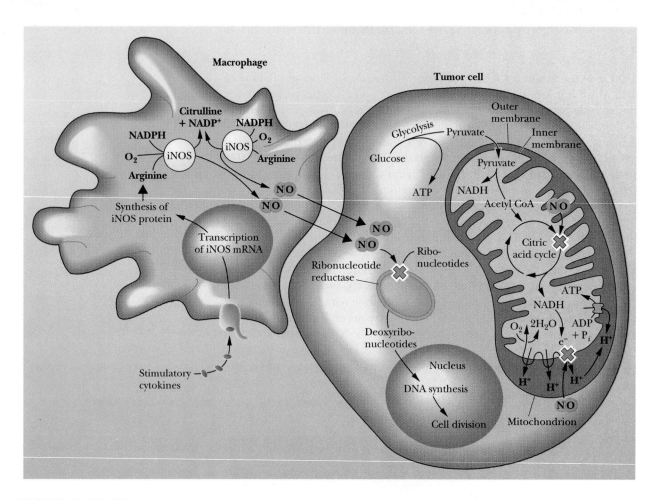

FIGURE 18.25

The role of nitric oxide in the immune system. The production of NO in the immune system takes place in macrophages (components of the immune system). When stimulatory cytokines bind to macrophages, they set off a chain of events that leads to synthesis of NOS. The DNA that encodes the gene for NOS is activated, producing the messenger RNA for the enzyme, which is then synthesized by the cell. An enzyme that is produced in this way is referred to as inducible. Consequently, this form of NOS is designated iNOS (i for inducible). When iNOS is formed, it diffuses to tumor cells close to the macrophage. There it interferes with a number of cellular processes, which occur at high rates in rapidly growing cancer cells. One site of action is inhibition of the enzyme aconitase, which catalyzes one of the early steps of the citric acid cycle. NO also interferes with Complex I of the electron transport chain. Another site of action is inhibition of ribonucleotide reductase, preventing conversion of ribonucleotides to deoxyribonucleotides and thus interfering with DNA synthesis in the cancer cell.

the synapse. Glutamate is the chemical signal (the neurotransmitter) that binds to a membrane receptor of the adjacent postsynaptic neuron. As a result of the binding, a channel opens in the receptor, allowing Ca^{2+} to flow into the cell and to bind to calmodulin. The calcium–calmodulin complex binds to the specific NOS found in brain cells, bNOS (b for brain), activating it so that it can catalyze the formation of NO. The NO thus formed can have one of several functions. One that has occasioned much speculation is that it diffuses through the membranes of brain cells to other neurons to reinforce connections in neural pathways. NO itself may serve as an intercellular signal or may trigger the formation of another signal molecule. This question is the subject of active investigation.

The production of NO in the immune system takes place in the cells known as macrophages. We saw these components of the immune system in the last section, in their role in the development of T cells. They play another part when they produce NO: that of attacking cancer cells. When stimulatory cytokines bind to macrophages, they set off a chain of events that leads to synthesis of NOS (Figure 18.25). The gene for NOS is activated, producing the messenger RNA for the enzyme, which is then synthesized by the cell. (This process is an example of the genetic control of enzyme synthesis that we will discuss in detail in Chapter 20, Section 20.9.) An enzyme that is produced in this way is referred to as *inducible*. Consequently, this form of NOS is designated iNOS (i for inducible).

When iNOS is formed, it diffuses to tumor cells close to the macrophage. There it interferes with a number of cellular processes, which occur at high rates in rapidly growing cancer cells. One site of action is inhibition of the enzyme aconitase, which catalyzes one of the early steps of the citric acid cycle (Chapter 12, Section 12.3). NO also interferes with Complex I of the electron transport chain (Chapter 13, Section 13.2). Another site of action is inhibition of ribonucleotide reductase, preventing conversion of ribonucleotides to deoxyribonucleotides and thus interfering with DNA synthesis in the cancer cell (Chapter 17, Section 17.10).

S U M M A R Y

All metabolic pathways are related. Some metabolites appear in several pathways. Furthermore, many reactions of metabolism can take place simultaneously. The citric acid cycle plays a central role in metabolism, in both catabolic and anabolic pathways. The breakdown products of sugars, fatty acids, and amino acids all enter the citric acid cycle.

While the citric acid cycle takes place in mitochondria, many anabolic reactions take place in the cytosol. Oxaloacetate, the starting material for gluconeogenesis, is a component of the citric acid cycle. Malate, but not oxaloacetate, can be transported across the mitochondrial membrane. Once malate from mitochondria is carried to the cytosol, it can be converted to oxaloacetate by malate dehydrogenase, an enzyme that requires NAD^+.

In the case of lipid anabolism, it is not definitely settled whether acetyl-CoA, the starting material, can be transported across the mitochondrial membrane. Malate, which does cross the mitochondrial membrane, plays a role in lipid anabolism, in a reaction in which malate is oxidatively decarboxylated to pyruvate by an enzyme that requires $NADP^+$, producing NADPH. This reaction is an important source of NADPH for lipid anabolism, with the pentose phosphate pathway the only other source.

The sources of substrates for catabolism and for anabolism are the nutrients derived from foodstuffs. In humans the choice of diet becomes important in the interest of obtaining enough of essential nutrients while avoiding excesses of others, such as fats, where excess is known to play a role in the development of health problems.

Sophisticated fine-tuning of metabolic processes in multicellular organisms is possible through the actions of hormones and second messengers. In humans a complex hormonal system has evolved that requires releasing factors (under the control of the hypothalamus), trophic hormones (under the control of the pituitary), and specific hormones for target organs (under the control of endocrine glands). Feedback control occurs at every level of the system. When a hormone binds to its receptor on the plasma membrane of a target cell, it sets off a cascade of reactions by which second messengers elicit the actual cellular response. Two of the most important second messengers, *cyclic AMP* (cAMP) and phosphatidylinositol 4,5-*bis*phosphate (PIP_2), activate protein kinases. Calcium ion is intimately involved in the action of PIP$_2$. Hormonal triggering can be added to other levels of control of metabolism, such as allosteric activation and covalent modification, to ensure an efficient response to the needs of the organism.

The immune system makes the distinction between self and nonself. The recognition process operates on both the cellular and molecular levels. Two kinds of lymphocytes (a class of white blood cells) are involved: T cells and B cells. Killer T cells attack cells infected by disease-causing pathogens, and helper T cells aid in the development of the B cells, which produce the proteins known as antibodies. The antibodies in turn circulate in the bloodstream and bind to antigens (the agents that trigger the immune response), marking them for destruction.

One of the most striking examples of the intimate connections within biochemical pathways and between biochemistry and physiological responses is found in the biochemical role of nitric oxide. This simple molecule has profound effects in a number of different kinds of cells. A complex enzyme is responsible for its production; the various forms of the enzyme have essentially the same structure in the cell types in which they are found, but must be activated to catalyze the production of NO. It is the mode of enzyme activation that makes the difference between one cell type and another, giving rise to the many different roles that NO can play.

E X E R C I S E S

(*Hint:* You may want to review material in Chapters 11 through 17.)

1. Immature rats are fed all the essential amino acids but one. Three hours later they are fed the missing amino acid. The rats fail to grow. Explain this observation.
2. NADH is an important coenzyme in catabolic processes, whereas NADPH appears in anabolic processes. Explain how an exchange of the two can be effected.
3. A cat named Lucullus is so spoiled that he will eat nothing but freshly opened canned tuna. Another cat, Griselda, is given only dry cat food by her far less indulgent owner. Canned tuna is essentially all protein, whereas dry cat food can be considered 70% carbohydrate and 30% protein. Assuming that these animals have no other sources of food, what can you say about the differences and similarities in their catabolic activities? (The pun is intended.)
4. Kwashiorkor is a protein-deficiency disease that occurs most commonly in small children, who characteristically have thin arms and legs and bloated, distended abdomens due to fluid imbalance. When such children are placed on adequate diets, they tend to lose weight at first. Explain this observation.
5. Recent recommendations on diet suggest that the sources of calories should be distributed as follows: 50 to 55% carbohydrate, 25 to 30% fats, and 20% protein. Suggest some reasons for these recommendations.
6. When PIP$_2$ is hydrolyzed, why does IP$_3$ diffuse into the cytosol while DAG remains in the membrane?
7. Briefly describe the series of events that takes place when cAMP acts as a second messenger.
8. How do the actions of the hypothalamus and pituitary affect the workings of endocrine glands?
9. For each of three hormones discussed in this chapter, give its source and chemical nature; also discuss the mode of action of each hormone.
10. What are the main structural features of antibody molecules? How are they related to the binding of antigens?
11. How can the observed enormous diversity of antibodies be accounted for by a relatively small number of genes?
12. Outline a procedure for production of monoclonal antibodies.

A N N O T A T E D   B I B L I O G R A P H Y

Alfin-Slater, R., and D. Kritchevsky, eds. *Cancer and Nutrition.* New York: Plenum Press, 1990. [The latest volume in a comprehensive series on human nutrition.]

Barinaga, M. Immune Mystery Revealed: How MHC Meets Antigen. *Science* **250,** 1657–1658 (1990). [The mechanism by which the major histocompatability complex (MHC) protein operates. MHC presents peptides derived from viruses that infect a cell to the organism's immune system.]

Boon, T. Teaching the Immune System To Fight Cancer. *Sci. Amer.* **268** (3), 82–89 (1993). [A report on a promising form of cancer therapy: developing antibodies to antigens on the surfaces of tumor cells.]

Cohen, J. New Protein Steals the Show as "Costimulator" of T Cells. *Science* **262**, 844–845 (1993). [A report on continuing research on the second of two signals needed for *T* cell growth and differentiation.]

Editors of *Science* and various authors. Breast Cancer Research: A Special Report. *Science* **259**, 616–638 (1993). [A series of articles on prevention and treatment.]

Editors of *Science* and various authors. AIDS: The Unanswered Questions. *Science* **260**, 1253–1293 (1993). [A report based on a survey of 150 investigators in AIDS research.]

Editors of *Scientific American* and various authors. Life, Death, and the Immune System. *Sci. Amer.* **269** (3), 52–144 (1993). [A special issue that appeared in September 1993. An excellent source of information about the immune system and its relationship to cancer, AIDS, and autoimmune diseases.]

Englehard, V. How Cells Process Antigens. *Sci. Amer.* **271** (2), 54–61 (1994). [The first stages of the body's reaction to viruses, bacteria, and parasites. Particularly informative about macrophages and the histocompatibility complex.]

Galione, A. Cyclic ADP-Ribose: A New Way To Control Calcium. *Science* **259**, 325–326 (1993). [A short review of new research results on the role of calcium as second messenger.]

Hiller, S. A Better Way To Make the Medicine Go Down. *Science* **253**, 1095–1096 (1991). [Prodrugs, compounds converted to active substances by metabolism, show promise for treating cancer and AIDS.]

Hoffman, M. Determining What Immune Cells See. *Science* **255**, 531–534 (1992). [A comparison of the two MHC classes.]

Hoffman, M. Antigen Processing: A New Pathway Discovered. *Science* **255**, 1214–1215 (1992). [A mechanism for antigen processing that involves transport of peptides across membranes without transporter proteins.]

Johnson, H., J. Russell, and C. Pontzer. Superantigens in Human Disease. *Sci. Amer.* **266** (4), 92–101 (1992). [A study of proteins that cause food poisoning and toxic shock and may be involved in arthritis and AIDS.]

Katch, F. I., and W. D. McArdle. *Nutrition, Weight Control, and Exercise.* 3rd ed. Philadelphia: Lea & Febiger, 1988. [A discussion of nutrition, exercise, and health, intended to substitute facts for myth on the subject.]

Kritchevsky, K., C. Bonfield, and J. Anderson, eds. *Dietary Fiber: Chemistry, Physiology, and Health Effects.* New York: Plenum Press, 1990. [A topic of considerable current interest, with explicit connections to the biochemistry of plant cell walls.]

Lancaster, J. R. Nitric Oxide in Cells. *American Scientist* **80**, 248–259 (1992). [A review of nitric oxide's ubiquitous activities in the body.]

Linder, M., and A. Gilman. G Proteins. *Sci. Amer.* **267** (1), 56–65 (1992). [A description of the role of an important class of membrane proteins in transmitting hormonal messages into the cell.]

Marx, J. Structure of MHC Protein Solved. *Science* **238**, 613–614 (1987). [The structure gives insight into important features of antigen processing.]

Marx, J. Clue Found to *T* Cell Loss in AIDS. *Science* **254**, 798–800 (1991). [The putative role of a superantigen in the destruction of the immune system by the AIDS virus.]

Rasmussen, H. The Cycling of Calcium as an Intracellular Messenger. *Sci. Amer.* **261** (4), 66–73 (1989). [An article on the role of calcium as a second messenger.]

Reitschel, E., and H. Brade. Bacterial Endotoxins. *Sci. Amer.* **267** (2), 54–61 (1992). [Toxins that are an integral part of disease-causing bacteria can be harnessed to enhance the immune system.]

Rennie, J. The Body Against Itself. *Sci. Amer.* **263** (6), 106–115 (1990). [A review of autoimmune disease.]

Rosen, O. After Insulin Binds. *Science* **237**, 1452–1458 (1987). [An account of the role of a protein kinase in the effect of insulin inside the cell.]

Schwartz, R. *T* Cell Anergy. *Sci. Amer.* **269** (2), 62–71 (1993). [How the body avoids autoimmune disease.]

Scrimshaw, N. Iron Deficiency. *Sci. Amer.* **265** (4), 46–52 (1991). [A description of a nutritional problem that occurs all over the world.]

Todorov, I. How Cells Maintain Stability. *Sci. Amer.* **263** (6), 66–75 (1990). [The concept of homeostasis applied to protein biosynthesis.]

Travis, J. Tracing the Immune System's Evolutionary History. *Science* **261**, 164–165 (1993). [Attempts to trace evolution of the immune system.]

Uvnas-Moberg, K. The Gastrointestinal Tract in Growth and Reproduction. *Sci. Amer.* **261** (1), 78–83 (1989). [How the endocrine and digestive systems cooperate to provide for the needs of the fetus and the newborn.]

Von Boehmer, H., and P. Kisielow. How the Immune System Learns About Self. *Sci. Amer.* **265** (4), 74–81 (1991). [The role of the thymus in the immune system.]

Willett, W., Diet and Health: What Should We Eat? *Science* **264**, 532–537 (1994). [An excellent summary of many aspects of a complex topic.]

Young, J., and Z. Cohn. How Killer Cells Kill. *Sci. Amer.* **258** (1), 38–44 (1988). [The method by which killer *T* cells produce the pores that cause target cells to leak and die.]

See also the bibliographies for Chapters 11 through 17.

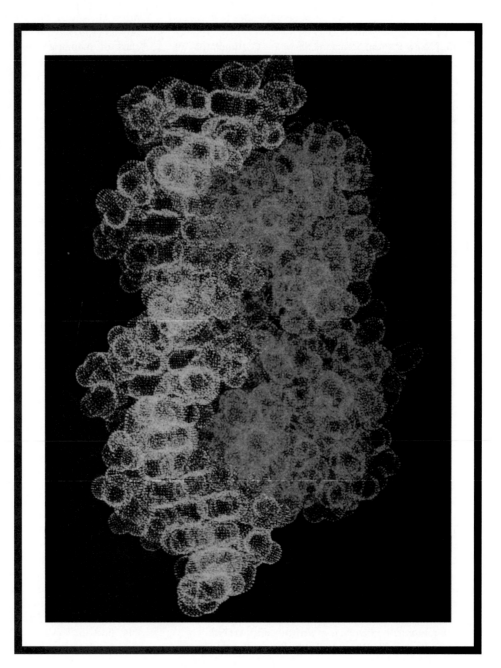

A computer-generated view of the DNA double helix (blue), with a peptide (red) bound in the major groove.

Part IV

Workings of the Genetic Code

Outline

Interview

Ponzy Lu

onzy Lu was born in Shanghai, China, and received his primary and secondary schooling in southern California and London, England. He is currently Professor of Chemistry and Chair of the Biochemistry Major Program at The University of Pennsylvania, where he often teaches undergraduate biochemistry courses. He received his bachelor of science degree (1964) in chemistry from the California Institute of Technology and his Ph.D. (1970) in biophysics from the Massachusetts Institute of Technology, followed by three postdoctoral years in Germany and Switzerland. The choice of Germany was motivated by a southern Californian fantasy of owning a Porsche and using it without speed limits. This dream was realized in Göttingen, where quantum physics and organic chemistry originated. In Switzerland he joined Jeffrey Miller, now at UCLA, to learn bacterial genetics firsthand, which also began a longstanding collaboration on work with the lactose operon repressor. He has been at The University of Pennsylvania since 1973, with one year at l'Université d'Aix-Marseille, France, in 1980. He has published over 60 articles and abstracts in the field of molecular biology. His research interests involve a combination of nuclear magnetic resonance (NMR) methods and genetics directed at repressor proteins. Professor Lu is often invited to lecture at universities and biochemical conferences, such as the international Conference on Magnetic Resonance in Biological Systems. Professor Lu's primary interests are biochemical: food, wine, and gene regulation.

How did you get interested in science and choose Caltech for your undergraduate training?

In high school my idea of fun was mechanical things: toys, model airplanes, cars, and explosions — firecrackers during Chinese New Year. I really didn't choose Caltech. Being a Californian and good at high school math, it was an obvious place to go. It wasn't until I arrived at Caltech's "Frosh Camp," a weekend with faculty in the San Bernardino Mountains before classes began, that I realized what science was all about. Science was everything (Caltech was not yet coed): individual challenge, self-aggrandizement, even human progress; but most important, the laboratory and computational equipment offered a vast range of toys.

Much of what we now call molecular biology was initiated at Caltech. Pauling's lectures in freshman chemistry on DNA, RNA, protein structure, and his theories on anesthesia with inert gases made me excited about studying biological science. Even today, learning about biochemical topics in freshman chemistry is unconventional.

How did graduate school at MIT affect your research interests?

MIT was never a serious consideration for graduate school; one institute of technology was enough. I visited several other graduate programs, including those of liberal arts schools in Massachusetts. At one, several disgruntled graduate students, especially a violin-playing geneticist, suggested that I visit MIT because of the breadth of research interests in the biology department.

During the time I was a graduate student, all of the central dogma — information flow from DNA to protein, the genetic code, and its implementation — was being discovered. As I was

594

finishing my Ph.D. at MIT, it was clear to me that there were two research directions I could take: macromolecular structure and interactions or cells and organisms. Given my simplistic, mechanical approach to biochemical research, cells and organisms seemed too complicated, so I chose structure and interactions. This was at a time before restriction enzymes and the whole array of methods using recombinant DNA that now allow dissection of any organism with Petri dishes and gels. I suspect that I chose to work on molecular structure because I could play with more research toys.

How did your postdoctoral experience shape your career in biochemistry?

Most of my postdoctoral work was done at the Max-Planck Institute for Biophysical Chemistry in Göttingen. The most important thing I learned was that you need pure, active proteins or nucleic acids in order to do structure analysis. No matter how fancy your x-ray or NMR instrumentation, if you put "garbage in" you get "garbage out." First, you use Petri dishes and gels to verify genetics, obtain expression, and purify macromolecules; then you can do biophysical studies.

What are your research interests in your labs at The University of Pennsylvania?

We use a combination of modern genetic techniques and high-field NMR spectroscopy to study the mechanism of gene expression. My research group is interested in the regulatory proteins that control gene expression. In particular, we are probing the structure and dynamics of repressors, activators, and RNA polymerase, both alone and bound to their specific DNA sites. Since we are isolating and manipulating both the DNA sites and the proteins at the gene level, all the techniques that we use are applicable

to any cloned structural gene. The Penn Cancer Center's DNA synthesis service is housed in our laboratory.

Can you explain how NMR works?

NMR detects differences in the energy levels of atomic nuclei when they are in magnetic fields. The individual atomic nucleus in a molecule, placed in a magnet, sees the field through its own and the neighboring atoms' electrons. Each nucleus in a molecule sees a different magnetic field and thus has its own set of energy levels. NMR spectrometers not only can measure these energy level differences but also can identify individual atoms in the molecule. These are the "peaks" and groups of "peaks" seen in NMR spectra presented in organic chemistry texts. In addition, the spectrometer can selectively send energy into one nucleus and affect the energy levels of a neighboring nucleus. These effects follow defined rules of nuclear physics, without changing any of the electronic structure that holds the molecules together. Some effects travel through chemical bonds (spin–spin coupling) and some through space (nuclear Overhauser effect, NOE). NMR determines the geometric relationship of the atomic nuclei, and we learn how the molecule looks.

How does NMR differ from MRI (magnetic resonance imaging), used in medical diagnosis?

NMR and MRI observe the same physical phenomenon. The molecular compositions of the different tissues, including H_2O, are quite different. If you place a whole person in a magnet, the same molecules in different tissues have different nuclear energy levels because they are in different molecular environments. The MRI instrument records not only the energy level differences but also the spatial position in the magnet and therefore in the organism. The detection electronics and computations of MRI are identi-

Professor Ponzy Lu making measurements on an NMR superconducting magnet.

cal to those of molecular NMR. The presentation of results shows up as a color picture with areas outlined by anatomical shape. NMR works in the same electromagnetic frequency range as broadcast radio and television (10^8 Hz), so the energies involved are 10^{-10} that of x-rays (10^{18} Hz) used for looking at bones or in computed tomographic (CT) scans. The name "MRI" was chosen to remove the word "nuclear," which many associate with radioactivity and thus danger.

Generally, what can physical methods like NMR tell us about the nature of biological macromolecules?

Biochemists, in large part, work on an assumption that the structure of a molecule is related to function. To see the three-dimensional model of a molecule is the first step in this approach. NMR and x-ray crystallography are the only methods that yield comprehensive three-dimensional information. Macromolecules are not static and homoge-

neous. They are more like a partially wet sponge; parts are soft and parts are firm, and they are constantly undergoing thermal motion. NMR in solution allows one to see this heterogeneity and how it changes as the pH, ionic strength, and temperature are varied.

Specifically, what can NMR tell us about how molecules interact with DNA?

Again, we assume we need to see the structure first. NMR will tell us the structure of the protein–DNA or DNA–ligand complex. DNA has considerable sequence-dependent variation from the average helical structure. NMR analysis can tell us how this variation is exploited by specific interactions of the DNA with proteins.

Will your research have any medical applications, such as in diagnosis or the development of new drugs?

We are back to the assumption that structure is important. Most drugs have been found by accident. The term "rational drug design" is now used a lot. To design a drug, we need to see where it has to go. First, we need to know the structure of the target biological macromolecule. This structure can be elucidated by using NMR. This target site can be an enzyme-active site, or it can be the surface between two proteins or a protein and nucleic acid.

Is NMR the best technique for determining protein structure?

NMR has limitations. Complete structure determination of proteins by NMR measurements beyond 200 amino acids, which are multimeric, is at the present time not possible. We have been interested in *lac* repressor, a tetramer of 360 amino acids, for some time. Our NMR experiments are being used to probe large proteins, RNA polymerases, which in *Escherichia coli*

have five subunits, each ranging from 300 to 1400 amino acids. We will not get detailed atomic structures, but useful overviews.

What would you most want to tell students who are interested in a career in biochemical research?

Don't let introductory science courses discourage you. They are usually boring; unfortunately, the course content has not changed very much in the past 50 years. However, learning science and doing it are very different. Take lab courses and do independent research with one of your professors. Chemistry and biochemistry can be fun; learn how to play with the toys in the lab. To test if you enjoy learning about science or biochemical science, read *Scientific American* or the Tuesday *New York Times*. Read business publications such as the *Wall Street Journal*, *Fortune,* or *Business Week,* since new products use new technology. Once you start reading regularly, you will not stop. New discoveries are being made constantly. There are many areas of science to participate in. For example, biochemical science is very broad, ranging through biomass conversion for fuel, agriculture, food processing, cosmetics, drug abuse, DNA forensics, medicine, and pharmaceuticals.

Interview

Jacqueline K. Barton

D r. Jacqueline K. Barton is a native New Yorker and was educated in that city. She received her B.A. degree from Barnard College in 1974 and her Ph.D. from Columbia University in 1979. While at Columbia she worked in an area of platinum chemistry closely related to the work done by Dr. Barnett Rosenberg. Following her Ph.D. work, Dr. Barton did further research at Yale University and Bell Laboratories and then joined the faculty of Hunter College. In 1983 she returned to Columbia University, where she rose rapidly to the rank of full professor, and in the fall of 1989 she assumed her present position as Professor of Chemistry at the California Institute of Technology.

Dr. Barton has done outstanding research in the field of biochemistry, particularly in the design of simple molecular probes to explore the variations in structure and conformation along the DNA helix. In spite of having been a research scientist for a relatively short time, she has done important new work and has received many honors. In 1985 she received the Alan T. Waterman Award of the National Science

Foundation as the outstanding young scientist in the United States. In 1987 she was the recipient of the American Chemical Society's Eli Lilly Award in Biological Chemistry, and the following year she received the Society's Award in Pure Chemistry. That same year, 1988, she also received the Mayor of New York's Award of Honor in Science and Technology.

Like many chemists, Dr. Barton is interested in art. She has a painting by the Spanish artist Miró on her office wall as well as a print by the French artist Vasarel-ey, and this interest in form and color in art carries over into her research.

Dr. Barton, it is always of interest to learn how scientists came to their chosen fields. Did you have a strong background in chemistry?

I never took chemistry in high school. Maybe one shouldn't publicize that, but it's the truth. However, I was always very interested in mathematics, so I took a lot of calculus when I was in high school. I also took a course in geometry, and that interest in geome-

try has carried over into my research, since the sort of science I do now is very much governed by structures and shapes.

When I went to college I thought that in addition to taking math I should take some science courses. I walked into the freshman chemistry class, and there were about 150 people there. However, there was also a small honors class with about ten students. Even though I hadn't had chemistry before, I thought I would try it — and I loved it. What chemistry allowed me to do was to combine the abstract and the real. I was very excited by it.

But it was really the experience of the laboratory that got me interested in chemistry. Like many who are involved in chemistry, I was fascinated by color changes in reactions and the significance of these observations. However, I was also interested in trying to predict what would happen in a reaction and, if my prediction was not correct, to try to explain this and then to do more experiments that would solve the puzzle. That's really what got me started in science.

In addition, I also had an inspirational teacher and role model, Bernice

597

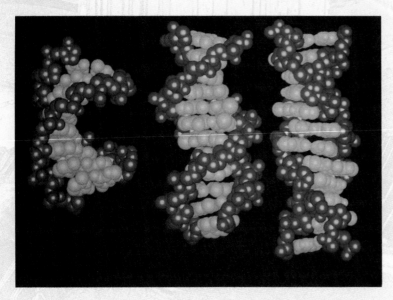

Double helical conformations of DNA: (left) A-DNA, (center) B-DNA, (right) Z-DNA.

Segal. She was an absolute inspiration to me. She gave a magnificent course, and was a tough lady who asked a lot of you — and you did it!

Your Ph.D. thesis research focused on compounds known as "platinum blues." Could you tell us more about these compounds?

Most platinum compounds are orange or red, and yet there are these magnificent blue complexes. What are they? What are their structures, and why are they blue? In fact, when Barney Rosenberg was looking at certain platinum compounds, he also found some 'platinum blues'; they were water-soluble, unlike cisplatin, and so people had hopes that they might even be better with respect to chemotherapy. But it turned out that was not the case. Nonetheless, that finding and others got Steve Lippard, my advisor at Columbia, interested in trying to work on the problem. Therefore, I made the type of complex that Rosen-

berg had made, and it was indeed blue. It's an absolutely beautiful molecule. We solved the structure of the molecule and found that it contains four platinum atoms in a line. We also found that it's mixed-valent, where the platinum has an average oxidation number of $2\frac{1}{4}$, a fact that helps to explain its blue color.[1]

You are currently doing research in bioinorganic chemistry, which deals with the role of metals in biological systems. You have received numerous awards for your work, indicating that the scientific community places great importance on the field and your contributions to it. Could you explain why this work has such significance?

The interest of my group is to exploit inorganic chemistry as a tool to ask questions of biological interest and to explore biological molecules. A lot of the work in bioinorganic chemistry thus far has been the exploration of

metal centers in biology. Why is blood red? Why does the iron [in heme] do what it does? That's just one example, but there are hundreds of others. Many enzymes and proteins within the body, in fact, contain metals, and the reason we've looked at blood and then the heme center within it has been because it's colored. An obvious tool that transition metal chemistry provides is color, and so things change color when reactions occur. That is one of the things that fascinated me in the first place.

Another wonderful thing about transition metal chemistry is that it allows us to build molecules that have interesting shapes and structures, depending upon the coordination geometry. In fact, you can create a wealth of different shapes, several of which are chiral, and that's something we take advantage of in particular. What we want to do is make a variety of molecules of different shapes, target these molecules to sites on a DNA strand, and then ask questions such as, 'Does DNA vary in its shape as a function of sequence?'[2] If we think about how proteins bind to DNA, do they also take advantage of shape recognition in binding to one site to activate one gene or turn off another gene? When scientists first wondered about these and other such problems, they would write down a one-dimensional sequence of DNA and would think about it in one-dimensional terms. How does the protein recognize a particular DNA sequence? DNA is clearly not one-dimensional. It has a three-dimensional structure, and different sequences of bases will generate different shapes and different forms. Therefore, we think we can build transition metal complexes of particular shapes, target them to particular sequences of bases in DNA, and then use these complexes to plot out the topology of DNA.[3] We can then ask how nature takes advantage of this topology. We

---

[1]Superconducting solids contain metal ions of different valences; that is, they are mixed-valent.

[2]Amino acids, proteins, nucleic acids such as DNA, and other aspects of biochemistry are discussed in the text.

[3]The results of such an experiment are seen in the structure of a complex formed between a ruthenium-based coordination compound and DNA.

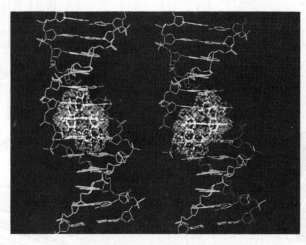

Tris *(phenanthroline) metal complexes (*Λ*, left, and *Δ*, right) are shown intercalated into right-handed, double-helical DNA.*

want to develop a true molecular understanding, a three-dimensional understanding, of the structure and the shapes of biologically important molecules such as DNA and RNA.

What are some recent developments in your field? Can you comment generally on the direction in which chemistry is moving?

I think our work may be an example of where chemistry is going in general. I think there has been a revolution in chemistry in the past ten years. The revolution is at the interface between chemistry and biology, where we can now ask chemical questions about biological molecules. First of all, we can make biological molecules that are pure. I can now go to a machine called a DNA synthesizer, and I can type in a sequence of DNA; from that sequence I can synthesize a pure material, with full knowledge of where all of the bonds are. Then I can run it through an HPLC and get it 100% pure.[4] Therefore, I can now talk about these biopolymers in chemical terms as molecules rather than as impure cellular extracts. I couldn't do that before.

Not only do we have the ability to prepare biological molecules in pure form, but we also have the techniques to characterize them in ways that chemists think about molecules. The development of new techniques allows us to make a bridge between chemistry and biology and ask chemical questions with molecular detail. It's an exciting time to be doing chemistry, and that is why I see it as a new frontier area.

Are you satisfied with the curriculum in science as it is, or would you make some changes?

It is chemists who are making new materials and making and exploring biological systems. It's the chemist who looks at questions of molecular detail and asks about structure and its relationship to function. Since this involves so many areas of chemistry, I believe that we are going to have to stop making divisions between inorganic, physical, analytical, and organic chemistry. We must all do a little bit of each. This is an attitude shared by many in chemical education today, and it means that we should perhaps rethink the curriculum in chemistry in

particular and science in general.

No matter what the curricular structure, however, I believe what is important in the education of scientists is to get across the excitement that now we can know what biologically important molecules look like. And, from knowing what they look like, we can manipulate them and change them a little. Then we can ask how those changes affect the function, so we can relate the structure of the molecule and its macroscopic function.

A protein molecule of average size is so small, you could put more than a billion billion of them on the head of a pin. We now know we can manipulate molecules that are of those dimensions and can know exactly what they look like. I can't imagine that we can't get people interested in chemistry if we can get across the excitement that comes from the realization that we are looking at things so small and yet can do surgery on them.

Could you describe those special attributes of chemistry that first attracted you and have kept you interested? In other words, how would you "sell" chemistry?

The bottom line is that chemistry is fun, it's addictive, and, if one has a sense of curiosity, it can be tremendously entertaining and appealing. And it is not so difficult. It's difficult when one thinks about it as rote memorization, which *is* difficult and boring. But that isn't what chemistry is. Chemistry is trying to understand the world around us in some detail. For example, we are interested in knowing such things as what makes skin soft, what makes things different in color, why sugar is sweet,[5] or why a particular pharmaceutical agent makes us feel better.

What is your perspective on the issue of women in science? Do you see

---

[4]An HPLC is a "high-pressure liquid chromatograph," an instrument capable of separating one type of molecule from another.

[5]See Box 3.1, "Aspartame, the Sweet Peptide," in Chapter 3.

special opportunities for women? Are there problems that women need to overcome or be aware of?

Because I am a woman, and there are so few women currently in professional positions in chemistry, I'm asked those questions often. First of all, I am not an expert on the subject. What I like to think my best contribution to women in chemistry can be is to do the best science I can, and to be recognized for my science, not for being a woman in science. I think that it is generally important when women go into science that they should appreciate that there are no special opportunities; that is, you will be treated like any other person doing science. But just as there should be no special opportunities in that respect, happily — maybe this is naive of me — I think there are also no special detriments or obstacles that one need consider in this day and age. One shouldn't think that 'because I am a woman I can't do it.' That's patently false. In fact, everyone is extremely supportive of women who do science. However, I remember talking to Bernice Segal, my former teacher at Barnard College, and having her explain to me that when she was a graduate student she had to do things behind a curtain, because the women weren't supposed to be doing chemistry. Mildred Cohn, another one of my role models, took over 20 years to have her own independent position as a professor, as opposed to being a laboratory assistant working for someone else. The bottom line is that I don't have a story like that to tell. That's the good news. In my generation there are few such stories of blatant discrimination. Now the world is a much better place for a woman to do science.

Dr. Barton's enthusiasm for her work and for chemistry in general is obvious. It is evident that she will continue to do some of the most important work in science and that her infectious enthusiasm for chemistry will bring many more young people into the profession.

Biosynthesis of Nucleic Acids: Replication and Transcription of the Genetic Code

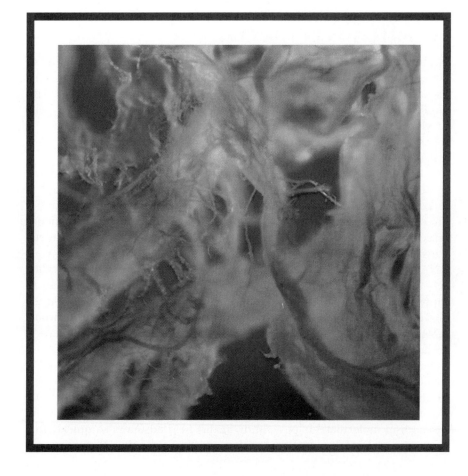

Fibers of purified DNA viewed under polarized light.

Before double helical DNA can be replicated, helical sections of DNA must be unwound so that the two parental strands can serve as templates for the synthesis of new daughter strands, thus making a precise copy of the original double helix. DNA polymerases synthesize DNA by joining nucleotides to the exposed single-stranded DNA template, with each new base bound to its complementary partner. The fidelity of DNA synthesis is of utmost importance, since errors of replication will be passed to future generations. The polymerases have "proofreading" powers capable of self-correcting. In the next stage of DNA processing, the sequence of DNA bases is *transcribed* into a complementary sequence of RNA bases called messenger RNA. The RNA message differs from DNA in one respect: the DNA base thymine (T) is replaced by the RNA base uracil (U). In eukaryotes, messenger RNA carries the genetic code from the nucleus to the ribosomes in the cytosol where the sequence of RNA bases is *translated* into the amino acid sequence of proteins. Copying the genetic message is a powerful way to amplify the production of protein molecules. Proteins in turn are the workhorses of the cell. They play a structural role, as well as serving as antibodies and receptors on membranes. Above all, they are catalysts, a function they share with only a few kinds of RNA.

19.1
THE FLOW OF GENETIC INFORMATION IN THE CELL

The sequence of bases in DNA contains the *genetic code*. The duplication of DNA, giving rise to a new DNA molecule with the same base sequence as the original, is necessary whenever a cell divides to produce daughter cells. This duplication process is called **replication.** The actual formation of gene products requires RNA; the production of RNA on a DNA template is called **transcription** of the genetic message. The base sequence of DNA is reflected in the base sequence of RNA.

Three kinds of RNA are involved in the biosynthesis of proteins; of the three, messenger RNA (mRNA) is of particular importance. A sequence of three bases in mRNA specifies the identity of one amino acid in a manner directed by the genetic code. The process by which the base sequence directs the amino acid sequence is called the **translation** of the genetic message. In nearly all organisms, the flow of genetic information is DNA $\longrightarrow$ RNA $\longrightarrow$ protein. The only major exceptions are some viruses (called retroviruses) in which RNA, rather than DNA, is the genetic material. In those viruses RNA can direct its own synthesis as well as that of DNA; the enzyme *reverse transcriptase* catalyzes this process. (Not all viruses in which RNA is the genetic material are retroviruses, but all retroviruses have a reverse transcriptase. See the article by Varmus listed in the bibliography at the end of this chapter.) In cases of infection by retroviruses, such as HIV, this enzyme is a target for drug design. Figure 19.1 shows ways in which information is transferred in the cell.

The nucleic acids are the focus of some of the most exciting current research in molecular biology. Study in this field advances so quickly that some discoveries are out of date within months of the time they appear in

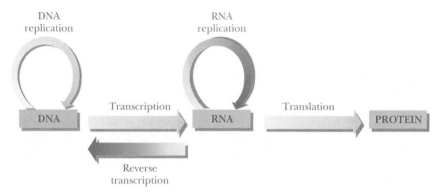

FIGURE 19.1

Mechanisms for transfer of information in the cell. The yellow arrows represent general cases, and the blue arrows represent special cases (mostly in RNA viruses).

print. The material in this chapter is well-established basic information on the subject. Considerably more needs to be learned, however; perhaps some of the students now reading this chapter will do research on nucleic acids in the future.

19.2
THE REPLICATION OF DNA: SOME GENERAL CONSIDERATIONS

Naturally occurring DNA exists in many forms. Single- and double-stranded DNAs are known, and both can exist in linear and circular forms. As a result, it is difficult to generalize about all possible cases of DNA replication. Since many DNAs are double-stranded, we can present some general features of the replication of double-stranded DNA, features that apply to both linear and circular DNA. Most of the details of the process that we shall discuss here were first investigated in prokaryotes, particularly *Escherichia coli*. We shall use information obtained by experiments on this organism throughout our discussion of the topic.

The process by which one double-helical DNA molecule is duplicated to produce two such double-stranded molecules is a complex one. The very complexity allows for a high degree of fine-tuning, which in turn ensures considerable fidelity in replication. The cell faces three important challenges in carrying out the necessary steps. The first challenge is how to *separate the two DNA strands*. In addition to achieving continuous unwinding of the double helix, the cell also has to protect the unwound portions of DNA from the action of nucleases that preferentially attack single-stranded DNA. The second task involves the *synthesis of DNA from the 5′ to the 3′ end*. Two antiparallel strands must be synthesized in the same direction on antiparallel templates. In other words, the template has one $5′ \longrightarrow 3′$ strand and one $3′ \longrightarrow 5′$ strand, as does the newly synthesized DNA. The third task is how to *guard against errors in replication,* ensuring that the correct base is added to the growing polynucleotide chain. The answers to these questions require the material in this and the three following sections.

Semiconservative Replication

DNA replication involves separation of the two original strands and production of two new strands with the original ones as templates. Each new DNA molecule contains one strand from the original DNA and one newly synthesized strand. This situation is what is called **semiconservative replication** (Figure 19.2). The details of the process differ in prokaryotes and eukaryotes, but the semiconservative nature of replication is observed in all organisms.

Semiconservative replication of DNA was established unequivocally in the late 1950s by experiments performed by Meselson and Stahl. *E. coli* bacteria were grown with $^{15}NH_4Cl$ as the sole nitrogen source, where ^{15}N is a heavy isotope of nitrogen. The usual form of nitrogen is ^{14}N. In such a medium, all newly formed nitrogen compounds, including purine and pyrimidine nucleobases, become labeled with ^{15}N. The ^{15}N-labeled DNA has a higher density than unlabeled DNA, which contains the usual isotope, ^{14}N. In this experiment, the ^{15}N-labeled cells were then transferred to a medium that contained only ^{14}N. The cells continued to grow in the new medium.

FIGURE 19.2

The labeling pattern of ^{15}N strands in semiconservative replication. (G_0 is original strands; G_1 is new strands after first generation; G_2 is new strands after second generation.)

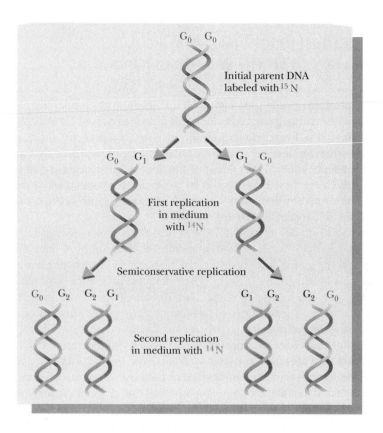

G_0 G_0

Initial parent DNA labeled with ^{15}N

G_0 G_1 G_1 G_0

First replication in medium with ^{14}N

Semiconservative replication

G_0 G_2 G_2 G_1 G_1 G_2 G_2 G_0

Second replication in medium with ^{14}N

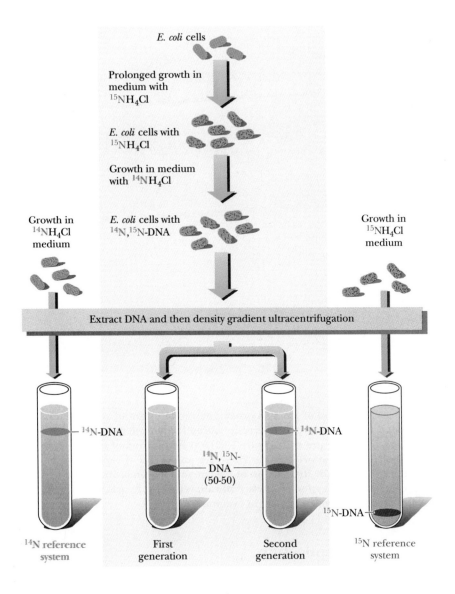

E. coli cells

Prolonged growth in medium with $^{15}NH_4Cl$

E. coli cells with $^{15}NH_4Cl$

Growth in medium with $^{14}NH_4Cl$

Growth in $^{14}NH_4Cl$ medium

E. coli cells with $^{14}N,^{15}N$-DNA

Growth in $^{15}NH_4Cl$ medium

Extract DNA and then density gradient ultracentrifugation

^{14}N-DNA

^{14}N-DNA

$^{14}N, ^{15}N$- DNA (50-50)

^{15}N-DNA

^{14}N reference system

First generation

Second generation

^{15}N reference system

FIGURE 19.3

The experimental evidence for semiconservative replication. Heavy DNA labeled with ^{15}N forms a band at the bottom of the tube, and light DNA with ^{14}N forms a band at the top. DNA that forms a band at an intermediate position has one heavy strand and one light strand.

With every new generation of growth, a sample of DNA was extracted and analyzed by the technique of **density-gradient centrifugation** (Figure 19.3). This technique depends on the fact that heavy ^{15}N DNA (DNA that contains ^{15}N alone) will form a band at the bottom of the tube; light ^{14}N DNA (^{14}N alone) will appear at the top of the tube. DNA containing a 50–50 mixture of ^{14}N and ^{15}N DNA will appear at a position halfway between the two bands. In the actual experiment this 50–50 hybrid DNA was observed after one generation, a result to be expected with semiconservative replication. After two generations in the lighter medium, half the DNA in the cells should be the 50–50 hybrid and half should be the lighter ^{14}N DNA. This prediction of the kind and amount of DNA observed is confirmed by the experiment.

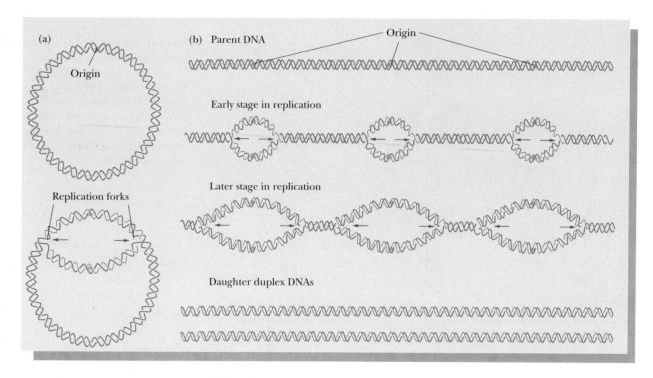

FIGURE 19.4

Bidirectional replication of DNA in prokaryotes (one origin of replication) and in eukaryotes (several origins). Bidirectional replication refers to overall synthesis (compare this figure with Figure 19.5). (a) Replication of the chromosome of *E. coli*, a typical prokaryote. There are one origin of replication and two replication forks. (b) Replication of a eukaryotic chromosome. There are several origins of replication and two replication forks for each origin. The "bubbles" that arise from each origin eventually coalesce.

Bidirectional Replication

During replication the DNA double helix unwinds at a specific point called the **origin of replication** (Ori). New polynucleotide chains are synthesized using each of the exposed strands as a template. Two possibilities exist for the growth of the new strands: synthesis can take place in both directions from the origin of replication or in only one direction. It has been established that DNA synthesis is bidirectional in most organisms, with the exception of a few viruses. For each origin of replication there are two points (**replication forks**) at which new polynucleotide chains are formed. A "bubble" of newly synthesized DNA between regions of the original DNA is a manifestation of the advance of the two replication forks in opposite directions. There is one such bubble (and one origin of replication) in the circular DNA of prokaryotes (Figure 19.4a). In eukaryotes several origins of replication and thus several bubbles exist (Figure 19.4b). The bubbles grow larger and eventually coalesce, giving rise to two complete daughter DNAs. This bidirectional growth of both new polynucleotide chains represents **net chain growth.** Both new polynucleotide chains are synthesized in the 5′-to-3′ direction.

19.3
THE DNA POLYMERASE REACTION

One Strand of DNA Is Synthesized Discontinuously

A major challenge for the cell in DNA replication is how to achieve $5' \rightarrow 3'$ polymerization in the opposite direction from the template strand, which is itself exposed from the $5' \rightarrow 3'$ direction. (There is no problem with the other strand, which is exposed by unwinding from the $3'$ to the $5'$ end.)

The problem is solved by different modes of polymerization for the two growing strands. One newly formed strand (the leading strand) is formed continuously from its $5'$ end to its $3'$ end at the replication fork on the exposed $3'$-to-$5'$ template strand. The other strand (the lagging strand) is formed discontinuously in small fragments (typically 1000 to 2000 nucleotides long), sometimes called **Okazaki fragments** after the scientist who first studied them (Figure 19.5). The $5'$ end of each of these fragments is closer to the replication fork than the $3'$ end. The fragments of the lagging strand are then linked enzymatically by an enzyme called **DNA ligase.**

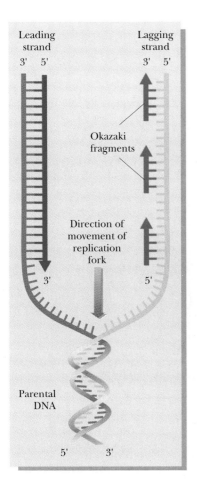

Leading strand — Lagging strand

Okazaki fragments

Direction of movement of replication fork

Parental DNA

FIGURE 19.5

Discontinuous and continuous synthesis of DNA strands at a replication fork. The leading strand of the DNA is synthesized continuously in the $5' \rightarrow 3'$ direction. In the lagging strand, short fragments (Okazaki fragments) are synthesized, also in the $5' \rightarrow 3'$ direction.

DNA Polymerase from *E. Coli*

The first DNA polymerase discovered was found in *E. coli*. A universal feature of DNA replication is that the nascent chain grows from the 5′ to the 3′ end; there is a 5′-phosphate on the sugar at one end and a free 3′-hydroxyl on the sugar at the other end. *DNA polymerase* catalyzes the successive addition of each new nucleotide to the growing chain. The 3′-hydroxyl group at the end of the growing chain is a nucleophile. It attacks the phosphorus adjacent to the sugar in the nucleotide to be added to the growing chain, leading to the elimination of the pyrophosphate and the formation of a new phosphodiester bond (Figure 19.6). We discussed nucleophilic attack by a hydroxyl group at length in the case of serine proteases (Chapter 5, Section 5.12); here we see another instance of this kind of mechanism.

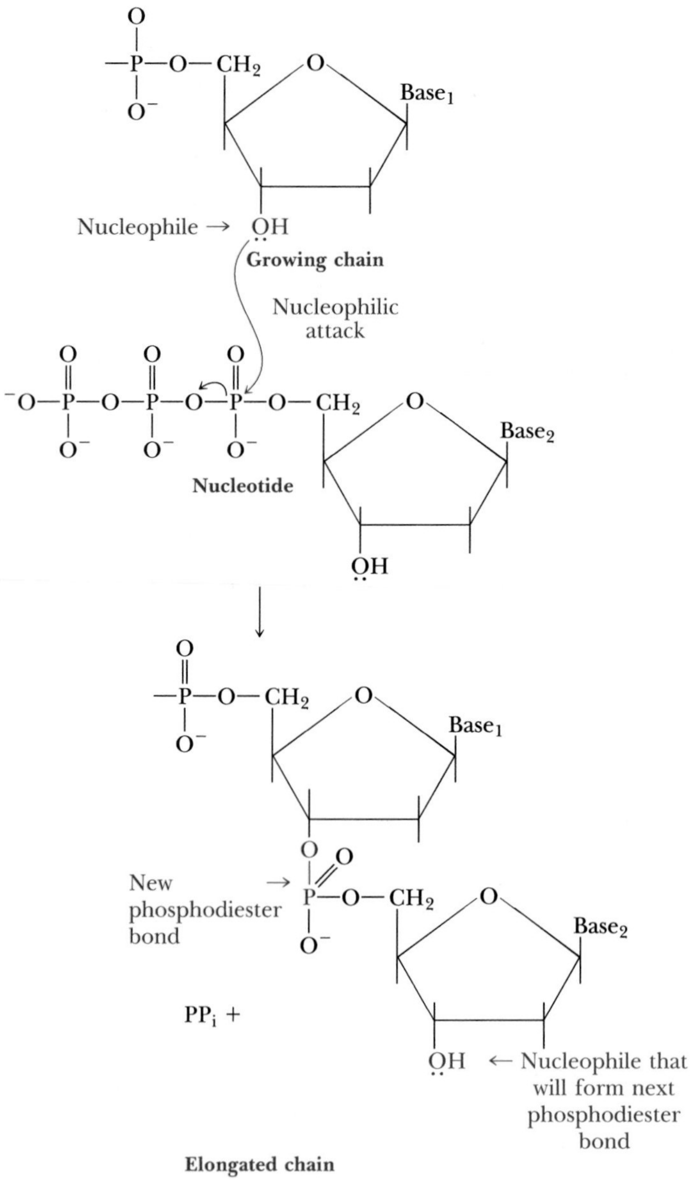

FIGURE 19.6

The addition of a nucleotide to a growing DNA chain. The 3′-hydroxyl group at the end of the growing DNA chain is a nucleophile. It attacks at the phosphorus adjacent to the sugar in the nucleotide, which will be added to the growing chain. Pyrophosphate is eliminated, and a new phosphodiester bond is formed.

Elongated chain

TABLE 19.1 Properties of DNA Polymerases of *E. coli*

	POLYMERASE I	POLYMERASE II	POLYMERASE III
Functions			
Polymerization: $5' \longrightarrow 3'$	+	+	+
Exonuclease $3' \longrightarrow 5'$	+	+	+
Exonuclease $5' \longrightarrow 3'$	+	−	+
Molecular weight	109,000	120,000	140,000
Molecules/cell	400	30	10–20
Number of nucleotides polymerized/min at 37° by one enzyme molecule	600		9,000

Data from A. Kornberg, 1980, *DNA Replication*, W. H. Freeman, New York, p. 169.

There are actually three DNA polymerases in *E. coli*; some of their properties are listed in Table 19.1. DNA polymerase I (Pol I) was discovered first, with the subsequent discovery of polymerases II (Pol II) and III (Pol III). Polymerases I and II consist of a single polypeptide chain, but polymerase III is a multisubunit protein (Figure 19.7). All these enzymes add nucleotides to a growing polynucleotide chain but have different roles in the overall replication process. All three enzymes require the presence of a **primer,** a short strand of RNA to which the growing polynucleotide chain is covalently attached in the early stages of replication.

The DNA polymerase reaction requires all four deoxyribonucleoside triphosphates: dTTP, dATP, dGTP, and dCTP (Figure 19.8). Mg^{2+} and DNA itself are also necessary. Because of the requirement for an RNA primer, all four ribonucleoside triphosphates–ATP, UTP, GTP, and CTP—are needed as well; they are incorporated into the primer. The primer (RNA) is hydrogen bonded to the template (DNA); the primer provides a stable framework on which the nascent chain can start to grow. The newly synthesized DNA strand begins to grow by forming a covalent linkage to the free 3′-hydroxyl group of the primer.

It is now known that DNA polymerase I has a specialized function in replication, that of repair and "patching" of DNA, and that DNA polymerase III is the enzyme primarily responsible for the polymerization of the newly formed DNA strand. The major function of DNA polymerase II is not known. The exonuclease activity listed in Table 19.1 is part of the proofreading and repair functions of DNA polymerases; it is a process by which

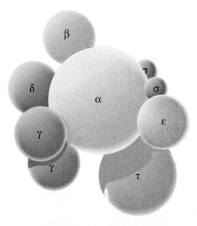

FIGURE 19.7

The subunit structure of DNA polymerase III of *E. coli*. The α subunit is the polymerase itself (the protein of molecular weight 140,000 listed in Table 19.1). The whole complex, called the DNA polymerase holoenzyme, has a molecular weight of 550,000. The β subunit is required for recognition of the parental DNA. Once the complex is bound to the DNA to be copied, the β subunit dissociates, leaving an active polymerase of molecular weight 400,000.

$$\text{dTTP, dATP, dGTP, dCTP} \xrightarrow{\text{DNA-dependent DNA polymerase}} \text{DNA (polymer)} + PP_i$$
(All four required)

FIGURE 19.8

The requirements for the DNA polymerase reaction. Template DNA, Mg^{2+}, and an RNA primer are also required. Because of the need for an RNA primer, there is also an implicit requirement for all four ribonucleoside triphosphates (ATP, UTP, GTP, and CTP) for formation of the primer.

incorrect nucleotides are removed from the polynucleotide so that the correct nucleotides can be incorporated. The $3' \longrightarrow 5'$ exonuclease activity, which all three polymerases possess, is part of the **proofreading** function; incorrect nucleotides are removed in the course of replication and replaced by the correct ones. Proofreading is done one nucleotide at a time. The $5' \longrightarrow 3'$ exonuclease activity clears away short stretches of nucleotides during **repair,** usually involving several nucleotides at a time.

19.4
DNA REPLICATION REQUIRES THE COMBINED ACTIONS OF SEVERAL ENZYMES

Unwinding the Double Helix

Two questions arise in separating the two strands of the original DNA so that it can be replicated. The first is how to achieve continuous unwinding of the double helix. This question is complicated by the fact that prokaryotic DNA exists in supercoiled, closed-circular form (see Tertiary Structure of DNA: Supercoiling, in Chapter 6, Section 6.3). The second, related question is how to protect single-stranded stretches of DNA that are exposed to intracellular nucleases as a result of the unwinding.

An enzyme called **DNA gyrase** catalyzes the conversion of relaxed, circular DNA with a nick in one strand to the supercoiled form with the nick sealed (Figure 19.9). A slight unwinding of the helix before the nick is sealed introduces the supercoiling. The energy required for the process is supplied by the hydrolysis of ATP. Some evidence exists that DNA gyrase causes a double-strand break in DNA in the process of converting the relaxed, circular form to the supercoiled form (a mode of action typical of Type II topoisomerases, the class of enzymes to which DNA gyrase belongs). In

FIGURE 19.9

DNA gyrase introduces super-twisting in circular DNA.

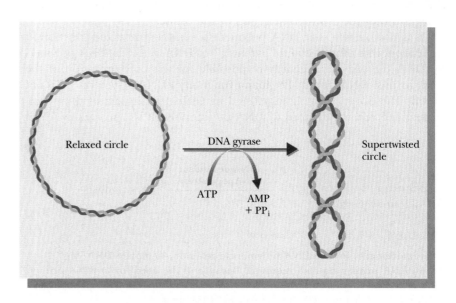

Relaxed circle   DNA gyrase   Supertwisted circle

ATP   AMP + PP$_i$

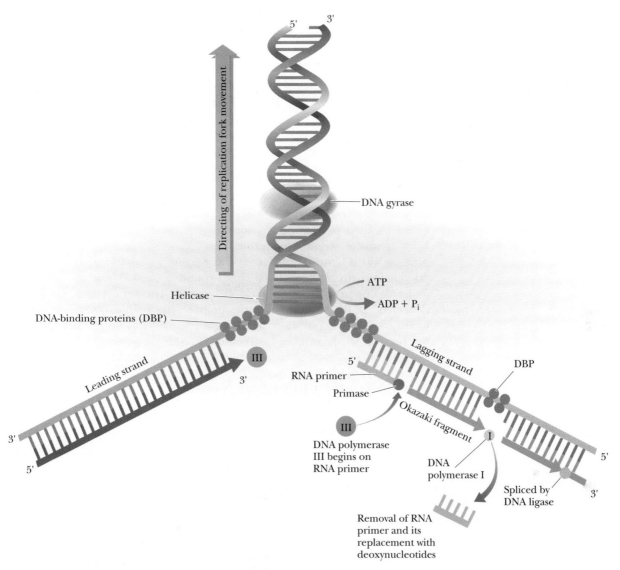

FIGURE 19.10

The steps in DNA replication. The exact site of the swivel point for unwinding of parental DNA and the site of action of DNA gyrase are not known with certainty. The DNA-binding protein is also called a single-strand binding protein.

replication, the role of the gyrase is somewhat different. It introduces a nick in supercoiled DNA; the reverse of the reaction that produces supercoiling gives rise to a **swivel point** in the DNA at the site of the nick. The gyrase opens and reseals the swivel point in advance of the replication fork (Figure 19.10). (The newly formed DNA automatically assumes the supercoiled form, since it does not have the nick at the swivel point.) A helix-destabilizing protein, also called a **helicase,** promotes unwinding by binding at the replication fork. This protein is also called the **rep protein.** Another protein, called SSB (single-strand binding) protein, stabilizes the single-stranded regions by binding tightly to these portions of the molecule. The presence of this DNA-binding protein protects the single-stranded regions from hydrolysis.

The Primase Reaction

One of the great surprises in studies of DNA replication was the discovery that *RNA serves as a primer in DNA replication.* In retrospect, it is not surprising at all, since RNA can be formed *de novo* without a primer, while DNA synthesis requires a primer. This finding lends support to theories of the origin of life in which RNA, rather than DNA, was the original genetic material. A primer in DNA replication must have a free 3'-hydroxyl to which the growing chain can attach, and both RNA and DNA can provide this group. The primer activity of RNA was first observed *in vivo.* In some of the original *in vitro* experiments, DNA was used as a primer, since a primer consisting of DNA was expected. Living organisms are, of course, far more complex than isolated molecular systems and, as a result, can be full of surprises for researchers. It has subsequently been found that a separate enzyme, called **primase,** is responsible for copying a short stretch of the DNA template strand to produce the RNA primer sequence. The first primase was discovered in *E. coli.* The enzyme consists of a single polypeptide chain, with a molecular weight around 60,000. There are 50 to 100 molecules of primase in a typical *E. coli* cell. The primer and the protein molecules at the replication fork compose the **primosome.** The general features of DNA replication, including the use of an RNA primer, appear to be common to all prokaryotes (Figure 19.10).

Synthesis and Linking of New DNA Strands

The synthesis of two new strands of DNA is begun by DNA polymerase III. The newly formed DNA is linked to the 3'-hydroxyl of the RNA primer, and synthesis proceeds from the 5' to the 3' end on both the leading and the

T A B L E   1 9 . 2   **A Summary of DNA Replication in Prokaryotes**

1. DNA synthesis is bidirectional. Two replication forks advance in opposite directions from an origin of replication.
2. The direction of DNA synthesis is from the 5' to the 3' end of the newly formed strand. One strand (the leading strand) is formed continuously, while the other strand (the lagging strand) is formed discontinuously. In the lagging strand, small fragments of DNA (Okazaki fragments) are subsequently linked together.
3. Three DNA polymerases have been found in *E. coli.* Polymerase III (Pol III) is primarily responsible for the synthesis of new strands. The first polymerase enzyme discovered, polymerase I (Pol I), is a repair enzyme. The function of polymerase II is unknown.
4. DNA gyrase introduces a swivel point in advance of the movement of the replication fork. A helix-destabilizing protein, helicase, binds at the replication fork and promotes unwinding. The exposed single-stranded regions of the template DNA are stabilized by a DNA-binding protein.
5. Primase catalyzes the synthesis of an RNA primer.
6. The synthesis of new strands is catalyzed by Pol III. The primer is removed by Pol I, which also replaces the primer with deoxynucleotides. DNA ligase seals the remaining nicks.

lagging strands. Two molecules of Pol III, one for the leading strand, one for the lagging strand, are physically linked to the primosome; the resulting multiprotein complex is called the **replisome.** As the replication fork moves away, the RNA primer is removed by polymerase I, using its exonuclease activity. The primer is replaced by deoxynucleotides, also by DNA polymerase I, using its polymerase activity. (The removal of the RNA primer and its replacement with the missing portions of the newly formed DNA strand by polymerase I are the repair function we mentioned earlier.) None of the DNA polymerases can seal the nicks that remain; DNA ligase is the enzyme responsible for the final linking of the new strand. Table 19.2 summarizes the main points of DNA replication in prokaryotes.

19.5
PROOFREADING AND REPAIR

On the average, DNA replication takes place only once each generation in each cell, unlike other processes, such as RNA and protein synthesis, which occur many times. It is essential that the fidelity of the replication process be as high as possible to prevent **mutations,** which are errors in replication. Most mutations are harmful, even lethal, to organisms. Nature has devised several ways to ensure that the base sequence of DNA is copied faithfully. Errors in replication occur spontaneously only once in every 10^9 to 10^{10} base pairs.

Proofreading removes incorrect nucleotides immediately after they are added to the growing DNA during the replication process. Errors in hydrogen bonding lead to the incorporation of an incorrect nucleotide into a growing DNA chain once in every 10^4 to 10^5 base pairs (Figure 19.11a).

FIGURE 19.11

Proofreading in DNA replication. (a) When an incorrect nucleotide has been incorporated into a growing polynucleotide chain, (b) DNA polymerase I uses its exonuclease activity to remove the mismatched nucleotide. (c) The polymerase then adds the correct nucleotide, and replication proceeds. (From M. Radman and R. Wagner, 1988, The High Fidelity of DNA Replication, *Sci. Amer.* 259 (2), 40–46.)

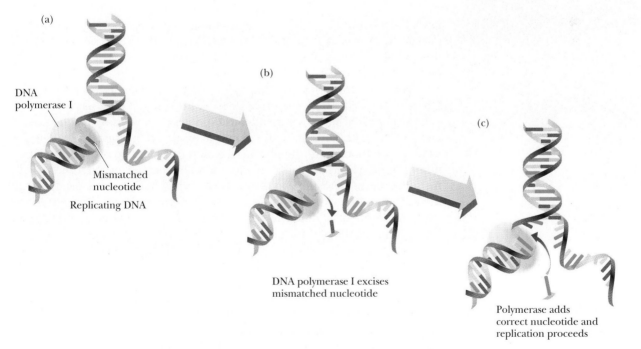

(a)

DNA polymerase I

Mismatched nucleotide

Replicating DNA

(b)

DNA polymerase I excises mismatched nucleotide

(c)

Polymerase adds correct nucleotide and replication proceeds

SITE-DIRECTED MUTAGENESIS

The introduction of mutations at specific points in a DNA sequence depends on the availability of chemically synthesized oligonucleotides. The sequence of these oligonucleotides, typically about 15 residues long, differs from that of a target DNA sequence by no more than a few residues. The oligonucleotide can anneal to the complementary sequence of a single-stranded DNA; the hydrogen-bonded synthetic oligonucleotide is then used as a primer for DNA synthesis. An altered strand has been introduced into the DNA sequence and can be cloned by the methods we discussed in Chapter 7. The mutated DNA can be used as a source of DNA, RNA, or a protein. This method has proved to be of great usefulness to biochemists who want to study the effect of changing a single selected residue in proteins. It eliminates the element of chance in looking for naturally occurring mutations by making it possible to produce "custom-made" alterations in a given protein. This method was developed by Michael Smith, a feat for which he shared the 1993 Nobel Prize in chemistry with the inventor of the polymerase chain reaction.

Mismatched primer

```
3'  C       GT        5'
GAA TACG GCACT
  | | |  | | | |     | | | | |
CTTTATGCAGCGTGA
5'                        3'
```

DNA polymerase → dATP + dCTP + dGTP + dTTP

```
3'    C        GT         5'
GAA TACG   GCACT
 | | | | | | | | | | | | | | |
CTTTATGCAGCGTGA
5'                           3'
```

Altered gene

Site-directed mutagenesis. A chemically synthesized oligonucleotide is annealed to the DNA of the gene to be altered. With the oligonucleotide as primer, the DNA polymerase reaction is allowed to proceed, giving rise to DNA with the altered gene. The mutated DNA can be cloned to give a desired DNA, RNA, or protein.

DNA polymerase I uses its 3′ exonuclease activity to remove the incorrect nucleotide (Figure 19.11b). Replication resumes when the correct nucleotide is added, also by DNA polymerase I (Figure 19.11c). The specificity of hydrogen-bonded base pairing accounts for one error in every 10^4 to 10^5 base pairs; the proofreading function of DNA polymerase improves the fidelity of replication to one error in every 10^9 to 10^{10} base pairs.

During replication, a **cut-and-patch** process catalyzed by polymerase I takes place. The cutting is the removal of the RNA primer by the 5′ exonuclease function of the polymerase, and the patching is the incorporation of the required deoxynucleotides, done by the polymerase function of the same enzyme. Existing DNA can also be repaired by polymerase I, using the cut-and-patch method, if one or more bases have been damaged by an external agent. In addition to experiencing those spontaneous mutations caused by misreading of the genetic code, organisms are frequently exposed to **mutagens,** agents that produce mutations. Common mutagens include ultraviolet light, ionizing radiation (radioactivity), and various chemical agents, all of which lead to changes in DNA over and above those produced by spontaneous mutation. In the repair of existing DNA, the process of cutting and patching is called **excision-repair;** the damaged portion of the DNA is removed, and the new, correct portion is substituted. Box 19.1 describes a situation in which mismatched bases are deliberately introduced by biochemists to produce desired mutations.

Defects in DNA repair mechanisms can have drastic consequences. One of the most remarkable examples is the disease **xeroderma pigmentosum.** Affected individuals develop numerous skin cancers at an early age because they do not have the repair system to correct damage caused by ultraviolet light. The endonuclease that nicks the damaged portion of the DNA is probably the missing enzyme. The cancerous lesions eventually spread throughout the body, causing death.

19.6
MODIFICATION OF DNA AFTER REPLICATION

Once DNA is formed by the replication mechanism of the cell, modifications can take place for specific purposes. The most important of these modifications is **methylation.** Specific adenine and thymine residues are altered in prokaryotes, but only cytosine residues are affected in eukaryotes. The products include N^6-methyladenine and 5-methylcytosine. S-Adenosylmethionine (see One-Carbon Transfers and the Serine Family in Chapter 17, Section 17.4) is the methyl group donor. The residues modified are frequently found in sequences that would otherwise be attacked by restriction endonucleases. These enzymes (see Box 6.2 in Chapter 6), which occur mainly in bacteria, produce breaks in specific places in both strands of DNA. The methylation protects the DNA of the bacterium from the restriction endonuclease (Figure 19.12), but DNA of an infecting bacteriophage does not have the methyl groups and is cleaved. The cleavage of the bacterio-

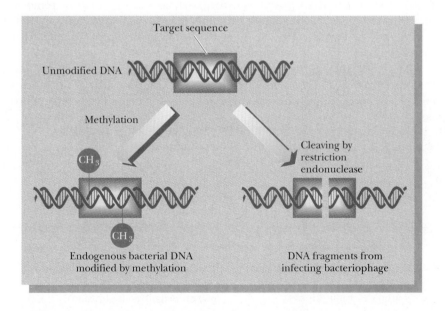

FIGURE 19.12

Methylation of endogenous DNA protects it from cleavage by its own restriction nucleases.

phage DNA restricts the growth of the bacteriophage in the infected cell, giving rise to the name "restriction nuclease."

19.7
THE REPLICATION OF DNA IN EUKARYOTES

Three different DNA polymerases have been isolated from animal systems. The use of animals rather than plants for study avoids the complication of any DNA synthesis in chloroplasts. The three different polymerases are called α, β, and γ. The α- and β-enzymes are found in the nucleus, and the γ form occurs in mitochondria. DNA polymerase α plays a role similar to that of DNA polymerase III in prokaryotes, in the sense that the α-enzyme is primarily responsible for the polymerization of DNA. DNA polymerase β appears to be a repair enzyme. DNA polymerase γ carries out DNA replication in mitochondria. None of the DNA polymerases isolated from animals acts as an exonuclease, and in this regard the animal enzymes differ from prokaryotic DNA polymerases. Separate exonucleolytic enzymes exist in animal cells.

The general features of DNA replication in eukaryotes are similar to those in prokaryotes but are not as extensively studied. Many replication origins are found in eukaryotes, rather than a single one as in prokaryotes, but the steps in the replication process are basically the same. As with prokaryotes, DNA replication in eukaryotes is semiconservative. There is a leading strand with continuous synthesis in the $5' \longrightarrow 3'$ direction and a lagging strand with discontinuous synthesis in the $5' \longrightarrow 3'$ direction. An RNA primer is formed by a specific enzyme in eukaryotic DNA replication, as is the case with prokaryotes. The formation of Okazaki fragments (typically 150 to 200 nucleotides long in eukaryotes) is catalyzed by DNA polymerase α. The RNA primer is hydrolyzed, and the gaps left by removal of the primer are filled in by a reaction that is also catalyzed by DNA polymerase α. Finally, DNA ligase seals the nicks that separate the fragments.

An important difference between DNA replication in prokaryotes and that in eukaryotes is that prokaryotic DNA exists in "naked" form, not complexed to proteins, whereas eukaryotic DNA is complexed to proteins,

T A B L E 1 9 . 3 **Differences in DNA Replication in Prokaryotes and Eukaryotes**

PROKARYOTES	EUKARYOTES
Three polymerases (I, II, III)	Three polymerases (α, β, γ)
Functions of polymerases:	Fuctions of polymerases:
I is a repair enzyme	α is main polymerizing enzyme
II — function unknown	β is a repair enzyme
III is main polymerizing enzyme	γ — mitochondrial DNA synthesis
Polymerases are also exonucleases	Separate enzymes are exonucleases
One origin of replication	Several origins of replication
Okazaki fragments 1000–2000 residues long	Okazaki fragments 150–200 residues long
No proteins complexed to DNA	Histones complexed to DNA

principally histones. Histone biosynthesis occurs at the same time and at the same rate as DNA biosynthesis. In eukaryotic replication, histones are associated with DNA as it is formed. Table 19.3 summarizes the differences between prokaryotes and eukaryotes in DNA replication.

19.8
RNA BIOSYNTHESIS: TRANSCRIPTION OF THE GENETIC MESSAGE

The details of RNA transcription differ somewhat in prokaryotes and eukaryotes. Most of the research on the subject has been done in prokaryotes, especially *E. coli,* but some general features are found in all organisms except RNA viruses. Table 19.4 summarizes the main features of the process.

RNA Polymerase from *Escherichia coli*

The most extensively studied RNA polymerase is that isolated from *E. coli.* The molecular weight of this enzyme is about 500,000, and it has a multisubunit structure. Four different types of subunits, designated α, β, β', and σ, have been identified. The actual composition of the enzyme is $\alpha_2\beta\beta'\sigma$. The σ subunit is rather loosely bound to the rest of the enzyme (the $\alpha_2\beta\beta'$ portion), which is called the **core enzyme.** The **holoenzyme** consists of all the subunits, including the σ subunit. The essential role of the σ subunit is recognition of the **promoter locus** (a DNA sequence that signals the start of RNA transcription). The loosely bound σ subunit is released after transcription begins.

The promoter region to which RNA polymerase binds is closer to the $3'$ end of the DNA than is the actual gene for the RNA to be synthesized. (The RNA is formed from the $5'$ to the $3'$ end, so the polymerase moves along the DNA from the $3'$ to the $5'$ end.) The binding site for the polymerase is said to lie *upstream* of the start of transcription. The base sequence of promoter regions has been determined for a number of prokaryotic genes, and a striking feature is that they contain many sequences in common (**consensus**

T A B L E 1 9 . 4   **General Features of RNA Synthesis**

1. All RNAs are synthesized on a DNA template; the enzyme that catalyzes the process is **DNA-dependent RNA polymerase.**
2. All four ribonucleoside triphosphates (ATP, GTP, CTP, and UTP) are required, as is Mg^{2+}.
3. A primer is not needed in RNA synthesis, but a DNA template is required.
4. As is the case with DNA biosynthesis, the RNA chain grows from the $5'$ to the $3'$ end (Figure 19.13). The nucleotide at the $5'$ end of the chain retains its triphosphate group (abbreviated ppp).
5. The enzyme uses one strand of the DNA (the sense strand) as the template for RNA synthesis. The base sequence of the DNA contains signals for initiation and termination of RNA synthesis. The enzyme binds to the sense strand and moves along it in the $3'$-to-$5'$ direction.
6. The template is unchanged (Figure 19.13).

sequences). Promoter regions are A–T-rich, with two hydrogen bonds per base pair, and thus more easily unwound than G–C-rich regions, with three hydrogen bonds per base pair. In prokaryotes these consensus regions frequently occur 35 base pairs (bp) and 10 bp upstream from the start of transcription. These two regions are called the **−35 box** and the **−10 box** (also called the **Pribnow box**). The consensus sequence for a prokaryotic promoter is

$$-35 \qquad\qquad -10$$
$$\text{TTGACA 17 bp TATAAT}$$

In eukaryotes, a similar sequence called the **TATA box** lies about 25 bp upstream from the start of transcription.

A good deal more can be said about the operation of the promoter region of genes, but we are going to need some information about protein synthesis for an effective discussion of the subject. The control of genetic expression is complex and important enough that we shall return to the topic in Chapter 20, Section 20.9, to look at it in light of new information.

FIGURE 19.13

The transcription of RNA on a DNA template. In *E. coli*, transcription begins when the RNA polymerase holoenzyme ($\alpha_2\beta\beta'\sigma$) binds to DNA. The σ subunit is rquired for binding but dissociates from the complex after initiation of transcription. Synthesis of RNA proceeds from the 5′ to the 3′ end. Termination of chain growth requires the ρ subunit.

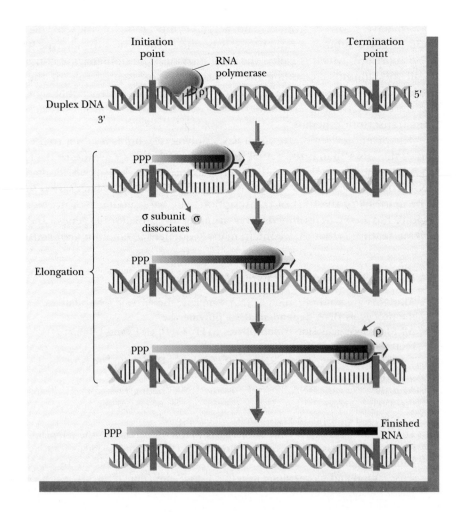

Termination of RNA transcription also involves specific sequences *downstream* of the actual gene for the RNA to be transcribed, and in prokaryotes a protein designated ρ (rho) is involved in the termination process. This **ρ-protein,** which is a tetramer with a molecular weight of approximately 200,000, binds to DNA, RNA, and RNA polymerase. The ρ-factor triggers the release of the completed RNA from the DNA–RNA polymerase complex (Figure 19.13). No ρ-protein has been detected in eukaryotic systems.

19.9
POSTTRANSCRIPTIONAL MODIFICATION OF RNA

The three kinds of RNA—transfer RNA (tRNA), ribosomal RNA (rRNA), and messenger RNA (mRNA)—are all modified enzymatically after transcription to give rise to the functional form of the RNA in question. The type of processing in prokaryotes can differ greatly from that in eukaryotes, especially in the case of mRNA. The initial size of the RNA transcripts is greater than the final size because of leader sequences at the 5′ end and trailer sequences at the 3′ end. The leader and trailer sequences must be removed, and other forms of **trimming** are possible as well. *Terminal sequences* can be added after transcription, and *base modification* is frequently observed, especially in tRNA.

Transfer RNA and Ribosomal RNA

The precursor of several tRNA molecules is frequently transcribed in one long polynucleotide sequence. All three types of modification—trimming, addition of terminal sequences, and base modification—take place in the transformation of the initial transcript to the mature tRNAs (Figure 19.14). (The enzyme that is responsible for generating the 5′ ends of all *E. coli* tRNAs, **RNase P,** consists of both RNA and protein. The RNA moiety is responsible for the catalytic activity. This was one of the first examples of catalytic RNA [Section 19.10].) Some base modifications take place before trimming, and some occur after. Methylation and substitution of sulfur for oxygen are two of the more usual types of base modification. (See Chapter 6, Section 6.2, and Transfer RNA in Section 6.5 for the structures of some of the modified bases.) One type of methylated nucleotide found only in eukaryotes contains a 2′-*O*-methylribosyl group (Figure 19.15).

The trimming and addition of terminal nucleotides produce tRNAs with the proper size and base sequence. Every tRNA contains a CCA sequence at the 3′ end. The presence of this portion of the molecule is of great importance in protein synthesis, since the 3′ end is the acceptor for amino acids to be added to a growing protein chain. Trimming of large precursors of eukaryotic tRNAs takes place in the nucleus, but most methylating enzymes occur in the cytosol.

The processing of rRNAs is primarily a matter of methylation and of trimming to the proper size. In prokaryotes there are three rRNAs in an intact ribosome, which has a sedimentation coefficient of 70S. (Recall that we discussed sedimentation coefficients and some aspects of ribosomal

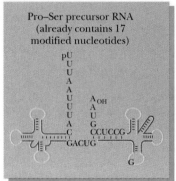

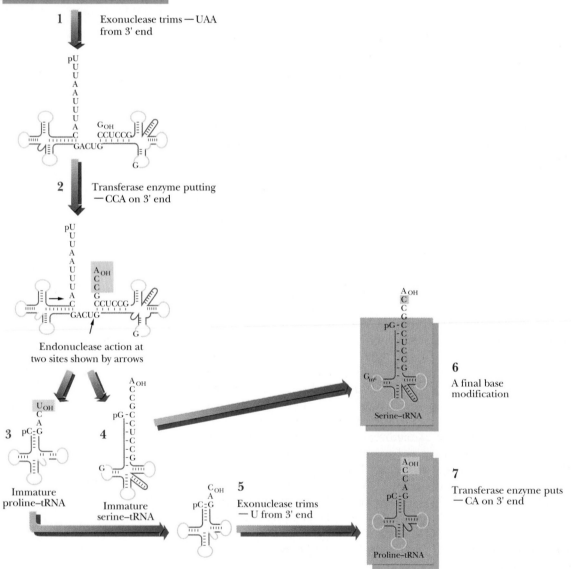

FIGURE 19.14

Posttranscriptional modification of a tRNA precursor. Dots represent hydrogen-bonded base pairs. The symbols G$_{OH}$, C$_{OH}$, A$_{OH}$, and U$_{OH}$ refer to a free 3′ end without a phosphate group; G$_m^2$ is a methylated guanine.

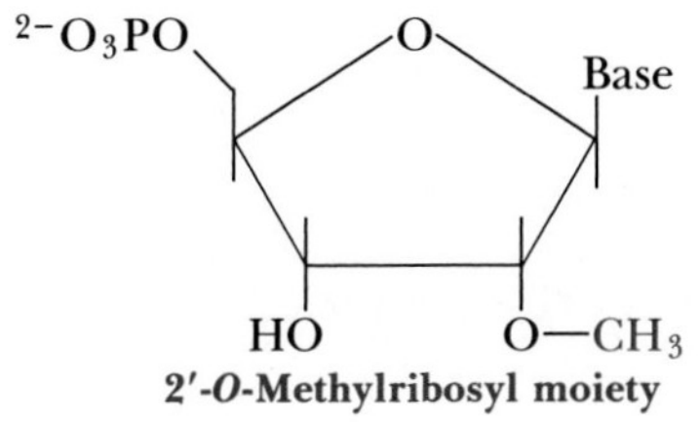

FIGURE 19.15

The structure of a nucleotide containing a 2'-O-methylribosyl group.

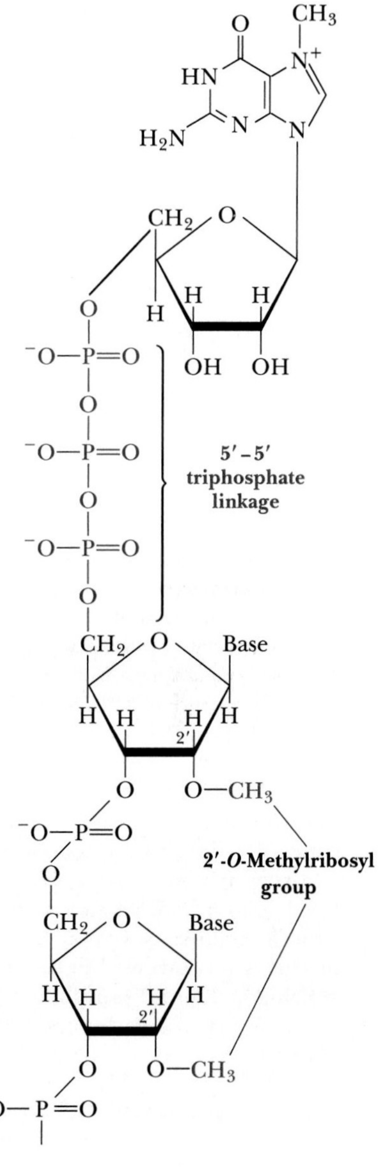

FIGURE 19.16

The structures of some typical mRNA caps.

structure in Ribosomal RNA, in Section 6.5.) In the smaller subunit, which has a sedimentation coefficient of 30S, there is one RNA molecule that has a sedimentation coefficient of 16S. The 50S subunit contains two kinds of RNA, with sedimentation coefficients of 5S and 23S. The ribosomes of eukaryotes have a sedimentation coefficient of 80S, with 40S and 60S subunits. The 40S subunit contains an 18S RNA, and the 60S subunit contains a 5S RNA, a 5.8S RNA, and a 28S RNA. Base modifications in both prokaryotic and eukaryotic rRNA are accomplished primarily by methylation.

Messenger RNA

Extensive processing takes place in eukaryotic mRNA. Modifications include *capping* of the 5' end, *polyadenylation* (addition of a poly A sequence) of the 3' end, and *splicing of coding sequences*. Such processing is not a feature of the synthesis of prokaryotic mRNA.

The cap at the 5' end of eukaryotic mRNA is a guanylate residue that is methylated at the N-7 position. This modified guanylate residue is attached to the neighboring residue by a $5' \longrightarrow 5'$ triphosphate linkage (Figure 19.16). The 2'-hydroxyl group of the ribosyl portion of the neighboring residue is frequently methylated, and sometimes that of the next nearest neighbor is, as well. The polyadenylate "tail" at the 3' end of a messenger (typically 100 to 200 nucleotides long) is added before the mRNA leaves the nucleus. It is thought that the presence of the tail serves to protect the mRNA from nucleases and phosphatases, which would degrade it. According to this point of view, the adenylate residues would be cleaved off before the portion of the molecule that contains the actual message is attacked. The presence of the 5' cap also protects the mRNA from exonuclease degradation.

The genes of prokaryotes are continuous; every base pair in a continuous prokaryotic gene is reflected in the base sequence of mRNA. The genes of eukaryotes are not necessarily continuous; eukaryotic genes frequently contain intervening sequences that do not appear in the final base sequence of the mRNA for that gene product. The DNA sequences that are expressed (the ones actually retained in the final product) are called **exons.** The intervening sequences, which are not expressed, are called **introns.** The β-

FIGURE 19.17

Intervening sequences (introns) in the β-globin gene.

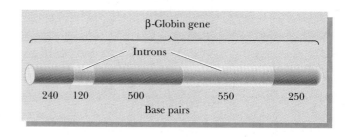

globin gene of the mouse, which codes for the β-chain of hemoglobin, is a well-known example. The actual gene is split into three parts, with two intervening sequences (Figure 19.17).

In eukaryotes the entire DNA sequence, both introns and exons, is transcribed to produce a precursor of the mature mRNA. In the processing of mRNA, the noncoding sequences, the introns, must be excised, and the coding sequences, the exons, must be spliced together. This process is done by the sequential actions of suitable nucleases and ligases. The number and size of the introns vary in different genes. There are 1 intron in the gene for the muscle protein actin, 2 for both the α- and β-chains of hemoglobin, 3 for lysozyme, and so on, up to as many as 50 introns in a single gene (the gene for one of the subunits of collagen).

FIGURE 19.18

The formation of a lariat structure as an intermediate in the splicing of mRNA. In the first step of the splicing process, one splice site is cut and attached to a binding site near the second splice site, forming a lariat. In the course of splicing, mRNA is bound to small nuclear ribonucleoproteins (snRNPs) (not shown here), which hold the splice sites in position. In the second step, the intron lariat is excised, and the mRNA exons are spliced together. A similar lariat mechanism is involved in the self-splicing of RNA. (From J. Steitz, 1988, "Snurps," *Sci. Amer.* 258 (6), 56–63.)

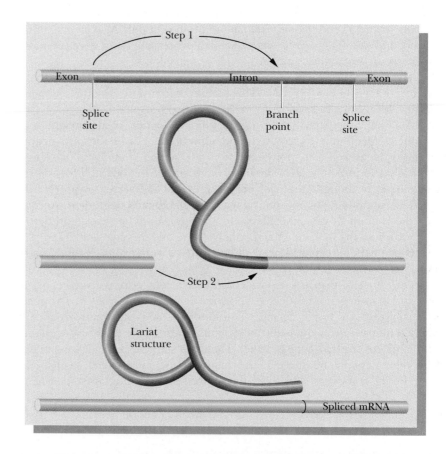

LUPUS: AN AUTOIMMUNE DISEASE INVOLVING RNA PROCESSING

Systemic lupus erythematosus is an autoimmune disease that can have fatal consequences. It starts, usually in late adolescence or early adulthood, with a rash on the forehead and cheekbones, giving the wolflike appearance from which the disease takes its name. (The word *lupus* means "wolf" in Latin.) Severe kidney damage follows, along with arthritis, accumulation of fluid around the heart, and inflammation of the lungs. About 90% of lupus patients are women. It has been established that this disease is of autoimmune origin, specifically from the production of antibodies to one of the snRNPs, **U1-snRNP.** This snRNP is so designated because it contains a uracil-rich RNA, U1-snRNA, which recognizes the 5′ splice junction of mRNA. Because the processing of mRNA affects every tissue and organ in the body, this disease affects widely dispersed target areas and can spread easily.

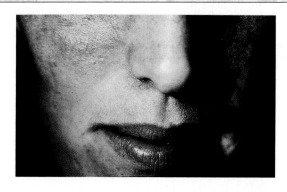

A characteristic rash is frequently seen on the cheekbones and foreheads of victims of systemic lupus erythematosus.

The removal of intervening sequences takes place in the nucleus, where small nuclear ribonucleoproteins, or snRNPs (pronounced "snurps"), mediate the process. The snRNPs, as their name implies, contain both RNA and proteins, such as the nucleases and ligases involved in the splicing process. The actual splicing involves a 50S to 60S particle called the **spliceosome.** Little is known about this particle except that it consists of pre-mRNA and an assemblage of snRNPs. It is now widely recognized that some RNAs can catalyze their own self-splicing; the present process involving ribonucleoproteins may well have evolved from the self-splicing of RNAs. An important similarity between the two processes is that both proceed via a lariat mechanism by which the splice sites are brought together (Figure 19.18). Not all details of the process are clear, but it is established that base pairing between the mRNA to be spliced and the RNA portion of the snRNPs plays a role in the recognition of the sequences to be spliced. (For additional information, see the article by Steitz listed in the bibliography at the end of this chapter.) Box 19.2 describes an autoimmune disease that develops when the body makes antibodies to one of these snRNPs.

19.10
RIBOZYMES: RNA AS AN ENZYME

There was a time when proteins were considered to be the only biological macromolecules capable of catalysis. The discovery of the catalytic activity of RNA has thus had a profound impact on the way biochemists think. The first catalytic RNAs (**ribozymes**) discovered included those that catalyze their

FIGURE 19.19

The self-splicing of pre-rRNA of the protozoan *Tetrahymena*, a Class I ribozyme. (a) A guanine nucleotide attacks at the splice site of the exon on the left, giving a free 3'-OH end. (b) The free 3'-OH end of the exon attacks the 5' end of the exon on the right, splicing the two exons and releasing the intron. (c) The free 3'-OH end of the intron then attacks a nucleotide 15 residues from the 5' end, cyclizing the intron and releasing a 5' terminal sequence.

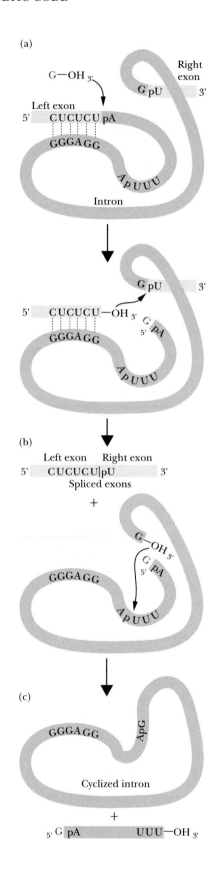

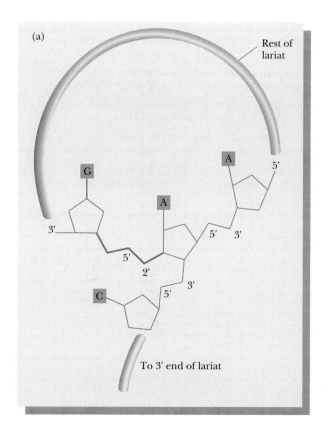

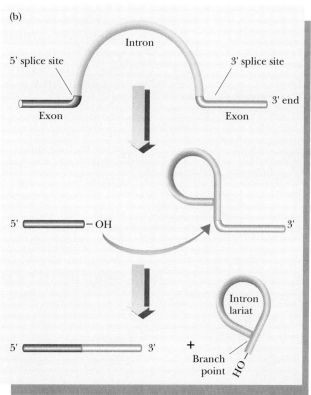

FIGURE 19.20

Self-splicing of a Class II ribozyme. (a) The backbone (yellow) of the lariat showing the 2', 5'-phosphodiester bond at the branch point (red). (b) The lariat is formed when the 2'-OH group of an adenine nucleotide residue forms a 2', 5'-phosphodiester bond. The RNA chain branches at that point. (From T. Cech, 1986, RNA as an Enzyme, *Sci. Amer.* 255 (5), 64–75.)

own self-splicing. It is easy to see a connection between this process and the splicing of mRNA by snRNPs. More recently, it has been shown that RNAs can catalyze reactions involved in protein synthesis.

Several groups of ribozymes are known to exist. In Group I ribozymes there is a requirement for an external guanosine, which becomes covalently bonded to the splice site in the course of excision. An example is the self-splicing that takes place in pre-rRNA of the protozoan *Tetrahymena* (Figure 19.19). The transesterification (of phosphoric acid esters) that takes place here releases one end of the intron. The free 3'-OH end of the exon attacks the 5' end of the other exon, splicing the two exons and releasing the intron. The free 3'-OH end of the intron then attacks a nucleotide 15 residues from the 5' end, cyclizing the intron and releasing a 5' terminal sequence. The precision of this sequence of reactions depends on the folded conformation of the RNA, which remains internally hydrogen bonded throughout the process.

Group II ribozymes display a lariat mechanism of operation that is strikingly similar to that of splicing of mRNA. There is no requirement for an external nucleotide; the 2'-OH of an internal adenosine attacks the phosphate at the 5' splice site. The exon has a free 3'-OH group that then attacks the 3' splice site. The exons are ligated and the intron is left as a lariat (Figure 19.20). Clearly, DNA cannot self-splice in this fashion because it does not have a 2'-OH.

The folding of the RNA is crucial to its catalytic activity, as is the case with protein catalysts. A divalent cation (Mg^{2+} or Mn^{2+}) is required; it is quite likely that metal ions stabilize the folded structure by neutralizing some of the negative charges on the phosphate groups of the RNA. The folding of RNA is such that large-scale conformational changes can take place with great precision. Similar large-scale changes take place in the ribosome in protein synthesis and in the spliceosome in the processing of mRNA. It is interesting to note that they remain RNA machines when proteins have taken over much of the catalytic function of the cell. The ability of RNA to undergo the requisite large-scale conformational changes may well play a role in the process.

A recently proposed clinical application of ribozymes has been suggested. If a ribozyme can be devised that can cleave the RNA genome of HIV, the virus that causes AIDS (Chapter 20, Section 20.10), it will be a great step forward in the treatment of this disease. Research on this topic is in progress in several laboratories. (See the 1993 article by Barinaga listed in the bibliography at the end of this chapter.)

S U M M A R Y

In all organisms except RNA viruses, the flow of genetic information is DNA $\longrightarrow$ RNA $\longrightarrow$ protein. The duplication of DNA is called replication, and the production of RNA on a DNA template is called transcription. Translation is the process of protein synthesis, in which the sequence of amino acids is directed by the sequence of bases in the genetic material.

Replication of DNA is semiconservative and bidirectional. Two replication forks advance in opposite directions from an origin of replication. Both new polynucleotide chains are synthesized in the 5′-to-3′ direction. One strand (the leading strand) is synthesized continuously, while the other (the lagging strand) is synthesized discontinuously, in fragments that are subsequently linked together. Two DNA polymerases play important roles in replication. Polymerase III (Pol III) is primarily responsible for the synthesis of new strands. The first polymerase enzyme discovered, polymerase I (Pol I), is mainly a repair enzyme. DNA gyrase introduces a swivel point in advance of the movement of the replication fork. A helix-destabilizing protein, helicase, binds at the replication fork and promotes unwinding. The exposed single-stranded regions of the template DNA are protected from nuclease digestion by a DNA-binding protein. Primase catalyzes the synthesis of an RNA primer. The synthesis of new strands linked to the primer is catalyzed by Pol III. The primer is removed by Pol I, which also replaces the primer with deoxynucleotides. DNA ligase seals any remaining nicks.

DNA replication takes place only once each generation in each cell. It is essential that the fidelity of the replication process be as high as possible to prevent mutations, which are errors in replication. Pol III does proofreading in the course of replication. In addition, Pol I carries out a cut-and-patch process, removing the RNA primer and replacing it with deoxyribonucleotides during replication. Pol I uses the same cut-and-patch process to repair existing DNA. Replication has been most extensively studied in prokaryotes. Replication in eukaryotes follows the same general outline, with the most important difference being the presence of proteins complexed to eukaryotic DNA.

RNA synthesis is the transcription of the base sequence of DNA to that of RNA. All RNAs are synthesized on a DNA template; the enzyme that catalyzes the process is DNA-dependent RNA polymerase. All four ribonucleoside triphosphates (ATP, GTP, CTP, and UTP) are required, as is Mg^{2+}. There is no need for a primer in RNA synthesis. As is the case with DNA biosynthesis, the RNA chain grows from the 5′ to the 3′ end. The enzyme uses one strand of the DNA (the sense strand) as the template for RNA synthesis. Posttranscriptional processing takes place in RNA. Base modification frequently occurs, as does trimming of long polynucleotide chains. In eukaryotes, the removal of intervening sequences, which reflect the base sequence of a portion of the DNA not expressed in the mature RNA, is an important step in the processing of RNA.

Some forms of RNA, called ribozymes, have catalytic activity. The mechanism by which such RNAs catalyze their own self-splicing may have an evolutionary relationship to the processing of eukaryotic mRNA.

EXERCISES

1. Is the following statement true or false? Why? "The flow of genetic information in the cell is always DNA ⟶ RNA ⟶ protein."
2. Define the terms "replication," "transcription," and "translation."
3. Why is the replication of DNA referred to as a semiconservative process? What is the experimental evidence for the semiconservative nature of the process? What experimental results would you expect if replication of DNA were a conservative process?
4. What is a replication fork, and why is it important in replication?
5. Compare and contrast the properties of the enzymes DNA polymerase I and polymerase III from *E. coli*.
6. List the substances required for replication of DNA catalyzed by DNA polymerase.
7. Describe the discontinuous synthesis of the lagging strand in DNA replication.
8. What are the functions of the gyrase, primase, and ligase enzymes in DNA replication?
9. How does proofreading take place in the process of DNA replication?

10. Describe the excision-repair process in DNA, using the excision of thymine dimers as an example.
11. How does DNA replication in eukaryotes differ from the process in prokaryotes?
12. List three important properties of RNA polymerase from *E. coli*.
13. Define the term "promoter region," and list three of its properties.
14. Why is a trimming process important in converting precursors of tRNA and rRNA to the active forms? List three molecular changes that take place in the processing of eukaryotic mRNA.
15. List three molecular changes that take place in the processing of eukaryotic mRNA.
16. Define the terms "exon" and "intron."
17. What are snRNPs, and what is their role in the processing of eukaryotic mRNAs?
18. What is site-directed mutagenesis? How can it be used to produce proteins that differ by one amino acid residue from those normally found in nature?
19. Outline a mechanism by which RNA can catalyze its own self-splicing.

ANNOTATED BIBLIOGRAPHY

Adams, R. L. P., J. T. Knowles, and D. P. Leader. *The Biochemistry of the Nucleic Acids*. 11th ed. New York: Chapman and Hall, 1992. [New authors have prepared this edition of a classic text originally written by J. N. Davidson.]

Barinaga, M. Dimers Direct Development. *Science* **251**, 1176—1177 (1991). [The "helix–loop–helix" structural motif in DNA-binding proteins plays a ubiquitous role in development.]

Barinaga, M. Ribozymes: Killing the Messenger. *Science* **262**, 1512–1514 (1993). [A report on research designed to use ribozymes to attack the RNA genome of HIV.]

Blackwood, E., and R. Eisenman. Max: A Helix–Loop–Helix Zipper Protein That Forms a Sequence-Specific DNA-Binding Complex with Myc. *Science* **251**, 1211–1217 (1991). [A leucine zipper protein binds to cancer-causing genes.]

Brenner, S., F. Jacob, and M. Meselson. An Unstable Intermediate Carrying Information from Genes to Ribosomes for Protein Synthesis. *Nature* **190**, 576–581 (1961). [One of the first descriptions of the concept of mRNA. Of historical interest.]

Buratowski, S. DNA Repair and Transcription: The Helicase Connection. *Science* **260**, 37–38 (1993). [How repair and transcription are coupled.]

Cech, T. R. RNA as an Enzyme. *Sci. Amer.* **255** (5), 64–75 (1986). [A description of the discovery that some RNAs can catalyze their own self-splicing. The author was a recipient of the 1989 Nobel Prize in chemistry for this work.]

Celander, D., and T. Cech. Visualizing the Higher Order Folding of a Catalytic RNA Molecule. *Science* **251,** 401–407 (1991). [The three-dimensional structure of catalytic RNA can give clues to its mode of operation.]

Darnell, J. E. RNA. *Sci. Amer.* **253** (4), 68–78 (1985). [An article on the function and processing of RNA.]

Echols, H. Multiple DNA–Protein Interactions Governing High-Precision DNA Transaction. *Science* **233,** 1050–1056 (1986). [A discussion of factors needed for faithful replication.]

Hoffman, M. RNA Editing: What's in a Mechanism? *Science* **253,** 136–138 (1991). [Similarities between editing and splicing of RNA, with possible therapeutic applications.]

Kassavetis, G., and E. P. Geiduschek. RNA Polymerase Marching Backward. *Science* **259,** 944–945 (1993). [Some questions (and a few answers) about transcription.]

Kornberg, A. *DNA Replication.* San Francisco: W. H. Freeman, 1980. [Most aspects of DNA biosynthesis are covered. The author received a Nobel Prize for his work in this field.]

McCorkle, G. M., and S. Altman. RNA's as Catalysts. *J. Chem. Ed.* **64,** 221–226 (1987). [An article on the mechanism of action of catalytic RNA. Altman shared the 1989 Nobel Prize in chemistry with T. Cech for work on the role of RNA as a catalyst.]

McKnight, S. Molecular Zippers in Gene Regulation. *Sci. Amer.* **264** (4), 54–64 (1991). [The role of leucine zipper proteins in gene regulation.]

Pyle, A. Ribozymes: A Distinct Class of Metalloenzymes. *Science* **261,** 709–714 (1993). [Emphasizes the role of metal ions in the structures and functions of ribozymes.]

Radman, M., and R. Wagner. The High Fidelity of DNA Duplication. *Sci. Amer.* **259** (1), 40–46 (1988). [A description of replication, concentrating on the mechanisms for minimizing errors.]

Rhodes, D., and A. Klug. Zinc Fingers. *Sci. Amer.* **268** (2), 56–65 (1993). [How the structure of these zinc-containing proteins enables them to play a role in regulating the activity of genes.]

Sharp, P., "Five Easy Pieces." *Science* **254,** 663 (1991). [A review by a Nobel laureate on some highly conserved RNAs that play a ubiquitous role in cellular processes.]

Simpson, L. RNA Editing — A Novel Genetic Phenomenon? *Science* **250,** 512–513 (1990). [A review of some of the less common aspects of RNA processing.]

Smith, L., J. Z. Sanders, R. J. Kaiser, *et al.* Fluorescence Detection in Automated DNA Sequence Analysis. *Nature* **321,** 674–679 (1986). [The first report of an automated method for sequencing of DNA.]

Steitz, J. A. Snurps. *Sci. Amer.* **258** (6), 56–63 (1988). [A discussion of the role of small nuclear ribonucleoproteins, or snRNPs (pronounced "snurps"), in the removal of introns from mRNA.]

Varmus, H. Reverse Transcription. *Sci. Amer.* **257** (3), 56–64 (1987). [A description of RNA-directed DNA synthesis. The author was one of the recipients of the 1989 Nobel Prize in medicine or physiology for his work on the role of reverse transcription in cancer. He is currently Director of the National Institutes of Health.]

Von Hippel, P., and T. Yager. The Elongation–Termination Decision in Transcription. *Science* **255,** 809 — 812 (1992). [Energy considerations in transcription.]

Wang, J.-F., W. Downs, and T. Cech. Movement of the Guide Sequence During RNA Catalysis by a Group I Ribozyme. *Science* **260,** 504–508 (1993). [RNA can guide large-scale, specific conformational changes in catalytic processes.]

Watson, J. D. *Molecular Biology of the Gene.* 4th ed. Menlo Park, CA: Benjamin/Cummings, 1987. [A highly readable book on molecular biology.]

Wong, I., and T. Lohman. Allosteric Effects of Nucleotide Cofactors on *Escherichia coli* Rep Helicase–DNA Binding. *Science* **256,** 350–355 (1992). [The mode of action of an allosteric enzyme that is important in DNA replication.]

Chapter 20

Protein Synthesis: Translation of the Genetic Message

Crystals of the protein lactate dehydrogenase. Like all proteins, it is synthesized on ribosomes.

After the DNA base sequence has been transcribed to RNA, the genetic code is needed to *translate* the RNA sequence into the amino acid sequence of a protein. In eukaryotic cells, DNA is transcribed in the nucleus but translated in the cytosol. The transcript is exported from the nucleus in the form of *messenger RNA,* which is read and translated at the ribosome. Molecules of transfer RNA (tRNA), one for each amino acid, are required to collect activated amino acids and deliver them, one at a time, to the ribosome. Here they are sequentially joined to synthesize the polypeptide chain of a protein. The sequence of amino acids, derived from the sequence of DNA bases, is specified by the genetic code, using the four RNA bases A, U, G, and C taken three at a time. In the triplet code there are 64 possible "code words," called codons, of which 3 are stop signals and 61 specify the 20 amino acids with considerable redundancy. In the actual mechanism of translation, messenger RNA is temporarily bonded to transfer RNA. As tRNA molecules deliver their amino acids to the ribosome in succession, peptide bonds covalently join the amino acids to form the growing polypeptide chain. A stop signal on the messenger RNA terminates the protein chain, which is then released from the ribosome. This whole mechanism is subject to outside takeover when a virus infects a cell, sometimes with drastic consequences.

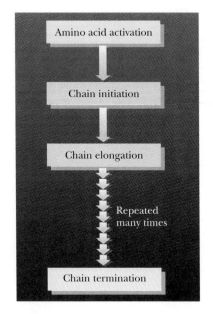

FIGURE 20.1

A flow chart showing the steps in protein biosynthesis.

20.1
THE PROCESS OF TRANSLATING THE GENETIC MESSAGE

Protein biosynthesis is a complex process requiring ribosomes, messenger RNA (mRNA), transfer RNA (tRNA), and a number of protein factors. The ribosome is the site of protein synthesis. The mRNA and tRNA, which are bound to the ribosome in the course of protein synthesis, are responsible for the correct order of amino acids in the growing protein chain.

Before an amino acid can be incorporated into a growing protein chain, it must first be **activated,** a process involving both tRNA and a specific enzyme of the class known as **aminoacyl-tRNA synthetases.** The amino acid is covalently bonded to the tRNA in the process, forming an aminoacyl-tRNA. The actual formation of the polypeptide chain occurs in three steps. In the **initiation** step, the first aminoacyl-tRNA is bound to the mRNA at the site that encodes the start of polypeptide synthesis. In this complex the mRNA and the ribosome are bound to each other. The next aminoacyl-tRNA forms a complex with the ribosome and with mRNA. The binding site for the second aminoacyl-tRNA is close to that for the first aminoacyl-tRNA. A peptide bond is formed between the amino acids; this step is called chain **elongation.** The chain elongation process repeats itself until the polypeptide chain is complete. Finally, chain **termination** takes place. Each of these steps has many distinguishing features (Figure 20.1), and we shall look at each of them in detail.

20.2
THE GENETIC CODE

Some of the most important features of the code can be specified by saying that the genetic message is contained in a *triplet, nonoverlapping, commaless, degenerate, universal code.* Each of these terms has a definite meaning that describes the way in which the code is translated.

A **triplet** code means that a sequence of three bases (called a **codon**) is needed to specify one amino acid. The term **nonoverlapping** indicates that no bases are shared between consecutive codons; the ribosome moves along the mRNA three bases at a time rather than one or two at a time (Figure 20.2). If the ribosome moved along the mRNA more than three bases at a time, this situation would be referred to as "the presence of commas in the code." Since no intervening bases exist between codons, the code is **commaless.** In a **degenerate** code, more than one triplet can code for the same amino acid. There are 64 (4 × 4 × 4) possible triplets of the four bases that occur in RNA, and all of them are used to code for 20 amino acids or for one of the three stop signals. A **universal** code is one that is the same in all organisms. The universality of the code has been observed in viruses, prokaryotes, and eukaryotes; the only exceptions are the differences in some codons seen in mitochondria. The evolutionary origin of these differences is not known at this writing.

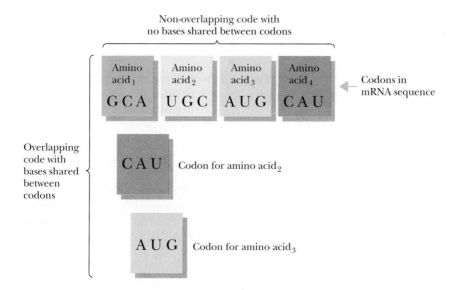

FIGURE 20.2

A comparison of nonoverlapping and overlapping codes. In a nonoverlapping code, adjacent codons do not share any nucleotides in the mRNA sequence. In an overlapping code, adjacent codons have one or more nucleotides in common.

TABLE 20.1   **The Assignment of the 64 Triplet Codons in the 5′ ⟶ 3′ Sequence of mRNA**

BASE AT 5′ END OF CODON ↓	MIDDLE BASE OF CODON				BASE AT 3′ END OF CODON ↓
	U	C	A	G	
U	phe (UUU)	ser	tyr	cys	U
	phe	ser	tyr	cys	C
	leu	ser	Termination	Termination	A
	leu	ser	Termination	trp	G
C	leu	pro	his	arg	U
	leu	pro	his	arg	C
	leu	pro	gln	arg	A
	leu	pro	gln	arg	G
A	ile	thr	asn	ser	U
	ile	thr	asn	ser	C
	ile	thr	lys	arg	A
	met (and initiation)	thr	lys	arg	G
G	val	ala	asp	gly	U
	val	ala	asp	gly	C
	val	ala	glu	gly	A
	val	ala	glu	gly	G

All 64 codons have been assigned meanings, with 61 of them coding for amino acids and the remaining 3 serving as the termination signals (Table 20.1).

Two amino acids, tryptophan and methionine, have only one codon each, but the rest have more than one. A single amino acid can have as many as six codons, as is the case with leucine and arginine. Multiple codons for a single amino acid are not randomly distributed in Table 20.1 but have one or two bases in common. The bases that are common to several codons are usually the first and second bases, with more room for variation in the third base, which is called the "wobble" base.

The assignment of triplets in the genetic code is based on several types of experiment. One of the most significant involves the use of synthetic polyribonucleotides as messengers. This approach can give some information about the nature of the code, but ambiguities remain. Unambiguous assignments require another important type of experiment that is based on binding studies involving tRNAs and triplets synthesized in the laboratory.

When homopolynucleotides (polyribonucleotides that contain only one type of base) are used as a **synthetic mRNA** in laboratory systems for polypeptide synthesis, homopolypeptides (polypeptides that contain only one kind of amino acid) are produced. When poly U is the messenger, the product is polyphenylalanine. With poly A as the messenger, polylysine is formed. The product for poly C is polyproline, and with poly G, polyglycine results. When an alternating copolymer (a polymer with an alternating sequence of two bases) is the messenger, the product is an alternating polypeptide (a polypeptide with an alternating sequence of two amino acids). For example, when the sequence of the polynucleotide is -ACACACACACACACACACACAC-, the polypeptide produced is poly(thr-his), with the alternating sequence threonine-histidine. There are two types of coding triplets in this polynucleotide, ACA and CAC, but this experiment cannot establish which one codes for threonine and which one for histidine. More information is needed for an unambiguous assignment, but it is interesting that this result proves that the code is a triplet code. If it were a doublet code, the product would be a mixture of two homopolymers, one specified by the codon AC and the other by the codon CA. (The terminology for the different ways of reading this message as a doublet is to say that they have different **reading frames,** /AC/AC/ and /CA/CA/. In a triplet code only one reading frame is possible, namely /ACA/CAC/ACA/CAC/, which gives rise to an alternating polypeptide.) Use of other synthetic polynucleotides can yield other coding assignments, but, as in our example here, many questions remain.

Other methods are needed to answer the remaining questions about codon assignment. One of the most useful methods is the **binding assay.** This technique depends on the fact that aminoacyl-tRNAs bind strongly to ribosomes in the presence of trinucleotides. In this situation, the trinucleotide plays the role of an mRNA codon. The possible trinucleotides are synthesized by chemical methods, and binding assays are carried out with each type of trinucleotide. Aminoacyl-tRNAs are tested for their ability to bind in the presence of a given trinucleotide. For example, if the aminoacyl-tRNA for isoleucine binds to the ribosome in the presence of the trinucleotide AUC, the sequence AUC is established as a codon for isoleucine. About 50 of the 64 codons have been identified by this method.

Codon–Anticodon Pairing and Wobble

A codon forms base pairs with a complementary anticodon of a tRNA when an amino acid is incorporated during protein synthesis. Some tRNAs bond to one codon exclusively, but many of them can recognize more than one codon because of variations in the allowed pattern of hydrogen bonding. This variation is called "wobble" (Figure 20.3), and it applies to the first base of an anticodon, the one at the 5′ end, but not to the second or the third base. Recall that mRNA is read from the 5′ to the 3′ end. The first (wobble) base of the anticodon hydrogen-bonds to the third base of the codon, the one at the 3′ end. The base in the wobble position of the

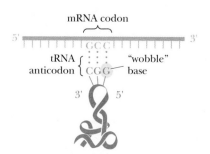

FIGURE 20.3

"Wobble" base pairing. The wobble base of the anticodon is the one at the 5′ end; it forms hydrogen bonds with the last base of the mRNA codon, the one at the 3′ end of the codon.

TABLE 20.2   **Base-Pairing Combinations in the Wobble Scheme**

BASE AT 5′ END OF ANTICODON	BASE AT 3′ END OF CODON
I*	A, C, or U
G	C or U
U	A or G
A	U
C	G

I* = hypoxanthine.
Note that there are no variations in base when the wobble position is occupied by A or C.

anticodon can base-pair with several different bases in the codon, not just the base specified by Watson–Crick base pairing (Table 20.2).

When the wobble base of the anticodon is uracil, it can base-pair not only with adenine, as expected, but also with guanine, the other purine base. When the wobble base is guanine, it can base-pair with cytosine, as expected, and also with uracil, the other pyrimidine base. The purine base hypoxanthine frequently occurs in the wobble position in many tRNAs, and it can base-pair with adenine, cytosine, and uracil in the codon (Figure 20.4). Adenine and cytosine do not form any base pairs other than the expected

FIGURE 20.4

Base pairing involving inosine. When inosine occupies the wobble position of the anticodon, it can form hydrogen-bonded base pairs with adenosine, uridine, and cytidine. (R indicates the point of attachment of the base to the ribose.)

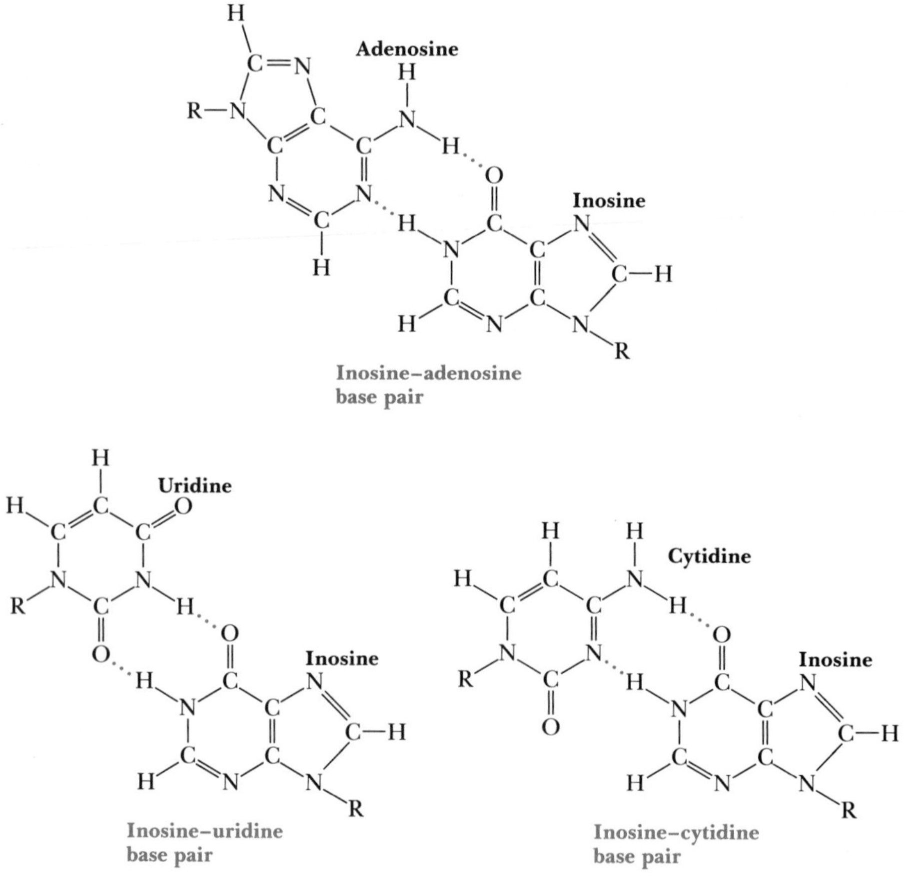

Inosine–adenosine base pair

Inosine–uridine base pair

Inosine–cytidine base pair

ones with uracil and guanine, respectively (Table 20.2). To summarize, when the wobble position is occupied by I (from inosine, the nucleoside made up of ribose and hypoxanthine), G, or U, variations in hydrogen bonding are allowed; when the wobble position is occupied by A or C, these variations do not occur.

The wobble hypothesis provides insight into some aspects of the degeneracy of the code. In many, but not all, cases the degenerate codons for a given amino acid differ in the third base, the one that pairs with the wobble base of the anticodon. Fewer different tRNAs are needed, since a given tRNA can base-pair with several codons. As a result, a cell would have to invest less energy in the synthesis of needed tRNAs. The existence of wobble also minimizes the damage that can be caused by misreading of the code. If, for example, a leucine codon, CUU, were to be misread as CUC, CUA, or CUG during transcription of mRNA, this codon would still be translated as leucine during protein synthesis; no damage to the organism would occur. We saw earlier that drastic consequences can result from misreading of the genetic code in other codon positions, but here we see that such effects are not inevitable.

20.3
AMINO ACID ACTIVATION

The activation of the amino acid and the formation of the aminoacyl-tRNA take place in two separate steps, both of which are catalyzed by the aminoacyl-tRNA synthetase (Figure 20.5). First, the amino acid forms a covalent bond to an adenine nucleotide, producing an aminoacyl-AMP. The free energy of hydrolysis of ATP provides energy for bond formation. The aminoacyl moiety is then transferred to tRNA, forming an aminoacyl-tRNA.

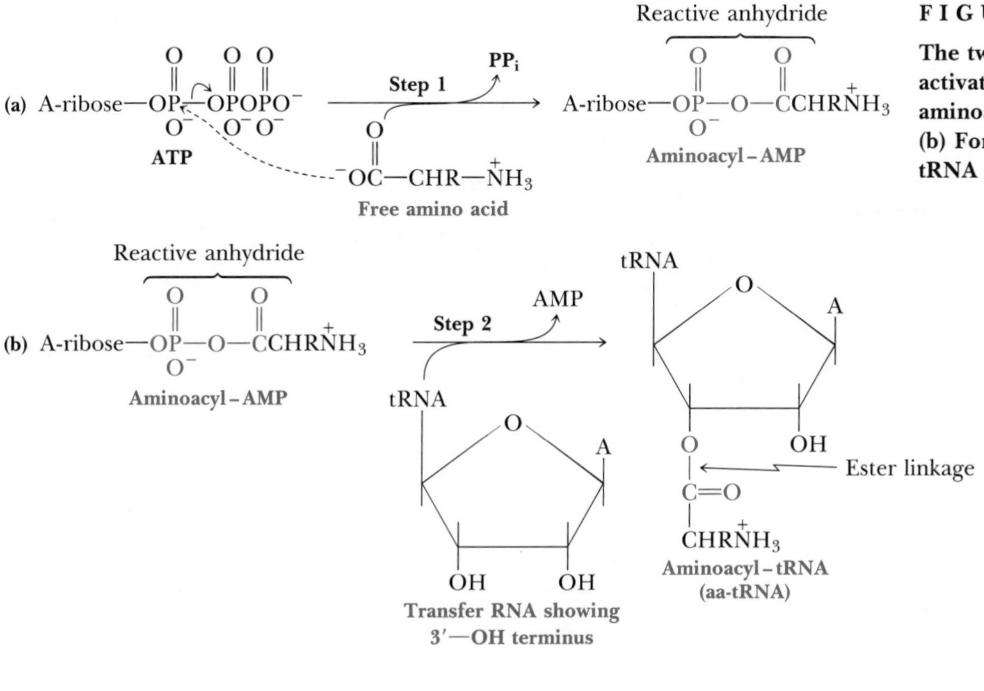

FIGURE 20.5

The two steps of amino acid activation. (a) Formation of aminoacyl-AMP intermediate. (b) Formation of aminoacyl-tRNA from aminoacyl-AMP.

$$\text{Amino acid} + \text{ATP} \longrightarrow \text{aminoacyl-AMP} + \text{PP}_i$$

$$\text{Aminoacyl-AMP} + \text{tRNA} \longrightarrow \text{aminoacyl-tRNA} + \text{AMP}$$

$$\overline{\text{Amino acid} + \text{ATP} + \text{tRNA} \longrightarrow \text{aminoacyl-tRNA} + \text{AMP} + \text{PP}_i}$$

Aminoacyl-AMP is a mixed anhydride of a carboxylic acid and a phosphoric acid. Since anhydrides are reactive compounds, the free energy change for the hydrolysis of aminoacyl-AMP favors the second step of the overall reaction. Another point that favors the process is the energy released when pyrophosphate (PP_i) is hydrolyzed to orthophosphate (P_i) to replenish the phosphate pool in the cell. The synthetase enzyme requires Mg^{2+} and is highly specific both for the amino acid and for the tRNA. A separate synthetase exists for each amino acid.

In the second part of the reaction, an ester linkage is formed between the amino acid and the 3'-hydroxyl end of the tRNA. Several tRNAs can exist for each amino acid, but a given tRNA will not bond to more than one amino acid. Each synthetase has a high degree of specificity for the correct tRNA and the correct amino acid. The specificity of the enzyme contributes to the accuracy of the translation process.

20.4
CHAIN INITIATION

The details of chain initiation differ somewhat in prokaryotes and eukaryotes. Like DNA and RNA synthesis, this process has been more thoroughly studied in prokaryotes. We shall use *Escherichia coli* as our principal example, since all aspects of protein synthesis have been most extensively studied in this bacterium. In all organisms the synthesis of polypeptide chains starts at the N-terminal end; the chain grows from the N-terminal to the C-terminal end.

In prokaryotes the initial N-terminal amino acid of all proteins is *N*-formylmethionine (fmet) (Figure 20.6). However, this residue can be, and

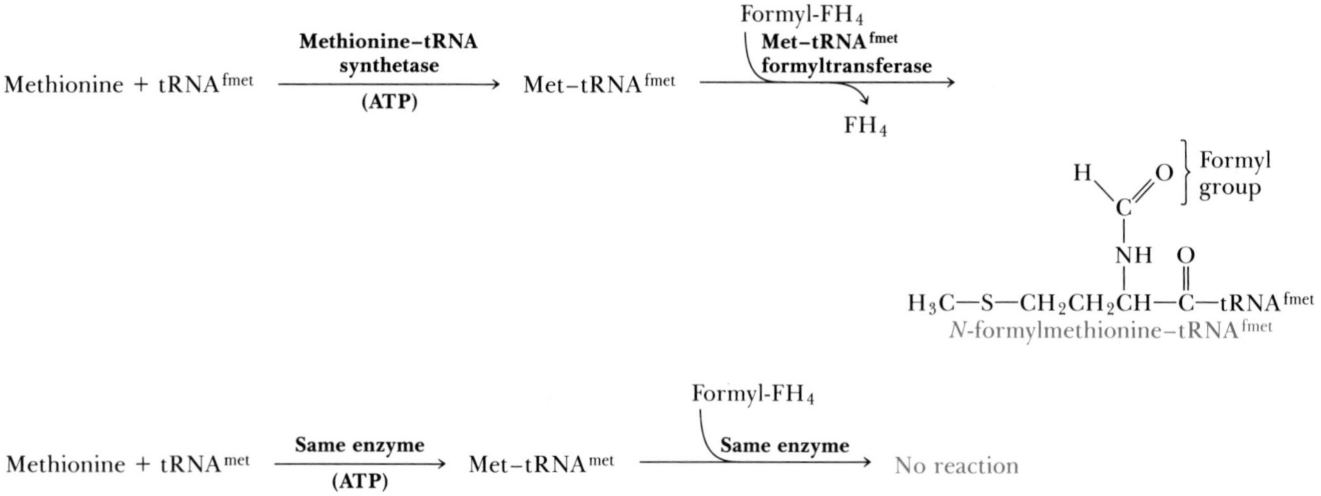

FIGURE 20.6

Formation of *N*-formylmethionine-tRNA^{fmet} (first reaction). Methionine must be bound to tRNA^{fmet} to be formylated. (FH$_4$ is tetrahydrofolate.) Methionine bound to tRNA^{met} is not formylated (second equation).

often is, removed by posttranslational processing after the polypeptide chain is synthesized. There are two different tRNAs for methionine in *E. coli*, one for unmodified methionine and one for *N*-formylmethionine. These two tRNAs are called tRNA^{met} and tRNA^{fmet}, respectively (the superscript identifies the tRNA). The aminoacyl-tRNAs that they form with methionine are called met-tRNA^{met} and met-tRNA^{fmet}, respectively (the prefix identifies the bound amino acid). In the case of met-tRNA^{fmet}, a formylation reaction takes place after methionine is bonded to the tRNA, producing **N-formylmethionine-tRNA^{fmet}** (fmet-tRNA^{fmet}). The source of the formyl group is N^{10}-formyltetrahydrofolate (see One-Carbon Transfers and the Serine Family in Chapter 17, Section 17.4). Methionine bound to tRNA^{met} is not formylated.

Both tRNAs (tRNA^{met} and tRNA^{fmet}) contain a specific sequence of three bases (a triplet), 3'-UAC-5', which base-pairs with the sequence 5'-AUG-3' in the mRNA sequence. The tRNA^{fmet} triplet in question, UAC, recognizes the AUG triplet, which is the start signal when it occurs at the beginning of the mRNA sequence that directs the synthesis of the polypeptide. The same UAC triplet in tRNA^{met} recognizes the AUG triplet when it is found in an internal position in the mRNA sequence. The mRNA triplet, the codon, specifies the nature of the amino acid to be added to the growing polypeptide chain. The tRNA triplet that hydrogen-bonds to the codon is called the **anticodon.** Note that the anticodon is antiparallel and complementary to the mRNA codon. (The list of amino acids specified by each triplet codon is given in Section 20.2.) The start signal is preceded by a leader segment of mRNA, the Shine–Dalgarno sequence (5'-GGAGGU-3'), which usually lies about ten nucleotides upstream of the AUG start signal (nearer the 5' end of the RNA). The genetic message of mRNA is read from the 5' to the 3' end. A portion of the mRNA leader segment binds to the 30S subunit of the ribosome by forming base pairs with the 3' portion of the 16S rRNA of the subunit.

The start of polypeptide synthesis requires formation of an **initiation complex.** At least eight components enter into the formation of the initiation complex, including mRNA, the 30S ribosomal subunit, fmet-tRNA^{fmet}, GTP, and three protein initiation factors, called IF-1, IF-2, and IF-3. The IF-3 protein facilitates the binding of mRNA to the 30S ribosomal subunit. It also appears to prevent premature binding of the 50S subunit, which takes place in a subsequent step of the initiation process. IF-2 binds GTP and aids in the selection of the initiator tRNA (fmet-tRNA^{fmet}) from all the other aminoacylated tRNAs available. The function of IF-1 is less clear; it appears to bind to IF-3 and to IF-2 and facilitates the action of both. The resulting combination of mRNA, the 30S ribosomal subunit, and fmet-tRNA^{fmet} is the **30S initiation complex** (Figure 20.7). A 50S ribosomal subunit binds to the 30S initiation complex to produce the **70S initiation complex.** The hydrolysis of GTP to GDP and P$_i$ favors the process by providing energy; the initiation factors are released at the same time.

The process of chain initiation in eukaryotes has not been studied to the extent it has been in bacteria, but several differences are known to exist. Methionine, not *N*-formylmethionine, is the amino acid used for initiation in eukaryotes. (Questions have been raised regarding whether eukaryotes use amino acids other than methionine for initiation. *N*-acetylated amino acids have been found in some proteins, and it is known that there is N-terminal

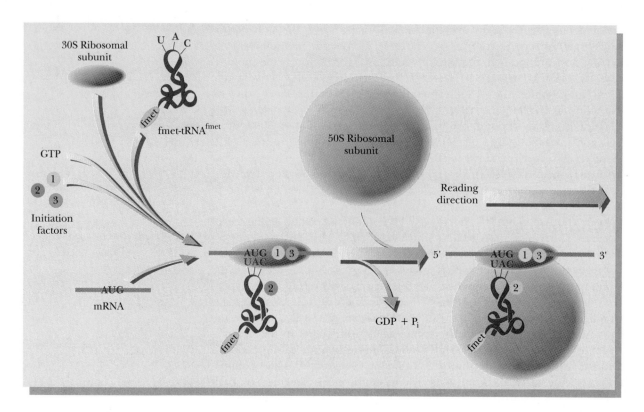

FIGURE 20.7

The formation of an initiation complex. The 30S ribosomal subunit binds to mRNA and fmet-tRNA^{fmet} in the presence of GTP and the three initiation factors, IF-1, -2, and -3, forming the 30S initiation complex. The 50S ribosomal subunit is added, forming the 70S initiation complex.

modification of proteins in eukaryotes.) Two tRNAs for methionine are found in eukaryotes, and with one of them, the formylase enzyme from *E. coli* can produce fmet-tRNA. Apparently eukaryotes have lost the ability to carry out the reaction because of the lack of the proper enzyme rather than because of the properties of the tRNA. There is no Shine–Dalgarno sequence in eukaryotes. There are at least eight initiation factors in eukaryotes; they are designated eIF-1, eIF-2, and so on, with the "e" (for eukaryotic) used to distinguish them from prokaryotic initiation factors. Considerably less is known about eukaryotic initiation factors than about prokaryotic ones, but the eukaryotic proteins are the subject of current research.

20.5
CHAIN ELONGATION

The elongation phase of prokaryotic protein synthesis (Figure 20.8) makes use of the fact that two binding sites for tRNA are present on the 50S subunit of the 70S ribosome. The two tRNA binding sites are called the P (peptidyl) site and the A (aminoacyl) site. The P site binds a tRNA that carries a peptide chain, and the A site binds an aminoacyl-tRNA. (More advanced discussions of chain elongation include a third site on the ribosome, the E [exit] site for deacylated tRNA. For more details, see the article by Nierhaus listed in the bibliography at the end of this chapter.) Chain elongation begins with the addition of the second amino acid

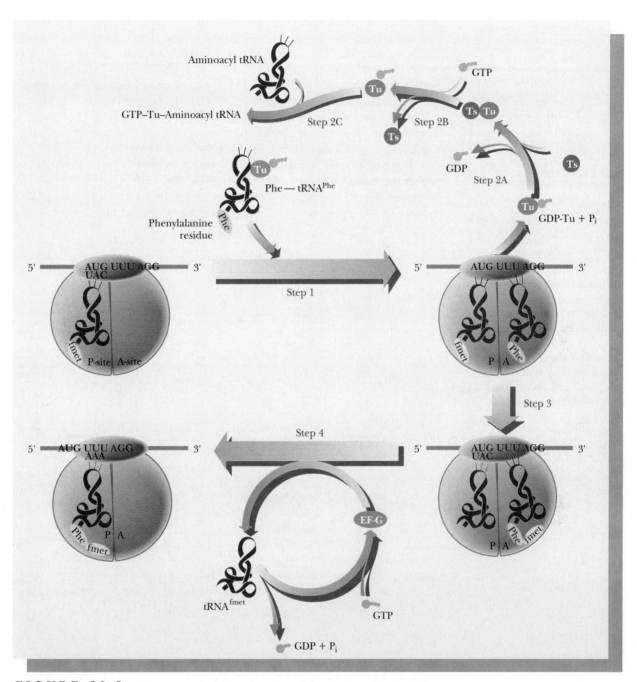

FIGURE 20.8

A summary of the steps in chain elongation. *Step 1.* An aminoacyl-tRNA is bound to the A site on the ribosome. Elongation factor EF-Tu (Tu) and GTP are required. The P side on the ribosome is already occupied. *Step 2.* Elongation factor EF-Tu is released from the ribosome and regenerated in a process requiring elongation factor EF-Ts (Ts) and GTP. *Step 3.* The peptide bond is formed, leaving an uncharged tRNA at the P site. *Step 4.* In the translocation step, the uncharged tRNA is released. The peptidyl-tRNA is translocated to the P site, leaving an empty A site. Elongation factor EF-G and GTP are required. (UUU is the codon for phenylalanine; AAA is the anticodon for phenylalanine.)

specified by the mRNA to the 70S initiation complex (Step 1). The P site on the ribosome is the one initially occupied by the fmet-tRNA^{fmet} in the 70S initiation complex. The second aminoacyl-tRNA binds at the A site. A triplet of tRNA bases (the anticodon AAA in our example) forms hydrogen bonds with a triplet of mRNA bases (UUU, the codon for phenylalanine, in this example). In addition, GTP and two protein elongation factors, EF-Tu and EF-Ts (temperature-unstable and temperature-stable elongation factors, respectively), are required for the binding of aminoacyl-tRNAs to the A site. GTP is hydrolyzed to GDP and P_i in the process (Step 2A).

A **peptide bond** is formed in a reaction catalyzed by *peptidyl transferase,* which is a part of the 50S subunit (Step 3). The carboxyl group of the *N*-formylmethionyl residue is transferred to the amino group of the amino acid bound to the tRNA at the A site. (See Figures 20.8 and 20.9 for the

FIGURE 20.9

The mode of action of puromycin. (a) A comparison of the structures of puromycin and the 3′ end of an aminoacyl-tRNA. (b) Formation of a peptide bond between a peptidyl-tRNA bound at the P site of a ribosome and puromycin bound at the A site. Protein synthesis cannot continue, and the product dissociates from the ribosome.

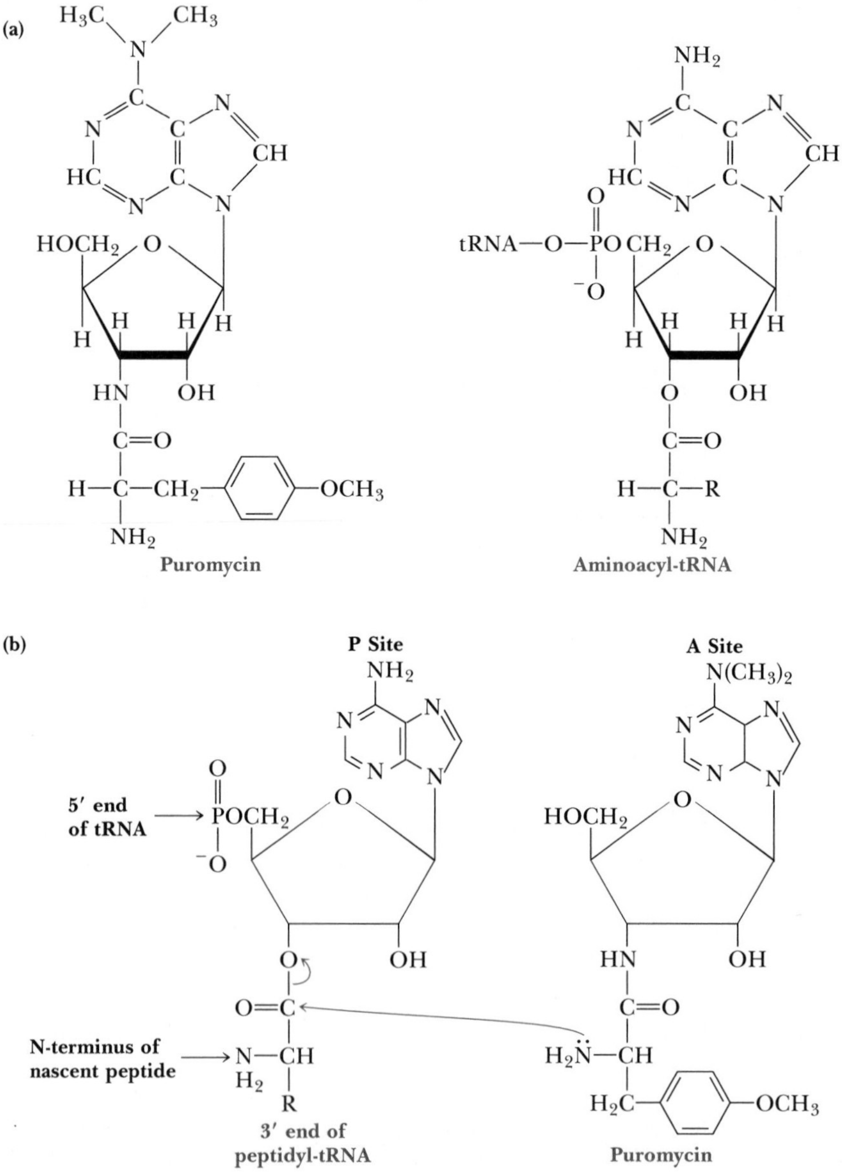

mechanism of this reaction.) There are now a dipeptidyl-tRNA at the A site and a tRNA with no amino acid attached (an uncharged tRNA) at the P site.

A **translocation** step then takes place before another amino acid can be added to the growing chain (Step 4). In the process the uncharged tRNA is released from the P site, and the peptidyl-tRNA moves from the A site to the vacated P site. In addition, the mRNA moves with respect to the ribosome. Another elongation factor, EF-G, also a protein, is required at this point, and once again GTP is hydrolyzed to GDP and P_i.

The three steps of the chain elongation process are aminoacyl-tRNA binding, peptide bond formation, and translocation (Steps 1, 3, and 4 in Figure 20.8). They are repeated for each amino acid specified by the genetic message of the mRNA until the stop signal is reached. Steps 2A through 2C in Figure 20.8 show the regeneration of aminoacyl-tRNA.

Much of the information about this phase of protein synthesis has been gained from the use of inhibitors. Puromycin is a structural analogue of the 3' end of an aminoacyl-tRNA, making it a useful probe to study chain elongation (Figure 20.9). In an experiment of this sort, puromycin binds to the A site, and a peptide bond is formed between the C-terminal of the growing polypeptide and the puromycin. The peptidyl puromycin is weakly bound to the ribosome and dissociates from it easily, resulting in premature termination and a defective protein. Puromycin also binds to the P site and blocks the translocation process, although it does not react with peptidyl-tRNA in this case. The existence of A and P sites was determined by these experiments with puromycin.

The main features of the elongation process are the same in prokaryotes and eukaryotes, but the details differ. These differences can be seen in the response to inhibitors of protein synthesis and to toxins. The antibiotic chloramphenicol (a trade name is Chloromycetin) binds to the A site and inhibits peptidyl transferase activity in prokaryotes, but not in eukaryotes. This property has made chloramphenicol useful in treating bacterial infections. In eukaryotes, diphtheria toxin is a protein that interferes with protein synthesis by decreasing the activity of the eukaryotic elongation factor eEF-2.

20.6
CHAIN TERMINATION

A stop signal is required for the termination of protein synthesis. The codons UAA, UAG, and UGA are the stop signals. One of two protein release factors (RF-1 or RF-2) is also required, as is GTP, which is bound to a third release factor, RF-3. RF-1 binds to UAA and UAG, and RF-2 binds to UAA and UGA. RF-3 does not bind to any codon, but it does facilitate the activity of the other two release factors. Either RF-1 or RF-2 is bound near the A site of the ribosome when one of the termination codons is reached. The release factor not only blocks the binding of a new aminoacyl-tRNA but also affects the activity of the peptidyl transferase, so that the bond between the carboxyl end of the peptide and the tRNA is hydrolyzed. GTP is hydrolyzed in the process. The whole complex dissociates, setting free release factors, tRNA, mRNA, and the 30S and 50S ribosomal subunits. All

T A B L E   2 0 . 3   **Components Required for Each Step
of Protein Synthesis in *Escherichia coli***

STEP	COMPONENTS
Amino acid activation	Amino acids
	tRNAs
	Aminoacyl-tRNA synthetases
	ATP, Mg^{2+}
Chain initiation	fmet-tRNA^{fmet}
	Initiation codon (AUG) of mRNA
	30S ribosomal subunit
	50S ribosomal subunit
	Initiation factors (IF-1, IF-2, and IF-3)
	GTP, Mg^{2+}
Chain elongation	70S ribosome
	Codons of mRNA
	Aminoacyl-tRNAs
	Elongation factors (EF-Tu, EF-Ts, and EF-G)
	GTP, Mg^{2+}
Chain termination	70S ribosome
	Termination codons (UAA, UAG, and UGA) of mRNA
	Release factors (RF-1, RF-2, and RF-3)
	GTP, Mg^{2+}

these components can be reused in further protein synthesis. Table 20.3 summarizes the steps in protein synthesis and the components required for each step.

20.7
POSTTRANSLATIONAL
MODIFICATION OF PROTEINS

Newly synthesized polypeptides are frequently processed before they reach the form in which they have biological activity. We have already mentioned the *N*-formylmethionine in prokaryotes is cleaved off. Specific bonds in precursors can be hydrolyzed, as in the cleavage of preproinsulin to proinsulin and of proinsulin to insulin (Figure 20.10). Proteins destined for export to specific parts of the cell or from the cell have leader sequences at their N-terminal ends. These leader sequences, which direct the proteins to their proper destinations, are recognized and removed by specific proteases associated with the endoplasmic reticulum. The finished protein then enters the Golgi apparatus, which directs it to its final destination.

In addition to the processing of proteins by breaking bonds, other substances can be linked to the newly formed polypeptide. Various cofactors such as heme groups are added, and disulfide bonds are formed (Figure 20.10). Some amino acid residues are also covalently modified, as in the conversion of proline to hydroxyproline. Other covalent modifications can take place, an example being the addition of carbohydrates to yield an active final form of the protein in question.

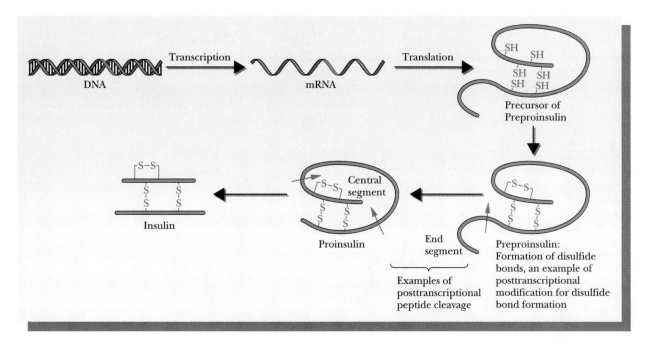

FIGURE 20.10

Some examples of posttranslational modification of proteins. After a precursor of preproinsulin is formed by the transcription–translation process, it is transformed into preproinsulin by formation of three disulfide bonds. Specific cleavage that removes an end segment converts preproinsulin to proinsulin. Finally, two further specific cleavages remove a central segment, with insulin as the end result.

20.8
POLYSOMES AND THE SIMULTANEOUS PRODUCTION OF SEVERAL COPIES OF THE SAME POLYPEPTIDE

In our description of protein synthesis, we have considered, up to now, the reactions that take place at one ribosome. It is, however, not only possible but quite usual for several ribosomes to be attached to the same mRNA. Each of these ribosomes will bear a polypeptide in one of various stages of completion, depending on the position of the ribosome as it moves along the mRNA (Figure 20.11). This complex of mRNA with several ribosomes is called a **polysome;** an alternative name is *polyribosome*.

In prokaryotes, translation begins very soon after mRNA transcription. It is possible for a molecule of mRNA that is still being transcribed to have a number of ribosomes attached to it that are in various stages of translating that mRNA. It is also possible for DNA to be in various stages of being transcribed. In this situation several molecules of RNA polymerase are attached to a single gene, giving rise to several mRNA molecules, each of which has a number of ribosomes attached to it. The prokaryotic gene is being simultaneously transcribed and translated. This process is called **coupled translation** (Figure 20.12); it is possible in prokaryotes because of

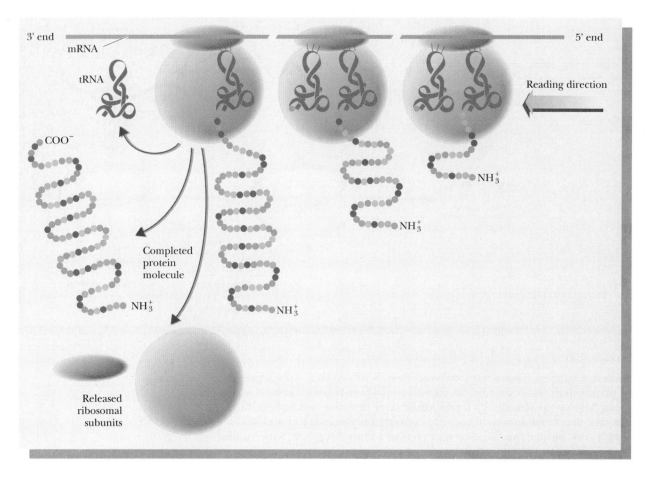

FIGURE 20.11

Simultaneous protein synthesis on polysomes. A single mRNA molecule is translated by several ribosomes simultaneously. Each ribosome produces one copy of the polypeptide chain specified by the mRNA. When the protein has been completed, the ribosome dissociates into subunits that are used in further rounds of protein synthesis.

FIGURE 20.12

Electron micrograph showing coupled translation. The dark spots are ribosomes, arranged in clusters on a strand of mRNA. Several mRNAs have been transcribed from one strand of DNA (diagonal line from center to upper right).

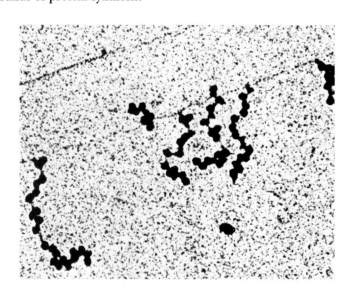

the lack of cell compartmentalization. In eukaryotes, mRNA is produced in the nucleus, and protein synthesis takes place in the cytosol.

20.9
GENETIC REGULATION OF TRANSCRIPTION AND TRANSLATION

Some proteins are not synthesized by cells at all times. Rather, the production of these proteins can be triggered by the presence of a suitable substance, called the **inducer.** This phenomenon is called **induction;** the process is under genetic control. A particularly well-studied example of an inducible protein is the enzyme **β-galactosidase** in *E. coli,* which we shall use as a case study.

The disaccharide **lactose** (a β-galactoside; Chapter 10, Section 10.3) is the substrate of β-galactosidase. The enzyme hydrolyzes the glycosidic linkage between galactose and glucose, the monosaccharides that are the component parts of lactose. *E. coli* can survive with lactose as its sole carbon source. To do so, the bacterium needs β-galactosidase to catalyze the first step in lactose degradation. The production of β-galactosidase takes place only in the presence of lactose, not in the presence of other carbon sources, such as glucose. Lactose is the inducer, and β-galactosidase is an **inducible enzyme.**

In 1961 Jacob and Monod proposed a theory to account for the experimental facts given in the last paragraph; the main features of the theory have been supported by further experimental results. According to this point of view, the actual production of an inducible protein such as β-galactosidase is under the control of a **structural gene** (Z) (Figure 20.13). The base sequence of the structural gene specifies the amino acid sequence of the protein. The expression of one or more structural genes is, in turn, under the control of a regulatory gene (*I*), and the mode of operation of the regulatory gene is the most important feature of the theory. The regulatory gene is responsible for the production of a protein, the **repressor.** As the name indicates, the repressor inhibits the expression of the structural gene. In the presence of an inducer, this inhibition is removed.

The repressor operates by binding to a portion of the DNA known as the **operator** (O) (Figure 20.13a). When a repressor is bound to the operator, RNA polymerase cannot bind to the adjacent **promoter region** (P), which facilitates the expression of the structural gene. The operator and promoter together constitute the **control sites.**

In induction, the inducer binds to the repressor, producing an inactive repressor that cannot bind to the operator (Figure 20.13b). Since the operator is no longer bound to the repressor, RNA polymerase can now bind to the promoter, and transcription and eventual translation of the structural gene (or genes) can take place. The whole assemblage of promoter, operator, and structural genes is called an **operon.** The control sites, the promoter and operator, are physically adjacent to the structural genes in the DNA sequence, but the regulatory gene can be quite far removed from the operon. (In *E. coli,* the regulatory gene for this operon is adjacent to the promoter).

When *E. coli* is presented with lactose as a carbon source, β-galactosidase is not the only protein induced. In other words, several

The *lac* operon

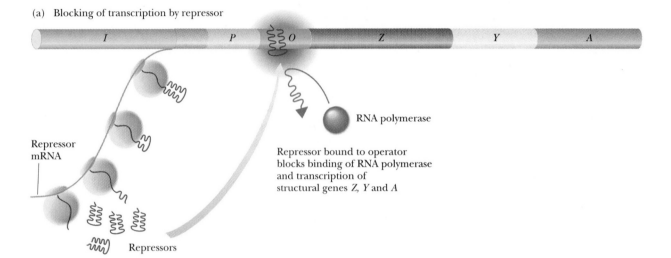

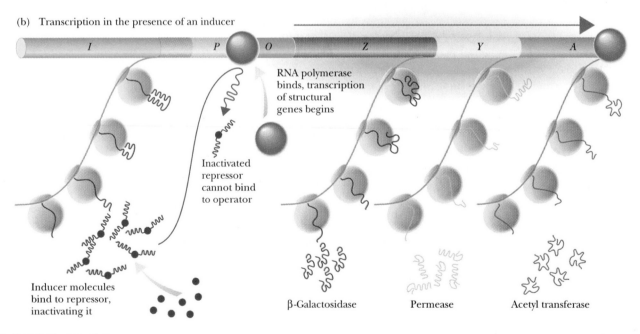

FIGURE 20.13

The *lac* operon in the absence and presence of inducer. (a) When no inducer is present, the regulatory gene (*I*) is transcribed, leading to the production of repressor. The repressor in turn binds to the operator gene (*O*), blocking transcription of the structural genes that code for β-galactosidase and the other proteins produced under the control of the *lac* operon. (b) In the presence of inducer (small blue circles), the repressor is still produced but becomes inactive when the inducer is bound to it. The inactive repressor does not bind to the operator, and the structural genes of the *luc* operon are expressed. Other regions of the DNA shown here are the promoter (*P*) and the genes for β-galactosidase (*Z*), permease (*Y*), and acetyl transferase (*A*).

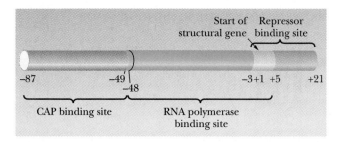

FIGURE 20.14

Binding sites in the *lac* operon. Numbering refers to base pairs. Negative numbers are assigned to base pairs in the regulatory sites. Positive numbers indicate the structural gene, starting with base pair +1. (There is some overlap between promoter and operator regions.)

structural genes are found in this operon, which is called the ***lac* operon.** In addition to β-galactosidase, the *lac* operon is responsible for the production of two other enzymes, a permease and an acetyltransferase. The permease mediates active transport of lactose into the cell, but the function of the acetyltransferase is not well understood. The structural genes of the *lac* operon are the *Z* gene, which codes for β-galactosidase; the *Y* gene, which codes for the permease; and the *A* gene, which codes for the acetyltransferase.

The *lac* operon is induced when *E. coli* has lactose, and no glucose, available to it as a carbon source. When both glucose and lactose are present, the cell does not make the *lac* proteins. The repression of the synthesis of *lac* proteins by glucose is called **catabolite repression.** The mechanism by which *E. coli* recognizes the presence of glucose involves the promoter gene. The promoter has two regions. One is the entry site for RNA polymerase, and the other is the binding site for another regulatory protein, the **catabolite activator protein** (CAP) (Figure 20.14). The binding site for RNA polymerase also overlaps the binding site for repressor in the operator region (the promoter and operator overlap).

The binding of the CAP protein to the promoter depends on the presence or absence of 3′, 5′-cyclic AMP (cAMP). When glucose is not present, cAMP is formed, serving as a "hunger signal" for the cell. CAP forms a complex with cAMP. The complex binds to the CAP site in the

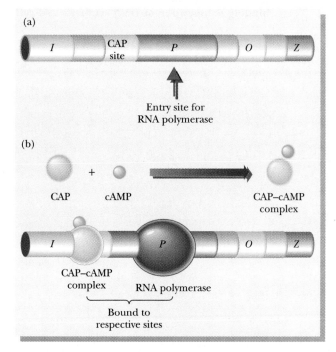

FIGURE 20.15

Catabolite repression. (a) The control sites of the *lac* operon. The CAP–cAMP complex, not CAP alone, binds to the CAP site of the *lac* promoter. When the CAP site on the promoter is not occupied, RNA polymerase cannot bind to the entry site for it on the promoter. (b) In the absence of glucose, cAMP forms a complex with CAP. The complex binds to the CAP site, allowing RNA polymerase to bind to the entry site on the promoter and transcribe the structural genes.

A GENETIC KEY TO CANCER: WHY IS THE *p53* PROTEIN A TUMOR SUPPRESSOR?

A 53-kilodalton protein designated p53 has become the focus of feverish activity in cancer research. Mutations in the gene that codes for this protein are found in about half of all human cancers. When the gene is operating normally, it acts as a tumor suppressor; when it is mutated, it is involved in a wide variety of cancers. By the end of 1993, *p53* mutations had been found in 51 types of human tumors.

The mode of action of *p53* depends on its role in the production of another protein that plays a key role in cell division. When the *p53* protein is made, it "turns on" the transcription and translation of a gene that encodes a 21-kilodalton protein that is a key regulator of DNA synthesis and thus of cell division. It has been shown that cell division depends on the activity of enzymes known as cyclin-dependent kinases (cdk), which, as their name implies, become active only when they associate with proteins called cyclins. The 21-kilodalton protein binds to the cyclin-dependent kinases and their associated cyclins. The 21-kilodalton protein is present in cdk–cyclin complexes in normal cells, but not in cancer cells. This understanding of the uncontrolled growth of cancer cells compared to the controlled growth of normal cells can be expected to lead to new forms of therapy. Research is continuing on this topic. Another protein, p16, may be a tumor suppressor as well. The protein p16 is an intrinsic part of the control mechanism for cell growth involving cyclins and cyclin-dependent kinases.

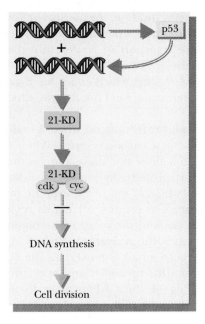

The *p53* protein turns on the production of a 21-kilodalton protein. This protein binds to complexes of cyclin-dependent kinases (cdk) and cyclins. The result of binding is inhibition of DNA synthesis and cell growth.

promoter region. When the complex is bound to the CAP site on the promoter, the RNA polymerase binds strongly to the entry site for it on the promoter. If, at the same time, lactose is present, the repressor–inducer complex forms, so the RNA polymerase can bind at the entry site available to it and proceed with transcription (Figure 20.15).

When the cell has an adequate supply of glucose, the level of cAMP is low. CAP binds to the promoter only when it is complexed to cAMP. The entry site is not available to the RNA polymerase when the CAP–cAMP complex is not bound to the promoter, and the *lac* operon proteins are not produced. Note that the *lac* operon is subject to positive control by the cAMP–CAP complex (allowing binding of RNA polymerase to the promoter) and to negative control by the repressor (steric blocking of RNA polymerase binding in the presence of the repressor).

There is another possible aspect to the control of transcription and translation by repressors. In some cases a corepressor binds to the repressor protein. The repressor–corepressor complex binds to the operator, whereas the repressor alone does not. The tryptophan operon of *E. coli* is an example of control involving a **corepressor.** This operon contains five structural genes for the enzymes that convert chorismate to tryptophan (see Chapter B, Section B.4). The corepressor of this operon is tryptophan itself. When tryptophan is abundant, the cell does not need to invest the energy needed to produce the mRNA for these five enzymes or to produce the enzymes themselves. When tryptophan is required, the cell forms it as well as the necessary enzymes and mRNAs.

Regulation of the production of inducible enzymes by the induction–repression mechanism allows for a high degree of fine-tuning of metabolism by organisms, even apparently simple ones, such as bacteria. Parenthetically, there are no known operons in eukaryotes. The simplest living cell is a highly complex and well-organized entity, far more than just a "bag of enzymes"; it is a great challenge for present and future biochemists to discover the details of metabolic processes and their control mechanisms. Box 20.1 describes a particularly striking example of genetic regulation of transcription and translation.

20.10
VIRUSES, CANCER, AND AIDS

Now that we know something about the processes involved in the replication, transcription, and translation of the genetic message of DNA, we can say something about how all these processes work together. The best understood examples of how these pathways interact are those of viral infection. The importance of this topic cannot be overemphasized. Viruses play a role in the development of cancer; one of the most notorious of all viruses, human immunodeficiency virus (HIV), is the causative agent of AIDS. In Box 20.2 we shall look at the structures of two viruses. Then we can go on to a discussion of the roles they play in cancer and AIDS.

The outcome of infection by SV40 depends on the organism infected. When simian cells are infected, the virus enters the cell and loses its protein

(Text continues on p. 653)

A TALE OF TWO VIRUSES: SV40 AND HIV

Simian virus 40 (SV40) is an example of a DNA virus. It appears to be spherical, but it is actually an icosahedron, a geometric figure with 20 faces that are equilateral triangles. The genome of this virus is a closed circle of double-stranded DNA, with genes that encode the amino acid sequences of five proteins. Three of the five proteins are coat proteins. Of the remaining two proteins, one, the large-T protein, is involved in the development of the virus when it infects a cell. The function of the fifth protein, the small-T protein, is not known. In the complete virus particle (called a **virion**), the coat proteins are packed around the DNA to give the observed icosahedral shape.

HIV is a more complex virus than SV40. Its genome is a single-stranded RNA that has a number of proteins packed around it, including the virus-specific reverse transcriptase and protease. There is a protein coat around the RNA–protein assemblage, giving the overall shape of a truncated cone. Finally there is a membrane envelope around the protein coat. The envelope consists of a phospholipid bilayer formed from the plasma membrane of cells infected earlier in the life cycle of the virus, as well as some specific glycoproteins, such as gp41 (glycoprotein 41) and gp120.

(a)
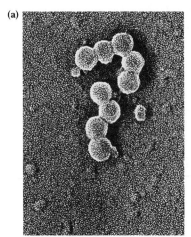
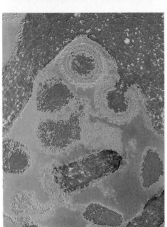

(b)
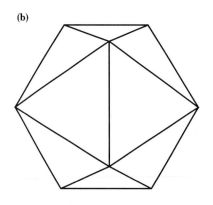

(c)
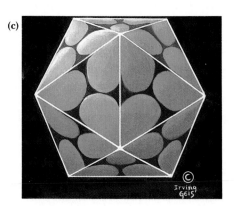

The architecture of simian virus 40 (SV40). (a) Virus particles appear almost spherical in electron micrographs but on closer examination can be seen to have an icosahedral shape. (b) The geometry of icosahedron. This regular polyhedron has 20 faces, all of which are equilateral triangles of identical size. (c) A drawing showing the arrangement of 60 subunits to form an icosahedron.

A TALE OF TWO VIRUSES: SV40 AND HIV *(continued)*

The genome of simian virus 40 is a single circle of double-stranded DNA. Note that some of the genes overlap, and that the gene for the large-T protein is split. The proteins encoded by genes VP_1, VP_2, and VP_3 are part of the protein coat of the virus.

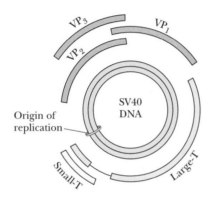

The architecture of HIV. (a) The RNA genome is surrounded by P7 nucleocapsid proteins and by several viral enzymes, namely reverse transcriptase, integrase, and protease. The truncated cone consists of P24 capsid protein subunits. The P17 matrix (another layer of protein) lies inside the envelope, which consists of a lipid bilayer and glycoproteins such as gp41 and gp120. (b) An electron micrograph shows both mature virus particles, in which the core (the truncated cone) is visible, and immature virus particles, in which it is not.

(a)

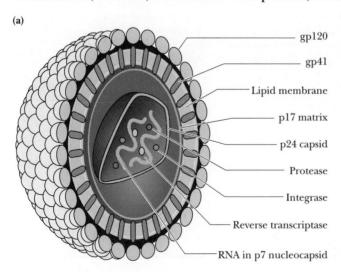

(b)

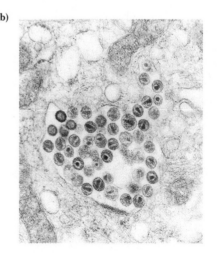

FIGURE 20.16

The outcome of infection of cells by simian virus 40 depends on the nature of the cells. When primate cells are infected, the large-T protein is produced and its presence ultimately leads to the production of new viral DNA and coat proteins. New virus particles are assembled and released; the death of the host cell takes place when the new virions are released.

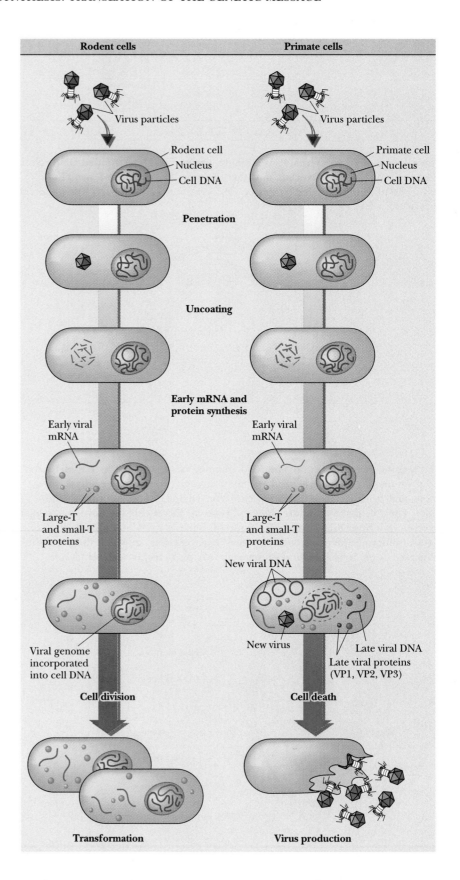

coat. The viral DNA is expressed as mRNAs and then as proteins. The large-T protein is the first made (Figure 20.16), triggering the replication of viral DNA, followed by viral coat proteins. The virus takes over the cellular machinery for both replication of DNA and protein synthesis. New virus particles are assembled, and eventually the infected cell bursts, releasing the new virus particles to infect other cells.

The results are different when SV40 infects rodent cells. The process is the same as far as the production of the large-T protein, but replication of the viral DNA does not take place. The SV40 DNA already present in the cell can be lost or can be integrated into the DNA of the host cell. If the SV40 DNA is lost, there is no apparent result of the infection. If it becomes integrated into the DNA of the host cell, the infected cell loses control of its own growth. As a result of accumulation of large-T protein, the infected cell behaves like a cancer cell. The large-T gene is an **oncogene,** one that causes cancer. Its mechanism is a subject of active research.

Oncogenes are transported from one cell to another by viruses, but they normally originate in cells that have been infected by the virus at some earlier time. *Analogues of these genes exist in normal cells in many different species, cells that have not been infected by a virus. When some triggering event takes place, they are transformed. As a result of the change in the gene, the cell becomes a cancer cell.* Viral oncogenes have been incorporated into the genomes of many viruses and are carried along with the genes needed specifically for the replication of the virus. The oncogenes of DNA viruses do not appear to have normal analogues in the cell, but the situation is different with RNA viruses.

Many of the viruses that transmit oncogenes are **retroviruses,** RNA viruses that make use of reverse transcriptase to copy their RNA genomes into DNA. RNA tumor viruses have been known since the discovery in 1911 of Rous sarcoma virus, a retrovirus that leads to sarcomas (connective tissue tumors) in chickens. Figure 20.17 shows the genomes of some retroviruses. When the viral DNA, produced by the action of reverse transcriptase, becomes integrated into the DNA of the host cell, the transformation into a cancer cell can take place. It is not known whether abnormal regulation of the oncogene or accumulation of its abnormal product, or both processes together, gives rise to a tumor. Research on this topic is a particularly active area.

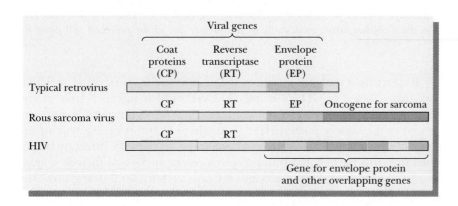

FIGURE 20.17

The RNA genomes of all retroviruses have genes for coat proteins (CP), for reverse transcriptase (RT), and for envelope protein (EP). In addition to these essential genes, the Rous sarcoma virus carries the sarcoma oncogene. The HIV genome is more complex, with a number of overlapping genes for envelope protein and other proteins.

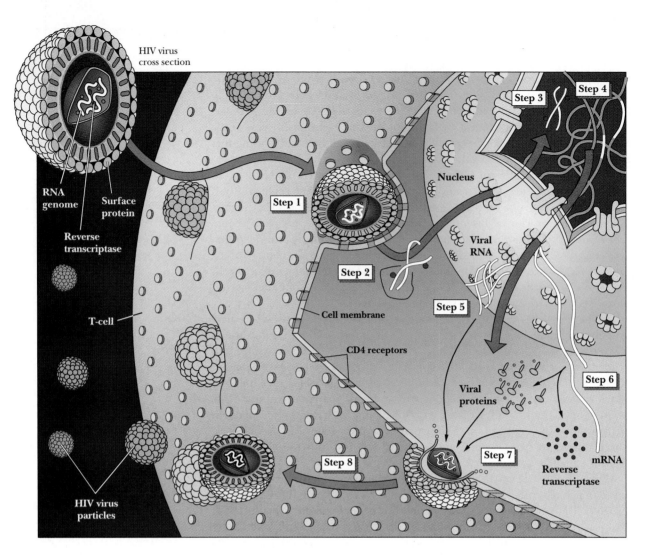

FIGURE 20.18

HIV infection begins when the virus particle binds to CD4 receptors on the surface of the cell (Step 1). The viral core is inserted into the cell and partially disintegrates (Step 2). The reverse transcriptase catalyzes the production of DNA from the viral RNA. The viral DNA is integrated into the DNA of the host cell (Step 3). The DNA, including the integrated viral DNA, is transcribed to RNA (Step 4). Smaller RNAs are produced first, specifying the amino acid sequence of viral regulatory proteins (Step 5). Larger RNAs, ones that specify the amino acid sequences of viral enzymes and coat proteins, are made next (Step 6). The viral protease assumes particular importance in the budding of new virus particles (Step 7). Both the viral RNA and viral proteins are included in the budding virus, as is some of the membrane of the infected cell (Step 8).

Retroviruses that cause cancer in humans are known; some forms of leukemia (caused by HTLV-I and HTLV-II, which infect *T* cells of the immune system) are well-known examples. Their fame, however, is obscured by that of the closely related HIV. Its mode of action is an example of the mode of operation of retroviruses. It is known that the HIV infection begins when the virus particle binds to receptors on the surface of the cell (Figure 20.18). The viral core is inserted into the cell and partially disintegrates. The reverse transcriptase catalyzes the production of DNA from the viral RNA.

The viral DNA is integrated into the DNA of the host cell. The DNA, including the integrated viral DNA, is transcribed to RNA. Smaller RNAs are produced first, specifying the amino acid sequences of viral regulatory proteins. Larger RNAs, ones that specify the amino acid sequences of viral enzymes and coat proteins, are made next. The viral protease (Chapter 5, Box 5.2) assumes particular importance in the budding of new virus particles. Both the viral RNA and viral proteins are included in the budding virus, as is some of the membrane of the infected cell.

The reverse transcriptase of HIV is not very accurate in replication. The result is rapid mutation of HIV, a situation that presents a considerable challenge to those who want to devise treatments for AIDS. It seems likely that effective treatment will have to include a battery of drugs that affect the virus at various stages of its life cycle.

S U M M A R Y

Protein biosynthesis requires ribosomes, messenger RNA (mRNA), transfer RNA (tRNA), and a number of protein factors. The ribosome is the site of protein synthesis. The mRNA and tRNA, which are bound to the ribosome in the course of protein synthesis, are responsible for the correct order of amino acids in the growing protein chain.

The genetic message is contained in a triplet, nonoverlapping, commaless, degenerate, universal code. A codon—in other words, a series of three bases adjacent to one another in sequence (nonoverlapping and commaless)—specifies a given amino acid. Several codons can and usually do specify the same amino acid (degeneracy of the code). The same code has been observed in viruses, prokaryotes, and eukaryotes (universality of the code); the only exceptions are the differences in some codons observed in mitochondria. All 64 possible codons have been assigned meanings, with 61 of them coding for amino acids and the remaining 3 serving as the termination signals.

Before an amino acid can be incorporated into a growing protein chain, it must first be activated. A covalent bond is formed between the amino acid and a tRNA, yielding an aminoacyl-tRNA. The actual formation of the polypeptide chain takes place in three steps. In the initiation step, the first aminoacyl-tRNA is bound to the ribosome and to mRNA. A second aminoacyl-tRNA forms a complex with the ribosome and with mRNA. The binding site for the second aminoacyl-tRNA is close to that for the first aminoacyl-tRNA. A peptide bond is formed between the amino acids (chain elongation). The chain elongation process—which involves translocation of the ribosome along the mRNA, in addition to peptide bond formation—repeats itself until the polypeptide chain is complete. Finally, chain termination takes place. Newly synthesized polypeptides frequently undergo post-translational modification to produce the final active form of the protein. In the actual translation process, it is usual for several ribosomes to be bound to the same mRNA. Such a complex is called a polysome. Each of the ribosomes in the polysome has a polypeptide in one of various stages of completion, depending on the position of the ribosome as it moves along the mRNA and transcribes the genetic message.

Not all proteins are synthesized by cells at all times. The process of induction, which triggers the production of such proteins in the presence of a suitable substance called the inducer, is under genetic control in prokaryotes. The actual synthesis of an inducible protein is regulated by a structural gene. The expression of one or more structural genes is, in turn, under the control of a regulatory gene. The regulatory gene is responsible for the production of a protein, the repressor, which binds to the operator site. When a repressor is bound to the operator, RNA polymerase cannot bind to the adjacent promoter site. The operator and promoter together constitute the control sites. The whole assemblage of promoter, operator, and structural genes is called an operon. In induction, the inducer binds to the repressor, producing an inactive repressor that cannot bind to the operator. Since the operator is no longer bound to the repressor, RNA polymerase can now bind to the promoter, and transcription and eventual translation of the structural gene (or genes) can take place. Other variations in operon activity, such as catabolite repression and the action of corepressors, are possible, allowing for several different ways to control protein synthesis.

The whole cellular mechanism of replication, transcription, and translation becomes subject to takeover during viral infection. The virus uses the cellular machinery to produce new virus particles. Not only can viruses cause diseases ranging from the common cold to influenza, they are involved in the development of tumors and in AIDS.

E X E R C I S E S

1. Prepare a flow chart showing the stages of protein synthesis.

2. What is the role of ATP in amino acid activation?

3. A friend tells you that she is starting a research project on aminoacyl esters. She asks you to describe the biological role of this class of compounds. What do you tell her?

4. *E. coli* has two tRNAs for methionine. What is the basis for the distinction between the two?

5. Describe the recognition process by which the tRNA for *N*-formylmethionine interacts with the portion of mRNA that specifies the start of transcription.

6. What are the components of the initiation complex in protein synthesis, and how do they interact with one another?

7. What are the A site and the P site? How are their roles in protein synthesis similar, and how do they differ?

8. Identify the following by describing their functions: EF-G, EF-Tu, EF-Ts, and peptidyl transferase.

9. How does puromycin function as an inhibitor of protein synthesis?

10. Describe the role of the stop signals in protein synthesis.

11. The amino acid hydroxyproline is found in collagen. There is no codon for hydroxyproline. Give an explanation for the occurrence of this amino acid in a common protein.

12. In the early days of research on protein synthesis, some scientists observed that their most highly purified ribosome preparations, containing almost exclusively single ribosomes, were less active than preparations that were less highly purified. Suggest an explanation for this observation.

13. A genetic code in which two bases code for a single amino acid is not adequate for protein synthesis. Give a reason why.

14. It is possible for the codons for a single amino acid to have the first two bases in common and to differ in the third base. Why is this experimental observation consistent with the concept of wobble?

15. The base hypoxanthine (the corresponding nucleoside is inosine) frequently occurs as the third base in codons. What role does hypoxanthine play in wobble base pairing?

16. Define the terms "inducer" and "inducible enzyme."

17. What are the component parts of an operon, and what roles do they play in inducing enzyme synthesis?

18. What is the distinction between a repressor and a corepressor?

19. Do our own genes play a role in the development of tumors? Give the reason for your answer.

20. Drugs used in the treatment of AIDS frequently target some of the viral enzymes such as the reverse transcriptase and the protease. Why are these enzymes chosen as targets?

A N N O T A T E D   B I B L I O G R A P H Y

Abraham, A. K., T. S. Eikhon, and I. F. Pryme, eds. *Protein Synthesis: Translational and Post-Translational Events.* Clifton, NJ: Humana Press, 1983. [A collection of articles on all aspects of protein synthesis, with extensive coverage of modification of proteins after synthesis.]

Adams, R. L. P., J. T. Knowles, and D. P. Leader. *The Biochemistry of the Nucleic Acids.* 11th ed. New York: Chapman and Hall, 1992. [New authors have prepared this edition of a classic text originally written by J. N. Davidson.]

Berg, P., and M. Singer. *Dealing with Genes: The Language of Heredity*. Mill Valley, CA: University Science Books, 1992. [An eminently readable account of molecular genetics written by two outstanding biochemists.]

Brenner, S., F. Jacob, and M. Meselson. An Unstable Intermediate Carrying Information from Genes to Ribosomes for Protein Synthesis. *Nature* **190**, 576–581 (1961). [One of the first descriptions of the concept of mRNA. Of historical interest.]

Crick, F. H. C. Codon–Anticodon Pairing: The Wobble Hypothesis. *J. Mol. Biol.* **19**, 548–555 (1966). [The first statement of the wobble hypothesis, and still one of the best.]

Greene, W. AIDS and the Immune System. *Sci. Amer.* **269** (3), 98–105 (1993). [A description of the HIV virus and its life cycle in *T* cells.]

Gualerzi, C., and C. Pon. Initiation of mRNA Translation in Prokaryotes. *Biochemistry* **29**, 588–589 (1990). [A review of the initiation step in protein synthesis.]

Jacob, F., and J. Monod. Genetic Regulatory Mechanisms in the Synthesis of Proteins. *J. Mol. Biol.* **3**, 318–356 (1961). [The original article in which the concept of repression was postulated. Also one of the first descriptions of mRNA. Mostly of historical interest but quite well written.]

Jaenicke, R. Protein Folding: Local Structures, Domains, Subunits, and Assemblies. *Biochemistry* **30**, 3147–3161 (1991). [A review of a topic that has been called "the second half of the genetic code."]

Jordan, S., and C. Pabo. Structure of the Lambda Complex at 2.5 Å Resolution: Details of the Repressor–Operator Interactions. *Science* **242**, 893–899 (1988). [Detailed information on repressor binding to DNA.]

Moldave, K., ed. *RNA and Protein Synthesis*. New York: Academic Press, 1981. [A collection of articles on all aspects of protein synthesis, with emphasis on the role of tRNA.]

Nierhaus, K. The Allosteric Three-Site Model for the Ribosomal Elongation Cycle: Features and Future. *Biochemistry* **29**, 4997–5008 (1990). [A review of the role of ribosomal binding sites for tRNA in protein chain elongation.]

Ptashne, M. *A Genetic Switch: Gene Control and Phage λ*. Palo Alto, CA: Cell Press/Blackwell Scientific, 1986. [An extensive discussion of induction and repression. The life cycle of a bacteriophage that infects *E. coli* is used as an example.]

Ptashne, M. How Gene Activators Work. *Sci. Amer.* **260** (1), 41–47 (1989). [An explanation of the mode of action of repressors in eukaryotic and prokaryotic systems.]

Ross, J. The Turnover of Messenger RNA. *Sci. Amer.* **260** (4), 48–55 (1989). [A description of the regulation of the rate at which mRNA is degraded in the cell.]

Schultz, S., G. Shields, and T. Steitz. Crystal Structure of a CAP–DNA Complex: The DNA Is Bent by 90°. *Science* **253**, 1001–1007 (1991). [The structure of a complex involved in transcriptional control of gene expression.]

Waldrop, M. Finding RNA Makes Proteins Gives "RNA World" a Big Boost. *Science* **256**, 1396–1397 (1992). [RNA can catalyze the formation of peptide bonds.]

Watson, J. D. *Molecular Biology of the Gene*. 4th ed. Menlo Park, CA: Benjamin/Cummings, 1987. [A highly readable book on molecular biology.]

Answers to Exercises

Chapter 1

1. A polymer is a very large molecule formed by the linking together of smaller units (monomers). A protein is a polymer formed by the linking together of amino acids. A nucleic acid is a polymer formed by the linking together of nucleotides. Catalysis is the process that increases the rate of a chemical reaction compared to the uncatalyzed reaction. Biological catalysts are proteins in almost all cases; the only exceptions are a few types of RNA that can catalyze some of the reactions of their own metabolism. The genetic code is the means by which the information for the structures and functions of all living things is passed from one generation to the next. The sequence of purines and pyrimidines in DNA carries the genetic code (RNA is the coding material in some viruses). Anabolism is the synthesis of important biomolecules from simpler compounds. Catabolism is the breakdown of biomolecules to release energy.

2. The theory that proteins arose first in the origins of life gives a good explanation of catalysis and metabolic pathways but is vague about the origin of coding. The theory that nucleic acids arose first gives prime importance to coding but does not address the problem of lack of stability of unprotected nucleic acids. The double-origin theory that life arose on the surface of clay particles suggests a stable coding system (the clay surface), which also served as a site for catalysis. Later, more efficient biomolecules replaced clay particles in life coding and catalysis for life processes.

3. With respect to coding, RNA has been produced from monomers in the absence of either a preexisting RNA to be copied or an enzyme to catalyze the process. The observation that some existing RNA molecules can catalyze their own processing suggests a role for RNA in catalysis. With this dual role, RNA may have been the original informational macromolecule in the origin of life.

4. It is unlikely that cells could have arisen as bare cytoplasm without plasma membranes. The presence of the membrane protects cellular components from the environment and prevents them from diffusing away from each other. The molecules within a cell can react more easily if they are closer to each other.

5. Five differences between prokaryotes and eukaryotes are as follows. (1) Prokaryotes do not have a well-defined nucleus, but eukaryotes have a nucleus marked off from the rest of the cell by a double membrane. (2) Prokaryotes have only a plasma (cell) membrane; eukaryotes have an extensive internal membrane system. (3) Eukaryotic cells contain membrane-bounded organelles, while prokaryotic cells do not. (4) Eukaryotic cells are normally larger than the cells of prokaryotes. (5) Prokaryotes are single-celled organisms, while eukaryotes can be multicellular as well as single-celled.

6. See Section 1.5 for the functions of the parts of an animal cell, which are shown in Figure 1.6a.

7. See Section 1.5 for the functions of the parts of a plant cell, which are shown in Figure 1.6b.

8. In green plants, photosynthesis takes place in the membrane systems of chloroplasts, which are large membrane-bounded organelles. In photosynthetic bacteria there are extensions of the plasma membrane into the interior of the cell called chromatophores, which are the sites of photosynthesis.

9. Nuclei, mitochondria, and chloroplasts are bounded by a double membrane.

10. Nuclei, mitochondria, and chloroplasts all contain DNA. The DNA found in mitochondria and in chloroplasts differs from that in the nucleus.

11. Mitochondria carry out a high percentage of the oxidation energy-releasing reactions of the cell. They are the primary sites of ATP synthesis.

12. Protein synthesis takes place on ribosomes in both prokaryotes and eukaryotes. In eukaryotes, ribosomes may be bound to the endoplasmic reticulum or found

free in the cytoplasm; in prokaryotes, ribosomes are only found free in the cytoplasm.

13. The Golgi apparatus is involved in carbohydrate metabolism and in the export of substances from the cell. Lysosomes contain hydrolytic enzymes, peroxisomes contain catalase (needed for the metabolism of peroxides), and glyoxysomes contain enzymes needed by plants for the glyoxylate cycle. Each of these organelles has the appearance of a flattened sac bounded by a single membrane.

14. It is unlikely that mitochondria would be found in bacteria. These eukaryotic organelles are bounded by a double membrane, and bacteria do not have an internal membrane system. The mitochondria found in eukaryotic cells are about the same size as most bacteria.

Chapter 2

1. The C—H bond is not sufficiently polar for greatly unequal distribution of electrons at its two ends. Also, there are no unshared pairs of electrons to serve as hydrogen bond acceptors.

2. In a hydrogen-bonded dimer of acetic acid, the —OH portion of the carboxyl group on molecule 1 is hydrogen-bonded to the —C=O portion of the carboxyl group on molecule 2, and vice versa.

$$CH_3 - \underset{OH\text{---}O}{\overset{O\text{---}HO}{\diamondsuit}} - CH_3$$

3. $\dfrac{(CH_3)_3NH^+ \text{ (conjugate acid)}}{(CH_3)_3N \text{ (conjugate base)}}$

$\dfrac{^+H_3N-CH_2-COOH \text{ (conjugate acid)}}{^+H_3N-CH_2-COO^- \text{ (conjugate base)}}$

$\dfrac{^+H_3N-CH_2-COO^- \text{ (conjugate acid)}}{H_2N-CH_2-COO^- \text{ (conjugate base)}}$

$\dfrac{^-OOC-CH_2-COOH \text{ (conjugate acid)}}{^-OOC-CH_2-COO^- \text{ (conjugate base)}}$

$\dfrac{^-OOC-CH_2-COOH \text{ (conjugate base)}}{HOOC-CH_2-COOH \text{ (conjugate acid)}}$

4.
Blood plasma, pH 7.4	$[H^+] = 4.0 \times 10^{-8}$ M
Orange juice, pH 3.5	$[H^+] = 3.2 \times 10^{-4}$ M
Human urine, pH 6.2	$[H^+] = 6.3 \times 10^{-7}$ M
Household ammonia, pH 11.5	$[H^+] = 3.2 \times 10^{-12}$ M
Gastric juice, pH 1.8	$[H^+] = 1.6 \times 10^{-2}$ M

5. In all cases the suitable buffer range covers a pH range of $pK_a \pm 1$ pH unit.
   Lactic acid ($pK_a = 3.86$) and its sodium salt, pH 2.86–4.86
   Acetic acid ($pK_a = 4.76$) and its sodium salt, pH 3.76–5.76
   TRIS (see Table 2.4; $pK_a = 8.3$) in its protonated form and its free amine form, pH 7.3–9.3
   HEPES (see Table 2.4; $pK_a = 7.55$) in its zwitterionic and anionic forms, pH 6.55–8.55

6. Use the Henderson-Hasselbalch equation:

$$pH = pK_a + \log\left(\frac{[CH_3COO^-]}{[CH_3COOH]}\right)$$

$$5.00 = 4.76 + \log\left(\frac{[CH_3COO^-]}{[CH_3COOH]}\right)$$

$$0.24 = \log\left(\frac{[CH_3COO^-]}{[CH_3COOH]}\right)$$

$$\left(\frac{[CH_3COO^-]}{[CH_3COOH]}\right) = \frac{1.7}{1}$$

7. At pH 7.5, the ratio of $[HPO_4^{2-}]/[H_2PO_4^-]$ is 2/1 (pK_a of $H_2PO_4^- = 7.2$), as calculated using the Henderson–Hasselbalch equation. K_2HPO_4 is a source of the base form, and HCl must be added to convert one third of it to the acid form, according to the 2/1 base/acid ratio. Weigh 8.7 grams of K_2HPO_4 (0.05 moles, based on a formula weight of 174 grams/mole), dissolve it in a small quantity of distilled water, add 16.7 mL of 1 M HCl (gives 1/3 of 0.05 moles of hydrogen ion, which converts 1/3 of the 0.05 moles of HPO_4^{2-} to $H_2PO_4^-$), and dilute the resulting mixture to one liter.

8. A 2/1 ratio of the base form to acid form is still needed, because the pH of the buffer is the same in both problems. NaH_2PO_4 is a source of the acid form, and NaOH must be added to convert two thirds of it to the base form. Weigh 6.0 grams of NaH_2PO_4 (0.05 moles, based on a formula weight of 120 grams/mole), dissolve it in a small quantity of distilled water, add 33.3 mL of 1 M NaOH (gives 2/3 of 0.05 moles of hydroxide ion, which converts 2/3 of the 0.05 moles of $H_2PO_4^-$ to HPO_4^{2-}), and dilute the resulting mixture to one liter.

9. At the equivalence point of the titration, a small amount of acetic acid remains because of the equilibrium $CH_3COOH \rightleftharpoons H^+ + CH_3COO^-$. There is a small, but nonzero, amount of acetic acid left.

10. Buffering capacity refers to the amounts of the acid and base forms present in the buffer solution. A solution with a high buffering capacity can react with a large amount of added acid or base without drastic changes

in pH. A solution with a low buffering capacity can react with only comparatively small amounts of acid or base before showing changes in pH. The more concentrated the buffer, the higher its buffering capacity. The first buffer listed here has 10 times less buffering capacity than the second, which in turn has 10 times less buffering capacity than the third. All three buffers have the same pH, since they all have the same relative amounts of the acid and base forms.

11. The only zwitterion is $^+H_3N-CH_2-COO^-$.

12. The solution is a buffer because it contains equal concentrations of TRIS in the acid and free amine forms. When the two solutions are mixed, the concentrations of the resulting solution (in the absence of reaction) are 0.05 M HCl and 0.1 M TRIS because of dilution. The HCl reacts with half the TRIS present, giving 0.05 M TRIS (protonated form) and 0.05 M TRIS (free amine form).

13. $[H^+] = 7.9 \times 10^{-3}$ M

14. Use the Henderson–Hasselbalch equation. [Acetate ion]/[acetic acid] = 2.3/1

15. A substance with a pK_a' of 3.9 has a buffer range of 2.9 to 4.9. It will not buffer effectively at pH 7.5.

16. Hypoventilation decreases the pH of blood.

17. Aspirin is electrically neutral at the pH of the stomach and can pass the membrane more easily there than in the small intestine.

18. The correct matches of functional group and compound containing that functional group are given in the following list.

Amino group	$CH_3CH_2NH_2$
Carbonyl group (ketone)	CH_3COCH_3
Hydroxyl group	CH_3OH
Carboxyl group	CH_3COOH
Carbonyl group (aldehyde)	CH_3CH_2CHO
Thiol group	CH_3SH
Ester linkage	$CH_3COOCH_2CH_3$
Double bond	$CH_3CH=CHCH_3$
Amide linkage	$CH_3CON(CH_3)_2$
Ether	$CH_3CH_2OCH_2CH_3$

19. The functional groups are identified in the compounds that follow.

Glucose

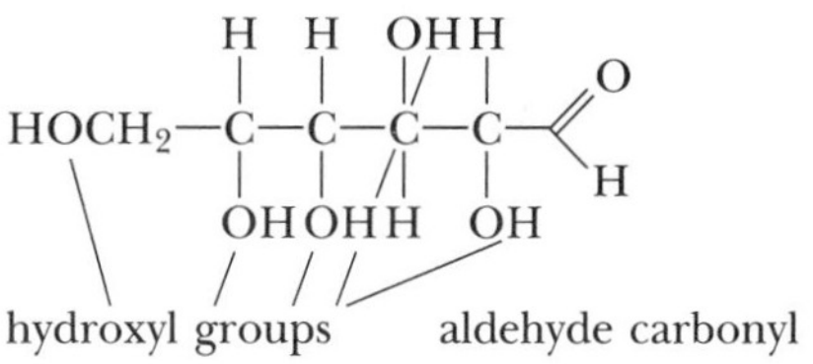

hydroxyl groups   aldehyde carbonyl

A triglyceride

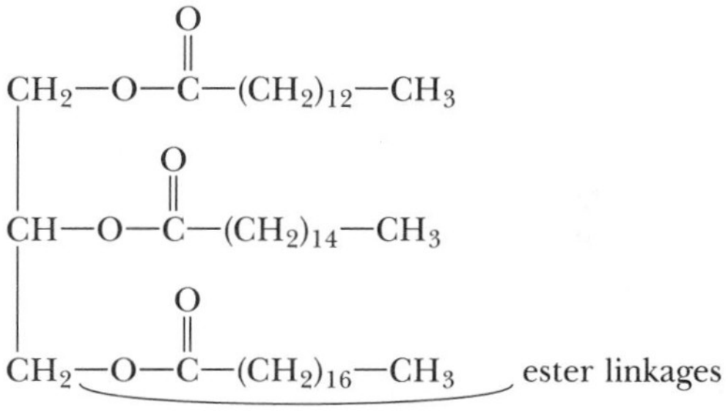

ester linkages

A peptide

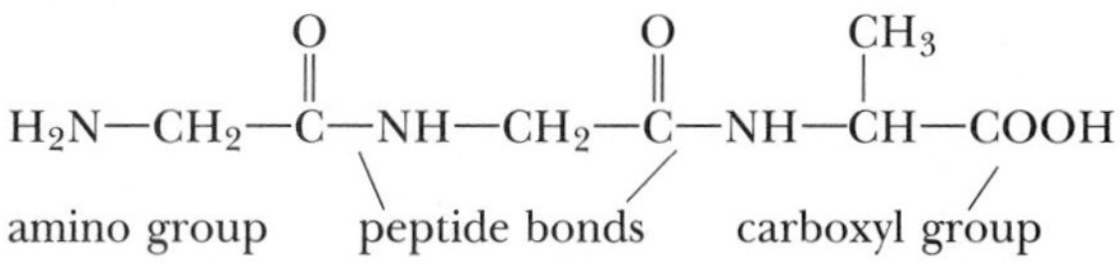

amino group   peptide bonds   carboxyl group

Vitamin A

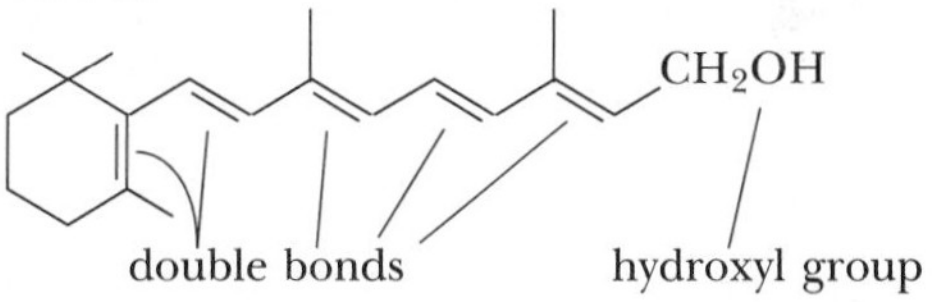

double bonds   hydroxyl group

20. Urea, like all organic compounds, has the same molecular structure whether it is produced by a living organism or not.

Chapter 3

1. The ionic dissociation reactions of the following amino
   acids: aspartic acid, valine, histidine, serine, and lysine

Aspartic acid

$$
\begin{array}{ccccccc}
\text{COOH} & & \text{COO}^{\ominus} & & \text{COO}^{\ominus} & & \text{COO}^{\ominus} \\
| & & | & & | & & | \\
\overset{\oplus}{\text{H}_3\text{N}}-\text{C}-\text{H} & \underset{\substack{\text{p}K_a' \\ 2.09}}{\rightleftharpoons} & \overset{\oplus}{\text{H}_3\text{N}}-\text{C}-\text{H} & \underset{\substack{\text{p}K_a' \\ 3.86}}{\rightleftharpoons} & \overset{\oplus}{\text{H}_3\text{N}}-\text{C}-\text{H} & \underset{\substack{\text{p}K_a' \\ 9.82}}{\rightleftharpoons} & \text{H}_2\text{N}-\text{C}-\text{H} \\
| & & | & & | & & | \\
\text{CH}_2 & & \text{CH}_2 & & \text{CH}_2 & & \text{CH}_2 \\
| & & | & & | & & | \\
\text{COOH} & & \text{COOH} & & \text{COO}^{\ominus} & & \text{COO}^{\ominus}
\end{array}
$$

+1 net charge   0 net charge   −1 net charge   −2 net charge

Valine

$$
\begin{array}{ccccc}
\text{COOH} & & \text{COO}^{\ominus} & & \text{COO}^{\ominus} \\
| & & | & & | \\
\overset{\oplus}{\text{H}_3\text{N}}-\text{C}-\text{H} & \underset{\substack{\text{p}K_a' \\ 2.32}}{\rightleftharpoons} & \overset{\oplus}{\text{H}_3\text{N}}-\text{C}-\text{H} & \underset{\substack{\text{p}K_a' \\ 9.62}}{\rightleftharpoons} & \text{H}_2\text{N}-\text{C}-\text{H} \\
| & & | & & | \\
\text{H}_3\text{C}-\text{C}-\text{H} & & \text{H}_3\text{C}-\text{C}-\text{H} & & \text{H}_3\text{C}-\text{C}-\text{H} \\
| & & | & & | \\
\text{CH}_3 & & \text{CH}_3 & & \text{CH}_3
\end{array}
$$

+1 net charge   0 net charge   −1 net charge

Histidine

$$
\begin{array}{ccccccc}
\text{COOH} & & \text{COO}^{\ominus} & & \text{COO}^{\ominus} & & \text{COO}^{\ominus} \\
| & & | & & | & & | \\
\overset{\oplus}{\text{H}_3\text{N}}-\text{C}-\text{H} & \underset{\substack{\text{p}K_a' \\ 1.83}}{\rightleftharpoons} & \overset{\oplus}{\text{H}_3\text{N}}-\text{C}-\text{H} & \underset{\substack{\text{p}K_a' \\ 6.0}}{\rightleftharpoons} & \overset{\oplus}{\text{H}_3\text{N}}-\text{C}-\text{H} & \underset{\substack{\text{p}K_a' \\ 9.2}}{\rightleftharpoons} & \text{H}_2\ddot{\text{N}}-\text{C}-\text{H} \\
| & & | & & | & & |
\end{array}
$$

with imidazole side chains:

+2 net charge   +1 net charge   0 net charge   −1 net charge

Serine

$$
\begin{array}{ccccc}
\text{COOH} & & \text{COO}^{\ominus} & & \text{COO}^{\ominus} \\
| & & | & & | \\
\overset{\oplus}{\text{H}_3\text{N}}-\text{C}-\text{H} & \underset{\substack{\text{p}K_a' \\ 2.21}}{\rightleftharpoons} & \overset{\oplus}{\text{H}_3\text{N}}-\text{C}-\text{H} & \underset{\substack{\text{p}K_a' \\ 9.15}}{\rightleftharpoons} & \text{H}_2\text{N}-\text{C}-\text{H} \\
| & & | & & | \\
\text{CH}_2\text{OH} & & \text{CH}_2\text{OH} & & \text{CH}_2\text{OH}
\end{array}
$$

+1 net charge   0 net charge   −1 net charge

Lysine

$$
\begin{array}{ccccccc}
\text{COOH} & & \text{COO}^{\ominus} & & \text{COO}^{\ominus} & & \text{COO}^{\ominus} \\
| & & | & & | & & | \\
\overset{\oplus}{\text{H}_3\text{N}}-\text{C}-\text{H} & \underset{\substack{\text{p}K_a' \\ 2.18}}{\rightleftharpoons} & \overset{\oplus}{\text{H}_3\text{N}}-\text{C}-\text{H} & \underset{\substack{\text{p}K_a' \\ 8.95}}{\rightleftharpoons} & \text{H}_2\text{N}-\text{C}-\text{H} & \underset{\substack{\text{p}K_a' \\ 10.53}}{\rightleftharpoons} & \text{H}_2\text{N}-\text{C}-\text{H} \\
| & & | & & | & & | \\
(\text{CH}_2)_4 & & (\text{CH}_2)_4 & & (\text{CH}_2)_4 & & (\text{CH}_2)_4 \\
| & & | & & | & & | \\
\overset{\oplus}{\text{NH}_3} & & \overset{\oplus}{\text{NH}_3} & & \overset{\oplus}{\text{NH}_3} & & \text{NH}_2
\end{array}
$$

+2 net charge   +1 net charge   0 net charge   −1 net charge

2. The ionized form of the following amino acids at pH 7: glutamic acid, leucine, threonine, histidine, and arginine

Glutamic acid

$$\overset{\oplus}{H_3N}-\overset{\displaystyle COO^{\ominus}}{\underset{\displaystyle CH_2}{\underset{\displaystyle \underset{\displaystyle COO^{\ominus}}{CH_2}}{\overset{\displaystyle |}{C}-H}}}$$

Leucine

$$\overset{\oplus}{H_3N}-\overset{\displaystyle COO^{\ominus}}{\underset{\displaystyle CH_2}{\underset{\displaystyle \underset{\displaystyle CH_3 \quad CH_3}{CH}}{\overset{\displaystyle |}{C}-H}}}$$

Threonine

$$\overset{\oplus}{H_3N}-\overset{\displaystyle COO^{\ominus}}{\underset{\displaystyle CHOH}{\underset{\displaystyle CH_3}{\overset{\displaystyle |}{C}-H}}}$$

pH 7

Histidine

$$\overset{\oplus}{H_3N}-\overset{\displaystyle COO^{\ominus}}{\underset{\displaystyle CH_2}{\overset{\displaystyle |}{C}-H}}$$

Arginine

$$\overset{\oplus}{H_3N}-\overset{\displaystyle COO^{\ominus}}{\underset{\displaystyle (CH_2)_3}{\underset{\displaystyle NH}{\underset{\displaystyle \overset{\oplus}{C}=NH_2}{\underset{\displaystyle NH_2}{\overset{\displaystyle |}{C}-H}}}}}$$

3. The pK'_a for the ionization of the thiol group of cysteine is 8.33, so this amino acid could serve as a buffer in the —SH and —S— forms over the pH range 7.33 to 9.33. The α-amino groups of asparagine and lysine have pK'_a values of 8.80 and 8.95, respectively; these are also possible buffers, but they are both near the end of their buffer ranges.

4. In the peptide, Val-Met-Ser-Ile-Phe-Arg-Cys-Tyr-Leu, the polar amino acids are Ser, Arg, Cys, and Tyr; the aromatic amino acids are Phe and Tyr; and the sulfur-containing amino acids are Met and Cys.

5. At pH 1 the charged groups are the N-terminal NH_3^+ on valine and the protonated guanidino group on arginine. The charged groups at pH 7 are the same as those at pH 1 with the addition of the carboxylate group on the C-terminal leucine.

6.
Ser-Leu-Phe	Leu-Ser-Phe	Phe-Ser-Leu
Ser-Phe-Leu	Leu-Phe-Ser	Phe-Leu-Ser

7.

Histidine

$$\overset{\oplus}{H_3N}-\overset{\displaystyle COO^{\ominus}}{\underset{\displaystyle CH_2}{\overset{\displaystyle |}{C}-H}}$$

Asparagine

$$\overset{\oplus}{H_3N}-\overset{\displaystyle COO^{\ominus}}{\underset{\displaystyle CH_2}{\underset{\displaystyle O=C}{\underset{\displaystyle NH_2}{\overset{\displaystyle |}{C}-H}}}}$$

Tryptophan

$$\overset{\oplus}{H_3N}-\overset{\displaystyle COO^{\ominus}}{\underset{\displaystyle CH_2}{\overset{\displaystyle |}{C}-H}}$$

pH 4

Proline

$$\overset{\displaystyle COO^{\ominus}}{\underset{\displaystyle CH}{\underset{\displaystyle H_2C \qquad \overset{\oplus}{N}H_2}{\underset{\displaystyle H_2C-CH_2}{}}}}$$

Tyrosine

$$\overset{\oplus}{H_3N}-\overset{\displaystyle COO^{\ominus}}{\underset{\displaystyle CH_2}{\overset{\displaystyle |}{C}-H}}$$

8. Both peptides, Phe-Glu-Ser-Met and Val-Trp-Cys-Leu, have charges of +1 at pH 1 because of the protonated N-terminal amino group. At pH 7, the peptide on the right has no net charge because of the protonated N-terminal amino group and the ionized C-terminal carboxylate negative charge. The peptide on the left has a net charge of −1 at pH 7 because of the side-chain carboxylate group on the glutamate in addition to the charges on the N-terminal and C-terminal groups.

9. Cysteine will have no net charge at pH 5.02 = $\dfrac{(1.71 + 8.33)}{2}$.

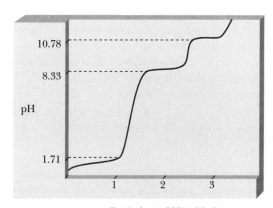

Equivalents OH⁻ added

10. The conjugate acid–base pair will act as a buffer in the pH range 1.09 to 3.09.

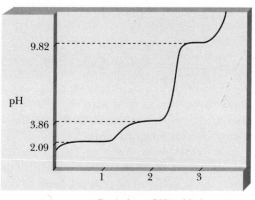

Equivalents OH⁻ added

11. The two peptides differ in amino acid sequence but not in composition. Consequently, they will have titration curves of the same shape. The pK'_a values of the α-amino and α-carboxyl groups will differ.

12. Oxytocin has an isoleucine at position 3 and a leucine at position 8; it stimulates smooth muscle contraction in the uterus during labor and in the mammary glands during lactation. Vasopressin has a phenylalanine at position 3 and an arginine at position 8; it stimulates resorption of water by the kidneys, thus raising blood pressure.

13. The reduced form of glutathione consists of three amino acids with a sulfhydryl group; the oxidized form consists of six amino acids and can be considered the result of linking two molecules of reduced glutathione by a disulfide bridge.

14. Gramicidin S is an antibiotic; its sequence is L-Val-L-Orn-L-Leu-D-Phe-L-Pro-L-Val-D-Orn-L-Leu-L-Phe-L-Pro.

15. The different stereochemistries of the two peptides lead to different binding with taste receptors and to the sweet taste for one and the bitter taste for the other.

16. See Figure 3.5 for the structures of modified amino acids. Hydroxyproline and hydroxylysine are found in collagen, and thyroxine is found in thyroglobulin.

17. See Figure 3.9. The resonance structures contribute to the planar arrangement by giving the C—N bond partial double-bond character.

Chapter 4

1. (a) (2); (b) (4); (c) (1); (d) (3)

2. Meat consists largely of animal proteins and fat. The temperatures involved in cooking meat are usually more than enough to denature the protein part of the meat.

3. The principal component of wool is the protein keratin, which is a classic example of α-helical structure. The principal component of silk is the protein fibroin, which is a classic example of β-pleated sheet structure. The statement is somewhat an oversimplification, but it is fundamentally valid.

4. The "random" portions of a protein do not contain structural motifs that are repeated within the protein, such as α-helix or β-pleated sheet, but three-dimensional features in these parts of the protein are repeated from one molecule to another. Thus, the term "random" is something of a misnomer.

5. (1) Backbone H-bonds, involving the CO and NH groups of the peptide chain   (2) Side-chain H-bonds, involving any possible hydrogen bond donor or acceptors on the side chains   (3) Hydrophobic interactions, involving the nonpolar groups on the protein   (4) Electrostatic interactions, involving any charged groups on the protein   (5) Metal ligation, involving coordination bonds between side chains and a metal ion.

6. When a protein is denatured, the interactions that determine secondary, tertiary, and any quaternary structure are overcome by the presence of the denaturing agent. Only the primary structure remains intact.

7. Similarities: both contain heme group; oxygen binding; secondary structure primarily α-helix. Differences: hemoglobin is a tetramer, whereas myoglobin is a monomer; oxygen binding is cooperative to hemoglobin, noncooperative to myoglobin.

8. The function of hemoglobin is oxygen transport; its sigmoidal binding curve reflects the fact that it can bind easily to oxygen at comparatively high pressures and release oxygen at lower pressures. The function of myoglobin is oxygen storage; as a result, it is easily saturated with oxygen at low pressures, as shown by its hyperbolic binding curve.

9. Deoxygenated hemoglobin is a weaker acid (has a higher $pK'a$) than oxygenated hemoglobin. In other words, deoxygenated hemoglobin binds more strongly to H^+ than does oxygenated hemoglobin. The binding of H^+ (and of CO_2) to hemoglobin favors the change in quaternary structure to the deoxygenated form of hemoglobin.

10. When a protein is covalently modified, its primary structure is changed. The primary structure determines the final three-dimensional structure of the protein. The modification disrupts the folding process.

11. The α-helix is not fully extended, and its hydrogen bonds are parallel to the protein fiber. The β-pleated sheet structure is almost fully extended, and its hydrogen bonds are perpendicular to the protein fiber.

12. The $\alpha\alpha$ unit, the $\beta\alpha\beta$ unit, the β-meander, the Greek key, the β-barrel

13. (a) Serine has a small side chain that can fit in any relatively polar environment. (b) Tryptophan has the largest side chain of any of the common amino acids, and it tends to require a nonpolar environment. (c) Lysine and arginine are both basic amino acids; exchanging one for the other would not affect the side-chain pK_a in a significant way. Similar reasoning applies to the substitution of a nonpolar isoleucine for a nonpolar leucine.

14. Persons with sickle-cell trait have some abnormal hemoglobin, impairing their capacity to transport oxygen in the bloodstream. At high altitudes, there is less oxygen and the decreased efficiency becomes more apparent.

15. In the presence of H^+ and CO_2, both of which bind to hemoglobin, the oxygen-binding capacity of hemoglobin decreases.

16. The geometry of the proline residue is such that it does not fit into the α-helix, but it does fit exactly for a reverse turn. See Figure 4.12c.

17. In the absence of 2,3-*bis*phosphoglycerate, the binding of oxygen by hemoglobin resembles that of myoglobin, characterized by lack of cooperativity. 2,3-*Bis*phosphoglycerate binds at the center of the hemoglobin molecule, increases cooperativity, and modulates the binding of oxygen so that it can easily be released in the capillaries.

18. Fetal hemoglobin binds oxygen more strongly than adult hemoglobin.

19. Hb S (sickle-cell anemia), Glu A3(6)β ⟶ Val; Hb E, Glu B8(26)β ⟶ Lys; Hb Savannah, Gly B6(24)β ⟶

Val; Hb Bibba, Leu H 19(136)α ⟶ Pro; HbM Iwate, His F8(87)α ⟶ Tyr; HbM Milwaukee, Val E11(67)β ⟶ Glu

20. In fetal hemoglobin, the subunit composition is $\alpha_2\gamma_2$ with replacement of the β-chains by the γ-chains. The sickle-cell mutation affects the β-chain, so the fetus homozygous for HbS has normal fetal hemoglobin.

21. Blood changes color from red to brown when the iron ions in it are oxidized from Fe(II) to Fe(III) in air. In methemoglobin (an abnormal hemoglobin), the iron is already Fe(III).

Interchapter A

1. Glutamic acid will be eluted from the column first. It will be necessary to raise the pH to elute lysine from the column.

2. Phenylalanine will have the largest R_f value; glutamic acid will have the smallest.

3. Glutamic acid will move fastest and phenylalanine will move most slowly — the reverse of the situation in Exercise 2.

4. Small particles, such as the ammonium and sulfate ions, enter the pores in the molecular sieve material, while large molecules, such as the protein, do not. The protein is eluted from the molecular sieve column before the salts.

5.

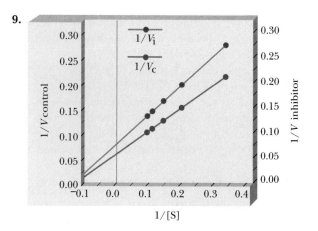

phenylthiohydantoin
derivative of leucine

6. Val-Leu-Gly-Met-Ser-Arg-Asn-Thr-Trp-Met-Ile-Lys-Gly-Tyr-Met-Gln-Phe

7. The sequence is Met-Val-Ser-Thr-Lys-Leu-Phe-Asn-Glu-Ser-Arg-Val-Ile-Trp-Thr-Leu-Met-Ile.

8. Molecular weights of newly isolated proteins can be estimated on a molecular sieve column by comparing elution volumes of known proteins with those of the unknown.

Chapter 5

1. The reaction of glucose with oxygen is thermodynamically favored, as shown by the negative free-energy change. The fact that glucose can be maintained in an oxygen atmosphere is a reflection of the kinetic aspects of the reaction, requiring the overcoming of an activation-energy barrier.

2. The reaction is first order with respect to A, first order with respect to B, and second order overall. The detailed mechanism of the reaction is likely to involve one molecule each of A and B.

3. In the lock-and-key model, the substrate fits into a comparatively rigid protein that has an active site with a well-defined shape. In the induced-fit model, the enzyme undergoes a conformational change on binding to substrate. The active site takes shape around the substrate.

4. See Figures 5.4 and 5.5.

5. The steady-state assumption is that the concentration of the enzyme–substrate complex does not change appreciably with time. The rate of appearance of the complex is equal to its rate of disappearance, simplifying the equations for enzyme kinetics.

6. Use equation 5.16. (a) $V = 0.5V_{max}$ (b) $V = 0.33V_{max}$ (c) $V = 0.09V_{max}$ (d) $V = 0.67V_{max}$ (e) $V = 0.91V_{max}$

7. Turnover number $= V_{max}/[E]_0$

8. In the case of competitive inhibition, the value of K_M increases; in noncompetitive inhibition, the value of K_M remains unchanged.

9. $K_M = 7.42$ mM; $V_{max} = 15.9$ mmol min^{-1}; noncompetitive inhibition

10. A competitive inhibitor binds to the active site of an enzyme, preventing binding of the substrate. A noncompetitive inhibitor binds at a site different from the active site, causing a conformational change that renders the active site less able to bind substrate and convert it to product.

11. The graph of rate against substrate concentration is sigmoidal for an allosteric enzyme but hyperbolic for an enzyme that obeys the Michaelis–Menten equation. Allosteric enzymes have multisubunit structures. Michaelis–Menten enzymes may be multisubunit or a single polypeptide chain.

12. In the concerted model, a conformational change caused by binding of an allosteric effector takes place simultaneously in all subunits. In the sequential model, a conformational change takes place in one subunit and is subsequently passed on to the other subunits.

13. Trypsin, chymotrypsin, fibrin

14. See Table 5.3.

15. False. The mechanisms of enzymic catalysis are the same as those encountered in organic chemistry, operating in a complex environment.

16. The results do not prove that the mechanism is correct, since results from different experiments could contradict the proposed mechanism. In that case, the mechanism would have to be modified to accommodate the new experimental results.

17. Metal ions can provide a "steering" effect by forming coordination bonds of specified geometry both to the substrate and to enzyme side chains. Metal ions can also be an aid to catalysis by acting as Lewis acids.

18. In the first step of the reaction, the serine hydroxyl is the nucleophile that attacks the substrate peptide bond. In the second step, water is the nucleophile that attacks the acyl-enzyme intermediate.

19. As a result of cleavage of the peptide bond, isoleucine 16 now has a free protonated amino group that forms an ionic bond with the carboxylate of aspartate 194. This linkage stabilizes the active form of the enzyme.

20. Instead of a phenylalanine moiety (similar to the usual substrates of chymotrypsin), use a nitrogen-containing basic group similar to the usual substrates of trypsin.

21. The easiest way to follow the rate of this reaction is to monitor the decrease in absorbance at 340 nm, reflecting the disappearance of NADH.

22. $V_{max} = 0.681$ mM min^{-1}, $K_M = 0.421$ M

23. Competitive inhibition, $K_M = 6.5 \times 10^{-4}$. The key point here is that the V_{max} is the same within the limits of error in this problem. Some of the concentrations are given to one significant figure.

Chapter 6

1. A-T base pairs have two hydrogen bonds, whereas G-C base pairs have three. It takes more energy and higher temperature to disrupt the structure of DNA rich in G-C base pairs.

2. Statements (c) and (d) are true; statements (a) and (b) are not.

3. True. There is room for binding and access to the base pairs in both the major and minor grooves of DNA.

4. Statement (c) is true. Statements (a) and (b) are false. Statement (d) is true for the B form of DNA but not for the A and Z forms.

5. Supercoiling is twists in DNA in addition to those of the double helix. Positive supercoiling is an extra twist in DNA caused by overwinding of the helix before sealing the ends to produce circular DNA. A topoisomerase is an enzyme that induces a single-strand break in supercoiled DNA, relaxes the supercoiling, and reseals the break. Negative supercoiling is the unwinding of the double helix before the ends are sealed to produce circular DNA.

6. Chromatin is the complex consisting of DNA and basic proteins found in eukaryotic nuclei (see Figure 6.12).

7. See Figures 6.16 and 6.21.

8. In the cleavage by *EcoRI* there are single-strand breaks between the G and A, but the two strands are held together by the A-T hydrogen bonds.

   G A-A-T-T-C

   C-T-T-A-A C

9. The sequence in Exercise 8 is an example of a palindrome. See the table in Box 6.2 for others.

10. The chi intermediate can give rise to either of the possibilities for recombinant DNA — single-strand or double-strand heterozygous — depending on how it is nicked, with subsequent resealing of the nicks.

11. A transposon is the portion of DNA that is moved from one place on a chromosome to another in the course of gene rearrangement. Homologous sequences in DNA may play a role in the looping out of transposons from the rest of the DNA.

12. More extensive hydrogen bonding occurs in tRNA than in mRNA. The folded structure of tRNA, which determines its binding to ribosomes in the course of protein synthesis, depends on its hydrogen-bonded arrangement of atoms. The coding sequences of mRNA must be accessible to direct the order of amino acids in proteins and should not be rendered inaccessible by hydrogen bonding.

13. Turnover of mRNA should be rapid to ensure that the cell can respond quickly when specific proteins are needed. Ribosomal subunits, including their rRNA component, can be recycled for many rounds of protein synthesis. As a result, mRNA is degraded more rapidly than rRNA.

14. Adenine–guanine base pairs occupy more space than is available in the interior of the double helix, whereas cytosine–thymine base pairs are too small to span the distance between the sites to which complementary bases are bonded. One would not normally expect to find such base pairs in DNA.

15. Four different kinds of bases — adenine, cytosine, guanine, and uracil — make up the preponderant majority of the bases found in RNA, but they are not the only ones. Modified bases occur to some extent, principally in tRNA.

16. The phosphate groups in DNA are negatively charged at physiological pH. If they were grouped together closely, as in the center of a long fiber, the result would be considerable electrostatic repulsion. Such a structure would be unstable.

Chapter 7

1. The use of restriction endonucleases with different specificities gives overlapping sequences that can be combined to give an overall sequence.

2. A portion of exogenous DNA is introduced into a suitable vector, frequently a bacterial plasmid, and many copies of the DNA are produced when the bacteria grow. Viruses are also commonly used as vectors.

3. X-ray film is used to detect the presence of radioactively labeled substances, which expose the film. The two principal applications are sequencing of DNA and selection of a specific DNA sequence out of a heterogeneous sample.

4. The polymerase chain reaction depends on repeated

cycles of separation of DNA strands followed by annealing of primers. The first step requires a significantly higher temperature than the second, giving rise to the requirement for strict temperature control.

5. The polymerase chain reaction can increase the amount of a desired DNA sample by a considerable factor. This makes possible definite identification of DNA samples that were too small to be characterized by other means. It can be used on hair and blood samples found at the scene of a crime to establish the presence of a suspect. This method can also be used to identify remains; the DNA samples from the Romanov skeletons described in Box 1.1 were amplified by PCR before comparative analysis.

6. If a DNA library is to represent the total genome of an organism, it must contain at least one clone for each DNA sequence. This requires several hundred thou-sand separate clones to ensure that every sequence is represented.

7. The restriction fragments of different sizes (restriction-fragment-length polymorphisms, or RFLPs) that result from different base sequences on paired chromosomes were used as genetic markers to determine the exact position of the cystic fibrosis gene on chromosome 7.

8. The DNA sequence to be inserted in the bacterial plasmid to direct the production of β-globin should be cDNA, which is a sequence complementary to the mRNA for β-globin. The cDNA can be produced on the mRNA template in a reaction catalyzed by reverse transcriptase.

9. Isolate the DNA that codes for the growth factor by means of suitable probes. Introduce the DNA into a bacterial genome. Allow the bacteria to grow and to produce human growth factor.

Chapter 8

1. In both types of lipids, glycerol is esterified to carboxylic acids, with three such ester linkages formed in triacylglycerols and two in phosphatidyl ethanolamines. The structural difference comes in the nature of the third ester linkage to glycerol. In phosphatidyl ethanolamines, the third hydroxyl group of glycerol is esterified not to a carboxylic acid but to phosphoric acid. The phosphoric acid moiety is esterified in turn to ethanolamine. (See Figures 8.2 and 8.5.)

2. Both sphingomyelins and phosphatidylcholines contain phosphoric acid esterified to an amino alcohol, which must be choline in the case of a phosphatidyl choline and may be choline in the case of a sphingomyelin. They differ in the second alcohol to which phosphoric acid is esterified. In phosphatidylcholines the second alcohol is glycerol, which has also formed ester bonds to two carboxylic acids. In sphingomyelins the second alcohol is another amino alcohol, sphingosine, which has formed an amide bond to a fatty acid. (See Figure 8.6.)

3. Triacylglycerols are not found in animal membranes.

4.

5. This lipid is a ceramide, which is one kind of sphingo-lipid.

6.

glycerol moiety

CH_2—O—C(=O)—$(CH_2)_{14}CH_3$ **Palmitic acid moiety**

CH—O—C(=O)—$(CH_2)_7CH=CH-CH_2-CH=CH(CH_2)_4CH_3$ **Linoleic acid moiety**

CH_2—O—C(=O)—$(CH_2)_7(CH=CHCH_2)_3CH_3$ **Linolenic acid moiety**

Any combination of fatty acids is possible.

7.

CH_2—O—C(=O)—$(CH_2)_{14}CH_3$

CH—O—C(=O)—$(CH_2)_7CH=CH-CH_2-CH=CH-(CH_2)_4CH_3$

CH_2—O—C(=O)—$(CH_2)_7-(CH=CH-CH_2)_3CH_3$

↓ **Aqueous NaOH**

CH_2OH $CH_3-(CH_2)_{14}-C(=O)-O^{\ominus}Na^{\oplus}$

CHOH   + $CH_3-(CH_2)_4-CH=CH-CH_2-CH=CH-(CH_2)_7-C(=O)-O^{\ominus}Na^{\oplus}$

CH_2OH $CH_3(CH_2-CH=CH)_3-(CH_2)_7-C(=O)-O^{\ominus}Na^{\oplus}$

8. Myelin is a multilayer sheath, consisting mainly of lipids (with some proteins), that insulates the axons of nerve cells, facilitating transmission of nerve impulses.

9. Steroids contain a characteristic fused-ring structure, which other lipids do not.

10. The *cis-trans* isomerization of retinal in rhodopsin triggers the transmission of an impulse to the optic nerve and is the primary photochemical event in vision.

11. Lipid-soluble vitamins accumulate in fatty tissue, leading to toxic effects. Water-soluble vitamins are excreted, drastically reducing the chances of an overdose.

12. Cholesterol is a precursor of vitamin D_3; the conversion reaction involves ring opening.

13. Vitamin E is an antioxidant.

14. Prostaglandins and leukotrienes are derived from arachidonic acid. They play roles in inflammation and in allergy and asthma attacks.

15. The transition temperature is lower in a lipid bilayer with mostly unsaturated fatty acids than in one with a high percentage of saturated fatty acids. The bilayer with the unsaturated fatty acids is already more disordered than the one with a high percentage of saturated fatty acids.

16. In a 100-gram sample of membrane, there are 50 grams of protein and 50 of phosphoglycerides.

$$50 \text{ g lipid} \times \frac{1 \text{ mol lipid}}{800 \text{ g lipid}} = 0.0625 \text{ mol lipid}$$

$$50 \text{ g protein} \times \frac{1 \text{ mol protein}}{50,000 \text{ g protein}} = 0.001 \text{ mol protein}$$

The molar ratio of lipid to protein is 0.0625/0.001 or 62.5/1.

17. Statements (c) and (d) are consistent with what is known about membranes. If there is any covalent bonding between lipids and proteins [statement (e)], it is rare. Proteins "float" in the lipid bilayers rather than being sandwiched between them [statement (a)]. Bulkier molecules tend to be found in the outer lipid layer [statement (b)].

18. Biological membranes are highly nonpolar environments. Charged ions tend to be excluded from such environments rather than dissolving in them, as they would have to do to pass the membrane by simple diffusion.

19. Statements (c) and (d) are correct. Transverse diffusion is normally not observed [statement (b)], and the term "mosaic" refers to the pattern of distribution of proteins in the lipid bilayer [statement (e)]. Peripheral proteins are also considered part of the membrane [statement (a)].

20. Statements (a) and (c) are correct; statement (b) is not correct since ions and larger molecules, especially polar ones, require channel proteins.

21. At the lower temperature, the membrane would tend to be less fluid. The presence of more unsaturated fatty acids would tend to compensate by increasing the fluidity of the membrane compared to one at the same temperature with a higher proportion of saturated fatty acids.

22. The binding site of the LDL receptor recognizes the protein portion of the LDL particle, specifically a protein exposed to the aqueous environment of the bloodstream. The LDL protein is likely to contain polar amino acids on its surface, as is the active site of the receptor.

23. Hydrophobic interactions among the hydrocarbon tails are the main thermodynamic driving force in the formation of lipid bilayers.

24. The higher percentage of unsaturated fatty acids in membranes in cold climates is an aid to membrane fluidity.

25. The waxy surface coating is a barrier that prevents loss of water.

26. The lecithin in the egg yolks serves as an emulsifying agent by forming closed vesicles. The lipids in the butter (frequently triacylglycerols) are retained in the vesicles and do not form a separate phase.

Chapter 9

1. The system is the nonpolar solute and water, which become more disordered when a solution is formed; ΔS_{sys} is positive but comparatively small. The ΔS_{surr} is negative and comparatively large, since it is a reflection of the unfavorable enthalpy change for forming the solution (ΔH_{sys}). Since $\Delta S_{univ} = \Delta S_{sys} + \Delta S_{surr}$, ΔS_{univ} is negative and does not favor the dissolution of nonpolar solutes in water.

2. Processes (a) and (b) are spontaneous, whereas processes (c) and (d) are not. The spontaneous processes represent an increase in disorder (increase in the entropy of the universe) and have a negative $\Delta G°$ at constant temperature and pressure. The opposite is true of the nonspontaneous processes.

3. In all cases there is an increase in entropy. In all cases the final state has more possible random arrangements than the initial state.

4. The first statement is true, but the second is not. The standard state of solutes is normally defined as unit activity (1 M for all but the most careful work). In biological systems the pH is frequently in the neutral range (i.e., H^+ is close to 10^{-7} M); the modification is a matter of convenience. Water is the solvent, not a solute, and its standard state is the pure liquid.

5. $\Delta G°' = \Delta H°' - T\Delta S°'$
   $\Delta S°' = 349 \text{ J mol}^{-1} \text{ K}^{-1} = 8.39 \text{ cal mol}^{-1} \text{ K}^{-1}$
   There are two particles on the reactant side of the equation and three on the product side, representing an increase in disorder.

6. Statements (a), (c), and (d) are correct. (a) The unfavorable entropy change for the water is reflected in the unfavorable heat of solution for nonpolar solutes in water. (c) This statement is a way of defining the enthalpy change. (d) Heat is a less useful form of energy than others, resulting from the degradation of ordered molecular motion into disordered motion. Statements (b) and (e) are incorrect. (b) The entropy of the universe reflects its randomness, which increases in spontaneous processes. (e) An endergonic reaction is one in which energy is taken up rather than given off, as is the case in spontaneous processes.

7. Nonpolar residues tend to be found in the interiors of proteins because of hydrophobic interactions. Polar residues tend to be found on the exteriors because of dipolar interactions with solvent water.

8. The local decrease in entropy associated with living organisms is balanced by the larger increase in the entropy of the surroundings caused by their presence.

9. Hydrophobic interactions stabilize the nonpolar interiors of membranes, with the charged and polar groups in contact with water.

10. The biosynthesis of proteins is endergonic and is accompanied by a large decrease in entropy.

11. The ATP constantly generated by living organisms is used as a source of chemical energy for endergonic processes. There is a good deal of turnover of molecules but no net change.

12. A large increase in entropy accompanies the hydrolysis of one molecule to five separate molecules.

13. The second half reaction (the one involving NADH) is oxidation, while the first half reaction (the one involving O_2) is reduction. The overall reaction is

$$\tfrac{1}{2} O_2 + NADH + H^+ \longrightarrow H_2O + NAD^+$$

O_2 is the oxidizing agent and NADH is the reducing agent.

14. Reaction (a) will not proceed as written; $\Delta G^{\circ\prime} = +12.6$ kJ. Reaction (b) will proceed as written; $\Delta G^{\circ\prime} = -20.8$ kJ. Reaction (c) will not proceed as written; $\Delta G^{\circ\prime} = +31.4$ kJ. Reaction (d) will proceed as written; $\Delta G^{\circ\prime} = -18.0$ kJ.

15. Sprints and similar short periods of exercise rely on anaerobic metabolism as a source of energy, producing lactic acid. Longer periods of exercise draw on aerobic metabolism as well.

16. Creatine phosphate + ADP $\longrightarrow$ creatine + ATP; $\Delta G^{\circ\prime} = -12.6$ kJ
ATP + glycerol $\longrightarrow$ ADP + glycerol 3-phosphate; $\Delta G^{\circ\prime} = -20.8$ kJ
Creatine phosphate + glycerol $\longrightarrow$ creatine + glycerol 3-phosphate

$$\Delta G^{\circ\prime} \text{ overall } = -33.4 \text{ kJ}$$

17. Glucose 1-phosphate $\longrightarrow$ glucose + P_i; $\Delta G^{\circ\prime} = -20.9$ kJ mol^{-1}
Glucose + P_i $\longrightarrow$ glucose 6-phosphate; $\Delta G^{\circ\prime} = +12.5$ kJ mol^{-1}
Glucose 1-phosphate $\longrightarrow$ glucose 6-phosphate; $\Delta G^{\circ\prime} = -8.4$ kJ mol^{-1}

18. In both pathways the overall reaction is

$$ATP + 2 H_2O \longrightarrow AMP + 2 P_i$$

Thermodynamic parameters such as energy are additive. The overall energy is the same, since the overall pathway is the same.

19. Phosphoarginine + ADP $\longrightarrow$ arginine + ATP; $\Delta G^{\circ\prime} = -1.7$ kJ
ATP + H_2O $\longrightarrow$ ADP + P_i; $\Delta G^{\circ\prime} = -30.5$ kJ
Phosphoarginine + H_2O $\longrightarrow$ arginine + P_i; $\Delta G^{\circ\prime} = -32.2$ kJ

20. Glucose 6-phosphate is oxidized, and NADP$^+$ is reduced. NADP$^+$ is the oxidizing agent, and glucose 6-phosphate is the reducing agent.

21. FAD is reduced, and succinate is oxidized. FAD is the oxidizing agent, and succinate is the reducing agent.

22. NAD$^+$, NADP$^+$, and FAD each contain an ADP moiety.

Chapter 10

1. Ester linkage. Repeating disaccharide of pectin:

Galacturonic acid (α form)

unmethylated   methylated

Repeating disaccharide

$\alpha(1 \rightarrow 4)$ α-anomeric end

2. Structure of gentibiose

$\beta(1 \rightarrow 6)$

3. To 2500, one place, 0.02%; to 1000, four places, 0.08%; to 200, 24 places, 0.48%

4. This polymer would be expected to have a structural role. The presence of the β-glycosidic linkage makes it useful as food only to animals such as termites and ruminants such as cows and horses; these animals harbor bacteria capable of attacking the β-linkage in their digestive tracts.

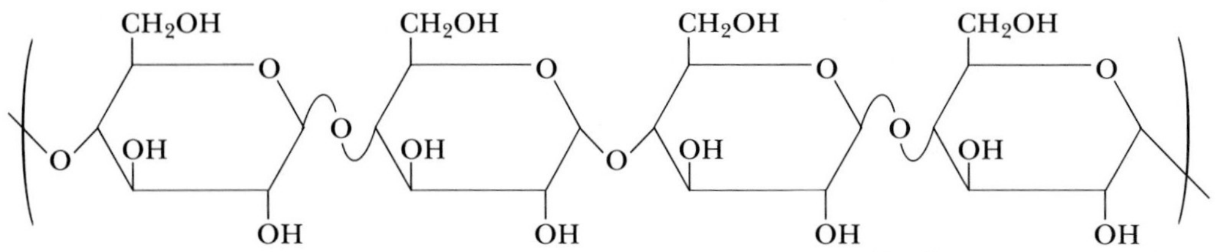

The structures at top with labels: CH₂OH groups, O, OH, OH

$\alpha(1\rightarrow4)$ $\beta(1\rightarrow4)$ $\alpha(1\rightarrow4)$ $\beta(1\rightarrow4)$ $\alpha(1\rightarrow4)$

5. A polysaccharide is a polymer of simple sugars, which are compounds that contain a single carbonyl group and several hydroxyl groups. A furanose is a cyclic sugar that contains a five-membered ring similar to that in furan.

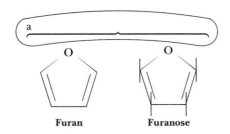

Furan   **Furanose**

A pyranose is a cyclic sugar that contains a six-membered ring similar to that in pyran.

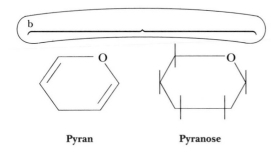

Pyran   **Pyranose**

An aldose is a sugar that contains an aldehyde group; a ketose is a sugar that contains a ketone group. A glycosidic bond is the acetal linkage that joins two sugars. An oligosaccharide is a compound formed by the linking of several simple sugars (monosaccharides) by glycosidic bonds. A glycoprotein is formed by the covalent bonding of sugars to a protein.

6. D-mannose and D-galactose are both epimers of D-glucose with inversion of configuration around carbon atoms 2 and 4, respectively; D-ribose has only five carbons, while the rest of the sugars named in this question have six.

7. All groups are aldose-ketose pairs. For example,

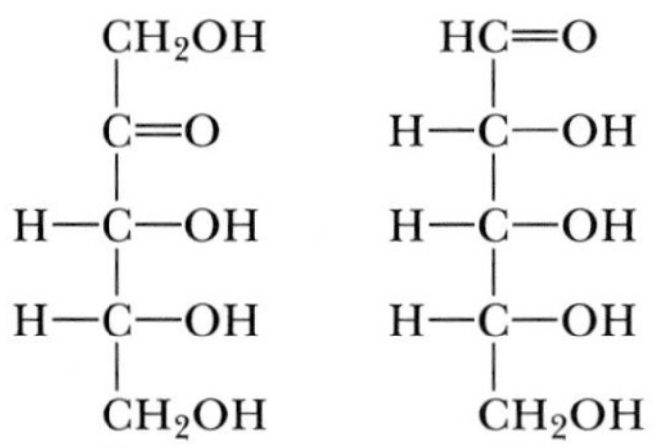

$$
\begin{array}{cc}
\text{CH}_2\text{OH} & \text{HC}=\text{O} \\
| & | \\
\text{C}=\text{O} & \text{H}-\text{C}-\text{OH} \\
| & | \\
\text{H}-\text{C}-\text{OH} & \text{H}-\text{C}-\text{OH} \\
| & | \\
\text{H}-\text{C}-\text{OH} & \text{H}-\text{C}-\text{OH} \\
| & | \\
\text{CH}_2\text{OH} & \text{CH}_2\text{OH} \\
\text{D-ribulose} & \text{D-ribose}
\end{array}
$$

8. In some cases the enzyme that degrades lactose (milk sugar) to its components, glucose and galactose, is missing. In other cases, the enzyme is the one that isomerizes galactose to glucose for further metabolic breakdown.

9. Enantiomers: a and f, b and d. Epimers: a and c, a and d, a and e, b and f. Five carbon sugars:

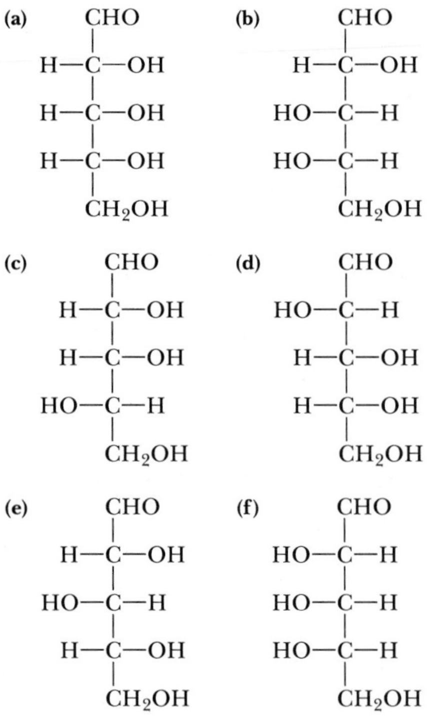

(a)
$$
\begin{array}{c}
\text{CHO} \\
| \\
\text{H}-\text{C}-\text{OH} \\
| \\
\text{H}-\text{C}-\text{OH} \\
| \\
\text{H}-\text{C}-\text{OH} \\
| \\
\text{CH}_2\text{OH}
\end{array}
$$

(b)
$$
\begin{array}{c}
\text{CHO} \\
| \\
\text{H}-\text{C}-\text{OH} \\
| \\
\text{HO}-\text{C}-\text{H} \\
| \\
\text{HO}-\text{C}-\text{H} \\
| \\
\text{CH}_2\text{OH}
\end{array}
$$

(c)
$$
\begin{array}{c}
\text{CHO} \\
| \\
\text{H}-\text{C}-\text{OH} \\
| \\
\text{H}-\text{C}-\text{OH} \\
| \\
\text{HO}-\text{C}-\text{H} \\
| \\
\text{CH}_2\text{OH}
\end{array}
$$

(d)
$$
\begin{array}{c}
\text{CHO} \\
| \\
\text{HO}-\text{C}-\text{H} \\
| \\
\text{H}-\text{C}-\text{OH} \\
| \\
\text{H}-\text{C}-\text{OH} \\
| \\
\text{CH}_2\text{OH}
\end{array}
$$

(e)
$$
\begin{array}{c}
\text{CHO} \\
| \\
\text{H}-\text{C}-\text{OH} \\
| \\
\text{HO}-\text{C}-\text{H} \\
| \\
\text{H}-\text{C}-\text{OH} \\
| \\
\text{CH}_2\text{OH}
\end{array}
$$

(f)
$$
\begin{array}{c}
\text{CHO} \\
| \\
\text{HO}-\text{C}-\text{H} \\
| \\
\text{HO}-\text{C}-\text{H} \\
| \\
\text{HO}-\text{C}-\text{H} \\
| \\
\text{CH}_2\text{OH}
\end{array}
$$

10. (a)

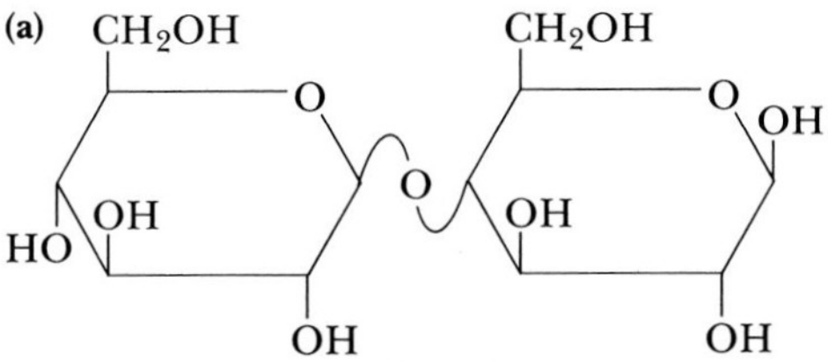

$\beta(1 \rightarrow 4)$

(b)

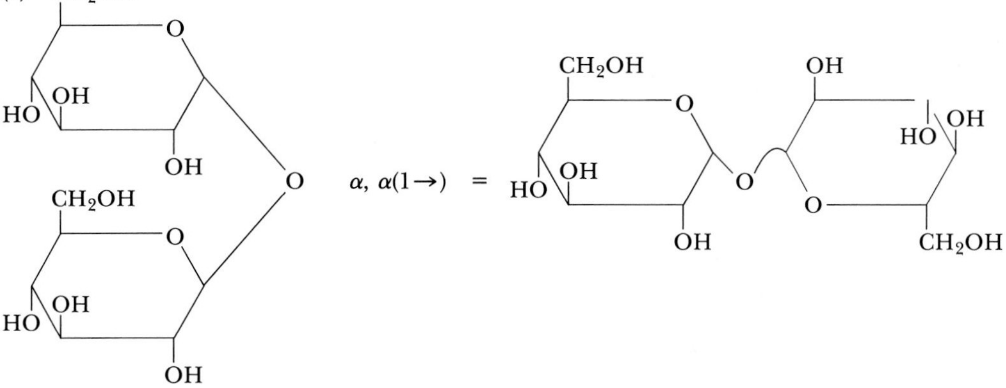

$\alpha, \alpha(1 \rightarrow)$ =

(c)

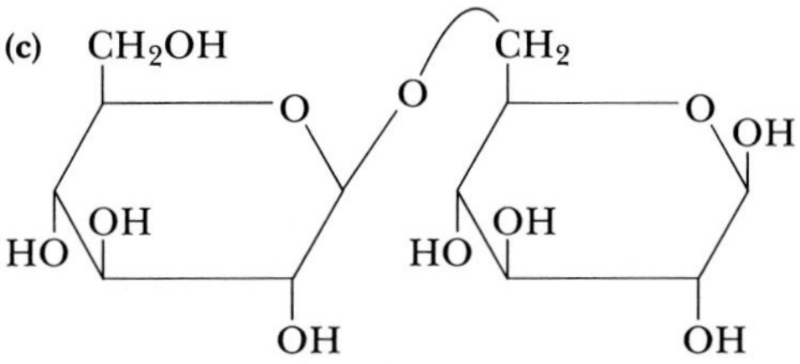

$\beta(1 \rightarrow 6)$

11. The cell walls of plants consist mainly of cellulose, while those of bacteria consist mainly of polysaccharides with peptide crosslinks.

12. Chitin is a polymer of *N*-acetyl-β-D-glucosamine, while cellulose is a polymer of D-glucose.

13. Glycogen and starch differ mainly in the degree of chain branching.

14. The enzyme β-amylase is an exoglycosidase, degrading polysaccharides from the ends. The enzyme α-amylase is an endoglycosidase, cleaving internal glycosidic bonds.

15. The sugar portions of the blood group glycoproteins are the source of the antigenic difference.

Chapter 11

1. The bubbles in beer are CO_2, produced by alcoholic fermentation. Tired and aching muscles are caused by a buildup of lactic acid, a product of anaerobic glycolysis.

2. Reactions that require ATP: phosphorylation of glucose to give glucose 6-phosphate, phosphorylation of fructose 6-phosphate to give fructose 1,6-*bis*phosphate

Reactions that produce ATP: transfer of phosphate group from 1,3-*bis*phosphoglycerate to ADP, transfer of phosphate group from phosphoenolpyruvate to ADP.

Enzymes that catalyze reactions requiring ATP: hexokinase, glucokinase, phosphofructokinase

Enzymes that catalyze reactions producing ATP: phosphoglycerate kinase, pyruvate kinase

3. Reactions that require NADH: reduction of pyruvate to lactate, reduction of acetaldehyde to ethanol

Reactions that require NAD^+: oxidation of glyceraldehyde 3-phosphate to give 1,3-diphosphoglycerate.

Enzymes that catalyze reactions requiring NADH: lactate dehydrogenase, alcohol dehydrogenase

Enzymes that catalyze reactions requiring NAD^+: glyceraldehyde 3-phosphate dehydrogenase

4. NADH-linked dehydrogenases: glyceraldehyde 3-phosphate dehydrogenase, lactate dehydrogenase, alcohol dehydrogenase

5. Aldolase catalyzes the reverse aldol condensation of fructose 1,6-diphosphate to glyceraldehyde 3-phosphate and dihydroxyacetone phosphate.

6. The energy released by all the reactions of glycolysis is 184.5 kJ per mol glucose. The energy released by glycolysis drives the phosphorylation of two ADP to ATP for each molecule of glucose, trapping 61.0 kJ per mol glucose. The estimate of 33 percent efficiency comes from the calculation $(61.0/184.5) \times 100 = 33$ percent.

7. Add the $\Delta G^{\circ\prime}$ mol^{-1} values for the reactions from glucose to glyceraldehyde 3-phosphate. The result is 2.5 kJ mol^{-1} = 0.6 kcal mol^{-1}.

8. Phosphoenolpyruvate $\longrightarrow$ pyruvate + P$_i$;
$\Delta G^{\circ\prime} = -61.9$ kJ mol^{-1} = -14.8 kcal mol
ADP + P$_i$ $\longrightarrow$ ATP;
$\Delta G^{\circ\prime} = 30.5$ kJ mol^{-1} = 7.3 kcal mol^{-1}
Phosphoenolpyruvate + ADP $\longrightarrow$ pyruvate + ATP;
$\Delta G^{\circ\prime} = -31.4$ kJ mol^{-1} = -7.5 kcal mol^{-1}

9. (a)

(b)

(c)

*For structures of NADH and NAD$^+$, see the answer for Exercise 11.9a.

10. There is a net gain of two ATP molecules per glucose molecule consumed in glycolysis.

11. Pyruvate can be converted to lactate, ethanol, or acetyl-CoA.

12. The reaction of 2-phosphoglycerate to phosphoenolpyruvate is a dehydration (loss of water) rather than a redox reaction.

13. The hexokinase molecule changes shape drastically on binding to substrate, consistent with the induced-fit theory of an enzyme adapting itself to its substrate.

14. ATP inhibits phosphofructokinase, consistent with the fact that ATP is produced by later reactions of glycolysis.

15. Phosphate ion, rather than ATP, is the source of phosphorus in substrate-level phosphorylation. An example is the conversion of glyceraldehyde-3-phosphate to 1,3-*bis*phosphoglycerate.

16. Isozymes are oligomeric enzymes that have slightly different amino acid compositions in different organs. Lactate dehydrogenase is an example, as is phosphofructokinase.

17. Thiamine pyrophosphate is a coenzyme in the transfer of two-carbon units. It is required for catalysis by pyruvate decarboxylase in alcoholic fermentation.

18. Thiamine pyrophosphate (TPP) is a coenzyme required in the reaction catalyzed by pyruvate carboxylase. Since this reaction is a part of the metabolism of ethanol, less TPP will be available to serve as a coenzyme in the reactions of other enzymes that require it.

19. Animals that have been run to death have accumulated large amounts of lactic acid in their muscle tissue, accounting for the sour taste of the meat.

20. The formation of fructose 1,6-*bis*phosphate is the committed step in the glycolytic pathway. The earlier components can play a part in other pathways, but not fructose 1,6-*bis*phosphate.

Chapter 12

1.

$$CH_2-COO^-$$
$$^-OOC-C-H$$
$$H-C-OH$$
$$COO^-$$

$$CH_2-COO^-$$
$$H-C-COO^-$$
$$H-C-OH$$
$$COO^-$$

$$CH_2-COO^-$$
$$^-OOC-C-H$$
$$HO-C-H$$
$$COO^-$$

2.

$$HO-\overset{..}{C}-\overset{-}{C}\overset{N^+}{\underset{S}{\diagdown}}$$
$$\underset{CH_3}{\big|}$$

3. The NADH and FADH$_2$ produced by the citric acid cycle are the electron donors in the electron transport chain linked to oxygen. Because of this connection, the citric acid cycle is considered part of aerobic metabolism.

4. There is an adenine nucleotide portion in the structure of NADH, with a specific binding site on NADH-linked dehydrogenases for this portion of NADH.

5. If the amount of ADP in a cell increases relative to the amount of ATP, the cell needs energy (ATP). This situation favors the reactions of the citric acid cycle, which release energy, activating isocitrate dehydrogenase. It also stimulates the formation of NADH and FADH$_2$ for ATP production by electron transport and oxidative phosphorylation.

6. If the amount of NADH in a cell increases relative to the amount of NAD$^+$, the cell has completed a number of energy-releasing reactions. There is less need for the citric acid cycle to be active, and as a result, the activity of pyruvate dehydrogenase is decreased.

7. The citric acid cycle is less active when a cell has high ATP/ADP and NADH/NAD$^+$ ratios. Both ratios signify a high "energy charge" in the cell, indicating less of a need for the energy-releasing reactions of the citric acid cycle.

8. (a) $CH_3-\overset{O}{\overset{\|}{C}}\overset{\overset{..}{O}:}{\underset{..}{C}}:\overset{..}{O}:^- + \text{CoA-SH} \longrightarrow CH_3-\overset{O}{\overset{\|}{C}}-\text{SCoA} + H^+ + :\overset{..}{O}::C::\overset{..}{O}: + 2e^-$

  Pyruvate   Acetyl CoA   Carbon dioxide

(b)

$$COO^-$$
$$CH_2$$
$$H-C-COO^-$$
$$:\overset{..}{O}:C:H$$
$$\overset{\overset{..}{H}}{\underset{}{\big|}}\ COO^-$$
Isocitrate

$\longrightarrow$

$$COO^-$$
$$CH_2$$
$$H-C-H$$
$$C::\overset{..}{O}:$$
$$COO^-$$
α-Ketoglutarate

$+ \ CO_2 + 2e^-$

(c)

$$\overset{..}{O}:$$
$$::$$
$$C:\overset{..}{O}:^-$$
$$C=O$$
$$CH_2 + \text{CoA-SH}$$
$$CH_2$$
$$COO^-$$
α-Ketoglutarate

$\longrightarrow$

$$S\,CoA$$
$$C=O$$
$$CH_2$$
$$CH_2$$
$$COO^-$$
Succinyl-CoA

$+ \ :\overset{..}{O}::C::\overset{..}{O}: + 2e^-$

(d)

$$COO^-$$
$$H:C-H$$
$$H-C:O$$
$$COO^-$$
Succinate

$\longrightarrow$

$$\underset{H}{\overset{COO^-}{\diagup}}\overset{\diagdown H}{C}$$
$$::$$
$$\underset{H}{\diagup}\overset{C}{\diagdown}_{COO^-}$$
Fumarate

$+ \ 2e^-$

(e)

$$
\begin{array}{c}
\text{COO}^- \\
| \\
\text{H}:\overset{..}{\underset{..}{\text{O}}}:\text{C}:\text{H} \\
| \\
\text{CH}_2 \\
| \\
\text{COO}^-
\end{array}
\qquad \longrightarrow \qquad
\begin{array}{c}
\text{COO}^- \\
| \\
\text{C}::\overset{..}{\underset{..}{\text{O}}}: \\
| \\
\text{CH}_2 \\
| \\
\text{COO}^-
\end{array}
\quad + \quad 2e^-
$$

Malate   Oxaloacetate

9. The conversion of fumarate to malate is a hydration reaction, not a redox reaction.

10. See Equation 12.3.

11. A condensation reaction is one in which a new carbon–carbon bond is formed. The reaction of acetyl-CoA and oxalacetate to produce citrate involves formation of such a carbon–carbon bond.

12. Thioesters are "high-energy" compounds that play a role in group-transfer reactions; consequently their $\Delta G^{\circ\prime}$ of hydrolysis is (a) large and negative to provide energy for the reaction.

13. The citric acid cycle takes place in the mitochondrial matrix, while glycolysis occurs in the cytosol.

14. NAD^+ and FAD are the primary electron acceptors of the citric acid cycle.

15. Lipoic acid plays a role both in redox and acetyl-transfer reactions.

16. In oxidative decarboxylation, the molecule that is oxidized loses a carboxyl group as carbon dioxide. Examples of oxidative decarboxylation include the conversion of pyruvate to acetyl-CoA, isocitrate to α-ketoglutarate, and α-ketoglutarate to succinyl-CoA.

17. Table 12.2 shows that the sum of the energies of the individual reactions is −44.3 kJ (−10.6 kcal) for each mole of acetyl-CoA that enters the cycle.

18. The glyoxylate cycle bypasses the two oxidative decarboxylations of the citric acid cycle by splitting isocitrate to glyoxylate and succinate. Glyoxylate reacts with acetyl-CoA to give malate, which is converted to oxaloacetate by reactions of the citric acid cycle.

Chapter 13

1. Electrons are passed from NADH to a flavin-containing protein to coenzyme Q. From coenzyme Q the electrons pass to cytochrome b, then to cytochrome c, via the Q cycle, followed by cytochromes a and a_3. From the cytochrome a/a_3 complex the electrons are finally passed to oxygen.

2. (a) 40; (b) 38; (c) 16; (d) 20; (e) 3; (f) 15

3. The half reaction of oxidation

$$\text{NADH} + \text{H}^+ \longrightarrow \text{NAD}^+ + 2\text{H}^+ + 2e^-$$

is strongly exergonic ($\Delta G^{\circ\prime} = -61.3$ kJ mol^{-1} = −14.8 kcal mol^{-1}), as is the overall reaction

$$\text{Pyruvate} + \text{NADH} + \text{H}^+ \longrightarrow \text{lactate} + \text{NAD}^+$$

($\Delta G^{\circ\prime} = -25.1$ kJ mol^{-1} = −6.0 kcal mol^{-1}).

4. The maximum yield of ATP, to the nearest whole number, is 3.

$$102.3 \text{ kJ released} \times \frac{1 \text{ ATP}}{30.5 \text{ kJ}} = 3.35 \text{ ATP}$$

One ATP is actually produced, so the efficiency of the process is

$$\frac{1 \text{ ATP}}{3 \text{ ATP}} \times 100 = 33.3\%$$

5. (a) Azide inhibits the transfer of electrons from cytochrome a/a_3 to oxygen. (b) Antimycin A inhibits the transfer of electrons from cytochrome b to coenzyme Q in the Q cycle. (c) Amytal inhibits the transfer of electrons from NADH reductase to coenzyme Q. (d) Rotenone inhibits the transfer of electrons from NADH reductase to coenzyme Q. (e) Dinitrophenol is an uncoupler of oxidative phosphorylation. (f) Gramicidin A is an uncoupler of oxidative phosphorylation. (g) Carbon monoxide inhibits the transfer of electrons from cytochrome a/a_3 to oxygen.

6. A P/O ratio of 2 can be expected because oxidation of succinate passes electrons to coenzyme Q via a flavoprotein intermediate, bypassing the first respiratory complex.

7. Succinate + $\frac{1}{2} \text{O}_2 \longrightarrow$ fumarate + H_2O

8. Cytochrome c is not tightly bound to the mitochondrial membrane and can easily be lost in the course of cell fractionation. This protein is so similar in most aerobic organisms that cytochrome c from one source can easily be substituted for that from another source.

9. The chemiosmotic coupling mechanism is based on the difference in hydrogen ion concentration between the intermembrane space and the matrix of actively respiring mitochondria. The hydrogen ion gradient is created by the proton pumping that accompanies the transfer of electrons. The flow of hydrogen ions back into the matrix through a channel in the ATP synthase is directly coupled to the phosphorylation of ADP.

10. The F_1 portion of the mitochondrial ATP synthase, which projects into the matrix, is the site of ATP synthesis.

11. The complete oxidation of glucose produces 36 molecules of ATP in muscle and brain and 38 ATP in liver, heart, and kidney. The underlying reason is the difference in shuttle mechanisms for transfer to mitochondria of electrons from the NADH produced in the cytosol by glycolysis.

12. $\Delta G^{\circ\prime} = -60$ kJ/mol

13. In all reactions, electrons are passed from the reduced form of one reactant to the oxidized form of the next reactant in the chain. The notation [Fe-S] refers to any one of a number of iron–sulfur proteins.

Reactions of Complex I

$$\left. \begin{array}{l} NADH + E\text{-}FMN \longrightarrow NAD^+ + E\text{-}FMNH_2 \\ E\text{-}FMNH_2 + [Fe\text{-}S]_{ox} \longrightarrow E\text{-}FMN + [Fe\text{-}S]_{red} \end{array} \right\} \text{1 ATP produced}$$

Transfer to coenzyme Q

$$[Fe\text{-}S]_{red} + CoQ \longrightarrow [Fe\text{-}S]_{ox} + CoQH_2$$

Reactions of Complex III

$$\left. \begin{array}{l} Q\text{-cycle reactions} \\ [Fe\text{-}S]_{red} + \text{cyt } c_{1ox} \longrightarrow [Fe\text{-}S]_{ox} + \text{cyt } c_{1red} \end{array} \right\} \text{1 ATP produced}$$

Transfer to cytochrome c

$$\text{cyt } c_{1red} + \text{cyt } c_{ox} \longrightarrow \text{cyt } c_{1ox} + \text{cyt } c_{red}$$

Reactions of Complex IV

$$\left. \begin{array}{l} \text{cyt } c_{red} + \text{cyt } a, a_{3ox} \longrightarrow \text{cyt } c_{ox} + \text{cyt } a, a_{3red} \\ \text{cyt } a, a_{3red} + \frac{1}{2} O_2 \longrightarrow \text{cyt } a, a_{3ox} + H_2O \end{array} \right\} \text{1 ATP produced}$$

There are three sites of oxidative phosphorylation in the electron transport chain, which results in a P/O ratio of 3/1.

14. When $FADH_2$ is the starting point for electron transport, electrons are passed from $FADH_2$ to coenzyme Q in a reaction carried out by Complex II that bypasses Complex I.

$$FADH_2 + [Fe\text{-}S]_{ox} \longrightarrow FAD + [Fe\text{-}S]_{red}$$

$$[Fe\text{-}S]_{red} + CoQ \longrightarrow [Fe\text{-}S]_{ox} + CoQH_2$$

There are only two sites of oxidative phosphorylation in the rest of the electron transport chain, giving rise to a P/O ratio of 2.

Chapter 14

1. Reactions that require ATP: formation of UDP-glucose from glucose 1-phosphate and UTP (indirect requirement, because ATP is needed to regenerate UTP), regeneration of UTP, carboxylation of pyruvate to oxaloacetate
   Reactions that produce ATP: none
   Enzymes that catalyze ATP-requiring reactions: UDP-glucose phosphorylase (indirect requirement), nucleoside phosphate kinase, pyruvate carboxylase
   Enzymes that catalyze ATP-producing reactions: none

2. NADPH has one more phosphate group than NADH (at the 2′ position of the ribose ring of the adenine nucleotide portion of the molecule). NADH is produced in oxidative reactions that give rise to ATP. NADPH is a reducing agent in biosynthesis. The enzymes that use NADH as a coenzyme are different from those that require NADPH.

3. Reactions that require acetyl-CoA: none
   Reactions that require biotin: carboxylation of pyruvate to oxaloacetate

4. Hemolytic anemia is caused by defective working of the pentose phosphate pathway. There is a deficiency of NADPH, which indirectly contributes to the integrity of the red blood cells. The pentose phosphate pathway is the only source of NADPH in red blood cells.

5. Three reactions of glycolysis are irreversible under physiological conditions. They are the production of pyruvate and ATP from phosphoenolpyruvate, the production of fructose 1,6-diphosphate from fructose 6-phosphate, and the production of glucose 6-phosphate from glucose. These are the reactions that are bypassed in gluconeogenesis; the reactions of gluconeogenesis differ from those of glycolysis at these points and are catalyzed by different enzymes.

6.

Glucose 6-phosphate   NADP$^+$ 6-Phosphoglucono lactone   NADPH

7. There is a net gain of three, rather than two, ATP when glycogen, not glucose, is the starting material of glycolysis.

8. Transketolase catalyzes the transfer of a two-carbon unit, whereas transaldolase catalyzes the transfer of a three-carbon unit.

9. Eating high-carbohydrate foods for several days prior to strenuous activity is intended to build up glycogen stores in the body. Glycogen will be available to supply required energy.

Chapter 15

1. Features in common: involvement of acetyl-CoA and thioesters; each round of breakdown or synthesis involves two-carbon units.

   Differences: malonyl-CoA is involved in biosynthesis, not in breakdown; thioesters involve CoA in breakdown, acyl carrier proteins in biosynthesis; biosynthesis occurs in the cytosol, breakdown in the mitochondrial matrix; breakdown is an oxidative process that requires NAD$^+$ and FAD and produces ATP by electron transport and oxidative phosphorylation, whereas biosynthesis is a reductive process that requires NADPH and ATP.

2. From seven cycles of β-oxidation: 8 acetyl-CoA, 7 FADH$_2$, 7 NADH

   From the processing of 8 acetyl-CoA in the citric acid cycle: 8 FADH$_2$, 24 NADH, 8 GTP

   From reoxidation of all FADH$_2$ and NADH: 30 ATP from 15 FADH2, 93 ATP from 31 NADH

   From 8 GTP: 8 ATP

   Subtotal: 131 ATP

   2 ATP equivalent used in activation step

   Grand total: 129 ATP

3. Seven carbon–carbon bonds are broken in the course of β-oxidation (see Figure 15.5).

4. Acyl groups are esterified to carnitine to cross the inner mitochondrial matrix. There are transesterification reactions from the acyl-CoA to carnitine and from acyl-carnitine to CoA (see Figure 15.3).

5. False. The oxidation of unsaturated fatty acids to acetyl-CoA requires a *cis-trans* isomerization and an epimerization (see Figure 15.7), reactions that are not found in the oxidation of saturated fatty acids.

6. The humps of camels contain lipids that can be degraded as a source of metabolic water rather than water as such.

7. Acetone is one of the "ketone bodies" produced by breakdown of lipids. Diabetics have impaired ability to metabolize carbohydrates and degrade an excessive amount of lipids as a result.

8. *Step 1:* Biotin is carboxylated using bicarbonate ion (HCO$_3^-$) as the source of the carboxyl group.

   Step 2: The carboxylated biotin is brought into proximity with enzyme-bound acetyl-CoA by a biotin carrier protein.

   Step 3: The carboxyl group is transferred to acetyl-CoA, forming malonyl-CoA.

9. Linoleate and linolenate cannot be synthesized by the body and must be obtained from dietary sources.

10. Acyl-CoA intermediates are essential in the conversion of fatty acids to other lipids.

11. In steroid biosynthesis, three acetyl-CoA molecules condense to form the six-carbon mevalonate, which then gives rise to a five-carbon isoprenoid unit. A second and then a third isoprenoid unit condense, giving rise to a 10-carbon unit and then a 15-carbon unit. Two of the 15-carbon units condense, forming a 30-carbon precursor of cholesterol.

12. See Figure 15.21.

13. Acetyl groups condense with oxaloacetate to form citrate, which can cross the mitochondrial membrane. Acetyl groups are regenerated in the cytosol by the reverse reaction.

14. All steroids have a characteristic fused-ring structure, implying a common biosynthetic origin.

15. Cholesterol is nonpolar and cannot dissolve in blood, which is an aqueous medium.

Chapter 16

1. In cyclic photophosphorylation, the excited chlorophyll of Photosystem I passes electrons directly to the elec-

tron transport chain that normally links Photosystem II to Photosystem I. This electron transport chain is coupled to ATP production (see Figure 16.8).

2. Electron transport and ATP production are coupled to one another by the same mechanism in mitochondria and chloroplasts. In both cases the coupling depends on the generation of a proton gradient across the inner mitochondrial membrane or across the thylakoid membrane, as the case may be.

3. From the standpoint of thermodynamics, the production of sugars in photosynthesis is the reverse of the complete oxidation of a sugar such as glucose to CO_2 and water. The complete oxidation reaction produces six moles of CO_2 for each mole of glucose oxidized. To get the energy change for the fixation of one mole of CO_2, change the sign of the energy for the complete oxidation of glucose and divide by 6.

4. In the fall, the chlorophyll in leaves is lost, and the red and yellow colors of the accessory pigments become visible, accounting for fall foliage colors.

5. The proton gradient is created by the operation of the electron transport chain that links the two photosystems in noncyclic photophosphorylation (see Exercise 1).

6. A prokaryotic organism that contains both chlorophyll *a* and chlorophyll *b* could be a relict of an evolutionary way station in the development of chloroplasts.

7. When light impinges on the reaction center of *Rhodopseudomonas*, the special pair of chlorophylls there is raised to an excited energy level. An electron is passed from the special pair to accessory pigments—first pheophytin, then menaquinone, and finally to ubiquinone. The electron lost by the special pair of chlorophylls is replaced by a soluble cytochrome, which diffuses away. The separation of charge represents stored energy (see Figure 16.5).

8. In Photosystem I and in Photosystem II, light energy is needed to raise the reaction-center chlorophylls to a higher energy level. Energy is needed to generate strong enough reducing agents to pass electrons to the next of the series of components in the pathway.

9. Glucose synthesized by photosynthesis is not uniformly labeled because only one molecule of CO_2 is incorporated into each molecule of ribulose 1,5-*bis*phosphate, which then goes on to give rise to sugars.

10. In tropical plants, the C_4 pathway is operative in addition to the Calvin cycle.

Chapter 17

1. Lysine is an essential amino acid that is frequently lacking in cereals.

2. See Figure 17.8.

3. Four α-amino acids—ornithine, citrulline, argininosuccinate, and arginine—participate in the urea cycle; of these, only arginine can be used for protein synthesis.

4. $H^+ + HCO_3^- + 2 NH_3 + 3 ATP \longrightarrow NH_2CONH_2 + 2 ADP + 2 P_i + AMP + PP_i + 2 H_2O$

The urea cycle is linked to the citric acid cycle by fumarate and by aspartate, which can be converted to malate by transamination (see Figure 17.15).

5. Fluorouracil substitutes for thymine in DNA synthesis. In rapidly dividing cells, such as cancer cells, the result is the production of defective DNA.

6. Conversion of homocysteine to methionine using *S*-adenosylmethionine as the methyl donor gives no net gain; one methionine is needed to produce another methionine.

7. Glutamate + α-keto acid $\longrightarrow$ α-ketoglutarate + α-amino acid

8. In both cases, the requirements are those given in Table 17.1.

9. In purine nucleotide biosynthesis the growing purine ring is covalently bonded to ribose, while in pyrimidine nucleotide biosynthesis the ribose is added after the ring is synthesized.

10. $N_2 \longrightarrow NH_4^+ \longrightarrow$ 3-phosphoserine $\longrightarrow$ serine $\longrightarrow$ glycine $\longrightarrow$ N-7

11. Sulfanilamide inhibits folic acid biosynthesis.

12. See the *S*-adenosylmethionine structure in Figure 17.12. The reactive methyl group is indicated.

13. Feedback control mechanisms slow down long biosynthetic pathways at or near their beginnings, saving energy for the organism.

14. A high-protein diet leads to increased production of urea. Drinking more water increases the volume of urine, ensuring elimination of the urea from the body with less strain on the kidneys than if urea were present at higher concentration.

15. The DNA of fast-growing cells, such as those of the hair follicles, is damaged by chemotherapeutic agents.

16. The purine salvage reaction that produces GMP requires the equivalent of 2 ATP. The pathway to IMP and then to GMP requires the equivalent of 8 ATP.

Chapter 18

1. All amino acids must be present at the same time for protein synthesis to occur. Newly synthesized proteins are necessary for growth in the immature rats.

2. The following series of reactions exchanges NADH for NADPH.

Oxaloacetate + NADH + $H^+ \longrightarrow$ malate + NAD^+

Malate + $NADP^+ \longrightarrow$ pyruvate + CO_2 + NADPH + H^+

3. Lucullus breaks down the protein in the tuna to amino acids, which in turn undergo the urea cycle and the breakdown of the carbon skeleton described in Chapter 17, eventually leading to the citric acid cycle and electron transport. In addition to protein catabolism, Griselda breaks down the carbohydrates to sugars, which then undergo glycolysis and enter the citric acid cycle. (Gratuitous information: Lucullus was a notorious Roman gourmand. In medieval literature, Griselda was the name usually given to a forbearing, long-suffering woman.)

4. The weight loss is due to correction of the bloating caused by retention of liquids.

5. Carbohydrates are the main energy source. Excess fat consumption can lead to the formation of "ketone bodies" and to atherosclerosis. Diets extremely high in protein can put a strain on the kidneys.

6. IP_3 is a polar compound and can dissolve in the aqueous environment of the cytosol, while DAG is nonpolar and interacts with the side chains of the membrane phospholipids.

7. When a stimulatory hormone binds to its receptor on a cell surface, it stimulates the action of adenylate cyclase, mediated by the G protein. The cAMP produced elicits the desired effect on the cell by stimulating a kinase that phosphorylates a target enzyme.

8. The hypothalamus secretes hormone-releasing factors. Under the influence of these factors, the pituitary secretes trophic hormones, which act on specific endocrine glands. Individual hormones are then released by the specific endocrine glands.

9. See Table 18.2.

10. Antibodies are Y-shaped molecules. The prongs of the Y bind to antigen, allowing for cross-linking. Several antibodies can bind to a single antigen, allowing for still more cross-linking.

11. The genes that encode the sequences of portions of antibody molecules are relatively small and undergo considerable rearrangement in the process of transcription and translation of the genetic message. There is also a high level of somatic mutation in cells that produce antibodies.

12. Lymphocytes that produce the desired antibody are allowed to fuse with mouse myeloma cells. The hybridoma (hybrid myeloma) cells thus produced can be cloned and grown continuously in culture medium. The unfused lymphocytes cannot grow continuously in culture.

Chapter 19

1. False. In retroviruses, the flow of information is RNA $\longrightarrow$ DNA.

2. Replication is the production of new DNA on a DNA template. Transcription is the production of RNA on a DNA template. Translation is the synthesis of proteins directed by mRNA, which reflects the base sequence of DNA.

3. The semiconservative replication of DNA means that a newly formed DNA molecule has one new strand and one strand from the original DNA. The experimental evidence for semiconservative replication comes from density-gradient centrifugation (Figure 19.3). If replication were a conservative process, the original DNA would have two heavy strands and all newly formed DNA would have light strands.

4. A replication fork is the site of formation of new DNA. The two strands of the original DNA separate, and a new strand is formed on each original strand.

5. DNA polymerase I is primarily a repair enzyme. DNA polymerase III is mainly responsible for the synthesis of new DNA. See Table 19.1.

6. All four deoxyribonucleoside triphosphates, template DNA, DNA polymerase, all four ribonucleoside triphosphates, primase, helicase, single-strand binding protein, DNA gyrase, DNA ligase

7. DNA is synthesized from the 5′ to the 3′ end, and the new strand is antiparallel to the template strand. One of the strands is exposed from the 5′ to the 3′ end as a result of unwinding. Small stretches of new DNA are synthesized, still in an antiparallel direction from the 5′ to the 3′ end, and are linked by DNA ligase. See Figure 19.5.

8. DNA gyrase introduces a swivel point in advance of the replication fork. Primase synthesizes the RNA primer. DNA ligase links small newly formed strands to produce longer ones.

9. When an incorrect nucleotide is introduced into a growing DNA chain as a result of mismatched base pairing, DNA polymerase acts as a 3′-exonuclease, removing the incorrect nucleotide. The same enzyme then incorporates the correct nucleotide.

10. An exonuclease nicks the DNA near the site of the thymine dimers. Polymerase I acts as a nuclease and excises the incorrect nucleotides, then acts as a polymerase to incorporate the correct ones. DNA ligase seals the nick.

11. The general features of DNA replication are similar in prokaryotes and eukaryotes. The main differences are that eukaryotic DNA polymerases do not have exonuclease activity. After synthesis, eukaryotic DNA is complexed with proteins, while prokaryotic DNA is not.

12. (1) RNA polymerase from *E. coli* has a molecular weight of about 500,000 and four different kinds of subunits. (2) It uses one strand of the DNA template to direct

RNA synthesis. (3) It catalyzes polymerization from the 5′ to the 3′ end.

13. The promoter region is the portion of DNA to which RNA polymerase binds at the start of transcription. This region lies upstream (nearer the 3′ end of the DNA) of the actual gene for the RNA. The promoter regions of DNA from many organisms have sequences in common (consensus sequences). The consensus sequences frequently lie 10 base pairs and 35 base pairs upstream of the start of transcription.

14. Trimming is necessary to obtain RNA transcripts of the proper size. Frequently, several tRNAs are transcribed in one long RNA molecule and must be trimmed to obtain active tRNAs.

15. Capping, polyadenylation, and splicing of coding sequences take place in the processing of eukaryotic mRNA.

16. Exons are the portions of DNA that are expressed, which means that they are reflected in the base sequence of the final mRNA product. Introns are the intervening sequences that do not appear in the final product but are removed during the splicing of mRNA.

17. The snRNPs are small nuclear ribonucleoprotein particles. They are the sites of mRNA splicing.

18. Site-directed mutagenesis is a method for producing specific changes in a few bases of a DNA sequence. Synthetic oligonucleotides with altered bases are allowed to hydrogen-bond to a target sequence of DNA, and replication is allowed to proceed with the synthetic oligomer as a primer. Alterations can be chosen to substitute the coding sequence of one amino acid for another.

19. Two mechanisms for RNA self-splicing are known. In Group I ribozymes, an external guanosine is covalently bonded at the splice site, releasing one end of the intron. The free end of the exon thus produced attacks the end of the other exon to splice the two. The intron cyclizes in the process. (See Figure 19.19.) Group II ribozymes display a lariat mechanism. The 2′-OH of an internal adenosine attacks the splice site. (See Figure 19.20.)

Chapter 20

1. See Figure 20.1.

2. The hydrolysis of ATP to AMP and PP_i provides the energy to drive the activation step.

3. The linkage of amino acids to tRNA is as an aminoacyl ester.

4. Methionine bound to $tRNA^{fmet}$ can be formylated, but methionine bonded to $tRNA^{met}$ cannot.

5. The methionine anticodon (UAC) on the tRNA base-pairs with the methionine codon AUG in the mRNA sequence that signals the start of protein synthesis.

6. The initiation complex in *E. coli* requires mRNA, the 30S ribosomal subunit, $fmet-tRNA^{fmet}$, GTP, and three protein initiation factors, called IF-1, IF-2, and IF-3. The IF-3 protein is needed for the binding of mRNA to the ribosomal subunit. The other two protein factors are required for the binding of $fmet-tRNA^{fmet}$ to the mRNA-30S complex.

7. The A site and the P site on the ribosome are both binding sites for charged tRNAs taking part in protein synthesis. The P (peptidyl) site binds a tRNA to which the growing polypeptide chain is bonded. The A (aminoacyl) site binds to an aminoacyl tRNA. The amino acid moiety will be the next added to the nascent protein.

8. Peptidyl transferase catalyzes the formation of a new peptide bond in protein synthesis. The elongation factors, EF-Tu and EF-Ts, are required for binding of aminoacyl tRNA to the A site. The third elongation factor, EF-G, is needed for the translocation step in which the mRNA moves with respect to the ribosome, exposing the codon for the next amino acid.

9. Puromycin terminates the growing polypeptide chain by forming a peptide bond with its C-terminus, which prevents the formation of new peptide bonds (see Figure 20.9).

10. The stop codons bind to release factors, proteins that block binding of aminoacyl tRNAs to the ribosome and release the newly formed protein.

11. Hydroxyproline is formed from proline, an amino acid for which there are four codons, by posttranslational modification of the collagen precursor.

12. The less highly purified ribosome preparations contained polysomes, which are more active in protein synthesis than single ribosomes.

13. A code in which two bases code for a single amino acid allows for only 16 (4 × 4) possible codons, not adequate to code for 20 amino acids.

14. The concept of wobble specifies that the first two bases of a codon remain the same, while there is room for variation in the third base. This is precisely what is observed experimentally.

15. Hypoxanthine is the most versatile of the wobble bases; it can base-pair with adenine, cytosine, or uracil.

16. An inducible enzyme is one that is not produced at all times by a cell. The synthesis of such an enzyme is triggered by the presence of a specific substance, the inducer.

17. An operon consists of an operator gene, a promoter gene, and structural genes. When a repressor is bound

to the operator, RNA polymerase cannot bind to the promoter to start transcription of the structural genes. When an inducer is present, it binds to the repressor, rendering it inactive. The inactive repressor can no longer bind to the operator. As a result, RNA polymerase can bind to the promoter, leading to the eventual transcription of the structural genes.

18. A repressor is a protein that binds to the operator gene. A corepressor is a substance that binds to the repressor.

Where there is a corepressor, the repressor–corepressor complex binds to the operator and the repressor alone does not.

19. When our genes are transformed by a viral infection, somatic mutation, or some other factor, tumors result.

20. The virus needs these enzymes to catalyze vital processes in its reproduction.

Photograph and Illustration Credits

Frontmatter: Photograph of Mary Campbell by Curtis Smith; photograph of Irving Geis by Sandy Geis; p. xv: NCI/CNRI/Phototake; p. xvi: P. Plailly/Photo Researchers, Inc.; p. xviii top: Electra/Dan/Phototake; p. xviii bottom: Dr. Dennis Kunkel/Phototake; p. xix: Dr. Dennis Kunkel/Phototake; p. xx: Don Fawcett/Visuals Unlimited; p. xxi: Dr. Dennis Kunkel/Phototake; p. xxii: Stanley Flegler/Visuals Unlimited; p. xxiii: Dr. Dennis Kunkel/Phototake; p. xxiv: Murti/Phototake.

Part I

Part Opener, p. xxvi: Dr. Dennis Kunkel/Phototake.

Chapter 1: Chapter opener, p. 9: NCI/CNRI/Phototake; unnum. fig. p. 13: NASA; unnum. fig. p. 24: Le Petit Journal/Mary Evans Picture Library, London; Fig. 1.14: Joe McDonald/Visuals Unlimited; Fig. 1.15: C.P. Vance/ Visuals Unlimited.

Chapter 2: Chapter opener, p. 36: Frans Lanting, Minden Pictures, Inc.

Part II

Part Opener, p. 62: Leonard Lessin/Peter Arnold, Inc.

Chapter 3: Chapter opener, p. 68: Herb C. Ohlmeyer/ Fran Heyl and Associates; Figs. 3.1, 3.3, 3.8, 3.12, and unnum. fig. p. 81: Leonard Lessin/Waldo Feng/Mt. Sinai CORE; unnum. fig. p. 66: courtesy of Agouron Pharmaceuticals; unnum. fig. p. 83: Edgar Berstein/Peter Arnold, Inc.

Chapter 4: Chapter opener, p. 86: Jane and David Richardson; Figs. 4.2, 4.3, 4.5, 4.12(a), 4.20, 4.23: © Irving Geis; Fig. 4.7: The National Archeological Museum of Athens/The Bridgeman Art Library, London; Fig. 4.10(b): P. Plailly/Photo Researchers, Inc.; Fig. 4.11(a): © Petsko, Ringe, Schlicting and Kutsube, Peter Arnold, Inc.; Fig. 4.11(b): courtesy of Dr. C.M. Dobson, University of Oxford; Figs. 4.12(b), 4.14, 4.18(b): Leonard Lessin/Waldo Feng/Mt. Sinai CORE; unnum. fig. p. 112 right: Walter Reinhart/CNRI/Phototake; unnum. fig. p. 112 left: Dr. Dennis Kunkel/Phototake.

Interchapter A: Chapter opener, p. 119: Professor T.L. Blundell, Dept. of Crystallography, Burbeck College/ Science Photo Library/Photo Researchers, Inc.; Fig. A.4: Michael Gabridge/Visuals Unlimited; Fig. A.5(b): courtesy of Patrick O'Farrell.

Chapter 5: Chapter opener, p. 136: © Irving Geis; Fig. 5.18: Abeles, Frey, Jencks: *Biochemistry,* © Boston: Jones and Bartlett Publisher, reprinted by permission.

Chapter 6: Chapter opener, p. 179: Geis/Stodola, Photo Researchers, Inc.; Fig. 6.9 photos: Robert Stodola, Fox Chase Cancer Research Center and Irving Geis; Figs. 6.9 line drawing, 6.19: © Irving Geis; Fig. 6.11: Phillip A. Harrington/Fran Heyl and Associates; Fig. 6.24(a): Arthur M. Siegelman/Visuals Unlimited; Fig. 6.24(b): James Gathany and the Centers for Disease Control; unnum. fig. p. 192: adapted from *Science,* Figure 1, vol. 252, 1991, p. 1375, Dervan, P.B., © 1991 by the AAAS.

Chapter 7: Chapter opener, p. 213: Murti/Phototake; Fig. 7.2: Science Source/Photo Researchers, Inc.; Figs. 7.4, 7.5, 7.6, 7.7, 7.8, 7.9, 7.10, 7.11, 7.12, 7.13, 7.15, 7.16, 7.17, 7.18, 7.19, 7.20, 7.21, 7.22, 7.23, 7.24, 7.25, 7.26: adapted from "Dealing with Genes: The Language of Heredity," by Paul Berg and Maxine Singer, © 1992 by University Science Books; Fig. 7.14: adapted from "The Unusual Origin of the Polymerase Chain Reaction," by Kary B. Mullis, illustration by Michael Goodman, *Scientific American,* April 1990; Fig. 7.27: courtesy of Calgene Corporation; unnum. fig. p. 221: Electra/Dan/Phototake; unnum. fig. p. 227: Wide World Photos; unnum. fig. p. 230: Dr. Dennis Kunkel/Phototake.

Chapter 8: Chapter opener, p. 242: J.J. Sullivan/Rainbow; Fig. 8.2: © Irving Geis; Fig. 8.24: courtesy of Anthony Tu, University of Colorado; unnum. fig. p. 250: Harry Redl/Black Star; unnum. fig. p. 275: J.S. Reid/Custom Medical Stock.

Part III

Part Opener, p. 278: Dr. Dennis Kunkel/Phototake.

Chapter 9: Chapter opener, p. 284: © David Muench; Figs. 9.9, 9.11: Leonard Lessin/Waldo Feng/Mt. Sinai

CORE; Figs. 9.12, 9.20: © Irving Geis; Fig. 9.17: courtesy of Dr. Andrew Staehelin, University of Colorado, Boulder; unnum. fig. p. 285 top: Elizabeth Weiland/Photo Researchers, Inc.,; unnum. fig. p. 285 middle: Mike Neumann/Photo Researchers, Inc.; unnum. fig. p. 285 bottom: Dick Rowan/Photo Researchers, Inc., unnum. figs. p. 286, 289: The Bettmann Archive.

Chapter 10: Chapter opener, p. 311: Phillip A. Harrington/Fran Heyl and Associates; Figs. 10.1(b.2) and (b.4), 10.7(c.3)L Leonard Lessin/Waldo Feng/Mt. Sinai CORE; Fig. 10.4(b): Tito Simboli; unnum. fig. p. 314: The Bettman Archive; unnum. fig. p. 331 top: Albert Grant/Photo Researchers, Inc.; unnum. fig. p. 331 bottom: William E. Weber/Visuals Unlimited; unnum. fig. p. 337 top: Stanley Flegler/Visuals Unlimited; unnum. fig. p. 337 bottom: Don Fawcett/Visuals Unlimited.

Chapter 11: Chapter opener, p. 341: D.A. Lesk/Laboratory of Molecular Biology/Photo Researchers, Inc.

Chapter 12: Chapter opener, p. 365: Mel Pollinger/Fran Heyl and Associates; unnum. fig. p. 377: Stephen Kraseman/Photo Researchers, Inc.

Chapter 13: Chapter opener, p. 391: Dr. Dennis Kunkel/Phototake; Fig. 13.8(b): Photo Researchers, Inc.

Chapter 14: Chapter opener, p. 418: Dr. Dennis Kunkel/Phototake; unnum. fig. p. 422 top: Matt Biondi/Photo Researchers, Inc.; unnum. fig. p. 422 bottom: Bruce Gaylord/Visuals Unlimited; unnum. fig. p. 433: Wide World Photos.

Chapter 15: Chapter opener, p. 443: Ulof Bjorg Christianson/Fran Heyl and Associates; unnum. fig. p. 445: Leonard Lee Rue II/Photo Researchers, Inc.; unnum. fig. p. 450 top: George Holton/Photo Researchers, Inc.; unnum. fig. p. 450 bottom: Tom McHugh/Photo Researchers, Inc.

Chapter 16: Chapter opener, p. 479: Arthur Gurmankin/Phototake; unnum. fig. p. 493: David Newman/Visuals Unlimited.

Chapter 17: Chapter opener, p. 510: Ulof Bjorg Christianson.

Interchapter B: Chapter opener, p. 540: Mel Pollinger/Fran Heyl and Associates; unnum. fig. p. 552: Renee Lynn/Photo Researchers, Inc.

Chapter 18: Chapter opener, p. 555: Mel Pollinger/Fran Heyl and Associates; unnum. fig. p. 565: USDA; unnum. fig. p. 577 top: Kevin Beebe/Custom Medical Stock; unnum. fig. p. 577 bottom: The Bettman Archive; Fig. 18.16: adapted from "How the Immune System Recognizes Invaders," by Charles A. Janeway, Jr., illustration by Ian Warpole, *Scientific American,* September 1993; Fig. 18.18: adapted from "How the Immune System Develops," by Irving L. Weissman and Max D. Cooper, illustration by Jared Schneidman, *Scientific American,* September 1993; Figs. 18.19 and 18.21: adapted from "How the Immune System Recognizes Invaders," by Charles A. Janeway, Jr., *Scientific American,* September 1993; Figs. 18.24 and 18.25: adapted with permission from "The Surprising Life of Nitric Oxide," by P.L. Feldman, O.W. Griffith and D.J. Stuehr, *Chemical and Engineering News, 71*(51), December 20, 1993, pp. 27 and 32, © 1993 American Chemical Society.

Part IV

Part Opener, p. 592: Leonard Lessin/Peter Arnold, Inc.; unnum. figs. pp. 594-596: Becca Gruliow.

Chapter 19: Chapter opener, p. 601: Phillip A. Harrington/Fran Heyl and Associates; unnum. fig. p. 623: Ken Greer/Visuals Unlimited.

Chapter 20: Chapter opener, p. 629: Al Lemme/Len/Phototake; unnum. fig. p. 648: adapted from *Science,* Figure 1, Vol. 262, 1993, p. 1644, by K. Sutliff, © 1993 by the Cancer Center/University of North Carolina; unnum. fig. p. 650, part (a) bottom: A.B. Dowcett/Photo Researchers, Inc.; unnum. fig. p. 650, part (c): © Irving Geis; unnum. fig. p. 651, part (a): reprinted from "AIDS and the Immune System," by Warner C. Greene, illustration by Kirk Muldoff, *Scientific American,* September 1993; unnum. fig. p. 651, part (b): Dr. Jan Orenstein; Figs. 20.16, 20.17, and unnum. fig. p. 651 top: adapted from "Dealing with Genes: The Language of Heredity," by Paul Berg and Maxine Singer, © 1992 by University Science Books; Fig. 20.18: adapted from "AIDS and the Immune System," by Warner C. Greene, illustration by Tomo Narachima, *Scientific American,* September 1993.

GLOSSARY

absolute specificity An enzyme's property of acting on one, and only one, substrate

accessory pigments Plant pigments other than chlorophyll that play roles in photosynthesis

acid dissociation constant A number that expresses the strength of an acid

acid strength The tendency of an acid to dissociate to a hydrogen ion and its conjugate base

acromegaly A disease caused by an excess of growth hormone after the skeleton has stopped growing, characterized by enlarged hands, feet, and facial features

actin A protein, the principal component of the thin filament of muscle fibers

activation energy The energy required to start a reaction

activation step The beginning of a multistep process, in which a substrate is converted to a more reactive compound

active site The part of an enzyme to which the substrate binds and at which the reaction takes place

active transport The energy-requiring process of moving substances into a cell against a concentration gradient

acyl-CoA synthetase The enzyme that catalyzes the activation step in lipid catabolism

Addison's disease A disease caused by a deficiency of steroid hormones

of the adrenal cortex, characterized by hypoglycemia, weakness, and increased susceptibility to stress

adenine One of the purine bases found in nucleic acids

adenylate cyclase The enzyme that catalyzes the production of cyclic AMP

ADP (adenosine diphosphate) A compound that can serve as an energy carrier when it is phosphorylated to form ATP

albinism The condition of depigmentation in an organism, caused by the lack of an enzyme responsible for the conversion of the amino acid tyrosine to melanin

alcohol dehydrogenase An NADH-linked redox enzyme; catalyzes the conversion of acetaldehyde to ethanol

alcoholic fermentation The anaerobic pathway that converts glucose to ethanol

aldolase In glycolysis, the enzyme that catalyzes the reverse aldol condensation of fructose 1,6-*bis*phosphate

aldose A sugar that contains an aldehyde group as part of its structure

alleles Corresponding genes on paired chromosomes

allosteric Describes multisubunit proteins in which a conformational change in one subunit induces a change in another subunit

allosteric effector A substance — substrate, inhibitor, or activator —

that binds to an allosteric enzyme and affects its activity

amino acid activation The formation of an ester bond between an amino acid and its specific tRNA, catalyzed by a suitable synthetase

amino acid analyzer An instrument that gives information on the number and kind of amino acids in a protein

aminoacyl-tRNA synthetases Enzymes that catalyze the formation of an ester linkage between an amino acid and tRNA

aminopeptidase An enzyme that removes amino acids from the N-terminal end of a protein

amphibolic Able to be a part of both anabolism and catabolism

amphiphilic Describes a molecule with one end that dissolves in water and another end that dissolves in nonpolar solvents

α-amylase An enzyme that hydrolyzes glycosidic linkages anywhere along a polysaccharide chain

β-amylase An enzyme that hydrolyzes glycosidic linkages starting at the end of a polysaccharide chain

amylopectin A form of starch, a branched-chain polymer of glucose

amylose A form of starch, a linear polymer of glucose

anabolism The synthesis of biomolecules from simpler compounds

anaerobic glycolysis The pathway of conversion of glucose to lactate, distinguished from glycolysis, which is the conversion of glucose to pyruvate

analytical ultracentrifugation   The technique for observing the motion of particles as they sediment in a centrifuge

anaplerotic   Describes a reaction that ensures an adequate supply of an important metabolite

androgens   Male sex hormones, steroids in chemical nature

anomer   One of the possible stereo-isomers formed when a sugar assumes the cyclic form

anomeric carbon   The chiral center created when a sugar cyclizes

antibody   A glycoprotein that binds to and immobilizes a substance that is recognized by the cell as foreign

anticodon   The sequence of three bases (triplet) in tRNA that hydrogen-bonds with the mRNA triplet that specifies a given amino acid

antigen   A substance that triggers an immune response

antigenic determinant   The portion of a molecule that is recognized by antibodies as foreign and to which they bind

antioxidant   A strong reducing agent, which is easily oxidized and thus prevents the oxidation of other substances

apoenzyme   An enzyme that consists of polypeptide chains alone, lacking required nonprotein cofactors

arachidonic acid   A fatty acid that contains 20 carbon atoms and four double bonds, the precursor of prostaglandins and leukotrienes

aspartate transcarbamoylase (ATCase)   A classic example of an allosteric enzyme, catalyzes an early reaction in pyrimidine biosynthesis

atherosclerosis   The blockage of arteries by cholesterol deposits

ATP (adenosine triphosphate)   A universal energy carrier

ATP synthase   The enzyme responsible for production of ATP in mitochondria

autoimmune diseases   Diseases in which the immune system attacks the body's own tissues

autoradiography   The technique of locating radioactively labeled substances by allowing them to expose photographic film

B-cells   A type of white blood cell that plays an important role in the immune system, involved in the production of antibodies

bacterial plasmid   A portion of circular DNA separate from the main genome of the bacterium

bacteriochlorophyll   The form of chlorophyll that occurs in bacteria that carry out photosynthesis not linked to oxygen

β-barrel   A β-pleated sheet extensive enough to fold back on itself

bidentate ligand   A substance that forms two bonds when it binds to another molecule

binding assay   An experimental method for selecting one molecule out of a number of possibilities by specific binding, used to determine the natures of many triplets of the genetic code

biotin   A CO_2 carrier molecule

blotting   A technique for transferring a portion of a sample for further analysis

Bohr effect   The decrease in oxygen binding by hemoglobin caused by binding of carbon dioxide and hydrogen ion

buffer solution   A solution that resists a change in pH on the addition of moderate amounts of strong acid or strong base

buffering capacity   A measure of the amount of acid or base that reacts with a given buffer solution

C-terminal   The end of a protein or peptide with a carboxyl group not bonded to another amino acid

Calvin cycle   The pathway of carbon dioxide fixation in photosynthesis

carboxypeptidase   An enzyme that removes amino acids from the C-terminal end of a protein

β-carotene   An unsaturated hydrocarbon, the precursor of vitamin A

carrier protein   A membrane protein

to which a substance binds in passive transport into the cell

cascade   A series of steps that take place in hormonal control of metabolism, affecting a series of enzymes and amplifying the effect of a small amount of hormone

catabolism   The breakdown of nutrients to provide energy

catabolite activator protein (CAP)   A protein that can bind to a promoter when complexed with cAMP, allowing RNA polymerase to bind to its entry site on the same promoter

catabolite repression   Repression of the synthesis of *lac* proteins by glucose

catalysis   The process of increasing the rate of chemical reactions

cDNA (complementary DNA)   A form of DNA synthesized on an RNA template

cell membrane   The outer membrane of the cell that separates it from the outside world

cell wall   The outer coating of bacterial and plant cells

cellulose   A polymer of glucose, an important structural material in plants

ceramide   A lipid that contains one fatty acid linked to sphingosine (*vide infra*) by an amide bond

cerebroside   A glycolipid that contains sphingosine and a fatty acid in addition to the sugar moiety

channel protein   A membrane protein that has a channel through which a substance passes without binding in passive transport into the cell

chemiosmotic coupling   The mechanism for coupling electron transport to oxidative phosphorylation, requires a proton gradient across the inner mitochondrial membrane

chimeric DNA   DNA from more than one species covalently linked together

chiral   Describes an object that is not superimposable on its mirror image

chlorophyll   The principal photosynthetic pigment, responsible for trapping light energy from the sun

chloroplast   The organelle that is the site of photosynthesis in green plants

cholera   A disease caused by the bacterium *Vibrio cholerae*, characterized by dehydration due to excessive Na^+ transport in epithelial cells

cholesterol   A steroid that occurs in cell membranes, the precursor of other steroids

chromatin   A complex of DNA and protein found in eukaryotic nuclei

chromatography   An experimental method for separating substances

chromosome   A linear structure that contains the genetic material

chromosome walking   A technique for locating the genes on a relatively long portion of a chromosome

chymotrypsin   A proteolytic enzyme that preferentially hydrolyzes amide bonds adjacent to aromatic amino acid residues

chymotrypsinogen   The precursor of chymotrypsin, converted to it by proteolytic cleavage of specific peptide bonds

***cis-trans* isomerase**   An enzyme that catalyzes a *cis-trans* isomerization in the catabolism of unsaturated fatty acids

citric acid cycle   A central metabolic pathway, part of aerobic metabolism

clonal selection   The process by which the immune system responds selectively to antibodies actually present in an organism

clone   A genetically identical population of organisms, cells, viruses, or DNA molecules

cloning of DNA   The introduction of a section of DNA into a genome in which it can be reproduced many times

clotting factors   Proteins that play a role in blood clotting

cloverleaf structure   The characteristic secondary structure of tRNA; includes hydrogen-bonded stems and loops without hydrogen bonds

codon   A sequence of three bases on mRNA that specifies a given amino acid

coenzyme   A nonprotein substance that takes part in an enzymatic reaction and is regenerated at the end of the reaction

coenzyme A   A carrier of carboxylic acids bound to its thiol group by a thioester linkage

coenzyme Q   An oxidation–reduction coenzyme in mitochondrial electron transport

column chromatography   A form of chromatography in which the stationary phase is packed in a column

committed step   In a metabolic pathway, the formation of a substance that can play no other role in metabolism but to undergo the rest of the reactions of the pathway

competitive inhibition   A decrease in enzymatic activity caused by binding of a substrate analogue to the active site

complementary   Refers to the specific hydrogen bonding of adenine with thymine (or uracil) and guanine with cytosine in nucleic acids

concerted model   A description of allosteric activity in which the conformations of all subunits change simultaneously

configuration   The three-dimensional arrangement of groups around a chiral carbon atom

conformational coupling   A mechanism for coupling of electron transport to oxidative phosphorylation that depends on a conformational change in the ATP synthase

conjugation   Contact between bacteria that allows exchange of DNA

consensus sequences   DNA sequences to which RNA polymerase binds, identical in many organisms

contact inhibition   The process by which cell growth stops when cells touch each other; absent in cancer cells

control sites   The operator and promoter genes that modulate the pro-

duction of proteins whose amino acid sequences are specified by the structural genes under their control

cooperative binding   Binding to several sites such that when the first ligand is bound, the binding of subsequent ones is easier

cooperative transition   A transition that takes place in an all-or-nothing fashion, such as the melting of a crystal

corepressor   A substance that binds to a repressor protein making it active and able to bind to an operator gene

Cori cycle   A pathway in carbohydrate metabolism that links glycolysis in liver with gluconeogenesis in liver

coupled translation   In prokaryotes, the situation in which a gene is simultaneously transcribed and translated

coupling   The process by which an exergonic reaction provides energy for an endergonic one

cut and patch   A mechanism for the repair of DNA by enzymatic removal of incorrect nucleotides and substitution of correct ones

cyanogen bromide   A reagent that cleaves proteins at internal methionine residues

cyanosis   A short supply of oxygen in the bloodstream, characterized by bluish skin

cyclic AMP   A nucleotide in which the same phosphate group is esterified to the 3' and 5' hydroxyl groups of a single adenosine; an important second messenger

cytochrome   Any one of a group of heme-containing proteins in the electron transport chain

cytokines   Soluble protein factors produced by one cell that affect another cell

cytosine   One of the pyrimidine bases found in nucleic acids

cytoskeleton   A lattice of fine strands, consisting mostly of protein, that pervades the cytosol (*see* microtrabecular lattice)

cytosol   The portion of the cell that lies outside the nucleus and other membrane-bounded organelles

debranching enzyme   An enzyme that hydrolyzes the linkages in a branched-chain polymer such as amylopectin

degenerate code   The coding of more than one triplet of bases for the same amino acid

denaturation   The unraveling of the three-dimensional structure of a macromolecule, caused by the breakdown of noncovalent interactions

denitrification   The process by which nitrates are broken down to molecular nitrogen

density-gradient centrifugation   The technique of separating substances in an ultracentrifuge by applying the sample to the top of a tube that contains a solution of varying densities

deoxy sugar   A sugar in which one of the hydroxyl groups has been reduced to a hydrogen

deoxyribonucleoside   A compound formed when a nucleobase and deoxyribose form a glycosidic bond

deoxyribose   A sugar that is part of the structure of DNA

diastereomers   Nonsuperimposable, nonmirror-image stereoisomers

dimer   An aggregate consisting of two subunits

dipole   A molecule with a positive end and a negative end due to uneven distribution of electrons in bonds

disaccharide   Two monosaccharides (monomeric sugars) linked by a glycosidic bond

DNA   Deoxyribonucleic acid, the molecule that contains the genetic code

DNA gyrase   An enzyme that introduces supercoiling into closed circular DNA

DNA library   A collection of clones that includes the total genome of an organism

DNA ligase   The enzyme that links separate stretches of DNA

DNA polymerase   The enzyme that forms DNA from deoxyribonucleotides on a DNA template

DNase (deoxyribonuclease)   An enzyme that specifically hydrolyzes DNA

domain   A portion of a polypeptide chain that folds independently of other portions of the chain

double-strand heterozygous   Describes a recombinant DNA in which both strands incorporate portions of the two original molecules

downstream   In transcription, describes a portion of the DNA sequence nearer the $5'$ end than the gene to be transcribed, where the DNA is read from the $3'$ to the $5'$ end and the RNA is formed from the $5'$ to the $3'$ end; in translation, nearer the $3'$ end of mRNA

dwarfism   A disease caused by a deficiency of growth hormone

Edman degradation   A method for determining the amino acid sequences of peptides and proteins

electron transport to oxygen   A series of oxidation–reduction reactions by which the electrons derived from oxidation of nutrients are passed to oxygen

electronegativity   A measure of the tendency of an atom to attract electrons to it in a chemical bond

electrophile   An electron-poor substance that tends to react with centers of negative charge or polarization

electrophoresis   A method for separating molecules on the basis of the ratio of charge to size

elongation step   In protein synthesis, the succession of reactions in which the peptide bonds are formed

enantiomers   Mirror-image, nonsuperimposable stereoisomers

endergonic   Energy-absorbing

endocrine system   The series of ductless glands that release hormones into the bloodstream

endocytosis   The process by which portions of a cell membrane are pinched off into the cell

endoglycosidase   An enzyme that hydrolyzes glycosidic linkages anywhere along a polysaccharide chain

endonuclease   An enzyme that hydrolyzes nucleic acids, attacking linkages in the middle of the polynucleotide chain

endoplasmic reticulum (ER)   A continuous single-membrane system throughout the cell

endosymbiosis   A relationship in which a smaller organism is completely contained in a larger organism

enthalpy   A thermodynamic quantity, measured as the heat of reaction at constant pressure

entropy   A thermodynamic quantity, a measure of the disorder of the universe

enzyme   A biological catalyst, usually a globular protein, with self-splicing RNA as the only exception

epimerase   An enzyme that catalyzes the inversion of configuration around a single carbon atom

epimers   Stereoisomers that differ only in configuration around one of several chiral carbon atoms

epitope   A binding site for an antibody on an antigen

equilibrium   The state in which a forward process and a reverse process occur at the same rate

essential amino acids   Amino acids that cannot be synthesized by the body and must be obtained in the diet

essential fatty acids   The polyunsaturated fatty acids (such as linoleic acid) that cannot be synthesized by the body and must be obtained from dietary sources

estradiol   A steroid sex hormone

estrogens   Female sex hormones, steroids in chemical nature

eukaryote   An organism in which the cells have well-defined nuclei and membrane-bounded organelles

excision repair   Repair of DNA by enzymatic removal of incorrect nucleotides and replacement by the correct ones

exergonic   Energy-releasing

exoglycosidase   An enzyme that hydrolyzes glycosidic linkages, starting at the end of a polysaccharide chain

exon A DNA sequence that is expressed in the sequence of mRNA

exonuclease An enzyme that hydrolyzes nucleic acids, starting at the end of the polynucleotide chain

facilitated diffusion A process by which substances enter a cell by binding to a carrier protein; does not require energy

familial hypercholesterolemia A disease characterized by high cholesterol levels in the bloodstream and early heart attacks

farnesyl pyrophosphate A 15-carbon intermediate in cholesterol biosynthesis

fatty acid A compound with a carboxyl group at one end and a long, normally unbranched hydrocarbon tail at the other; the hydrocarbon tail may be saturated or unsaturated

feedback inhibition The process by which the final product of a series of reactions inhibits the first reaction in the series

fibrin An insoluble protein formed as the final stage in the blood clotting process

fibrinogen A soluble protein transformed into fibrin by selective hydrolysis of peptide bonds

fibrous protein A protein whose overall shape is that of a long, narrow rod

first order Describes a reaction whose rate depends on the first power of the concentration of a single reactant

Fischer projection A two-dimensional representation of the stereochemistry of three-dimensional molecules

fluid mosaic model The model for membrane structure in which proteins and a lipid bilayer exist side by side without covalent bonds between the proteins and lipids

folate reductase The enzyme that reduces dihydrofolate to tetrahydrofolate, a target for cancer chemotherapy

free energy A thermodynamic quantity, diagnostic for the spontaneity of a reaction at constant temperature

free radical A molecule that contains one or more unpaired electrons

functional group One of the groups of atoms that give rise to the characteristic reactions of organic compounds

Fungi One of the five kingdoms used to classify living organisms; includes molds and mushrooms

furanose A cyclic sugar with a six-membered ring, named for its resemblance to the ring system in furan

furanoside A glycoside involving a furanose

G-protein A membrane-bound protein that mediates the action of adenylate cyclase

β-galactosidase The enzyme that hydrolyzes lactose to galactose and glucose; the classic example of an inducible enzyme

gated channels Proteins that permit transient passage of external substances into a cell when suitably triggered

gel electrophoresis A method for separating molecules on the basis of charge-to-size ratio, using a gel as a support and sieving material

gene The individual unit of inheritance

gene therapy A method for treating genetic disease by introducing a good copy of the defective gene

general acid–base catalysis A form of catalysis that depends on transfer of protons

genetic code The information for the structure and function of all living organisms

genetic map A schematic representation of the order of genes on a chromosome

genetic recombination The combining of two different DNA molecules to produce a third molecule that is different from either of the original ones

genome The total DNA of the cell

geranyl pyrophosphate A ten-carbon intermediate in cholesterol biosynthesis

giantism A disease caused by overproduction of growth hormone before the skeleton has stopped growing

globular protein A protein whose overall shape is more or less spherical

glucocorticoid A kind of steroid hormone involved in the metabolism of sugars

glucogenic amino acid An amino acid that has pyruvate or oxaloacetate as a catabolic breakdown product

gluconeogenesis The pathway of synthesis of glucose from lactate

glucose A monosaccharide, a ubiquitous metabolite

glyceraldehyde The simplest carbohydrate that contains a chiral carbon, the starting point of a system of describing optical isomers

glyceraldehyde 3-phosphate A key intermediate in the reactions of sugars

glycerol phosphate shuttle A mechanism for transferring electrons from NADH in the cytosol to $FADH_2$ in the mitochondrion

glycogen A polymer of glucose, an important energy storage molecule in animals

glycolipid A lipid to which a sugar moiety is bonded

glycolysis The anaerobic breakdown of glucose to three-carbon compounds

glycoside A compound in which one or more sugars are involved in a linkage to another molecule

glyoxysomes Membrane-bounded organelles that contain the enzymes of the glyoxalate cycle

Golgi apparatus A system of flattened membranous sacs, usually involved in secretion of proteins

gout A disease characterized by painful deposits of uric acid in the joints of the fingers and toes

grana Bodies within the chloroplast that contain the thylakoid disks, the site of photosynthesis

Greek key A form of polypeptide

chain folding that resembles a motif found on ancient pottery

guanine   One of the purine bases found in nucleic acids

half reaction   An equation that shows either the oxidative or the reductive part of an oxidation–reduction reaction

Haworth projection formulas   Perspective representations of the cyclic forms of sugars

helicase (rep protein)   A protein that unwinds the double helix of DNA in the process of replication

α-helix   One of the most frequently encountered folding patterns in the protein backbone

heme   An iron-containing cyclic compound, found in cytochromes, hemoglobin, and myoglobin

hemiacetal   A compound formed by reaction of an aldehyde with an alcohol, found in the cyclic structure of sugars

hemiketal   A compound formed by reaction of a ketone with an alcohol, found in the cyclic structure of sugars

hemolytic anemia   A disease characterized by destruction of red blood cells

hemophilia   A molecular disease characterized by uncontrollable bleeding

Henderson–Hasselbalch equation   A mathematical relationship between the $pK'a$ of an acid and the pH of a solution containing the acid and its conjugate base

heteropolysaccharide   A polysaccharide that contains more than one kind of sugar monomer

heterotropic effects   Allosteric effects that occur when different substances are bound to a protein

heterozygous   Exhibiting differences in a given gene on one chromosome and the corresponding gene on the paired chromosome

hexose monophosphate shunt   A synonym for the pentose phosphate pathway, in which glucose is converted to five-carbon sugars with concomitant production of NADPH

histones   Basic proteins found complexed to eukaryotic DNA

holoenzyme   An enzyme that has all component parts, including coenzymes and all subunits

homeostasis   The balance of biological activities in the body

homologous sequences   Regions of different macromolecules that have the same order of monomers

homopolysaccharide   A polysaccharide that contains only one kind of sugar monomer

homotropic effects   Allosteric effects that occur when several identical molecules are bound to a protein

homozygous   Exhibiting no differences in a given gene on one chromosome and the corresponding gene on the paired chromosome

hormone   A substance produced by endocrine glands and delivered by the bloodstream to target cells, producing a desired effect

HPLC (high-performance liquid chromatography)   A form of column chromatography

hyaluronic acid   A polysaccharide that occurs in the lubricating fluid of joints

hydrazine   A reagent for labeling the C-terminal ends of proteins

hydride ion transfer   The transfer of a proton (H^+) and two electrons that occurs in many biological redox reactions

hydrogen bonding   A noncovalent association formed between a hydrogen atom covalently bonded to one electronegative atom and a lone pair of electrons on another electronegative atom

hydrophilic   Tending to dissolve in water

hydrophobic   Tending not to dissolve in water

β - hydroxy - β - methylglutaryl - CoA   An intermediate in the biosynthesis of cholesterol

hyperbolic   Rising quickly and then leveling off; used for a curve on a graph

hyperchromicity   The increase in absorption of ultraviolet light that accompanies the denaturation of DNA

hyperglycemia   The condition of elevated blood glucose levels

hypoglycemia   The condition of low blood glucose levels

hypothalamus   The portion of the brain that controls much of the workings of the endocrine system

induced-fit model   A description of substrate binding to an enzyme in which the conformation of the enzyme changes to accommodate the shape of the substrate

inducible enzyme   An enzyme whose synthesis can be triggered by the presence of some substance, the inducer

induction of enzyme synthesis   The triggering of the production of an enzyme by the presence of a specific inducer

initial rate   The rate of a reaction immediately after it starts, before any significant accumulation of product

initiation complex   The aggregate of mRNA, N-formylmethione tRNA, ribosomal subunits, and initiation factors needed at the start of protein synthesis

initiation step   The start of protein synthesis; the formation of the initiation complex

integral protein   A protein that is embedded in a membrane

interleukins   Proteins that play a role in the immune system

intermembrane space   The region between the inner and outer mitochondrial membranes

intron   An intervening sequence in DNA that does not appear in the final sequence of mRNA

ion-exchange chromatography   A method for separating substances on the basis of charge

ion product constant for water   A measure of the tendency of water to dissociate to give hydrogen ion and hydroxide ion

ionophore A peptide or protein that serves as a channel through membranes for ions

irreversible inhibition Covalent binding of an inhibitor to an enzyme, causing permanent inactivation

isoelectric focusing A method for separating substances on the basis of their isoelectric points

isoelectric point The pH at which a molecule has no net charge

isoprene A five-carbon unsaturated group, part of the structures of many lipids

ketogenic amino acid An amino acid that has acetyl-CoA or acetoacetyl-CoA as a catabolic breakdown product

α-ketoglutarate dehydrogenase complex One of the enzymes of the citric acid cycle; catalyzes the conversion of α-ketoglutarate to succinyl-CoA

ketose A sugar that contains a ketone group as part of its structure

Krebs cycle An alternative name for the citric acid cycle

kwashiorkor A disease caused by serious protein deficiency

L and D amino acids Amino acids whose stereochemistry is the same as the stereochemical standards L- and D-glyceraldehyde, respectively

labeling Covalent modification of a specific residue on an enzyme

lac **operon** The promoter, operator, and structural genes involved in the induction of β-galactosidase and related proteins

lactate dehydrogenase An NADH-linked dehydrogenase; catalyzes the conversion of pyruvate to lactate

lactone A cyclic ester

lagging strand In DNA replication, the strand that is formed in small fragments subsequently joined by DNA ligase

lanosterol A precursor of cholesterol

leading strand In DNA replication, the strand that is continuously formed in one long stretch

Lesch–Nyhan syndrome A metabolic disease characterized by severe retardation and a tendency to self-mutilate, caused by a deficiency of an enzyme of the purine salvage pathway

leukotriene A substance derived from leukocytes (white blood cells) that has three double bonds; of pharmaceutical importance

ligand-gated Describes a channel protein that opens transiently on binding a specific molecule

"light up" To select a sample of interest out of a mixture by binding of a radioactive tracer

lignin A polymer of coniferyl alcohol; a structural material found in woody plants

Lineweaver–Burk double-reciprocal plot A graphical method for analyzing the kinetics of enzyme-catalyzed reactions

lipase An enzyme that hydrolyzes lipids

lipid A compound insoluble in water and soluble in organic solvents

lipid bilayers Aggregates in which the polar head groups are in contact with water and the hydrophobic parts are not

lipoic acid A coenzyme that can function either in redox reactions or as an acyl transfer agent

liposome An aggregate of lipids arranged so that the polar ends are in contact with water and the nonpolar tails are sequestered from water

lock-and-key model A description of the binding of substrate to an enzyme such that the active site and the substrate exactly match each other in shape

lymphocyte A type of white blood cell; a major component of the immune system

lymphokines Soluble protein factors produced by one lymphocyte that affect another cell

lysosomes Membrane-bounded organelles that contain hydrolytic enzymes

macronutrients Nutrients needed in large amounts, such as proteins, carbohydrates, or fats

malate–aspartate shuttle A mechanism for transferring electrons from NADH in the cytosol to NADH in the mitochondrion

malonyl-CoA A three-carbon intermediate that is important in the biosynthesis of fatty acids

matrix (mitochondrial) The part of a mitochondrion enclosed within the inner mitochondrial membrane

melting The denaturation of DNA by heat

mercaptoethanol A reagent for reducing the disulfide bonds in proteins to sulfhydryl groups

metabolic water The water produced as a result of complete oxidation of nutrients; sometimes the only water source of desert-dwelling organisms

metabolism The sum total of all biochemical reactions that take place in an organism

metal-ion catalysis (Lewis acid–base catalysis) A form of catalysis that depends on the Lewis definition of an acid as an electron–pair acceptor and a base as an electron pair acceptor

methemoglobin An abnormal form of hemoglobin in which the iron is Fe(III) rather than the normal Fe(II)

micelle An aggregate formed by amphiphilic molecules such that their polar ends are in contact with water and their nonpolar portions are on the interior

Michaelis constant A numerical value for the strength of binding of a substrate to an enzyme; an important parameter in enzyme kinetics

micronutrients Vitamins and minerals, needed in small amounts

microspheres Spherical aggregates of artificially synthesized proteinoids

microtrabecular lattice (cytoskeleton) The network of microtubules that pervades the cell

microtubules Filaments made up of the protein tubulin, the material of the cytoskeleton

mineralocorticoid   A kind of steroid hormone involved in the regulation of levels of inorganic ions ("minerals")

minerals   In nutrition, inorganic substances required as the ions or free elements

mitochondrion   An organelle that contains the apparatus responsible for aerobic oxidation of nutrients

mobile phase (eluent)   In chromatography, the portion of the system in which the mixture to be separated moves

molecular sieve chromatography   A method for separating molecules on the basis of size

Monera   One of the five kingdoms used to classify living organisms; includes prokaryotes

monoclonal antibodies   Antibodies produced from the progeny of a single cell and specific for a single antigen

monomer   A small molecule that bonds to many others to form a macromolecule

monosaccharide   A compound that contains a single carbonyl group and two or more hydroxyl groups

mRNA   The kind of RNA that specifies the order of amino acids in a protein

mucopolysaccharide   A polysaccharide that has a gelatinous consistency

multifunctional enzyme   An enzyme in which a single protein catalyzes several reactions

multiple sclerosis   A disease in which the lipid sheaths of nerve cells are progressively destroyed

muscular dystrophy   A disease characterized by muscle weakness

mutagen   An agent that brings about a mutation; includes radiation and chemical substances that alter DNA

mutation   A change in DNA, causing subsequent changes in the organism that can be transmitted genetically

myelin   The lipid-rich sheath of nerve cells

myofibrils   Muscle fibers

myosin   A protein, the principal component of the thick filament of muscle fibers

N-formylmethionine-tRNA^{fmet} An essential factor for the start of protein synthesis, the amino acid methionine formylated at the amino group and covalently bonded to its specific tRNA

N-terminal   The end of a protein or polypeptide with its amino group not linked to another amino acid by a peptide bond

native conformation   A three-dimensional shape of a protein with biological activity

nitrification   The conversion of nitrates to ammonia

nitrogen fixation   The conversion of molecular nitrogen to ammonia

nitrogenase   The enzyme complex that catalyzes nitrogen fixation

noncompetitive inhibition   A form of enzyme inactivation in which a substance binds to a place other than the active site but distorts the active site so that the reaction is inhibited

nonheme (iron–sulfur) protein   A protein that contains iron and sulfur but no heme group

nonoverlapping, commaless code   The sequences of three bases (triplets) that specify an amino acid with no shared or intervening bases

nonpolar bond   A bond in which two atoms share electrons evenly

nuclear magnetic resonance (NMR) spectroscopy   A method for determining the three-dimensional shapes of proteins in solution

nuclear region   The portion of a prokaryotic cell that contains the DNA

nucleic acid   A macromolecule formed by polymerization of nucleotides; carries the genetic message

nucleic-acid base (nucleobase)   One of the nitrogen-containing aromatic compounds; makes up the coding portion of a nucleic acid

nucleolus   A portion of the nucleus rich in RNA

nucleophile   An electron-rich substance that tends to react with sites of positive charge or polarization

nucleophilic substitution reaction   A reaction in which one functional group is replaced by another as the result of nucleophilic attack

nucleoside   A purine or pyrimidine base bonded to a sugar (ribose or deoxyribose)

nucleosome   A structure in which DNA is wrapped around an aggregate of histones

nucleotide   A purine or pyrimidine base bonded to a sugar (ribose or deoxyribose), which in turn is bonded to a phosphate group

nucleus   The organelle that contains the main genetic apparatus in eukaryotes

Okazaki fragments   Short stretches of DNA formed in the lagging strand in replication and subsequently linked by DNA ligase

oligomer   An aggregate of several smaller units (monomers); bonding may be covalent or noncovalent

oligosaccharide   A few sugars linked by glycosidic bonds

oncogene   A gene that causes cancer when a triggering event takes place

one-carbon transfer   A reaction in which the transfer usually involves carbon dioxide, a methyl group, or a formyl group

operator   The gene to which a repressor of protein synthesis binds

operon   A group of operator, promoter, and structural genes

opsin   A protein in the rod and cone cells of the retina; plays a crucial role in vision

optical isomers   *See* stereoisomers.

order of a reaction   The experimentally determined dependence of the rate on substrate concentrations

organelle   A membrane-bounded portion of a cell with a specific function

organic chemistry   The study of compounds of carbon, especially of carbon and hydrogen and their derivatives

origin   In chromatography, the place to which the mixture to be separated is applied

origin of replication   The point at which the DNA double helix begins to unwind at the start of replication

oxidation   The loss of electrons

β-oxidation   The main pathway of catabolism of fatty acids

oxidative decarboxylation   Loss of carbon dioxide accompanied by oxidation

oxidative phosphorylation   A process for generating ATP; depends on the creation of a pH gradient within the mitochondrion as a result of electron transport

oxidizing agent   A substance that accepts electrons from other substances

palindrome   A message that reads the same from left to right as from right to left

palmitate   A 16-carbon saturated fatty acid, the end product of fatty acid biosynthesis in mammals

partial double-bond character   The character of a bond normally written as a single bond that can also be written as a double bond by a simple shift of electrons

passive transport   The process by which a substance enters a cell without expenditure of energy by the cell

pectin   A polymer of galacturonic acid; occurs in plant cell walls

pentose phosphate pathway   A pathway in sugar metabolism that gives rise to five-carbon sugars and NADPH

peptide bond   An amide bond between amino acids in a protein

peptides   Molecules formed by linking two to several dozen amino acids by amide bonds

peptidoglycan   A polysaccharide that contains peptide cross-links, found in bacterial cell walls

peptidyl transferase   In protein synthesis, the enzyme that catalyzes formation of the peptide bond; part of the 50S ribosomal subunit

peripheral proteins   Proteins loosely bound to the outside of a membrane

peroxisomes   Membrane-bounded sacs that contain enzymes involved in the metabolism of hydrogen peroxide (H_2O_2)

pH   A measure of the acidity of a solution

phenylketonuria   A disease characterized by mental retardation in developing children, caused by a lack of the enzyme that converts phenylalanine to tyrosine

phosphatidic acid   A compound in which two fatty acids and phosphoric acid are esterified to the three hydroxyl groups of glycerol

phosphatidylinositol *bis*phosphate (PIP$_2$)   A membrane-bound substance that mediates the action of Ca^{2+} as a second messenger

phosphoaclyglycerol (phosphoglyceride)   A phosphatidic acid (*vide supra*) with another alcohol esterified to the phosphoric acid moiety

3′,5′-phosphodiester bond   A covalent linkage in which phosphoric acid is esterified to the 3′ hydroxyl of one nucleoside and the 5′ hydroxyl of another nucleoside; forms the backbone of nucleic acids

phosphofructokinase   The key allosteric control enzyme in glycolysis; catalyzes the phosphorylation of fructose 6-phosphate

phospholipase   An enzyme that hydrolyzes phospholipids

photophosphorylation   The synthesis of ATP coupled to photosynthesis

photorespiration   The process by which plants oxidize carbohydrates aerobically in the light

photosynthesis   The process of using light energy from the sun to drive the synthesis of carbohydrates

photosynthetic unit   The assemblage of chlorophylls that includes light-harvesting molecules and the special

pair that actually carries out the reaction

Photosystem I   The portion of the photosynthetic apparatus responsible for the production of NADPH

Photosystem II   The portion of the photosynthetic apparatus responsible for the splitting of water to oxygen

pituitary   The gland that releases trophic hormones to specific endocrine glands, under the control of the hypothalamus

plasma membrane   Another name for the cell membrane, the outer boundary of the cell

plastocyanin   A copper-containing protein, part of the electron transport chain that links the two photosystems in photosynthesis

plastoquinone   A substance similar to coenzyme Q, part of the electron transport chain that links the two photosystems in photosynthesis

β-pleated sheet   One of the most important types of secondary structure, in which the protein backbone is almost fully extended

P/O ratio   The ratio of ATP produced by oxidative phosphorylation to oxygen atoms consumed in electron transport

polar bond   A bond in which two atoms have unequal shares in the bonding electrons

polyacrylamide gel electrophoresis (PAGE)   A form of electrophoresis in which a polyacrylamide gel serves as both a sieve and a supporting medium

polymer   A macromolecule formed by the bonding of smaller units

polymerase chain reaction   A method of amplifying the amount of DNA based on the reaction of isolated enzymes rather than on cloning

polypeptide chain   The backbone of a protein, formed by the linking of amino acids by peptide (amide) bonds

polysaccharide   A polymer of sugars

polysome   The assemblage of several ribosomes bound to one mRNA

porphyrins   Large-ring compounds formed by linking four pyrrole rings; combine with iron ions to form the heme group

power stroke   In muscle contraction, the stage in the process that causes the muscle fibers to slide past one another

pregnenelone   A steroid sex hormone

primary structure   The order in which the amino acids in a protein are linked by peptide bonds

primer   In DNA replication, a short stretch of RNA hydrogen-bonded to the template DNA to which the growing DNA strand is bonded at the start of synthesis

primosome   The complex at the replication fork in DNA synthesis; consists of the RNA primer, primase, and helicase

probe   A radioactively labeled strand of a nucleic acid, used for selecting a complementary strand out of a mixture

progesterone   A steroid sex hormone

prokaryote   An organism without a well-defined nucleus

promoter   The portion of DNA to which RNA polymerase binds at the start of transcription

proofreading   The process of removing incorrect nucleotides during DNA replication

prostaglandin   One of a group of derivatives of arachidonic acid, containing a five-membered ring; of pharmaceutical importance

prosthetic group   A portion of a protein that does not consist of amino acids

protease   An enzyme that hydrolyzes proteins

protein   A macromolecule formed by polymerization of amino acids

proteolysis   The hydrolysis of proteins

prothrombin   A protein that plays a part in blood clotting after activation

by selective hydrolysis to form thrombin

Protista   One of the five kingdoms used to classify living organisms, includes single-celled eukaryotes

proton gradient   The difference between the hydrogen ion concentrations in the mitochondrial matrix and the intermembrane space; the basis of coupling between oxidation and phosphorylation

purine   A nitrogen-containing aromatic compound that contains a six-membered ring fused to a five-membered ring; the parent compound of two nucleobases

pyranose   A cyclic form of a sugar containing a five-membered ring, named for its resemblance to pyran

pyranoside   A glycoside involving a pyranose

pyrimidine   A nitrogen-containing aromatic compound that contains a six-membered ring; the parent compound of several nucleobases

pyrrole ring   A five-membered ring that contains one nitrogen atom; part of the structures of porphyrins and heme

pyruvate carboxylase   The enzyme that catalyzes the conversion of pyruvate to oxaloacetate

pyruvate dehydrogenase complex   The enzyme that catalyzes the conversion of pyruvate to acetyl-CoA and carbon dioxide

quaternary structure   The interaction of several polypeptide chains in a multisubunit protein

R group   The side chain of an amino acid that determines its identity

rate constant   A proportionality constant in the equation that describes the rate of a reaction

rate-limiting step   The slowest step in a reaction mechanism; determines the maximum velocity of the reaction

reaction center   The site of the special pair of chlorophylls responsible for trapping light energy from the sun

reading frame   The starting point for the reading of a genetic message

receptor protein   A protein on a cell membrane with specific binding sites for extracellular substances

reducing agent   A substance that gives up electrons to other substances

reducing sugar   A sugar that has a free carbonyl group, one that can react with an oxidizing agent

reduction   The loss of electrons

regulatory gene   A gene that directs the synthesis of a repressor protein

relative specificity   An enzyme's property of acting on several related substrates

repair   The enzymatic removal of incorrect nucleotides from DNA and their replacement by correct ones

replication   The process of duplication of DNA

replication fork   In DNA replication, the point at which new DNA strands are formed

replisome   A complex of DNA polymerase, the RNA primer, primase, and helicase at the replication fork

repressor   A protein that binds to an operator gene, blocking the transcription and eventual translation of structural genes under the control of that operator

residues   The portion of a monomer unit included in a polymer after splitting out of water between the monomers

resonance structures   Structural formulas that differ from each other only in the positions of electrons

respiratory complexes   The multienzyme systems in the inner mitochondrial membrane that carry out the reactions of electron transport

restriction nuclease   An enzyme that catalyzes a double-strand hydrolysis of DNA at a defined point in a specific sequence

retinal   The aldehyde form of vitamin A

retrovirus   A virus in which the base

sequence of RNA directs the synthesis of DNA

reverse turn   A part of a protein where the polypeptide chain folds back on itself

reversible inhibitor   An inhibitor that is not covalently bound to an enzyme; can be removed with restoration of activity

rhodopsin   A molecule crucial in vision, formed by the reaction of retinal and opsin

ribonucleoside   A compound formed when a nucleobase forms a glycosidic bond with ribose

ribose   A sugar that is part of the structure of RNA

ribosome   The site of protein synthesis in all organisms, consisting of RNA and protein

ribozyme   catalytic RNA

ribulose 1,5-*bis*phosphate   A key intermediate in the production of sugars in photosynthesis

rickets   A disease characterized by skeletal deformities, caused by a deficiency of vitamin D

RNA   Ribonucleic acid

RNA polymerase   The enzyme that catalyzes the production of RNA on a DNA template

RNase P   One of the first examples of catalytic RNA

rRNA   The kind of RNA found in ribosomes

S-adenosylmethionine   A carrier of methyl groups

salvage reactions   Reactions that reuse compounds, such as purines, that require a large amount of energy to produce

saponification   The reaction of a triacylglycerol with a base to produce glycerol and three molecules of fatty acid

sarcomere   A repeating unit in muscle fibers

saturated   Having all carbon–carbon bonds as single bonds

Schiff base   A linkage of a carbonyl-containing substrate to an amino group on an enzyme

sclerotic plaque   The damage to the myelin sheath of nerve cells in multiple sclerosis

second messenger   A substance produced or released by a cell in response to hormone binding to a receptor on the cell surface; elicits the actual response in the cell

secondary structure   The arrangement in space of the backbone atoms in a polypeptide chain

sedimentation coefficient   The number expressing the rate at which particles move to the bottom of the tube during centrifugation

self-assembly of ribosomes   The reversible aggregation of ribosomal subunits in the presence of Mg^{2+}

semiconservative replication   The mode in which DNA reproduces itself such that one strand comes from parent DNA and the other strand is newly formed

sequential model   A description of the action of allosteric proteins in which a conformational change in one subunit is passed along to the other subunits

serine protease   A proteolytic enzyme in which a serine hydroxyl plays an essential role in catalysis

severe combined immune deficiency (SCID)   A genetic disease that affects DNA synthesis in the cells of the immune system

sickle-cell anemia   A disease caused by the change of one amino acid in two of the four polypeptide chains of hemoglobin, characterized by blockage of blood vessels

side-chain group   The portion of an amino acid that determines its identity

sigmoidal   An S-shaped curve on a graph, characteristic of cooperative interactions

simple diffusion   The process of passing through a pore or opening in a membrane with no requirement for a carrier or for energy

single-strand-binding protein (SSB)   In DNA replication, a protein that protects exposed single-strand sections of DNA from nucleases

single-strand heterozygous   A form of recombinant DNA in which one strand comes entirely from one of the original DNA molecules and the other contains sequences from both the original molecules

sodium–potassium ion pump   The export of sodium ion from the cell with simultaneous inflow of potassium ion, both against concentration gradients

sphingolipid   A lipid whose structure is based on sphingosine

sphingosine   A long-chain amino alcohol, the basis of the structures of a number of lipids

spin labeling   The tagging of a part of a macromolecule with a group that has an unpaired electron that can be detected by magnetic measurements

spontaneous   In thermodynamics, taking place without outside intervention; used for a reaction or process

standard state   The standard set of conditions used for comparison of chemical reactions

starch   A polymer of glucose that plays an energy-storage role in plants

stationary phase   In chromatography, the substance that selectively retards the flow of the sample, effecting the separation

steady state   The condition in which the concentration of an enzyme–substrate complex remains constant in spite of continuous turnover

stereochemistry   The three-dimensional shape of a molecule

stereoisomers (optical isomers)   Molecules that differ from each other only in configuration (three-dimensional shape)

stereospecific   Able to distinguish between stereoisomers

steroid   A lipid with a characteristic fused-ring structure

sticky ends   Short, single-stranded stretches at the ends of double-stranded DNA; can overlap and provide sites at which DNA molecules can be linked

stroma A portion of the chloroplast equivalent to the mitochondrial matrix; the site of production of sugars in photosynthesis

structural gene A gene that directs the synthesis of a protein under the control of some regulatory gene

subunits The individual parts of a larger molecule, e.g., the individual polypeptide chains that make up a complete protein

sugar–phosphate backbone The series of ester bonds between phosphoric acid and deoxyribose (in DNA) or ribose (in RNA)

supercoiling The presence of extra twists (in addition to those of the double helix) in closed circular DNA

supersecondary structure Specific clusters of secondary structure motifs in proteins

substrate cycling The control process in which opposing reactions are catalyzed by different enzymes

substrate-level phosphorylation A reaction in which the source of phosphorus is inorganic phosphate ion, not ATP

Svedberg unit (S) The unit of the sedimentation coefficient, a number that characterizes the motion of a particle in an ultracentrifuge tube

swivel point In DNA replication, the nick at which DNA unwinds before new strands can be formed

synapse The space between nerve cells through which a signal is transmitted

T-cell One of two kinds of white blood cells important in the immune system—killer T-cells, which destroy infected cells, and helper T-cells, which are involved in the process of B-cell maturation

Tay–Sachs disease A disease caused by the lack of an enzyme that breaks down gangliosides, characterized by retardation and early death

termination step In protein synthesis, the point at which the stop signal is reached, releasing the newly formed protein from the ribosome

tertiary structure The arrangement in space of all the atoms in a protein

testosterone A steroid sex hormone

tetrahydrofolate The metabolically active form of the vitamin folic acid, a carrier of one-carbon groups

tetramer An aggregate consisting of four subunits

thiamine pyrophosphate A coenzyme involved in the transfer of two-carbon units

thioester A sulfur-containing analogue of an ester

thrombin A protein that catalyzes the last stage in blood clotting

thylakoid disks The sites of the light-trapping reaction in chloroplasts

thylakoid space The portion of the chloroplast between the thylakoid disks

thymidylate synthetase The enzyme that catalyzes the production of thymine nucleotides needed for DNA synthesis; a target for cancer chemotherapy

thymine One of the pyrimidine bases found in nucleic acids

thymine dimers A defect in DNA structure caused by the action of ultraviolet light

titration An experiment in which a measured amount of base is added to an acid

α-tocopherol The most active form of vitamin E

topoisomerase An enzyme that relaxes supercoiling in closed circular DNA

torr A unit of pressure; that exerted by a column of mercury 1 mm high at 0°C

transaldolase An enzyme that transfers a two-carbon unit in reactions of sugars

transamination The transfer of amino groups from one molecule to another, an important process in the anabolism and catabolism of amino acids

transcription The process of formation of RNA on a DNA template

transgene A gene transferred from the genome of one organism to another

transgenic organism An organism that carries a gene introduced by genetic engineering

transition state The intermediate stage in a reaction in which old bonds break and new bonds are formed

transketolase An enzyme that transfers a three-carbon unit in reactions of sugars

translation The process of protein synthesis in which the amino acid sequence of the protein reflects the sequence of bases in the gene that codes for that protein

translocation In protein synthesis, the motion of the ribosome along the mRNA as the genetic message is being read

transport protein A component of a membrane that mediates the entry of specific substances into a cell

transposable elements Genes that can change their positions on a chromosome

triacylglycerol (triglyceride) A lipid formed by esterification of three fatty acids to glycerol

tricarboxylic acid cycle The citric acid cycle

trimer An aggregate consisting of three subunits

triosephosphate isomerase The enzyme that catalyzes the conversion of dihydroxyacetone phosphate to glyceraldehyde 3-phosphate

triplet code A sequence of three bases (a triplet) in mRNA that specifies one amino acid in a protein

tRNA The kind of RNA to which amino acids are bonded, preliminary to being incorporated into a growing polypeptide chain

trophic hormones   Hormones produced by the pituitary gland under the direction of the hypothalamus; in turn, they cause the release of specific hormones by individual endocrine glands

tropocollagen   A protein consisting of three polypeptide chains in a helical conformation, found in connective tissue; a classic example of a fibrous protein

trypsin   A proteolytic enzyme specific for basic amino acid residues as the site of hydrolysis

turnover number   The number of moles of substrate that react per second per mole of enzyme

uncoupler   A substance that overcomes the proton gradient in mitochondria, allowing electron transport to proceed in the absence of phosphorylation

universal code   The genetic code that is the same in all organisms

unsaturated   Having some carbon–carbon double bonds

upstream   In transcription, describes a portion of the DNA sequence nearer the $3'$ end than the gene to be transcribed, where the DNA is read from the $3'$ to the $5'$ end and the RNA is formed from the $5'$ to the $3'$ end; in translation, nearer the $5'$ end of the mRNA

uracil   One of the pyrimidine bases found in nucleic acids

urea cycle   A pathway that leads to excretion of waste products of nitrogen metabolism, especially that of amino acids

uric acid   A product of catabolism of nitrogen-containing compounds, especially purines; in humans, accumulation in joints causes gout

vacuoles   Membrane-bounded sacs that tend to be found in plant rather than animal cells; storage locations for toxic waste materials

van der Waals bond   A noncovalent association based on the attraction of transient dipoles for one another

vector   A carrier molecule for transfer of genes in DNA recombination

virion   A complete virus particle consisting of nucleic acid and coat protein

voltage-gated   Describes a kind of channel protein transiently opened by changes in the membrane voltage

wobble   The possible variation in the third base of a codon allowed by several acceptable forms of base pairing between mRNA and tRNA

x-ray crystallography   An experimental method for determining the tertiary structures of proteins

zero-order   Describes a reaction that proceeds at a constant rate, independent of the concentration of reactant

zwitterion   A molecule that has both a positive charge and a negative charge

zymogen   An inactive protein that can be activated by specific hydrolysis of peptide bonds

Index

Page numbers in **boldface** refer to a major discussion of the entry. F after a page number refers to a figure or structural formula. T after a page number refers to a table. Positional and configurational designations in chemical names (e.g., 3-, N-, a-) are ignored in alphabetizing.